ANNALS OF THE NEW YORK ACADEMY OF SCIENCES

Volume 604

EDITORIAL STAFF
Executive Editor
BILL BOLAND
Managing Editor
JUSTINE CULLINAN
Associate Editor
SHEILA K. TREITLER

The New York Academy of Sciences
2 East 63rd Street
New York, New York 10021

PRESYNAPTIC RECEPTORS AND THE QUESTION OF AUTOREGULATION OF NEUROTRANSMITTER RELEASE

PRESYNAPTIC RECEPTORS AND THE QUESTION OF AUTOREGULATION OF NEUROTRANSMITTER RELEASE

Edited by Stanley Kalsner and Thomas C. Westfall

The New York Academy of Sciences
New York, New York
1990

Library of Congress Cataloging-in Publication Data

Presynaptic receptors and the question of autoregulation of
  neurotransmitter release / edited by Stanley Kalsner and Thomas C.
  Westfall.
      p.   cm. — (Annals of the New York Academy of Sciences ; ISSN
  0077-8923 ; v. 604)
      Papers presented at a conference held in Philadelphia, Pa., Dec.
  4–6, 1989.
      Includes bibliographical references (p.   ) and index.
      ISBN 0-89766-613-5 (alk. paper). — ISBN 0-89766-614-3 (pbk. :
  alk. paper)
      1. Autoreceptors—Congresses.   2. Presynaptic receptors—
  —Congresses.   I. Kalsner, Stanley, 1936–   .   II. Westfall, Thomas
  C. III. New York Academy of Sciences.   IV. Series.
  Q11.N5   vol. 604
  [QP364.7]
  500 s—dc20   90-13208
  [599′.0188]   CIP

SP
Printed in the United States of America
ISBN 0-89766-613-5 (cloth)
ISBN 0-89766-614-3 (paper)
ISSN 0077-8923

Volume 604
August 31, 1990

PRESYNAPTIC RECEPTORS AND THE QUESTION OF AUTOREGULATION OF NEUROTRANSMITTER RELEASE[a]

Editors and Conference Organizers
STANLEY KALSNER and THOMAS C. WESTFALL

CONTENTS

[a]This volume is the result of a conference entitled Presynaptic Receptors and the Question of Autoregulation of Neurotransmitter Release held from December 4 to December 6, 1989 in Philadelphia, Pennsylvania and sponsored by The New York Academy of Sciences.

Part III. Transduction Mechanisms and Linkages in the Activation of Presynaptic Receptors I

Part IV. Transduction Mechanisms in the Activation of Presynaptic Receptors II

Conference Supporter:
- NATIONAL INSTITUTE OF MENTAL HEALTH/NIH

Conference Contributors:
- ABBOTT LABORATORIES
- BRISTOL MYERS
- BURROUGHS WELLCOME
- CIBA-GEIGY
- E.I. DUPONT DE NEMOURS & COMPANY
- FISONS PHARMACEUTICALS
- GLAXO RESEARCH
- HOECHST ROUSSEL PHARMACEUTICALS
- HOFFMANN-LA ROCHE
- ICI PHARMACEUTICALS
- LEDERLE LABORATORIES/AMERICAN CYANAMID COMPANY
- MERCK SHARP & DOHME
- MILES INC./BAYER AG
- SMITHKLINE BEECHAM
- UPJOHN & COMPANY
- WYETH AYERST

Preface

STANLEY KALSNER

Department of Physiology
City University of New York Medical School
138th Street and Convent Avenue
New York, New York 10031

Autoreceptors are a special kind of presynaptic receptor, located at nerve terminals, through which the neurone's own transmitter might, under appropriate conditions, modify transmitter release. The existence of autoreceptors was postulated by several groups independently, and by 1971 had been proposed for noradrenergic, central cholinergic, and GABA (gamma-aminobutyric acid) neurones. Acting as a target for the neurone's own transmitter the receptors are thought to mediate positive or negative feedback. Presynaptic facilitatory nicotine receptors at cholinergic axon terminals, for instance, are thought by some to be links in a positive feedback in which released acetylcholine enhances its own release. Presynaptic inhibitory alpha-adrenoreceptors at noradrenergic terminal axons may be links in a negative feedback in which released noradrenaline reduces its own subsequent release.

Presynaptic autoreceptors have been a focal point of neurobiological research since they were first postulated, almost 20 years ago. They introduce an intriguing element of self-regulation into the activity of a neurone; self-regulation of the amplitude of the neurone's chemical signal. The theory has had much international and interdisciplinary appeal as it has considerable explanatory power, and potential relevance to diverse disease processes, in particular cardiovascular diseases and mental illness. The theory purports to explain many previously puzzling findings, and has been integral to such developments as the $alpha_2$ and histamine H_3-receptors, for which the respective presynaptic receptors were the prototypes.

For all of this interest, presynaptic autoreceptors are a subject of considerable uncertainty and controversy. First, there is disagreement about the transduction mechanisms behind the receptors. In the case of presynaptic alpha-autoreceptors (noradrenergic neurones), for instance, some authors postulate that the receptors, when activated, block the propagation of the action potential in the terminal region, whereas others suggest that propagation is unimpaired but electrosecretory coupling in the individual varicosities is depressed. Moreover, whereas some maintain that an inhibition of voltage-dependent Ca^{++} channels is the primary ionic mechanism, others believe that the primary action is the increase of a presynaptic K^+ conductance. There is also uncertainty as to the involvement of a presynaptic adenylate cyclase, as opposed to a direct coupling of the receptors to ion channels independent of any intraaxonal second messenger. In the case of presynaptic D_2-autoreceptors (dopamine neurones), the same questions can be raised, and in addition an increase in protein carboxymethylation has been suggested to be the mechanism of action.

Second, there is much discussion about the physiological importance of presynaptic receptors. The feedback mechanisms mentioned above have by no means remained undisputed. One fundamental finding supporting their operation is that agonists at the respective receptors exert an effect opposite to the effect of agonists. Curare alkaloids, for instance, depress the release of acetylcholine, and alpha-adrenoceptor antagonists increase the release of noradrenaline, suggesting that in the one case a positive, and in the other case a negative, feedback system is interrupted. Some workers, however, think that these antagonists effects are not due to blockade of autoreceptors and interruption of a physiological presynaptic feedback but to other mechanisms, and a body of evidence is now available that questions the interpretation of data purporting to sustain the authenticity of autoreceptor operation. Experiments are described in this symposium showing a seeming lack of correlation between theoretical expectations and physiological performance, in a variety of peripheral and central tissues. Very often the relationship of the antagonist and agonist effect and the intensity of nerve stimulation has not been as predicted by the autoreceptor hypothesis.

Third, the most fundamental contradiction has been that receptors at nerve terminals for the neurone's own transmitter do not exist at all, at least in a functional sense, as reflected in the operation of agonists and antagonists. Evidence is provided that enhancement by antagonists may be attributable to ion channel activation and unrelated to alpha-adrenoceptors. Another difficulty encountered is in attempts to detect autoreceptors in radioligand binding experiments. Moreover, if there are autoreceptors, they should be transported down from the perikarya, where they are synthesized, to the terminal axons, and such axonal transport has not yet been clearly demonstrated.

All of these uncertainties and controversies have developed over the last decade. This symposium has brought together experts from around the world, working in interdependent fields, or holding different views. Autoreceptors have not been the subject of any previous conference, although there have been small meetings and perhaps one large European conference on presynaptic receptors in general, over 10 years ago. An additional incentive for this particular symposium is that United States scientists have not had convenient exposure to the very active group of European scientists who are currently in the forefront in this area.

A central function of the proposed symposium is to encourage broad expression of current research and thinking on presynaptic receptors from an international group of participants, and to have as its axis a fundamental and extensive examination of the question of autoreceptors and their physiological relevance.

Since there are two basically opposing views on presynaptic autoreceptors and on their physiological relevance, the symposium has been organized and co-chaired by proponents of these two positions. It should be emphasized that the co-chairmen have decidedly different perspectives about the concept of autoreceptors, and on the central question of their physiological relevance. There is much closer agreement, however, about heteroreceptors, and about possible transduction mechanisms. It is our joint hope that the symposium has helped to crystallize differences of opinion and point the way towards their experimental resolution.

The selection of participants was aimed at bringing together both new and established investigators in the diverse disciplines that comprise research on presynaptic receptors. These disciplines encompass molecular biology, biochemistry, neuroscience, physiology, pharmacology, and clinical medicine. It was felt that the diverse representation with respect to speakers would be matched by the different backgrounds of those attending the conference. And we believe this ambition was met.

We thank the Academy, and in particular Dr. Charles Nicholson, for their unflagging support of the symposium, and Dr. Klaus Starke for his early contributions to the development of this successful meeting.

Heteroreceptors, Autoreceptors, and Other Terminal Sites

STANLEY KALSNER

Department of Physiology
City University of New York Medical School
138th Street and Convent Avenue
New York, New York 10031

The concept of presynaptic receptors and the idea that neuronal action can be facilitated or suppressed through their activation predate even the concept of chemical transmission. In the late 19th and early 20th centuries nicotine, acetylcholine, and even epinephrine were all thought to act on nerve terminals to alter the electrical conductance of neurones, and these terminals were thought to remain intact even after nerve section and degeneration (see Dale for a first-hand account of some of the early history).[1]

With the development of our understanding of chemical transmission and the characteristics of the synaptic space, the receptors for neurotransmitters were all assigned postsynaptic locations. But even in the 1930s, what we now call neuronal "autoreceptors" were recognized.

For example, in 1939, Marrazzi described inhibition by epinephrine of transmission at autonomic sympathetic ganglia,[2] through receptors now named alpha 2.[3] In the 1940s, 1950s, and 1960s Burn, Muscholl, and others wrote frequently about sites for acetylcholine and for sympathomimetics on sympathetic nerve terminals (see Fozard).[4] In 1961, George Koelle put forward perhaps the first formal proposal for presynaptic autoreceptors and feedback, namely, nicotinic facilitatory receptors at the terminals of peripheral cholinergic nerves.[5] Presynaptic autoreceptors and the concept of negative feedback of neurotransmitter release, the idea most familiar to us today, was probably first raised by Polak in 1967.[6,7] He described the effects of atropine to enhance the output of acetylcholine in cortical slices from rat brain. He employed the term "negative feedback mechanism" and commented that released acetylcholine inhibits further release of acetylcholine and that this is antagonized by atropine. In 1971 several groups of workers independently enunciated the concept of negative feedback and applied it to adrenergic mechanisms. Among them were Kirpekar and Puig,[8] Langer and associates,[9] Farnebo and Hamberger,[10] and Starke,[11–13] and they all contributed a good deal to our understanding of presynaptic receptors.

Our current terminology about neuronal receptors, including their location, was summarized by Starke and Langer.[14] It can be described as follows. All receptors at the terminals, even if they are postsynaptic to another neuron's transmitter, are termed "presynaptic receptors" reflecting an action on the secretory portion of the neuron. At the soma-dendritic, impulse-generating region of the neuron, definitions are more complex. If the region receives formal presynaptic input, the receptors are called "postsynaptic;" otherwise they are called soma-dendritic.

The above definitions lead to some difficulties. For example, in the periphery an autonomic neurone may release acetylcholine (Ach) and have muscarinic and nicotinic receptors both at its terminals and at the soma-dendritic region. The neuron may, in turn, receive formal preganglionic input at the cell body region from another cholinergic neuron. Are the cholinergic receptors in the cell body region to be described only

TABLE 1. Presynaptic Receptors

A. Current Terminology
Heteroreceptors—Can be activated by some agonists but not readily by neurone's own transmitter.
Autoreceptors—Can be activated readily by neurone's own transmitter (assumption: endogenous activation as part of feedback link).
(Auto = "self"—"Denotes action or operation")
B. Suggested Expansion
1. Function not implicit
   Heteroreceptors—Can be activated by some agonists but not readily by neurone's own transmitter.
   Homoreceptors—Can be activated by neurone's own transmitter when delivered from an "external" source.
2. Function implicit
   Autoreceptors—Can be activated by neurone's own transmitter as part of a feedback loop in neurosecretion

as postsynaptic receptors or also as soma-dendritic autoreceptors? These questions are intensified in the central nervous system (CNS). Recurrent collaterals which release transmitter and dendritic release onto the neurons own cell body region illustrate some of the complexity. Further, short interneurons confined to particular cortical regions may be suffused by foreign transmitter (e.g., Ach, norepinephrine) equally at its terminals and at its dendrites and soma. Are the points of contact for such transmitter at the terminals described as presynaptic and those at the soma as postsynaptic? To sometimes label receptors at the soma-dendritic portion of the neuron "presynaptic" would not do away with the confusion. We would then have to modify the nomenclature correspondingly at the terminal sites to acknowledge formal "postsynaptic contacts." Formal synaptic input to the terminal region would be mediated through "postsynaptic" rather than "presynaptic" receptor contacts. The current topographical categorizations should probably be kept, at least for the time being, despite the occasional difficulties.

The terminology used to describe presynaptic receptors is very restricted. As shown in TABLE 1A, heteroreceptors are those terminal sites that recognize agonists other than the neuron's own transmitter whereas autoreceptors respond to the neuron's own transmitter. The word "auto," however, not only means self, but when applied to scientific terminology implies function and operation (see *Oxford English Dictionary*);[15] in this instance endogenous feedback regulation. I suggest a clarifying expansion of the existing terminology (TABLE 1B). The term "homoreceptors" would be introduced as an appropriate match to the existing term "heteroreceptors" (see *Oxford English Dictionary* "hetero-homo")[15] (TABLE 1B). Heteroreceptors are sites that respond to agonists other than the neurone's own transmitter whereas homoreceptors respond to the neurone's own transmitter when added from an external source. Neither homo- nor heteroreceptors imply endogenous activity or function. Should an investigator decide that the evidence is sufficient to indicate that feedback through endogenous transmitter is operative in the system under study, then the term "autoreceptors" would be employed. Autoreceptors would represent a subset of homoreceptors to which a particular endogenous function, namely, feedback, is assigned.

For me, this expansion in terminology serves an obvious need. Whereas I have no problem with heteroreceptors and homoreceptors, I have difficulty with the available evidence for the operation of autoreceptors. In principle, terminal regulation of transmitter release is an exciting idea, and feedback may operate under some circumstances and with some transmitters. It is not possible, however, to sort out systems where feedback operates by relying simply on qualitative increases or decreases in transmitter release produced by agonists and antagonists, which may be due to other causes.[16]

Work that describes a neuronal system as having "functional autoreceptors" simply because the neurotransmitter added from outside enhances or reduces release should be viewed with caution. There is almost equal difficulty when the enhancement by an antagonist such as yohimbine is used, as it has been, to define the relative amount of ongoing feedback, such as in dopaminergic, adrenergic, or noradrenergic systems in the hypothalamus; or when the lack of enhancement of norepinephrine release by yohimbine in some hearts is used, as it has been, to decide that feedback is lost in old age.

Each of us interprets to some extent differently the existing body of data and as a result has different views about feedback. In my opinion this comes out of the highly indirect character of the available evidence, allowing much room for subjective interpretation. As shown in TABLE 2, looking at the same body of published data, feedback has been described in various and discrepant ways. It has been viewed by some as a system operating over the range of physiological frequencies;[12] by others as a system restricted in its operation to very low, or negligible, levels of neuronal activity,[17] and overridden at higher levels;[13] or as a system not active at the lowest levels but only once a critical threshold is reached, whereupon further activity is braked.[18-20] Still

TABLE 2. Autoreceptors—Some Views

1. Feedback regulation of transmitter release over the physiological range (Starke[12] & numerous).	11. Regulates distal spread of impulses along terminal axon fibers.[24,27,35]
2. To curtail low level "stray" stimuli during quiet periods.[17]	12. It conserves transmitter for later use by modulating short-term release.[32]
3. A system active at very low stimulation parameters and overridden at higher levels.[13]	13. Function and physiological relevance is unknown.[13]
4. Synaptic transmitter level must rise until a critical threshold is reached—shuts off further liberation of the transmitter.[18-20]	14. Effects of agonists and antagonists are not interpretable in terms of feedback/autoreceptors (Kalsner[16,37] & others).
5. A "last resort" mechanism that operates when the level of efferent activity is very high.[21]	15. In CNS—to achieve "waterspray effect" and modulation of neuronal activity at various contact points (e.g., adrenaline or noradrenaline or dopamine on alpha 2 sites) (Kuffler[31] & others).
6. Prevents sympathetic activity from increasing to a pathological range.[22]	16. To receive input from remote sources (e.g., periphery and adrenal medulla).[27-29]
7. A "modulatory, marginal system for fine functional adjustment."[24]	17. Terminal receptors incidental to soma-dendritic receptors (Groves[34] & others).
8. A system that smoothes out the junctional concentration of transmitter throughout an excitation-secretion cycle.[25]	18. Initiates long-term structural changes in size and conductivity of terminals (habit, facilitation, learning).[33]
9. It curbs competing process of facilitation of transmitter release over a range of stimulation parameters (Blakeley[23] & numerous).	19. Evolutionary debris (several).
10. Shuts down individual varicosities to ensure minimal overlap of released transmitter.[26]	20. To be determined (Kalsner[16] & others).

others perceive it as a "last resort" mechanism that operates only when neuronal activity is high[21] or entering into the pathological range.[22]

Feedback has been described as operating broadly over the physiological range to curb facilitation of release[23] or, oppositely, as a "marginal" system,[24] or perhaps a means of fine tuning neurosecretion.[25] Even its locus of operation is not agreed upon. Bevan *et al.* describe it as a system shutting down individual varicosities,[26] whereas Stjarne perceives it as acting downstream, shutting down terminal portions of neurones encompassing many release sites rather than individual varicosities.[24,27]

Some workers seem to see these presynaptic terminal sites as operating in a way more consonant with the proposed definition for homoreceptors than for autoreceptors. Presynaptic beta receptors on sympathetic nerves in the periphery, for example, may respond to circulating catecholamines rather than to neuronally released transmitter.[27-29] Similarly, the diffuse release of transmitter in the CNS[30] may allow for activation of terminal sites by mediator from distant sources,[31] and these activities would not represent autoreceptor but, instead, homoreceptor activity (e.g., activation of alpha 2 sites on noradrenergic terminals by adrenaline released from another set of adrenergic neurones).

A novel interpretation by Mcafee *et al.* sees negative feedback as a system for holding down or conserving transmitter over the short term for use over the long term,[32] and in this way facilitating transmitter release. Kandel and Schwartz[33] and others see terminal receptors not only as acting to modify transmitter release but, perhaps more important centrally, as serving to mediate alterations in the shape and conductivity of neurone terminals over the long term. The relevance to learning, habituation, and memory is apparent. Still others see terminal receptors as incidental to functional soma-dendritic receptors[34] or simply as evolutionary debris. My own work puts forward the view that feedback regulation of transmitter release does not routinely operate in most systems either in the periphery or centrally, and that many of the supporting data need to be reinterpreted.[33,35-37]

Should the evidence for "feedback" deserve reinterpretation it means that the explanation for the effects of antagonists to alter release, at least in some systems, needs to be sought elsewhere. This search for the mechanism of "antagonist" action, which may involve direct actions to alter transmitter release, could provide an entirely new and positive approach to theory and to the therapy of diseases as diverse as hypertension and schizophrenia. You don't need functional autoreceptors to have a satisfying explanation for the accumulated body of data.

REFERENCES

1. DALE, H. 1953. Adventures in Physiology. William Clowes and Sons Limited. London, England.
2. MARRAZZI, A. S. 1939. Electrical studies on the pharmacology of autonomic synapses. II. The action of a sympathomimetic drug (epinephrine) on sympathetic ganglia. J. Pharmacol. Exp. Ther. **65:** 395–404.
3. COLE, A. E. & P. SHINNICK-GALLAGHER. 1981. Comparison of the receptors mediating the catecholamine hyperpolarization and slow inhibitory postsynaptic potential in sympathetic ganglia. J. Pharmacol. Exp. Ther. **217:** 440–444.
4. FOZARD, J. 1979. Cholinergic mechanisms in adrenergic function. *In* Trends in Autonomic Pharmacology. S. Kalsner, Ed. Urban & Schwarzenberg. Baltimore, Md.
5. KOELLE, G. B. 1961. A proposed dual neurohumoral role of acetylcholine: its functions at the pre- and post-synaptic sites. Nature London **190:** 208–211.
6. POLAK, R. L. 1967. The influence of antimuscarinic drugs on the synthesis and release of acetylcholine by the isolated cerebral cortex of the rat. J. Physiol. London **191:** 34–35.

7. POLAK, R. L. 1971. Stimulating action of atropine on the release of acetylcholine by rat cerebral cortex in vitro. Br. J. Pharmacol. **41:** 600–606.

8. KIRPEKAR, S. M. & M. PUIG. 1971. Effect of flow-stop on noradrenaline release from normal spleens and spleens treated with cocaine, phentolamine or phenoxybenzamine. Br. J. Pharmacol. **43:** 359–369.

9. LANGER, S. Z., E. ADLER, M. A. ENERO & F. J. E. STEFANO. 1971. The role of the alpha receptor in regulating noradrenaline overflow by nerve stimulation. *In* Proceedings of the 25th International Congress on Physiological Science: 335. The German Physiological Society. Munich, FRG.

10. FARNEBO, L. O. & B. HAMBERGER. 1971. Drug-induced changes in the release of [^{3}H]-noradrenaline from field stimulated rat iris. Br. J. Pharmacol. **43:** 97–106.

11. STARKE, K. 1971. Influence of α receptor stimulants on noradrenaline release. Naturwissenschaften **58:** 420.

12. STARKE, K. 1977. Regulation of noradrenaline release by presynaptic receptor systems. Rev. Physiol. Biochem. Pharmacol. **77:** 1–124.

13. STARKE, K. 1987. Presynaptic α-autoreceptors. Rev. Physiol. Biochem. Pharmacol. **107:** 73–146.

14. STARKE, K. & S. Z. LANGER. 1979. A note on terminology for presynaptic receptors. *In* Presynaptic Receptors. S. Z. Langer & M. L. Dubocovich, Eds.: 1–3 Oxford Press & Pergamon. Oxford & New York.

15. ONIONS, C. T., Ed. 1973. The Shorter Oxford English Dictionary: on Historical Principles. Oxford University Press. Oxford, England.

16. KALSNER, S. 1985. Is there feedback regulation of neurotransmitter release by autoreceptors? Biol. Chem. **34:** 4085–4097.

17. VON EULER, V. 1979. General views on the relevance of presynaptic receptor systems. *In* Presynaptic Receptors. O. O. Langer & O. O. Dubocovich, Eds.: 5–9. Oxford Press & Pergamon. Oxford & New York.

18. WESTFALL, T. C. 1977. Local regulation of adrenergic neurotransmission. Physiol. Rev. **57:** 659–728.

19. RAND, M. J., M. W. MCCULLOCH & D. F. STORY. 1982. Feedback modulation of noradrenergic transmission. Trends Pharmacol. Sci. **3:** 8–11.

20. LANGER, S. Z. 1981. Presynaptic regulation of the release of catecholamines. Pharmacol. Rev. **32:** 337–362.

21. ANGUS, J. A., P. I. KORNER, G. P. JACKMAN, *et al.* 1984. Role of autoinhibitory feedback in cardiac sympathetic transmission. Clin. Exp. Hypertens. Part A Theory Pract. **A6**(1 & 2): 371–385.

22. KATO, E., K. KOKETSU, K. KUBA & E. KUMAMOTO. 1985. The mechanism of the inhibitory action of adrenaline on transmitter release in bullfrog sympathetic ganglia: independence of cyclic AMP and calcium ions. Br. J. Pharmacol. **84:** 435–443.

23. BLAKELEY, A. G. H., A. MATHIE & S. A. PETERSEN. 1984. Facilitation at single release sites of a sympathetic neuroeffector junction in the mouse. J. Physiol. **349:** 57–71.

24. STJARNE, L. 1985. Scope and mechanisms of control of stimulus secretion coupling in single varicosities of sympathetic nerve. Clin. Sci. **68** (Suppl): 77s–81s.

25. SHEPHERD, J. T. & P. M. VANHOUTTE. 1981. Local modulation of adrenergic neurotransmission. Circulation **64:** 655–666.

26. BEVAN, J. A., F. M. TAYO, R. A. ROWAN & R. D. BEVAN. 1984. Presynaptic α-receptor control of adrenergic transmitter release in blood vessels. Fed. Proc. **43:** 1365–1370.

27. STJARNE, L. 1989. Basic mechanisms and local modulation of nerve impulse–induced secretion of neurotransmitters from individual sympathetic nerve varicosities. Rev. Physiol. Biochem. Pharmacol. **112:** 1–137.

28. WESTFALL, T. C., M. J. PEACH & V. TITTERMARY. 1979. Enhancement of electrically-induced release of norepinephrine from the rat portal vein. Euro. J. Pharmacol. **58:** 67–74.

29. KALSNER, S. 1982. Positive feedback regulation of noradrenaline release from sympathetic nerves: a questionable hypothesis. Can. J. Physiol. Pharmacol. **60:** 737–743.

30. MOBLEY, P. & P. GREENGARD. 1985. Evidence for widespread effects of noradrenaline on axon terminals in the rat frontal cortex. Proc. Nat. Acad. Sci. USA **82:** 945–947.

31. KUFFLER, S. W., J. G. NICHOLLS & A. R. MARTIN. 1984. From Neuron to Brain. Sinauer
     Associates Inc. Sunderland, Mass.
32. MCAFEE, D. A., B. K. HENON, J. P. HORN & P. YAROWSKY. 1981. Calcium currents
     modulated by adrenergic receptors in sympathetic neurons. Fed. Proc. **40:** 2246–2249.
33. KANDEL, E. & J. H. SCHWARTZ. 1985. Principles of Neural Science. Elsevier. New York,
     N.Y.
34. GROVES, P. M., C. J. WILSON, S. J. YOUNG & G. V. REBEC. 1975. Self-inhibition by
     dopaminergic neurons. Science **190:** 522–529.
35. KALSNER, S. 1982. Feedback regulation of neurotransmitter release through adrenergic
     presynaptic receptors: time for a reassessment. *In* Trends in Autonomic Pharmacology. S.
     Kalsner, Ed. **2:** 385–425. Urban & Schwarzenberg, Inc. Baltimore, Md.
36. KALSNER, S. 1979. Single pulse stimulation of guinea-pig vas deferens and the presynaptic
     receptor hypothesis. Br. J. Pharmacol. **66:** 343–349.
37. KALSNER, S. 1984. Limitations of presynaptic theory: no support for feedback control of
     autonomic effectors. Fed. Proc. **43:** 1358–1364.

Presynaptic Receptors on Peripheral Noradrenergic Neurons

S. Z. LANGER AND S. ARBILLA

Department of Biology
Synthélabo Research (LERS)
58, rue de la Glacière
75013 Paris, France

INTRODUCTION

The general view that neurotransmitters can regulate their own release through presynaptic autoreceptors represents a new concept in the field of neurotransmission which has been demonstrated for noradrenaline, dopamine, serotonin, acetylcholine, γ-aminobutyric acid (GABA) and can probably be extended for other transmitters as well.[1-4]

In addition to presynaptic autoreceptors, through which neurotransmitters regulate their own release, many nerve terminals possess presynaptic receptors sensitive to endogenous compounds other than the neuron's own transmitter. This second group of presynaptic receptors are referred to as presynaptic heteroreceptors and are acted upon by cotransmitter neuropeptides, by transmitters released from adjacent terminals, or by locally produced or blood-borne substances that either facilitate or inhibit the calcium-dependent release of the neurotransmitter.[1-4] The relevance of presynaptic release modulating receptors has been demonstrated under both *in vitro* and *in vivo* experimental conditions.[1,4]

In the noradrenergic system, the concept of presynaptic modulation of transmitter release developed in parallel with the pharmacological evidence for two subtypes of α-adrenoceptors as defined by a different profile of affinity and relative order of potencies for agonists as well as for antagonists.[1]

In blood vessels, both the α_1- and α_2-adrenoceptor subtypes can be identified postjunctionally where they mediate vasoconstriction.[5,9] The α_1-adrenoceptor predominates in most vascular smooth muscles and it is the preferentially innervated subtype. The presynaptic adrenoceptor that mediates the inhibition of the release of noradrenaline corresponds to the α_2-adrenoceptor subtype.[1,10-12]

In addition to presynaptic α_2-inhibitory autoreceptors, the noradrenergic varicosity possesses inhibitory receptors sensitive to other neurotransmitters.[13-16] Two types of facilitatory receptors have been reported to be present on noradrenergic nerve terminals, β_2-adrenoceptors[10,17-19] and angiotensin II receptors.[20,21]

The present article examines the physiological and pharmacological relevance of presynaptic receptors on peripheral noradrenergic neurons.

PRESYNAPTIC INHIBITORY AUTORECEPTORS ON NORADRENERGIC NEUROTRANSMISSION

In support of the view that presynaptic α-adrenoceptors regulate the release of norepinephrine through a negative feedback mechanism, it has been demonstrated under *in vitro* as well as under *in vivo* conditions that α-adrenoceptor agonists inhibit, whereas α-adrenoceptor antagonists enhance, the release of norepinephrine elicited by nerve stimulation. The effects of α-adrenoceptor agonists and antagonists on noradrenergic neurotransmission are obtained regardless of the α or β type of the postsynaptic adrenoceptor that mediates the response of the effector organ (FIGURE 1).

Before the discovery of presynaptic, release-modulating α-adrenoceptors, it was generally accepted that the α-adrenoceptors represented a single homogeneous class of receptors. Experimental evidence suggesting differences between pre- and postsynaptic α-adrenoceptors was first obtained in the perfused cat spleen.[22] Subsequently, it was

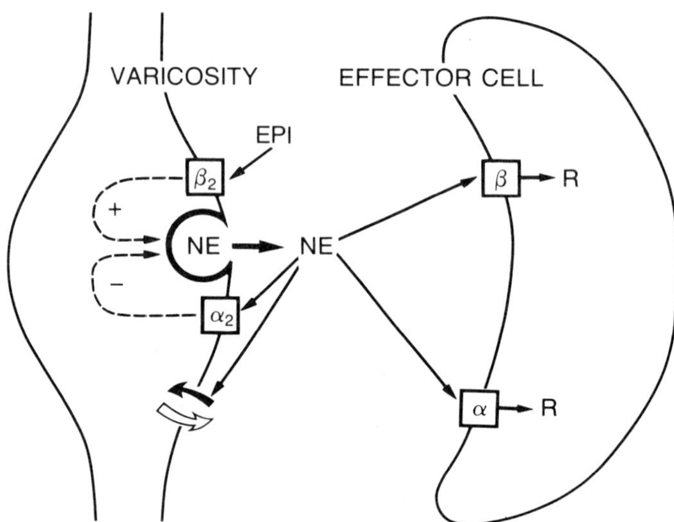

FIGURE 1. Role of the presynaptic α- and β-adrenoceptors in the regulation of norepinephrine release during nerve stimulation. During norepinephrine (NE) release at low frequencies of nerve stimulation (when the concentration of the released transmitter in the synaptic cleft is rather low), circulating or locally released epinephrine (EPI) can activate presynaptic β-adrenoceptors, of the β_2 subtype, leading to an increase in transmitter release. As the concentration of released NE increases, a threshold is reached at which the negative feedback mechanism mediated by presynaptic α-adrenoceptors of the α_2 subtype is triggered, leading to inhibition of transmitter release. Both presynaptic receptor mechanisms are present in nerves, irrespective of the α or β nature of the postsynaptic receptors that mediate the response (R) of the effector organ. As shown in the figure, NE is a substrate for the presynaptic transporter which represents the main inactivating mechanism for the transmitter.

shown that phenoxybenzamine is nearly 100 times more potent in blocking the postsynaptic α-adrenoceptors than it is in blocking the presynaptic α-adrenoceptors.[23,24] These results led to the proposal that the α-adrenoceptors should be subclassified into α_1 and α_2 subtypes.[25] Subsequent studies provided the additional pharmacological evidence in terms of differences in the relative order of potencies of α-adrenoceptors agonists and antagonists that is essential for the classification of α-adrenoceptors into α_1 and α_2 subtypes.[1,26] At present it is well established that the second messengers are different for α_1- and α_2-adrenoceptor subtypes and in addition both α_1- and α_2-receptor subtypes have now been cloned, and appear to exist in the form of multiple subtypes for each receptor.[27]

The α-adrenoceptor subtype located presynaptically on noradrenergic nerve terminals corresponds to the α_2-subtype and it is linked to the modulation of transmitter release. Activation of α_2-adrenoceptors by exogenous agonists reduces transmitter release, while the increase in norepinephrine release obtained in the presence of α_2-adrenoceptor antagonists depends upon the frequency of nerve impulses and duration of the period of depolarization.[12] These results support the view that there is an operational negative feedback mechanism through which the released transmitter can regulate its own release once a threshold concentration of norepinephrine is achieved in the synaptic cleft.[1,3,12] Certain aspects of this concept have, however, recently been challenged.[28] It is of interest to note that in addition to the largely predominating α_2-autoreceptors, presynaptic autoreceptors of the α_1 subtype have also been reported in some tissues.[29]

PRESYNAPTIC HETERORECEPTORS MODULATING NORADRENERGIC NEUROTRANSMISSION

Inhibitory Heteroreceptors

As already discussed, the autoreceptor control of the release of norepinephrine is mediated by presynaptic α_2-adrenoceptors which are involved in a negative feedback mechanism whereby the synaptic concentration of released norepinephrine can modulate further release of the neurotransmitter.

Several other receptor-mediated inhibitory mechanisms exist at the level of the release of norepinephrine (FIGURE 2). The inhibition of peripheral noradrenergic neurotransmission by presynaptic muscarinic receptors was mainly studied at the level of the heart,[30,31] where it is likely to play a physiological role, since both cholinergic and noradrenergic pathways innervate cardiac muscle.[32] The close proximity of cholinergic and noradrenergic terminals in the heart is compatible with the view that the muscarinic receptors on the sympathetic terminals in atria can be activated by acetylcholine released from the parasympathetic nerves.

Stimulation of presynaptic dopamine receptors of the D2 subtype by dopamine and other dopamine-receptor agonists mediate an inhibition of norepinephrine release. These effects have been demonstrated under *in vitro* as well under *in vivo* conditions.[13–17] The reduction of norepinephrine release, elicited by dopamine-receptor agonists, is accompanied by a decrease in the magnitude of the end-organ responses to sympathetic nerve stimulation.[13,14,16] The presynaptic inhibitory dopamine receptors are therefore a target for exogenously applied agonists (e.g., pergolide) which inhibit norepinephrine release and can reduce blood pressure and heart rate.[13,14] In contrast to presynaptic α_2-adrenoceptors, which play a physiological role in modulating noradrenergic neurotransmission, the presynaptic inhibitory dopamine receptors, although pharmacologically relevant, do not play a physiologically significant role since blockade of presynaptic D2 receptors (e.g., sulpiride) fails to enhance the stimulation-evoked release of norepinephrine, although this D2 receptor antagonist effectively blocks the inhibitory actions of dopamine-receptor agonists.[13,14,33]

The presynaptic inhibitory D2 receptors on noradrenergic nerve terminals differ pharmacologically from the postsynaptic D1 receptors, which mediate vasodilatation in the mesenteric and renal vascular beds.[16]

Among the other types of presynaptic receptors that can modify the release of norepinephrine, transynaptic mechanisms may involve the formation at the level of the effector organ of mediators that can act presynaptically to inhibit the release of norepinephrine. The inhibition of peripheral noradrenergic transmission by prostaglandins of the E series is well documented.[34,35] The stimulus that triggers the local release of prostaglandins is probably related to the activation by norepinephrine of the postsynaptic adrenoceptor.

Another transynaptic mechanism may involve the inhibition of the release of norepinephrine through the activation of adenosine receptors.[36–38] The possible cotransmitter role for ATP in noradrenergic neurons[39–42] may be relevant to the presynaptic

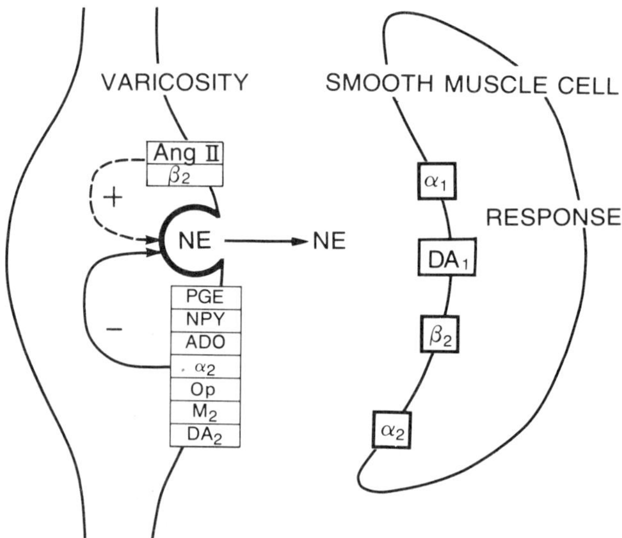

FIGURE 2. Schematic representation of pre- and postsynaptic receptors in a noradrenergic neuroeffector junction in the peripheral nervous system. Noradrenergic varicosity: activation of angiotensin II (Ang II) and β-adrenoceptors of the β_2 subtype enhances norepinephrine release. On the other hand, receptors for prostaglandins of the E series (PGE), for neuropeptide Y (NPY), adenosine (ADO), norepinephrine (α_2), opioids (Op), acetylcholine (M$_2$), and dopamine (DA$_2$) trigger a decrease in norepinephrine release. At the postjunctional level (smooth muscle cell), α-adrenoceptors (α_1) mediate contraction while dopamine (DA$_1$) and β_2-adrenoceptors (β_2) mediate relaxation. In the heart the β-adrenoceptor is of β_1 subtype (not shown in the figure). Adrenergic receptors of the α_2 subtype which also mediate contraction are predominantly located extrajunctionally, although under certain conditions they can be stimulated by norepinephrine release from the varicosity.

inhibitory effects of adenosine on the release of norepinephrine. The purinergic receptor involved in this inhibitory mechanism has been identified as the P$_2$ subtype,[43–45] although more recently a new P$_3$ purinergic inhibitory receptor has been proposed in the rat caudal artery.[46]

Opioid peptides, vasopressin, somatostatin, and neuropeptide Y, have all been proposed to act on specific receptors on peripheral noradrenergic nerve terminals, leading to changes in transmitter release.

Presynaptic inhibitory opioid receptors modulating norepinephrine release are present in several peripheral tissues like the cat nictitating membrane.[47] At present there is no evidence as to whether vasopressin may activate presynaptic receptors to modulate noradrenergic transmission and the question remains open also for somatostatin as far as peripheral noradrenergic neurons is concerned. On the other hand, it has been shown that neuropeptide Y inhibits the release of norepinephrine in some tissues.[48–50]

FACILITATORY HETERORECEPTORS

Receptor-mediated mechanisms that enhance presynaptically the release of norepinephrine appear to be particularly relevant in the cardiovascular system. There is now considerable evidence for the presence of presynaptic facilitatory β-adrenoceptors on peripheral noradrenergic nerve terminals.[10,17–19] Exposure to low concentrations of isoproterenol or epinephrine enhance the stimulation-evoked release of norepinephrine and this effect is blocked by propranolol,[1,19] but not by betaxolol[2] indicating that the β-adrenoceptor involved in the facilitation of noradrenergic neurotransmission corresponds to the β_2 subtype (FIGURE 1). The fact that epinephrine preferentially stimulates the presynaptic facilitatory β_2-adrenoceptor was proposed as associated to the pathogenesis of some forms of hypertension. Nevertheless, it remains to be established whether the clinical efficacy of β-adrenoceptor-blocking drugs in hypertension involves to some extent the blockade of presynaptic facilitatory β-adrenoceptors.[18,5]

More recently a facilitatory role of purinergic receptors on noradrenergic transmission has been postulated in the rabbit ear artery. The receptor involved in this particular case appears to be of the P_2 subtype.[52] Another facilitatory mechanism at the level of sympathetic nerve endings involves angiotensin II,[3,20,21] and it is of interest that inhibitors of the angiotensin converting enzyme (ACE) such as captopril, which have antihypertensive properties, can prevent the angiotensin II–mediated facilitation of noradrenergic neurotransmission.[20,21] There are indications for the existence of subtypes of angiotensin II receptors but it is not known which one corresponds to the presynaptic receptor that mediates facilitation of norepinephrine release.

PRESYNAPTIC SITES OF NEURONAL UPTAKE IN NORADRENERGIC NEURONS

The termination of the action of norepinephrine involves the process of neuronal uptake, namely, the active sodium-dependent transport of the neurotransmitter across the membrane of the nerve terminal. In analogy with presynaptic receptors that modulate the release of norepinephrine, the transporter for the neuronal uptake of the transmitter possesses many of the properties of a pharmacological receptor.[53–55] This transporter site in peripheral noradrenergic neurons (FIGURE 1) was labeled with ^{3}H-desipramine, an inhibitor of neuronal uptake of the transmitter.[53–55] More recently a high-affinity recognition site associated with the epinephrine transporter has also been labeled with ^{3}H-desipramine in the frog heart in which epinephrine is the transmitter.[56]

EVIDENCE FOR THE PRESENCE OF PRESYNAPTIC RECEPTORS IN HUMAN PERIPHERAL SYMPATHETIC NEURONS

The evidence in support of the presence of presynaptic α_2-adrenoceptors mediating the negative feedback control of norepinephrine release in human vasoconstrictor nerves was reported in isolated superfused field-stimulated biopsy specimens of human peripheral arteries and veins.[57,58] An enhancement of norepinephrine release was demonstrated in the presence of α-adrenoceptor antagonists, while an inhibition of the release induced by exposure to exogenous norepinephrine was observed.[57,58] More recent studies have confirmed and extended these observations.[59]

In agreement with the existence of facilitatory β-adrenoceptors in peripheral human noradrenergic nerve endings, it was shown that isoprenaline increases the release of norepinephrine during nerve stimulation at low frequency in the human oviduct.[60] Both isoprenaline and low concentration of norepinephrine increase the release of norepinephrine elicited by nerve stimulation from strips of human omental arteries and veins.[58] On the basis of these observations it was postulated that presynaptic β-adrenoceptors can be activated by circulating catecholamines and may subserve the function of enhancing the release of sympathetic transmitter during conditions of increased secretion of adrenomedullary hormone.[58]

OPEN QUESTIONS IN THE FIELD OF PRESYNAPTIC MODULATION OF NOREPINEPHRINE RELEASE

In addition to the physiological relevance of presynaptic α_2-adrenoceptors modulating norepinephrine release, there are other questions that remain to be explored in future research. The first one involves the presynaptic effects of the peptide cotransmitters like NPY, enkephalins, and galanine, which may be released simultaneously with norepinephrine. The second one involves the existence of subtypes of α_2-adrenoceptors and of possible pharmacological differences between presynaptic and postsynaptic α_2-adrenoceptors.[61] Third is the existence of imidazoline-specific sites as targets of the action of α_2-adrenoceptor agonists like clonidine[62] and finally there are the reports on the presence of a low molecular weight clonidine-displacing substance (CDS) in the bovine brain which is noncatecholaminergic in nature and which competes for ^{3}H-clonidine and ^{3}H-rauwolscine binding.[63,64]

CONCLUSIONS

In addition to the classical receptors that mediate the responses of the effector organ, it is now well established that presynaptic autoreceptors as well as receptors to other mediators are located in the sympathetic nerve terminals and are associated with the modulation of norepinephrine release. A separate receptor-mediated process, different from the autoreceptor that modulates transmitter release, involves the presynaptic transporter for norepinephrine reponsible for the termination of action of the released transmitter.

The discovery of presynaptic receptors has contributed to the understanding of the physiology and pharmacology of the sympathetic nerve ending. The fact that modulation of transmitter release through α_2-adrenoceptors occurs predominantly at low and intermediate frequencies of stimulation and that blockade of these receptors results in the enhancement of transmitter release points to the physiological relevance of this autoregulatory mechanism.

From the pharmacological point of view, the stimulation or blockade by drugs of presynaptic receptors in noradrenergic nerve endings may offer a novel approach to modulate sympathetic neurotransmission. Presynaptic receptors have been demonstrated in many species including man where they represent a target of action for drugs that act through the modulation of sympathetic tone.

REFERENCES

1. LANGER, S. Z. 1981. Presynaptic regulation of the release of catecholamines. Pharmacol. Rev. **32:** 337–362.
2. LANGER, S. Z. & A. M. GALZIN. 1983. Beta-adrenoceptors and noradrenergic neurotransmission: effects of betaxolol on the stimulation evoked release of ^{3}H-noradrenaline in isolated rat atria and rabbit hypothalamic slices. *In* Betaxolol: a New β_1-Adrenoceptor Antagonist. P. L. Morselli, J. R. Kilborn, I. Cavero, D. C. Harrison & S. Z. Langer. Eds.: **1:** 21–30. Raven Press. New York, N.Y.
3. WESTFALL, T. C. 1977. Local regulation of adrenergic neurotransmission. Physiol. Rev. **57:** 659–728.
4. STARKE, K., M. GÖTHERT & H. KILBINGER. 1989. Modulation of neurotransmitter release by presynaptic autoreceptors. Physiol. Rev. **69:** 864–989.
5. FLAVAHAN, N. A., T. J. RIMELE & J. P. COOKE. 1984. Characterization of postjunctional alpha$_1$ and alpha$_2$-adrenoceptors activated by exogenous or nerve released norepinephrine in the canine saphenous vein. J. Pharmacol. Exp. Ther. **230:** 699–705.
6. HICKS, P. E., I. C. MEDGETT & S. Z. LANGER. 1984. Postsynaptic alpha$_2$-adrenoceptors mediated vasoconstriction in SHR tail arteries in-vitro. Hypertension **6:** 12–18.
7. LANGER, S. Z. & N. B. SHEPPERSON. 1982. Recent developments in vascular smooth muscle pharmacology—the postsynaptic alpha$_2$-adrenoceptor. Trends Pharmacol. Sci. **3:** 440–444.
8. MEDGETT, I. C. & S. Z. LANGER. 1984. Heterogeneity of smooth muscle alpha-adrenoceptors in rat tail artery in-vitro. J. Pharmacol. Exp. Ther. **229:** 823–830.
9. MEDGETT, I. C., P. E. HICKS & S. Z. LANGER. 1984. Smooth muscle alpha$_2$-adrenoceptors mediate vasoconstrictor responses to exogenous norepinephrine and to sympathetic stimulation to a greater extent in SHR than in WKY rat tail arteries. J. Pharmacol. Exp. Ther. **231:** 159–165.
10. LANGER, S. Z. 1976. The role of alpha and beta-presynaptic receptors in the regulation of noradrenaline release elicited by nerve stimulation. Clin. Sci. Mol. Med. **51:** 423s–426s.
11. LANGER, S. Z. 1981. Presence and physiological role of presynaptic inhibitory alpha$_2$-adrenoceptors in guinea-pig atria. Nature **294.** 671–672.
12. STORY, D. F., M. W. McCULLOCH, M. J. RAND & C. A. STANFORD STARR. 1981. Conditions required for the inhibitory feedback loop in noradrenergic transmission. Nature **293:** 62–65.
13. CAVERO, I., F. LEFEVRE-BORG & R. GOMENI. 1981. Blood pressure lowering effects of *N*,*N*-di-*n*-propyldopamine in rats: evidence for stimulation of peripheral dopamine receptors leading to inhibition of sympathetic vasculature. J. Pharmacol. Exp. Ther. **218:** 515–524.
14. CAVERO, I., F. LEFEVRE-BORG, F. LHOSTE, F. SABATIER, C. RICHER & J. F. GIUDICELLI. 1984. Pharmacological haemodynamic and autonomic nervous system mechanisms responsible for the blood pressure and heart rate lowering effects of pergolide in rats. J. Pharmacol. Exp. Ther. **228:** 779–791.
15. ENERO, M. A. & S. Z. LANGER. 1975. Inhibition by dopamine of ^{3}H-noradrenaline release elicited by nerve stimulation in the isolated cat's nictitating membrane. Naunyn Schmiedebergs Arch. Pharmacol. **289:** 179–203.
16. SHEPPERSON, N. B., N. DUVAL, R. MASSINGHAM & S. Z. LANGER. 1982. Differential blocking effects of several dopamine receptor antagonists for peripheral pre- and postsynaptic dopamine receptors in the anaesthetized dog. J. Pharmacol. Exp. Ther. **221:** 753–761.
17. ADLER-GRASCHINSKY, E. & S. Z. LANGER. 1975. Possible role of a beta adrenoceptor in the regulation of noradrenaline release by nerve stimulation. Br. J. Pharmacol. **53:** 43–50.
18. MISU, Y. & T. KUBO. 1983. Presynaptic beta-adrenoceptors. Trends Pharmacol. Sci. **4:** 506–508.
19. STÄRJNE, L. & J. BRUNDIN. 1976. Beta$_2$-adrenoceptors facilitating noradrenaline secretion from human vasoconstrictor nerves. Acta Physiol. Scand. **94:** 88–93.
20. ANTONACCIO, M. J. & L. KERWIN. 1981. Pre- and post-junctional inhibition of vascular sympathetic function by captopril in SHR. Hypertension **3:** 154–162.

21. CLOUGH, D. P., R. HATTON, J. R. KEDDIE & M. G. COLLIS. 1982. Hypotensive action of captopril in spontaneously hypertensive and normotensive rats. Interference with neurogenic vasoconstriction. Hypertension **4:** 764–772.

22. LANGER, S. Z. 1973. The regulation of transmitter release elicited by nerve stimulation through a presynaptic feedback mechanism. *In* Frontiers in Catecholamines Research. E. Usdin & S. Snyder, Eds.: 543–549. Pergamon Press. New York, N.Y.

23. CUBEDDU, L. X., E. M. BARNES, S. Z. LANGER & N. WEINER. 1974. Release of norepinephrine and dopamine-β-hydroxylase by nerve stimulation. I. Role of neuronal and extraneuronal uptake and of alpha presynaptic receptors. J. Pharmacol. Exp. Ther. **190:** 431–450.

24. DUBOCOVICH, M. L. & S. Z. LANGER. 1974. Negative feed-back regulation of noradrenaline release by nerve stimulation in the perfused cat's spleen: differences in potency of phenoxybenzamine in blocking the pre- and post-synaptic adrenergic receptors. J. Physiol. London **237:** 505–519.

25. LANGER, S. Z. 1974. Presynaptic regulation of catecholamine release. Biochem. Pharmacol. **23:** 793–1800.

26. STARKE, K. & S. Z. LANGER. 1979. A note on terminology for presynaptic receptors. *In* Presynaptic Receptors. S. Z. Langer, K. Starke & M. L. Dubocovich, Eds.: 1–4. Pergamon Press. New York, N.Y.

27. KOBILKA, B. K., H. MATSUI, T. S. KOBILKA, T. L. YANG-FENG, U. FRANCKE, M. G. CARON, R. J. LEFKOWITZ & J. W. REGAN. 1987. Cloning, sequencing and expression of the gene coding for the human platelet α_2-adrenergic receptors. Science **238:** 650–656.

28. ANGUS, J. A., A. BOBIK, G. P. JACKMAN, I. J. KOPIN & P. KORNER. 1984. Role of auto-inhibitory feedback in cardiac sympathetic transmission assessed by simultaneous changes in ^{3}H-efflux and atrial rate in guinea pig atrium. Br. J. Pharmacol. **81:** 201–214.

29. HICKS, P. E., M. NAJAR, M. VIDAL & S. Z. LANGER. 1986. Possible involvement of presynaptic α_1-adrenoceptors in the effects of idazoxan and prazosin on ^{3}H-noradrenaline release from tail arteries of SHR. Naunyn Schmiedebergs Arch. Pharmacol. **333:** 354–361.

30. LÖFFELHOLZ, K. & E. MUSCHOLL. 1970. Inhibition by parasympathetic nerve stimulation of the release of the adrenergic transmitter. Arch. Exp. Pathol. Pharmacol. **267:** 181–184.

31. FOZARD, J. R. & E. MUSCHOLL. 1972. Effects of several muscarinic agonists on cardiac performance and the release of noradrenaline from sympathetic nerves of the perfused rabbit heart. Br. J. Pharmacol. **45:** 616–629.

32. MUSCHOLL, E. 1979. Presynaptic muscarinic receptors and inhibition of release, *In* The Release of Catecholamines from Adrenergic Neurons. D. M. Paton, Ed.: 87–110. Pergamon Press. Oxford, England.

33. MASSINGHAM, R., M. L. DUBOCOVICH & S. Z. LANGER. 1980. The role of presynaptic receptors in the cardiovascular actions of *N,N*-di-*n*-propyldopamine in the cat and dog. Naunyn Schmiedebergs Arch. Pharmacol. **314:** 17–28.

34. DUBOCOVICH, M. L. & S. Z. LANGER. 1975. Evidence against a physiological role of prostaglandins in the regulation of noradrenaline release in the cat spleen. J. Physiol. London **251:** 737–762.

35. HEDQVIST, P. 1976. Further evidence that prostaglandins inhibit the release of noradrenaline from adrenergic nerve terminals by restriction of availability of calcium. Br. J. Pharmacol. **58:** 599–603.

36. LANGER, S. Z. 1977. Presynaptic receptors and their role in the regulation of transmitter release. Sixth Gaddum Memorial Lecture. Br. J. Pharmacol. **60:** 481–497.

37. LUCHELLI-FORTIS, M. A., B. B. FREDHOLM & S. Z. LANGER. 1979. Release of radioactive purines from cat nictitating membrane labeled with ^{3}H-adenine. Eur. J. Pharmacol. **58:** 389–397.

38. LUCHELLI-FORTIS, M. A., B. B. FREDHOLM & S. Z. LANGER. 1981. Evidence against the presence of presynaptic inhibitory adenosine receptors in the cat nictitating membrane. J. Pharmacol. Exp. Ther. **219:** 235–242.

39. DUVAL, N., P. E. HICKS & S. Z. LANGER. 1985. Inhibitory effects of alpha, beta-methylene ATP on nerve-mediated contractions of the nictitating membrane in reserpinised cats. Eur. J. Pharmacol. **110:** 373–377.

40. SNEDDON, P. & G. BURNSTOCK. 1984. Inhibition of excitatory junctional potentials in

guinea-pig vas deferens by alpha, beta-methylene ATP: further evidence for ATP and noradrenaline as co-transmitters. Eur. J. Pharmacol. **100:** 85–90.

41. VIDAL, M., P. E. HICKS & S. Z. LANGER. 1986. Effects of alpha-beta-methylene-ATP on responses to nerve stimulation in SHR and WKY tail arteries. Naunyn Schmiedebergs Arch. Pharmacol. **332:** 384–390.

42. BULLOCH, J. M. & K. STARKE. 1990. Presynaptic α_2-autoinhibition in the vascular neuroeffector junction where ATP and noradrenaline act as co-transmitters. Br. J. Pharmacol. **99:** 279–284.

43. FUJIOKA, M. & D. M. CHEUNG. 1987. Autoregulation of neuromuscular transmission in the guinea-pig saphenous artery. Eur. J. Pharmacol. **139:** 147–153.

44. LUNDBERG, J. M., A. RUDEHILL, A. SOLLEVY & B. HAMBERGER. 1989. Evidence for cotransmitter role of neuropeptide Y in the pig spleen. Br. J. Pharmacol. **96:** 675–687.

45. STJARNE, L. & P. ASTRAND. 1985. Relative pre- and post junctional roles of noradrenaline and adenosine 5-triphosphate as neurotransmitters of the sympathetic nerves of guinea-pig and mouse vas deferens. Neuroscience **14:** 929–946.

46. SHINOZUKA, K., R. A. BJUR & D. P. WESTFALL. 1986. Characterization of prejunctional purinoceptors on adrenergic nerves of the rat caudal artery. Naunyn Schmiedebergs Arch. Pharmacol. **338:** 221–227.

47. DUBOCOVICH, M. L. & S. Z. LANGER. 1980. Pharmacological differentiation of presynaptic inhibitory alpha-adrenoceptors and opiate receptors in the cat nictitating membrane. Br. J. Pharmacol. **70:** 383–393.

48. DÄHLOF, C. P., K. DÄHLOF, K. TATEMOTO & J. M. LUNDBERG. 1985. Neuropeptide Y (NPY) reduces field stimulation–evoked release of noradrenaline and enhances force of contraction in the rat portal vein. Naunyn Schmiedebergs Arch. Pharmacol. **328:** 327–330.

49. HAASS, M., B. CHENG, G. RICHARDT & E. LANG. 1989. Characterization and presynaptic modulation of stimulation-evoked exocytotic co-release of noradrenaline and neuropeptide Y in guinea-pig heart. Naunyn Schmiedebergs Arch. Pharmacol. **339:** 71–78.

50. STJARNE, L., J. M. LUNDBERG & P. ASTRAND. 1986. Neuropeptide Y—a cotransmitter with noradrenaline and adenosine 5-triphosphates in the sympathetic nerves of the mouse vas deferens. A biochemical physiological and electropharmacological study. Neuroscience. **18:** 151–166.

51. RAND, M. J., H. MAJEWSKI, M. W. McCULLOCH & D. F. STORY. 1978. An adrenaline mediated positive feedback loop in sympathetic transmission and its possible role in hypertension. *In* Presynaptic Receptors. S. Z. Langer, K. Starke & M. L. Dubocovich, Eds.: 263. Oxford University Press. Oxford, England.

52. MIYAHARA, H. & H. S SUZUKI. 1987. Pre and postjunctional effects of adenosine triphosphate on noradrenergic transmission in the rabbit ear artery. J. Physiol. London **389:** 423–440.

53. RAISMAN, R., M. SETTE, C. PIMOULE, M. S. BRILEY & S. Z. LANGER. 1982. High affinity ^{3}H desipramine binding in the peripheral and central nervous system: a specific site associated with the neuronal uptake of noradrenaline. Eur. J. Pharmacol. **78:** 345–351.

54. LANGER, S. Z., L. TAHRAOUI, R. RAISMAN, S. ARBILLA, M. NAJAR & J. DEDEK. 1984. ^{3}H-Desipramine labels a site associated with the neuronal uptake of noradrenaline in the peripheral and central nervous system. *In* Neuronal and Extraneuronal Events. W. W. Fleming, K. H. Graefe, S. Z. Langer & N. Weiner, Eds.: 37–49. Raven Press. New York, N.Y.

55. LANGER, S. Z. & S. ARBILLA. 1988. Different presynaptic receptors modulate uptake and release of catecholamines. *In* Progress in Catecholamine Research. Part A. Basic Aspects and Peripheral Mechanisms. A. Dahlström, R. H. Belmaker & M. Sandler, Eds.: 7–12. Alan R. Liss Inc. New York, N.Y.

56. PIMOULE, C., H. SCHOEMAKER & S. Z. LANGER. 1987. ^{3}H-Desipramine labels with high affinity the neuronal transporter for adrenaline in the frog heart. Eur. J. Pharmacol. **137:** 277–280.

57. STJARNE, L. & K. GRIPE. 1973. Prostaglandin-dependent and -independent feedback control of noradrenaline secretion in vasoconstrictor nerves of normotensive human subjects. A preliminary report. Naunyn Schmiedebergs Arch. Pharmacol. **280:** 441–446.

58. STJARNE, L. & J. BRUNDIN. 1975. Dual adrenoceptor-mediated control of noradrenaline secretion from human vasoconstrictor nerves: facilitation by β-receptors and inhibition by α-receptors. Acta Physiol. Scand. **94:** 139–141.
59. JIE, K., P. VAN BRUMMELEN, P. VERMEY, P. B. M. W. M. TIMMERMANS & P. A. VAN ZWIETEN. 1987. Modulation of noradrenaline release by peripheral presynaptic α_2-adrenoceptors in humans. J. Cardiovasc. Pharmacol. **9:** 407–413.
60. HEDQVIST, P. & A. MOAWAD. 1975. Presynaptic α and β-adrenoceptor mediated control of noradrenaline release in human oviduct. Acta Physiol. Scand. **95:** 494–496.
61. LANGER, S. Z. & H. SCHOEMAKER. 1989. Alpha-adrenoceptor subtypes in blood vessels: physiology and pharmacology. Clin. Exp. Hypertension **A11**(Suppl. 1): 21–30.
62. BOUSQUET, P. & J. SCHWARTZ. 1983. Alpha adrenergic drugs: pharmacological tools for the study of the central vasomotor control. Biochem. Pharmacol. **32:** 1459–1465.
63. ATLAS, D. & Y. BURNSTEIN. 1984. Isolation and partial purification of a clonidine-displacing endogenous brain substance. Eur. J. Biochem. **144:** 287–293.
64. ATLAS, D., S. DIAMANT, H. M. FALES & L. PANNELL. 1987. The brain's own clonidine: purification and characterization of endogenous clonidine displacing substance from brain. J. Cardiovasc. Pharmacol. **10:** S122–S127.

Presynaptic Regulation of Dopamine Release

Implications for the Functional Organization of the Basal Ganglia

M-F. CHESSELET

Department of Pharmacology
University of Pennsylvania
John Morgan Building
36 and Hamilton Walk
Philadelphia, Pennsylvania 19104

The hypothesis that neurotransmitter release can be regulated presynaptically in the central nervous system has elicited widespread interest since the description of such regulation in the peripheral nervous system.[1-3] Because of the anatomical organization of the central nervous system, presynaptic regulation is particularly difficult to demonstrate convincingly in this tissue. The technical difficulties encountered led to numerous controversies, but data obtained during the past few years strongly argue for the existence of central presynaptic regulation. Its functional significance, however, is still a matter of speculation.

The goal of this brief review is not to survey all evidence for or against the existence of presynaptic regulation in the central nervous system, but to highlight a few examples in which research on presynaptic regulation led to the development of novel concepts regarding the control of neurotransmitter release in the brain. These examples will be taken from studies on the dopaminergic nigrostriatal pathway, a system that has been extensively studied with respect to presynaptic regulation of neurotransmitter release (see Reference 4 for review). This is largely due to the fact that these neurons, which have their cell bodies in the pars compacta of the substantia nigra, terminate in the striatum (caudate-putamen), an area of the brain that does not contain intrinsic catecholaminergic neurons. Consequently, regulation of dopamine release observed in striatal sections or synaptosomal preparations is most certainly operating through presynaptic mechanisms. These can be either direct, involving receptors located on the dopaminergic nerve endings, or indirect, mediated through interneurons or other nerve terminals which, in turn, modulate dopamine release through an action on presynaptic receptors.

Earlier studies of the regulation of dopamine release in the striatum relied upon the use of radioisotopes in order to reach sufficient sensitivity to detect the small amounts of dopamine released. We used a method based on the addition of ^{3}H-tyrosine to the superfusion medium of striatal slices. The ^{3}H-dopamine endogenously formed from the labeled tyrosine was isolated from precursors and metabolites with ion-exchange and alumina columns.[5] An additional advantage of this method is the possibility of obtaining simultaneously an index of dopamine synthesis by measuring the amount of ^{3}H-water formed during the hydroxylation of ^{3}H-tyrosine.[6]

With this approach, several neurotransmitters normally present in the striatum, or agonist of their receptors, were shown to modulate dopamine release when applied to striatal slices (for review see Reference 4). These agents include dopamine itself, which reduces its own release when applied in the presence of an uptake inhibitor, acetylcholine, gamma-aminobutyric acid (GABA), glycine, beta-adrenergic agonists, glutamate, somatostatin, angiotensin, D-Ala$_2$, met-enkephalin, and agonists of delta opiate receptors. Substance P has also recently been shown to increase the release of newly synthesized dopamine in striatal slices.[7] These effects are calcium dependent, and, when available, are blocked by antagonists of the corresponding receptors. The present review will address some issues related to the effects of opioid peptides, acetylcholine, beta-adrenergic agonists, and glutamate on striatal dopamine release. These examples were chosen because they relate to the morphological basis for presynaptic regulation in the central nervous system, the biochemical mechanisms involved in this phenomenon and its functional significance.

MORPHOLOGICAL EVIDENCE FOR PRESYNAPTIC RECEPTORS IN THE STRIATUM

The peptidase-resistant analogue of enkephalin, D-Ala$_2$, met-enkephalinamide, increased the release of newly synthesized dopamine in rat striatal slices in a dose-dependent manner.[8] This effect was reproduced by selective agonists of delta but not mu opiate receptors[8,9] and, as expected from an action on delta opiate receptors, was blocked by high concentrations of naloxone. The effect of enkephalin analogues on striatal dopamine release *in vitro* was still observed in the presence of tetrodotoxin, suggesting, but not proving, that it is mediated through an action on opiate receptors located on dopaminergic terminals. This hypothesis is supported by more recent evidence by E. Hamel and A. Beaudet who showed that a substantial number of opiate binding sites in the striatum are associated with axon terminals.[10]

This work provided the first morphological evidence that receptors may be associated with presynaptic terminals in the caudate-putamen. Whether these receptors are accessible to endogenous neurotransmitters under physiological conditions is still unclear. Experimental observations that pharmacological blockade of receptors believed to be associated with dopaminergic terminals modifies dopamine release *in vivo* support the hypothesis that presynaptic control does occur *in vivo*.[11,12] There is, however, no evidence that axoaxonic synapses exist in the rat or primate striatum.[13] It is conceivable that neurotransmitter released from nearby nerve endings can act upon presynaptic receptors without the occurrence of classical synapses. This is substantiated by electron microscopic evidence that close appositions between axon terminals exist in the striatum.[14] These could constitute the morphological basis for regulation involving presynaptic heteroreceptors in the caudate-putamen.

Direct morphological evidence of the presynaptic location of other types of receptors in the striatum is still missing. Other receptors likely to be associated with dopaminergic nerve terminals in the rat, however, include those mediating the effects of acetylcholine on dopamine release. Pharmacological evidence suggests that both muscarinic and nicotinic receptors are involved in the action of acetylcholine in rat striatal slices.[5,15] The effects of muscarinic and nicotinic agonists on striatal dopamine release *in vitro* persist in the presence of tetrodotoxin and are also observed in synaptosomal preparations,[16,17] suggesting that both muscarinic and nicotinic receptors are present on dopaminergic nerve endings.

The presence of nicotinic receptors in the brain has long been controversial, in part because their pharmacological characteristics differ from those of nicotinic receptors in the periphery.[18] Recent data, however, show that mRNAs encoding various subunits of the nicotinic receptor are present in the brain and, in particular, can be detected by *in situ* hybridization histochemistry in neurons of the substantia nigra pars compacta.[19]

This is compatible with the hypothesis that nicotinic receptors are indeed synthesized by dopaminergic neurons. In addition, ligand binding studies suggest that these receptors are transported to the nerve terminals in the striatum because 6-hydroxydopamine-induced lesions of the dopaminergic nigrostriatal neurons result in a decrease in the number of nicotinic binding sites in the striatum ipsilateral to the lesion.[20] A functional role for these receptors is further suggested by the observation that nicotinic antagonists decrease dopamine turnover *in vivo,* suggesting that dopamine release is tonically regulated by nicotinic mechanisms.[12]

PRESYNAPTIC REGULATION AND SECOND MESSENGER SYSTEMS

Little is known yet about the second messenger systems associated with presynaptic receptors. An interesting possibility is that they may either have different pharmacological properties, or be coupled to different second messenger systems than postsynaptic receptors are. The beta-adrenergic agonist isoproterenol increased the release of dopamine in slices of the rat striatum.[13] This effect was tetrodotoxin resistant and antagonized by propranolol, but not by the alpha antagonist phentolamine. Despite the fact that many beta-adrenergic actions are mediated by an increased level of intracellular cAMP resulting from an activation of adenylyl cyclase, isoproterenol-induced increase in striatal dopamine release appears to be independent of the activation of adenylyl cyclase by this compound.[21] In view of recent data showing that stimulation of beta-adrenergic receptors can lead to a direct modulation of ion channels, unrelated to its effects on cAMP levels, it is conceivable that the presynaptic actions of beta-adrenergic agonists on dopamine release in the striatum may be mediated through such mechanisms.[22]

FUNCTIONAL IMPLICATIONS OF PRESYNAPTIC REGULATION OF STRIATAL DOPAMINE RELEASE

Glutamate, a major excitatory amino acid neurotransmitter in the central nervous system increases dopamine release in rat striatal slices. The effect is tetrodotoxin insensitive, and not affected by cholinergic antagonists.[23] Recent studies have shown that several types of glutamate receptors (NMDA, quisqualate, and kainate) are involved in the effects of glutamate on striatal dopamine release[24-26] (Glowinski *et al,* personal communication). Much evidence suggests that glutamate or aspartate is the neurotransmitter of corticostriatal neurons.[27-29] *In vivo* studies have shown that stimulation of the cerebral cortex results in an increase in striatal dopamine release, compatible with a presynaptic effect of glutamate released from corticostriatal neurons.[30,31] In addition, the increase in striatal dopamine release elicited by local application of GABA in thalamic motor nuclei[32] is blocked by application of glutamate antagonists into the caudate nucleus and by unilateral lesion of the sensory-motor cortex.[33] This supports the hypothesis that a presynaptic regulation of dopamine release by corticostriatal afferents mediates the effects of alterations in other neuronal systems on striatal dopamine. It has been recently postulated that this presynaptic modulation by glutamate is essential for the regulation of tonic dopamine release in the striatum, and that a selective attenuation of tonic dopamine release secondary to a decrease in corticostriatal activity may play a role in the etiology of schizophrenia.[34]

In conclusion, a number of *in vivo* studies suggest that presynaptic regulation of dopamine release is part of the physiological control of dopaminergic neurotransmission in the striatum. The existence of modulations of dopamine release occurring within the striatum, at the level of dopaminergic nerve terminals, implies that inputs to the striatum, in addition to those that impinge on the dopaminergic cell bodies and dendrites in the substantia nigra, may play a critical role in the regulation of striatal dopaminergic neurotransmission and that modulation of firing activity is not the sole factor involved in the regulation of the release of this neurotransmitter.[34-36]

In view of the complex anatomical organization of the striatum, which is comprised of two intermingled compartments, the striosomes and the extrastriosomal matrix,[37] topographical differences in presynaptic regulation could exist within this region. For example, those dopaminergic neurons that terminate in the extrastriosomal matrix of the striatum[38,39] are in a position to be regulated by somatostatin,[40] a neuropeptide present in much greater abundance in the matrix than in the striosomes.[41,42] The hypothesis that regionally different presynaptic regulation exists in the striatum is further substantiated by recent evidence that the modalities of cholinergic regulation of dopamine release differ in the striosomes and the extrastriosomal matrix of the cat caudate nucleus.[43] Thus presynaptic regulation could contribute to an array of local modulations of the transmitter release related to the anatomical specificity of neuronal organization in the striatum.

Taken together, the numerous data documenting local effects of neurotransmitters, or agonists of their receptors, on the release of dopamine and other brain neurotransmitters,[4] and the more recent evidence from *in vivo* experiments that have brought strong support for a physiological role of this phenomenon, suggest that the occurrence of presynaptic regulation must be taken into account for models of neurotransmitter interaction in the brain. A better understanding of its role may shed new light into some pathological alterations of brain function.[34]

REFERENCES

1. WESTFALL, T. C. 1977. Local regulation of adrenergic neurotransmission. Physiol. Rev. **57:** 659–728.
2. STARKE, K. 1979. Presynaptic regulation of release in the central nervous system. *In* The Release of Catecholamine from Adrenergic Neurons. D. M. Paton, Ed.: 143–183. Pergamon Press. Oxford, England.
3. ERULKAR, S. D. 1983. The modulation of neurotransmitter release at synaptic junctions. Rev. Physiol. Biochem. Pharmacol. **98:** 64–175.
4. CHESSELET, M-F. 1984. Presynaptic regulation of neurotransmitter release in the brain: facts and hypothesis. Neuroscience **12:** 347–375.
5. GIORGUIEFF, M-F., M-L. LE FLOC'H, J. GLOWINSKI & M-J. BESSON. 1977. Involvement of cholinergic presynaptic receptors of nicotinic and muscarinic types in the control of the spontaneous release of dopamine from striatal dopaminergic terminals in the rat. J. Pharmacol. Exp. Ther. **200:** 535–544.
6. WESTFALL, T. C., M. J. BESSON, M-F. GIORGUIEFF & J. GLOWINSKI. 1976. The role of presynaptic receptors in the release and synthesis of ³H-dopamine by slices of rat striatum. Naunyn Schmiedebergs Arch. Pharmacol. **292:** 279–287.
7. PETIT, F. & J. GLOWINSKI. 1986. Stimulatory effect of substance P on the spontaneous release of newly synthesized ³H-dopamine from rat striatal slices: a tetrodotoxin-sensitive process. Neuropharmacol. **25:** 1015–1021.
8. LUBETZKI, C., M-F. CHESSELET & J. GLOWINSKI. 1982. Modulation of dopamine release in rat striatal slices by delta opiate agonists. J. Pharmacol. Exp. Ther. **222:** 435–440.
9. PETIT, F., M. HAMON, M-C. FOURNIE-ZALUSKI, B. ROQUES & J. GLOWINSKI. 1986. Further evidence for a role of delta-opiate receptors in the presynaptic regulation of newly synthesized dopamine release. Eur. J. Pharmacol. **126:** 1–10.
10. HAMEL, E. & A. BEAUDET. 1984. Localization of opioid binding sites in rat brain by electron microscopic autoradiography. J. Electron Microscopy Tech. **1:** 317–329.

11. REISINE, T. D., M-F. CHESSELET & J. GLOWINSKI. 1982. A role for striatal beta-adrenergic receptors in the regulation of dopamine release. Brain Res. **241:** 123–130.

12. HAIKALA, H. & L. AHTEE. 1988. Antagonism of the nicotine-induced changes of the striatal dopamine metabolism in mice by mecamylamine and pempidine. Naunyn Schmiedebergs Arch. Pharmacol. **338:** 169–173.

13. KEMP, J. M. & T. P. S. POWELL. 1971. The site of termination of afferent fibers in the caudate nucleus. Philos. Trans. R. Soc. London Ser. B **262:** 413–427.

14. BOUYER, J. J., D. H. PARK, T. H. JOH & V. M. PICKEL. 1984. Chemical and structural analysis of the relation between cortical inputs and tyrosine hydroxylase–containing terminals in rat neostriatum. Brain Res. **302:** 267–275.

15. GIORGUIEFF-CHESSELET, M-F., M-L. KEMEL, D. WANDSCHEER & J. GLOWINSKI. 1977. Regulation of dopamine release by nicotinic receptors in rat striatal slices: effect of nicotine in a low concentration. Life Sci. **25:** 1257–1262.

16. DE BELLEROCHE, J. & H. BRADFORD. 1978. Biochemical evidence for the presence of presynaptic receptors on dopamine nerve terminals. Brain Res. **142:** 53–68.

17. RAITERI, M., M. MARCHI & F. MAURA. 1982. Presynaptic muscarinic receptors increase striatal dopamine release evoked by quasi-physiological depolarization. Eur. J. Pharmacol. **83:** 127–129.

18. SCHULTZ, D. W. & R. E. ZIGMOND. 1989. Neuronal bungarotoxin blocks the nicotinic stimulation of endogenous dopamine release from rat striatum. Neurosci. Lett. **98:** 310–316.

19. WADA, E., K. WADA, J. BOULTER, E. DENERIS, S. HEINEMANN, J. PATRICK & L. SWANSON. 1989. Distribution of alpha2, alpha3, alpha4 and beta2 neuronal nicotinic receptor subunit mRNAs in the central nervous system: a hybridization histochemical study in the rat. J. Comp. Neurol. **284:** 314–335.

20. CLARKE, P. B. & A. PERT. 1985. Autoradiographic evidence for nicotinic receptors on nigrostriatal and mesolimbic dopaminergic neurons. Brain Res. **348:** 355–358.

21. REISINE, T. D., M-F. CHESSELET & J. GLOWINSKI. 1982. Striatal dopamine release in vitro: a beta-adrenergic regulated response not mediated through cyclic AMP. J. Neurochem. **39:** 976–981.

22. SCHUBERT, B., A. M. J. VANDONGEN, G. E. KIRSCH & A. M. BROWN. 1989. Beta-adrenergic inhibition of cardiac sodium channels by dual G-protein pathways. Science **245:** 516–519.

23. GIORGUIEFF, M-F., M-L. KEMEL & J. GLOWINSKI. 1977. Presynaptic effect of L-glutamic acid on the release of dopamine in rat striatal slices. Brain Res. **139:** 115–130.

24. ROBERTS, P. J. & A. D. ANDERSON. 1979. Stimulatory effect of L-glutamate and related amino acids on ^{3}H-dopamine release from rat striatum: an in vitro model for glutamate actions. J. Neurochem. **32:** 1539–1545.

25. CHERAMY, A., R. ROMO, G. GODEHEU, P. BARUCH & J. GLOWINSKI. 1986. In vivo presynaptic control of dopamine release in the cat caudate nucleus. II. Facilitatory or inhibitory influence of L-glutamate. Neuroscience **19:** 1081–1090.

26. KREBS, M-O., M-L. KEMEL, C. GAUCHY, M. DESBAN & J. GLOWINSKI. 1989. Glycine potentiates the NMDA-induced release of dopamine through a strychnine-insensitive site in the rat striatum. Eur. J. Pharmacol. **166:** 567–570.

27. MCGEER, P. L., E. G. MCGEER, U. SCHERER & K. SINGH. 1977. A glutamatergic corticostriatal path? Brain Res. **128:** 369–373.

28. HASSLER, R., P. HAUG, C. NITSCH, J. S. KIM & K. PAIK. 1982. Effect of motor and premotor cortex ablation on concentrations of amino-acids, monoamines and acetylcholine and on the ultrastructure in rat striatum. A confirmation of glutamate as the specific corticostriatal transmitter. J. Neurochem. **38:** 1087–1098.

29. GIRAULT, J. A., L. BARBEITO, U. SPAMPITANO, H. GOZLAN, J. GLOWINSKI & M. J. BESSON. 1986. In vivo release of endogenous amino acids from the rat striatum: further evidence for a role of glutamate and aspartate in corticostriatal neurotransmission. J. Neurochem. **47:** 98–106.

30. NIEOULLON, A., A. CHERAMY & J. GLOWINSKI. 1978. Release of dopamine evoked by electrical stimulation of motor and visual areas of the cerebral cortex in both caudate nuclei and the substantia nigra in the cat. Brain Res. **145:** 69–83.

31. BARBEITO, L., J. A. GIRAULT, G. GODEHEU, A. PITTALUGA, J. GLOWINSKI & A. CHERAMY. 1989. Activation of the bilateral corticostriatal glutamatergic projection by infusion of GABA into thalamic motor nuclei in the cat: an in vivo release study. Neuroscience **28:** 365–374.

32. CHESSELET, M-F., A. CHERAMY, R. ROMO, M. DESBAN & J. GLOWINSKI. 1983. GABA in the thalamic motor nuclei modulates dopamine release from the two dopaminergic nigro-striatal pathways in the cat. Exp. Brain Res. **51:** 275–282.

33. ROMO, R., A. CHERAMY, G. GODEHEU & J. GLOWINSKI. 1986. In vivo presynaptic control of dopamine release in the cat caudate nucleus. III. Further evidence for the implication of cortico-striatal neurons. Neuroscience **19:** 1091–1099.

34. GRACE, A. Phasic versus tonic dopamine release and the modulation of dopamine system responsivity: a hypothesis for the etiology of schizophrenia. Biol. Psychiatry. (In press.)

35. DELONG, M. R., M. D. CRUTCHER & A. P. GEORGOPOULOS. 1983. Relations between movement and single cell discharge in the substantia nigra of the behaving monkey. J. Neurosci. **3:** 1599–1606.

36. ROMO, R., A. CHERAMY, G. GODEHEU & J. GLOWINSKI. 1986. In vivo presynaptic control of dopamine release in the cat caudate nucleus. I. Opposite changes in neuronal activity and release evoked from thalamic motor nuclei. Neuroscience **19:** 1067–1079.

37. GRAYBIEL, A. M. 1984. Neurochemically specified subsystems in the basal ganglia. Ciba Found. Symp. **107:** 114–149.

38. JIMENEZ-CASTELLANOS, J. & A. M. GRAYBIEL. 1987. Subdivisions of the dopamine-containing A8-A9-A10 complex identified by their meso-striatal innervation of striosomes and extrastriosomal matrix. Neuroscience **23:** 223–242.

39. GERFEN, C. R., M. HERKENHAM & J. THIBAULT. 1987. The neo-striatal mosaic. II. Patch- and matrix-directed mesostriatal dopaminergic and non-dopaminergic systems. J. Neurosci. **7:** 3915–3934.

40. CHESSELET, M-F. & T. D. REISINE. 1983. Somatostatin regulates dopamine release in rat striatal slices and cat caudate nuclei. J. Neurosci. **3:** 232–237.

41. GERFEN, C. R. 1984. The neostriatal mosaic: compartmentalization of corticostriatal input and striatonigral output systems. Nature London **311:** 461–464.

42. CHESSELET, M-F. & A. M. GRAYBIEL. 1986. Striatal neurons expressing somatostatin-like immunoreactivity: evidence for a peptidergic interneuronal system in the cat. Neuroscience **17:** 547–571.

43. KEMEL, M-L., M. DESBAN, J. GLOWINSKI & C. GAUCHY. 1989. Distinct presynaptic control of dopamine release in striosomal and matrix areas of the cat caudate nucleus. Proc. Nat. Acad. Sci. USA **86:** 9005–9010.

Presynaptic Receptors and Autonomic Effectors

STANLEY KALSNER

Department of Physiology
City University of New York Medical School
138th Street and Convent Avenue
New York, New York 10031

INTRODUCTION

Whereas homoreceptors are defined simply by their responsivity to transmitter added from outside the system under study, the appropriate utilization of the term "autoreceptors" is somewhat more complex. Since the descriptive term "autoreceptors" signifies a system engaged in feedback regulation, it should be expected that several criteria be met. These center around (1) the actions of antagonists, (2) the actions of agonists, and (3) the characteristics of transmitter release in the absence of release-altering drugs. This study explores the extent to which the latter two criteria for negative feedback regulation are met in the periphery. Although the work described here was done with norepinephrine, primarily in autonomic noradrenergic systems, the findings should be considered as generally applicable to other peripheral and also to central neurotransmitter systems.

METHODS

Tissue preparations were done as described previously, and dependent on species and organ.[1,2] For example, renal arteries were removed from beef cattle at the slaughter house and taken promptly to the laboratory in cold previously oxygenated Krebs-Henseleit solution [composition (millimolar concentrations): NaCl, 115.3; KCl, 4.6; $CaCl_2$, 2.3; $MgSO_4$, 1.1; $NaHCO_3$, 22.1; KH_2PO_4, 1.1; glucose, 7.8; and disodium EDTA, 0.03]. After trimming of fat and connective tissue the vessels were cut into spiral strips of about 4 mm × 30 mm.

Guinea pigs and rabbits of either sex were killed by cervical dislocation or by barbiturate overdosage (sodium pentobarbital) and the appropriate tissues were immediately removed and placed in oxygenated (95% O_2 and 5% CO_2) Krebs-Henseleit (Krebs') solution at pH 7.4. The left atria of guinea pigs were bisected along the axis of base to apex, as previously described,[3,4] and ureters were desheathed and bisected into proximal and distal halves.[5] The rabbit thoracic aorta was trimmed of adherent fat and connective tissue and cut into spiral strips, with approximate dimensions of 2.0 × 0.2 cm.

All tissues were maintained for approximately 30 minutes in oxygenated Krebs' solution before incubation with tritiated norepinephrine. The desired preparations were then incubated for 60 minutes in 4.0 ml of oxygenated Krebs' solution containing 1-[7,8-^{3}H]norepinephrine (10 μCi/ml; 7.0–10 × 10^{-7} M) at 37°C. The tissues were then dipped in fresh Krebs' solution, suspended under 1 or 2 g of tension in a pump-driven superfusion apparatus, and superfused continuously at 4 ml/minute with warmed (37°C) and oxygenated Krebs' solution containing cocaine (1 or 3 × 10^{-5} M) or desmethylimipramine (1.1×10^{-6} M), to inhibit neuronal reuptake, and normetanephrine (1×10^{-5} M), to inhibit extraneuronal uptake of released amine.[4]

Stimulation Parameters

The tissues were mounted between platinum wire electrodes fixed vertically (or for heart tissue, horizontally) on opposite sides and stimulated transmurally. Biphasic pulses, usually of 0.5 msecond duration at supramaximal voltage (50–70 V) were delivered at the desired frequency using Grass stimulators. A 90-minute equilibration period under superfusion conditions was allowed for all tissues before the onset of stimulation protocols. After the equilibration period, which included one primer stimulation (10 pulses at 1 Hz given at 60 minutes), each tissue received, at 9-minute intervals, a set of 100 pulses of 0.5 msecond duration, twice at each of the test frequencies, e.g., 2 and 5 Hz. In a typical protocol, after completion of the initial set of stimulations (S_1) the experimental tissues, but not the controls, were exposed to the desired agonist or antagonists, dissolved in the superfusate for the indicated time, and the identical stimulation cycle was repeated (S_2) in both the untreated and treated tissues. Other stimulation conditions are as described under Results.

Drugs and Radiochemicals

The drugs used and their sources were as follows: cocaine hydrochloride (May & Baker Ltd., Manchester, England), normetanephrine hydrochloride (Calbiochem Biochemicals, San Diego, Calif.), desmethylimipramine (Merrell Dow Pharmaceuticals, Cincinnati, Ohio), phenoxybenzamine hydrochloride (Smith Kline & French, Philadelphia, Pa.), yohimbine hydrochloride (Sigma Chemical Co., St. Louis, Mo.), l-norepinephrine bitartrate (Calbiochem), carbamyl choline chloride (carbachol) (Aldrich). The radioisotope, l-[7,8-^{3}H]noradrenaline hydrochloride (specific activity about 15 Ci/mmol) was usually obtained from Amersham International (Bucks, England). It was diluted to a stock concentration of 100 μCi/ml (7.0–10.0×10^{-6} M) in ascorbic acid (50 μg/ml) and stored at 4°C in 10-ml aliquots, under nitrogen gas. To obtain a final concentration of 10 μCi/ml (7.0–10.0×10^{-7} M) in the incubation medium, 0.4 ml of this stock solution was added to 3.6 ml of Krebs' solution.

Efflux of Transmitter

The basal and stimulation-induced efflux of tritium from the tissues was determined by assaying 1.0-ml aliquots of the superfusate, collected in vials by a fraction collector which rotates at 3-minute intervals. The aliquots are subsequently transferred to vials containing 10-ml of Aqueous Counting Scintillant (Amersham/Searle Corp., Des Plaines, Ill.) and counted to a 1% error in a Wallace Rackbeta LKB scintillation system with automatic external standardization to determine efficiency or in a Beckman L5-5801. Basal efflux is expressed as disintegrations per minute and is referred to as the total radioactivity detected in the 3-minute sample collected immediately before each stimulation. Stimulation-evoked efflux is expressed as disintegrations per minute and was determined by subtracting the basal efflux from the total disintegrations per minute detected in the 3-minute sample collected during, and immediately after, stimulation. Transmural stimulation, when given, was always administered at the onset of a collection period, and collection of superfusate was continued until tritium levels returned to basal levels.[1]

Data Evaluation

Data are routinely expressed either as absolute disintegrations per minute collected for individual tissues during a stimulation period (minus basal efflux) or as the ratio of stimulation-induced efflux during a second period of stimulation (usually in the

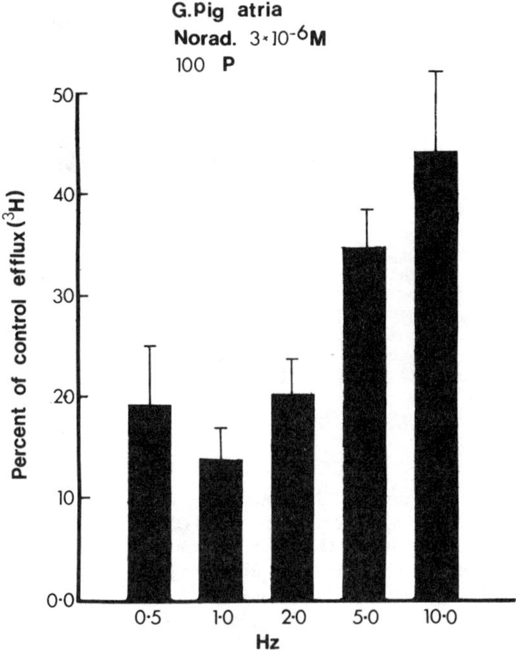

FIGURE 1. Effect of norepinephrine (noradrenaline) on stimulation-induced efflux of ^{3}H-norepinephrine in guinea pig atria. Tissues were exposed to norepinephrine (3×10^{-6} M) after the initial period of stimulation with 100 pulses at each test frequency followed 20 minutes later, in the presence of the catecholamine, by a repetition of the stimulations (0.5 to 10 Hz). See Methods for details.

presence of the desired test drug) to that obtained in the first period of stimulation (usually in the absence of drug).[1] This latter procedure eliminates any differences between preparations or animals in the absolute amounts of tritium released because the effect of a drug is determined by a same-tissue (intratissue) comparison of stimulation-induced efflux, first in the absence and then in the presence of drug, and a ratio is expressed. A matching group of control tissues taken from the same animals and treated identically except for the drug under study serves to correct and account for (normalize) any changes in efflux due to time alone. This is achieved by comparison of S_2/S_1 or other desired ratio in treated strips with that of the comparable ratio in untreated preparations. Additionally, it should be noted that no material decline in

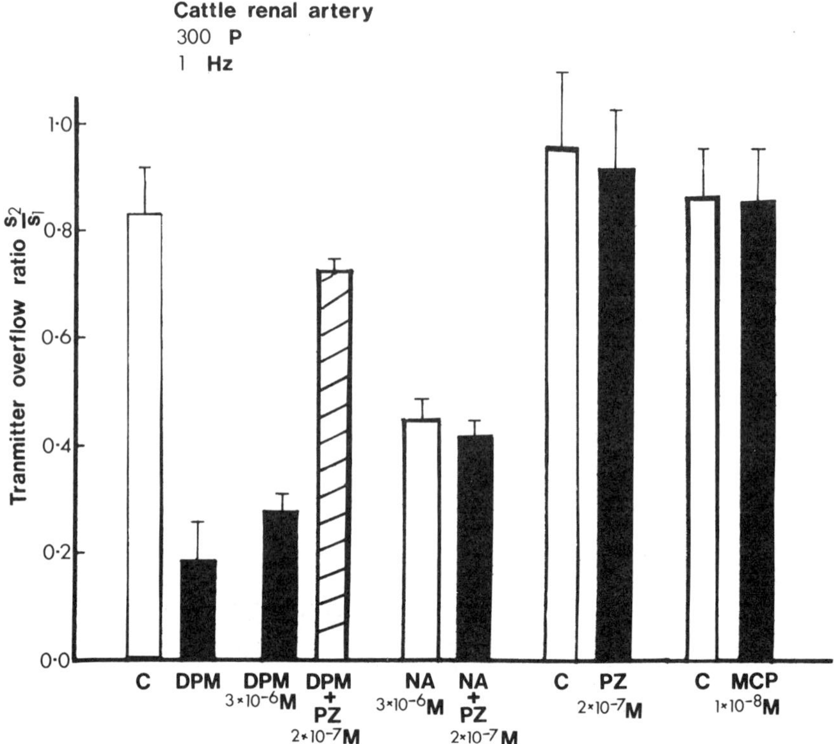

FIGURE 2. The effects of dopamine and dopamine antagonists on the stimulation-induced efflux of ^{3}H-norepinephrine in cattle renal artery. Dopamine (DPM), norepinephrine (NA), pimozide (PZ), and metoclopromide are used at the indicated concentrations. The transmitter overflow ratio (second period:first period) with 300 pulses at 1 Hz is shown for groups of strips treated prior to the second period of stimulation with the indicated drug(s) for 20 minutes. C indicates preparations not exposed to drugs in the interval between the first and second stimulation periods.

stimulation-induced efflux was noted in control preparations, with any of the tissues used here, when second vs. first stimulation runs were compared.

The above-described procedure, now widely used, is a precise and desirable way of recording efflux events and avoids a correction related to total tissue content of tritium, a factor of which the full implications are not yet clear.[6] A preliminary study, however, confirmed that data expressed by the fractional release method[7] and the ratio method[1] give essentially similar results.

A procedure even more sensitive than the intergroup comparison of ratios was also used, as deemed appropriate. The ratio of the absolute output of tritium in the second vs. the first (or third or fourth vs. the first) period of stimulation or any given treated tissue was calculated as a percentage of the similarly obtained ratio for the matching control (untreated) tissue, taken from the same animal. Thus, e.g., a ratio of 1.2 for a control tissue and 0.9 for its drug-treated mate yields a value of 75.0% of the control output of tritium, for that particular experiment. Mean percentage values for each treatment condition then allow a precise assessment of the comparative effects of a given drug on the stimulation-induced efflux of tritium by nulling out the sometimes encountered spontaneous variations in the absolute output of radioactivity between first and subsequent stimulation periods, which are reflected in ratio determinations.

Data on efflux are presented as mean ± standard error (SE). Student's paired t test was used for all intrastrip comparisons with the unpaired test used for comparisons between groups; p values of <0.05 were considered significant.[4]

RESULTS

Competition between Endogenous and Exogenous Norepinephrine

When there is competition between endogenous and exogenous norepinephrine for the occupancy of presynaptic alpha receptor sites, a fixed quantity of exogenous transmitter becomes an increasingly smaller proportion of the total transmitter incident on the membrane receptors as the intensity of stimulation increases. Consequently, a constant concentration of added norepinephrine should have less of an inhibitory effect on stimulation-induced efflux of transmitter as the frequency of stimulation increases. As shown in FIGURE 1, exogenous norepinephrine becomes increasingly less effective in inhibiting stimulation-induced transmitter release in the guinea pig atria as the frequency of stimulation with 100 pulses increases, in seeming

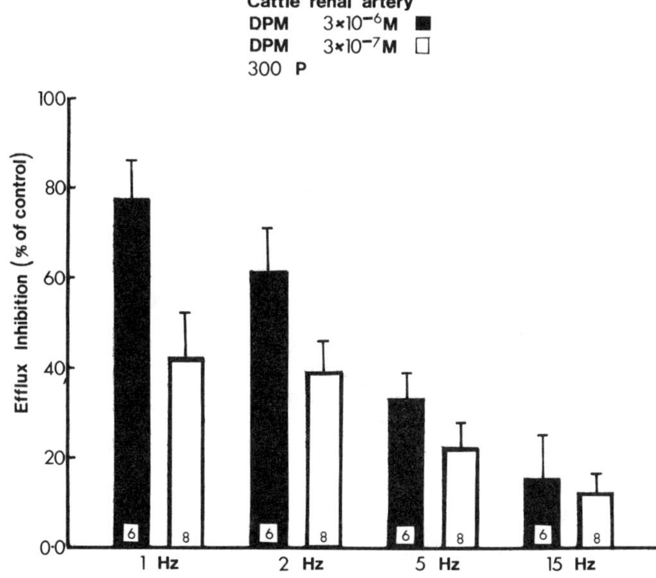

FIGURE 3. The effects of two concentrations of dopamine on the stimulation-induced efflux of ^{3}H-norepinephrine in cattle renal artery. Dopamine was administered at the indicated concentrations for 20 minutes in the interval between the first and second stimulation periods with 300 pulses, and the vessels were stimulated in its presence. See Methods for details. The correlation coefficients (r) between the length of the stimulus interval and the mean values for percentage of inhibition of transmitter efflux were determined. The correlation coefficient in the presence of the lower concentration of dopamine was 0.904 ($p < 0.02$) and with the higher concentration it was 0.957 ($p < 0.01$).

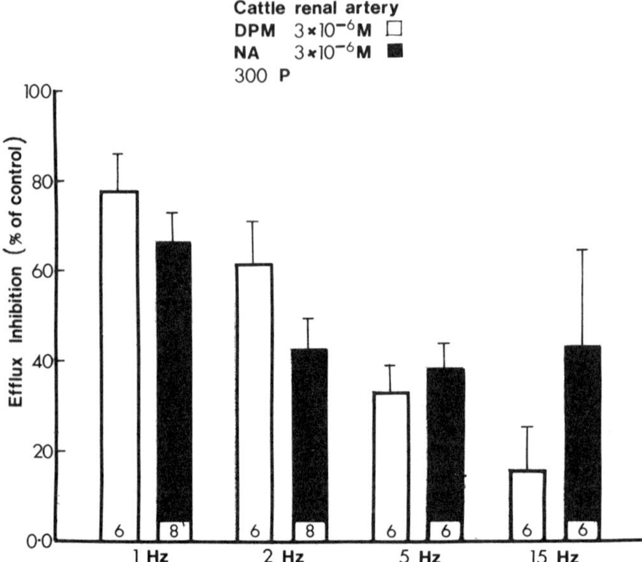

FIGURE 4. Comparison of the effects of dopamine (DPM) and norepinephrine (NA) on the stimulation-induced efflux of ^{3}H-norepinephrine in cattle renal artery preparations. Dopamine or norepinephrine was administered at the indicated concentration for 20 minutes in the interval between the first and second stimulation periods with 300 pulses at 1 to 15 Hz and the vessels were stimulated in its presence. See Methods for details.

accordance with these theoretical expectations. However, for a pattern of declining agonist efficacy to be reliably interpreted as providing evidence of competition between foreign and local transmitter for neuronal receptors, it must be shown that activation of heteroreceptors, which does not involve competition with endogenous transmitter, initiates a decidedly different profile of effect. In the case of heteroreceptor activation, a reasonably uniform amount of inhibition of transmitter release by the test agonist should be anticipated, regardless of stimulation intensity, at least over the physiological range. This was easily put to the test.

Cattle renal artery preparations respond to dopamine with inhibition of stimulation-induced ^{3}H-norepinephrine release. Inhibition of release by dopamine is selectively blocked by the dopamine antagonists pimozide (shown in FIGURE 2) and metoclopromide (not shown), at concentrations that do not alter the inhibitory effect of exogenous norepinephrine on ^{3}H-norepinephrine release.[8] As shown in FIGURE 2, neither of the dopamine antagonists, by themselves, altered the stimulation-induced release of transmitter, confirming the absence of an ongoing dopaminergic inhibitory feedback loop in the test preparation. This observation is not surprising since dopamine is a precursor of the transmitter norepinephrine in peripheral sympathetic nerves, and comprises at best a few percent of the total catecholamine resident in the nerve terminals. Despite these observations, dopamine, in two test concentrations, diminished the release of norepinephrine, and did so in inverse proportion to the intensity of the applied stimulation, over the physiological range (FIGURE 3).

Surprisingly, exogenous norepinephrine inhibited S-I ^{3}H-norepinephrine efflux in the renal artery preparation less adequately, in terms of theoretical expectations for competition between exogenous and endogenous amine, than did dopamine, which does not compete materially (FIGURE 4). These findings cannot be explained by exhaustion of a common second messenger system since the lower concentration of dopamine, with its more modest effects on release, also produced a pattern of declining efficacy with increasing stimulation frequency (FIGURE 3). Further, the inhibitory effect of norepinephrine actually did not decline at all over most of the test frequency range (FIGURE 4). Similar results with heteroreceptor activation were obtained with the cholinergic agonist carbamylcholine in rabbit ureter (FIGURE 5). A declining inhibition of stimulation-induced ^{3}H-norepinephrine release by the cholinergic compound was seen as the frequency increased from 2 to 5 Hz, even though the lack of enhancement of release by the muscarinic antagonist atropine confirmed the absence of ongoing inhibition by endogenous acetylcholine. It is clear that a pattern of agonist inhibition of neurotransmitter release that shows a decline with increasing stimulation intensity cannot be interpreted as the outcome of competition with endogenous transmitter for receptor sites, nor does it certify the operation of a negative feedback system.

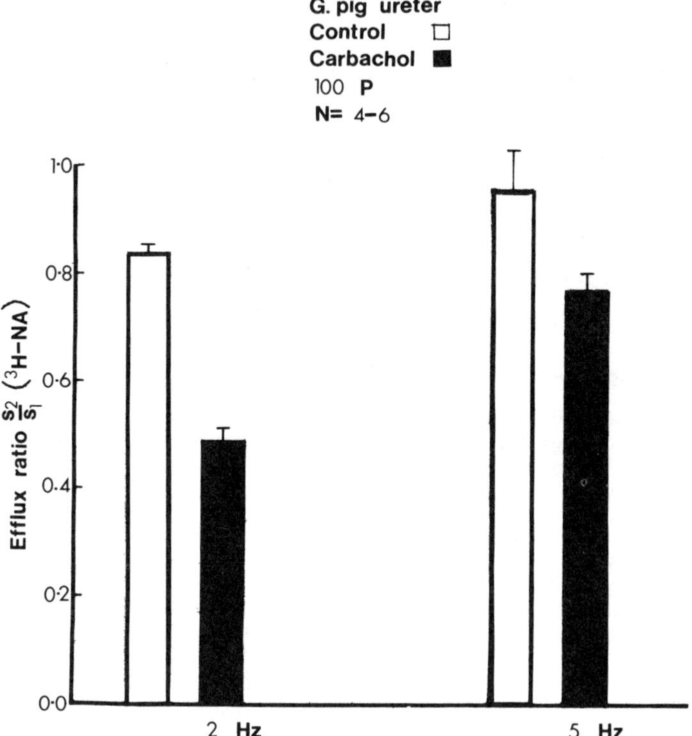

FIGURE 5. Effects of carbamylcholine (carbachol) on stimulation-induced efflux of ^{3}H-norepinephrine in guinea pig ureter. Carbachol (5.5×10^{-6} M), when used, was administered for 60 minutes in the interval between S_1 and S_2. Control ureter preparations were not treated with carbachol between S_1 and S_2. See Methods for details.

Relationship of Agonist and Antagonist Effects on Transmitter Efflux

In those instances where the pattern of declining agonist efficacy with increasing intensity does indeed derive from competition with endogenous transmitter for inhibitory receptor sites, certain other considerations must follow. As the agonist moves from being a highly effective to a substantially less effective inhibitor of transmitter release, during progressive increases in stimulation intensity, the antagonist should become increasingly more effective in increasing release (FIGURE 6, top). This is because the

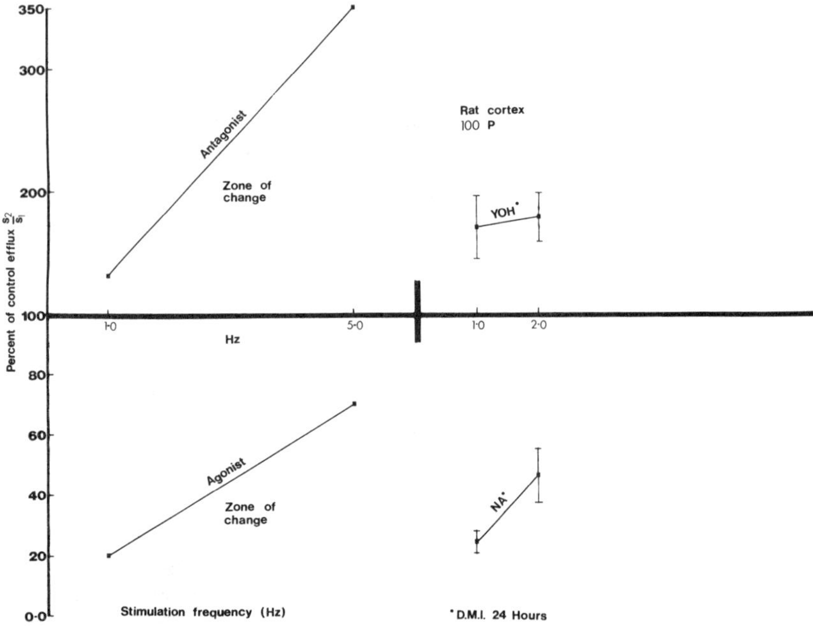

FIGURE 6. Theoretically expected profiles for agonist and antagonist effects on transmitter efflux during the operation of a negative feedback system. **Left:** During feedback regulation with increasing stimulation frequency the effectiveness of the agonist to inhibit efflux of transmitter declines while that of the antagonist to elevate efflux increases. **Right:** The effect of yohimbine (Yoh) and norepinephrine (NA) on the stimulation-induced efflux of ^{3}H-norepinephrine during stimulation with 100 pulses at 1 Hz and 2 Hz in slices of rat occipital cortex. See Methods for details.

agonist and antagonist patterns are presumed to be the simultaneous outcome of increasing levels of endogenous transmitter around the receptors. As the transmitter concentration in the biophase increases with stimulation frequency, the amount of feedback inhibition increases and enhancement of transmitter release by an antagonist of feedback inhibition should also increase (see FIGURE 6). The profile of agonist inhibition, however, does not satisfactorily correlate with that of antagonist performance, confirming the unreliability of the agonist as an index of feedback. This is seen in rat cortex, in which the effectiveness of yohimbine did not increase during the frequency range in which the agonist became decidedly less effective (FIGURE 6).

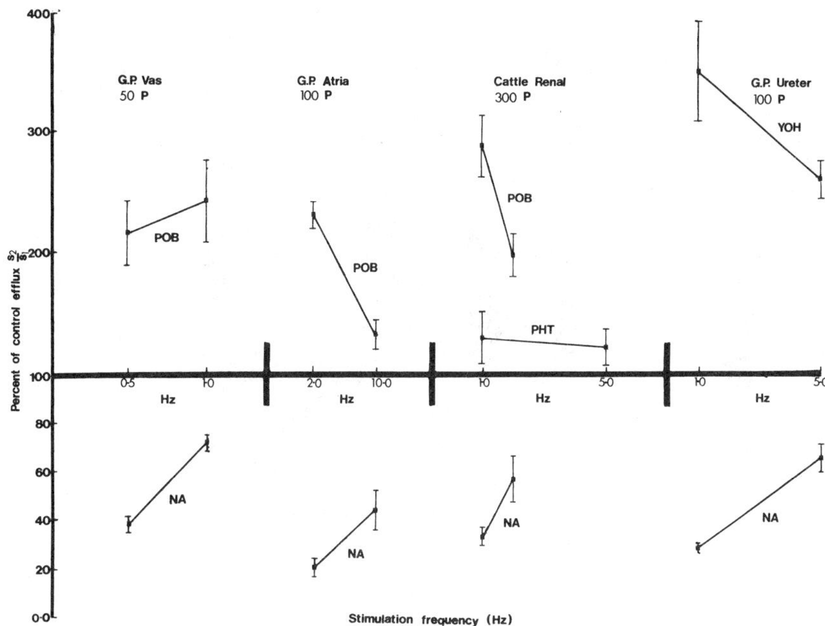

FIGURE 7. Profile of agonist and antagonist effects on transmitter efflux in several preparations during conditions of increasing stimulation frequency. The effects of the antagonists, phenoxybenzamine (POB) (1×10^{-5} M), phentolamine (3×10^{-7} M), and yohimbine (Yoh) (3×10^{-7} M), are shown along with that of the agonist norepinephrine (NA) (1.8×10^{-7} M or 1.8×10^{-6} M). The concentrations of antagonists used were sufficient to produce marked receptor blockade and maximal or near-maximal enhancement of transmitter (^{3}H-norepinephrine) efflux.

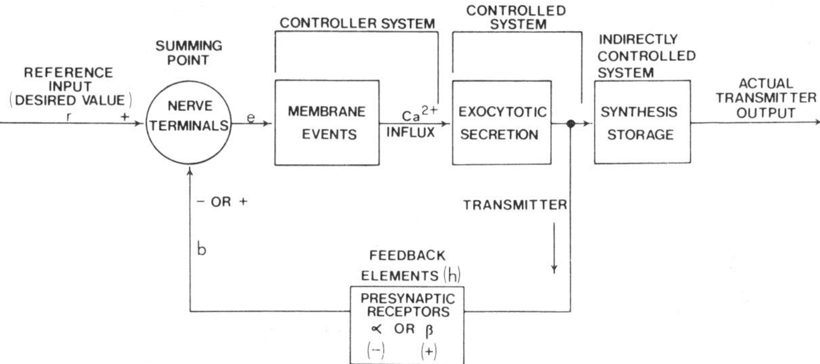

FIGURE 8. A block diagram of neurotransmitter release involving negative and positive feedback: r, reference input; e, activating signal; h, feedback transfer function; b, primary feedback. The control signal (e) is the difference between the input and output signal in negative feedback.

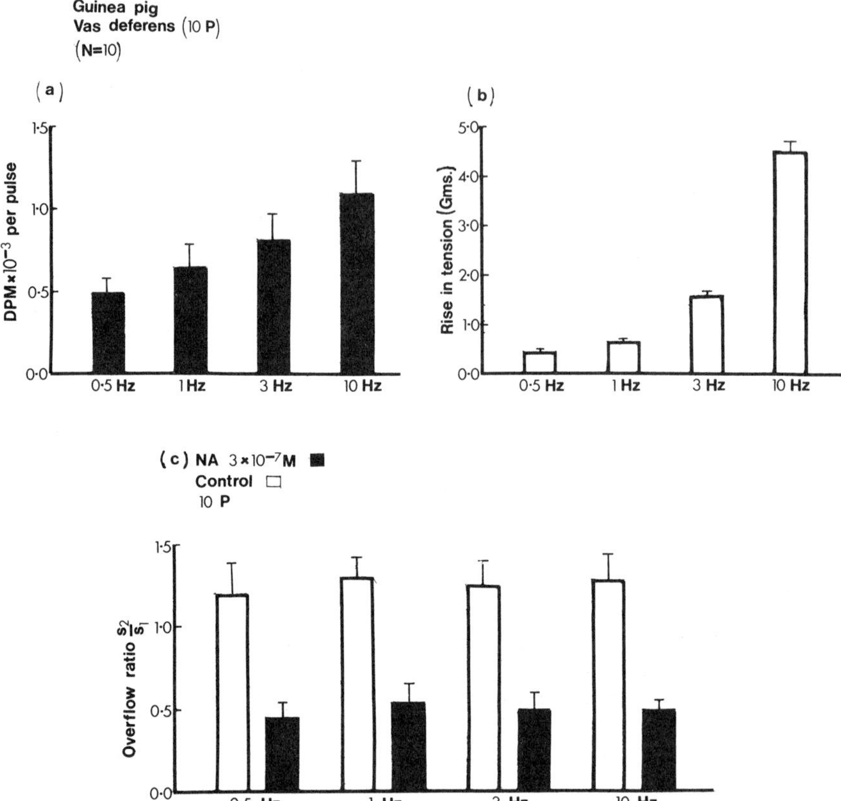

FIGURE 9. Transmitter efflux, mechanical response, and the effects of exogenous norepinephrine in guinea pig vas deferens. (**a**) Per pulse efflux of ^{3}H-norepinephrine over a range of frequencies with 10 pulses at 0.5 msecond duration. (**b**) Contractions to field stimulation of guinea pig vas with 10 pulses at the indicated range of frequencies. (**c**) Efflux ratios of ^{3}H-norepinephrine in control and norepinephrine-treated preparations stimulated at the indicated frequencies during S_1 and S_2. Tissues were exposed to norepinephrine (3×10^{-7} M) for 20 minutes in the interval between S_1 and S_2 (filled bars), and stimulation was repeated in its presence. Control preparations (open bars) were not exposed to norepinephrine. See Methods for details.

Similar results were obtained in a variety of other tissues and with several antagonists, including yohimbine, phentolamine, and phenoxybenzamine (FIGURE 7), and cannot be attributed to inadequate blocking capacity. For example, the haloalkylamine compound, phenoxybenzamine, generates a highly reactive ethyleimminium intermediate which forms a covalent chemical link with a part of the receptor, and this is not susceptible to agonist competition. In some tissues, e.g., ureter, atria, both the agonist and the antagonist become simultaneously less effective in altering transmitter release as the intensity of stimulation increases (FIGURE 7), decidedly not in keeping with the assumed meaning of agonist inhibition.

Cybernetic Considerations

A model of feedback regulation of neurotransmitter release described in cybernetic terms is shown in FIGURE 8. Effective regulation of neurotransmitter release through feedback inhibition, perhaps like blood sugar or blood pressure, would assure a biophase level of transmitter that only marginally increases with increasing stimulation intensity. Such regulation of the perineuronal concentration of transmitter would be achieved by negative feedback reduction of the per pulse release of transmitter, throughout the neuronal terminal field. However, the consequence of such precise and striking regulation of transmitter release would be an effector response that approximates a single monotonic value. Of course, this would be physiologically inappropriate and does not occur. If feedback inhibition operates to curb output of transmitter, even only moderately, it must still be achieved through a reduction in the average per pulse output throughout the stimulated tissue, and even this does not occur. In fact, the per pulse release of transmitter does not usually decline at all when stimulation intensity is increased over the physiological range. The average per pulse output of transmitter increases with increasing frequency over the physiological range, through a process of facilitation of transmitter release[9,10] (e.g., FIGURE 9a). The possibility has been suggested that negative feedback operates primarily to brake facilitation. However, in the presence of an adrenergic antagonist the process of facilitation is not "unbraked" but is often diminished. This is because the antagonist usually has its greatest effects at the lowest stimulation frequencies (FIGURE 10).[2]

Inhibition by Endogenous and Exogenous Agonists

Exogenous norepinephrine can shut down stimulation-induced release of norepinephrine almost completely, in a variety of effector systems. For example, in guinea pig ureter the catecholamine reduces release to less than five percent of control values

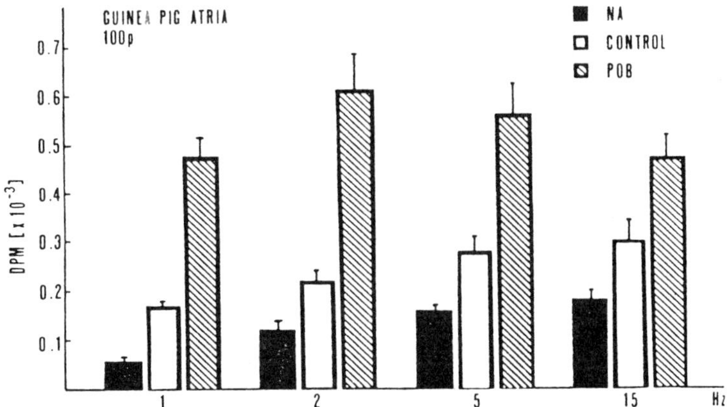

FIGURE 10. Relationship between stimulus frequency and per pulse efflux of ^{3}H-norepinephrine in guinea pig atria. Preparations were stimulated with 100 pulses at 1, 2, 5, and 15 Hz in untreated (open bars), norepinephrine (3 × 10^{-6} M) treated (filled bars), and phenoxybenzamine-treated (10 μm) (hatched bars). See Methods for details.

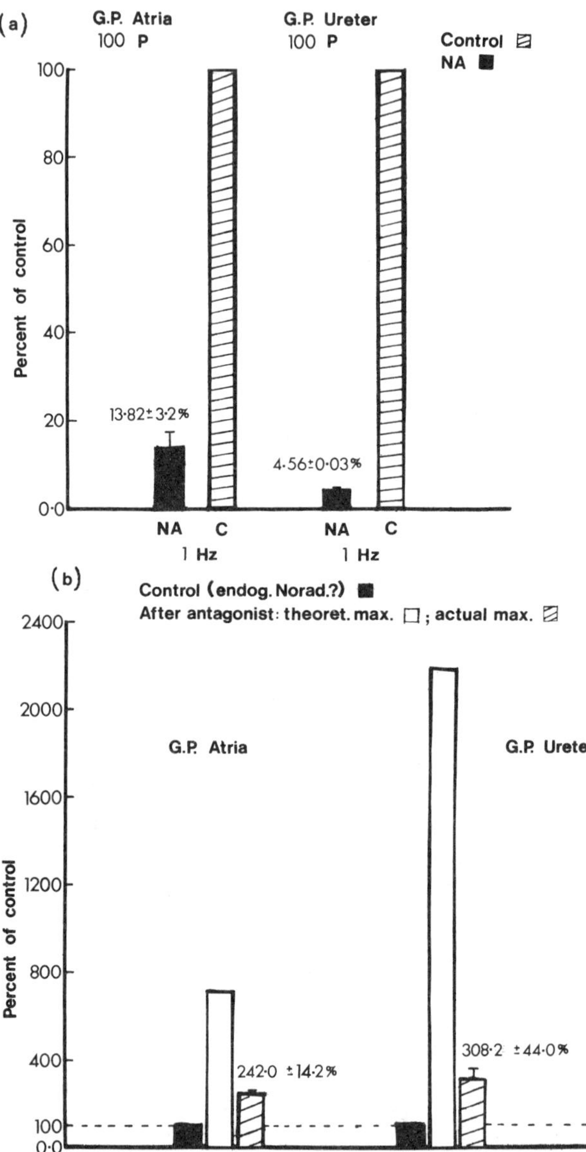

FIGURE 11. Theoretical and observed effects of norepinephrine (NA) and an antagonist (yohimbine) on efflux of ^{3}H-norepinephrine in guinea pig atria and ureter. See text for details.

(1/20th of control values) during low frequency stimulation (FIGURE 11a and b). If endogenous norepinephrine is capable of inhibiting transmitter release to a comparable extent as exogenous amine, then antagonist blockade should reveal this through a 20-fold enhancement of transmitter release. The maximal enhancement by an antagonist recorded in my laboratory with guinea pig ureter, examined under a very broad range of stimulation conditions, was only four to five times control values of stimulation-induced transmitter release (FIGURE 11b). A similar disparity between the extent of inhibition by exogenous agonist and the amount of enhancement by antagonist has been seen with other tissues (e.g., guinea pig atria FIGURE 11). A likely explanation is that endogenously released transmitter is not capable of activating as large a population of alpha receptors as exogenous amine.

Additional support for the likelihood that endogenous and exogenous transmitter do not compete for a common population of receptors during ordinary neurotransmission is provided by examining the response characteristics of some of the test preparations (e.g., FIGURE 12). Cattle renal arteries respond with progressively increasing contraction magnitudes as the frequency of stimulation is increased from 1 Hz to 15 Hz, reflecting the increasing concentration of norepinephrine in the biophase.[11] However, over most of this stimulation range the extent of inhibition of transmitter output by exogenous norepinephrine remains constant. This is not in keeping with expectations for competition between endogenous and exogenous amine for a common pool of receptors. Although a profile of declining agonist efficacy with increasing intensity

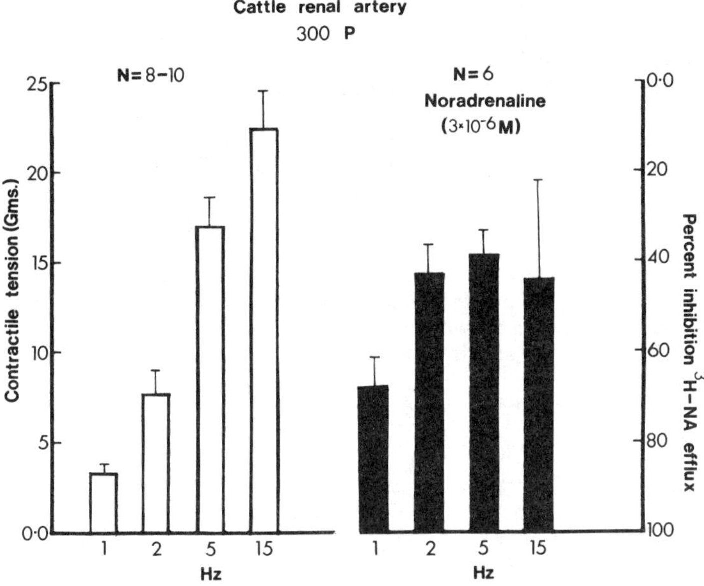

FIGURE 12. Stimulation-induced contractor responses and the inhibition of transmitter release by exogenous norepinephrine in cattle renal artery preparations. **Left:** Renal artery preparations were stimulated transmurally with 300 pulses at 1, 2, 5, and 15 Hz and contraction magnitudes recorded. **Right:** Preparations were exposed to norepinephrine (3×10^{-6} M) for 20 minutes in the interval between S_1 and S_2, and the extent of inhibition of transmitter release at each of the test frequencies was recorded. Details are given in Methods.

appears to be due primarily to causes other than competition, the lack of such a profile speaks overtly against competition. A similar discordancy between the effectiveness of exogenous norepinephrine as an inhibitor of stimulation-induced transmitter release, and the size of the postsynaptic response, has been noted in other tissues, e.g., guinea pig vas deferens (FIGURE 9b and c).[11,12] Although progressively larger contractions were elicited in the vas with 10 pulses by increasing the frequency of stimulation between 1 and 15 Hz, the inhibition of transmitter release by exogenous norepinephrine remained an unvarying value (FIGURE 9c).

Locus of Agonist Inhibition

Guinea pig ureter preparations were incubated for 20 minutes with the sodium channel blocker tetrodotoxin (3×10^{-7} or 1×10^{-6}) and then stimulated with 100 pulses at 2 and 5 Hz. As shown in FIGURE 13, the per pulse output of ^{3}H-norepinephrine was severely reduced by the toxin, signifying that field stimulation was exciting conducting portions of the terminal postganglionic fibers, and not directly the varicosities, which utilize Ca^{++} channels for neurosecretion. The capacity of norepinephrine to inhibit stimulation-induced transmitter release was almost abolished by the toxin at the lower of the two test frequencies, and much reduced at the higher frequency (FIGURE 13). These data support the conclusion that a moderate concentration of exogenous transmitter, distributed under "equilibrium" conditions, acts substantially on the conducting portions of the neuron and not only on the varicosities. These intervaricose segments likely are not easily accessible to endogenous transmitter liberated as pulses from point sources (varicosities) because the initial peak concentrations rapidly decline with distance through the process of diffusion.

What function could such a distribution of alpha$_2$ receptors on postganglionic sympathetic nerves serve? During conditions of exercise and stress the plasma levels of epinephrine and norepinephrine rise considerably.[13] Epinephrine is released into the blood by the adrenal medulla and norepinephrine diffuses into the blood from diverse peripheral adrenergic terminals, and this circulating pool of amine may provide a clue to function. Is it possible that elevated plasma levels of catecholamines curb effector organ activity in the periphery by inhibition of sympathetic discharge? This would dilate the vasculature, and thereby increase blood flow to certain organs, e.g., skeletal muscle. In this context, the motor activity of some autonomic effectors may thereby be beneficially reduced. As can be seen in FIGURE 14 (top), it appears that this may indeed be the case. Epinephrine was effective in inhibiting transmitter release in guinea pig aorta, at surprisingly low concentrations (3×10^{-9} g/ml), and Abrahamsen and Nedergaard have made similar observations with rabbit vascular tissue.[14] In my experiments, stimulation-induced release of norepinephrine was inhibited substantially in preparations of guinea pig ureter and in uterus by very low concentrations of epinephrine (even 3×10^{-10} g/ml), as well as by norepinephrine (FIGURE 14 bottom). The catecholamine concentrations used for these experiments are in keeping with blood levels for these amines encountered during certain physiological situations.[15]

SUMMARY

Agonist interactions with release processes do not validate autoreceptor operation. Instead, the data suggest that autoreceptors function as homoreceptors. Declining efficacy of agonist inhibition of transmitter release with increasing stimulation "intensity" is not assignable to competition with endogenous transmitter for a finite population of receptors. Heteroreceptor activation also reveals a pattern of declining efficacy with increasing intensity. Inhibition of stimulation-induced release by agonists also does not correlate inversely with that of antagonist effectiveness, as it should if both effects are linked to biophase levels of transmitter. Further, per pulse release of transmitter, in the absence of drugs, does not comply with expectations for autoinhibition. Experiments with tetrodotoxin, and study of the magnitudes of agonist and antagonist effects suggest that in the periphery putative autoreceptors may actually be homoreceptors, e.g., sympathetic nerve terminal receptors responsive to circulating catecholamines. In the central nervous system the paracrine secretion of transmitter

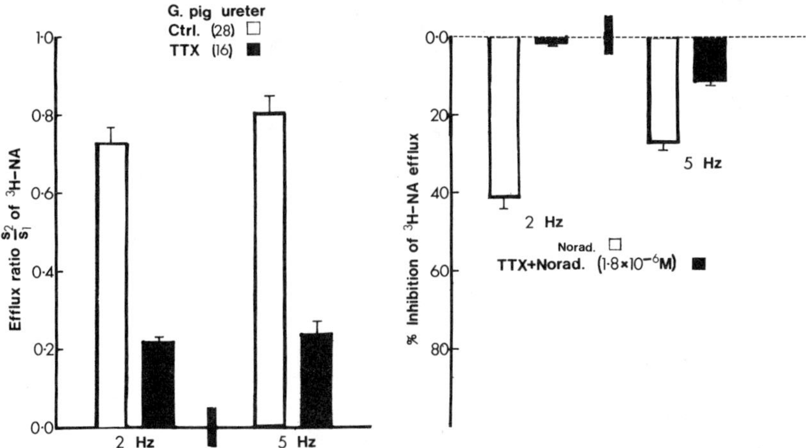

FIGURE 13. Effect of tetrodotoxin on the efflux of ^{3}H-norepinephrine and on norepinephrine-induced inhibition of release in guinea pig ureter. **Left:** Reduction of stimulation-induced transmitter efflux by tetrodotoxin (TTX). Tetrodotoxin (3×10^{-7} or 1×10^{-6}) was administered for 20 minutes in the interval between S_1 and S_2 and stimulation repeated in its presence. Tissues were stimulated twice with 100 pulses at 2 Hz and at 5 Hz (S_1) and after treatment the stimulation was repeated (S_2). Control tissues were not exposed to TTX. **Right:** Diminution of norepinephrine effectiveness after treatment with tetrodotoxin. Inhibition by norepinephrine of preparations exposed to norepinephrine alone (open bar) or to norepinephrine plus tetrodotoxin (filled bar) in the interval between S_1 and S_2.

may invoke homoreceptor activation rather than autoreceptors. Such activation may include, for example, the release of adrenaline from one set of fibers in the hypothalamus to act on dopaminergic or noradrenergic fibers or the release of noradrenaline from the terminals of some noradrenergic fibers in the cortex to activate alpha$_2$ receptors on other cortical noradrenergic fibers, the latter with a somewhat different function in the same brain region. The action of antagonist drugs to enhance transmitter release may be direct on nerve membranes in particular on sodium channels, and often unrelated to feedback regulation. This possibility is discussed by me elsewhere in this volume. It is shown that yohimbine and a low concentration of veratridine have similar and nonadditive effects on transmitter release.

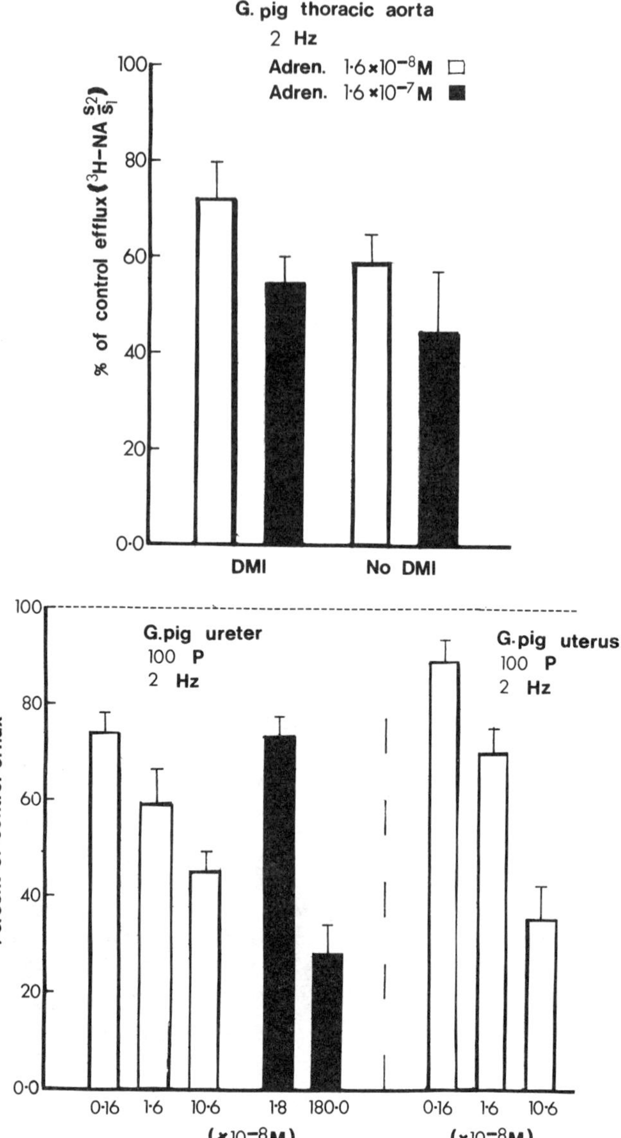

FIGURE 14. Top: Effects of epinephrine (adrenaline) on the stimulation-induced efflux of ³H-norepinephrine in guinea pig aorta stimulated with 100 pulses at 2 Hz. Tissues were exposed to the indicated concentration of epinephrine for 20 minutes in the interval between S₁ and S₂, and stimulations (S₂) were repeated in its presence. Desmethylimipramine (3 × 10⁻⁷ M), to block neuronal uptake, was present in the experiment depicted on the left, but not on the right. **Bottom:** Effects of epinephrine on the stimulation-induced efflux of ³H-norepinephrine in guinea pig ureter and uterus. Epinephrine (open bars) and norepinephrine (closed bars) were administered at the indicated concentrations in the interval between S₁ and S₂, and stimulation repeated in the presence of the amine. See Methods for details.

REFERENCES

1. KALSNER, S. 1979. Limitations of presynaptic adrenoceptor theory: The characteristics of the effects of noradrenaline and phenoxybenzamine on stimulation-induced efflux of [^{3}H]noradrenaline in vas deferens. J. Pharmacol. Exp. Ther. **212:** 232–239.
2. KALSNER, S. 1985. Is there feedback regulation for neurotransmitter release by autoreceptors? Biochem. Pharmacol. **34:** 4085–4097.
3. FURCHGOTT, R. F., P. S. GARCIA, A. R. WAKADE & P. CERVONI. 1971. Interactions of bretylium and other drugs on guinea-pig atria: evidence for inhibition of neuronal monamine oxidase by bretylium. J. Pharmacol. Exp. Ther. **179:** 171–185.
4. KALSNER, S. 1980. The effects of (+) and (−) propranolol on ^{3}H-transmitter efflux in guinea pig atria and the presynaptic α adrenoceptor hypothesis. Br. J. Pharmacol. **70:** 491–498.
5. KALSNER, S. 1982. Evidence against the unitary hypothesis of agonist and antagonist action at presynaptic adrenoceptors. Br. J. Pharmacol. **77:** 375–380.
6. HUGHES, J. & R. H. ROTH. 1974. Variation in noradrenaline output with changes in stimulus frequency and train length: role of different noradrenaline pools. Br. J. Pharmacol. **51:** 373–381.
7. STARKE, K., T. ENDO & H. D. TAUBE. 1975. Relative pre- and postsynaptic potencies of adrenoceptor agonists in the rabbit pulmonary artery. Naunyn-Schmiedeberg's Arch. Pharmacol. **291:** 55–78.
8. KALSNER, S. & C. C. CHAN. 1979. Adrenergic antagonists and the presynaptic receptor hypothesis in vascular tissue. J. Pharmacol. Exp. Ther. **211:** 257–264.
9. STJARNE, L. 1981. On sites and mechanisms of presynaptic control of noradrenaline secretion. *In* Presynaptic Receptors. L. Stjarne, Ed.: 257–272. Academic Press. London, England.
10. KALSNER, S. 1984. Limitations of presynaptic theory: no support for feedback control of autonomic effectors. Fed. Proc. **43:** 1358–1364.
11. CHAN, C. C. & S. KALSNER. 1979. An examination of the negative feedback function of presynaptic adrenoceptors in a vascular tissue. Br. J. Pharmacol. **67:** 401–407.
12. KALSNER, S. 1980. Limitations of presynaptic adrenoceptor theory: the characteristics of the effects of noradrenaline and phenoxybenzamine on stimulation-induced efflux of [^{3}H]noradrenaline in vas deferens. J. Pharmacol. Exp. Ther. **212:** 232–239.
13. WEST, J. B., Ed. Best and Taylor's Physiological Basis of Medical Practice. 12th edit. Williams & Wilkins. Baltimore, Md. (In press.)
14. ABRAHAMSEN, J. & O. A. NEDERGAARD. 1989. Presynaptic action of adrenaline on adrenoceptors modulating stimulation-evoked ^{3}H-noradrenaline release from rabbit isolated aorta. Naunyn-Schmiedeberg's Arch Pharmacol. **339:** 281–287.
15. ESLER, M., G. JENNINGS, P. KORNER, P. BLOMBRY, N. SACHARIAS & P. LEONARD. 1984. Measurement of total and organ-specific norepinephrine kinetics in humans. Am. J. Physiol. Endocrinol. Metab. **247**(10): E21–E28.

Modulation of Histamine Synthesis and Release in Brain via Presynaptic Autoreceptors and Heteroreceptors

JEAN-CHARLES SCHWARTZ, JEAN-MICHEL ARRANG,
MONIQUE GARBARG, CHRISTIANE GULAT-MARNAY,
AND HÉLÈNE POLLARD

Neurobiology and Pharmacology Unit
Paul Broca Center
INSERM
2ter rue d'Alésia
75014 Paris, France

The role of histamine (HA) as a neurotransmitter in the central nervous system (CNS) was already clear by the mid-70s when the biochemical machinery of putative histaminergic neurons was unraveled and the two classes (H_1 and H_2) of HA receptors, first identified in peripheral tissues, were characterized in the mammalian brain.[1]

Histaminergic neurons are now known to be located in the tuberomammillary nucleus of the posterior hypothalamus and to project in a widespread manner to most areas of the neuroaxis (reviewed by Schwartz *et al.*)[2] This anatomical disposition as well as various other neurochemical or neurophysiological features suggest that histaminergic neurons have functions similar to those of other monoaminergic neurons.

Until 1983, however, little information was available regarding the control of HA synthesis and release via presynaptic mechanisms that had already been evidenced in other aminergic systems. This was then studied on a model developed earlier in our laboratory which consisted of labeling the endogenous pool of neuronal HA in brain slices using the [^{3}H] precursor, i.e., [^{3}H]L-histidine, and isolating neosynthetized [^{3}H]HA by ion-exchange chromatography.[3] Using this model, it appeared that exogenous HA exerted a "braking" effect on the K^+-evoked release of the neosynthetized [^{3}H]amine and that this response involved the stimulation of a non-H_1, non-H_2 receptor, leading to pharmacological definition of a third (H_3) HA receptor subtype.[4] The H_3 receptor was then shown to represent a true presynaptic receptor and to control not only release but also synthesis of the amine. With the design of highly selective and potent H_3-receptor ligands,[5] the role played by this receptor *in vivo* and its localization in brain as well as peripheral tissues could be established. More recently the influence of agents acting at other presynaptic receptor sybtypes (heteroreceptors) on HA synthesis and release was investigated.

These studies have already contributed to improving our knowledge of the factors controlling the activity of HA neurons in brain as well as of the drugs that modify it.

H_3-RECEPTOR-MEDIATED CONTROL OF HISTAMINE
RELEASE *IN VITRO*

In slices from several regions of rat brain the release of neosynthesized [^{3}H]HA induced by either K^+,[4–8] veratridine,[4,6,7] or field electrical stimulation[9–11] is strongly inhibited upon addition of exogenous HA in the external medium. In overflow experiments, the maximal inhibition of release may be as high as 80% with 30 mM K^+ stimulations[12] or nearly total with electrical stimulation.[9,10] Inhibition is more marked for depolarizing stimuli of low intensity.[7,9,11] The autoinhibitory effect of HA appears to be a receptor-mediated effect since it displays saturability (the EC_{50} of HA is ~ 0.1 μM), reversibility (it is suppressed by washing out the excess of exogenous HA from the preparation), and high pharmacological specificity, being mimicked or antagonized in a competitive manner by several agents displaying low if any activity at H_1 and H_2 receptors. This last feature has led to the pharmacological definition of H_3 receptors.[4,5]

This model was used to design in collaboration with the laboratory of Prof. W. Schunack (University of Berlin, FRG) a potent and selective H_3 receptor chiral agonist: (R) α-methylhistamine is approximately 15 times more potent than HA itself at the H_3 receptor whereas its potency relative to HA at the H_1 or H_2 receptor is only about 1% (FIGURE 1); (S) α-methylhistamine is approximately 100 times less potent that the R isomer. In parallel thioperamide, an imidazolylpiperidine derivative, was designed in collaboration with the laboratory of Prof. M. Robba (University of Caen, France) and shown to represent a selective, potent, and competitive H_3-receptor antagonist.[5] Finally the drug 4-[2-(1-pyrrolidinyl)ethyl] imidazole was shown to represent a partial agonist at the H_3 receptor (FIGURE 2)

That H_3 receptors are directly located on histaminergic terminals was shown by various data: (1) the autoinhibitory effect of HA persists when the propagation of action potentials in the brain slice is blocked by tetrodotoxin; (2) it also persists in slices from kainate-injected striatum; (3) it is also found with a synaptosomal preparation.[6,7] The extent of the autoinhibition can be modulated in a complex manner by changes in extracellular Ca^{++}, suggesting that H_3 receptors regulate HA release via a control of Ca^{++} entry.[7]

Various H_3-receptor antgonists (impromidine, burimamide, or the selective agent thioperamide) enhance the depolarization-induced release of $[^3H]HA$[5,7] (FIGURE 1). This effect is particularly marked when slices are previously loaded with HA by preincubation with $[^3H]$histidine in high concentrations: the increased $[^3H]HA$ synthesis occurring under these circumstances is attributable to the fact that, due to its K_M higher than the tissue histidine level, L-histidine decarboxylase is not saturated by its substrate under basal conditions.

The autoinhibitory effect of exogeneous HA was found on slices from various regions known to contain terminals of extrinsic histaminergic neurons, suggesting that all these terminals are endowed with H_3 receptors.

H_3-RECEPTOR-MEDIATED CONTROL OF HISTAMINE SYNTHESIS *IN VITRO*

The regulation of HA synthesis via H_3 receptors was evidenced on rat brain slices or snyaptosomes labeled with $[^3H]$L-histidine.[5,13] On this model, depolarization by increased extracellular K^+ concentration enhances by about twofold the $[^3H]HA$ formation in slices of hypothalamus[3] or cerebral cortex.[5,6,13] This stimulation is also observed, although to a lesser extent, in cortical synaptosomes or slices from the posterior hypothalamus where histaminergic cell bodies are located,[14-16] suggesting that it may occur in perikarya as well as in nerve endings. In the presence of exogenous HA, the K^+-induced stimulation of synthesis is reduced by up to 60–70% with an EC_{50} of 0.3 μM, similar to that found at H_3 receptors controlling $[^3H]$histamine release. Furthermore, the effect of HA is mimicked by the selective and potent H_3 agonist (R)α-methylhistamine and antagonized in a selective and competitive manner by various H_3 antagonists including thioperamide. These effects occur in slices of cerebral

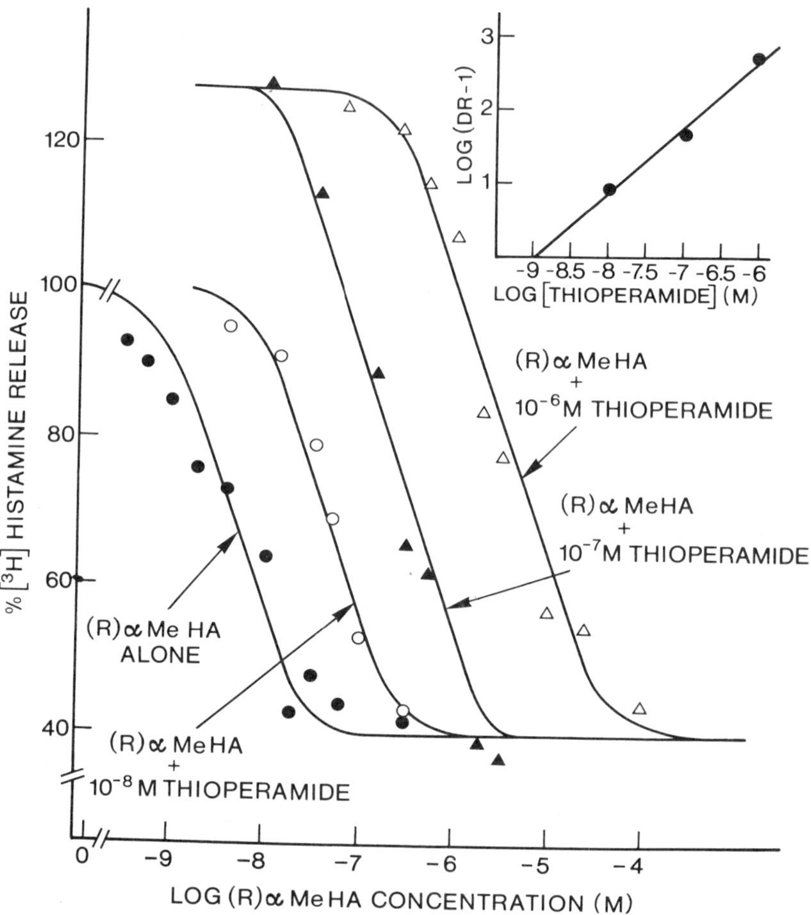

FIGURE 1. Effects of (R) α-methylhistamine [(R) α-MeHA], a selective H_3-receptor agonist, and thioperamide, a selective H_3-receptor antagonist, on [³H]histamine release from slices of rat cerebral cortex. Slices were prelabeled in the presence of [³H]L-histidine, and the release (expressed as percent of total [³H]histamine released over spontaneous efflux) was elicited in the presence of 30 mM K^+. **Inset:** Schild plot analysis of the competitive antagonism by thioperamide which leads to a pA_2 of 8.96, neglecting the influence of endogenous histamine. Note the overstimulation of efflux in the presence of thioperamide in high concentrations. (From Arrang *et al.*[5] by permission of *Nature*, London.)

cortex or posterior hypothalamus as well as in cortical synaptosomes. In addition, even in the absence of added HA or agonists, H_3 antagonists enhance the depolarization-induced stimulation of synthesis, indicating a participation of released HA in the control process.[5,13]

It can be concluded that H_3 receptors control not only release but also synthesis at the level of nerve endings and, possibly, perikarya. A relationship between the two regulatory processes seems likely and may occur via intracellular calcium, but the mechanism remains to be investigated.

RADIOLABELING OF H_3 RECEPTORS IN CEREBRAL MEMBRANES

The highly potent agonist [^{3}H] (R)α-methylhistamine (α-MeHA) constitutes a suitable probe for the selective labeling of the H_3 receptor.[5] In the absence of divalent cations, [^{3}H] (R) α-MeHA binds in a saturable manner to an apparently homogeneous population of sites in membranes of rat cerebral cortex. The binding is reversible, as indicated by the similar dissociation constants derived from either association/dissociation kinetics (K_D = 0.3 nM) or saturation kinetics at equilibrium (K_D = 0.4 nM). These binding sites were pharmacologically identified as H_3 receptors by competition studies performed with various H_1-, H_2- or H_3-histaminergic drugs.[5] The relative potencies of various agonists and affinity constants of various antagonists are highly correlated with the corresponding values obtained at functional H_3 autoreceptors regulating HA release or synthesis. For instance, the (R) isomer of α-MeHA is about 10 times as potent as HA itself and (S) α-MeHA is about 100 times less potent than (R) α-MeHA. This confirms that enantiomers corresponding to S-configured L-histidine are highly preferred at H_3 receptors, whereas enantiomers corresponding to D-histidine are more potent at H_2 receptors, no difference being observed at H_1 receptors.[5,17] The potent and specific H_3-receptor antagonist thioperamide, the H_1-receptor agonist betahistine,[18] the H_2-receptor agonist impromidine,[4] and the psychoactive drug phencyclidine[19] inhibit [^{3}H] (R)α-MeHA binding with the same affinity as that displayed when tested as antagonists at functional H_3 autoreceptors.

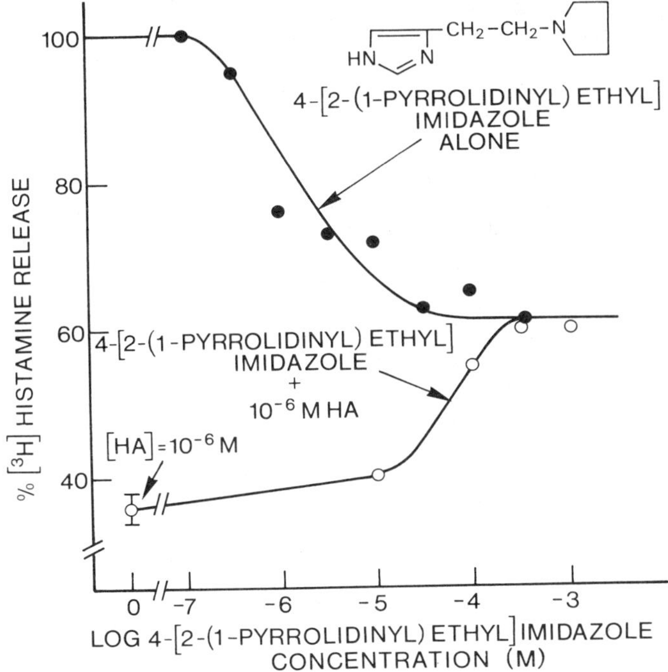

FIGURE 2. Effects of histamine and 4-[2-(1-pyrrolidinyl)ethyl] imidazole, a partial agonist at H_3 receptors, on [^{3}H]histamine release from slices of rat cerebral cortex. The maximal inhibition elicited by 4-[2-(1-pyrrolidinyl)ethyl]imidazole alone was 36 ± 2% as compared to 64 ± 2% for histamine alone. (From Arrang *et al.*[5] by permission of *Nature*, London.)

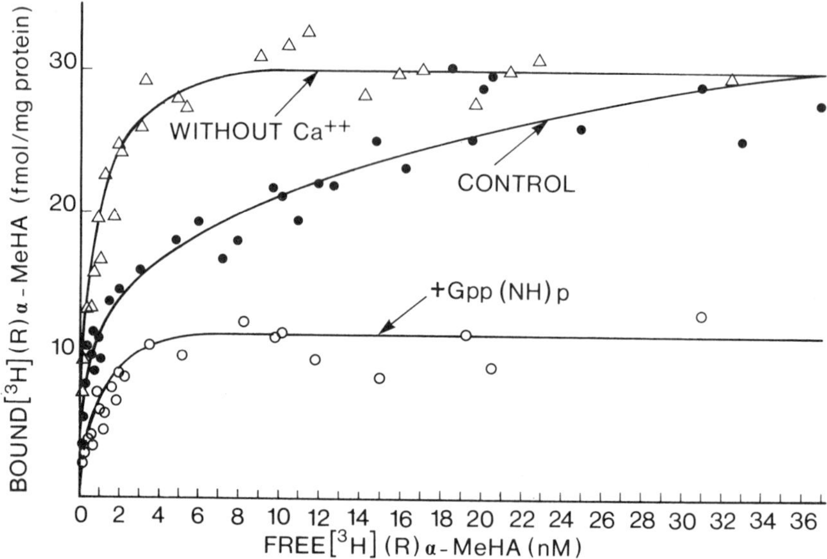

FIGURE 3. Effects of a guanylnucleotide and calcium ions on [³H] (R) α-methylhistamine specific binding to H_3 receptors of rat cerebral cortex membranes. Membranes were incubated for 60 minutes at 25°C in a modified Krebs medium containing 2.6 mM $CaCl_2$ (control). Specific binding was defined as that inhibited by 3 μM thioperamide. When required 100 μM Gpp(NHp) was added or calcium omitted.

Two affinity states of the H_3 receptor are observed when the binding of [³H] (R) α-MeHA is studied in the presence of Ca^{++} in physiological concentrations, an effect that is not reproduced by Mg^{++} and apparently arises from the conversion of a major fraction of H_3 receptors to a lower affinity state (K_D = 16 nM) (FIGURE 3). The binding of the ³H ligand to this low-affinity component is entirely and specifically inhibited in the presence of guanylnucleotides, indicating that the H_3 receptor is coupled to its (so far unknown) effector system via a G protein. However, at numerous other receptor types, guanylnucleotides affect binding to the high- and not to the low-affinity component. The low-affinity component may correspond to the functionally active receptor since the K_D value of (R) α-MeHA at this component is close to its EC_{50} values at H_3 autoreceptors regulating HA release and synthesis. However, the functional involvement of the high-affinity population of sites cannot be entirely excluded, since it also appears to be (although very partially) GTP regulated. The 10 times higher affinity of agonists to this high affinity component, as compared with their potency in functional studies, suggests that these sites may represent the H_3 receptor in a modified (desensitized?) state.

AUTORADIOGRAPHIC LOCALIZATION OF H_3 RECEPTORS IN BRAIN

The mean density of cerebral H_3 receptors in the rat is rather low, i.e., about 30 fmol/mg membrane protein. It is even lower in the guinea pig brain, suggesting that the number of H_3 receptors, as that of H_1 receptors, varies among species. H_3 receptors mediate the autoinhibition of HA release in human brain with a pharmacology apparently similar to that of corresponding receptors in rodents.[20]

The distribution of H_3 receptors in rat brain, established from either membrane binding studies or autoradiographic studies, is highly heterogeneous (FIGURES 4 and 5).

In cerebral cortex, where they are rather dense, H_3 receptors are found in all areas and layers with, however, a higher abundance in rostral areas and laminae IV Rnd V. In the hippocampal formation, they are moderately to highly abundant, their density being the highest in the dentate gyrus, moderate in subiculum, and very low in the fimbria. In the amygdaloid complex, high densities are found in central, lateral, and basolateral nuclei as well as in the bed nucleus of the stria terminalis, which contains a dense histaminergic innervation.[21]

In the basal forebrain, numerous H_3 receptors are present in anterior olfactory nuclei, nucleus accumbens, olfactory tubercles, as well as in striatum, particularly in its dorsomedial part; they are less numerous in the globus pallidus and even less in the septum. In the thalamus, H_3 receptors are mainly detected in various midline, intralaminar, and lateral nuclei. In the hypothalamus, their moderate density contrasts with the high density of HA axons but they are detectable at the level of the tuberomammillary nucleus, where they may reside on perikarya or dendrites (FIGURE 5). In the mesencephalon, they are numerous in the substantia nigra, particularly in its pars reticulata, the ventral tegmental area, and superior colliculi. In cerebellum, low densities are present in all layers. In brain stem, they are mainly present in pontine nuclei, around the fourth ventricle, in locus coeruleus, and in the dorsal tegmental

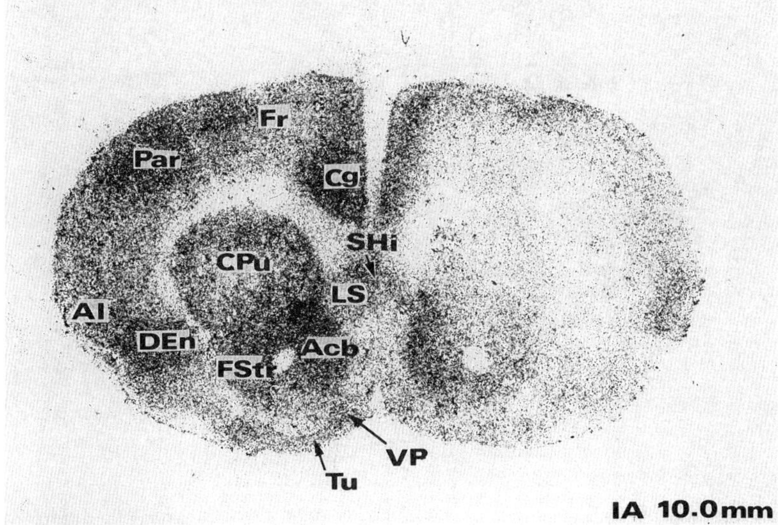

FIGURE 4. Autoradiographic localization of H_3 receptors in a frontal section of rat brain and the effect of intrastriatal kainate. Animals were killed four weeks after receiving 2.5 µg of kainic acid into the right caudate putamen, and H_3 receptors were visualized using 1 nM [^{3}H] (R) α-methylhistamine. Abbreviations: Acb, nucleus accumbens; AI, agranular insular cortex; Cg, cingulum; CPu, caudate putamen; Fr, frontal cortex; FStr, fundus striati; LS, lateral septum; Par, parietal cortex; Tu, olfactory tubercle; VP, ventral pallidum.

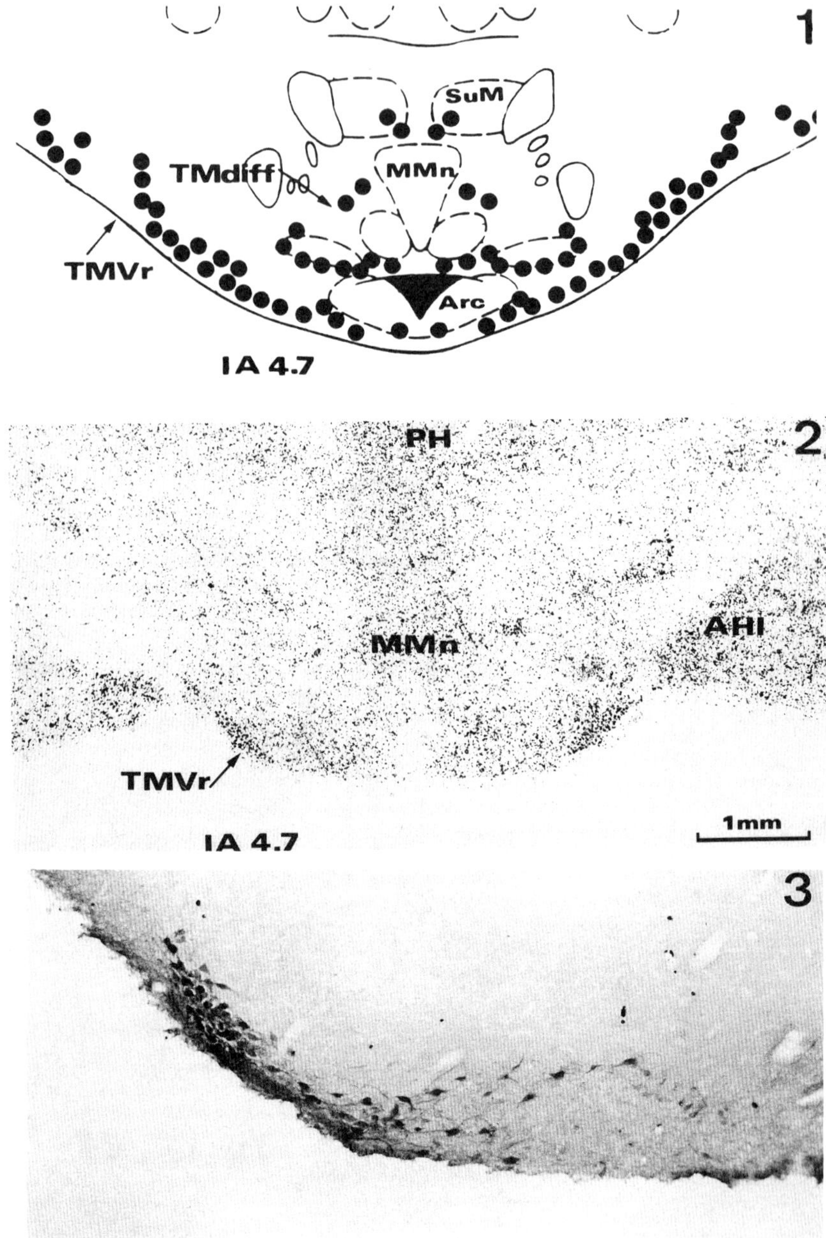

FIGURE 5. Comparative localization of H_3 receptors and histamine-immunoreactive neurons at the level of the posterior hypothalamus of the rat. (1) Schematic localization of histaminergic perikarya in the tuberomammillary nucleus on a section represented according to the atlas of Paxinos and Watson at the level of the caudal hypothalamus. (2) Autoradiographic visualization of H_3 receptors at the level of the ventral tuberomammillary nucleus (rostral part). (3) The ventral tuberomammillary nucleus (rostral part) labeled with a polyclonal antihistamine antibody kindly provided by Dr. P. Panula. Abbreviations: AHi, amygdalo hippocampal area; Arc, arcuate nucleus; MMn, medial mammillary nucleus, medial part; PH, posterior hypothalamic nucleus; SuM, supramammillary nucleus; TM diff, tuberomammillary nucleus, diffuse part; TMVr, ventral tuberomammillary nucleus subgroup, rostral part.

nucleus. In the spinal cord, a low density is present, mainly in external layers of the dorsal horn.

This distribution of H_3 receptors, not strictly parallel to that of histaminergic axons, suggests that they are not restricted to the latter. This is confirmed by the identification of H_3 receptors on serotoninergic nerve terminals in the cerebral cortex[22] and on a cerebral vessel.[23] Their decrease elicited in striatum by local administration of the neurotoxin kainate (FIGURE 4) is consistent with a major neuronal localization (Pollard *et al.*, in preparation).

H_3 RECEPTORS SELECTIVELY CONTROL CEREBRAL HISTAMINE METABOLISM *IN VIVO*

The turnover of cerebral amines is under a feedback control either via long neuronal loops involving postsynaptic receptors or via presynaptic autoreceptors.

The first process does not appear to regulate the activity of histaminergic neurons in rat brain since blockade of H_1 and/or H_2 receptors does not result in any significant change in histamine turnover[24–26] (TABLE 1).

By contrast either stimulation or blockade of H_3 receptors elicited by systemic administration of the agonist (R) α-methylhistamine or the antagonist thioperamide respectively induces in rats[26] or mice[27] marked and opposite changes in various indices of cerebral histamine turnover (FIGURES 6 and 7).

Hence in rats thioperamide, given at doses in the low mg/kg range (intraperitoneal or per oral) diminishes the steady-state histamine level, accelerates the α-FMH-induced depletion of endogenous histamine[28] (TABLE 1), the rate of [^{3}H]histamine synthesis, or the steady-state level of t-methylhistamine, a major HA metabolite in brain[26] (FIGURE 7). In mice a similar rise in t-methylhistamine level accompanied by a slight decrease in histamine level was reported.[27] This suggests that H_3 receptors selectively regulate the activity of histaminergic neurons *in vivo* and that these receptors are tonically activated by endogenous histamine. It is perhaps relevant that

TABLE 1. Effects of H_1-, H_2-, and H_3-Receptor Antagonists on Histamine Turnover in Rat Cerebral Cortex[a]

Treatment	Histamine Level	
	ng/g Protein	Change
Saline	347 ± 17	—
α-Fluoromethylhistidine		
Alone	222 ± 17	-36%
+ mepyramine	207 ± 25^b	-40%
+ zolantidine	206 ± 29^b	-41%
+ zolantidine + mepyramine	222 ± 20^b	-36%
+ thioperamide	133 ± 5^c	-62%

[a]The depletion of synaptosomal histamine induced by α-fluoromethylhistidine, an irreversible L-histidine decarboxylase inhibitor, was taken as an index of histamine turnover. The animals received mepyramine (3 mg/kg, intraperitoneal), an H_1-receptor antagonist, and/or zolantidine (30 mg/kg, intraperitoneal), an H_2-receptor antagonist, or thioperamide, an H_3-receptor antagonist (3 mg/kg, intraperitoneal), 20 minutes before α-fluoromethylhistidine (50 mg/kg, intraperitoneal) and were killed 40 minutes later.

[b]Nonsignificant as compared with α-fluoromethylhistidine alone.

[c]$p < 0.01$ as compared with α-fluoromethylhistidine alone.

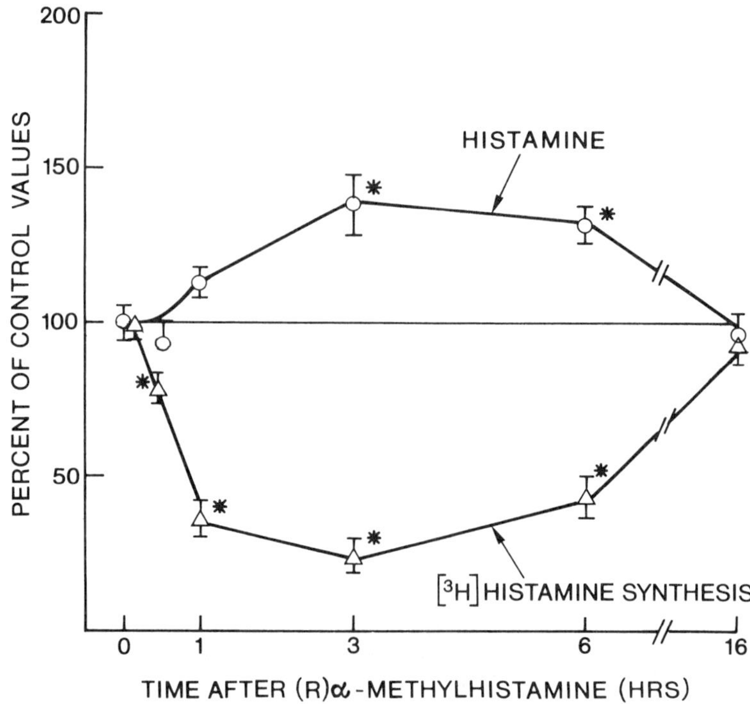

FIGURE 6. Effects of (R) α-methylhistamine on histamine level and [³H]histamine synthesis in rat cerebral cortex. The rats were killed at various times after oral administration of the H₃-receptor agonist (10 mg/kg). Histamine was measured by a radioenzymatic assay in a crude P₂ synaptosòmal fraction. Other groups of rats received [³H]L-histidine (250 μCi, intravenous) 10 minutes before being killed, and [³H]histamine synthetized was isolated by ion-exchange chromatography. (From Garbarg et al.[26] by permission of the *European Journal of Pharmacology*.)

the concentration of the amine in the cerebrospinal fluid (CSF), and therefore presumably in the interstitial fluid, is large enough to stimulate H₃ receptors.[29]

On the same tests (R) α-MeHA elicits with an ED₅₀ of ~5 mg/kg (per oral) an opposite, although generally less marked, response (FIGURE 6), indicating that, *in vivo,* H₃-receptors are not maximally activated by the endogenous amine under basal conditions.

From this, it appears that the autoreceptor-mediated control of histaminergic neuron activity constitutes a major regulatory mechanism under physiological conditions. Furthermore both types of H₃ ligands represent useful tools for investigating behavioral and other roles of these neurons, particularly the antagonists that constitute the first agents able to markedly enhance histaminergic transmissions in brain after systemic administration.

For example, the role of histaminergic neurons in arousal proposed earlier[30] was recently confirmed in cats treated with H₃-receptor ligands (Lin, Sakai, Arrang, Garbarg, Schwartz and Jouvet, submitted). Oral administration of (R) α-methylhistamine significantly enhanced deep slow-wave sleep (S₂), mainly at the expense of wakefulness and superficial slow-wave sleep. The effect was reversed by thioperamide which, administered alone, elicited a marked arousal effect. Thioperamide actions were blocked not only by (R) α-MeHA but also by mepyramine, an H_1 antagonist. These data suggest that the arousing effects of the H_3 antagonist were, indeed, mediated by an enhanced release of HA which acted via H_1 receptors presumably located in the posterior hypothalamus.[31,32]

A "sedative" action mediated by H_3 receptors was also evidenced in rats in which infusion of (R) α-MeHA into the nucleus accumbens depressed the locomotor activity, an action reversed by thioperamide.[33]

HETERORECEPTOR-MEDIATED MODULATION OF HISTAMINE SYNTHESIS AND RELEASE IN BRAIN

Using the same *in vitro* models as those on which H_3 autoreceptors were evidenced, heteroreceptor-mediated controls of HA synthesis and release were recently identified. Three pharmacologically defined receptor subtypes seem to subserve such functions as α_2-adrenergic receptors,[8,34] M_1-muscarinic receptors,[35] and κ-opioid receptors.[36] These

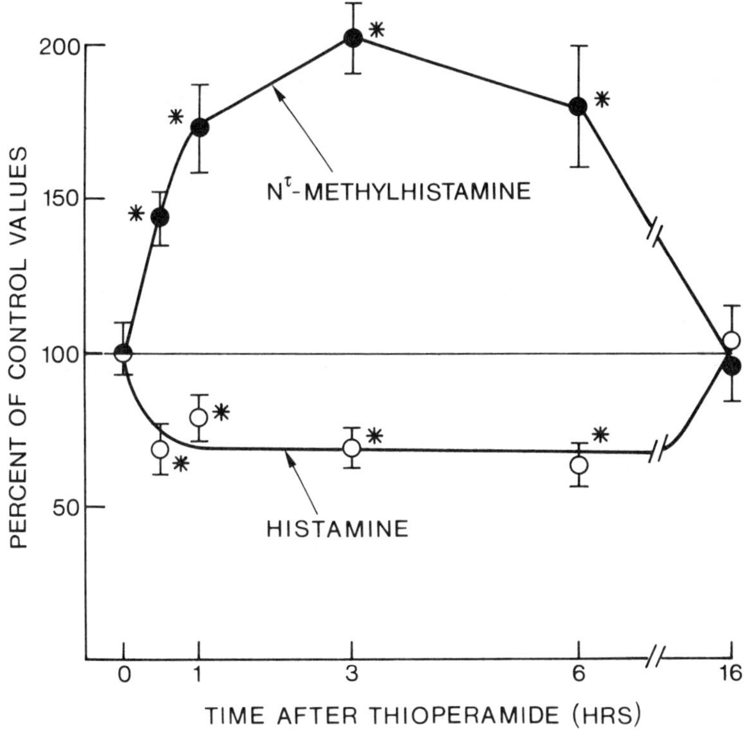

FIGURE 7. Effects of thioperamide (5 mg/kg, per oral), an H_3-receptor antagonist, on histamine and N$^\tau$-methylhistamine levels in rat cerebral cortex. Histamine was measured by a radioenzymatic assay on a crude P_2 synaptosomal fraction, and its methylated metabolite was radioimmunoassayed on a perchloric acid extract. (From Garbarg *et al.*[26] by permission of the *European Journal of Pharmacology*.)

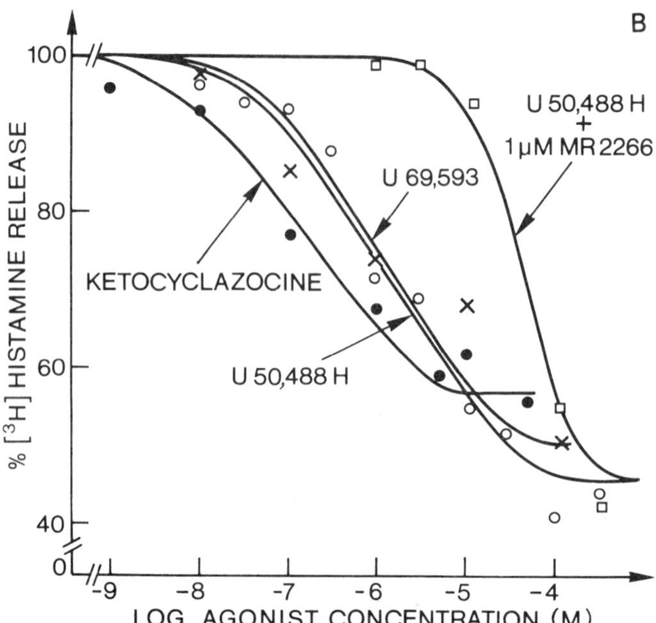

FIGURE 8. Effects of various kappa opioid receptor agonists and an antagonist on [³H]histamine release from slices of rat cerebral cortex. The effect of U50,488 H was inhibited by MR 2266, a kappa receptor antagonist, with an apparent K_i of 36 ± 15 nM (data shown) as well as by naloxone (K_i = 170 ± 80 nM) or norbinaltorphimine (K_i = 2.6 ± 1.1 nM) (not shown).

actions are illustrated in the last case by the effects of several preferential κ-agonists on [³H]HA release from slices of cerebral cortex and the competitive antagonism by MR 2266 (FIGURE 8).

The responses mediated by these heteroreceptors share several common features. *Firstly,* the maximal inhibitions of [³H]HA release from brain slices are of similar amplitudes as those mediated by H_3 receptors (TABLE 2). It is noteworthy, however, that, on this model, several agents behaved as partial agonists, such as clonidine acting at α_2-adrenergic receptors,[8,34] oxotremorine at muscarinic M_1 receptors,[35] or dynorphin

TABLE 2. Inhibition of Histamine Release in Slices of Rat Cerebral Cortex by Auto- and Heteroreceptors

Agonist	Maximal Inhibition	Antagonist (K_i)	Receptor
(R) α-methylhistamine	60%	Thioperamide (4 nM)	H_3 histamine
Carbachol	60%	Pirenzepine (30 nM)	M_1 muscarinic
Noradrenaline	50%	Yohimbine (30 nM)	α_2 adrenergic
U 50,488	50%	Norbinaltorphimine (3 nM)	κ opioid

(1–13) at κ-receptors.[36] *Secondly* these heteroreceptors seem to represent true presynaptic receptors since they were shown, at least in the case of M_1-muscarinic and α-adrenergic receptors, to exert inhibitory actions on [³H]HA release from synaptosomes. *Thirdly* as in the case of H_3 autoreceptors, they seem to control not only [³H]HA release but also [³H]HA synthesis in depolarized brain slices.[35] Inhibition of [³H]HA synthesis can also be observed *in vivo* after administration of clonidine, an effect blocked by yohimbine,[34] or by oxotremorine, an effect blocked by scopolamine.[35]

However, one major difference between H_3 autoreceptors and the three heteroreceptor subtypes is that, at the latter, antagonists failed to enhance [³H]HA release from depolarized slices or [³H]HA synthesis *in vivo*.[34–36] This indicates that, under basal conditions, these heteroreceptors of histaminergic axons are not activated by the endogenous neurotransmitters. Nevertheless, this may not be true under all physiological conditions since, in the case of M_1-muscarinic receptors, neostigmine, an acetylcholinesterase inhibitor, inhibited [³H]HA release in a pirenzepine-reversible manner, suggesting the participation of endogenous acetylcholine.[35]

Among a large series of agents tested on the same models, none was found to modify [³H]HA release significantly (TABLE 3). These conclusions should be under-

TABLE 3. Agents that Failed to Significantly Affect Histamine Release from Slices of Rat Cerebral Cortex at the Indicated Concentration (μM)

A. Putative cotransmitters of histaminergic neurons			
• GABA	(100)	• Galanin	(10)
• Adenosine	(100)	• Substance P	(10)
• Enkephalins	(1)	• TRH	(20)
B. Other agents			
• Serotonin	(10)	• Cholecystokinin-8	(4)
• Nicotine	(100)	• Bombesine	(10)
• Dopamine	(1)	• Neurotensin	(10)
• Apomorphine	(10)	• Somatostatin	(10)
• Aspartate	(100)	• Neuropeptide Y	(10)
• Glutamate	(100)		

lined in the case of the numerous neurotransmitters (or putative neurotransmitters) that immunohistochemical studies have shown (or suggested) to coexist with HA in tuberomammillary neurons, i.e., γ-aminobutyric acid (GABA), adenosine, substance P, galanin, enkephalins, or thyrotropin releasing hormone (TRH) (reviewed in Schwartz *et al.*).[2] Assuming that these substances are coreleased with HA (which remains to be established), this suggests that they do not participate in a "heterofeedback" control of the amine release. It would be interesting to assess whether the reverse is also true, i.e., whether H_3 receptors participate in the control of these cotransmitters' release, but, in view of the relatively low number of histaminergic neurons in brain, this presents some technical difficulties.

CONCLUSIONS

Most of the observations regarding the operation of autoreceptors and heteroreceptors on histaminergic neurons are clearly reminiscent of those of the corresponding receptors on other monoaminergic systems in brain. Furthermore, the molecular and intracellular mechanisms leading to the observed responses have been often investigated with more detail in the case of other presynaptic receptors.

We believe that, at this stage, the major interest of our studies lies in the fact that they have enabled a better understanding of the role of histaminergic systems and led to the development of novel pharmacological tools that can be used to affect these systems in experimental studies and, possibly, in therapeutics. For instance, the H_3-receptor antagonists crossing the blood-brain barrier represent the first class of agents able to promote histaminergic neurotransmission in the CNS and they have already allowed us to confirm the role of HA as a "waking" amine. This role may as well account, at least partly, for the "sedative" properties of agents inhibiting HA release via activation of presynaptic α_2-adrenergic or κ-opioid receptors. In addition, the presence of H_3 receptors in peripheral tissues such as the lung, where they inhibit not only HA synthesis[5] but also apparently the release of peptidergic transmitters controlling various airway functions,[37,38] may offer new therapeutic opportunities for H_3-receptor agonists.

REFERENCES

1. SCHWARTZ, J. C. 1975. Histamine as a transmitter in brain. Life Sci. **17:** 503–518.
2. SCHWARTZ, J. C., J. M. ARRANG, M. GARBARG, H. POLLARD & M. RUAT. Histaminergic transmission in the mammalian brain. Physiol. Rev. (In press.)
3. VERDIÈRE, M., C. ROSE & J. C. SCHWARTZ. 1975. Synthesis and release of histamine studied on slices from rat hypothalamus. Eur. J. Pharmacol. **34:** 157–168.
4. ARRANG, J. M., M. GARBARG & J. C. SCHWARTZ. 1983. Autoinhibition of brain histamine release mediated by a novel class (H_3) of histamine receptor. Nature London **302:** 832–837.
5. ARRANG, J. M., M. GARBARG, J. C. LANCELOT, J. M. LECOMTE, H. POLLARD, M. ROBBA, W. SCHUNACK & J. C. SCHWARTZ. 1987. Highly potent and selective ligands for histamine H_3-receptors. Nature London **327:** 117–123.
6. ARRANG, J. M., M. GARBARG & J. C. SCHWARTZ. 1985. Histamine synthesis and release in CNS: control by autoreceptors (H_3). In Frontiers in Histamine Research. C. R. Ganellin & J. C. Schwartz, Eds.: 143–153. Pergamon. Oxford, England.
7. ARRANG, J. M., M. GARBARG & J. C. SCHWARTZ. 1985. Autoregulation of histamine release in brain by presynaptic H_3-receptors. Neuroscience **15:** 553–562.
8. HILL, S. J. & R. M. STRAW. 1988. α 2-Adrenoceptor-mediated inhibition of histamine release from rat cerebral cortical slices. Br. J. Pharmacol. **95:** 1213–1219.
9. VAN DER WERF, J. F., A. BAST, G. J. BIJLOO, A. VAN DER VLIET & H. TIMMERMAN. 1987. HA autoreceptor assay with superfused slices of rat brain cortex and electrical stimulation. Eur. J. Pharmacol. **138:** 199–206.
10. VAN DER WERF, J. F., G. J. BIJLOO, A. VAN DER VLIET, A. BAST & H. TIMMERMAN. 1987. H_3 receptor assay in electrically stimulated superfused slices of rat brain cortex; effects of N^x-alkylated histamines and impromidine analogues. Agents Actions **20:** 239–243.
11. VAN DER VLIET, A., J. F. VAN DER WERF, A. BAST & H. TIMMERMAN. 1988. Frequency-dependent autoinhibition of histamine release from rat cortical slices: a possible role for H_3 receptor reserve. J. Pharm. Pharmacol. **40:** 577–579.
12. SCHWARTZ, J. C., M. GARBARG & H. POLLARD. 1986. Histaminergic transmission in the brain. In Handbook of Physiology. The Nervous System. F. E. Bloom, V. B. Mountcastle & S. R. Geiger, Eds. **4:** 257–316. American Physiological Society. Bethesda, Md.
13. ARRANG, J. M., M. GARBARG & J. C. SCHWARTZ. 1987. Autoinhibition of histamine synthesis mediated by presynaptic H_3-receptors. Neuroscience **23:** 149–157.
14. PANULA, P., H. Y. T. YANG & E. COSTA. 1984. Histamine-containing neurons in the rat hypothalamus. Proc. Nat. Acad. Sci. USA **81:** 2572–2576.
15. WATANABE, T., Y. TAGUCHI, H. HAYASHI, H. WADA, H. KUBOTA, Y. TERANO, J. TANAKA,

S. SHIOSAKA & M. TOHYAMA. 1983. Evidence for the presence of a histaminergic neuron system in the rat brain: an immunochemical analysis. Neurosci. Lett. **39:** 249–254.

16. POLLARD, H., I. PACHOT, P. LEGRAIN, G. BUTTIN & J. C. SCHWARTZ. 1985. Development of a monoclonal antibody against L-histidine decarboxylase as a selective tool for the localisation of histamine synthetising cells. *In* Frontiers in Histamine Research. C. R. Ganellin & J. C. Schwartz, Eds.: 103–118. Pergamon. Oxford, England.

17. ARRANG, J. M., J. C. SCHWARTZ & W. SCHUNACK. 1985. Stereoselectivity of the histamine H_3-presynaptic autoreceptor. Eur. J. Pharmacol. **117:** 109–114.

18. ARRANG, J. M., M. GARBARG, T. T. QUACH, M. DAM TRUNG TUONG, E. YERAMIAN & J. C. SCHWARTZ. 1985. Actions of betahistine at histamine receptors in the brain. Eur. J. Pharmacol. **111:** 73–84.

19. ARRANG, J. M., N. DEFONTAINE & J. C. SCHWARTZ. 1988. Phencyclidine blocks histamine H_3-receptors in rat brain. Eur. J. Pharmacol. **157:** 31–35.

20. ARRANG, J. M., B. DEVAUX, J. P. CHODKIEWICZ & J. C. SCHWARTZ. 1988. H_3-receptors control histamine release in human brain. J. Neurochem. **51:** 105–108.

21. BEN-ARI, Y., G. LE GAL LA SALLE, G. BARBIN, J. C. SCHWARTZ & M. GARBARG. 1977. Histamine synthesizing afferents within the amygdaloid complex and bed nucleus of the stria terminalis of the rat. Brain Res. **138:** 285–294.

22. SCHLICKER, E., R. BETZ & M. GOTHERT. 1988. Histamine H_3-receptor-mediated inhibition of serotonin release in the rat brain cortex. Naunyn-Schmiedeberg's Arch. Pharmacol. **337:** 588–590.

23. EA KIM, L. & N. OUDART. 1988. A highly potent and selective H_3 agonist relaxes rabbit middle cerebral artery, in vitro. Eur. J. Pharmacol. **150:** 393–396.

24. POLLARD, H., S. BISCHOFF & J. C. SCHWARTZ. 1973. Modifications of brain histamine metabolism induced by antihistamines. Agents Actions **3:** 190.

25. HOUGH, L. B., S. JACKOWSKI, N. EBERLE, K. R. GOGAS, N. A. CAMAROTA & D. CUE. 1989. Actions of the brain penetrating H_2-antagonist zolantidine on histamine dynamics and metabolism in rat brain. Biochem. Pharmacol. **37:** 4707–4711.

26. GARBARG, M., M. D. TRUNG TUONG, C. GROS & J. C. SCHWARTZ. 1989. Effects of histamine H_3-receptor ligands on various biochemical indices of histaminergic neuron activity in rat brain. Eur. J. Pharmacol. **164:** 1–11.

27. OISHI, R., Y. ITOH, M. NISHIBORI & K. SAEKI. 1989. Effects of the histamine H_3-agonist (R) α-methylhistamine and the antagonist thioperamide on histamine metabolism in the mouse and rat brain. J. Neurochem. **52:** 1388–1392.

28. GARBARG, M., G. BARBIN, E. RODERGAS & J. C. SCHWARTZ. 1980. Inhibition of histamine synthesis in brain by α-fluoromethylhistidine, a new irreversible inhibitor: in vitro and in vivo studies. J. Neurochem. **35:** 1045–1052.

29. KHANDELWAL, J. K., L. B. HOUGH, A. M. MORRISHOW & J. P. GREEN. 1982. Measurement of tele-methylhistamine and histamine in human cerebrospinal fluid, urine, and plasma. Agents Actions **12:** 583–590.

30. SCHWARTZ, J. C. 1977. Histaminergic mechanisms in brain. Annu. Rev. Pharmacol. Toxicol. **17:** 325–339.

31. LIN, J. S., K. SAKAI & M. JOUVET. 1988. Evidence for histaminergic arousal mechanisms in the hypothalamus of cat. Neuropharmacology **27:** 111–122.

32. BOUTHENET, M. L., M. RUAT, N. SALÈS, M. GARBARG & J. C. SCHWARTZ. 1988. A detailed mapping of histamine H_1-receptors in guinea-pig central nervous system established by autoradiography with [^{125}I]iodobolpyramine. Neuroscience **26:** 553–600.

33. BRISTOW, L. J. & G. W. BENNETT. 1988. Biphasic effects of intra-accumbens histamine administration on spontaneous motor activity in the rat; a role for central histamine receptors. Br. J. Pharmacol. **95:** 1292–1302.

34. GULAT-MARNAY, C., A. LAFITTE, J. M. ARRANG & J. C. SCHWARTZ. 1989. Modulation of histamine release and synthesis in the brain mediated by α_2-adrenoceptors. J. Neurochem. **53:** 519–524.

35. GULAT-MARNAY, C., A. LAFITTE, J. M. ARRANG & J. C. SCHWARTZ. 1989. Regulation of histamine release and synthesis in the brain by muscarinic receptors. J. Neurochem. **52:** 248–254.

36. GULAT-MARNAY, C., A. LAFITTE, J. M. ARRANG & J. C. SCHWARTZ. Modulation of histamine release in the rat brain by kappa opioid receptors. J. Neurochem. (In press.)
37. ICHINOSE, M. & P. J. BARNES. 1989. Histamine H_3-receptors modulate nonadrenergic noncholinergic bronchoconstriction in guinea pig in vivo. Eur. J. Pharmacol. **174:** 49–56.
38. ICHINOSE, M., M. G. BELVISI & P. J. BARNES. 1990. Histamine H^3-receptors inhibit neurogenic microvascular leakage in airways. J. Appl. Physiol. **68:** 21–25.

Estimation of the Role of Presynaptic α_2-Adrenoceptors in the Circulation

Influence of Neuronal Uptake

JAMES A. ANGUS,[a] ANN C. DYKE,
AND PAUL I. KORNER

Baker Medical Research Institute
Melbourne, Victoria, Australia

INTRODUCTION

Activation of presynaptic α_2-adrenoceptors located on sympathetic varicosities causes a reduction in transmitter release and postjunctional response. The existence of these receptors can be readily demonstrated pharmacologically by applying the α_2-adrenoceptor agonist clonidine. But the more difficult question to answer is whether these receptors are activated by transmitter norepinephrine released during sympathetic activity to initiate so-called autoinhibitory feedback, leading to subsequent reduction of transmitter release. Under most circumstances there may be important reasons why these receptors are *not* activated. Here we review (1) our contribution to the debate of the role of autoinhibitory feedback in sympathetic transmission; (2) the importance of neuronal uptake and its interaction with α_2-adrenoceptor blockade; (3) the variability of autoinhibitory feedback between rat and guinea pig atria; and (4) the development of neuronal uptake and prejunctional α_2-adrenoceptors in SHR and WKY rat atria and in mesenteric resistance arteries.

EVIDENCE AGAINST AUTOINHIBITORY FEEDBACK IN GUINEA PIG RIGHT ATRIA

Transmitter released from sympathetic nerves cannot be measured directly. As an approximation, either ^{3}H-labeled (or unlabeled) norepinephrine and its metabolites can be measured in the surrounding bathing medium or we can assay the effector response to the released transmitter. Each method has advantages and limitations.

In the guinea pig isolated right atrium in the presence of atropine (1 μM), a single 2 msecond electrical field pulse delivered from adjacent field electrodes will cause a reproducible tachycardia and a fall in atrial period of about 30 mseconds measured from the atrial surface electrogram (FIGURE 1). Four field pulses delivered one per four consecutive refractory periods gives a marked peak tachycardia that is about 50% of the maximum tissue response that can be achieved to β-adrenoceptor stimulation by isoproterenol. This method of refractory period field stimulation of sympathetic nerves is dependent on the atrial rate which is between 120 and 180 beats/minute (i.e. 2–3 Hz). If there is autoinhibitory feedback, then the α_2-adrenoceptor antagonist phentolamine (1–10 μM) should enhance the tachycardia in response to four field pulses. We did not observe this in our experiments (FIGURE 1), calling into question the general

[a] Address correspondence to Dr. Angus at the Baker Medical Research Institute, Commercial Road, Prahran, Victoria 3181, Australia.

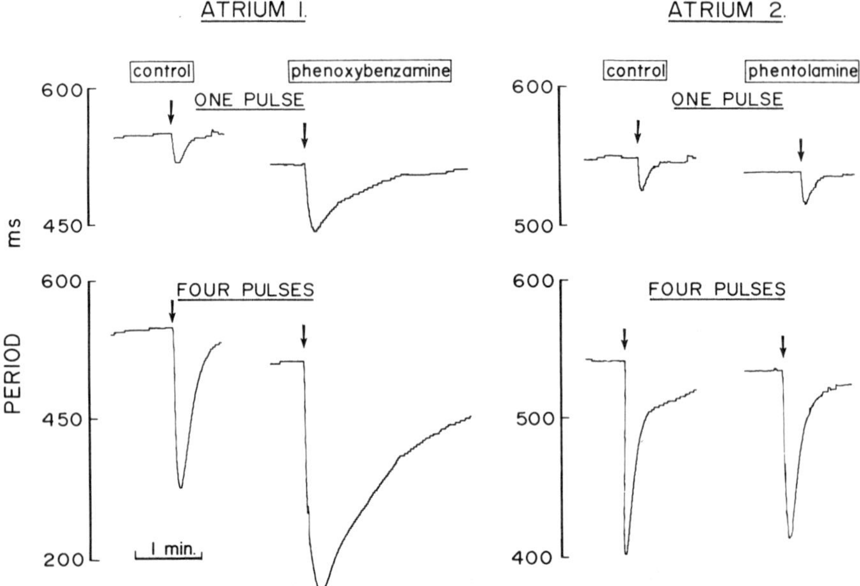

FIGURE 1. Chart records of the atrial period from two guinea pig right atria in response to one (top traces) and four (bottom traces) electrical field pulses delivered as one per refractory period. In atrium 1 the responses were increased and prolonged by phenoxybenzamine (10 μM) pretreatment for 15 minutes followed by 30 minutes wash. In contrast the responses were little affected by phentolamine (10 μM) pretreatment for 30 minutes. Atropine (1 μM) was present throughout. (Reproduced from Reference 1 with permission from Macmillan Press Ltd.)

role of autoinhibitory feedback.[1] Phenoxybenzamine which blocks neuronal uptake and α_2-adrenoceptors markedly enhanced the peak and duration of the tachycardia to 1 and 4 field pulses; phentolamine (10 μM) and desipramine (0.1 μM) together also caused a similar increase in response to one and four field pulses. We concluded that at least in the guinea pig right atrium, these studies provided evidence against a physiologically important role of presynaptic α_2-adrenoceptors.[1]

At the time these findings were controversial and prompted others to reexamine the role of autoinhibitory feedback. Story and colleagues proposed that there were two major conditions that determined the operation of autoinhibitory feedback in atria; one that the field pulses in a train had to be sufficiently frequent ($\geq$4 seconds, i.e., 0.25 Hz) to enable the feedback to be effective and that the train had to be of adequate length >1.5 seconds at higher frequencies (2 Hz).[2] In addition, debate flourished over the use of ^{3}H efflux and tissue response (tachycardia) as the appropriate measure of transmitter release. We performed a wide ranging study where tachycardia and ^{3}H efflux could be measured simultaneously with the aid of a flow cell.[3] This brought to light a number of technical problems with the then available chemical methods of norepinephrine release. First, the release of total norepinephrine (6–12 fmol/field pulse) was low and beyond the sensitivity of electrochemical detectors used with high performance liquid chromatography (HPLC). Second, that ^{3}H-norepinephrine labeling did not trace label the releasable neuronal norepinephrine pool since 32% of this norepinephrine was labeled. Third, we found that a large amount of ^{3}H efflux came from the plastic tissue support and electrodes if the same bath was used for the ^{3}H labeling procedure and for the experiment. In response to these problems we developed a flow cell in which to run the experiment after the tissue had been ^{3}H labeled in a separate bath with the high specific activity, ring-labeled $(-)$norepinephrine (47.7 Ci/mmol, 0.1 μM).[3] In the absence of desipramine, phentolamine (1 μM) had no effect on either the ^{3}H efflux or tachycardia in response to 4–12 field pulses at 0.25 Hz (FIGURE 2 top panels). But at 2 Hz both measures of transmitter release indicated a significant enhancement for 12 pulses after phentolamine (FIGURE 3, top panels). Thus under normal conditions, in the presence of neuronal uptake, autoinhibitory feedback occurred only at high frequency,

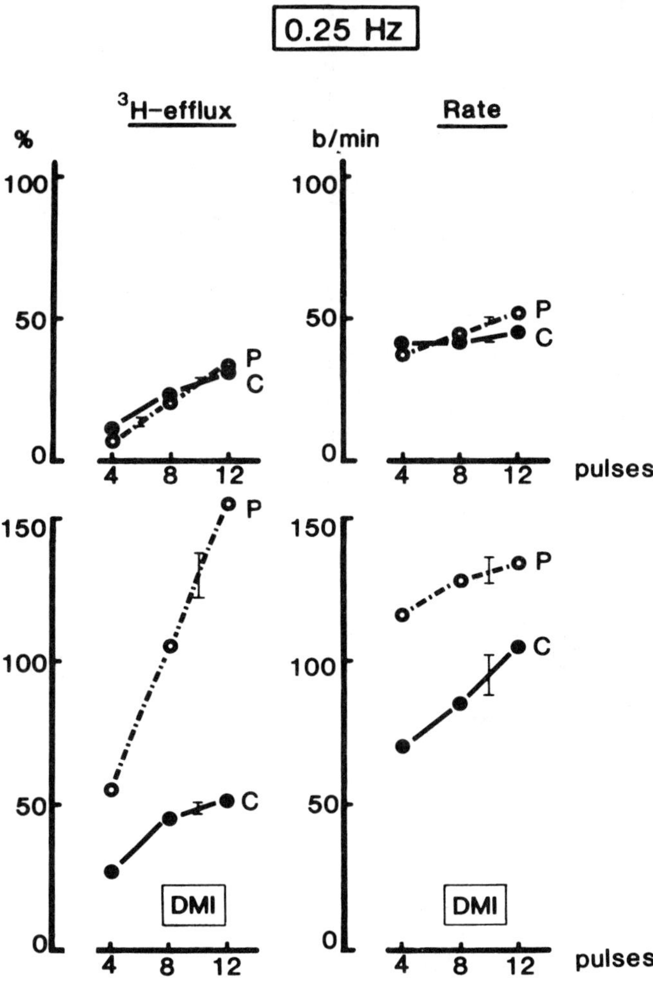

FIGURE 2. Relationship between ^{3}H efflux (% increase) (left panels) and increase in atrial rate (beats/minute) (right panels) with the number of field pulses (abscissae) at low frequency (0.25 Hz) in guinea pig right atria. Solid symbols and lines are normal atria (C, $n = 5$) open symbols and broken lines are separate atria ($n = 5$) equilibrated with phentolamine (P, 1 μM). Lower panels were atria pretreated with desipramine (DMI, 0.1 μM). (Reproduced from Reference 3 with permission from Macmillan Press Ltd.)

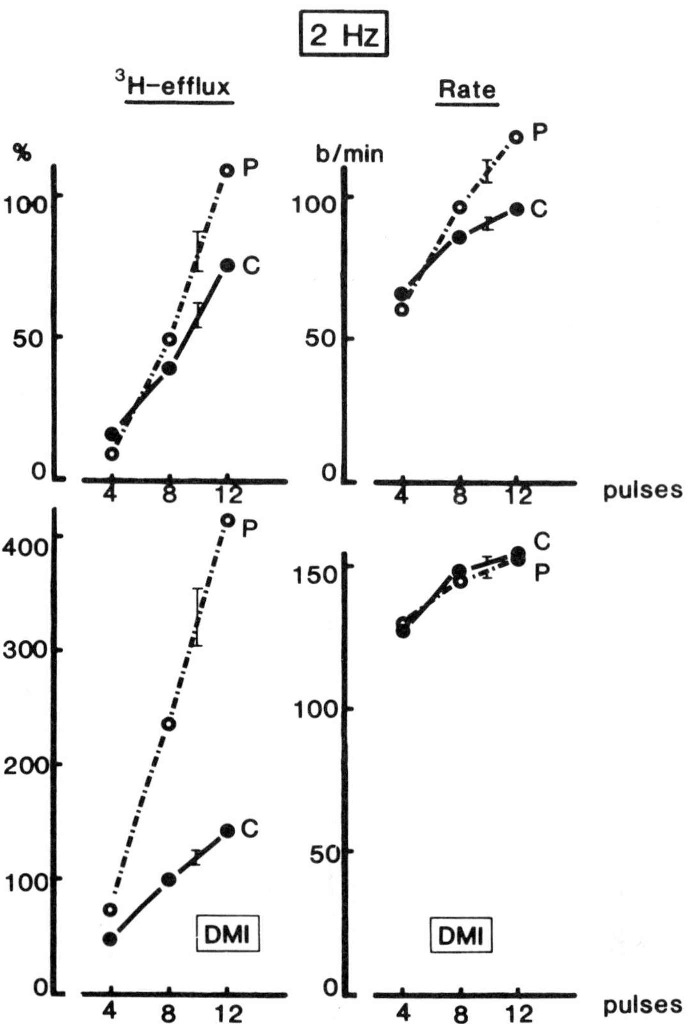

FIGURE 3. See legend for FIGURE 2, except that the field pulses (4–12) were applied at the higher frequency of 2 Hz.

and with prolonged pulse trains that gave near maximum tachycardia. This assumes, of course, that phentolamine at 1 μM acted solely to inhibit prejunctional α_2-adrenoceptors.

Repeating these experiments in the presence of inhibition of neuronal uptake was most revealing. Even at 4 pulses at 0.25 Hz there was marked enhancement of both efflux and tachycardia to 4, 8, or 12 field pulses. At 2 Hz, despite the marked increases in ^{3}H efflux of as much as 400% for 12 pulses there was no increase in response indicating that the biological assay was at tissue maximum. The important point here is that neuronal uptake normally limits the transmitter available to activate the autoinhibitory feedback. But if neuronal uptake is inhibited, or high-frequency stimulation is applied for a long train, then the autoinhibitory feedback can and does reduce transmitter release in the atrium. These studies suggest that the chemical assay of transmitter release alone gives information on the *capacity* of the release process while the biological response provides the scale of what conditions of stimulation are relevant to the effector system.

PHARMACOLOGICAL EVIDENCE OF PRESYNAPTIC α_2-ADRENOCEPTORS

Tachycardia in response to refractory period field stimulation of 1–4 field pulses in guinea pig right atrium was inhibited by clonidine in a concentration-dependent manner (FIGURE 4). Evidence that this was a presynaptic effect was that tachycardia to exogenously applied norepinephrine was unaffected by clonidine (100 nM). Moreover, a low concentration of phentolamine (0.1 μM) antagonized the action of clonidine in an apparently competitive fashion[4] (FIGURE 4, right). Thus pharmacological evidence for α_2-adrenoceptors located on these cardiac sympathetic nerve terminals is easily demonstrated. Moreover, the use of phentolamine at 1–10 μM is more than sufficient to test a role of autoinhibitory feedback given the interaction of phentolamine 0.1 μM with clonidine at this α_2-adrenoceptor.

INTERACTION BETWEEN AUTOINHIBITORY FEEDBACK AND NEURONAL UPTAKE: RAT AND GUINEA PIG ATRIA

The peak tachycardia, the integrated tachycardia trace (area) above baseline, and the half-response time ($t_{1/2}$) are also useful measures of transmitter release.[1,5] The $t_{1/2}$ was measured as the time taken for the atrial period to pass from resting before the onset of stimulation through the peak response and returned to a point halfway between the peak and resting periods. This measure was used to compare the effects of neuronal uptake block and α_2-adrenoceptor blockade. To block prejunctional α_2-

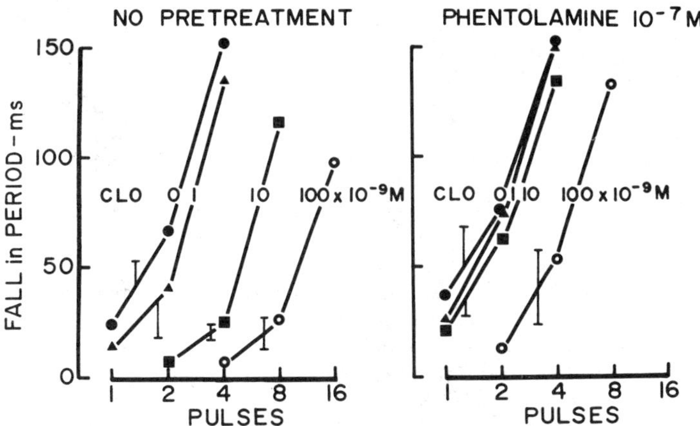

FIGURE 4. Relationship between the tachycardia (fall in atrial period) and number of electrical field pulses applied one per successive refractory period in guinea pig atria. Left, effect of clonidine (1–100 nM in 4 atria). Right, effect of clonidine in the presence of phentolamine (0.1 μM).

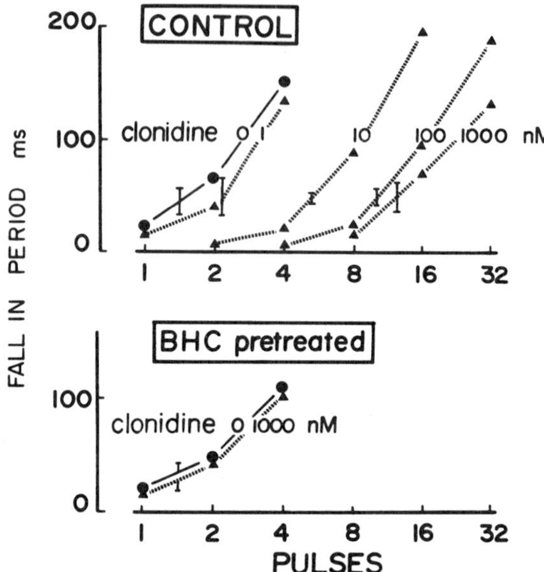

FIGURE 5. Relationship between tachycardia (fall in atrial period) and number of electrical field pulses (one per successive refractory period) in the presence of increasing concentrations of clonidine (upper panel). In separate atria pretreated with benextramine (1 μM for one hour and washing for one hour) the response to clonidine (1 μM) was completely blocked (lower panel).

adrenoceptors irreversibly, the atria were treated with benextramine 1 μM for 60 minutes followed by 60 minutes in drug-free solution. This treatment leaves the atria with a fully functional neuronal uptake system but completely unresponsive to the inhibitory effects of clonidine (FIGURE 5) (and see Reference 6).

In guinea pig and rat right atria, field pulses (1–8) delivered at 1 Hz gave reproducible $t_{1/2}$ values over four runs, each run separated by 30 minutes (FIGURES 6 and 7, left panels). Neuronal uptake inhibition (desipramine 0.01-1μM) increased $t_{1/2}$ substantially more in rat atria than in guinea pig atria (FIGURES 6 and 7, middle panels). In separate atria pretreated first with benextramine but in the absence of desipramine, there was very little increase in $t_{1/2}$ except at 8 pulses. But uptake blockade after benextramine caused a marked increase in $t_{1/2}$ in the guinea pig compared with uptake blockade alone. This suggests that autoinhibitory feedback was active, particularly with 4 field pulses, in the presence of uptake inhibition and limited the transmitter release and thus $t_{1/2}$. Thus the effect of either α_2-adrenoceptor blockade or uptake inhibition alone gives an inadequate measure of the function of each modulatory system due to the marked interactive effect of the alternative system. In the rat atrium, however, the neuronal uptake process seems to be the most important modulatory system. Unlike the guinea pig, the enhanced neural cleft concentration of norepinephrine in the presence of neuronal uptake block after 4–8 field pulses is *not* apparently sufficient to significantly activate the prejunctional α_2-adrenoceptors in the rat. Overall then, these comparative experiments show important species differences in the role of autoinhibitory feedback, being much greater in the guinea pig than in the rat right atrium. These experiments also indicate again the major importance of neuronal uptake in limiting and often preventing the action of autoinhibitory feedback.

We have speculated that the location of the prejunctional α_2-adrenoceptors may be somewhat distant (even extrasynaptic) from the site of transmitter release.[4] Under these circumstances, diffusion from a point source and neuronal uptake would lower the concentration of transmitter as it diffused towards the α_2-adrenoceptors (FIGURE 8).

DO DEVELOPMENT AND GENETIC HYPERTENSION ALTER NEURONAL UPTAKE OR α_2-ADRENOCEPTORS?

In the spontaneously hypertensive rat (SHR) there is increased transmitter turnover in the heart and some vascular beds particularly in the early phase of development of hypertension.[7] A decrease in neuronal uptake or in autoinhibitory activity could contribute to the increase in cardiac norepinephrine turnover. We first tested the postjunctional tissue response to β-adrenoceptor stimulation. Isoproterenol concentra-

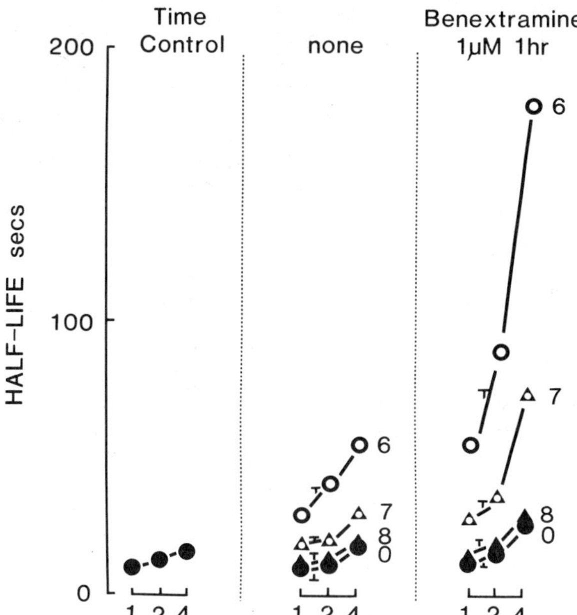

FIGURE 6. Effect of desipramine (DMI) on half-response time ($t_{1/2}$, half-life) of atrial period of trains of 1–4 field pulses in guinea pig atria. **Left panel:** time Control of 4 successive runs of 1–4 field pulses in the absence of DMI. The points for runs two, three, and four are superimposed on the first run. **Center panel:** Half-life responses to successive runs of field pulses in the absence (o) and presence of DMI 10^{-8} M (8); 10^{-7} M (7); 10^{-6} M (6). **Right panel:** As for center panel but after benextramine pretreatment (1 μM, 1 hour and washout). Data are average of 6 atria, error bar is the average standard error of the mean (SEM). (Reproduced from Reference 5 with permission from Blackwell Scientific Publications Ltd.)

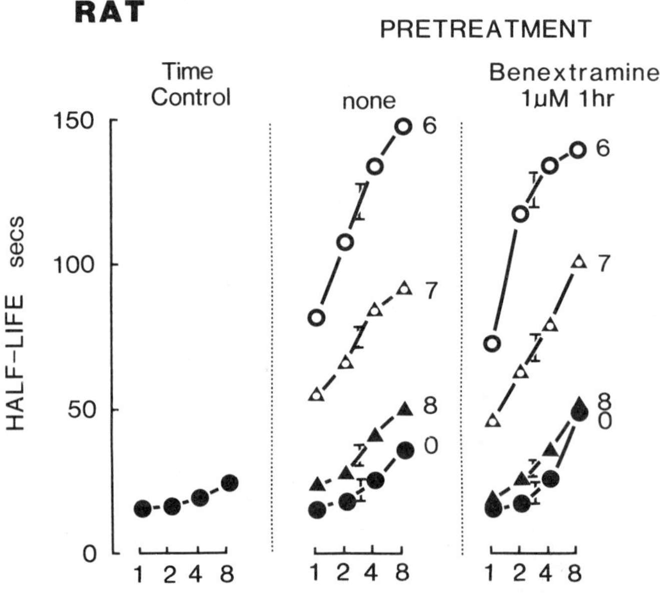

FIGURE 7. Effect of desipramine (DMI) on half-response time ($t_{1/2}$, half-life) of atrial period to trains of 1–8 field pulses in rat atria. Other details, see legend FIGURE 6. (Reproduced from Reference 5 with permission from Blackwell Scientific Publications.)

tion–tachycardia curves in atria from SHR and normotensive (WKY) rats had similar sensitivities (EC_{50}) across all ages (4–50 weeks), but there was an age-related fall in resting atrial rate in both strains. To test the role of neuronal uptake in both strains, field pulses (1–32 at 1 Hz) were applied in the absence and presence of desipramine (0.01–1 µM). The time-dependent recovery ($t_{1/2}$) of the tachycardia was increased by desipramine in a concentration-dependent manner that was similar in magnitude at all ages and in both strains.[8] This suggests that neuronal uptake at least in this cardiac sympathetic synapse was not the cause of increased norepinephrine turnover in SHR.

We also examined the activity of α_2-adrenoceptor stimulation by applying clonidine (0.1 µM). This concentration caused a marked rightward shift in the tachycardia–number of field pulses relationship in the guinea pig (see FIGURE 5). In the mature rat (20 weeks), clonidine was most effective in inhibiting transmitter release (FIGURE 9). The degree of presynaptic inhibition or "clonidine activity" was measured as a field pulse ratio by taking the ratio of field pulses giving equal tachycardia in the presence/absence of clonidine. At 4 weeks of age, the "clonidine activity" field pulse ratio was markedly reduced in both SHR and WKY atria to only 30–50% of that in the mature rats (FIGURE 9). In addition, over the entire age range the clonidine activity in SHR was less than in WKY atria (FIGURES 9 and 10). Given the previous findings that autoinhibitory feedback was relatively poorly developed in the rat atrium (FIGURE 7) even in the presence of neuronal uptake, these findings with clonidine are probably of little consequence in the SHR to explain the sympathetic overactivity. The interesting effect of age suggests that potentially the autoinhibitory α_2-adrenoceptor modulation is slow to develop in the rat heart.

ROLE OF AUTOINHIBITORY FEEDBACK IN RESISTANCE ARTERIES

Functional response of increased transmitter release is easily measured provided the pharmacological tools used to enhance release do not interfere with the postjunctional response. In the heart we made extensive use of the β-adrenoceptor-mediated response while manipulating the prejunctional α_2-adrenoceptors. The small resistance arteries removed from the rat mesentery and mounted in a Mulvany-Halpern

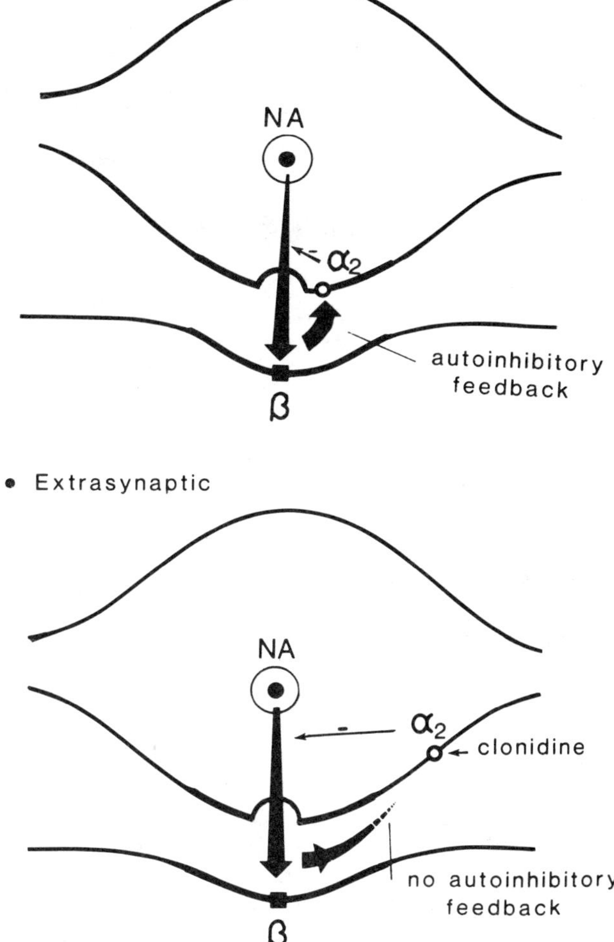

FIGURE 8. Schematic diagram of proposed location of α_2-adrenoceptors in guinea pig or rat atria. **Top:** Presynaptic location on sympathetic varicosities where the α_2-adrenoceptor is intrasynaptic or extrasynaptic (bottom). An extrasynaptic location would allow neuronal uptake and diffusion processes to lower transmitter concentration to below the threshold required for activation of autoinhibitory feedback. Clonidine can still inhibit transmitter release.

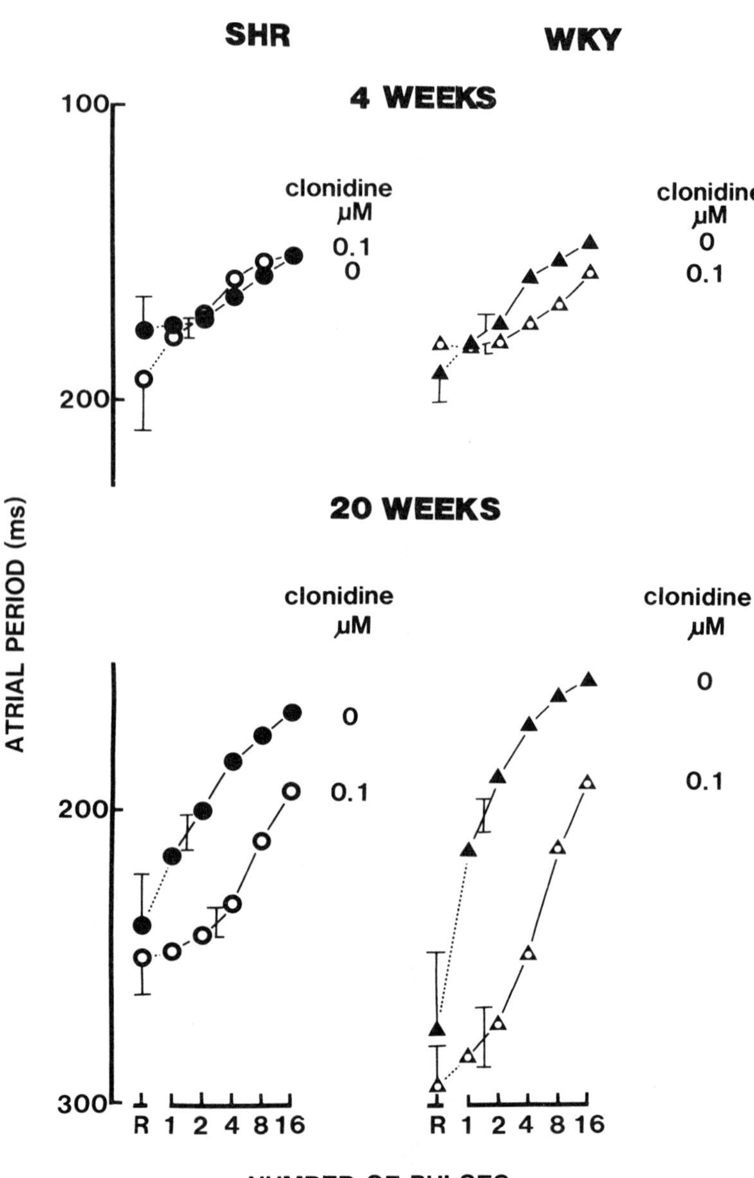

FIGURE 9. Atrial period in response to electrical field pulses (1–16) in right atria from SHR and WKY rats of 4 (top) and 20 (bottom) weeks old. Clonidine (0.1 µM) caused only a small right shift in the stimulus-response lines in 4 week atria compared with the mature atria (n = 5–6 in each group). (Reproduced from Reference 8 with permission from Current Science Ltd.)

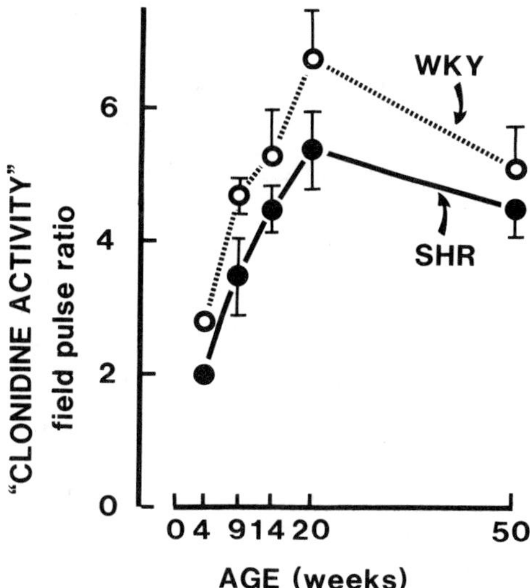

FIGURE 10. Effect of clonidine on the tachycardia in response to electrical field stimulation in SHR and WKY rat atria. The right shift in the stimulus-response curves (i.e., FIGURE 9) were calculated as clonidine activity–ratio of field pulse number in the presence/absence of clonidine (0.1 μM). (Error bars are ± 1 SEM, $n = 5$–7 atria per point). (Reproduced from Reference 8 with permission from Current Science Ltd.)

myograph[9,10] represent a convenient preparation to study the modulatory mechanisms surrounding sympathetic innervation of the resistance arteries. In these vessels (internal diameter 150–250 μm) the excitatory junction potential (ejp) is probably mediated by ATP while the major contractile response and more sustained depolarization is mediated by norepinephrine acting at postsynaptic α_1-adrenoceptors.[10] Electrical field pulses of 4–24 Hz for trains of 3 seconds cause increases in isometric wall tension (ΔT) that can be equated with increases in transmural pressure ΔP where ($\Delta P = \Delta T/$ internal radius, Laplace relationship). The response to 24 Hz, 3 seconds is about 60% of the maximum response to added norepinephrine. Here the response to electrical stimulation in control experiments (no added drugs) is greater in SHR vessels than in WKY vessels (FIGURE 11) since there is some increase in media thickness in SHR, and thus increased range of contraction to K^+ depolarization or norepinephrine concentration–response curves (Angus and Dyke, unpublished). We have examined the role of neuronal uptake and α_2-adrenoceptors at 9 weeks in SHR and WKY rats. Pretreatment of the vessels for 30 minutes with desipramine 0.1 μM—a concentration that markedly inhibited neuronal uptake in rat atria (FIGURE 7)—had no significant α_1-blocking activity against methoxamine concentration contraction curves in these vessels (Angus, unpublished). Surprisingly, this concentration of desipramine also had no significant effect on the response to field stimulation in either rat strain (FIGURE 11). To block the autoinhibitory α_2-adrenoceptors we used an α_1-adrenoceptor protection regime. Here prazosin 0.1 μM was equilibrated first with the vessels for 5 minutes before adding benextramine 3 μM for 5 minutes. The bathing solution was replaced with drug-free solution to allow the prazosin and benextramine to be removed leaving postjunctional α_1-adrenoceptors free to be activated by transmitter norepinephrine and the α_2-adrenoceptors irreversibly blocked by the covalently bound benextramine.[10] Restimulation showed a marked enhancement of response at all frequencies and in

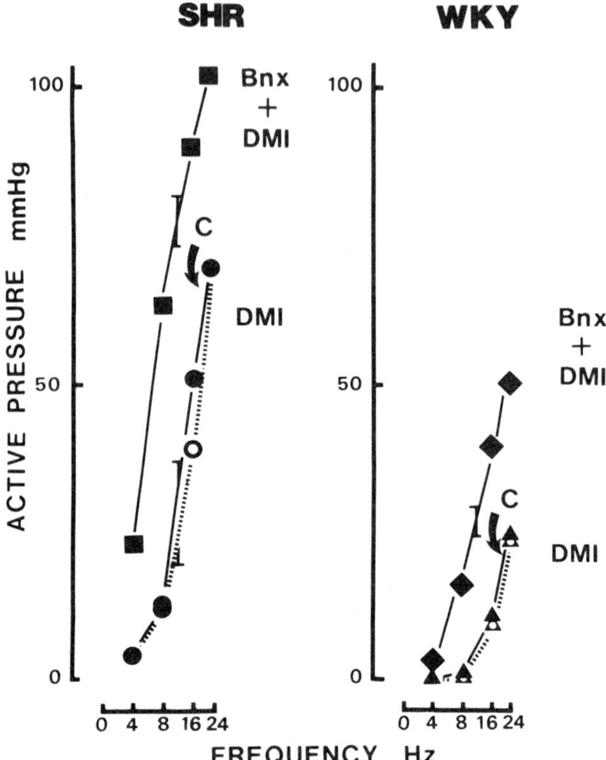

FIGURE 11. Average contraction responses of SHR ($n = 5$) and WKY ($n = 5$) rat isolated mesenteric resistance arteries to electrical field stimulation (4–24 Hz, 3 second train). Response was calculated as increase in pressure where $\Delta P = \Delta\text{tension}/\text{radius}$. Three stimulus-response lines were constructed in each vessel. First, in the absence of any inhibitors. Second, in the presence of desipramine (DMI, 0.1 μM), and third with DMI (0.1 μM) after pretreatment with benextramine (3 μM) to irreversibly block α_2-adrenoceptors.

both strains (FIGURE 11). In a separate group of vessels the benextramine treatment was applied first before the desipramine (FIGURE 12). The stimulus-response curves were left-shifted by benextramine alone, and desipramine treatment added nothing more. From this study, it would appear that the blood vessel neuroeffector junction has a poorly developed neuronal uptake; the major modulatory mechanism being provided by autoinhibitory feedback. Examination of the age-dependent development of this modulatory mechanism indicated that it was present at 4 weeks of age, as large as in mature rats, and similar in SHR and WKY rats (Angus and Dyke, unpublished).

CONCLUSIONS

Modulation of transmitter release by autoinhibitory feedback at sympathetic varicosities does not occur under physiological conditions under all circumstances and not to a uniform degree at the different synapses. Our studies suggest that there are three major variants of the control of transmitter release: (a) in guinea pig right atria where autoinhibitory feedback is of little importance unless intense stimulation is applied or neuronal uptake is inhibited; (b) in rat right atria where neuronal uptake is the most important system with a very minor role of autoinhibitory feedback even though the presynaptic α_2-adrenoceptors are present; (c) in rat mesenteric arteries where autoinhibitory feedback is of major importance and neuronal uptake of no consequence.

In vitro experiments do not simulate normal conditions of neural activity, but probably define optimum conditions under which autoinhibitory feedback could occur. Thus, under physiological neural activity, action potentials may not invade every varicosity.[13] On the other hand, field stimulation *in vitro* probably gives the "best chance" for autoinhibitory feedback to play a role since all varicosities are depolarized synchronously by each field pulse.[11,12] In our view, autoinhibitory feedback should not be considered as a general homeostatic mechanism. The variation in its importance may be related to the geometry of the synaptic cleft, the siting of the α_2-adrenoceptors, intra- or extrasynaptically, and the profusion of neuronal uptake.

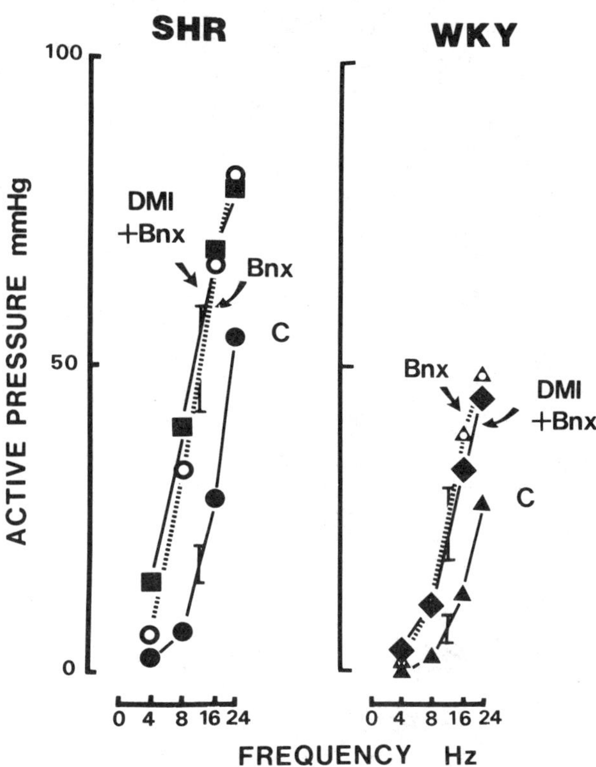

FIGURE 12. Contraction responses of SHR ($n = 5$) and WKY rat ($n = 5$) isolated mesenteric resistance arteries to electrical field stimulation (4–24 Hz, 3 second train). The order of treatment was, nil control (C); benextramine (Bnx), and Bnx + desipramine (DMI, 0.1 μM) (see legend FIGURE 11 for details).

REFERENCES

1. ANGUS, J. A. & P. I. KORNER. 1980. Nature **286**: 288–291.
2. STORY, D. F., M. W. McCULLOCH, M. J. RAND & C. A. STANDFORD-STARR. 1981. Nature **293**: 62–65.
3. ANGUS, J. A., A. BOBIK, G. P. JACKMAN, I. J. KOPIN & P. I. KORNER. 1984. Br. J. Pharmacol. **81**: 201–214.
4. KORNER, P. I., J. A. ANGUS, M. J. LEW & B. G. J. HEINZOW. 1983. Chest **83S**: 345S–349S.
5. DYKE, A. & J. A. ANGUS. 1988. J. Auton. Pharmacol. **8**: 219–228.
6. LEW, M. J. & J. A. ANGUS. 1984. Eur. J. Pharmacol. **98**: 27–34.
7. ADAMS, M., A. BOBIK & P. I. KORNER. 1989. Hypertension **14**: 191–202.
8. DYKE, A. C., J. A. ANGUS & P. I. KORNER. 1989. J. Hypertension **7**: 345–353.
9. MULVANY, M. J. & W. HALPERN. 1977. Circ. Res. **41**: 19–21.
10. ANGUS, J. A., A. BROUGHTON & M. J. MULVANY. 1988. J. Physiol. **403**: 495–510.
11. ANGUS, J. A. & P. I. KORNER. 1981. Nature **294**: 671–672.
12. SPERELAKIS, N. 1975. In Methods in Pharmacology. E. DANIEL & D. M. PATON, Eds. **3**: Chapter 16. Plenum. New York, N.Y.
13. BLAKELY, A. G. H. & T. C. CUNNANE. 1988. J. Physiol. **280**: 30P–31P.

Presynaptic Receptors in the
Neuromuscular Junction

W. C. BOWMAN, C. PRIOR, AND I. G. MARSHALL

Department of Physiology & Pharmacology
University of Strathclyde
Glasgow, G1 1XW, Scotland

INTRODUCTION

The suggestion that motor nerve endings in skeletal muscle possess cholinoceptors, though still controversial, is not new. Early evidence in support of their presence at this site notably includes that of Masland and Wigton,[1] and Riker, Hubbard, and Blaber, and their coworkers. Much of their early work has been described in reviews.[2–6] The question of whether such prejunctional cholinoceptors, assuming they exist, play a part in the normal transmission process or merely reflect some general, perhaps vestigeal, tendency for nonmyelinated neuronal membranes to respond pharmacologically to acetylcholine is also controversial. However, several workers,[2,3,6,7–13] including ourselves, have ascribed autoreceptor functions, at least to some of them, and have incorporated them into postulated physiological mechanisms for controlling evoked transmitter release.

PREJUNCTIONAL NICOTINIC RECEPTORS

The available evidence suggests to us[6,12] that there are two separate populations of prejunctional nicotinic receptors, each with different characteristics and each mediating different effects at the nerve endings. One such population may be involved in a physiological control mechanism over transmitter availability, whereas the other mediates a pharmacological effect, of excess acetylcholine or of other nicotinic agonist, that results in depolarization of some site at the terminal membrane. Although there is controversy over this interpretation, it is convenient to discuss prejunctional nicotinic receptors under headings that imply the existence of the subpopulations.

Nicotinic Autoreceptors that Mediate Enhanced Availability of Transmitter

It is well known that tubocurarine and related neuromuscular blocking drugs produce two effects on tetanic contractions of skeletal muscle evoked by stimulating the motor nerve: they depress peak tension, and they cause a rapid waning of tension to the resting level (i.e., tetanic fade) despite continuing stimulation. There is ample evidence that these two effects are consequences of separate and independent mechanisms of action since, for the same degree of tension depression, different neuromuscular blocking drugs produce different degrees of tetanic fade,[9,14,15] and even with the same neuromuscular blocking drug, the occurrence of tetanic fade, but not of depression of peak tension, is dependent on the route of injection.[6] Obviously, if both effects were a consequence of a common mechanism, they would hold a constant relationship to each other, and this is clearly not so.

69

It is not necessary to stimulate the nerve with tetanic frequencies in order to observe the two effects and their independence one from another. The pattern of stimulation often used by anesthesiologists and known as train-of-four stimulation[16] is adequate. In this pattern of stimulation, the motor nerve is stimulated with groups of 4 stimuli delivered at 2 Hz. There is an interval of 30 seconds between groups. Neuromuscular blocking drugs both depress amplitude and cause rundown of the twitches within each group (train-of-four fade).

The electrical counterpart of tetanic fade is the well-known rundown, to a plateau, in trains of end plate potentials (epps)[17] or end plate currents (epcs).[18] Glavinovic showed that in the absence of tubocurarine, there was relatively little rundown in trains of end plate currents, but when tubocurarine was added, not only was overall amplitude depressed, but rundown within each train became pronounced.[18]

Depression of peak amplitude produced by tubocurarine and related drugs is mainly attributed to block of postjunctional acetylcholine receptors, and there is abundant evidence that this is the explanation. It remains then to determine the mechanism underlying fade. Fade is of course a use-dependent phenomenon, and attempts have been made to explain it in terms of another use-dependent effect produced by tubocurarine and related drugs, that is, occlusion of the postjunctional acetylcholine receptor-operated ion channels.[19] However, examination of the evidence makes this explanation untenable.[20] Rundown in trains of epcs produced by tubo-curarine is not affected by changes in membrane potential,[21,22] yet ion-channel occlusion is strongly voltage dependent.[23] Furthermore, it has been calculated from a knowledge of the rate constants for association with and dissociation from open channels that far too few channels could be occluded by tubocurarine to contribute to rundown of epcs.[21,22] Receptor desensitization is also excluded as a factor in rundown, since this phenomenon too is voltage dependent, yet rundown is not. Gibb and Marshall effectively excluded any kind of postjunctional mechanism from underlying rundown produced by tubocurarine in experiments in which they recorded trains of epcs evoked by repetitive nerve stimulation at 50 Hz and compared them with trains evoked by repetitive jets of ionophoretically released acetylcholine at the same frequency.[22] Tubocurarine produced both depression of initial neurally evoked epc amplitude and rundown to a plateau of the neurally evoked trains of epcs, but it caused only a uniform depression of amplitude, without rundown, in the trains of responses evoked by ionophoretically applied acetylcholine. The quantal content of epcs of the plateau was reduced when compared with the initial response of the train.[24]

FIGURE 1 illustrates analogous experiments[25] with train-of-four stimulation and with another neuromuscular blocking drug, vecuronium. Depression of the first response and subsequent rundown are evident in the neurally evoked mechanical responses and neurally evoked epcs in the presence of vecuronium, but only a uniform depression of amplitude is seen in the train-of-four currents evoked by ionophoretic application of acetylcholine. Similar results to those obtained with tubocurarine and vecuronium were obtained with several other blocking drugs, including hexa-methonium,[24,26–28] but with erabutoxin b, only a uniform depression of amplitude of epcs was produced whether they were evoked by nerve stimulation or by ionophoreti-cally applied acetylcholine; i.e., there was no rundown. When a drug that acts at the neuromuscular junction primarily by occluding open receptor-activated ion channels rather than by binding to the acetylcholine binding sites was tested (i.e., tri-metmethaphan),[29] rundown was evident in trains of both neurally evoked and ionophoret-ically evoked epcs (FIGURE 2). These results show beyond reasonable doubt that, in addition to blocking postjunctional acetylcholine receptors, tubocurarine and related drugs act on the nerve endings to impair a component of evoked acetylcholine release in a use-dependent manner.

Experiments in which evoked transmitter release was measured in nerve-muscle preparations previously loaded with tritium-labeled choline[30,31] broadly confirmed the electrophysiological experiments, showing that tubocurarine and related drugs, including hexamethonium (FIGURE 3), produce a frequency-dependent reduction in evoked transmitter output.

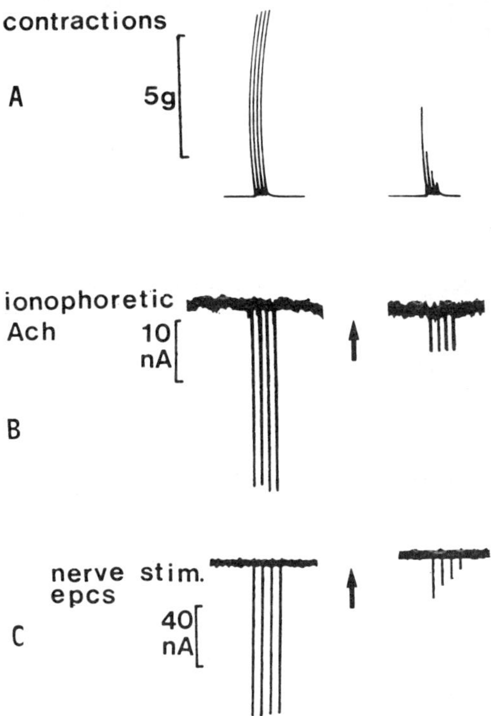

FIGURE 1. Isolated phrenic nerve–hemidiaphragm preparations of rats. All responses are to train-of-four stimulation (2 Hz for 1.9 seconds) before (left) and in the presence of 5 μM vecuronium (right). A: Twitches evoked by nerve stimulation. B: End plate current responses, recorded from a cut muscle fiber clamped at −80 mV, evoked by jets of acetylcholine applied ionophoretically. C: End plate currents (epcs), recorded from a cut muscle fiber clamped at −60 mV, evoked by stimulation of the motor nerve. Rundown, as distinct from depression of overall amplitude, was present only when the nerve was stimulated (A and C). From the same experiments as those published in Reference 25.

These experiments provide convincing evidence that tubocurarine and related quaternary ammonium compounds act on the nerve endings, but they do not provide any evidence that the prejunctional site concerned is an acetylcholine receptor as distinct from, for example, a voltage-operated ion channel. If an acetylcholine receptor is involved, then acetylcholine and other nicotinic agonists should exert the opposite effect to tubocurarine, and should antagonize the inhibitory effect of tubocurarine, assuming it blocks competitively. Here the electrophysiological evidence leaves more room for doubt, since in a large number of experiments in which acetylcholine,

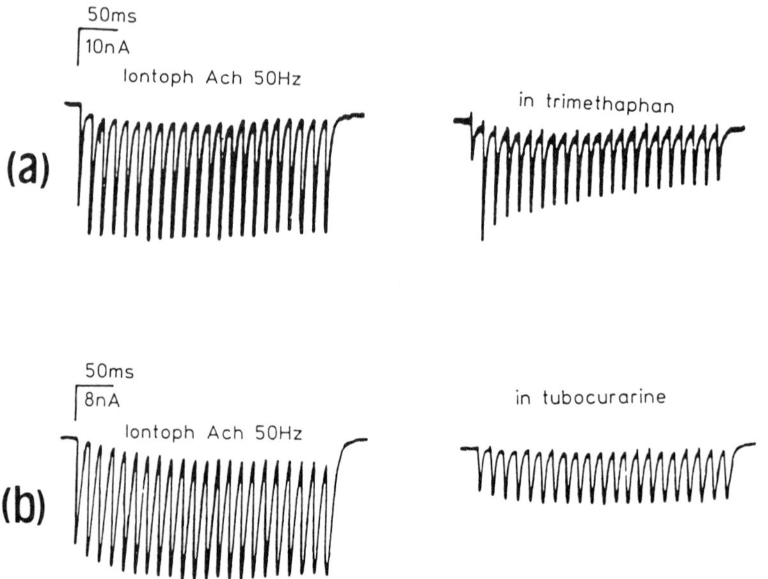

FIGURE 2. Cut rat diaphragm fibers, voltage clamped at −60 mV. End plate currents generated by ionophoretically applied jets of acetylcholine at a frequency of 50 Hz before (on the left) and in the presence of (on the right) either (a) trimethaphan (25 μM) or (b) tubocurarine (0.25 μM). Note the rundown of responses in the presence of trimethaphan (a channel blocker), but uniform depression with absence of rundown in the presence of tubocurarine. (a) and (b) are from different experiments of the series described in Reference 29.

carbachol, or dimethylphenylpiperazinium (DMPP) was tested, we have found it difficult to demonstrate any enhancement of epc amplitude in the absence of tubocurarine, and reversal of tubocurarine-induced rundown was weak and inconsistent. At best, and only at 37°C but not at room temperature, acetylcholine produced no more than a 25–30% reversal of tubocurarine-induced rundown;[26] sometimes no reversal is detectable. Possibly this difficulty arises because of the abnormal way that the nicotinic agonist is necessarily applied to the nerve terminals. The whole junction becomes flooded, and it may be that the receptors rapidly desensitize. Alternatively it might be that tubocurarine, hexamethonium, and related drugs bind most favorably, not with the acetylcholine recognition site of a prejunctional receptor, but instead with its associated ion channel, after the manner that they are proposed to do in some autonomic ganglia.[32] The results of experiments in which labeled choline output is estimated[33,34] are more convincing than the electrophysiological experiments with regard to the facilitatory effects of nicotinic agonists on transmitter release (FIGURE 3). Under these conditions, a clear increase in evoked output was produced by brief application of acetylcholine, nicotine, or DMPP,[30,31,35,36] although with prolonged application the opposite effect occurred and this was attributed by the authors to receptor desensitization.[35,37] Similarly, during depressed output produced by the continued presence of tubocurarine or hexamethonium, the nicotinic agonists restored output towards control.

Until a suitable irreversible ligand, analogous to α-bungarotoxin at the postjunctional receptor, is discovered and the receptors are biochemically isolated and cloned, it will probably not be possible to accept the presence of prejunctional receptors with absolute conviction. α-Bungarotoxin itself is not suitable for binding irreversibly to nerve terminals,[38] and, generally speaking, this and other α-neurotoxins (cobra toxin, erabutoxin b), even when they can be shown to produce prejunctional effects,[39,40] appear to be readily removed by washing. κ-Bungarotoxin and κ-flavitoxin do not produce fade or rundown,[41] and therefore presumably do not bind to prejunctional cholinoceptors.

Nevertheless, despite these reservations, it seems to us that the circumstantial evidence is in favor of the presence of prejunctional nicotinic receptors that function in a positive feedback mechanism to facilitate transmitter mobilization within the terminal axoplasm, so that availability for release keeps pace with the demands of high frequencies of nerve impulses. Block of the postjunctional receptors depresses only the amplitude of the responses, whereas block of the prejunctional receptors causes the various "fade" phenomena (tetanic fade, train-of-four fade, rundown to a plateau in trains of epps and epcs) and accounts for the frequency dependence of neuromuscular block produced by tubocurarine and related drugs.[20,42–44] It is therefore appropriate to consider what subtype of nicotinic cholinoceptor might mediate the presynaptic effect, what component or components of the mobilization system may be involved, and what second messenger system might couple receptor activation to the mobilization process.

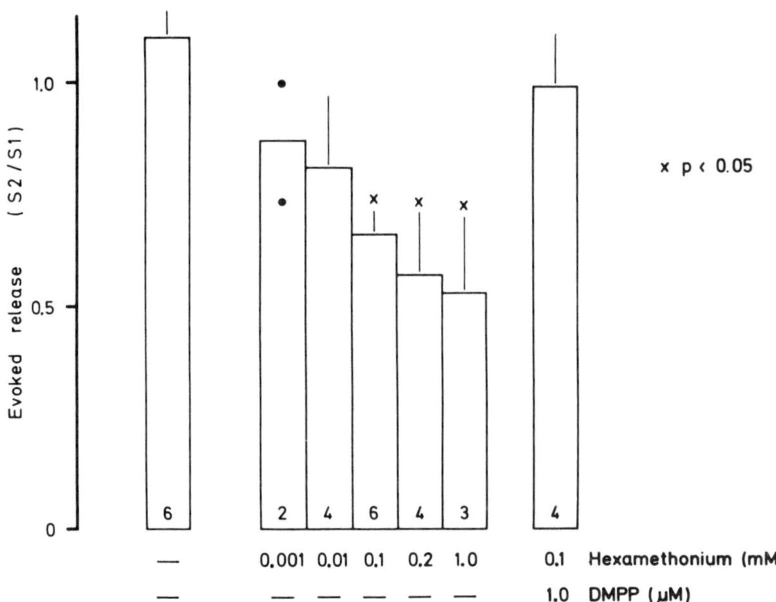

FIGURE 3. Inhibition by hexamethonium of [^{3}H]acetylcholine release. After labeling with 1 μM [^{3}H]choline and a 60 minute washout, tritium efflux was measured every 3 minutes. S1 and S2 were carried out at 5 Hz (100 pulses). Drugs were added to the incubation medium 15 minutes before S2. The effects on evoked [^{3}H]acetylcholine release are expressed by the ratios S2/S1. The concentrations of the drugs are indicated below each column. Each column indicates the mean ratio in the number of experiments noted. The vertical lines are the standard errors of the mean (SEM). In control experiments, evoked [^{3}H]acetylcholine release during S1 was 20,000 ± 3,600 dpm/g ($n = 6$). (From Reference 30 with permission.)

At the present time, and until it becomes possible to determine dissociation constants and make proper binding studies, it would be premature to characterize the prejunctional receptors as definitely of a different nicotinic subtype from the postjunctional receptors. However, insofar as the evidence goes, the prejunctional receptors appear to differ both from the postjunctional motor end plate receptors and from ganglionic receptors.[35,45] Thus, hexamethonium is more potent in producing the various fade phenomena and depressing transmitter release than it is in blocking postjunctional receptors,[9,28,30] but it is not as potent as it is in blocking autonomic ganglia. Trimetaphan, a powerful ganglion blocking agent, has little ability to produce fade or rundown by a prejunctional mechanism, and acts postjunctionally to block open end plate ion channels rather than receptor recognition sites.[29] There is some controversy about the prejunctional effects of snake α-toxins. Certainly α-bungarotoxin, α-cobratoxin, and erabutoxin b can be shown to produce virtually complete twitch block and pronounced depression of tetanic tension without producing tetanic fade or train-of-four rundown, thereby demonstrating the separate origins of the two compo-

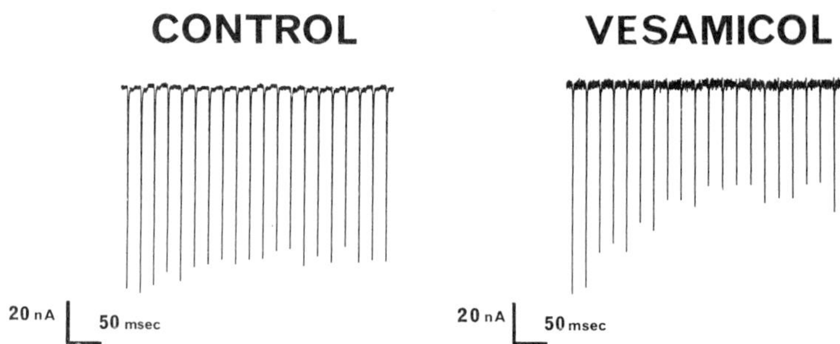

FIGURE 4. End plate currents in cut rat diaphragm fibers evoked by stimulating the motor nerve at a frequency of 50 Hz. The fibers were voltage clamped at -50 mV. A control response is shown on the left. The record on the right was taken after bathing the nonstimulated preparation for 30 minutes in 1 μM of $(-)$-vesamicol. Note that, in contrast to the effect of tubocurarine, the increased rundown in the presence of vesamicol is not associated with a decrease in the amplitude of the first response of the train. The records are from two different representative cells.

nents of action.[39,40,46,47] However, when small concentrations are allowed to act for a long time, fade and depression of transmitter output may begin to appear,[13,39,40] although even under these circumstances Wessler failed to detect any effect of α-cobratoxin on transmitter output,[13] and Bradley and his coworkers failed to demonstrate tetanic fade with α-bungarotoxin.[39] It may be that the prejunctional receptors are sensitive to the large toxin molecules but diffusion barriers slow down their rate of access to them. Be that as it may, there is sufficient evidence at least to suggest that the prejunctional receptors belong to a third nicotinic subtype, and accordingly they have been labeled N_3 in the diagram of FIGURE 6 (N_1 being the postjunctional motor end plate subtype, and N_2 those of autonomic ganglia). In fact, the prejunctional nicotinic receptors appear to resemble some of those in the central nervous system (CNS).[48,49] Several different subtypes of nicotinic receptor have been detected in the chick CNS.[50]

Recent experiments with the compound called vesamicol (AH 5183)[51] have shown that, in concentrations below those that block postjunctional acetylcholine receptor-operated ion channels or that exert any other postjunctional action, it produces rundown in trains of epcs (FIGURE 4), an effect that superficially resembles that of tubocurarine. Vesamicol inhibits the process responsible for the loading of vesicles with acetylcholine,[52] and as far as we are aware the drug has no other actions in these concentrations. In the presence of a sufficiently high concentration of tubocurarine (0.5 μM) to produce its maximal effect on rundown in a train of epcs, vesamicol produced no further effect, suggesting that the tubocurarine had already blocked the system upon which vesamicol acts. It is therefore possible that rapid reloading of the vesicles in the readily releasable store is one component of the mobilization process that is stimulated when prejunctional nicotinic receptors are activated. Alternative, or additional, mechanisms that might be involved include the synapsin I mechanism and its possible role in vesicular docking at the active zones,[53] enhanced fusion of vesicles with the terminal membrane, stimulated vesicular recycling, or increased availability of release sites. Nothing is yet known about the linking mechanism between receptor activation and the mobilization process. Other nicotinic cholinoceptors are linked directly to cation channels, and it might be expected that this would also be so at the motor nerve terminals, but as yet this possibility has not been tested.

Presynaptic Nicotinic Receptors that Mediate Inhibition of Acetylcholine Release

When junctional acetylcholinesterase is inhibited by anticholinesterase drugs, tubocurarine no longer decreases acetylcholine release, but in fact may enhance it.[30,54,55] α-Bungarotoxin shares the ability of tubocurarine to enhance transmitter release in the presence of an anticholinesterase drug.[56,57] These results suggest that, under these conditions, transmitter acetylcholine, acting on presynaptic nicotinic receptors, may be exerting a negative modulatory effect on its own release. This idea is supported by electrophysiological experiments which demonstrate a marked fall in the quantal content of the epp during high-frequency nerve stimulation in the presence of neostigmine or physostigmine,[55,58] and by biochemical experiments in which the release of tritium-labeled acetylcholine is estimated in the presence of the same anticholinesterase agents.[55] It is supposed that the transmitter, persisting in the junctional cleft in the absence of functional cholinesterase, acts on the nerve endings to depress release. Stable nicotinic agonists, such as nicotine, carbachol, and DMPP, applied either in higher concentrations than those necessary to activate the positive feedback mechanism or in lower concentrations for a prolonged period, inhibit the evoked release of transmitter, whether or not cholinesterase is inhibited.[55,59]

The inhibitory effect on acetylcholine release produced either by stable nicotinic agonists or by the transmitter itself, persisting in the presence of inhibited cholinesterase, is blocked by tubocurarine or α-bungarotoxin.[35,60,61] FIGURE 5 illustrates experiments in which tritium-labeled acetylcholine was collected from the phrenic nerve of the isolated diaphragm of the rat in the presence of neostigmine. The motor nerve was stimulated at 50 Hz four times at regular intervals. Under these conditions, acetylcholine output fell off during successive stimulation periods to about 60% of its initial amount. However, when tubocurarine was present in the bath, the falloff in output was less (to about 85%). Falloff to this extent is similar to that in the absence of neostigmine. Hence, neostigmine enhanced the falloff in output, and tubocurarine prevented that extra falloff, restoring it towards the control level. Even at low frequencies of stimulation in the presence of an anticholinesterase, there may be sufficient transmitter persisting (evoked release plus both quantal and nonquantal spontaneous release) to activate the negative modulatory effect. By blocking this inhibitory effect, tubocurarine and other nicotinic antagonists give the impression that

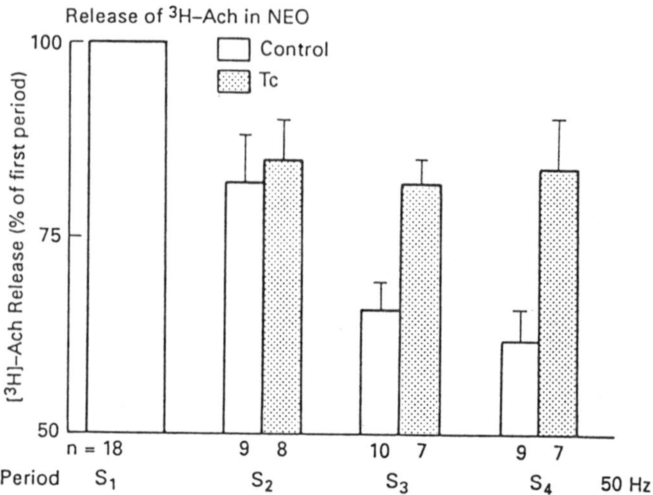

FIGURE 5. Rat phrenic nerve–diaphragm preparations. Release of radiolabeled acetylcholine in the presence of neostigmine (3×10^{-5} M) but absence of hemicholinium during stimulation of the nerve to each preparation (50 Hz for 7.5 minutes) on four occasions (S1, S2, S3, and S4). Acetylcholine output fell with successive stimulation periods. However, tubocurarine (10^{-5} M) to a substantial extent prevented the fall in acetylcholine output, thereby giving the impression that it increased output, especially in S4. The method used was the same as that described in Reference 84. (The figure was supplied by Dr. D. Joseph from unpublished experiments.)

they increase release, but in reality they may merely be restoring an inhibited release towards normal.

The inhibitory effect of persisting transmitter or of stable nicotinic agonists has been attributed to removal, by receptor desensitization, of the otherwise continually active positive modulatory effect on transmitter mobilization described above.[35,39] However, while there is evidence to suggest that these "positive feedback" receptors do desensitize, it is difficult to attribute the negative modulatory effect simply to such desensitization. Tubocurarine prevents the negative modulatory effect, presumably by blocking the receptors involved. Even if tubocurarine were to remove the desensitization, the receptors would presumably remain inactive because of the tubocurarine block, and hence it is difficult to see how the positive feedback mechanism could be reinstated by this mechanism. We are inclined to the view that a different population of nicotinic receptors, mediating a different action, is involved in the negative feedback mechanism, and this view is reinforced by the observation that α-bungarotoxin readily blocks these receptors, whereas it interacts only with difficulty or not at all with the receptors involved in positive feedback. The fact that α-bungarotoxin is active at the site involved in negative feedback suggests that the receptors may be of the same subtype as those of the postjunctional membrane of the motor end plate (FIGURE 6). The fact that α-bungarotoxin interacts at this site is at variance with results that fail to demonstrate binding of the toxin at motor nerve endings.[38] However, there is controversy over this matter; other workers have described clear binding.[62,63]

There is evidence that acetylcholine may depolarize many nonmyelinated neuronal membranes even when no physiological function can be ascribed to the effect: for example, nodes of Ranvier,[64] sensory nerve terminals,[65] crustacean motor nerve fibers,[66] mammalian C fibers,[67] adrenergic nerve terminals,[68] locust motor nerve terminals;[69] so it might be surprising if motor nerve endings in skeletal muscle were an exception and did not respond similarly. In fact, there is evidence that a site on such motor nerve endings is depolarized by acetylcholine acting on tubocurarine-sensitive nicotinic receptors.[70] Depolarization of motor nerve endings by cathodal currents reduces the evoked release of transmitter.[71] Hence, a possible explanation of the negative modulatory effect of stable nicotinic agonists or of transmitter acetylcholine persisting in the presence of an anticholinesterase is that they depolarize some membrane site near the terminal. Since the effect is produced only by pharmacological agents (nicotine, DMPP, carbachol) or by transmitter acetylcholine only when cholinesterase is inhibited, it seems to have pharmacological significance but no obvious physiological role in the transmission mechanism. It may be that one of the functions of junctional cholinesterase, especially that component associated with the nerve ending membrane, is to protect the sensitive site on the nerve endings from the depolarizing action of the transmitter. Depolarization of the nerve endings during tetanic stimulation of the motor nerve when cholinesterase is inhibited could account for the prejunctional component of the transmission block that is produced,[58] and also for the excess leakage of transmitter that occurs simultaneously but independently of nerve impulses, the so-called regenerative release of transmitter.[72]

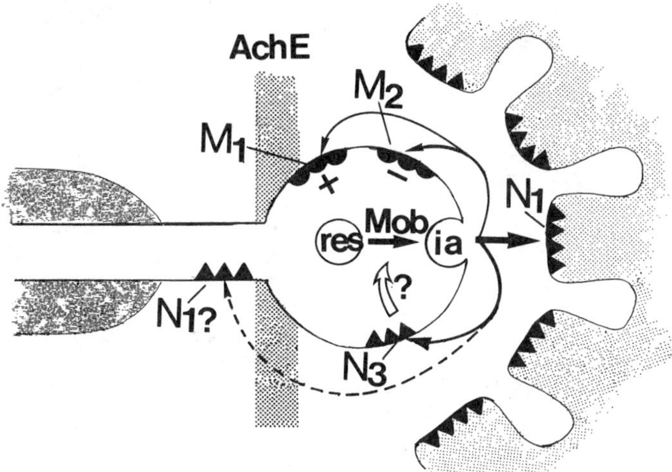

FIGURE 6. Postulated sites of cholinoceptors at the neuromuscular junction. The postjunctional nicotinic receptors (N_1) mediate the epp and transmission. The prejunctional nicotinic receptors (N_3) act to facilitate mobilization (mob) which includes all the processes referred to in the text, and not simply the movement of reserve (res) vesicles into the immediately available (ia) store as implied by the diagram. (By this scheme, autonomic ganglionic nicotinic receptors would be designated N_2). The nerve endings appear also to possess another population of nicotinic receptors which seem to resemble the postjunctional receptors (N_1?) and which mediate terminal depolarization, and hence repetitive nerve activity or conduction block, depending on the degree of exposure to the agonist. These receptors are stimulated by transmitter acetylcholine only when cholinesterase is inhibited. It is possible therefore that the acetylcholinesterase (AchE) of the basement membrane and of the nerve terminal membrane serves as a barrier to protect the nerve endings from depolarization by the transmitter. There is also evidence for the existence of muscarinic receptors that modify the release process. The M_1 (?) subtype may facilitate release whereas the M_2 (?) subtype may inhibit release. The nomenclature of the nicotinic receptor subtypes (N_1, N_2, N_3) has no status other than to indicate that there appear to be at least three subtypes.

At very low frequencies of stimulation (0.1–0.4 Hz) in the presence of an anticholinesterase, transient depolarization of the same site in the terminals, by the acetylcholine released by each nerve impulse, may serve as a generator potential that triggers the repetitive antidromic nerve firing that follows each orthodromic nerve impulse.[3]

PREJUNCTIONAL MUSCARINIC RECEPTORS

Ligand-binding studies applied to subcellular fractions of the *Torpedo* electroplaque have shown that muscarinic receptors in this tissue are associated with the synaptosomal or nerve terminal fraction.[73–75] Oxotremorine, a muscarinic agonist, inhibited acetylcholine release from *Torpedo* synaptosomes, suggesting that the muscarinic receptors have an inhibitory role in transmission.[76,77]

Some workers obtained evidence suggesting that the nerve terminal muscarinic receptors facilitated acetylcholine release from the rat phrenic nerve.[78,79] However, other workers were unable to repeat these observations.[80,81] Some studies have shown the opposite. At the frog neuromuscular junction, it was found that carbachol, oxotremorine, or anticholinesterase agents produced reductions in miniature end plate potential frequency and in the quantal content of the epp. These effects were blocked by atropine but not by tubocurarine.[82,83] The authors concluded that muscarinic receptors mediate a negative feedback control of transmitter release. From experiments in which the output of tritiated transmitter from the rat phrenic nerve was estimated, Abbs and Joseph also deduced that muscarinic receptors mediated inhibition of acetylcholine release.[84] They found that atropine enhanced the evoked release of labeled acetylcholine, suggesting that transmitter release might have been under continuous muscarinic receptor-mediated inhibitory modulation. The muscarinic agonist oxotremorine antagonized the facilitatory effects of atropine, although by itself it did not inhibit acetylcholine release. Possibly, under the conditions of their experiments, which included inhibition of cholinesterase, the extent of muscarinic receptor-mediated inhibitory modulation by the transmitter itself was already maximal, so that oxotremorine could not produce any additional effect, except in the presence of atropine. Similar experiments have been carried out on the mouse phrenic nerve–diaphragm preparation.[36] The authors measured the release of tritiated choline, initially in the absence of anticholinesterase drugs, and they confirmed the observation that atropine enhanced transmitter release. They also found that physostigmine, presumably through accumulating acetylcholine, also produced a pronounced increase in transmitter release which was blocked by tubocurarine and which they attributed to stimulation of the positive feedback nicotinic receptors described above. In the presence of physostigmine, atropine did not produce any further increase in transmitter release. However, when tubocurarine was added to abolish the enhancing effect of physostigmine, then the enhancing effects of atropine reappeared. The authors concluded that in the mouse preparation, the positive nicotinic receptor-mediated automodulation was dominant and that when it was fully operative, the muscarinic receptor-mediated negative feedback was much less effective, possibly because of competition at the level of second messenger systems.

Wessler's group to some extent settled the controversy concerning the direction of release modulation via muscarinic receptors by showing that both facilitation and inhibition can occur under different circumstances.[35,85,86] They assayed the release of tritiated choline from the rat phrenic nerve and showed that in low concentrations (10 nmol/l) oxotremorine enhanced and in high concentration (1 μmol/l) it reduced the evoked release of transmitter. Both effects were blocked by the muscarinic antagonist hyoscine. In the absence of oxotremorine, hyoscine enhanced transmitter release evoked by 100 pulses at 5 Hz, but reduced the release evoked by 1500 pulses at 5 or 25 Hz. Thus, it appeared that two muscarinic systems may operate—a negative modulation of release during short stimulation periods and a facilitatory system during long stimulation periods. The authors further showed that the receptor that mediates the facilitatory effect is blocked by low concentrations of pirenzepine, which has some selectivity for the M_1 subclass of muscarinic receptors. They proposed that the M_1 facilitatory receptors are situated more proximally on the motor nerve endings than are the M_2 (?) inhibitory receptors.

The physiological roles of the muscarinic receptors, if any, are not known, nor are the coupling mechanisms between receptor activation and modulation of the transmitter release mechanism. However, some speculation is possible. Muscarinic receptors of the M_1 subtype are usually coupled to K^+ M channels.[87] Coupling between the M_1 receptor and the K^+ channel is thought to involve diacylglycerol and protein kinase C as second messengers,[88] and these agents have been shown to enhance acetylcholine release.[89] It might therefore be that K^+ efflux through M channels plays a part in terminating the depolarization of the nerve endings that triggers transmitter release. If this is so, it is possible that under some circumstances transmitter acetylcholine might act back on the nerve endings to close the M channels by stimulating the terminal M_1 receptors. This would have the effect of prolonging the terminal depolarization and thereby enhancing the evoked transmitter release. In this connection it is worthy of note that there is some evidence[90] that neostigmine prolongs the duration of the depolarization in the nerve endings. Possibly this arises through preserved acetylcholine stimulating the terminal M_1 receptors.

The M_2 subtype of muscarinic receptor often mediates inhibition of adenylate cyclase and therefore a fall in the concentration of cyclic AMP. A fall in cyclic AMP concentration may oppose transmitter mobilization,[91] and it is feasible that M_2 receptor activation inhibits acetylcholine release in this way.

CONCLUSION

Thus, there is pharmacological evidence for the existence of two populations of nicotinic autoreceptors and two populations of muscarinic autoreceptors on motor nerve endings. Activation of these receptors modifies transmitter release in various ways, as summarized in FIGURE 6. Whether all of these receptor types have physiological roles in modulating transmitter release is uncertain. The evidence for a physiological role seems to be greatest for the nicotinic autoreceptors postulated to function in a positive feedback mechanism that enhances transmitter mobilization.

REFERENCES

1. MASLAND, R. L. & R. S. WIGTON. 1940. J. Neurophysiol. **3**: 269–275.
2. BOWMAN, W. C. & S. N. WEBB. 1972. *In* International Encyclopedia of Pharmacology and Therapeutics. J. Cheymol, Ed. **2**(Section 14): 427–502. Pergamon Press. Oxford, England.
3. RIKER, W. F. 1975. *In* Muscle Relaxants, Monographs in Anesthesiology. R. L. Katz, Ed. **3**: 59–102. North Holland Publishing Co. Amsterdam, the Netherlands.
4. MIYAMOTO, M. D. 1988. Pharmacol. Rev. **29**: 221–247.
5. FOLDES, F. F. & E. S. VIZI. *In* Modulation of Neurochemical Transmission. E. S. Vizi, Ed.: 335–382. Pergamon Press. Oxford, England.
6. BOWMAN, W. C., A. J. GIBB, A. L. HARVEY & I. G. MARSHALL. *In* Handbook of

Experimental Pharmacology. New Neuromuscular Blocking Agents. D. A. Kharkevich, Ed. **79:** 141–170. Springer-Verlag. Berlin, FRG.

7. LILLEHEIL, G. & K. NAESS. 1961. Acta Physiol. Scand. **52:** 120–136.
8. KOELLE, G. B. 1962. J. Pharm. Pharmacol. **14:** 65–90.
9. BOWMAN, W. C. & S. N. WEBB. 1976. Clin. Exp. Physiol. Pharmacol. **3:** 545–555.
10. BOWMAN, W. C. 1980. Anesth. & Anal. **59:** 935–943.
11. BOWMAN, W. C., I. G. MARSHALL & A. J. GIBB. 1984. Semin. Anaesth. **3:** 275–283.
12. BOWMAN, W. C., I. G. MARSHALL, A. J. GIBB & A. J. HARBORNE. 1988. Trends Pharmacol. Sci. **9:** 16–20.
13. WESSLER, I. 1989. Trends Pharmacol. Sci. **10:** 110–114.
14. WILLIAMS, N. E., S. N. WEBB. & T. N. CALVEY. 1980. Br. J. Anaesth. **52:** 1111–1115.
15. STANEC, A. & T. BAKER. 1984. Br. J. Anaesth. **56:** 607–611.
16. ALI, H. H. 1984. Semin. Anesth. **3:** 284–292.
17. LILEY, A. W. 1956. J. Physiol. **133:** 571–587.
18. GLAVINOVIC, A. M. 1979. J. Physiol. **290:** 499–506.
19. DREYER, F. 1982. Br. J. Anaesth. **54:** 115–130.
20. STORELLA, R. J. & G. G. BIERKAMPER. 1986. Eur. J. Pharmacol. **124:** 143–148.
21. MAGLEBY, K. L., B. S. PALOTTA & D. A. TERRAR. 1981. J. Physiol. **312:** 97–113.
22. GIBB, A. J. & I. G. MARSHALL. 1984. J. Physiol. **351:** 275–297.
23. COLQUHOUN, D., F. DREYER & R. E. SHERIDAN. 1979. J. Physiol. **293:** 247–284.
24. HARBORNE, A. J., W. C. BOWMAN & I. G. MARSHALL. 1988. Clin. Exp. Pharmacol. Physiol. **15:** 479–490.
25. BOWMAN, W. C., A. J. GIBB & I. G. MARSHALL. 1983. *In* Clinical Experiences with Norcuron. S. Agoston, W. C. Bowman, R. D. Miller & J. Viby-Mogensen, Eds.: 26–32. Elsevier. Amsterdam, the Netherlands.
26. GIBB, A. J., I. G. MARSHALL & W. C. BOWMAN. 1984. Proc. Aust. Physiol. Pharmacol. Soc. **15:** 86P.
27. GIBB, A. J. & I. G. MARSHALL. 1986. Br. J. Pharmacol. **89:** 619–624.
28. GIBB, A. J. & I. G. MARSHALL. 1987. Br. J. Pharmacol. **90:** 511–521.
29. GIBB, A. J. & I. G. MARSHALL. 1982. Br. J. Pharmacol. **76:** 187P.
30. WESSLER, I., M. HALANK, J. RASBACH & H. KILBINGER. 1986. Naunyn-Schmiedeberg's Arch. Pharmacol. **334:** 365–372.
31. VIZI, E. S., G. T. SOMOGYI, H. NAGASHIMA, D. DUNCALF, D. V. M. CHAUDHRY & F. F. FOLDES. 1987. Br. J. Anaesth. **59:** 226–231.
32. RANG, H. P. 1982. Br. J. Pharmacol. **75:** 151–168.
33. FOLDES, F. F., G. T. SOMOGYI, I. A. CHAUDRY, H. NAGASHIMA & D. DUNCALF. 1984. Anesthesiology **61:** A395.
34. WESSLER, I. & H. KILBINGER. 1986. Naunyn-Schmiedeberg's Arch. Pharmacol. **334:** 357–364.
35. WESSLER, I. 1989. Trends Pharmacol. Sci. **10:** 110–114.
36. VIZI, E. S. & G. T. SOMOGYI. 1989. Br. J. Pharmacol. **97:** 65–70.
37. WESSLER, I., M. J. RASBACH, B. SCHEUER, U. HILLEN & H. KILBINGER. 1987. Naunyn-Schmiedeberg's Arch. Pharmacol. **335:** 496–501.
38. JONES, S. W. & M. M. SALPETER. 1983. J. Neurosci. **3:** 326–331.
39. BRADLEY, R. J., M. K. PAGALA & M. T. EDGE. 1987. FEBS Lett. **224:** 277–282.
40. CHANG, C. C. & S. J. HONG. 1987. Exp. Neurol. **98:** 509–517.
41. MARSHALL, I. G. & V. A. CHIAPPINELLI. 1987. Unpublished experiments.
42. PRESTON, J. B & F. F. VAN MAANEN. 1953. J. Pharmacol. Exp. Ther. **107:** 165–171.
43. BLACKMAN, J. G. 1963. Br. J. Pharmacol. **20:** 5–16.
44. LEE, C. & R. L. KATZ. 1977. Anesth. Anal. **56:** 271.
45. BOWMAN, W. C., A. J. GIBB, A. J. HARBORNE & I. G. MARSHALL. 1987. *In* Pharmacology. M. J. Rand & C. Raper, Eds.: 91–94. Elsevier Amsterdam, the Netherlands.
46. LEE, C., D. CHEN & R. L. KATZ. 1977. Can. Anaesth. Soc. J. **24:** 212–219.
47. CHEAH, L. S. & M. C. E. GWEE. 1988. Clin. Exp. Pharmacol. Physiol. **15:** 937–943.
48. CLARKE, P. B. S. 1987. Trends Pharmacol. Sci. **8:** 32–34.
49. BLAKE, J. F., R. H. EVANS & D. A. S. SMITH. 1987. Br. J. Pharmacol. **90:** 167–173.
50. WOLF, K. M., A. CIARLEGLIO & V. A. CHIAPPINELLI. 1988. Brain Res. **439:** 249–258.

51. PRIOR, C., T. SEARL & I. G. MARSHALL. 1989. Br. J. Pharmacol. **98:** 826P.
52. MARSHALL, I. G. & S. M. PARSONS. 1987. Trends Neurosci. **10:** 174–177.
53. DE CAMILLI, P. & P. GREENGARD. 1986. Biochem. Pharmacol. **35:** 4349–4357.
54. FLETCHER, P. & T. FORRESTER. 1975. J. Physiol. **251:** 131–144.
55. BIERKAMPER, G. G., E. AIZENMAN & W. R. MILLINGTON. 1986. *In* Dynamics of Cholinergic Function. I. Hanin, Ed.: 447–457. Plenum Publishing Corp. New York, N.Y.
56. MILEDI, R., P. C. MOLENAAR & R. L. POLAK. 1978. Nature **272:** 641–643.
57. MILEDI, R., P. C. MOLENAAR & R. L. POLAK. 1983. J. Physiol. **334:** 245–254.
58. CHANG, C. C., S. J. HONG & J. L. KO. 1986. Br. J. Pharmacol. **87:** 757–762.
59. WESSLER, I., B. SCHEUER & H. KILBINGER. 1987. Eur. J. Pharmacol. **135:** 85–87.
60. BIERKAMPER, G. G. & E. AIZENMAN. 1984. Proc. West. Pharmacol. Soc. **27:** 353–356.
61. CHANG, C. C. 1987. Personal communication.
62. LENTZ, T. L., J. E. MAZURKIEWICZ & J. ROSENTHAL. 1977. Brain Res. **132:** 423–442.
63. TSUJITA, T. & H. OKUDA. 1983. Eur. J. Biochem. **133:** 215–220.
64. DETTBARN, W. D. 1960. Nature **186:** 891–892.
65. DOUGLAS, W. W. & J. M. RITCHIE. 1960. J. Physiol. **150:** 501–514.
66. DETTBARN, W. D. 1967. Ann. N.Y. Acad. Sci. **144:** 483–503.
67. RITCHIE, J. M. 1967. Ann. N.Y. Acad. Sci. **144:** 504–516.
68. KOPIN, I. J. 1967. Ann. N.Y. Acad. Sci. **144:** 558–570.
69. FULTON, B. P. & P. N. R. USHERWOOD. 1977. Neuropharmacology **16:** 877–880.
70. HUBBARD, J. I., R. F. SCHMIDT & T. YOKOTA. 1965. J. Physiol. **181:** 810–829.
71. HUBBARD, J. I. & W. D. WILLIS. 1968. J. Physiol. **194:** 381–405.
72. CHANG, C. C. & S. J. HONG. 1986. Neurosci. Lett. **69:** 203–207.
73. KLOOG, Y., D. M. MICHAELSON & M. SOKOLOVSKY. 1980. Brain Res. **194:** 97–115.
74. STRANGE, P. G., M. J. DOWDALL, P. R. GOLDS & M. R. PICKARD. 1980. FEBS Lett. **122:** 293–296.
75. DOWDALL, M. J., P. R. GOLDS & P. G. STRANGE. 1981. Biochem. Soc. Trans. **9:** 412–413.
76. MICHAELSON, D. M., S. ARISSAR, Y. KLOOG & M. SOKOLOVSKY. 1979. Proc. Nat. Acad. Sci. USA **76:** 6336–6340.
77. MICHAELSON, D. M., S. ARISSAR, I. OPHIR, I. PINCHASI, I. ANGEL, Y. KLOOG & M. SOKOLOVSKI. 1980. J. Physiol Paris **76:** 505–511.
78. DAS, M., D. K. GANGULY & J. R. VEDASIROMONI. 1978. Br. J. Pharmacol. **62:** 195–198.
79. GANGULY, D. K. & M. DAS. Nature **278:** 645–646.
80. GUNDERSON, C. B. & D. J. JENDEN. 1980. Br. J. Pharmacol. **70:** 8–10.
81. HAGGBLAD, J. & E. HEILBRONN. 1983. Br. J. Pharmacol. **80:** 471–476.
82. DUNCAN, C. J. & S. J. PUBLICOVER. 1979. J. Physiol. **294:** 91–103.
83. ARENSON, M. S. 1989. Neuroscience **30:** 827–836.
84. ABBS, E. T. & D. N. JOSEPH. 1981. Br. J. Pharmacol. **73:** 481–483.
85. WESSLER, I., M. KARL, M. MAI & A. DIENER. 1987. Naunyn-Schmiedeberg's Arch. Pharmacol. **335:** 605–612.
86. WESSLER, I., A. DIENER & M. OFFERMANN. 1988. Naunyn-Schmiedeberg's Arch. Pharmacol. **338:** 138–142.
87. BROWN, D. A., N. V. MARRION & T. G. SMART. 1989. J. Physiol. **413:** 469–488.
88. BROWN, D. A., B. H. GAHWILER, S. J. MARSH & A. A. SELYANKO. 1986. Trends Pharmacol. Sci. Suppl.: 66–71.
89. MURPHY, R. L. W. & M. E. SMITH. 1987. Br. J. Pharmacol. **90:** 327–334.
90. HUBBARD, J. I. & R. F. SCHMIDT. 1961. Nature **191:** 1103–1104.
91. DRYDEN, W. F., Y. N. SINGH, T. GORDON & G. LAZARENKO. 1988. Can. J. Physiol. Pharmacol. **66:** 207–212.

Presynaptic Receptors in the Visual System[a]

MARGARITA L. DUBOCOVICH

Department of Pharmacology
Northwestern University Medical School
303 East Chicago Avenue
Chicago, Illinois 60611

INTRODUCTION

The retina is considered to be part of the central nervous system. The main neuronal elements of the mammalian retina are rod and cone photoreceptor cells, bipolar cells, horizontal cells, amacrine and interamacrine cells, and ganglion cells. Most classical neurotransmitters and neuromodulators have been localized in neurons of the inner retina, i.e., amacrine and interamacrine cells. This article broadly reviews the presynaptic receptor systems involved in the modulation of dopamine and acetylcholine release in the mammalian retina.

Dopamine and acetylcholine fulfill all the criteria for functional neurotransmitters in the mammalian retina. The dopaminergic neurons in the rabbit retina are a population of large amacrine cells localized in the inner plexiform layer that contact other amacrine neurons which may use acetylcholine, gamma-aminobutyric acid (GABA), glycine, or indoleamines as neurotransmitters.[1,2] Recently, dopaminergic amacrine cell processes have been shown to be postsynaptic to bipolar cells at ribbon synapses.[3] Dopamine is synthesized in the retina from the amino acid tyrosine by the sequential action of the rate-limiting synthetic enzyme tyrosine hydroxylase and aromatic L-amino acid decarboxylase.[4] The retina possesses the degradative enzymes as well as a high-affinity uptake system for dopamine.[5,6] Dopamine is released from the retina in response to photic and electrical stimulation, and potassium-induced depolarization.[5–9] In the retina dopamine activates the D-1 and D-2 dopamine receptor subtypes.[10,11] These dopamine receptors have been localized by receptor autoradiography in the inner and outer layers of the mammalian retina.[12,13] D-1 dopamine receptors located postsynaptically to dopamine-containing neurons are linked to the stimulation of adenylate cyclase. Activation of D-1 dopamine receptors in retina uncouples gap junctions between horizontal cells,[14,15] regulates the activity and responsiveness of ganglion cells to light stimuli,[16] and evokes calcium-dependent release of acetylcholine.[17] The D-2 dopamine receptors of the vertebrate retina, which are possibly linked to the inhibition of adenylate cyclase,[4] modulate the synthesis, release, metabolism, and turnover of dopamine.[4,18] Presynaptic D-2 dopamine autoreceptors on dopamine-containing amacrine cells regulate dopamine release.[6] Activation of D-2 dopamine receptors in rod outer segments increases the activity of cyclic GMP phosphodiesterase in a manner analogous to light suggesting coupling of D-2 dopamine receptors to a

[a]Recent work was supported by U.S. Public Health Service grant MH 42922.

GTP binding protein.[19] In retinas of lower vertebrates activation of D-2 dopamine receptors inhibits the activity of serotonin *N*-acetyltransferase, the rate-limiting enzyme involved in the synthesis of melatonin, and induces light-adaptive cone contraction and pigment epithelium dispersion.[4,20,21] In this review I will focus the discussion on the presynaptic D-2 autoreceptors and postsynaptic D-1 dopamine receptors involved in the modulation of acetylcholine and dopamine release.

The mediator(s) of photoperiodic changes that affect the activity of dopamine neurons in retina are not known. Dopamine-containing amacrine cells receive synaptic inputs from retinal cells localized in the inner plexiform layer, and certainly have no direct synaptic contacts with photoreceptor cells.[2] Therefore the activation of dopamine-containing amacrine cells by light may be mediated indirectly either through the release of neurotransmitter(s) from retinal cells synapsing with dopamine neurons, e.g., GABA, or the release of neuromodulator(s) by retinal cells distant from dopamine neurons, e.g., melatonin. Presynaptic heterorecptors for neurotransmitter and neuro-modulator substances that inhibit dopaminergic activity include alpha$_2$-adrenoceptors, GABA, and opiate and melatonin receptors.[6,16,23–27] The role of these receptors in modulating dopaminergic activity is reviewed below.

Acethylcholine functions as a neurotransmitter of a subpopulation of amacrine cells in rabbit retina.[29,30] The cholinergic markers choline acetyltransferase, required for the synthesis of acetylcholine from choline, and acetylcholinesterase, the enzyme responsible for degrading acetylcholine, are present in retina and localized to the inner plexiform layer.[31] Light and potassium stimulate the calcium-dependent release of acetylcholine from the rabbit retina.[32] Studies on light-evoked release of acetylcholine from rabbit retina *in vivo* suggest that the activity of cholinergic neurons is modulated by the putative neurotransmitters GABA, glycine, and taurine.[29,30] Dopamine increases acetylcholine release from retinal neurons at synapses they form with striated muscle cells in culture, suggesting a direct dopaminergic input into these neurons.[33] A review of the presynaptic dopamine and GABA receptors modulating acetylcholine release from rabbit retina is presented below.

D-2 DOPAMINE AUTORECEPTOR

In the rabbit retina the calcium-dependent release of dopamine is modulated through activation of presynaptic D-2 dopamine autoreceptors. This was first reported by Dubocovich and Weiner using the rabbit retina labeled *in vitro* with ³H-dopamine,[6] and subsequently confirmed by others.[8,34] From this preparation calcium-dependent release of ³H-dopamine can be elicited by depolarization with electrical field, potassium, or light stimulation.[6–8,11,34–35] FIGURE 1 shows the efflux of radioactivity from control rabbit retina tissue labeled with ³H-dopamine. In this study the calcium-dependent and tetrodotoxin sensitive release of ³H-dopamine was elicited by electrical field stimulation at 3 Hz during 2 minutes, twice during the same experiment. Using this protocol we found that dopamine at nanomolar concentrations inhibited in a concentration-dependent manner the release of ³H-dopamine from rabbit retina (FIGURE 2A). This effect of dopamine was mimicked by dopamine agonists such as bromocriptine, apomorphine, pergolide, and quinperole, which are known to activate dopamine D-2 receptors.[11,36] The D-2 dopamine antagonist S-sulpiride completely antagonized the inhibition of release elicited by dopamine (FIGURE 2A). The potentiation by S-sulpiride (FIGURE 3A), spiperone, and fluphenazine of the calcium-dependent release of dopamine from rabbit retina indicates that these antagonists block receptor sites endogenously activated by dopamine.[6,11] The D-2 dopamine

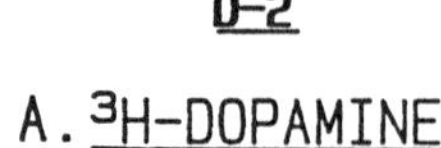

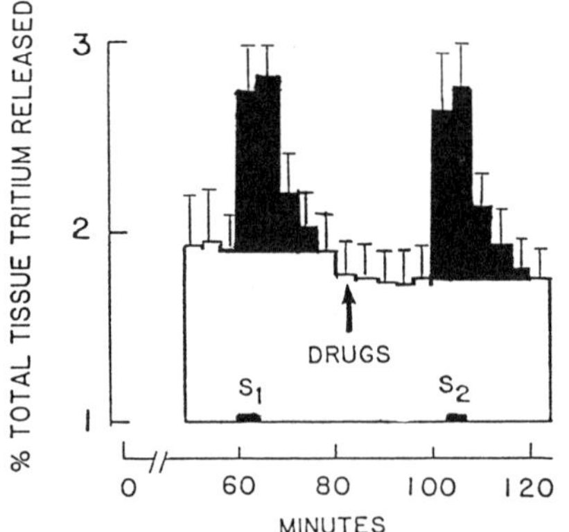

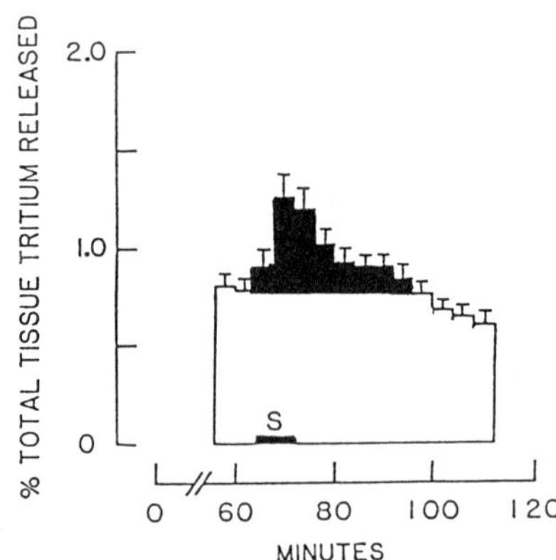

FIGURE 1. Efflux of radioactivity from rabbit retina labeled *in vitro* with [³H]dopamine or [³H]acetylcholine. **(A)** Superfusate samples were collected in 4-minute fractions beginning 56 minutes after the end of the incubation with [³H]dopamine. Ordinate: percentage of total tissue tritium released in each 4-minute fraction. The calcium-dependent release of [³H]dopamine (dark areas: [³H]dopamine overflow, S) was evoked by field stimulation (3 Hz, 2 minutes, 20 mA, 2 mseconds) above the spontaneous levels of tritium release (light area: spontaneous overflow, S_p) at 60 minutes (S_1) and 100 minutes (S_2) of superfusion. (Redrawn from Reference 6.) **(B)** Rabbit retinas were labeled with [³H]acetylcholine. Superfusate samples were collected in 4-minute fractions beginning 50 minutes after the end of the incubation with [³H]acetylcholine. Ordinate: percentage total tissue tritium release in each 4-minute fraction. The calcium-dependent release of [³H]acetylcholine (dark areas: [³H]acetylcholine overflow) was evoked by superfusion with dopamine (100 μM for one 8-minute period) above the spontaneous levels of tritium release (light area: spontaneous outflow). (Redrawn from Reference 58.) Shown are mean values ± standard errors of the mean (SEM).

autoreceptors of rabbit retina demonstrate stereoselectivity for the isomers of sulpiride, and of the D-2 dopamine agonist LY 141865 (i.e., LY 171555 or quinperole and LY 181990).[6,36] The pharamcological characteristics of the presynaptic D-2 dopamine autoreceptor of the rabbit retina are pharmacologically distinct from the postsynaptic D-1 dopamine receptors.[17,37]

The potency of dopamine agonists to decrease and of dopamine antagonists to increase the calcium-dependent release of dopamine from rabbit retina depends on the endogenous levels of neurotransmitter in the synaptic gap.[6,11,36] Increases in the synaptic concentration of dopamine, by either the frequency of stimulation from 1 to 6 Hz keeping the total number of pulses constant or inhibiting the neuronal uptake of dopamine by nomifensine, reduce the inhibitory effects of D-2 dopamine agonists and potentiate the effects of D-2 dopamine antagonists.[36] These studies suggest that D-2 dopamine autoreceptors may be involved in the physiological modulation of dopamine release from retina.

MODULATION OF DOPAMINERGIC ACTIVITY BY
PRESYNAPTIC HETERORECEPTORS

Alpha₂ Presynaptic Heteroreceptors

Various presynaptic heteroreceptors are involved in the modulation of dopamine release from the mammalian retina. Activation of alpha₂-adrenergic receptor by agents such as norepinephrine, epinephrine, and the alpha₂-adrenergic agonist clonidine inhibits the calcium-dependent release of dopamine. The inhibitory effect of alpha agonists is antagonized by phentolamine or the alpha₂-adrenergic receptor antagonist yohimbine, but not the alpha₁ antagonist prazosin, suggesting that the alpha-adrenergic presynaptic heteroreceptor is of the alpha₂ subtype.[24,25] Alpha₂-adrenergic receptors in the mammalian retina have also been implicated in the regulation of

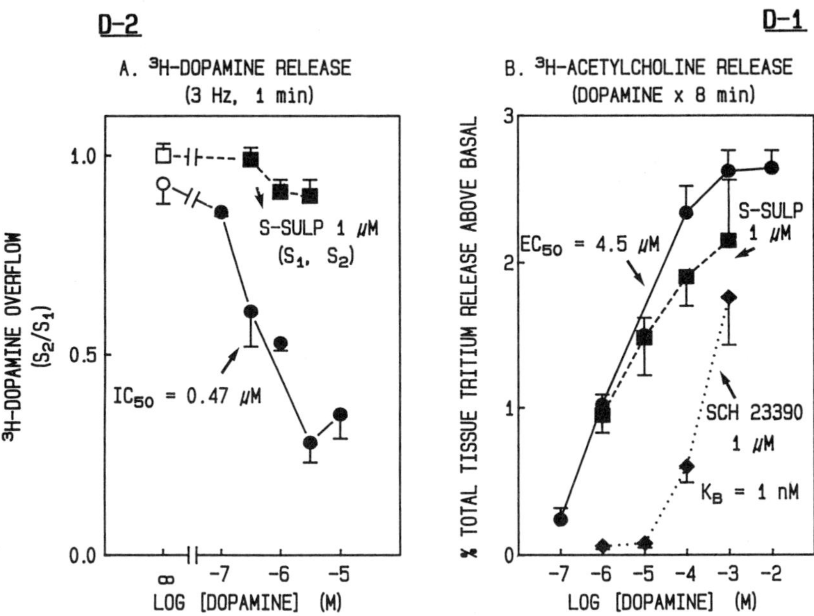

FIGURE 2. Modulation of [³H]dopamine and [³H]acetylcholine release from rabbit retina. (**A**) Ordinate:[³H]dopamine overflow; the percentage of total tissue radioactivity released by field stimulation (3 Hz, 2 minutes, 20 mA, 2 mseconds) above the spontaneous levels of release. Results are expressed as the ratio obtained between the second (S₂) and first (S₁) stimulation periods within the same experiment. The abscissa represents the molar concentration of dopamine (logarithmic scale) added 20 minutes before the second period of stimulation (S₂) (filled symbols). Control (open circle): S-sulpiride (squares) at a concentration of 1 μM was added 40 minutes before the first period of stimulation (S₁) and remained present throughout. (Redrawn from Reference 6.) (**B**) Ordinate: percentage of the total tissue tritium released above basal. In A the calcium-dependent release of [³H]acetylcholine was elicited by superfusion with dopamine (0.1 μM–10 mM) for one 8-minute period. S-sulpiride (1 μM) and SCH 23390 (1 μM) were added to the superfusion medium 48 minutes before the period of stimulation and remained throughout the experiment. Circles, dopamine (8 minutes); squares, dopamine (8 minutes) in the presence of S-sulpiride (0.1 μM); diamonds, dopamine (8 minutes) in the presence of SCH 23390 (1 μM). (Redrawn from Reference 17.)

dopamine turnover, and the activity of tyrosine hydroxylase and aromatic amino acid decarboxylase, two enzymes involved in the synthesis of dopamine.[38–40]

Opiate and GABA Presynaptic Heteroreceptors

In addition to catecholamines many other putative neurotransmitters and neuro-modulators have been found in specific retinal neurons, where they exert a variety of

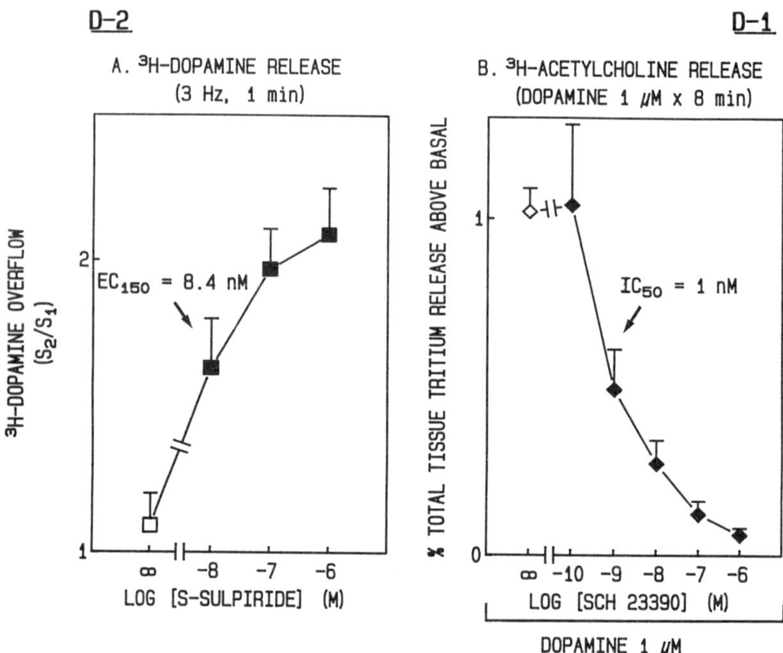

FIGURE 3. Modulation of [³H]dopamine and [³H]acetylcholine release from rabbit retina. (A) Ordinate: [³H]dopamine overflow; percentage of total tissue radioactivity released expressed as the ratio obtained between the second (S_2) and the first (S_1) stimulation periods within the same experiment. Release was elicited by a 1-minute period of electrical stimulation at 3 Hz (20 mA, 2 mseconds). Abscissa: concentration of the drugs (logarithmic scale). S-sulpiride was added 20 minutes before S_2. (Redrawn from Reference 6.) (B) Ordinate: percentage of the total tissue tritium released above basal. The calcium-dependent release of [³H]acetylcholine was elicited by superfusion with dopamine (1 µM, 8 minutes). SCH 23390 was added to the superfusion medium 40 minutes before the period of stimulation. Abscissa: molar contraction of dopamine receptor antagonists (logarithmic scale). (Redrawn from Reference 37.) Shown are mean values ± SEM.

functions. In the rabbit retina stereoselective presynaptic inhibitory opiate receptors modulate the release of dopamine.[23] These sites may be activated by the endogenous opiate peptides release from retinal cells.[41]

Inhibitory GABA receptors mediate inhibition of light-evoked increase in dopamine turnover and tyrosine hydroxylase activity in rat retina.[28,4] However, in the rabbit retina neither GABA nor muscimol inhibit the calcium-dependent release of dopamine. The increase in spontaneous and evoked release of radioactivity from rabbit retina labeled by ^{3}H-dopamine appears not to be GABA receptor mediated since it was not antagonized by picrotoxin (Dubocovich, unpublished).

Melatonin Presynaptic Heteroreceptors

Serotonin has been suggested to play a neurotransmitter role in the retina,[42] however this indoleamine may not be directly involved in the modulation of dopaminergic activity.[26] In mammalian retina where the levels of serotonin are low, this neurotransmitter may function as a precursor for other indoleamines, such as *N*-acetylserotonin and melatonin,[43] both of which are very potent modulators of dopaminergic activity.[26]

Melatonin (5-methoxy-*N*-acetyltryptamine) is synthesized from serotonin following a two-step biochemical sequence involving the enzymes *N*-acetyltransferase and hydroxyindole-*O*-methyltransferase.[4] A system for melatonin synthesis exists in retina, where the activity of serotonin *N*-acetyltransferase varies with changes in light intensity following a diurnal pattern.[44–46] Melatonin and melatoninlike immunoreactivity have been detected in retina.[45,47] In rabbit retina, the highest density of 2-[^{125}I]-iodomelatonin binding sites was localized over the inner plexiform layer where the dopamine-containing amacrine cells are localized.[48] Since the activity of dopaminergic neurons follows a circadian cycle with higher levels during the light period,[49] it was suggested that presynaptic melatonin heteroreceptors on dopaminergic neurons may mediate these changes.

Picomolar concentrations of melatonin inhibit the calcium-dependent release of dopamine from retina elicited by electrical field stimulation (FIGURE 4),[26] by potassium,[35] by light,[9,50] and by glutamate (Nicholson and Dubocovich, unpublished). The calcium-independent release of dopamine elicited by the indirect amine tyramine is not affected by melatonin, nor does it modify the spontaneous outflow of radioactivity.[26] FIGURE 4 shows a concentration-dependent inhibition of the field-stimulation-evoked release of dopamine by melatonin in rabbit retina. The concentration of melatonin inhibiting the release of dopamine by 50% (IC$_{50}$) was 25 pM, with a maximal inhibitory effect at 1 nM.[26,27] Melatonin is about 1000 times more potent than its precursor *N*-acetylserotonin to inhibit dopamine release, while serotonin was inactive.[26] These results suggest that melatonin inhibits the release of dopamine through activation of a site that is pharmacologically different from the presynaptic serotonin receptor.[26] This effect of melatonin was mimicked by other *N*-acetyltryptamines possessing a 5-methoxy group on the indole ring, such as 2-iodomelatonin, 6-chloromelatonin, 6-hydroxymelatonin, and 6-methoxymelatonin.[37]

The existence of a presynaptic melatonin receptor site is further documented by studies using the new melatonin receptor antagonist luzindole. Luzindole is an *N*-acetyltryptamine with a novel chemical structure that was recently found to be a competitive melatonin receptor antagonist.[51,52] Luzindole when added alone does not modify either the spontaneous or the stimulation-evoked overflow of dopamine in experiments conducted during the light period, when the levels of melatonin are low. FIGURE 4 shows that luzindole (1 μM) produced a parallel shift of the concentration-effect curve of melatonin to the right, suggesting that luzindole competitively antagonized the inhibitory effect of melatonin.[52]

Although several lines of evidence suggest that light may convey information to dopamine neurons indirectly through the release of excitatory neurotransmitters by neurons linking photoreceptors and dopamine-containing retinal cells, a role for the hormone melatonin in modulating dopaminergic activity is becoming evident.[26] In darkness depolarized photoreceptors produce and secrete melatonin, while exposure to light rapidly inactivates N-acetyltransferase, leading to a decrease in melatonin synthesis and secretion by the depolarized photoreceptors.[26] Several lines of evidence suggest that during the dark period the elevated concentration of melatonin in retina exerts an inhibitory action on the activity of retinal dopamine neurons. In experiments in which we mimicked the effect of dark by superfusing the rabbit retina labeled *in*

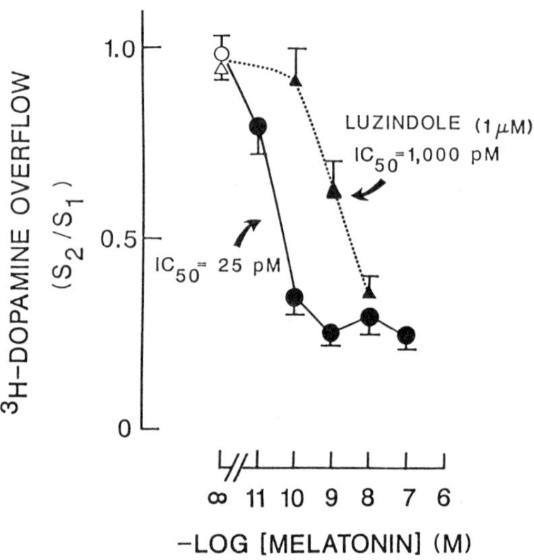

FIGURE 4. Luzindole antagonizes the melatonin-induced inhibition of [³H]dopamine release from rabbit retina. Ordinate: [³H]dopamine overflow; the percentage of total tissue radioactivity released by field stimulation (3 Hz, 2 minutes, 20 mA, 2 mseconds) above the spontaneous levels of release. Results are expressed as the ratio obtained between the second (S_2) and first (S_1) stimulation periods within the same experiment. The abscissa represents the molar concentration of melatonin (logarithmic scale), added 20 minutes before the second period of stimulation (S_2) (filled symbols). Control (open circle); luzindole (triangles) at a concentration of 1 μM was added 40 minutes before the first period of stimulation (S_1) and remained present throughout. Shown are mean values ± SEM. (Redrawn from Reference 52.)

vitro with ³H-dopamine with exogenous melatonin, we demonstrated that the melatonin antagonist luzindole increased the calcium-dependent release of dopamine when added alone (M.L. Dubocovich, unpublished). These experiments suggested that endogenous melatonin may modulate the release of dopamine during the dark period through a negative feedback mechanism. In contrast, during the light period the low levels of melatonin might disinhibit the dopamine-containing amacrine cells of the retina, leading to an increase in tyrosine hydroxylase activity and therefore an increase in the rate of dopamine release, synthesis, and turnover.[36,4]

DOPAMINE AND MELATONIN INTERACTIONS IN RETINA

The nature of the stimulatory input that leads to the activation of dopaminergic cells in retina following stimulation with light is not known.[22] It has been suggested that the light-induced release of alpha-melanocyte stimulating hormone (α-MSH), excitatory amino acids, or glycine may stimulate dopaminergic activity. α-MSH as well as glutamate stimulates the release of dopamine from retina [53,54] (Dubocovich and Nicholson, unpublished). The activity of dopaminergic neurons in retina is higher during the light period when the levels of melatonin are low.[9,34,45,46] Since the dopamine and melatonin systems in retina follow opposite diurnal rhythms, melatonin is a likely candidate to mediate the effects of light and dark on dopaminergic transmission.[4,26] The role of retinal melatonin receptors in modulating physiological functions is discussed below.

Interactions between the dopamine and melatonin systems in retina regulate physiological processes that follow a diurnal rhythm.[1] Dopamine synthesized in retinal amacrine cells is secreted on exposure to light. Light exposure also increases the biosynthesis, release, and metabolism of dopamine in vertebrate retinas.[4,6–8] The activation of D-2 dopamine receptors by dopamine released in response to light stimulation mediates, in part, the light-induced suppression of the nocturnal increase in retinal N-acetyltransferase activity.[55] In addition dopamine promotes light-adaptative movements of photoreceptors and melanin granules, and inhibition of disc shedding.[20,21,56]

Melatonin synthesized and secreted by retinal photoreceptors or other cells appears to modulate the activity of dopaminergic neurons. Presynaptic melatonin receptors modulate the calcium-dependent release of dopamine through activation of ML-1 melatonin receptors.[26,34,50] The release of melatonin during the dark period activates melatonin receptor sites and may directly or indirectly regulate several retinal functions through inhibition of dopamine release. In the guinea pig eye melatonin induces pigment aggregation in the retinal pigment epithelium and choroid,[57] while in *Xenopus laevis* retina it causes cone elongation, thus mimicking the effect of darkness.[55] On the other hand, unlike dopamine, melatonin activates rod photoreceptor disc shedding.[55]

D-1 POSTSYNAPTIC DOPAMINE RECEPTORS

A postsynaptic dopamine receptor with the pharmacological and biochemical characteristics of the D-1 dopamine subtype mediates release of acetylcholine from rabbit retina. In rabbit retina pieces labeled with ^{3}H-choline exposure to dopamine (0.1 μM–10 mM) evoked, in a concentration-dependent manner, the release of ^{3}H-acetylcholine (FIGURES 1B and 2B).[17,37] This release was abolished by omission of calcium, and not affected by the presence of the sodium channel blocker tetrodotoxin.[17] Endogenous dopamine, released from dopaminergic amacrine neurons by the indirect amines tyramine or D-amphetamine, also evoked the calcium-dependent release of ^{3}H-acetylcholine from rabbit retina.[58] The concentration of dopamine necessary to induce 50% of the maximal release of tritium above spontaneous levels (EC_{50}) was 4.5 μM (FIGURE 2B). This correlates with the potency of dopamine to activate adenylate cyclase (EC_{50} = 2.46 μM) in rabbit retina. The stimulatory effect of exogenous and endogenous dopamine is competitively antagonized by the selective D-1 dopamine receptor antagonist SCH 23390, but not by the selective D-2 dopamine receptor antagonist S-sulpiride (FIGURE 2B). The equilibrium dissociation constant (K_B) for SCH 23390 calculated using the Schild equation was 1 nM.[37] The effect of dopamine was stereoselectively antagonized by the isomers of the dopamine receptor antagonist flupenthixol. Activation or blockade of D-2 dopamine, α_2-, or β-adrenergic receptors did not stimulate or attenuate the spontaneous or dopamine-stimulated release of acetylcholine from rabbit retina.[17,37]

The potencies of dopamine receptor agonist and antagonists at the dopamine receptor mediating ^{3}H-acetylcholine release is characteristic of the D-1 dopamine receptor. Dopamine receptor agonists evoked the release of ^{3}H-acetylcholine with the following order of potency: apomorphine $\geq$ SKF(R)82626 > SKF 85174 > SKF(R)38393 $\geq$ pergolide $\geq$ dopamine (EC$_{50}$ = 4.5 μM) > SKF(S)82526 $\geq$ SFK(S)38393. Dopamine receptor antagonists inhibited the dopamine-evoked release of ^{3}H-acetylcholine with the following order of potency SCH 23390 (IC$_{50}$ = 1 nM) > (+)-butaclamol $\geq$ α-flupenthixol > fluphenazine > perphenazine > β-flupentixol > R-sulpiride. These potencies were correlated with the potencies of dopamine receptor agonists and antagonists at the D-1 dopamine receptor in rabbit retina as labeled by ^{3}H-SCH 23390, or as determined by adenylate cyclase activity.[37]

The release of acetylcholine evoked by activation of the D-1 dopamine receptor appears to be mediated through increases in intracellular cyclic AMP. In support of this hypothesis we demonstrated that increases in the levels of cyclic AMP by administration of the analogue 8-bromo-cyclic AMP, activation of adenylate cyclase by forskolin, or inhibition of phosphodiesterase elicited acetylcholine release in a calcium-dependent manner.[17]

The basal and calcium-dependent release of ^{3}H-acetylcholine evoked by exogenous dopamine is modulated by GABA agonists and antagonists. GABA and the agonist muscimol inhibit, while the antagonists picrotoxin and muscimol increase, the spontaneous release of radioactivity and the dopamine-evoked release of ^{3}H-acetylcholine from rabbit retina labeled with ^{3}H-choline.[58] These results support the existence of a tonic GABA-ergic inhibition of cholinergic amacrine cells in retina.[29,30] Dopamine-evoked acetylcholine release from rabbit retina may be physiologically important as D-1 receptor–mediated increases in acetylcholine release may be mediated by endogenous dopamine release by either tyramine or amphetamine.[58]

PRESYNAPTIC D-2 AUTORECEPTORS AND POSTSYNAPTIC D-1 DOPAMINE RECEPTORS

The established identity of and location of cholinergic and dopaminergic neurons in rabbit retina make this tissue ideal for the study of interactions between these two neurotransmitter systems. In particular, the rabbit retina provides a model system in which to determine the selectivity of dopamine receptor agonists and antagonists for the presynaptic D-2 and the postsynaptic D-1 dopamine receptors modulating the release of ^{3}H-dopamine and ^{3}H-acetylcholine, respectively. The calcium-dependent release of ^{3}H-dopamine was elicited by electrical stimulation (3Hz, 20 mA, 2 mseconds) and ^{3}H-acetylcholine release was elicited by superfusion with dopamine or dopamine agonists for an 8 minute period according to the experimental protocol showed in FIGURE 1A and B. The potency of agonists on the presynaptic D-2 dopamine autoreceptor was expressed as the concentration inhibiting the release of ^{3}H-dopamine by 50% (IC$_{50}$), and on the postsynaptic D-1 dopamine receptor as the concentration increasing ^{3}H-acetylcholine release by 50% (EC$_{50}$).

FIGURE 2 shows the potencies of dopamine to inhibit the calcium-dependent release of ^{3}H-dopamine (D-2) and to induce calcium-dependent release of ^{3}H-acetylcholine (D-1). Using this model we have calculated the potency ratio of the dopamine agonists as the IC$_{50}$ agonist/IC$_{50}$ for dopamine, where the IC$_{50}$ for dopamine was 0.47 μM.[17,59]

The potency ratios for N-0437 (0.008), pergolide (0.013), bromocriptine (0.034), apomorphine (0.047), and LY 171555 (0.74) reflect their high selectivity for the D-2 dopamine autoreceptor rather than for the D-1 postsynaptic dopamine receptor. On the other hand the potency ratios for SKF (R) 82526 (0.056), SKF 85174 (0.14), SKF (R) 38393 (0.5), and pergolide (0.59) indicate their selectivity for the D-1 dopamine receptor.[11,17,57]

FIGURE 3A shows the potency of S-sulpiride to increase the calcium-dependent release of ^{3}H-dopamine, while FIGURE 3B shows the potency of SCH 23390 to antagonize the dopamine-evoked increases in ^{3}H-acetylcholine release. Using this model we have demonstrated that the dopamine antagonists (EC_{150}, nM) S-sulpiride (8.4), domperidone (14 nM), spiperone (14 nM), and metoclopramide (22 nM) increased the release of ^{3}H-dopamine (D-2) but did not antagonize the dopamine-evoked (dopamine = 1 μM, 8 minutes) increase of ^{3}H-acetylcholine release (D-1). Fluphenazine blocked the D-1 and D-2 receptors with the same potency. The most potent D-1 dopamine receptor agonist was SCH 23390. The potency of SCH 23390 to antagonize dopamine on the D-1 postsynaptic receptor was 1 nM, and on the D-2 presynaptic autoreceptor was 430 nM.[11,17,57]

The selectivity of dopamine agonists and antagonists for the D-1 and D-2 receptor subtypes is as follows:

D-2	N-0437 > bromocriptine > 1Y171555 > pergolide
D-2 ≥ D-1	apomorphine ≥ dopamine
D-1	SKF (R) 82626 > SKF 85174 > SKF (S) 38393
D-2	S-sulpiride > domperidone > spiperone > metoclopramide
D-2 ≥ D-1	fluphenazine ≥ (+)-butaclamol ≥ alpha-flupenthixol
D-1	SCH 23390

The rabbit retina provides a model with which to determine the potency of dopamine agonists and antagonists on the D-2 presynaptic autoreceptor and the D-1 postsynaptic dopamine receptors in the central nervous system.

CONCLUSIONS

In the retina, dopamine function as a neurotransmitter and a neuromodulator, regulating a variety of retinal functions through activation of D-1 and D-2 dopamine receptors. Dopamine activates D-2 dopamine autoreceptors involved in the modulation of dopamine release through a negative feedback mechanism (FIGURE 5), and D-1 dopamine receptors that evoke the calcium-dependent release of acetylcholine. GABA inhibitory presynaptic receptors appear to modulate the release of acetylcholine elicited by D-1 dopamine receptor activation (FIGURE 5). Presynaptic inhibitory heteroreceptors for neurotransmitters and neuromodulators secreted by other retinal neurons appear to control dopaminergic activity, as does α-adrenergic, opiate, GABA, and melatonin receptors, which are inhibitory in nature (FIGURE 5). In addition to light, α-MSH, glycine, and glutamate increase the release of dopamine.

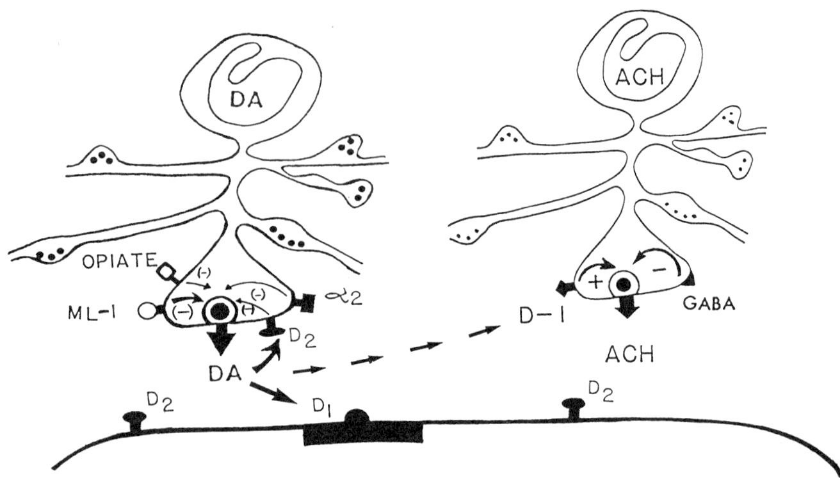

FIGURE 5. Schematic representation of dopamine- and acetylcholine-containing retinal amacrine neurons in mammalian retina. D-1 and D-2 dopamine receptors located on retinal cells postsynaptic to dopamine neurons modulate a variety of retinal functions. Activation of D-1 dopamine receptors evoked calcium-dependent release of acetylcholine. This release is modulated through activation of inhibitory GABA receptors. D-2 dopamine autoreceptors modulate the calcium-dependent release of dopamine through a negative feedback mechanism. Presynaptic and postsynaptic D-1 and D-2 dopamine receptors are not necessarily located on the same synapse. Activation of presynaptic alpha$_2$-adrenergic, opiate, and melatonin (ML-1) heteroreceptors inhibits the calcium-dependent release of dopamine.

ACKNOWLEDGMENTS

The author would like to thank Mr. Kevin McMasters for the library work and helping in editing the manuscript.

REFERENCES

1. DOWLING, J. E. & B. EHINGER. 1978. Synaptic organization of the dopaminergic neurons in the rabbit retina. J. Comp. Neurol. **180:** 203–220.
2. HOLMGREN-TAYLOR, I. 1982. Ultrastructure and synapses of the [^{3}H]dopamine-accumulating neurons in the retina of the rabbit. Exp. Eye Res. **35:** 555–572.
3. HOKOC, J. N. & A. P. MARIANI. 1988. Synapses from bipolar cells onto dopaminergic amacrine cells in cat and rabbit retinas. Brain Res. **461:** 17–26.
4. IUVONE, P. M. 1986. Neurotransmitters and neuromodulators in the retina: regulation, interactions, and cellular effects. *In* The Retina: a Model for Cell Biology Studies. R. Adler & D. Farber, Eds. **2:** 1–72. Academic Press. London, England.
5. THOMAS, T. N., T. C. CLEMENT-CORMIER & D. A. REDBURN. 1978. Uptake and release of [^{3}H]dopamine and dopamine-sensitive adenylate cyclase activity in retinal synaptosomal fractions. Brain Res. **155:** 391–396.
6. DUBOCOVICH, M. L. & N. WEINER. 1981. Modulation of the stimulation-evoked release of [^{3}H]dopamine in the rabbit retina. J. Pharmacol. Exp. Ther. **219:** 701–707.
7. BAUER, B., B. EHINGER & L. ABERG. 1980. ^{3}H-Dopamine release from rabbit retina. Albrecht von Graefs Arch. klin. exp. Ophthalmol. **215:** 71–78.

8. READING, H. W. 1983. Dopaminergic receptors in bovine retina and their interaction with thyrotropin-releasing hormone. J. Neurochem. **41:** 1587–1595.

9. NOWAK, J. Z. & E. ZURAWSKA. 1989. Dopamine in the rabbit retina and striatum: diurnal rhythm and effect of light stimulation. J. Neural. Transm. **75:** 201–212.

10. STOOF, J. C. & J. W. KEBABIAN. 1984. Two dopamine receptors: biochemistry, physiology and pharmacology. Life Sci. **35:** 2281–2296.

11. DUBOCOVICH, M. L. & N. WEINER. 1985. Pharmacological differences between the dopamine receptors of the D-1 and D-2 subtype in the rabbit retina. J. Pharmacol. Exp. Ther. **233:** 747–754.

12. ZARBIN, M. A., J. K. WAMSLEY, J. M. PALACIOS & M. J. KUHAR. 1986. Autoradiographic localization of high affinity GABA, benzodiazepine, dopaminergic, adrenergic and muscarinic cholinergic receptors in the rat, monkey, and human retina. Brain Res. **374:**75–92.

13. ELENA, P. P., P. DENIS, M. KOSINA-BOIX & P. LAPALUS. 1989. Dopamine receptors in rabbit and rat eye: characterization and localization of DA_1 and DA_2 binding sites. Curr. Eye Res. **8:** 75–83.

14. PICCOLINO, M., J. NEYTON & H. M. GERSCHENFELD. 1984. Decrease of gap junction permeability induced by dopamine and cyclic adenosine 3′:5′-monophosphate in horizontal cells of turtle retina. J. Neurosci. **4:** 2477–2488.

15. LASATER, E. M. & J. E. DOWLING. 1985. Dopamine decreases conductance of the electrical junctions between cultural retinal horizontal cells. Proc. Nat. Acad. Sci. USA **82:** 3025–3029.

16. JENSON, R. J. & N. W. DAW. 1986. Effects of dopamine and its agonists and antagonists on the receptive field properties of ganglion in the rabbit retina. Neuroscience **17:** 837–855.

17. HENSLER, J. G. & M. L. DUBOCOVICH. 1986. D-1 dopamine receptor activation mediates ^{3}H-acetylcholine release from rabbit retina. Brain Res. **398:**407–412.

18. MORGAN, W. W. 1982. Dopamine neurons in the retina: a new pharmacological model. *In* Cell Biology of the Eye. D. C. McDevitt, Ed.: 533–553. Academic Press. New York, N.Y.

19. BRANN, M. R. & C. L. JELSEMA. 1985. Dopamine receptors on photoreceptor membrane couple to a GTP-binding protein which is sensitive to both pertussis and cholera toxin. Biochem. Biophys. Res. Commun. **133:** 222–227.

20. DEARRY, A. & B. BURNSIDE. 1986. Dopaminergic regulation of cone retinomotor movement in isolated teleost retinas: I. Induction of cone contraction is mediated by D2 receptors. J. Neurochem. **46:** 1006–1021.

21. PIERCE, M. E. & J. C. BESHARSE. 1985. Circadian regulation of retinomotor movements. I. Interaction of melatonin and dopamine in the control of cone length. J. Gen. Physiol. **86:** 671–689.

22. EHINGER, B. 1983. Functional role of dopamine in the retina. *In* Progress in Retinal Research. N. Osborne & G. Chader, Eds. **2:** 213–232.

23. DUBOCOVICH, M. L. & N. WEINER. 1983. Enkephalins modulate [^{3}H]dopamine release from rabbit retina in vitro. J. Pharmacol. Exp. Ther. **224:** 634–639.

24. DUBOCOVICH, M. L. 1984. Alpha-2 adrenoceptors modulate ^{3}H-dopamine release from rabbit retina. J. Pharmacol. Exp. Ther. **230:** 149–155.

25. DUBOCOVICH, M. L. 1984. Presynaptic alpha-adrenoceptors in the central nervous system. Ann N.Y. Acad. Sci. **430:** 7–25.

26. DUBOCOVICH, M. L. 1983. Melatonin is a potent modulator of dopamine release in the retina. Nature **306:** 782–784.

27. DUBOCOVICH, M. L. 1985. Characterization of a retinal melatonin receptor. J. Pharmacol. Exp. Ther. **234:** 395–401.

28. MARSHBURN, P. B. & P. M. IUVONE. 1981. The role of GABA in the regulation of the dopamine/tyrosine hydroxylase-containing neurons of the rat retina. Brain Res. **214:** 335–347.

29. CUNNINGHAM, J. R. & M. J. NEAL. 1983. Effect of gamma-aminobutyric acid agonists, glycine taurine and neuropeptides on acetylcholine release from the rabbit retina. J. Physiol. London **336:** 563–577.

30. MASSEY, S. C. & D. A. REDBURN. 1983. A tonic gamma-aminobutyric acid-mediated inhibition of cholinergic amacrine cells in rabbit retina. J. Neurosci. **2:** 1633–1643.

31. ROSS, C. D. & D. B. MCDOUGALL, JR. 1976. The distribution of choline acetyltransferase activity in retina. J. Neurochem. **26:** 521–526.

32. MASLAND, R. H., J. W. MILLS & C. CASSIDY. 1984. The functions of acetylcholine in the rabbit retina. Proc. R. Soc. London Ser. B **223:** 121–139.

33. TSAI, W.-H. & D. G. PURO. 1986. Dopamine modulates evoked transmission at cholinergic synapses formed by rat retinal neurons with muscle cells. Brain Res. **380:** 375–378.

34. NOWAK, J. Z. 1987. The retina as a model neural tissue: comparative studies on retinal and brain aminergic mechanisms. Pol. J. Pharmacol. Pharm. **39:** 451–482.

35. DUBOCOVICH, M. L., J. T. EELLS & J. G. HENSLER. 1985. Release of ^{3}H-dopamine from rabbit retina by different depolarization stimuli: effect of melatonin. Br. J. Pharmacol. **86:** 533P.

36. DUBOCOVICH, M. L. & J. G. HENSLER. 1986. Modulation of ^{3}H-dopamine release by different frequencies of stimulation from rabbit retina. Br. J. Pharmacol. **88:**51–61.

37. HENSLER, J. G., D. J. COTTERELL & M. L. DUBOCOVICH. 1987. Pharmacological and biochemical characterization of the D-1 receptor mediating acetylcholine release in rabbit retina. J. Pharmacol. Exp. Ther. **243:** 857–867.

38. HADJICONSTANTINOU, M., J. COHEN, J. R. RUBENSTEIN & N. H. NEFF. 1984. An endogenous ligand modulates dopamine-containing neurons in retina via alpha-2 adrenoceptors. J. Pharmacol. Exp. Ther. **229:** 381–385.

39. HADJICONSTANTINOU, M., Z. ROSETTI & N. H. NEFF. 1986. Aromatic L-amino acid decarboxylase is modulated by alpha-2 adrenoceptors in rat retina. Fed. Proc. **28:** 135.

40. IUVONE, P. M. & A. L. RAUCH. 1983. Alpha-2 adrenergic receptors influence tyrosine hydroxylase activity in retinal dopamine neurons. Life Sci. **33:**2455–2466.

41. STELL, W., D. MARSHAK, T. YAMADA, N. BRECHA & H. KARTEN. 1980. Peptides are in the eye of the beholder. Trends Neurosci. **3:** 292–295.

42. REDBURN, D. A. 1985. Serotonin neurotransmitter systems in vertebrate retina. *In* Retinal Transmitters and Modulators: Models for the Brain. W. W. Morgan, Ed. **2:** 107–122. CRC Press. Boca Raton, Fla.

43. REDBURN, D. A. & C. K. MITCHELL. 1989. Darkness stimulates rapid synthesis and release of melatonin in rat retina. Visual Neurosci. **3:** 391–403.

44. BINKLEY, S., M. HRYSHCHYSHYN & L. REILLY. 1979. *N*-Acetyltransferase activity responds to environmental lighting in the eye as well as in the pineal gland. Nature **281:** 479–481.

45. DUBOCOVICH, M. L., R. C. LUCAS & J. S. TAKAHASHI. 1985. Light-dependent regulation of dopamine receptors in mammalian retina. Brain Res. **335:** 321–325.

46. NOWAK, J. Z., E. ZURAWSKA & J. ZAWILSKA. 1989. Melatonin and its generating system in vertebrate retina: circadian rhythm, effect of environmental lighting and interaction with dopamine. Neurochem. Int. **14:** 397–406.

47. HALL, F., C. TENGERDY, M. MORITA & E. PAULTER. 1985. Determination of bovine retinal melatonin by HPLC-EC. Curr. Eye. Res. **4:** 847–850.

48. BLAZYNSKI, C. & M. L. DUBOCOVICH. Localization of 2-[^{125}I]-iodomelatonin binding sites in mammalian retina. J. Neurochem. (Submitted.)

49. WIRZ-JUSTICE, A., M. DA PRADA & C. REME. 1984. Circadian rhythm in rat retinal dopamine. Neurosci. Lett. **45:** 21–25.

50. BOATRIGHT, J. H. & P. M. IUVONE. 1989. Melatonin suppresses the light-evoked release of endogenous dopamine from retinas of frogs (*Xenopus laevis*). Soc. Neurosci. Abstr. **15:** 1395.

51. DUBOCOVICH, M. L. 1988. Pharmacology and function of melatonin receptors. FASEB J. **2:** 2765–2773.

52. DUBOCOVICH, M. L. 1988. Luzindole (N-0774): a novel melatonin receptor antagonist. J. Pharmacol. Exp. Ther. **246:** 902–910.

53. BAUER, B. & B. EHINGER. 1980. Action of α-MSH on the release of neurotransmitters from the retina. Acta Physiol. Scand. **108:** 105–107.

54. KAMP, C. W. & W. W. MORGAN. 1983. Effects of excitatory amino acids on dopamine synthesis in the rat retina. Eur. J. Pharmacol. **92:** 139–142.
55. IUVONE, P. M., J. H. BOATRIGHT & M. M. BLOOM. 1987. Dopamine mediates the light-evoked suppression of serotonin N-acetyltransferase activity in retina. Brain Res. **418:** 314–324.
56. BESHARSE, J. C. & D. A. DUNIS. 1983. Methoxyindoles and photoreceptor metabolism: activation of rod shedding. Science **219:** 1341–1343.
57. PANG, S. F. & D. T. YEW. 1979. Pigment aggregation by melatonin in the retinal pigment epithelium and choroid of guinea pigs, *Cavia porcellus*. Experientia **35:** 231–233.
58. HENSLER, J. G. & M. L. DUBOCOVICH. 1988. GABA-ergic modulation of D-1 dopamine receptor-mediated ^{3}H-acetylcholine release from rabbit retina. Naunyn Schmiedebergs Arch. Pharmacol. **337:** 661–668.
59. DUBOCOVICH, M. L. & J. G. HENSLER. 1988. Functional presynaptic D-2 and postsynaptic D-1 dopamine receptors in the central nervous system. *In* Pharmacology and Functional Regulation of Dopaminergic Neurons. P. M. Beart, G. N. Woodruff & D. M. Jackson, Eds.: 175–177. Macmillan Press, London, England.

The Ionic Basis of Inhibitory Presynaptic Modulation and Substance B[a]

JACK R. COOPER AND GABOR ZOLTAY

Department of Pharmacology
Yale University
School of Medicine
New Haven, Connecticut 06510

For several years, our laboratory has been involved in identifying presynaptic receptors using synaptosomes prepared from the myenteric plexus of the guinea pig ileum.[1-4] With this preparation we have also discovered a factor in brain, tentatively referred to as substance B, which has no effect on the evoked release of acetylcholine (ACh) but which reverses inhibitory modulated release.[5,6] We recently turned our attention to investigating the mechanism involved in inhibitory presynaptic modulation and for this study we utilized cerebral rat cortical synaptosomes. Our hypothesis is that, regardless of the involvement of G proteins or of a second messenger system, the ultimate effect of presynaptic, receptor-activated inhibitory modulation is the opening of a specific K channel. This effect would hyperpolarize the terminal and therefore reduce the calcium uptake, and the result would be a diminished release of transmitter. Utilizing both KCl and veratridine evoked ^{86}Rb efflux from prelabeled synaptosomes as a measure of K fluxes and the release of [^{3}H]ACh in parallel experiments, this paper demonstrates that four inhibitory presynaptic modulators, 2-chloroadenosine, carbamylcholine, clonidine, and morphine, all increase K conductance and concomitantly inhibit ACh release. In addition, preliminary experiments suggest that substance B blocks K conductance.

Synaptosomes were prepared from the cerebral cortices of Sprague Dawley rats (Charles River Labs, Wilmington, Mass.) by the procedure of Nagy and Delgado-Escueta.[7] The preparation was incubated for 30 minutes at 37°C under 95% O_2:5% CO_2 in Krebs Ringer bicarbonate (KRB) buffer pH7.4 containing either ^{86}Rb (15–30 μCi/ml, and 0.1 mM unlabeled RbCl) or [^{3}H]choline (30 μCi/ml and 1.0 μM unlabeled choline chloride). The labeled preparation was washed twice with KRB with centrifugations performed at room temperature and subsequently resuspended in KRB and then equilibrated at 30°C for 20 minutes with 95% O_2:5% CO_2 gassing. Finally the preparation was again centrifuged at room temperature and, when 2 rat brains were used, resuspended at room temperature in 3 to 4 ml KRB buffer at a final protein concentration of 10 to 14 mg/ml. With synaptosomes that were prelabeled with [^{3}H]choline, after the labeling, eserine (20 μM) was routinely added to the KRB buffer. The composition of KRB buffer was NaCl, 118 mM; KCl, 4.7 mM; MgSO$_4$, 1.2 mM; CaCl$_2$, 2.5 mM; NaH$_2$PO$_4$, 1.2 mM; NaHCO$_3$, 20 mM, and glucose, 10 mM. The buffer was gassed with the O_2:CO_2 mixture and the pH adjusted to 7.4.

[a]This work was supported by the National Science Foundation, BNS-86-17115.

PROCEDURE FOR ^{86}Rb EFFLUX

In test tubes in a final volume of 200 μl were added KRB ± veratridine (8 μM) ± KCl (20 mM) and ± modulating agent (10^{-5}–10^{-6} M). The reaction was initiated by the addition of the labeled synaptosomes (100 μl) and tubes were incubated for 5 seconds at 30°C. The Rb efflux was terminated by the addition of 1 ml of the ice-cold "stop solution" of Bartschat and Blaustein,[8] containing tetraethylammonium chloride (145 mM), tetrabutylammonium chloride (1 mM), RbCl (5 mM), $MgCl_2$ (5 mM), $NiCl_2$ (10 mM), and HEPES buffer pH 7.4 (20 mM). The tubes were immediately centrifuged and 800 μl of the supernatant was placed in scintillation vials and the radioactivity determined by Cerenkov radiation. A zero time blank in ice was subtracted from all samples.

PROCEDURE FOR [³H]ACh RELEASE

The incubation media containing [³H]ACh-labeled synaptosomes was as described above but with the addition of eserine (20 μM) and choline oxidase (100 munits).

TABLE 1. Increase in Evoked [86]Rb Efflux in the Presence of Modulators[a]

	[86]Rb Efflux (% Increase)	
Modulator	KCl (20 mM)	Veratridine (8 μM)
2-Chloroadenosine (10^{-5} M)	130.4	95.0
Carbamylcholine (10^{-5} M)	23.0	29.9
Clonidine (10^{-5} M)	41.4	24.8
Morphine (10^{-6} M)	82.6	12.9

[a]Synaptosomes, preloaded with ^{86}Rb, were incubated in the presence and absence of reagents noted above for 5 seconds at 30°C. Efflux was terminated by the "stop solution" of Bartschat and Blaustein,[8] and after centrifugation, an aliquot of the radioactive supernatant was counted by Cerenkov radiation. Results are the mean of four experiments, each done in triplicate.

Choline oxidase was added in order to convert any released choline to betaine thus preventing a reuptake of choline and to obviate the separation of choline and ACh with the final extraction into tetraphenylboron in butyronitrile since betaine does not extract.[9]

After the 5 second incubation, the tubes were placed in ice, centrifuged, and 100 μl aliquots were pipetted into tubes containing 1 ml of tetraphenylboron in butyronitrile (10 mg/ml). After vortexing and centrifuging, 800 μl of the solvent was added to scintillation vials containing 5 ml Opti-fluor and the radioactivity determined. Again, zero time blanks were subtracted from the samples.

RESULTS AND DISCUSSION

Rb Efflux

As shown in TABLE 1, the presence of all four modulators increased ^{86}Rb efflux suggesting the opening of a K channel. It should be noted that suboptimal depolarizing concentrations of veratridine and KCl were employed in order to permit a further

TABLE 2. Decreased Evoked Release of [³H]ACh in the Presence of Modulators[a]

Modulator	[³H]ACh Release (% Decrease)	
	KCl	Veratridine
2-Chloroadenosine (10^{-5} M)	33.7	66.2
Carbamylcholine (10^{-5} M)	48.4	68.2
Morphine (10^{-6} M)	31.8	43.4
Clonidine (10^{-5} M)	31.7	73.9

[a]Synaptosomes, preloaded with [³H]choline in order to generate[³H]ACh, were incubated along with choline oxidase (100 mU) and eserine (20 μM) in the presence and absence of reagents noted above for 5 seconds at 30°C. Incubation was terminated by placing the tubes in ice water and centrifugation. An aliquot was removed, extracted with tetraphenylboron in butyronitrile (10 mg/ml), and after vortexing and centrifugation, the radioactivity in an aliquot of the organic phase was determined by liquid scintillation spectrometry.

increase in K conductance. A 5 second incubation at 30°C was chosen since in preliminary experiments when efflux was determined at 5, 15, and 30 seconds, the efflux curve tailed off at the 15 and 30 second time points. According to Bartschat and Blaustein who determined ^{86}Rb efflux from synaptosomes at 1 second intervals,[8] initial velocities are observed up to 5 seconds. Stimulated synaptosomes had an efflux rate of 1.5 to 3% of the total ^{86}Rb/second. We used ^{86}Rb rather than ^{42}K as a tracer because of its considerably longer half-life. The electrophysiological activity of Rb has been shown to be virtually identical to K both in electrophysiological experiments[10–12] and in synaptosomal preparations.[13,8]

ACh Release

As shown in TABLE 2, and as would be predicted, all modulators decreased the evoked release of ACh: as with Rb efflux experiments, this effect could also be observed in the absence of either veratridine or KCl as depolarizing agents. These results in addition disclose the presence in cortical synaptosomes of four inhibitory presynaptic receptors, purinergic, muscarinic (the carbamylcholine effect is blocked by atropine, 1 μM), α_2-adrenergic, and opioid.

One puzzling but consistent observation we made during the course of these experiments on ACh release (but not on Rb efflux) is the effect of the modulators in zero time control tubes. As shown in TABLE 3, when synaptosomes were added to the control tubes containing modulators as well as depolarizing agents kept in ice, an *increase* in ACh release was noted. (Both a depolarizing agent and a modulator had to be present to produce this increase.) The extent of this increase varied with the agents,

TABLE 3. Effect of Modulators on Zero Time Controls

Agent	Increase in [³H]ACh[a]
2-Chloroadenosine	52.8 ± 4.4
Clonidine	52.7 ± 16
Carbamylcholine	17.2 ± 22
Morphine	8.4 ± 9.4

[a]Control blanks ranged from 1274 to 3264 cpm in four experiments done in triplicate. Values are mean ± standard deviation.

with 2-chloroadenosine and clonidine giving the greatest release and carbamylcholine and morphine the least. That this increase released radioactivity was [^{3}H]ACh was shown by ion exchange chromatography using a Dowex CG 50 column. It thus appears that at a low temperature, the attachment of an inhibitory modulator to its presynaptic receptor somehow produces a membrane change that permits the release of ACh.

As shown in TABLE 4, substance B reverses the increased ^{86}Rb efflux produced by 2-chloroadenosine. For a reason that is as yet unclear, substance B also has an inhibitory effect on the veratridine-evoked efflux. This inhibitory effect is not observed in the electrically stimulated myenteric plexus–longitudinal muscle preparation. Here substance B only reverses the effects of the inhibitory modulators. Also unknown is whether this factor directly closes K channels or does so indirectly by an effect on a G protein or second messenger system.

Although these neurochemical experiments suggesting that inhibitory modulators operate by opening a K channel are new for synaptosomal preparations, there are in the literature many electrophysiological experiments that support our contention. Thus it is well documented that adenosine inhibits neuronal activity through the activation of a potassium conductance,[14–17] muscarinic activation of M_2 receptors in the heart activates a K channel,[18–20] and in locus ceruleus slices, opioids, α_2-adrenergic agonists, and

TABLE 4. Reversal of 2-Chloroadenosine Increased Efflux of ^{86}Rb in the Presence of Substance B[a]

Condition	^{86}Rb Efflux (cpm/mg protein)
Veratridine	4693
Veratridine plus 2-chloroadenosine	5536
Veratridine plus 2-chloroadenosine plus substance B	3770

[a]Experimental procedure as in legend to TABLE 1. The volume of substance B that was added was equivalent to the volume that completely reversed the effects of 2-chloroadenosine in the guinea pig ileal bioassay.

somatostatin increased a potassium conductance and hyperpolarization.[21–23] In addition, Andrade *et al.* have shown that baclofen, an agonist acting on primarily presynaptic γ-aminobutyric acid (GABA$_B$) receptors, activates a potassium current.[24]

As can be noted from the data presented in TABLES 1 and 2, a causal relationship between ^{86}Rb efflux and ACh release is not seen. Thus, for example, with KCl depolarization, carbamylcholine has the smallest effect on efflux but the largest in inhibiting ACh release whereas morphine has the reverse effect. This result is not surprising since the modulators would presumably hyperpolarize all the terminals and interneurons that are present in cortical synaptosomes and we are only measuring the release of one transmitter.

In some of these electrophysiological studies in mammalian systems, the evidence, though not conclusive, suggests that a receptor-activated G protein is directly coupled to a K channel without the intervention of second messenger systems. In the reverse situation however, protein kinase C activation appears to be involved in blocking a K channel and increasing transmitter release.[25–27] Whether in mammalian neurons protein phosphorylation is required for increasing transmitter release and a phosphoprotein phosphatase is not involved in decreasing release is still an open question.

Also to be determined is which K channel is implicated with presynaptic receptor activation. In Bartschat and Blaustein's investigations on K channels in synaptosomes, four different K channels were identified.[8,28] Using relatively specific blocking agents, our current aim is to identify the channel that is implicated in modulated release and perhaps a different channel that operates in the evoked release of transmitters.

SUMMARY

We have investigated the possibility that, regardless of the involvement of a second messenger system, the ultimate effect of presynaptic, receptor-activated inhibitory modulation is the opening of a K channel. This possibility was explored utilizing rat cortical synaptosomes that were prelabeled with either ^{86}Rb or [^{3}H]acetylcholine, depolarizing with either K^+ or veratridine, and measuring either efflux of ^{86}Rb or release of [^{3}H]acetylcholine in the presence or absence of inhibitory presynaptic modulators. The modulating agents used were 2-chloroadenosine, carbamylcholine, clonidine, and morphine. In all instances, these agents promoted an increased efflux of ^{86}Rb, indicating hyperpolarization, and decreased release of acetylcholine. These results support our contention that an increase in K conductance may be responsible for presynaptic inhibition of the release of neurotransmitters. We have also found that substance B, a compound that reverses presynaptic modulation, appears to act by closing K channels.

REFERENCES

1. BRIGGS, C. A. & J. R. COOPER. 1982. Cholinergic modulation of the release of [^{3}H] acetylcholine from synaptosomes of the myenteric plexus. J. Neurochem. **38:** 501–508.
2. REESE, J. H. & J. R. COOPER. 1982. Modulation of the release of acetylcholine from ileal synaptosomes by adenosine and adenosine 5′-triphosphate. J. Pharmacol. Exp. Ther. **223:** 612–616.
3. REESE, J. H. & J. R. COOPER. 1984. Noradrenergic inhibition of the nicotinically stimulated release of acetylcholine from guinea-pig ileal synaptosomes. Biochem. Pharmacol. **33:** 1145–1147.
4. REESE, J. H. & J. R. COOPER. 1984. Stimulation of acetylcholine release from guinea-pig ileal synaptosomes by cyclic nucleotides and forskolin. Biochem. Pharmacol. **33:** 3007–3011.
5. PEARCE, L. B., C. B. BENISHIN & J. R. COOPER. 1986. Substance B: an endogenous brain factor that reverses presynaptic inhibition of acetylcholine release. Proc. Nat. Acad. Sci. USA **83:** 7979–7983.
6. BENISHIN, C. B., L. B. PEARCE & J. R. COOPER. 1986. Isolation of a factor (substance B) that antagonizes presynaptic modulation: pharmacological properties. J. Pharmacol. Exp. Ther. **239:** 185–191.
7. NAGY, A. & A. V. DELGADO-ESCUETA. 1985. Rapid preparation of synaptosomes from mammalian brain using nontoxic isoosmotic gradient material (Percoll). J. Neurochem. **43:** 1114–1123.
8. BARTSCHAT, D. K. & M. P. BLAUSTEIN. 1985. Potassium channels in isolated presynaptic nerve terminals from rat brain. J. Physiol. **361:** 419–440.
9. COOPER, J. R. 1989. A simple method to determine released acetylcholine in the presence of choline. Life Sci. **45:** 2041–2042.
10. HILLE, B. 1975. Ionic selectivity of Na and K channels of nerve membranes. *In* Membranes: a Series of Advances **3:** 255–323. Sinaver Associates Inc. Sunderland, Mass.
11. REUTER, H. & C. F. STEVENS. 1980. Ion conductance and ion selectivity of potassium channels in small neurons. J. Membr. Biol. **57:** 103–118.
12. GORMAN, A. L. F., J. C. WOOLAM & M. C. CORNWALL. 1982. Selectivity of the calcium-activated and light-dependent K^+ channels for monovalent cations. Biophys. J. **38:** 319–332.

13. ADAM-VIZI, V. & E. LIGETI. 1984. Release of acetylcholine from rat brain synaptosomes by various agents in the absence of external calcium ions. J. Physiol. **353:** 505–521.
14. GREEN, R. W. & H. L. HAAS. 1985. Adenosine action on CA1 pyramidal neurones in rat hippocampal slices. J. Physiol. London **366:** 119–127.
15. TRUSSELL, L. O. & M. B. JACKSON. 1985. Adenosine-activated potassium conductance in cultured striatal neurones. Proc. Nat. Acad. Sci. USA **82:** 4857–4861.
16. TRUSSELL, L. O. & M. B. JACKSON. 1987. Dependence of an adenosine-activated potassium current on a GTP-binding protein in mammalian central neurons. J. Neurosci. **7:** 3306–3316.
17. MICHAELIS, M. L., K. K. JOHE, B. MOGHADAM & R. N. ADAMS. 1988. Studies on the ionic mechanism for the neuromodulatory actions of adenosine in the brain. Brain Res. **473:** 249–260.
18. SOEJIMA, M. & A. NOMA. 1984. Mode of regulation of the ACh-sensitive K-channel by the muscarinic receptor in rabbit atrial cells. Pflugers Arch. ges. Physiol. **400:** 424–431.
19. CODINA, J., A. YATANI, D. GRENET, A. M. BROWN & L. BIRNBAUMER. 1987. The α subunit of the GTP binding protein G_k opens atrial potassium channels. Science **236:** 442–445.
20. KURACHI, Y., T. NAKAJIMA & T. SUGIMOTO. 1986. Arachidonic acid metabolites as intracellular modulators of the G protein–gated cardiac K^+ channel. Pflugers. Arch. ges. Physiol. **407:** 2634–274.
21. AGHAJANIAN, G. K. & C. P. VANDER MAELEN. 1982. α_2-Adrenoreceptor-mediated hyperpolarization of locus coeruleus neurons: intracellular studies in vivo. Science **215:** 1394–1396.
22. AGHAJANIAN, G. K. & Y.-Y. WANG. 1987. Common α_2- and opiate effector mechanisms in the locus coeruleus: intracellular studies in brain slices. Neuropharmacology **26:** 793–799.
23. MIYAKE, M., J. C. MACDONALD & R. A. NORTH. 1989. Single potassium channels opened by opioids in rat locus ceruleus neurons. Proc. Nat. Acad. Sci. USA **86:** 3419–3422.
24. ANDRADE, R., R. C. MALENKA & R. A. NICOLL. 1986. A G protein couples serotonin and $GABA_B$ receptors to the same channels in hippocampus. Science **243:** 1261–1265.
25. BARABAN, J. M., S. H. SNYDER & B. E. ALGER. 1985. Protein kinase C regulates ionic conductance in hippocampal pyramidal neurons: electrophysiological effects of phorbol esters. Proc. Nat. Acad. Sci. USA **82:** 2538–2542.
26. HARVEY, A. L. & E. KARLSSON. 1980. Dendrotoxin from the venom of the green mamba, *Dendroaspis augusticeps*. A neurotoxin that enhances acetylcholine release at neuromuscular junctions. Naunyn Schmeideberg's Arch. Pharmacol. **312:** 1–6.
27. COLBY, K. A. & M. P. BLAUSTEIN. 1988. Inhibition of voltage-gated K channels in synaptosomes by sn-1,2-dioctanoylglycerol, an activator of protein kinase C. J. Neurosci. **8:** 4685–4692.
28. BARTSCHAT, D. K. & M. P. BLAUSTEIN. 1985. Calcium activated potassium channels in isolated presynaptic nerve terminals from rat brain. J. Physiol. **361:** 441–457.

Presynaptic Serotonin Receptors in the Central Nervous System

MANFRED GÖTHERT[a]

Institute of Pharmacology and Toxicology
University of Bonn
Reuterstrasse 2b
D-5300 Bonn 1, Federal Republic of Germany

INTRODUCTION

As with other presynaptic receptor systems, experiments on slices prepared from certain brain regions and the spinal cord were carried out in order to obtain basic evidence for the existence of presynaptic serotonin (5-hydroxytryptamine, 5-HT) auto- and heteroreceptors. The stimulation-evoked overflow of the relevant endogenous or tritiated neurotransmitter (previously taken up or newly synthesized from its ^{3}H precursor) was determined; under appropriate conditions the overflow reflects transmitter release from the respective neurone.[1,2] Stimulation was carried out either by electrical impulses or by an increase in the K^+ concentration in the superfusion (in a few cases incubation) medium or by reintroduction of Ca^{2+} after superfusion with Ca^{2+}-free, K^+-rich solution.[3–7] For example, it was expected (and found, see below) in experiments designed to identify inhibitory presynaptic 5-HT autoreceptors that the stimulation-evoked 5-HT release should be inhibited by a suitable 5-HT receptor agonist and disinhibited, i.e., facilitated, by an antagonist blocking the relevant 5-HT receptor class.

However, a major problem that had first to be overcome for both the identification and the classification of presynaptic 5-HT receptors was the lack of pharmacological tools. This reflected the need for a generally accepted classification scheme of 5-HT receptors. During the last decade criteria for the identification of a given receptor as belonging to a certain 5-HT receptor class or subclass have been elaborated.[8–10]

In the following section, the current classification scheme of 5-HT receptors will be very briefly outlined. This provides a framework for the subsequent description of experiments designed to obtain basic evidence for the existence of presynaptic 5-HT receptors and for their classification as derived from their pharmacological properties. This report will mainly focus on presynaptic 5-HT autoreceptors which have been studied most extensively; the literature will be quoted selectively rather than exhaustively, since several review articles on this topic are already available.[3–7] In the final section, presynaptic 5-HT heteroreceptors mediating inhibition or facilitation of 5-HT release will be briefly touched upon.

HETEROGENEITY AND CLASSIFICATION OF 5-HT RECEPTORS

By means of pharmacological, biochemical, electrophysiological, and molecular biological techniques at least three main classes of 5-HT receptors, termed 5-HT$_1$,

[a]The author's own work on presynaptic 5-HT receptors was supported by the Deutsche Forschungsgemeinschaft.

5-HT$_2$, and 5-HT$_3$, have been distinguished. The 5-HT$_1$ class does not represent a homogeneous population, since it has been subdivided into at least four subtypes named 1A, 1B, 1C, and 1D.[8-12] The identification of a given receptor as one of the various 5-HT$_1$ receptor subtypes is complicated by the fact that no selective antagonists are available.[9,12] The compounds that have been applied for this purpose block either 5-HT$_2$ receptors or even entirely different receptors as well. For instance, this holds true for spiperone which exhibits preference for the 5-HT$_{1A}$ subclass within the 5-HT$_1$ "family" but which also acts antagonistically not only at dopamine but also at 5-HT$_2$ receptors. Similarly, propranolol, pindolol, and cyanopindolol are antagonists at 5-HT$_{1A}$ and 5-HT$_{1B}$ receptors leaving 5-HT$_{1C}$ and 5-HT$_{1D}$ receptors unaffected, but, above all, they are β-adrenoceptor antagonists. Mesulergine, which is a suitable tool to distinguish 5-HT$_{1C}$ from other 5-HT$_1$ receptors, also counteracts 5-HT$_2$ receptor–mediated effects. It is a common feature of all 5-HT$_1$ receptor subtypes that they can be competitively blocked by metitepine (proposed generic name, frequently also termed methiothepine); however, this drug also acts antagonistically at 5-HT$_2$ receptors. Therefore, in order to classify a given receptor as belonging to the 5-HT$_1$ category, it is, in addition, necessary to apply a selective 5-HT$_2$ receptor antagonist, thus excluding the involvement of the latter class in the effect under study.

With few exceptions, no agonists are available that are so selective in activating a certain 5-HT$_1$ receptor subclass that their effectiveness per se represents reliable evidence for the involvement of the respective subclass. The drugs that are exceptional in this respect are, e.g., 8-hydroxy-di(n-propylamino)tetralin (8-OH-DPAT), ipsapirone, and urapidil (which in addition, is an α_1-adrenoceptor antagonist); they preferentially act either as full or as partial agonists at 5-HT$_{1A}$ receptors. However, it should be noted that 8-OH-DPAT in the higher nanomolar/low micromolar range also activates 5-HT$_{1D}$ receptors. 5-Carboxamidotryptamine is considered as a key drug for the classification of a given receptor as belonging to the 5-HT$_1$ receptor family,[8] but it should be taken into account that this drug is not very potent in activating 5-HT$_{1C}$ receptors. Thus, in spite of the progress made during the last decade in the classification of the 5-HT receptors, a rather large number of agonists and antagonists with an overlapping profile of preference for certain 5-HT$_1$ receptor subtypes have to be applied in order to be able to suggest that a given effect is mediated via a certain 5-HT receptor subclass. A reliable conclusion can, on the rule, be drawn if the potency of a sufficient number of drugs in producing a certain effect is compared with their affinity for the various 5-HT$_1$ binding sites.

In contrast to the effects mediated by 5-HT$_1$ receptors, it is rather easy to obtain evidence for the involvement of 5-HT$_2$ or 5-HT$_3$ receptors in a given action. This has been made possible by the development of relatively selective competitive antagonists. Thus, ketanserin, which (besides its affinity for α_1-adrenoceptors) only blocks 5-HT$_2$ but no other 5-HT receptors, played a key role in the identification of 5-HT$_2$ receptors.[11] The pharmacological characterization of the 5-HT$_3$ recognition sites has been facilitated by the development of highly selective antagonists, such as MDL 72222 (1αH, 3α, 5αH-tropan-3-yl-3,5-dichlorobenzoate) and ICS 205-930 [(3α-tropanyl)1H-indole-3-carboxylic acid ester]. 2-Methyl-5-hydroxytryptamine (2-CH$_3$-5-HT) is a preferential agonist at the 5-HT$_3$ receptors.[13]

BASIC EVIDENCE FOR THE EXISTENCE OF PRESYNAPTIC 5-HT AUTORECEPTORS

The ability of the nonselective 5-HT receptor agonist lysergide (LSD) to inhibit the electrically evoked ^{3}H-5-HT release in rat brain cortex slices was first described in

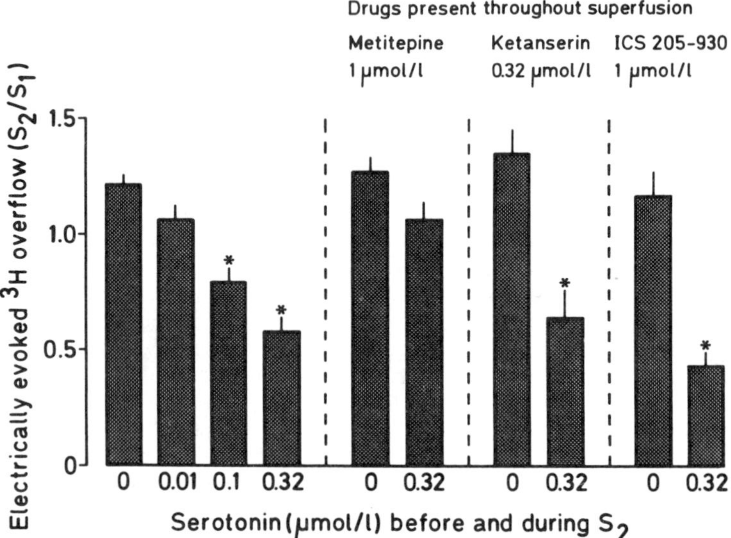

FIGURE 1. Effect of serotonin on the electrically evoked ^{3}H overflow from superfused pig brain cortex slices preincubated with ^{3}H-5-HT and interaction with either the mixed 5-HT$_1$/5-HT$_2$ receptor antagonist metitepine or the 5-HT$_2$ receptor antagonist ketanserin or the 5-HT$_3$ receptor antagonist ICS 205-930. Citalopram 10 μmol/l, phentolamine 3.2 μmol/l, and, when relevant, the 5-HT receptor antagonist under study were present throughout superfusion. ^{3}H overflow was evoked twice (S$_1$, S$_2$), and the ratio of the overflow evoked by S$_2$ over that evoked by S$_1$ is given. Serotonin was added 28 minutes before and during S$_2$. Means $\pm$ standard errors of the mean (SEM) of 4–12 experiments. The figure is based on data given by Fink et al.[23] (by permission).

1971 by Farnebo and Hamberger,[14] but it was 3 more years before it was suggested that the effect is receptor mediated. This proposal was based on the finding that not only LSD but also other 5-HT receptor agonists inhibit ^{3}H-5-HT release from rat brain coronal, hippocampal, or striatal slices, whereas metitepine increases it.[15,16] In 1979, it was shown in the rat hypothalamus[17] and cortex[18] that 5-HT itself is capable of inhibiting its own stimulation-evoked release. The inhibitory effect of 5-HT was competitively antagonized by metitepine,[1] suggesting that both compounds compete for the same site, namely, a release-modulating 5-HT receptor. Accordingly, the facilitatory effect of metitepine alone, observed in slices, is probably due to the prevention of the autoinhibitory effect of released 5-HT, i.e., to a disinhibition of the release process.

The possibility that the receptors involved in autoregulation of 5-HT release are located on the somadendritic part of 5-HT neurones is ruled out by the fact that they were identified in isolated preparations of brain areas that do not contain 5-HT cell bodies. Most of the latter occur in the raphe nuclei of the midbrain from where they project into almost all brain regions and the spinal cord. One experimental approach designed to obtain further evidence for the location of the release-modulating autoreceptors on the 5-HT nerve terminals themselves was to stimulate 5-HT release in the presence of tetrodotoxin. Even when this neurotoxin was present in the superfusion fluid of rat brain cortical slices, i.e., under a condition in which no impulse traffic occurs along the nerve axons of interneurones, serotonin inhibits its own release induced by high K$^+$ or by reintroduction of Ca^{2+} after superfusion with K$^+$-rich, Ca^{2+}-free solution.[1] Experiments on synaptosomes provided even more direct evidence that these 5-HT receptors are actually located presynaptically on the 5-HT terminals: 5-HT inhibited 5-HT release, and this effect was susceptible to blockade by metitepine.[17] Although the synaptosomal fraction may contain fragments of the adjacent postsynaptic membrane in addition to the resealed pinched-off nerve terminals, it is not very probable that transsynaptic modulation occurs. Transmitters released from neighboring nonserotoninergic synaptosomes also cannot be involved in the inhibitory effect of the 5-HT receptor agonists, since in superfused synaptosomes any transmitter released is, on the rule, removed very effectively by the superfusion fluid. Accordingly, endogenous transmitters are not present in the biophase of the synaptosomes at concentrations high enough to activate synaptosomal receptors. Since this holds also true for the autoreceptor, the 5-HT receptor antagonist metitepine, given alone, does not increase 5-HT release in superfused synaptosomes.[17]

As will be outlined in more detail below, further experiments revealed that the presynaptic 5-HT autoreceptor in the rat brain belongs to the 5-HT_{1B} subclass,[19] which, however, could not be identified by radioligand binding in, e.g., the human and porcine brain.[20,21] Accordingly, it became uncertain whether presynaptic 5-HT autoreceptors also exist in the brain of these species and, hence, appropriate experiments had to be carried out in order to clarify this matter. Basic evidence for the existence of such receptors in the porcine brain were of interest, since the latter might represent a model of the human one. In fact, it was found in both human[22] and porcine brain cortical slices[23] that serotonin inhibits the electrically evoked ^{3}H-5-HT release in a manner sensitive to blockade by metitepine (FIGURE 1); the latter given alone increased, i.e., disinhibited, release (FIGURE 2). These results indicate that autoregulation of seroto-

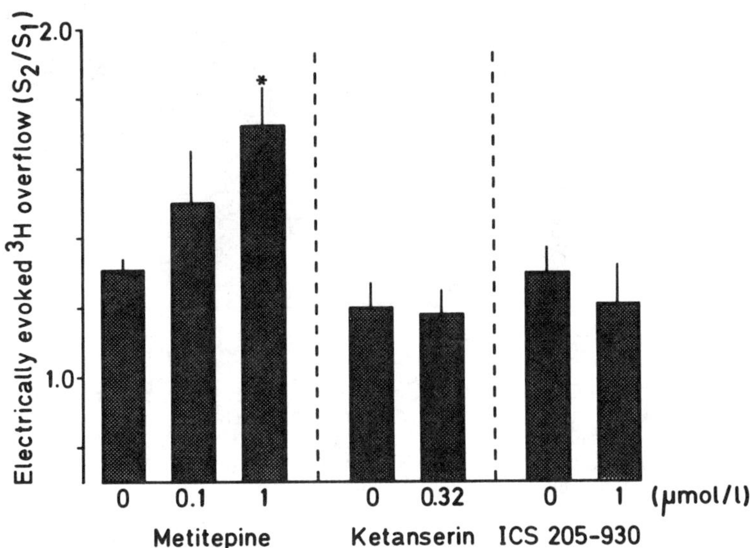

FIGURE 2. Effects of the 5-HT receptor antagonists metitepine, ketanserin, and ICS 205-930 on the electrically evoked ^{3}H overflow from superfused pig brain cortex slices preincubated with ^{3}H-5-HT. Citalopram 10 µmol/l and phentolamine 3.2 µmol/l were present throughout superfusion, and the drug under study was added 28 minutes before and during S_2. For further details, see FIGURE 1. Means ± SEM of 4–12 experiments. The figure is based on data given by Fink *et al.*[23] (by permission).

nin release also occurs in these species. In the pig brain cortex, the presynaptic location on the 5-HT nerve terminals themselves could be confirmed by experiments on synaptosomes: 5-HT and 5-methoxytryptamine inhibited K^+-evoked 5-HT release, and metitepine antagonized this inhibitory effect (FIGURE 3). In the meantime, it could be shown by application of the techniques outlined so far that presynaptic 5-HT

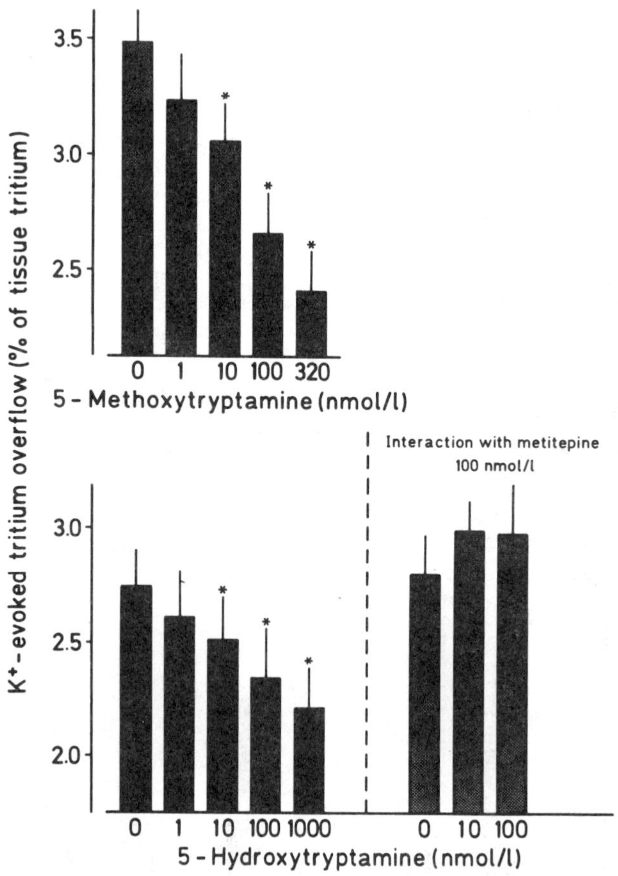

FIGURE 3. Effect of 5-hydroxytryptamine in the absence or presence of metitepine and of 5-methoxytryptamine on the K^+ (25 mmol/l) evoked ^3H overflow from superfused pig brain cortex synaptosomes preincubated with ^3H-5-HT. Superfusion with 5-HT was carried out in the presence of the 5-HT uptake inhibitor citalopram (10 μmol/l). Means ± SEM of 6 experiments in quadruplicate. *p < 0.05. (From Fink et al.,[23] by permission.)

autoreceptors occur in a great variety of brain regions and in the spinal cord of several species (TABLE 1). In fact, they have been identified in any region of the mammalian brain that has been investigated for this purpose.

Basic evidence for the existence of presynaptic 5-HT autoreceptors can also be derived from experiments in which the concentration of endogenous 5-HT is manipulated by, e.g., inhibition of neuronal 5-HT uptake[51] (TABLE 2). Under this condition, an

TABLE 1. Occurrence of Presynaptic 5-HT Autoreceptors

Rat	Rabbit
Cerebral cortex[1,2,18,19,24–31]	Cerebral cortex[48,49]
Hippocampus[26,30,32]	Hippocampus[50,51]
Nucleus accumbens[33]	Hypothalamus[52]
Corpus striatum[16,26]	Superior colliculus[53]
Hypothalamus[17,30,31,34–38]	
Cerebellum[39]	
Medulla oblongata[31]	
Spinal cord[40–45]	**Pig**
	Cerebral cortex[23,54]
Mouse	
Cerebellum[46]	
	Man
	Cerebral cortex[22,55]
Guinea pig	
Cerebral cortex[47]	

exogenous 5-HT receptor ligand competes with a high amount of endogenous 5-HT. In the case of a full agonist, its apparent potency and intrinsic activity are decreased. The decrease in apparent intrinsic activity is particularly evident with a partial agonist, which may no longer produce any inhibition or may even facilitate ^{3}H-5-HT release due to the manifestation of its antagonistic property. Finally, the release-increasing effect of a pure autoreceptor antagonist devoid of intrinsic activity is, under appropriate conditions, strongly reinforced.[51] Similar alterations in the effects of autoreceptor ligands can be found when the concentration of endogenous 5-HT in the biophase of the autoreceptor is enhanced by, e.g., increasing the frequency of electrical stimulation. In contrast, opposite modifications of the effects of autoreceptor ligands can be observed when the biophase concentration of endogenous 5-HT is decreased. If a single pulse is applied to a brain slice, either no or, at the most, minimum concentrations of

TABLE 2. Normal Effects (*Italics*) of Autoreceptor Ligands and Modifications ($\rightarrow$) by Increasing the Concentration of Endogenous 5-HT[a] in the Autoreceptor Biophase

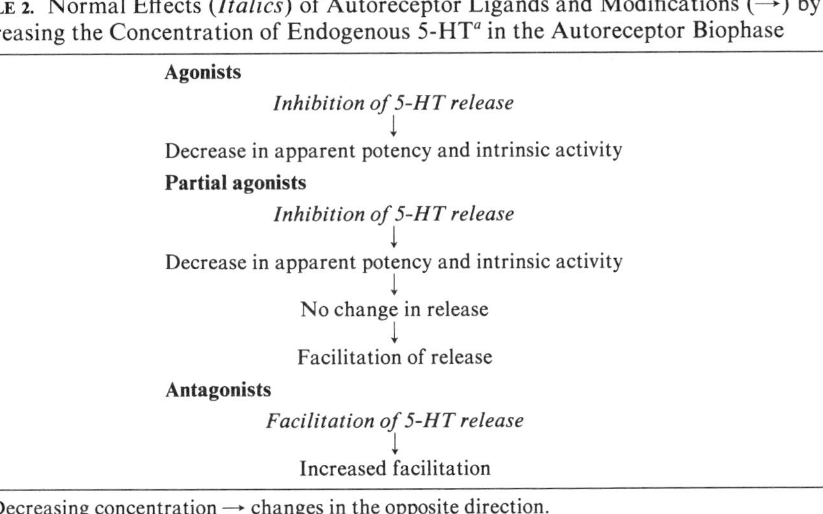

Agonists

Inhibition of 5-HT release
↓
Decrease in apparent potency and intrinsic activity

Partial agonists

Inhibition of 5-HT release
↓
Decrease in apparent potency and intrinsic activity
↓
No change in release
↓
Facilitation of release

Antagonists

Facilitation of 5-HT release
↓
Increased facilitation

[a]Decreasing concentration $\rightarrow$ changes in the opposite direction.

endogenous 5-HT resulting from spontaneous neuronal release should be present in the autoreceptor biophase. Under these conditions, the "true" potencies of exogenous agonists and antagonists can be determined. An autoreceptor antagonist given alone should, at best, result in a minimum increase in 5-HT release, which in fact was found.[49]

All of these consequences resulting from the manipulations of the biophase concentration of endogenous 5-HT not only provide basic evidence for the existence of presynaptic 5-HT autoreceptors but they also support the suggestion that these receptors play a physiological role in the regulation of 5-HT release. The latter view is also substantiated by the finding that the presynaptic 5-HT autoreceptors are operative *in vivo*.[56–58]

CLASSIFICATION OF PRESYNAPTIC 5-HT AUTORECEPTORS

In any species investigated so far, the presynaptic 5-HT autoreceptor belongs to the 5-HT_1 "family." This conclusion can be based on experiments with only few antagonists, as exemplified in FIGURES 1 and 2 for the pig brain cortex: the inhibitory effect of 5-HT on 5-HT release was antagonized by the mixed $5\text{-HT}_1/5\text{-HT}_2$ receptor antagonist metitepine but not by the selective 5-HT_2 and 5-HT_3 receptor antagonists ketanserin and ICS 205-930 respectively (FIGURE 1); accordingly, only metitepine, but not the other two antagonists, given alone, increased 5-HT release[23] (FIGURE 2). Similar results had already been obtained in rat brain slices, in which the first attempts to subclassify the presynaptic 5-HT autoreceptors have been made as well.[19,26]

Comparison of the potencies of a large number of 5-HT receptor agonists and antagonists in activating and blocking the autoreceptors, respectively, with their affinities for the various 5-HT binding sites revealed significant correlations with the affinities for 5-HT_{1A} and 5-HT_{1B} binding sites only.[19] Since 8-OH-DPAT and ipsapirone were inactive as agonists and since spiperone failed to antagonize the release-inhibiting effect of 5-HT, it could be concluded that the presynaptic 5-HT autoreceptor in the rat brain belongs to the 5-HT_{1B} subclass.

However, 5-HT_{1B} binding sites are absent in, e.g., the calf, pig, and human brain. Therefore, the question arose as to which class of recognition sites might be involved in presynaptic autoregulation of 5-HT release in the brain of the latter species. Since 5-HT_{1D} binding sites could recently be identified as a new subclass of the 5-HT_1 family in the bovine, porcine, and human brain,[59,60] it was an attractive hypothesis to assume that this recognition site might mediate the same function in these species as the 5-HT_{1B} receptor in the rat, i.e., presynaptic autoinhibition of 5-HT release. For the porcine brain this hypothesis was recently examined in our laboratory. In cerebral cortex slices, the potencies of 9 serotonin receptor agonists in inhibiting $^3\text{H-5-HT}$ release were significantly correlated with their affinities for 5-HT_{1C} and 5-HT_{1D} binding sites in membranes of the pig choroid plexus and caudate nucleus, respectively, but not with their affinities for 5-HT_{1A} and 5-HT_{1B} sites in cerebral cortical membranes of pig and rat, respectively (the selective 5-HT_{1A} receptor agonists ipsapirone and urapidil were inactive). The involvement of 5-HT_{1C} recognition sites in presynaptic autoregulation of 5-HT release was ruled out by the failure of mesulergine and the very low potency of mianserin to antagonize the inhibitory effect of 5-methoxytryptamine. Hence, the pharmacological properties of the presynaptic 5-HT autoreceptors in the pig brain cortex conform to the 5-HT_{1D} receptor subtype,[54] and similar conclusions have recently been drawn for the guinea pig brain cortex.[61] In view of the marked pharmacological similarities of the pig and human 5-HT_{1D} binding sites,[60] the pig brain may be assumed to represent an appropriate model of the human one with respect to presynaptic 5-HT autoreceptors.

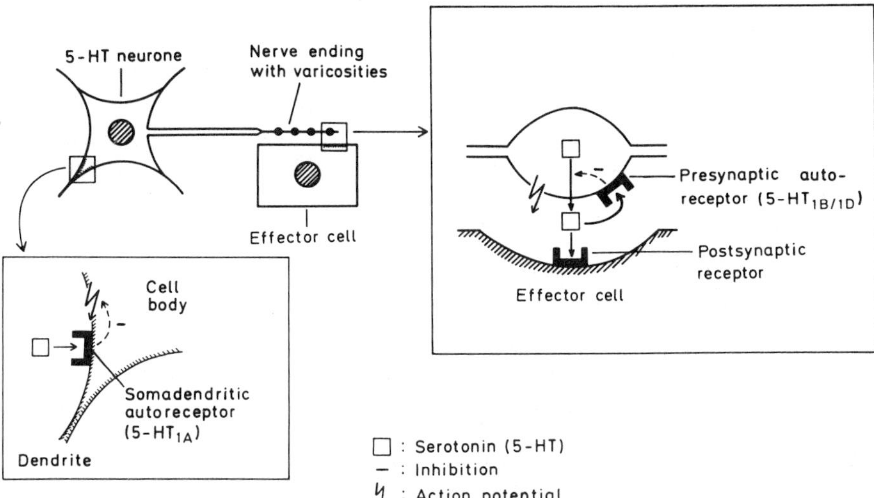

FIGURE 4. Schematic representation of a 5-HT neurone and its somadendritic and presynaptic 5-HT autoreceptors. The latter belong to the 5-HT$_{1B}$ subtype in the rat and mouse and to the 5-HT$_{1D}$ subclass in the pig, guinea pig, and, probably, man.

Taken together, the presynaptic 5-HT autoreceptors do not represent a homogeneous population; in rats and mice they belong to the 5-HT$_{1B}$ subtype; in the other species investigated so far they are of the 5-HT$_{1D}$ subtype (FIGURE 4). In this context it is of interest to note that the somadendritic 5-HT autoreceptors in the raphe nuclei do not fit to the pharmacological properties of either of these subclasses but belongs to the 5-HT$_{1A}$ subtype.[62,63]

INHIBITORY AND FACILITATORY 5-HT HETERORECEPTORS: BASIC EVIDENCE AND CLASSIFICATION

Slices and synaptosomes prepared from various rat brain regions have also been used to identify inhibitory and stimulatory 5-HT heteroreceptors. TABLE 3 summarizes results obtained in such experiments.

TABLE 3. Classification of Presynaptic 5-HT Heteroreceptors in the Rat Brain

Function	Neurone	Brain Region	5-HT Receptor Class or Subclass
Inhibition of release	DA neurone	Striatum[a,64,65]	Unknown
		Nucleus accumbens[b,66]	
	ACh neurone	Striatum[a,67,68]	5-HT$_1$[c]
		Hippocampus[b,69]	5-HT$_{1B}$
	GLU neurone	Cerebellum[b,70]	5-HT$_1$
Facilitation of release	DA neurone	Striatum[a,71]	5-HT$_3$

[a]Slices.
[b]Synaptosomes.
[c]Not yet subclassified; results compatible with the involvement of the 5-HT$_{1B}$ subtype.

The dopamine (DA) nerve terminals of the striatum and nucleus accumbens are endowed with inhibitory 5-HT receptors, at which 5-HT possesses only low affinity. This conclusion can be derived from the fact that high concentrations of 5-HT (in the μmolar range) were necessary to inhibit the K^+ or electrically evoked ^{3}H-DA release from rat striatal slices[64,65] and the K^+-evoked release of ^{3}H-DA from synaptosomes of nucleus accumbens.[66] The inhibitory effect of 5-HT in striatal slices was antagonized by methysergide or metitepine. Further experimental work is necessary to classify this receptor.

Inhibitory 5-HT heteroreceptors have also been identified on acetylcholine (ACh) nerve terminals. Thus, in striatal slices, 5-HT and other 5-HT receptor agonists inhibited the ^{3}H-ACh release evoked by K^+ (in the absence or presence of tetrodotoxin)[67] or ouabain.[68] Metitepine and methysergide, but not spiperone, counteracted the inhibitory effect, suggesting the involvement of a 5-HT$_1$ receptor subclass that does not conform to the pharmacological properties of the 5-HT$_{1A}$ subtype. Inhibitory effects of 5-HT and a mixed 5-HT$_{1A}$/5-HT$_{1B}$ receptor agonist, but not 8-OH-DPAT, were also observed in rat hippocampal synaptosomes; the response to 5-HT was antagonized by metitepine and $(-)$-propranolol, but not by ketanserin and spiperone.[69] These data suggest that the release-inhibiting 5-HT receptors on the hippocampal ACh nerve terminals are of the 5-HT$_{1B}$ subtype.

Furthermore, inhibitory 5-HT heteroreceptors are present on the glutamate (GLU) nerve terminals of the cerebellum, since the K^+-evoked release of endogenous GLU from synaptosomes of this brain region was decreased by 5-HT in a manner susceptible to antagonism by metitepine, but not $(-)$-propranolol, ketanserin, methysergide, or spiperone. 8-OH-DPAT at rather high concentration also acted inhibitorily.[70] These receptors seem to represent a subtype of the 5-HT$_1$ class, but their pharmacological properties do not conform to the characteristics of the 5-HT$_{1A}$, 5-HT$_{1B}$, or 5-HT$_{1C}$ recognition sites.

The DA nerve terminals of the striatum are endowed not only with inhibitory 5-HT receptors but also with 5-HT recognition sites mediating stimulation of DA release. In the experiments designed to identify these receptors, the overflow of endogenous DA from slices was determined. Both 5-HT and the 5-HT$_3$ receptor agonist 2-CH$_3$-5-HT produced a direct stimulation of DA release, and the response to 5-HT was antagonized by ICS 205-930.[71] These results strongly suggest that the receptor involved belongs to the 5-HT$_3$ class, a result that is compatible with the general functional property of 5-HT$_3$ receptors, namely, to produce neuronal excitation.[13]

SUMMARY AND CONCLUSIONS

Presynaptic 5-HT *autoreceptors* have been identified in any region of the mammalian CNS containing 5-HT nerve terminals that has been investigated for this purpose. They belong to the 5-HT$_{1B}$ receptor subclass in the rat and to the 5-HT$_{1D}$ subclass in the pig, guinea pig, and probably man. The presence and operation of presynaptic 5-HT autoreceptors have been proven by the ability of 5-HT receptor agonists to inhibit 5-HT release and of 5-HT receptor antagonists not only to competitively antagonize this effect but also to disclose the autoinhibitory effect of endogenous 5-HT by blocking the autoreceptor, thus interrupting the negative feedback loop. There is evidence that presynaptic 5-HT autoreceptors are operative *in vivo*. Presynaptic *inhibitory* 5-HT *heteroreceptors* have also been identified in various brain regions of the rat. DA nerve terminals in the striatum and nucleus accumbens as well as GLU nerve terminals in the cerebellum are endowed with such receptors, which were either not yet classified (DA neurone) or represent a not yet specified 5-HT$_1$ subtype (GLU neurone). Release-inhibiting 5-HT receptors on the acetylcholine nerve terminals in the hippocampus are of the 5-HT$_{1B}$ subtype, and those in the striatum were not yet classified in detail. A 5-HT heteroreceptor mediating *stimulation* of release occurs on rat striatal DA nerve terminals; it belongs to the 5-HT$_3$ class. Thus, presynaptic inhibitory 5-HT auto- and heteroreceptors as well as presynaptic excitatory 5-HT heteroreceptors are involved in the regulation of transmitter release in the brain.

ACKNOWLEDGMENT

The expert secretarial assistance of Mrs. S. Rudat is gratefully acknowledged.

REFERENCES

1. GÖTHERT, M. 1980. Naunyn-Schmiedeberg's Arch. Pharmacol. **314:** 223–230.
2. GÖTHERT, M. & E. SCHLICKER. 1983. Life Sci. **32:** 1183–1191.
3. GÖTHERT, M. 1982. Trends Pharmacol. Sci. **3:** 437–440.
4. MORET, C. 1985. *In* Neuropharmacology of Serotonin. A. R. Green, Ed.: 21–49. Oxford University Press. Oxford, England.
5. TIMMERMANS, P. B. M. W. M. & M. J. M. C. THOOLEN. 1987. Med. Res. Rev. **7:** 307–332.
6. MIDDLEMISS, D. N. 1988. *In* The Serotonin Receptors. E. Sanders-Bush, Ed.: 201–224. Humana. Clifton, N.J.
7. STARKE, K., M. GÖTHERT & H. KILBINGER. 1989. Physiol. Rev. **69:** 864–989.
8. BRADLEY, P. B., G. ENGEL, W. FENIUK, J. R. FOZARD, P. P. A. HUMPHREY, D. N. MIDDLEMISS, E. J. MYLECHARANE, B. P. RICHARDSON & P. R. SAXENA. 1986. Neuropharmacology **25:** 563–576.
9. PEROUTKA, S. J. 1988. Annu. Rev. Neurosci. **11:** 45–60.
10. LEFF, P. & G. R. MARTIN. 1988. Med. Res. Rev. **8:** 187–202.
11. GÖTHERT, M. *In* Serotonin and 5-HT$_2$ Receptor Blockade in the Cardiovascular System. M. Göthert, Ed. Progr. Pharmacol. Fischer. Stuttgart, FRG. (In press.)
12. GÖTHERT, M. & E. SCHLICKER. J. Cardiovasc. Pharmacol. (In press.)
13. RICHARDSON, B. P. & G. ENGEL. 1986. Trends Neurosci. **9:** 424–428.
14. FARNEBO, L.-O. & B. HAMBERGER. 1971. Acta Physiol. Scand. **371**(Suppl.): 35–44.
15. FARNEBO, L.-O. & B. HAMBERGER. 1974. J. Pharm. Pharmacol. **26:** 642–644.
16. HAMON, M., S. BOURGOIN, J. JAGGER & J. GLOWINSKI. 1974. Brain Res. **69:** 265–280.
17. CERRITO, F. & M. RAITERI. 1979. Eur. J. Pharmacol. **57:** 427–430.
18. GÖTHERT, M. & G. WEINHEIMER. 1979. Naunyn-Schmiedeberg's Arch. Pharmacol. **310:** 93–96.
19. ENGEL, G., M. GÖTHERT, D. HOYER, E. SCHLICKER & K. HILLENBRAND. 1986. Naunyn-Schmiedeberg's Arch. Pharmacol. **332:** 1–7.
20. HOYER, D., G. ENGEL & H. O. KALKMAN. 1985. Eur. J. Pharmacol. **118:** 13–23.
21. HOYER, D., A. PAZOS, A. PROBST & J. M. PALACIOS. 1986. Brain Res. **376:** 85–96.
22. SCHLICKER, E., F. BRANDT, K. CLASSEN & M. GÖTHERT. 1985. Brain Res. **331:** 337–341.
23. FINK, K., E. SCHLICKER, R. BETZ & M. GÖTHERT. 1988. Naunyn-Schmiedeberg's Arch. Pharmacol. **338:** 14–18.
24. BAUMANN, P. A. & P. C. WALDMEIER. 1981. Naunyn-Schmiedeberg's Arch. Pharmacol. **317:** 36–43.
25. ENGEL, G., M. GÖTHERT, E. MÜLLER-SCHWEINITZER, E. SCHLICKER, L. SISTONEN & P. A. STADLER. 1983. Naunyn-Schmiedeberg's Arch. Pharmacol. **324:** 116–124.
26. MIDDLEMISS, D. N. 1984. Naunyn-Schmiedeberg's Arch. Pharmacol. **327:** 18–22.
27. MIDDLEMISS, D. N. 1984. Eur. J. Pharmacol. **101:** 289–293.
28. BONANNO, G. & M. RAITERI. 1987. Naunyn-Schmiedeberg's Arch. Pharmacol. **335:** 219–225.
29. GÖTHERT, M., E. SCHLICKER, K. FINK & K. CLASSEN. 1987. Arch. Int. Pharmacodyn. Ther. **288:** 31–42.
30. FRIEDMAN, E. & H.-Y. WANG. 1988. J. Neurochem. **50:** 195–201.
31. SCHLICKER, E., K. CLASSEN & M. GÖTHERT. 1988. J. Cardiovasc. Pharmacol. **11:** 518–528.

32. WANG, H. Y. & E. FRIEDMAN. 1988. Psychopharmacology **94:** 312–314.
33. DRESCHER, K. & L. HETEY. 1988. Neuropharmacology **27:** 31–36.
34. COX, B. & C. ENNIS. 1982. J. Pharm. Pharmacol. **34:** 438–441.
35. LANGER, S. Z. & C. MORET. 1982. J. Pharmacol. Exp. Ther. **222:** 220–226.
36. MARTIN, L. L. & E. SANDERS-BUSH. 1982. Neuropharmacology **21:** 445–450.
37. RICHARDS, M. H. 1985. Naunyn-Schmiedeberg's Arch. Pharmacol. **329:** 359–366.
38. BLIER, P., R. RAMDINE, A.-M. GALZIN & S. Z. LANGER. 1989. Naunyn-Schmiedeberg's Arch. Pharmacol. **339:** 60–64.
39. BONANNO, G., G. MAURA & M. RAITERI. 1986. Eur. J. Pharmacol. **126:** 317–321.
40. MITCHELL, R. & S. FLEETWOOD-WALKER. 1981. Eur. J. Pharmacol. **76:** 119–120.
41. MONROE, P. J. & D. J. SMITH. 1985. J. Neurochem. **45:** 1886–1894.
42. MONROE, P. J., K. MICHAUX & D. J. SMITH. 1986. Neuropharmacology **25:** 261–265.
43. STAUDERMAN, K. A. & D. J. JONES. 1986. Eur. J. Pharmacol. **120:** 107–109.
44. BROWN, L., J. AMEDRO, G. WILLIAMS & D. SMITH. 1988. Eur. J. Pharmacol. **145:** 163–171.
45. MURPHY, R. M. & F. P. ZEMLAN. 1988. Neuropharmacology **27:** 37–42.
46. FIGUEROA, H. R., P. B. YÜRGENS, D. K. NEWTON & T. R. HALL. 1985. Gen. Pharmacol. **16:** 103–108.
47. MIDDLEMISS, D. N., M. E. BREMER & S. M. SMITH. 1988. Eur. J. Pharmacol. **157:** 101–107.
48. LIMBERGER, N., G. BONANNO, L. SPÄTH & K. STARKE. 1986. Naunyn-Schmiedeberg's Arch. Pharmacol. **332:** 324–331.
49. LIMBERGER, N., M. R. G. FISCHER, T. WICHMANN & K. STARKE. 1989. Naunyn-Schmiedeberg's Arch. Pharmacol. **340:** 52–61.
50. FEUERSTEIN, T. J., C. ALLGAIER & G. HERTTING. 1987. Eur. J. Pharmacol. **139:** 267–272.
51. FEUERSTEIN, T. J., A. LUPP & G. HERTTING. 1987. Neuropharmacology **26:** 1071–1080.
52. VERBEUREN, T. J., E. P. COEN, A. SCHOUPS, R. VAN DE VELDE, R. BAEYENS & W. P. DE POTTER. 1984. Naunyn-Schmiedeberg's Arch. Pharmacol. **327:** 102–196.
53. WICHMANN, T., N. LIMBERGER & K. STARKE. 1989. Neuroscience **32:** 141–151.
54. SCHLICKER, E., K. FINK, M. GÖTHERT, D. HOYER, G. MOLDERINGS, I. ROSCHKE & P. SCHOEFFTER. 1989. Naunyn-Schmiedeberg's Arch. Pharmacol. **340:** 45–51.
55. GALZIN, A.-M., J. P. CHODKIEWICZ, M.-F. POIRIER, H. LOO, F. X. ROUX, A. REDONDO, A. LISTA, R. RAMDINE, P. BLIER & S. Z. LANGER. 1988. Br. J. Pharmacol. **93:** 14P.
56. HÉRY, F., G. SIMONNET, S. BOURGOIN, P. SOUBRIÉ, F. ARTRAUD, M. HAMON & J. GLOWINSKI. 1979. Brain Res. **169:** 317–334.
57. BAUMANN, P. A. & P. C. WALDMEIER. 1984. Neuroscience **11:** 195–204.
58. CHAPUT, Y., P. BLIER & C. DE MONTIGNY. 1986. J. Neurosci. **6:** 2796–2801.
59. HEURING, R. E. & S. J. PEROUTKA. 1987. J. Neurosci. **7:** 894–903.
60. WAEBER, C., P. SCHOEFFTER, J. M. PALACIOS & D. HOYER. 1988. Naunyn-Schmiedeberg's Arch. Pharmacol. **337:** 595–601.
61. HOYER, D. & D. MIDDLEMISS. 1989. Tr. Pharmacol. Sci. **10:** 130–132.
62. DE MONTIGNY, C., P. BLIER & Y. CHAPUT. 1984. Neuropharmacology **23:** 1511–1520.
63. SPROUSE, J. S. & G. K. AGHAJANIAN. 1986. Eur. J. Pharmacol. **128:** 295–298.
64. ENNIS, C., J. D. KEMP & B. COX. 1981. J. Neurochem. **36:** 1515–1520.
65. WESTFALL, T. C. & V. TITTERMARY. 1982. Neurosci. Lett. **28:** 205–209.
66. HETEY, L. & K. DRESCHER. 1986. Neuropharmacology **25:** 1103–1109.
67. GILLET, G., S. AMMOR & G. FILLION. 1985. J. Neurochem. **45:** 1687–1691.
68. VIZI, E. S., L. G. HÁRSING & G. ZSILLA. 1981. Brain Res. **212:** 89–99.
69. MAURA, G. & M. RAITERI. 1986. Eur. J. Pharmacol. **129:** 333–337.
70. RAITERI, M., G. MAURA, G. BONANNO & A. PITTALUGA. 1986. J. Pharmacol. Exp. Ther. **237:** 644–648.
71. BLANDINA, P., J. GOLDFARB & J. P. GREEN. 1988. Eur. J. Pharmacol. **155:** 349–350.

Presynaptic Muscarinic Receptors in the Central Nervous System[a]

MAURIZIO RAITERI, MARIO MARCHI,
AND PAOLO PAUDICE

Institute of Pharmacology and Pharmacognosy
University of Genoa
Viale Cembrano 4
16148 Genoa, Italy

INTRODUCTION

Muscarinic receptors mediating regulation of neurotransmitter release exist on nerve terminals of the central nervous system (CNS). Due to their anatomical localization, these receptors have been termed presynaptic receptors and, according to their neuronal localization, divided into muscarinic autoreceptors and muscarinic heteroreceptors. Muscarinic autoreceptors are presynaptic receptors on which acetylcholine (ACh) is thought to act to regulate (in general to inhibit) its own release. Muscarinic heteroreceptors are presynaptic receptors on which ACh acts to mediate regulation of the release of a different transmitter.

Muscarinic autoreceptors and a number of muscarinic heteroreceptors have been characterized in various laboratories, including our own, using slices or nerve endings isolated from different areas of the mammalian brain. Some of the recent results are summarized in this chapter, in which particular emphasis has been given to the pharmacological aspects of the receptors, in other words to their characterization in terms of subtypes of the muscarinic receptor.

PRESYNAPTIC MUSCARINIC AUTORECEPTORS

The existence of an inhibitory muscarinic regulation of the release of ACh was first demonstrated *in vitro* by using brain slices (see the recent review by Starke *et al.*).[1]

The anatomical location of the receptors involved was sometimes difficult to demonstrate since the brain tissues utilized (slices of the corpus striatum, for instance) contained both nerve terminals and cell bodies. Thus the autocontrol of release could occur not only at the terminals but also at the soma-dendritic level. Some *in vitro* preparations did not contain cell bodies, however, and therefore the modulation of the release of ACh could not involve soma-dendritic receptors but most likely autoreceptors sited on the nerve terminals. The results obtained using isolated nerve endings (synaptosomes) clearly indicated that the autoinhibitory mechanism of ACh release occurs through the activation of muscarinic receptors located on cholinergic nerve terminals.[2,3] As shown in FIGURE 1, ACh or oxotremorine produced a concentration-dependent inhibition of the depolarization-evoked ^{3}H-ACh release from synaptosomes prepared from rat hippocampus and prelabeled with ^{3}H-choline. The inhibitory

[a]This work was supported by grants from the Italian Ministry of Education and from the Italian National Research Council.

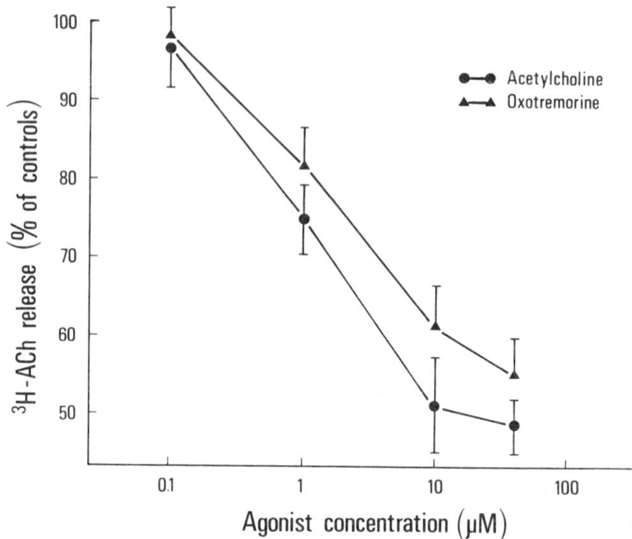

FIGURE 1. Effects of acetylcholine (ACh) and oxotremorine on ³H-ACh release in synaptosomes from rat hippocampus.

activity of extracellular ACh was totally counteracted by the muscarinic antagonist atropine but unaffected by the nicotinic antagonist mecamylamine (FIGURE 2).

It is well known that the sensitivity of a neutrotransmitter receptor can be modified after long-term activation or blockade by specific drugs. Desensitization occurs, in general, after long-lasting exposure to agonists, whereas reduced supply of the transmit-

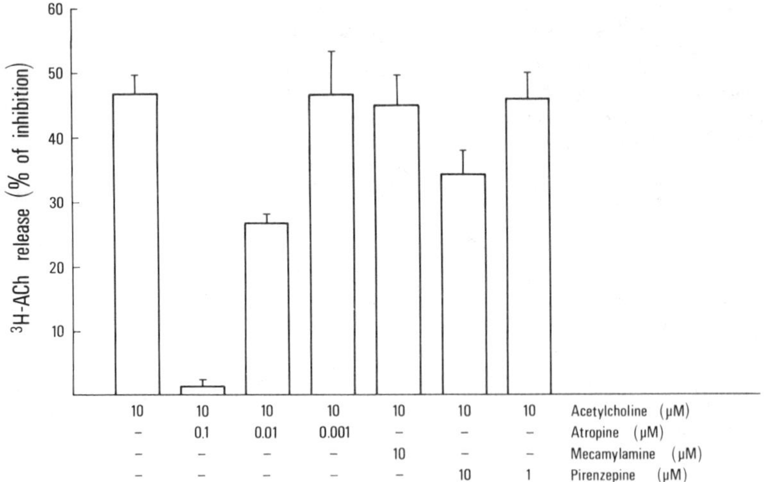

10	10	10	10	10	10	10	Acetylcholine (µM)
–	0.1	0.01	0.001	–	–	–	Atropine (µM)
–	–	–	–	10	–	–	Mecamylamine (µM)
–	–	–	–	–	10	1	Pirenzepine (µM)

FIGURE 2. Antagonism by different antimuscarinic drugs of the effect of exogenous acetylcholine on the release of ³H-ACh from synaptosomes of rat hippocampus.

ter to its receptors, as in the case of a chronic treatment with antagonist drugs, appears to cause receptor supersensitivity. Since it is conceivable that these modifications reflect adaptive changes at both pre- and postsynaptic receptors, we have investigated whether long-term treatment with paraoxon or scopolamine would induce super- or subsensitivity of presynaptic ACh autoreceptors, supporting the idea that these receptors play an *in vivo* role in the homeostatic regulation of neurotransmitter release.

Rats were treated chronically (14 days) with paraoxon, scopolamine, or saline. FIGURE 3 shows that, in synaptosomes from the animals treated chronically (but not acutely, not shown) with paraoxon or scopolamine, exogenous ACh inhibited ACh release, respectively, less and more efficiently than in synaptosomes prepared from animals treated with saline. These results indicate that cholinergic muscarinic autoreceptors undergo adaptive changes following long-term activation or blockade and suggest that a tonic activation of autoreceptors occurs in the living brain.[4]

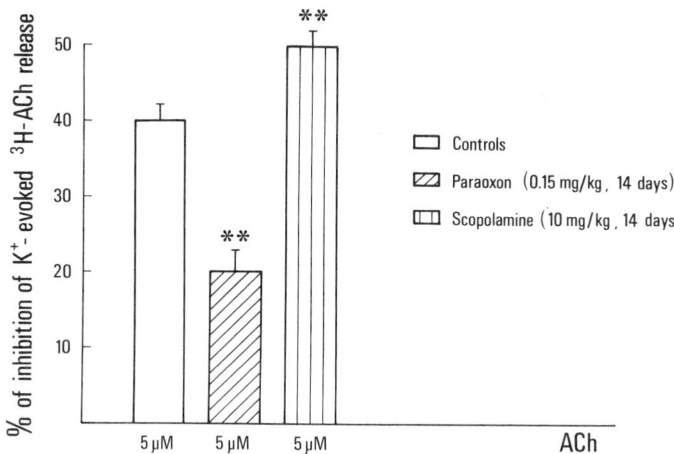

FIGURE 3. Effect of long-term treatment with paraoxon or scopolamine on the sensitivity of presynaptic muscarinic autoreceptors from rat hippocampus.

PRESYNAPTIC MUSCARINIC HETERORECEPTORS

Muscarinic Receptors Regulating Dopamine Release

Acetylcholine and dopamine (DA) interact reciprocally in the corpus striatum. While numerous studies allow us to conclude that DA plays an inhibitory role on striatal cholinergic interneurons, the effects of ACh on the dopaminergic system are less clear. After some initial reports in which a muscarinic-mediated inhibition of DA release was proposed,[5,6] different laboratories, including our own, have consistently reported the opposite effect.[7–9] A muscarinic receptor–mediated potentiation of DA release was also found in the frontal cortex of the rat.[10]

FIGURE 4 shows that the release of ³H-DA induced by 15 mM KCl from superfused striatal synaptosomes was increased in a concentration-dependent manner by exogenous ACh or oxotremorine. Depolarization with 15 mM KCl also stimulated the release of endogenous DA[11] (FIGURE 5). The K⁺-evoked release of endogenous DA was totally Ca^{2+} dependent. Similarly to that observed with the radioactive catecholamine,

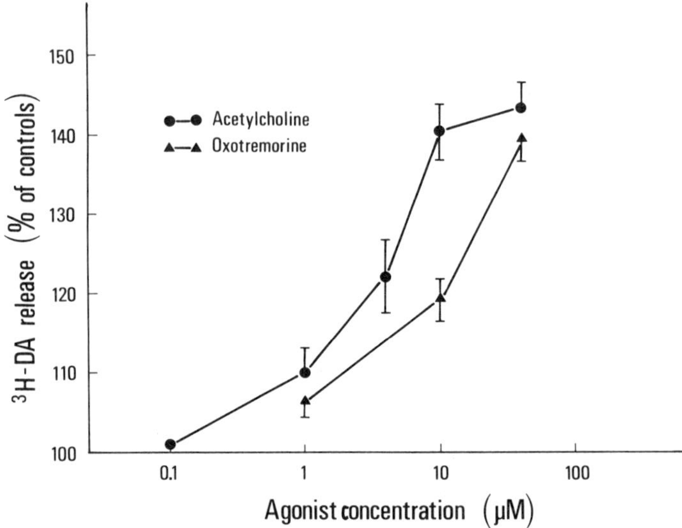

FIGURE 4. Effects of acetylcholine and oxotremorine on ³H-DA release in synaptosomes from rat corpus striatum.

ACh added to the superfusion medium produced potentiation (about 40%) of the K^+-evoked release of endogenous DA (FIGURE 5). The effects of ACh were antagonized by atropine suggesting the involvement, as in the case of the autoreceptors, of presynaptic receptors of the muscarinic type (FIGURES 5 and 6).

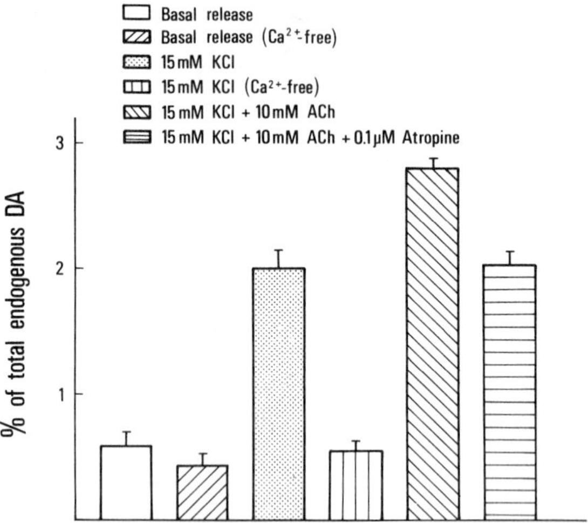

FIGURE 5. Release of endogenous dopamine from superfused rat striatal synaptosomes: effect of high K^+, Ca^{2+} ions and muscarinic agents.

When these experiments were performed it became evident that muscarinic receptors existed as subtypes in the CNS (for a review see Hirschowitz *et al.*).[12] Therefore, we investigated the possible pharmacological differences between the presynaptic muscarinic receptors present on DA terminals and the presynaptic muscarinic autoreceptors located on ACh terminals. As already mentioned, the actions of ACh (inhibition of ^{3}H-ACh release and potentiation of ^{3}H-DA or endogenous DA release) were counteracted by atropine which showed identical potency at the two muscarinic receptors (cf. FIGURES 2 and 6). However, other muscarinic antagonists behave differently. Pirenzepine strongly antagonized the receptors located on DA neurons (FIGURE 6) but was unable to counteract ACh at the autoreceptors (FIGURE 2). Dicyclomine, similarly to pirenzepine, was ineffective at the autoreceptors but antagonized the action of ACh on DA release.[13]

Studies on the regional distribution of ^{3}H-pirenzepine binding in the rat brain had provided evidence for distinct high-affinity (also termed M-1) and low-affinity musca-

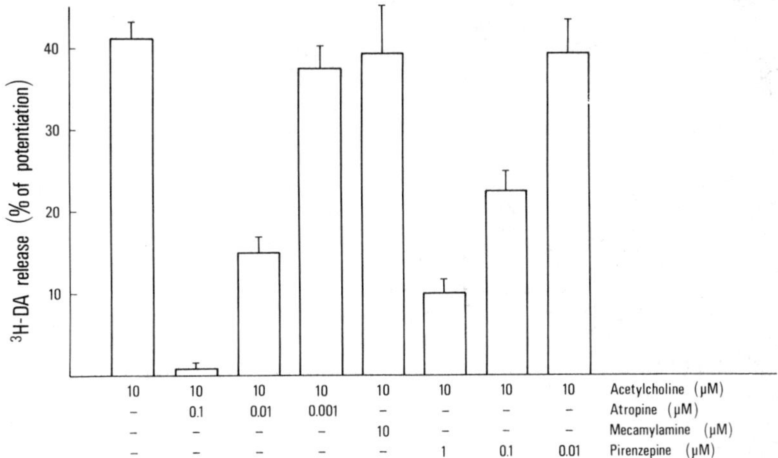

FIGURE 6. Antagonism by different antimuscarinic drugs of the effect of exogenous ACh on the release of ^{3}H-ACh from synaptosomes of rat corpus striatum.

rinic binding sites in the central nervous system.[12] From the results of our study, it can be concluded that the muscarinic heteroreceptors which potentiate DA release in striatal nerve endings (high affinity for pirenzepine) differ pharmacologically from the muscarinic autoreceptors inhibiting ACh release. A further characterization of the latter receptors has been done more recently using AF-DX 116 and the results will be presented in another part of this paper.

The above-mentioned experiments, performed with nerve endings isolated from corpus striatum, strongly indicate that M-1 muscarinic receptors are sited directly on the dopaminergic axon terminals. We thought that it would be of importance to provide an estimate of the number of the receptors in question. To this aim one could lesion the nigrostriatal pathways and look for a decrease in the number of binding sites possibly related to that presynaptic receptor within the corpus striatum. In several other instances concerning different nervous pathways, such an approach has been disappointing, however, particularly in investigations of autoreceptors. In fact, the possible denervation supersensitivity of the postsynaptic receptors[14] may obscure the loss of the presynaptic autoreceptors because the latter are often limited in number so that their disappearance may not be detected.

In our case the situation appeared to be particularly favorable for the following reasons: (a) the data obtained with synaptosomes (see above) suggested that the receptors in question (muscarinic) are located on *dopaminergic* nerve endings and therefore a degeneration of these endings would eventually lead to supersensitivity of *dopaminergic* postsynaptic receptors; (b) previous studies had demonstrated that the muscarinic binding sites in rat corpus striatum were heterogeneous[15,16] and that only about 50% belong to the subtype sensitive to pirenzepine;[17] (c) the muscarinic receptors regulating ^{3}H-DA release from striatal synaptosomes were sensitive to pirenzepine.[10] We thought that these considerations, together with the fact that the striatum is particularly rich in DA nerve endings, increased the probability of detecting a significant loss of muscarinic binding sites in a DA denervated corpus striatum using ^{3}H-pirenzepine (^{3}H-PZ) as a ligand.

TABLE 1 shows the data on endogenous DA content, ^{3}H-DA uptake, and ^{3}H-PZ binding measured in both the lesioned and the contralateral striata 8 days after the hemitransection of the nigrostriatal pathways. As expected, the lesion caused a significant decrease of endogenous DA content (-66%) and ^{3}H-DA uptake (-63%) in the lesioned striatum compared to the contralateral side.

The hemitransection of the nigrostriatal pathways reduced specific ^{3}H-PZ binding to striatal membranes (by 42%) and this decrease reflected a loss of maximal binding sites (B_{max}; from 1.077 ± 0.11 to 0.63 ± 0.18 pmol $\times$ mg protein^{-1}; $n = 12$) without any significant change in the equilibrium dissociation constant.

The nigrostriatal hemitransection performed under our experimental conditions led only to a partial decrease of striatal DA content suggesting that the lesion was not complete. One could therefore assume that in the presence of a more extensive lesion, the muscarinic binding sites might have diminished by more than 50%. Interestingly, intrastriatal kainic acid which is known to selectively destroy intrinsic cell bodies, sparing axons and terminals of extrinsic neurons,[18] caused a 40%–50% decrease of the binding of ^{3}H-QNB (a nonselective muscarinic ligand) to striatal membranes.[19,20] Subsequently, Luthin and Wolfe estimated the density of ^{3}H-PZ binding sites to be approximately half the density of ^{3}H-QNB binding in rat corpus striatum.[15] Thus, on the basis of these considerations, it is tempting to speculate that the muscarinic M-1 receptors sited on DA terminals represent about one-fourth of the total muscarinic receptors present in the rat striatum. It is not unreasonable to think that in the numerous studies carried out on the DA-ACh interactions in the corpus striatum, the importance of these novel receptors has been underevaluated.

Another aspect of the relations between striatal ACh-releasing structures and DA axon terminals has been investigated in order to establish a possible *in vivo* function of the muscarinic receptors located on these terminals. We thought that a destruction of the intrastriatal cholinergic neurons, the likely source of the ACh activating *in vivo* the

TABLE 1. Effect of Nigrostriatal Hemitransection on Some Dopaminergic Parameters 8 Days after the Lesion

Nigrostriatal Hemitransection	Unlesioned Side	Lesioned Side	Percent Decrease
Endogenous DA content (nmol/mg protein)	7.2 ± 1.2	2.5 ± 0.8	-66
^{3}H-DA uptake (pmol/mg protein per 2 minutes)	8.2 ± 0.6	3.0 ± 0.4	-64
^{3}H-PZ binding K_d (nM)	16.36 ± 1.6	15.72 ± 3.3	-4
B_{max}	1.07 ± 0.1	0.63 ± 0.2	-40

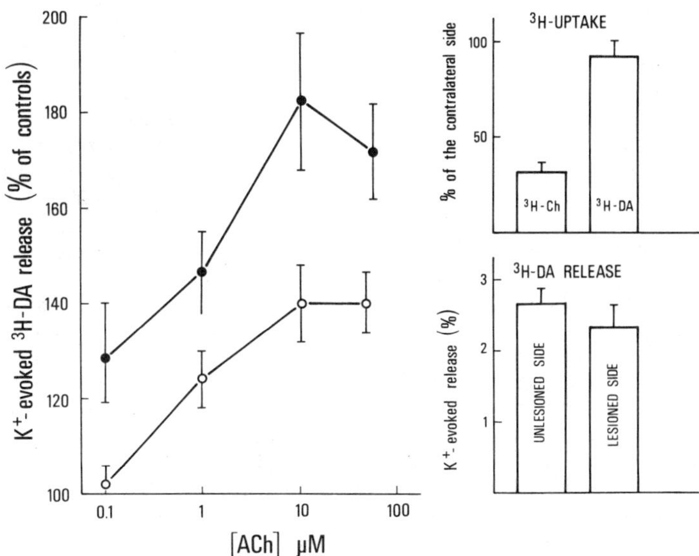

FIGURE 7. Effect of kainic acid treatment on ^{3}H-DA uptake, ^{3}H-DA release, and on the modulation by ACh of ^{3}H-DA release from synaptosomes of rat corpus striatum.

muscarinic receptors located on DA nerve endings, followed by analysis of the sensitivity to ACh of these receptors, could reveal whether or not the latter are *in vivo* targets for striatal ACh.

FIGURE 7 shows that intrastriatal kainic acid (KA) injection caused a strong decrease of ^{3}H-choline uptake in the injected striatum while the uptake of ^{3}H-DA and the K$^+$-evoked ^{3}H-DA release were not significantly diminished. Thus, according to expectation, a large proportion of striatal cholinergic interneurons were lesioned by KA while the DA nerve terminal remained intact. Interestingly, in the nerve endings prepared from KA-lesioned striata, ACh was much more effective in enhancing the K$^+$-evoked ^{3}H-DA release than in synaptosomes prepared from the contralateral unlesioned striata (FIGURE 7).

The most straightforward interpretation of this receptor supersensitivity is that the destruction of cholinergic interneurons had the effect of a denervation of the muscarinic receptors regulating DA release. Less likely seems the possibility that KA directly causes modifications of DA nerve terminals resulting in enhanced response to ACh.

As to the anatomical relationships between cholinergic interneurons and DA nerve terminals in the corpus striatum, the observed supersensitivity of the muscarinic receptors located on the DA nerve endings would be compatible with the existence of a classical synapse between ACh and DA axon terminals. However, electron microscopic studies have revealed few axoaxonic synapses between terminals in the corpus striatum.[21] Therefore, a dendritic release[22] of ACh onto DA terminals or a communication at distance[23,24] through ACh traveling in the extracellular space may represent alternatives to a classic synapse.

The possible physiological importance of the muscarinic receptors located on DA nerve endings deserves some comments. Lehmann and Langer[25] and Helmreich et al.[7] observed enhancement of ^{3}H-DA release by endogenous ACh only when acetylcholinesterase inhibitors were present in the caudate slices during electrical stimulation. This might lead to the conclusion that a muscarinic-mediated modulation of DA release does not occur under physiological conditions. However, it should be noted that the stimulation of the muscarinic receptors might have characteristics that do not allow it to be detected under the "acute" experimental conditions used by these authors, whereas it becomes evident only during prolonged reduction of ACh release.

In conclusion: (1) acetylcholine stimulates the K^+-evoked release of both previously accumulated radioactive DA and endogenous DA from isolated nerve endings; (2) this potentiation of the K^+-evoked release of DA occurs through the activation of muscarinic receptors located on dopaminergic nerve terminals; (3) these receptors are pirenzepine sensitive and pharmacologically different from muscarinic autoreceptors; (4) their presynaptic location is also demonstrated by the fact that after hemitransection of the nigrostriatal pathways the number of binding sites for ^{3}H-PZ in striatal membranes decreases significantly in parallel with the decrease of endogenous DA content and ^{3}H-DA uptake in striatal synaptosomes; (5) data obtained after KA lesion support the idea that these muscarinic receptors play a physiological role in the modulation of DA release in rat corpus striatum. If the above receptors existed also in the human striatum, a gradual loss of ^{3}H-PZ binding may be expected in the Parkinsonian brain. Therefore the availability of selective markers for these muscarinic receptors sited on DA terminals, together with that of the modern noninvasive techniques of brain imaging, might open new diagnostic possibilities in Parkinson's disease.

MUSCARINIC RECEPTORS REGULATING GLUTAMATE RELEASE

The neuroanatomical organization of both the cholinergic and the glutamatergic systems in the hippocampus is reasonably well known. The major component of the cholinergic system is extrinsic and mostly represented by the cholinergic projections coming from the septum.[26,27] Glutamic acid (GLU) seems to be a major neutrotransmitter of the neuronal pathways that originate in the etorhinal cortex and terminate in the hippocampal area dentata as well as of the intrinsic excitatory projections of the granule cells.[27–29]

The possible interaction between these two transmitter systems in the hippocampus has been, however, little investigated although both the neurotransmitter systems seem to be implicated in the process of learning and memory. Evidence has been accumulating that the cholinergic system is involved in these processes and it has been proposed recently that GLU, which serves as an excitatory neurotransmitter in the central nervous system,[30–32] also plays an important role in the processes of learning and memory in the hippocampal region.[33,34]

FIGURE 8 shows that exogenous ACh, added to the medium superfusing rat hippocampal synaptosomes depolarized with 15 mM K^+, inhibited in a concentration-dependent way the release of endogenous GLU. The inhibition by exogenous ACh (10 μM) of the release of GLU was fully antagonized by 0.1 μM atropine, whereas mecamylamine, tested at 10 μM, was ineffective. The finding that pirenzepine (1 μM) was ineffective as an antagonist of ACh at the muscarine receptor mediating inhibition of GLU release points to the classification of this receptor as a pirenzepine-insensitive type.

As previously mentioned it has been proposed that the muscarinic receptors sited on ACh terminals (autoreceptors) also are pirenzepine insensitive. The availability of the antimuscarinic AF-DX 116, which is able to distinguish between two different subtypes of pirenzepine-insensitive receptors, offered the opportunity of comparing the

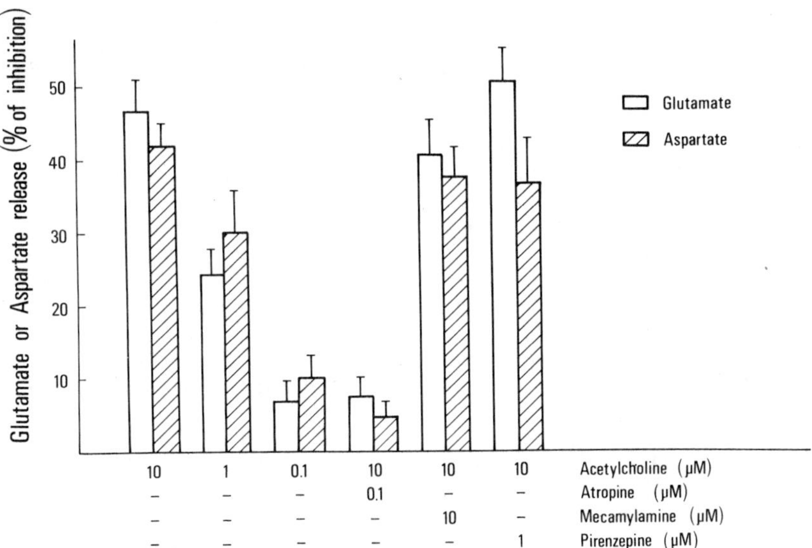

FIGURE 8. Antagonism by different antimuscarinic drugs of the effect of exogenous ACh on the release of glutamate and aspartate from rat hippocampal nerve endings.

muscarinic receptors located respectively on ACh (autoreceptors) or GLU nerve endings. The results shown in FIGURE 9 clearly show that the effectiveness of AF-DX 116 as an antagonist of ACh at the autoreceptors was much lower than that at the heteroreceptors regulating GLU release. These data suggest that the presynaptic

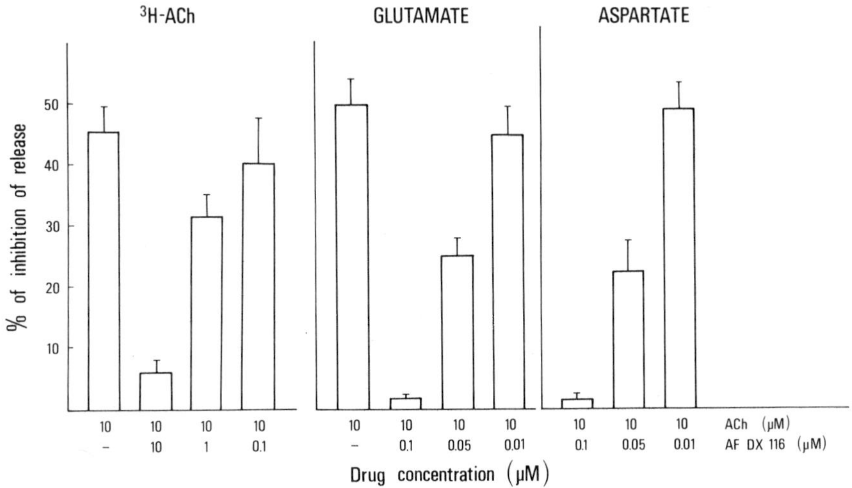

FIGURE 9. Antagonism by AF-DX 116 of the effect of exogenous ACh on the release of 3H-ACh, glutamate, and aspartate from hippocampal nerve endings.

autoreceptors are more similar to the so-called glandular or M-3 subtype, whereas the receptors located on GLU nerve endings resemble most the so-called cardiac or M-2 subtype.[35] Of course other factors such as ligand-induced conformational changes or the receptor microenvironment should be considered as alternative explanations for the differential antagonism by AF-DX 116 of the ACh effect at GLU or at ACh nerve endings, respectively. However, it has to be considered that the analysis of human and rat genomic clones indicates that there are at least four functional muscarinic receptor genes in the brain.[36,37] Thus, assuming that the ACh autoreceptors and the muscarinic heteroreceptors sited on GLU terminals differ pharmacologically from each other, it will be possible to design selective drugs able to modulate the ACh-GLU interaction in the hippocampus.

What may be the functional significance of an inhibitory muscarinic receptor located on GLU nerve endings can only be matter of speculation, particularly if one considers that both transmitters have been proposed to be positively involved in memory processes. L-Glutamate is probably the most widely used excitatory transmitter in the mammalian brain, capable of activating several receptor subtypes.[38,39] These diverse receptor subtypes certainly allow GLU to carry on multiple effects, including those related to memory. However, in addition to these physiological effects, it has been shown that administration of GLU directly into the brain may be highly toxic and cause destruction of neurons whose cell bodies lie in the administration site.[40–42] Such neutrotoxic effects possibly play an important role in certain pathological conditions. For instance, it has been proposed that the neuronal damage associated with cerebral ischemia, Huntington's chorea, Alzheimer's disease, and epilepsy may be the consequence of an excessive release of GLU onto neurons resulting in their damage.[41]

In recent work, GLU terminals have been shown to possess presynaptic receptors activation of which leads to potent inhibition of the release of the amino acid. For instance, cerebellar GLU nerve endings are endowed with inhibitory receptors of the $5-HT_1$ subtype,[43] while GLU terminals in the striatum possess inhibitory receptors of the D_2 subtype.[44] The presence of inhibitory muscarinic receptors on GLU terminals in the hippocampus might therefore have the function of preventing the release of abnormal quantities of the excitatory amino acid. Such an emergency control would not be incompatible with the widely accepted idea that both cholinergic and glutamatergic transmission activate cognitive processes. In fact, the hippocampal region appears to contain different glutamatergic pathways; only some may be involved in cognitive processes, and not all GLU terminals need to carry inhibitory muscarinic receptors.

MUSCARINIC RECEPTORS REGULATING ASPARTATE RELEASE

Although glutamate is probably the main excitatory transmitter in the mammalian central nervous system, experimental evidence suggests that aspartic acid (ASP) also may play a role in excitatory neurotransmission.[32,45–46]

GLU and ASP are structurally similar amino acids and share the same uptake system in synaptic membranes.[47,48] In general their functional responses appear to be sensitive to the same receptor antagonists.[45–49] Furthermore, both amino acids are important and ubiquitous components of intermediate metabolism. Because of these similarities, it has not been easy to demonstrate the existence of pure glutamatergic or aspartatergic neuronal pathways.

In fact, lesions in afferent excitatory projections to various regions have often been found to cause parallel reduction in the content and release of both GLU and ASP.[50–53] Only in a few cases has evidence been provided that some pathways use GLU or ASP exclusively. For instance, Nadler *et al.* proposed that, in the hippocampus, the commissural and ipsilateral projections to the fascia dentata are aspartatergic, whereas the transmitter of the perforant path fibers to the fascia dentata is GLU.[28] It should be noted, however, that the presence of large "metabolic pools" together with the "transmitter pools" makes it difficult to discriminate glutamatergic terminals from aspartatergic terminals. In conclusion, if the evidence in favor of GLU as an excitatory transmitter is convincing, the role of ASP has yet to be defined.

Interestingly, in experiments aimed at determining the calcium-dependency of GLU and ASP release as a criterion for their identification as putative neurotransmitters, the depolarization-evoked release of GLU from slices or synaptosomes has been found to be largely or totally calcium dependent, while conflicting results have been reported in the case of ASP.[54–58]

As amply reported in this volume, the release of neurotransmitters is often modulated by presynaptic receptors sited on the releasing nerve endings. The presence of release-modulating receptors may therefore be included as an additional criterion in the characterization of a putative transmitter.

FIGURE 8 shows that exogenous ACh added to the superfusion fluid concomitantly with high K^+ inhibited in a concentration-dependent manner the evoked-overflow of endogenous ASP from rat hippocampal synaptosomes. The release of ASP from isolated nerve endings depolarized by 15 mM KCl was totally calcium dependent as was the release of GLU.[59]

Similarly to what was observed when studying ACh on GLU release, the inhibition of the release of endogenous ASP by ACh was totally blocked by atropine (FIGURE 8). The nicotinic receptor antagonist mecamylamine was ineffective up to 10 μM. The effect of ACh was not counteracted by 1 μM pirenzepine. However, AF-DX 116 did antagonize ACh in a concentration-dependent way. The effect of 10 μM ACh was totally blocked by 0.1 μM AF-DX 116 and halved by 0.05 μM of the muscarinic antagonist (FIGURE 9). The release of both GLU and ASP can be therefore inhibited through the activation of pharmacologically identical presynaptic muscarinic receptors which appear to belong to the "cardiac" M-2 subtype. The complete calcium dependency of the release of ASP evoked by depolarization, together with the existence of presynaptic receptors regulating ASP release, strengthens the idea that, similarly to glutamic acid, ASP also is a neurotransmitter in the hippocampus. It was proposed on the basis of lesion studies that in rat hippocampus, GLU is the transmitter of the perforant path fibers while ASP is the transmitter of the commissural fibers.[28] This would imply that GLU and ASP are released from different boutons. However, the striking similarities of the effects of cholinergic drugs on GLU and ASP release are also compatible with the idea that the two amino acids coexist and are coreleased from a same excitatory projection. Corelease of GLU and ASP has been proposed to occur at the lobster neuromuscular junction[60] and in the rat hippocampus.[61]

PRESYNAPTIC MUSCARINIC AUTORECEPTORS IN HUMAN BRAIN

A muscarinic receptor–mediated mechanism for the regulation of ACh release similar to that present in the rodent was proposed to exist also in human brain. Using slices prepared from surgical specimens of human cerebral cortex, it was found that muscarinic receptor antagonists stimulated the synthesis[62] and the release[63] of ACh. Subsequently, Nilsson *et al.* reported that, in brain slices from human autopsies collected after a short postmortem delay, physostigmine reduced the release of ^{3}H-ACh.[64] However, a direct effect of cholinergic agonists and, in particular, of the natural transmitter ACh has so far not been reported. As previously discussed, muscarinic autoreceptors have been characterized pharmacologically in the rat and they seem to belong to the subtype having low affinity for pirenzepine and for AF-DX 116 (M-3 subtype).

It is obvious that receptor subtypes that have been shown to be present in the rodent brain may or may not exist in human brain. For instance, release-regulating autoreceptors of the γ-aminobutyric acid (GABA$_B$) type have been found both in rat[65,66] and in human neocortex.[67] However, 5-HT autoreceptors are likely to differ between the two species. In fact, these autoreceptors are 5-HT$_{1B}$ subtype in the rat brain,[68,69] while in human brain 5-HT$_{1B}$ binding sites do not seem to exist.[70] Thus it is essential to verify whether receptors present in the experimental animal exist and play the same role also in the human brain.

FIGURE 10 shows that the natural transmitter ACh inhibited in a concentration-dependent manner the release of ^{3}H-ACh from human neocortex nerve endings. When studying release-regulating receptors it is important to use the natural transmitter as an agonist. As previously mentioned, the existence of ACh autoreceptors in human brain was proposed only on the basis of indirect evidence. In fact, agonists were not used and the presence of muscarinic autoreceptors was inferred by observing that muscarinic antagonists increased the synthesis.[62] or the release[63] of ACh, probably by relieving the inhibition caused by the transmitter present in the autoreceptor biophase. Differently from slices, the system used in our experiments (synaptosomes superfused in a thin layer) allows the use of ACh even in the absence of acetylcholinesterase inhibitors. Moreover, this experimental setup makes interactions between transmitters very unlikely. Thus the observed inhibition of ^{3}H-ACh release by ACh was probably due to a direct action of ACh on the releasing particles. These particles are likely to be cholinergic nerve terminals since they were able to release ^{3}H-ACh newly synthesized from ^{3}H-Ch. The finding that the inhibitory effect of ACh was blocked by atropine (FIGURE 11) points to the involvement of muscarinic receptors. Altogether the data obtained allow us to conclude that cholinergic nerve terminals in human neocortex are endowed with muscarinic autoreceptors, activation of which brings about inhibition of ACh release. This autoregulatory system appears to be quantitatively similar to that present in rat cortical synaptosomes.[10] The maximal effect of ACh amounted to about 40% in both rat and human nerve endings. The EC$_{50}$ values for ACh as an inhibitor of

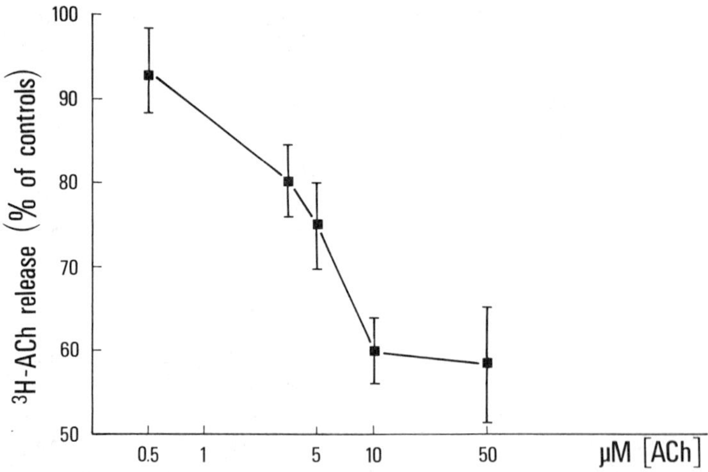

FIGURE 10. Effect of ACh on ^{3}H-ACh release in synaptosomes from human cortex.

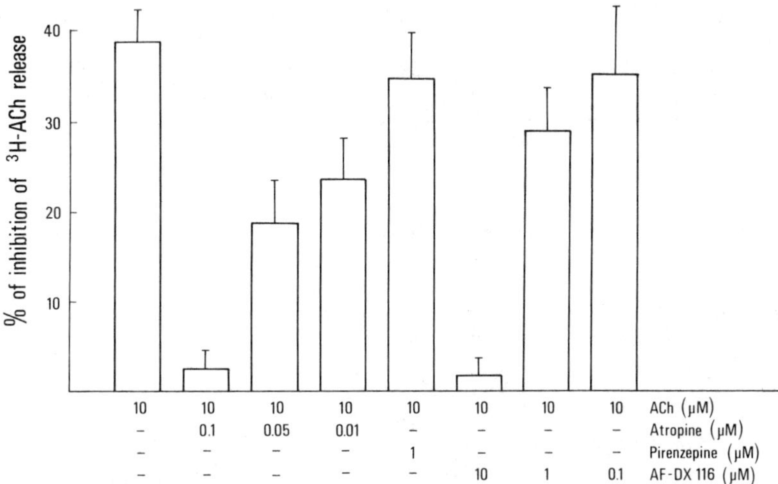

FIGURE 11. Antagonism by different antimuscarinic drugs of the effect of exogenous ACh on the release of ^{3}H-ACh from synaptosomes of human cortex.

^{3}H-ACh release were 1.5 μM and 1.7 μM in human and in rat cortical synaptosomes, respectively. Interestingly, ACh displayed similar potencies in decreasing the K^+-evoked release of ^{3}H-ACh in the four human cortex subregions tested (frontal, temporal, parietal, and occipital) (data not shown).

Figure 11 shows that 1 μM pirenzepine did not affect the inhibition of ^{3}H-ACh release caused by 10 μM ACh, whereas atropine already significantly antagonized 10 μM ACh at 0.01 μM. Thus pirenzepine appeared to be at least 100 times less effective than atropine. Similar results had been obtained in rat cortex synaptosomes.[10] Considering that the affinity of the M-1 pirenzepine-sensitive receptors for both pirenzepine and atropine was reported to fall in the nanomolar range,[17,71,72] to define the muscarinic receptors as pirenzepine insensitive appears to be justified.

FIGURE 11 also shows that the non–M-1 antagonist AF-DX 116 was able to antagonize in a concentration-dependent manner the inhibitory effect of 10 μM ACh. The potency of the drug on human autoreceptors was very similar to that shown on rat autoreceptors (cf. FIGURES 9 and 11), indicating that the human muscarinic autoreceptor seems to be pharmacologically similar to that of the rat and possibly belongs to a subtype having low sensitivity for pirenzepine as well as for AF-DX 116 (M-3 subtype).

In conclusion, cholinergic nerve endings in human cerebral cortex take up ^{3}H-Ch and, upon depolarization, release newly synthesized ^{3}H-ACh in a calcium-dependent manner. The depolarization-evoked ^{3}H-ACh release can be inhibited by ACh through the activation of muscarinic autoreceptors sited on the cholinergic terminals themselves. The autoreceptors are insensitive to pirenzepine and moderately sensitive to AF-DX 116. These results are very similar to those obtained in the rat[35] which therefore appears to represent a useful model for studying the properties of muscarinic autoreceptors in human brain and for testing new cholinergic drugs of potential therapeutic use.

The findings that in the human cortex the presynaptic muscarinic autoreceptors are pirenzepine insensitive whereas the muscarinic postsynaptic receptors are predominantly of the M-1 pirenzepine-sensitive subtype[73,74] open the possibility of some specific pharmacological approaches. For instance, it should be possible to improve cholinergic transmission with selective, centrally active autoreceptor antagonists which should increase ACh release without affecting the postsynaptic muscarinic receptors, or with selective postsynaptic muscarinic agonists incapable of activating the autoreceptors, or with an association of both.

CONCLUDING REMARKS

Release appears to be a powerful tool to study receptors and interaction between transmitters. The data presented clearly show that muscarinic receptors (but this can also be the case for other receptor systems) can be functionally investigated with

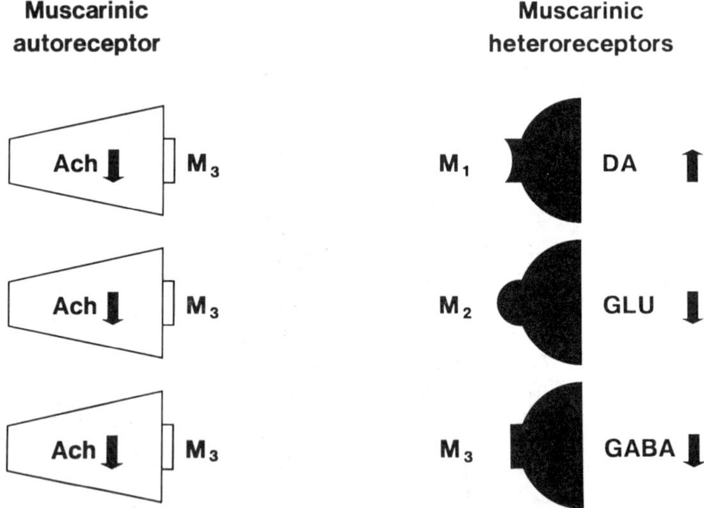

FIGURE 12. Possible interactions between the cholinergic system and other neurotransmitter systems with the involvement of different muscarinic receptor subtypes.

release techniques. By using superfused synaptosomes the neuronal system affected by the activation of a release-modulating receptor can be easily identified and the precise anatomical location of the receptor involved determined. The finding that activation of terminal presynaptic receptors affects transmitter release from superfused synaptosomes tends to exclude the idea that these receptors block the impulse propagation along the terminal axons and strongly supports the contention that they impair the stimulus-secretion coupling of the release process.

The changes of sensitivity of the muscarinic receptors observed in this study favor the idea that these receptors play a physiological role. The data already available on the pharmacology of the muscarinic receptors investigated in our studies appear to indicate that release-regulating muscarinic receptors may represent all or most of the proposed subtypes of muscarinic receptors. If this were the case, release would represent an important tool to test new and more selective drugs of potential therapeutical value. Of particular interest are the observations that muscarinic autoreceptors exist also in the human cerebral cortex and seem to belong to the same pharmacological subtype as those present in the rodent brain.[75]

The muscarinic release-regulating receptors so far identified allow us to draw the scheme reported in FIGURE 12 in which the ACh released from cholinergic terminals activates a variety of muscarinic receptors, including the autoreceptors, to modulate the release of various transmitters. One could easily envisage how muscarinic agonists or antagonists selective for one of the various receptor subtypes shown in the scheme should be able to produce important changes in the interplay between ACh and other transmitters.

REFERENCES

1. STARKE, K., M. GÖTHERT & H. KILBINGER. 1989. Physiol. Rev. **69:** 864–989.
2. NORDSTRÖM, O. & T. BARTFAI. 1980. Acta Physiol. Scand. **108:** 347–353.
3. MARCHI, M., P. PAUDICE & M. RAITERI. 1981. Eur. J. Pharmacol. **73:** 75–79.
4. RAITERI, M., M. MARCHI & P. PAUDICE. 1981. Eur. J. Pharmacol. **74:** 109–110.
5. WESTFALL, T. C. 1974. Neuropharmacology **13:** 693–700.
6. DE BELLEROCHE, J. & H. F. BRADFORD. 1978. Brain Res. **142:** 53–68.
7. HELMREICH, I., W. REIMANN, G. HERTTING & K. STARKE. 1982. Neuroscience **7:** 1559–1566.
8. LEHMANN, J. & S. Z. LANGER. 1983. Neuroscience **10:** 1105–1120.
9. RAITERI, M., M. MARCHI & G. MAURA. 1982. Eur. J. Pharmacol. **83:** 127–129.
10. MARCHI, M. & M. RAITERI. 1985. J. Pharmacol. Exp. Ther. **235:** 230–233.
11. BONANNO, G., M. MARCHI & M. RAITERI. 1985. Neuropharmacology **24:** 261–264.
12. HIRSCHOWITZ, B. I., R. HAMMER, A. GIACHETTI, J. J. KEIRNS & R. R. LEVINE, Eds. 1984. Subtypes of Muscarinic Receptors. Elsevier, North Holland. Amsterdam, the Netherlands.
13. RAITERI, M., R. LEARDI & M. MARCHI. 1984. J. Pharmacol. Exp. Ther. **228:** 209–214.
14. WESTLIND, A., M. GRYNFARB, B. HEDLUND, T. BARTFAI & K. FUXE. 1981. Brain Res. **225:** 131–141.
15. LUTHIN, G. R. & B. B. WOLFE. 1984. J. Pharmacol. Exp. Ther. **228:** 648–655.
16. WATSON, M., H. I. YAMAMURA & W. R. ROESKE. 1983. Life Sci. **32:** 3001–3011.
17. HAMMER, R., C. P. BERRIE, N. J. M. BIRDSALL, A. S. V. BURGEN & E. C. HULME. 1980. Nature London **283:** 90–92.
18. COYLE, J. T., E. G. MCGEER, P. L. MCGEER & R. SCHWARCZ. 1978. Neostriatal injections: a model for Huntington's chorea. *In* Kainic Acid as a Tool in Neurobiology. E. G. McGeer, J. W. Olney, P. L. McGeer, Eds.: 139–159. Raven Press. New York, N.Y.
19. HRUSKA, R. E., R. SCHWARCZ, J. T. COYLE & H. I. YAMAMURA. 1978. Brain Res. **152:** 620–625.
20. BRIGGS, R. S., P. REDGRAVE & S. R. NAHORSKI. 1981. Brain Res. **206:** 451–456.
21. KORNHUBER, J. & M. E. KORNHUBER. 1983. Eur. Neurol. **22:** 433–436.
22. CHERAMY, A., V. LEVIEL & J. GLOWINSKI. 1981. Nature London **289:** 537–542.
23. SCHMITT, F. O. 1984. Neuroscience **13:** 991–1001.
24. VIZI, E. S., Ed. 1984. Non-Synaptic Interactions between Neurons: Modulation of Neurochemical Transmission. John Wiley. New York, N.Y.
25. LEHMANN, J. & S. Z. LANGER. 1982. Brain Res. **248:** 61–69.
26. LEWIS, P. R. & C. C. D. SHUTE. 1967. Brain **90:** 521–540.
27. STORM-MATHISEN, J. 1978. Localization of putative transmitters in the hippocampal formation. Ciba Found. Symp. **58:** 49–86.
28. NADLER, J. V., K. W. VACA, W. F. WHITE, G. S. LYNCH & C. W. COTMAN. 1976. Nature London **260:** 538–540.
29. FONNUM, F. & I. WALAAS. 1978. J. Neurochem. **31:** 1173–1181.
30. REUBI, J. C. & M. CUÉNOD. 1979. Brain Res. **176:** 185–188.

31. COTMAN, C. W., A. FOSTER & T. LANTHORN. 1981. An overview of glutamate as a neurotransmitter. *In* Glutamate as a Neurotransmitter. G. Di Chiara & G. L. Gessa, Eds.: 1–27. Raven Press. New York, N.Y.
32. FONNUM, F. 1984. J. Neurochem. **31:** 1173–1181.
33. LYNCH, G. & M. BAUDRY. 1984. Science **224:** 1057–1063.
34. GUSTAFSSON, B. & H. WIGSTROM. 1988. Trends Neurosci. **11:** 156–162.
35. MARCHI, M. & M. RAITERI. 1989. J. Pharmacol. Exp. Ther. **248:** 1255–1260.
36. FUKUDA, K., H. HIGASHIDA, T. KUBO, A MAEDA, I. AKIBA, H. BUJO, M. MISHINA & S. NUMA. 1988. Nature London **335:** 355–358.
37. PERALTA, E. G., A. ASHKENAZI, J. W. WINSLOW, J. RAMACHANDRAN & D. J. KAPON. 1988. Nature London **334:** 434–437.
38. FOSTER, A. C. & G. E. FAGG. 1984. Brain Res. **319:** 103–164.
39. NICOLETTI, F., J. L. MEEK, M. J. IADAROLA, D. M. CHUANG, B. L. ROTH & E. COSTA. 1986. J. Neurochem. **46:** 40–46.
40. OLNEY, J. W. 1969. Science **164:** 719–721.
41. SCHWARCZ, R., G. S. BRUSH, A. C. FOSTER & E. D. FRENCH. 1984. Exp. Neurol. **84:** 1–17.
42. ROTHMAN, S. M. 1983. Science **200:** 536–537.
43. RAITERI, M., G. MAURA, G. BONANNO & A. PITTALUGA. 1986. J. Pharmacol. Exp. Ther. **237:** 644–649.
44. MAURA, G., A. GIARDI & M. RAITERI. 1988. J. Pharmacol. Exp. Ther. **247:** 680–684.
45. WATKINS, J. C. & R. H. EVANS. 1981. Annu. Rev. Pharmacol. Toxicol. **21:** 165–204.
46. FAGG, G. E. & A. C. FOSTER. 1983. Neuroscience. **9:** 701–719.
47. BALCAR, V. J. & G. A. R. JOHNSTON. 1972. J. Neurochem. **19:** 2657–2666.
48. LOGAN, W. J. & S. H. SNYDER. 1972. Brain Res. **42:** 413–431.
49. MCLENNAN, H. 1983. Prog. Neurobiol. **20:** 251–271.
50. WENTHOLD, R. J. 1979. Brain Res. **162:** 338–343.
51. FLINT, R. S., M. A. REA & W. J. MCBRIDE. 1981. J. Neurochem. **37:** 1425–1430.
52. SCHOLFIELD, C. N., F. MORONI, R. CORRADETTI & G. PEPEU. 1983. J. Neurochem. **41:** 135–138.
53. GIRAULT, J. A., L. BARBEITO, U. SPAMPINATO, H. GOZLAN, J. GLOWINSKI & M.-J. BESSON. 1986. J. Neurochem. **47:** 98–106.
54. LEVI, G., R. D. GORDON, V. GALLO, G. P. WILKIN & R. BALAZS. 1982. Brain Res. **239:** 425–445.
55. BARNES, S., G. E. LEIGHTON & J. A. DAVIES. 1988. J. Neurosci. Methods **23:** 57–61.
56. BURKE, S. P. & J. V. NADLER. 1988. J. Neurochem. **51:** 1541–1549.
57. SZERB, J. C. 1988. J. Neurochem. **50:** 219–224.
58. WILKINSON, R. & D. G. NICHOLLS. 1988. Biochem. Soc. Trans. **16:** 879–884.
59. RAITERI, M., M. MARCHI, A. COSTI & G. VOLPE. 1990. Eur. J. Pharmacol. **177:** 181–187.
60. FREEMAN, A. R., R. P. SHANK, J. KEPHART, M. DEKIN & M. WANG. 1981. A model for excitatory transmission at a glutamate synapse. *In* Glutamate as a Neurotransmitter. G. Di Chiara & G. L. Gessa, Eds.: 227–243. Raven Press. New York, N.Y.
61. CORRADETTI, R., G. MONETI, F. MORONI, G. PEPEU & A. WIERASZKO. 1983. J. Neurochem. **41:** 1518–1525.
62. MAREK, K. L., D. M. BOWEN, N. R. SIMS & A. N. DAVISON. 1982. Life Sci. **30:** 1517–1524.
63. NORSTRÖM, O., A. WESTLIND, A. UNDEN, B. MEYERSON, C. SACHS & T. BARTFAI. 1982. Brain Res. **234:** 287–297.
64. NILSSON, L., A. NORDBERG, J. HARDY, P. WESTER & B. WINBLAD. 1987. J. Neural Transm. **67:** 275–286.
65. PITTALUGA, A., D. ASARO, G. PELLERGRINI & M. RAITERI. 1987. Eur. J. Pharmacol. **144:** 45–52.
66. WALDMEIER, P. C., P. WICKI, J.-J. FELDTRAUER & P. A. BAUMANN. 1988. Naunyn-Schmiedeberg's Arch. Pharmacol. **337:** 289–295.
67. BONANNO, G., P. CAVAZZANI, G. C. ANDRIOLI, D. ASARO, G. PELLEGRINI & M. RAITERI. 1989. Br. J. Pharmacol. **96:** 341–346.
68. ENGEL, G., M. GÖTHERT, D. HOYER, E. SCHLICKER & K. HILLENBRAND. 1986. Naunyn-Schmiedeberg's Arch. Pharmacol. **332:** 1–7.

69. MAURA, G., E. ROCCATAGLIATA & M. RAITERI. 1986. Naunyn-Schmiedeberg's Arch. Pharmacol. **334:** 323–326.
70. HOYER, D., A. PAZOS, A. PROBST & J. M. PALACIOS. 1986. Brain Res. **376:** 85–96.
71. GIL, D. W. & B. B. WOLFE. 1985. J. Pharmacol. Exp. Ther. **232:** 608–616.
72. WATSON, M., W. R. ROESKE & H. I. YAMAMURA. 1982. Life Sci. **31:** 2019–2023.
73. MASH, D. C., D. D. FLYNN & L. T. POTTER. 1985. Science **228:** 115–117.
74. VOGT, B. A. 1988. TIPS (February suppl): 49–53.
75. MARCHI, M., A. RUELLE, G. C. ANDRIOLI & M. RAITERI. Brain Res. (In press.)

Presynaptic Purine Receptors[a]

DAVID P. WESTFALL, KAZUMASA SHINOZUKA,
KARYN M. FORSYTH, AND RICHARD A. BJUR

Department of Pharmacology
University of Nevada School of Medicine
Reno, Nevada 89557

The evidence is mounting that adenosine may act prejunctionally to limit the release of norepinephrine (NE) from postganglionic sympathetic nerves.[1–4] For example, adenosine has been shown to reduce the release of ^{3}H-NE from a variety of tissues including the vas deferens, salivary gland, ventricle and heart of the rat; guinea pig atria, heart, and vas deferens; and canine adipose tissue.[5–9] Similar observations have been made with blood vessels including the rat portal vein; the aorta, portal vein, and pulmonary artery of the rabbit; and the dog saphenous vein.[7,10–14] It has been believed generally that this action of adenosine is mediated via prejunctional purinoceptors of the P_1 class.

Interestingly, ATP also reduces the nerve stimulation–induced release of NE in a number of tissues.[2,4] Because nucleotides are found to be poor P_1-purinoceptor agonists,[15] the general notion has been that this effect of ATP results from its metabolism by ectoenzymes to adenosine, which then acts via prejunctional P_1-purinoceptors rather than ATP acting directly on P_2-purinoceptors. Indeed, in schematic representations of the role of adenine nucleotides and nucleosides in adrenergic transmission prejunctional modulation of transmitter release is usually depicted as occurring solely via P_1-purinoceptors.[4,16]

We believe this idea is not compatible with a number of observations. If the effect of ATP were due to its metabolism to adenosine, one might expect ATP or other nucleotides to be less potent than adenosine because the metabolism is not likely to be instantaneous or complete. We found it curious therefore that in those situations where the effects of both adenosine and ATP on the release of ^{3}H-NE have been compared, the agents are equipotent. This occurs in the rat vas deferens,[6] the rat portal vein,[10,13] the rat heart,[8] and the dog saphenous vein.[11]

The modulation of transmitter release by adenine nucleotides and nucleosides is not limited to adrenergic neuroeffector junctions. There is evidence that ATP and adenosine reduce the nerve stimulation–induced release of acetylcholine (ACh) at the neuromuscular junction and at the cholinergic–smooth muscle junction. As with adrenergic systems, ATP and adenosine seem to be of equal potency.[17,18]

Results such as these suggested to us that perhaps both P_1- and P_2-purpinoceptors mediate an inhibition of the release of NE or that the prejunctional purinoceptor is not a pure P_1- or P_2-receptor, but rather possesses some characteristics of both, i.e., a hybrid purinoceptor. We review in this paper findings from our laboratory[19,20] and discuss additional data that indicate the prejunctional purinoceptors may represent a distinct class, which we call P_3-receptors.

A critical point upon which this issue depends is whether nucleotides need to be converted to adenosine to produce effects or whether the nucleotides act per se. Two types of results indicate to us that the nucleotides can act without conversion to adenosine. The first is based on the relative order of potency of a series of nucleotide

[a]Work done in the authors' laboratory was supported by National Institutes of Health grant HL 38126.

and nucleoside compounds and the second derives from the use of adenosine uptake inhibitors.

In the rat caudal artery the relative order of potency of a series of nucleotides and nucleosides for inhibiting the nerve stimulation–induced release of NE is 2-chloroadenosine > beta, gamma-methylene ATP > ATP = adenosine.[19] This order of potency seems incompatible with the notion that the nucleotides need to be metabolized to adenosine to reduce NE release in that ATP and adenosine were equipotent and furthermore that beta, gamma-methylene ATP, a compound that is less rapidly hydrolyzed to adenosine than is ATP,[21] was more potent than either ATP or adenosine.

More recent studies in our laboratory have extended this type of analysis to purine modulation of NE release in the rat vas deferens. Shown in FIGURE 1 are concentration-

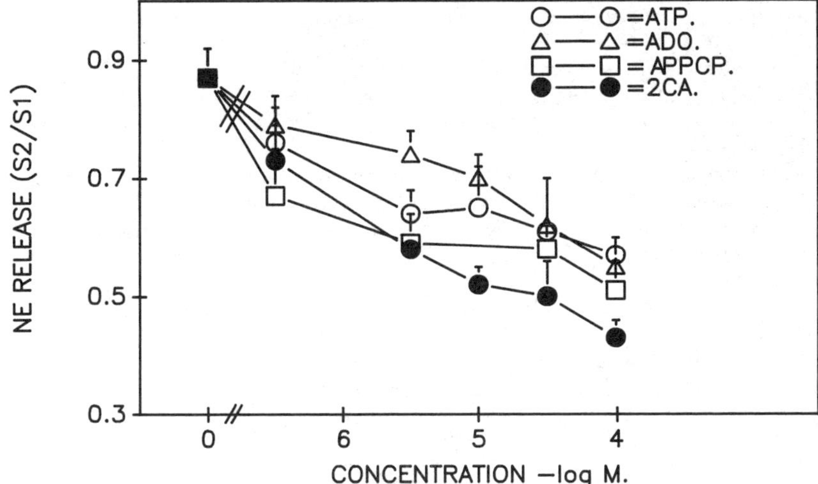

FIGURE 1. Concentration-response curves for several adenine nucleotides and nucleosides. The response being measured is the inhibition of NE release from the electrically stimulated rat vas deferens. The purines were 2-chloroadenosine (2CA); beta, gamma-methylene ATP (APPCP); ATP; and adenosine (ADO). Vasa were stimulated twice with an interval of 30 minutes between the 2 stimulations. Agonists were added 3 minutes before the second stimulation period (S_2). The effect of drugs on NE release was quantified by calculating the ratio of NE release between the second and first release period, i.e., S_2/S_1. NE was measured by high performance liquid chromatography–electrochemical detection techniques.[19]

response curves for 2-chloroadenosine, adenosine, ATP, and beta, gamma-methylene ATP in vas deferens. As was the situation with the rat caudal artery, beta, gamma-methylene ATP, the less rapidly hydrolyzed isostere of ATP, was more potent than ATP. Furthermore, ATP and adenosine exhibited similar potency.

These results in caudal artery and vas deferens support the notion that the nucleotides do not need to be converted to adenosine to be active. Additional support comes from the work of Lukacsko and Blumberg[22] and Husted and Nedergaard[23] with blood vessels. Lukacsko and Blumberg, for example, investigated the effects of a series of nucleotides and nucleosides for their ability to inhibit nerve stimulation–induced vasoconstriction in the perfused rat mesentery.[22] Inhibition of vasoconstriction in response to nerve stimulation was interpreted to occur as a result of the inhibition of transmitter release. These investigators found that beta, gamma-methylene ATP was more potent than adenosine. Similar studies with beta, gamma-methylene ATP in the rabbit pulmonary artery by Husted and Nedergaard led them to suggest that nucleotides could act per se without degradation to adenosine.[23]

The second type of result that argues for an action of adenine nucleotides that is independent of metabolism to adenosine derives from studies with adenosine uptake inhibitors such as S-*p*-nitrobenzyl-6-thioguanosine (NBTG). We found that NBTG potentiates the inhibition of NE release by adenosine in rat caudal artery, presumably by blocking its uptake into cells and thereby potentiating its extracellular effects. However, NBTG did not potentiate the effects of adenine nucleotides, thereby indicating that the actions of the nucleotides are independent of adenosine formation.[19,20]

A curious feature of purine modulation of NE release is that inhibition of release by both P_1-purinoceptor agonists, such as adenosine and 2-chloroadenosine, and P_2-purinoceptor agonists, such as ATP and beta, gamma-methylene ATP, is antagonized by P_1-receptor antagonists. This has been shown to occur in the rat tail artery and vas deferens with the P_1 antagonist 8-sulfophenyl-theophylline (8-SPT).[19,20,24] Similar findings occur with the perfused rat mesentery where theophylline was shown to antagonize the inhibition of nerve-induced vasoconstriction by beta, gamma-methylene ATP and adenosine.[22]

Another result that raises suspicions about the nature of the prejunctional purinoceptor derives from studies with the compound alpha, beta-methylene ATP. This isostere of ATP, similar to beta, gamma-methylene ATP, is less rapidly hydrolyzed than ATP. Alpha, beta-methylene ATP when administered acutely acts as an agonist at P_2-purinoceptors.[25] Furthermore upon repeated exposure to or in the continued presence of alpha, beta-methylene ATP, there is a desensitization of the P_2-purinoceptors.[26] Unlike other P_2-receptor agonists however, alpha, beta-methylene ATP does not inhibit nerve stimulation–induced release of NE from adrenergic nerves.[20] In spite of not being an agonist at these prejunctional purinoceptors, alpha, beta-methylene ATP does interact with them in that it reversed the ability of 2-chloroadenosine and beta, gamma-methylene ATP to reduce the release of NE.[20] Thus alpha, beta-methylene appears to be an antagonist not an agonist at the prejunctional purinoceptors, a finding that is not compatible with the receptors being either P_1 or P_2.

One possibility to explain these unusual results is to imagine that the prejunctional purinoceptors are not either P_1 or P_2 but rather are a third type of receptor. We refer to this receptor as P_3.[19,20] Some of the features that distinguish among these three receptors are listed below. (1) Whereas the P_1-receptor can be viewed primarily as a nucleoside receptor and the P_2-receptor as a nucleotide receptor,[15] the P_3-receptor recognizes the structure of both nucleosides and nucleotides. Indeed adenosine and ATP seem to be nearly equipotent at P_3-receptors. (2) Agents such as 8-SPT and other alkyl xanthines which antagonize responses mediated by P_1-receptors, but not P_2-receptors, also antagonize responses mediated by P_3 receptors. (3) Alpha, beta-methylene ATP is inactive at P_1-receptors, is an agonist and a desensitizing agent at P_2-receptors, but an antagonist at P_3-receptors.

Evidence for a unique prejunctional purinoceptor is not restricted to adrenergic nerves. In 1985, Gustafsson and colleagues reported results concerning the regulation of ACh release by adenine nucleosides and nucleotides[27] that are similar in many respects to what we subsequently observed concerning the regulation of NE release. Their studies were conducted with the isolated longitudinal smooth muscle–myenteric plexus preparation from the guinea pig. In this preparation, ATP and adenosine were equipotent in inhibiting the nerve-induced twitch contraction of the longitudinal smooth muscle presumably by decreasing the release of ACh. Beta, gamma-methylene

ATP and beta, gamma-imido ATP, both known to be less rapidly hydrolyzed to adenosine than ATP, also caused an inhibition of ACh release. These investigators provided evidence that the actions of both the nucleotides and the nucleosides were occurring via the same receptors because they were antagonized by 8-SPT. Furthermore the pA_2 values of 8-SPT versus the various nucleotides and nucleosides were very similar. As with the rat tail artery, alpha, beta-methylene ATP did not exert an agonistic effect, that is, it did not decrease the release of ACh. Gustafsson and colleagues concluded that the adenine nucleotides can inhibit neurotransmission per se via a prejunctional receptor common to nucleosides and nucleotides.[18,27] Unlike us, these investigators did not suggest a new subtype of purinoceptor but rather suggested that the criteria for classification of this receptor as a P_1-purinoceptor should be changed.[18]

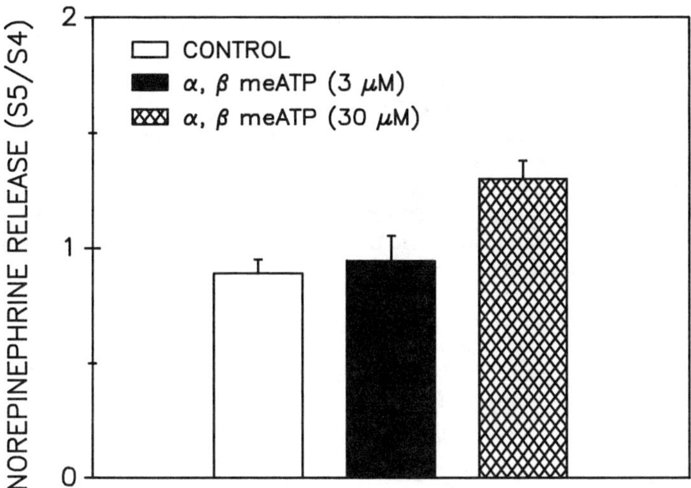

FIGURE 2. Influence of pretretement with alpha, beta-methylene ATP at 3 μM and 30 μM on the release of endogenous NE from the rat caudal artery stimulated at 8 Hz. Caudal arteries were stimulated 5 times at 30 minute intervals. Alpha, beta-methylene ATP was added 15 minutes prior to S_5 and the results are presented as the S_5/S_4 ratio. Each bar represents the mean of 4 experiments. NE release in the presence of 30 μM alpha, beta-methylene was significantly different from control and 3 μM ($p < 0.05$).

The evidence reviewed above indicates the presence of prejunctional purinoceptors, which when stimulated by adenine nucleotides and nucleosides, mediate an inhibition of transmitter releases. We wondered whether these receptors play any role in the normal physiological modulation of the release of NE, that is, are there sufficient quantities of extracellular adenine nucleotides and nucleosides to modulate transmitter release? If so, then drugs known to antagonize prejunctional P_3-receptors, such as alpha, beta-methylene ATP or 8-SPT, might be expected to enhance the release of NE in much the same manner as alpha$_2$-adrenoceptor antagonists enhance the release of NE.

To test this notion we examined the effect of alpha, beta-methylene ATP on the nerve stimulation–induced release of NE in the rat caudal artery. At a frequency of stimulation of 1 Hz, alpha, beta-methylene ATP did not significantly alter the release of NE. However, as shown in FIGURE 2, at a frequency of stimulation of 8 Hz, the release of NE was significantly enhanced by alpha, beta-methylene ATP. This effect was concentration dependent in that it was produced by 30 μM, but not 3 μM, of alpha, beta-methylene ATP.

The results of FIGURE 2, at first glance, would appear to be in conflict with a number of studies that have reported that alpha, beta-methylene ATP does not alter the nerve stimulation–induced release of NE.[28–33] In these studies, however, the release of NE was evoked by low frequencies of stimulation and the concentration of alpha, beta-methylene ATP was less than 10 μM. Our results with the caudal artery and those of Stjärne and Åstrand with the vas deferens[34] indicate that at somewhat higher frequencies and with concentrations of 10 μM or more of alpha, beta-methylene ATP, there is a significant increase in the release of NE.

We believe the ability of alpha, beta-methylene ATP to increase the release of NE at certain frequencies indicates that endogenous adenine nucleosides and nucleotides contribute to a prejunctional modulation of transmitter release. By producing antagonism of P_3 purinoceptors with alpha, beta-methylene ATP, the normal inhibitory effect of the purines would be obtunded and transmitter release would be increased. Previous studies in our laboratory have shown that large amounts of ATP and other nucleotides and nucleosides are released upon transmural nerve stimulation of the caudal artery at 8 Hz.[35]

The frequency of stimulation at which one can observe an effect of endogenous adenine nucleotides and nucleosides may depend upon the particular neuroeffector junction. Preliminary studies with the rat vas deferens indicate that 8-SPT enhances the release of NE evoked by stimulation at 2 Hz.[24]

It is tempting to speculate that because NE and ATP are thought to cocontribute to neuroeffector transmission as cotransmitters,[32,36,37] they may also cocontribute to feedback neuromodulation. Perhaps NE, acting at alpha$_2$-prejunctional receptors, is more important as a feedback modulator at low frequencies while ATP and/or its degradation products, acting at P_3-prejunctional receptors, are more important as a feedback modulator at higher frequencies.

REFERENCES

1. PATON, D. M. 1979. *In* Physiological and Regulatory Functions of Adenosine and Adenine Nucleotides. H. Baer and G. I. Dummond, Eds.: 69–77. Raven Press. New York, N.Y.
2. PATON, D. M. 1981. *In* Purinergic Receptors. G. Burnstock, Ed.: 199–219. Chapman and Hall. London, England.
3. FREDHOLM, B. B. & P. HEDQVIST. 1980. Biochem. Pharmacol. **29:** 1635–1643.
4. SU, C. 1983. Pharmacol. Rev. **23:** 397–411.
5. HEDQVIST, P. & B. B. FREDHOLM. 1976. Naunyn-Schmiedebergs Arch. Pharmacol. **283:** 217–223.
6. CLANACHAN, A. S., A. JOHNS & D. M. PATON. 1977. Neuroscience **2:** 597–602.
7. WAKADE, A. R. & T. D. WAKADE. 1978. J. Physiol. **282:** 35–49.
8. KHAN, M. T. & K. U. MALIK. 1980. J. Pharmacol. Exp. Ther. **68:** 551–561.
9. FREDHOLM, B. B., G. FRIED & P. HEDQVIST. 1982. Eur. J. Pharmacol. **79:** 233–243.
10. ENERO, M. A. & B. Q. SAIDMAN. 1977. Naunyn-Schmiedebergs Arch. Pharmacol. **297:** 39–46.
11. VERHAEGHE, R. H., P. M. VANHOUTTE & J. T. SHEPHERD. 1977. Circ. Res. **40:** 208–215.
12. SU, C. 1978. J. Pharmacol. Exp. Ther. **204:** 351–361.
13. MOYLAN, R. D. & T. C. WESTFALL. 1979. Blood Vessels **16:** 302–310.
14. HUSTED, S. E. & O. A. NEDERGAARD. 1981. Acta. Pharmacol. Toxicol. **49:** 334–353.
15. BURNSTOCK, G. 1978. *In* Cell Membrane Receptors for Drugs and Hormones: A Multidisciplinary Approach. L. Bolis & R. W. Staub, Eds.: 107–118. Rauch Press. New York, N.Y.

16. BURNSTOCK, G. 1985. Experientia **41:** 869–874.
17. RIBEIRO, J. A. & J. WALKER. 1975. Br. J. Pharmacol. **54:** 213–218.
18. WIKLUND, N. P., L. E. GUSTAFSSON & J. LUNDIN. 1985. Acta Physiol. Scand. **125:** 681–691.
19. SHINOZUKA, K., R. A. BJUR & D. P. WESTFALL. 1988. Naunyn-Schmiedebergs Arch. Pharmacol. **338:** 221–227.
20. WESTFALL, D. P., K. SHINOZUKA & R. A. BJUR. *In* Purines in Cellular Signalling: Targets for New Drugs. K. A. Jacobson, J. W. Daly & V. Manganiello, Eds. Springer Verlag. Amsterdam, the Netherlands (In press.)
21. MOODY, C. J. & G. BURNSTOCK. 1982. Eur. J. Pharmacol. **77:** 1–9.
22. LUKACSKO, P. & A. BLUMBERG. 1982. J. Pharmacol. Exp. Ther. **222:** 344–349.
23. HUSTED, S. E. & O. A. NEDERGAARD. 1985. Acta Pharmacol. Toxicol. **57:** 204–213.
24. FORSYTH, K., K. SHINOZUKA, R. A. BJUR & D. P. WESTFALL. *In* Biological Actions of Extracellular ATP. Ann. N.Y. Acad. Sci. **603.** (In press.)
25. FEDAN, J. S., G. K. HOGABOOM, D. P. WESTFALL & J. P. O'DONNELL. 1982. Eur. J. Pharmacol. **81:** 193–204.
26. KASAKOV, L & G. BURNSTOCK. 1983. Eur. J. Pharamcol. **86:** 291–294.
27. WIKLUND, N. P. & L. E. GUSTAFSSON. 1986. Acta Physiol. Scand. **126:** 217–233.
28. ALLCORN, R. J., T. C. CUNNANE & K. KIRKPATRICK. 1986. Br. J. Pharmacol. **89:** 647–659.
29. KASAKOV, L., J. ELLIS, K. KIRKPATRICK, P. MILLNER & G. BURNSTOCK. 1988. J. Auton. Nerv. Syst. **22:** 75–82.
30. MURAMATSU, I. & S. KIGOSHI. 1987. Br. J. Pharmacol. **92:** 901–908.
31. ISHIKAWA, S. 1985. Br. J. Pharamcol. **86:** 777–787.
32. KÜGELGEN, I. V. & K. STARKE. 1985. J. Physiol. **367:** 435–455.
33. MIYAHARA, H. & H. SUZUKI. 1987. J. Physiol. **389:** 423–440.
34. STJÄRNE, L. & P. ÅSTRAND. 1985. Neuroscience **14:** 929–946.
35. WESTFALL, D. P., K. SEDAA & R. A. BJUR. 1987. Blood Vessels **24:** 125–127.
36. FEDAN, J. S., G. K. HOGABOOM, J. P. O'DONNELL, J. COLBY & D. P. WESTFALL. 1981. Eur. J. Pharmacol. **69:** 41–53.
37. SNEDDON, P. & D. P. WESTFALL. 1984. J. Physiol. **347:** 561–580.

Presynaptic GABA Receptors

PETER C. WALDMEIER AND PETER A. BAUMANN

Research Department
Pharmaceuticals Division
Ciba-Geigy AG
CH-4002 Basel, Switzerland

The first reports about the existence of a γ-aminobutyric acid (GABA) autoreceptor which regulates the release of this amino acid from nerve endings, in a manner analogous to what was known of aminergic and cholinergic neuronal systems, were published more than 10 years ago. These early reports suggested that these autoreceptors were of the $GABA_A$ type. Thus, the potassium-stimulated release of radiolabeled GABA from slices of the rat cortex,[1] striatum,[2] or substantia nigra,[3] or synaptosomes from the rat cortex,[4,5] median eminence, and pituitary neurointermediate lobe,[6] was reported to be inhibited by muscimol in a bicuculline- or picrotoxin-sensitive manner. Similar results were observed with other $GABA_A$ agonists such as ∂-aminolaevulinic acid,[2,5] piperidine-4-sulfonic acid, 4,5,6,7-tetrahydroisoxazolo[5,4-C]pyridin-3(2H)-one (THIP), isoguvacine,[7] and imidazoleacetic acid.[2] On the other hand, Arbilla *et al.* were unable to find an inhibition by muscimol or exogenous GABA of [^{3}H]GABA release from rat occipital cortical slices,[3] and Limberger *et al.* found muscimol ineffective to inhibit the electrically evoked release of labeled GABA from rabbit striatal slices.[8]

The GABA analogue baclofen, which does not interact with classical muscimol- and bicuculline-sensitive GABA receptors, was found like GABA to inhibit the release of noradrenaline[9,10] and acetylcholine[11] from peripheral nerve endings, and later that of noradrenaline,[12] dopamine,[12,13] serotonin,[12,14,15] and glutamate[16] in the central nervous system (CNS). This inhibitory effect was thought to be mediated by a novel type of presynaptic heteroreceptor, which was termed $GABA_B$, in contrast to the classical GABA receptor which was designated $GABA_A$.[17] Although it might have seemed obvious that the release of GABA itself could be regulated by a receptor of the same type, either no such effects of baclofen were initially found at all, or they were not interpreted as indicating the existence of $GABA_B$ autoreceptors. Potashner did not find an effect of baclofen on the electrically induced (100 Hz) release of GABA formed from glucose in guinea pig cortical slices.[16] Johnston *et al.* observed no effect of this agent on the K^+-evoked release of [^{3}H]GABA from cortical and spinal cord slices.[18] Collins *et al.* described a reduction by 5 and 25 μM (−)-baclofen of the release of endogenous GABA from slices of the rat olfactory cortex elicited by very low frequency stimulation (0.083 Hz) of the lateral olfactory tract.[19] They thought it to be due to a reduction of the excitatory (glutamatergic and/or aspartergic) input to GABA-containing interneurons, because baclofen reduced the evoked field potential (P wave), and rejected the possibility of an effect on autoreceptors mainly because baclofen failed to affect GABA release evoked by 50 mM K^+ in their own experiments and by electrical stimulation at 100 Hz in those of Potashner.[16] Limberger *et al.* observed a small inhibitory effect of baclofen on the electrically induced (5 Hz) release of [^{3}H]GABA from rabbit striatal slices,[8] but did not consider it to indicate the existence of $GABA_B$ autoreceptors because exogenous GABA failed to have an effect in the presence of the uptake inhibitor nipecotic acid.

136

Anderson and Mitchell were the first to implicate $GABA_B$ autoreceptors in the control of GABA release.[20] They showed baclofen to inhibit the release of [³H]GABA elicited by 15 mM K^+ from synaptosomes of the median eminence, but not of the pituitary neurointermediate lobe. More recently, the evidence that the GABA autoreceptor is of the $GABA_B$ type was strengthened by new findings, which will be summarized in the following.

EVIDENCE THAT GABA INHIBITS ITS OWN RELEASE

Pittaluga *et al.* showed 10 μM GABA to inhibit the release of preloaded [³H]GABA elicited by 15 μM K^+ from superfused rat cortical synaptosomes by about 40%.[21] The threshold concentration was about 1 μM. Similar results were obtained in human cortical synaptosomes[22] and in rat cortical slices stimulated electrically at 5 Hz.[23] These experiments were carried out in the presence of SK&F 89976, a GABA uptake inhibitor more potent than nipecotic acid, which unlike the latter is not a substrate of the uptake carrier and does not displace GABA from its stores.[24,25] These shortcomings of nipecotic acid may explain the failure of Limberger *et al.* to observe a corresponding inhibitory effect of exogenous GABA in rabbit caudate slices stimulated electrically at 5 Hz.[8]

More evidence for an inhibitory effect of GABA on its own release comes from experiments with electrically stimulated brain slices. In a slice model, an inhibitory effect of GABA should result in a decrease of release with increasing frequency of stimulation, because the transmitter progressively accumulates in the extracellular space due to insufficient clearance by the superfusion medium, in particular if reuptake is blocked. This is not true in synaptosomal preparations, in which superfusion effectively removes transmitter once it is released. As expected, FIGURE 1 shows a clear-cut decrease of the stimulated fractional release per pulse of both endogenous and [³H]GABA in the range of 0.125–2 Hz. At higher frequencies, GABA release increased again (for a discussion of the characteristics of GABA release at these higher frequencies see References 26 and 27). The fact that the fractional release of endogenous GABA was about fourfold lower than that of [³H]GABA might suggest that the latter preferentially enters the pool(s) from which GABA is released under these conditions.

EVIDENCE THAT BACLOFEN INHIBITS GABA RELEASE

(−)-Baclofen, but not the (+)-enantiomer, was shown to inhibit the K^+-evoked release of [³H]- as well as of endogenous GABA from rat cortical and human synaptosomes and from rat cortical slices with a potency very similar to that of exogenous GABA.[21–23] The electrically evoked release of endogenous GABA from rat cortical slices was inhibited in a frequency-dependent manner, i.e., the percentage of inhibition decreased with increasing frequency of stimulation (TABLE 1). The order of potency of (−)-baclofen was comparable to that reported by Raiteri's group.

FIGURE 2 shows why the percentage of inhibition decreased with increasing frequency. The dependence of [³H]GABA release on the stimulation frequency is given in the presence or absence of 10 μM (−)-baclofen. This high concentration of the $GABA_B$ agonist suppressed the evoked release of [³H]GABA to a low and constant level between 0.125 and 4 Hz. Interestingly, (−)-baclofen became progressively ineffective when the frequency was raised from 4 to 16 Hz, corroborating the result

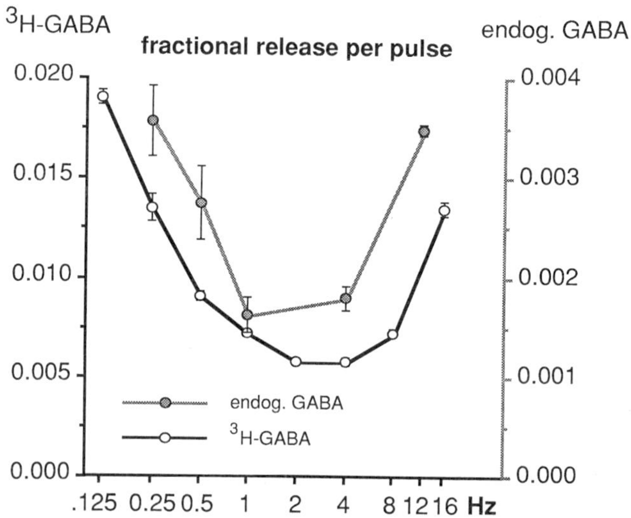

FIGURE 1. Frequency dependence of the electrically stimulated release of [³H]GABA and endogenous GABA from rat cortical slices. Cross-chopped rat cortical slices (1000 × 350 × 350 μm) were suspended in physiological buffer containing 2.6 mM Ca²⁺ and allowed to settle. For the experiments with radiolabeled GABA, they were incubated for 5 minutes with 100 nM [2, 3-³H(N)]GABA (25 Ci/mmol). Twenty-five–50 μl of slice suspension (1–2 mg protein) was transferred to stimulation chambers and superfused at 0.4 ml/minute with physiological buffer gassed with 5% CO₂ in O₂, containing 10 μM SK&F 89976 and, in the radioactive experiments, 50 μM of the GABA transaminase inhibitor aminooxyacetic acid. Fractions of 6 minutes were collected, beginning 50 minutes after starting the superfusion. The slices were stimulated at different frequencies by a constant number of pulses (2 mseconds, 40 mA; 60 pulses if [³H]GABA and 240 if endogenous GABA was measured). Endogenous GABA was determined by high performance liquid chromatography (HPLC) with electrochemical detection after derivatization with 2,4,6-trinitrobenzenesulphonic acid, by a modification of the procedure of Yamamoto *et al.*,[48] as described previously.[47] Radioactivity in the superfusion medium and in the slices (after solubilization in Irgasolv® and subsequent neutralization) was counted after addition of Irgascint A300® scintillator. Data are means of the stimulated fractional release per pulse ± standard errors of the mean (SEM) (left ordinate for [³H]GABA, right ordinate for endogenous GABA). For experimental details see Waldmeier *et al.*[32,47] and Baumann *et al.*[37]

obtained with endogenous release at 24 Hz (TABLE 1). There is a parallel to this frequency dependence of the inhibitory effect of (−)-baclofen in experiments with K⁺-stimulation. The ability of the GABA_B agonist to inhibit the release of [³H]GABA diminished with increasing K⁺ concentration, becoming virtually ineffective at 35 mM.[28] A similar result was obtained with endogenous GABA (TABLE 1).

(−)-Baclofen suppressed [³H]GABA release at 0.125 Hz to about 10% of the control level at 10μM and above. The IC₅₀ value of 0.37 μM (FIGURE 3) indicates that the compound is more effective and more potent in electrically stimulated slices than in synaptosomes stimulated by K⁺ concentrations of 9 or 15 mM.[21,22,28,29] This suggests that the effect of (−)-baclofen is the more clear-cut the milder the stimulus. The IC₅₀ of (+)-baclofen was about 500-fold higher than that of the (−)-enantiomer. The effect may be due to traces of the former; an impurity of 0.2% would be sufficient.

TABLE 1. Effects of the Enantiomers of Baclofen on the Release of Endogenous GABA from Cortical Slices[a]

	Concentration	Stimulation	Percent Inhibition
$(-)$-Baclofen	1 μM	0.25 Hz	41.4 ± 3.5[b]
$(-)$-Baclofen	1 μM	0.5 Hz	25.2 ± 1.4[b]
$(-)$-Baclofen	10 μM	0.5 Hz	67.7 ± 7.5[b]
$(-)$-Baclofen	10 μM	4 Hz	24.1 ± 1.6[b]
$(-)$-Baclofen	10 μM	24 Hz	0.0 ± 7.0
$(+)$-Baclofen	10 μM	0.5 Hz	-3.9 ± 4.8
$(-)$-Baclofen	10 μM	15 mM K^+	59.3 ± 3.1[b]
$(-)$-Baclofen	10 μM	30 mM K^+	-14.0 ± 7.5

[a]Data are means $\pm$ standard errors of the mean (SEM) from groups of 3–4 slices. For experimental details see Waldmeier *et al.*[26,32,47]

[b]$p > 0.01$ (Student's t-test)

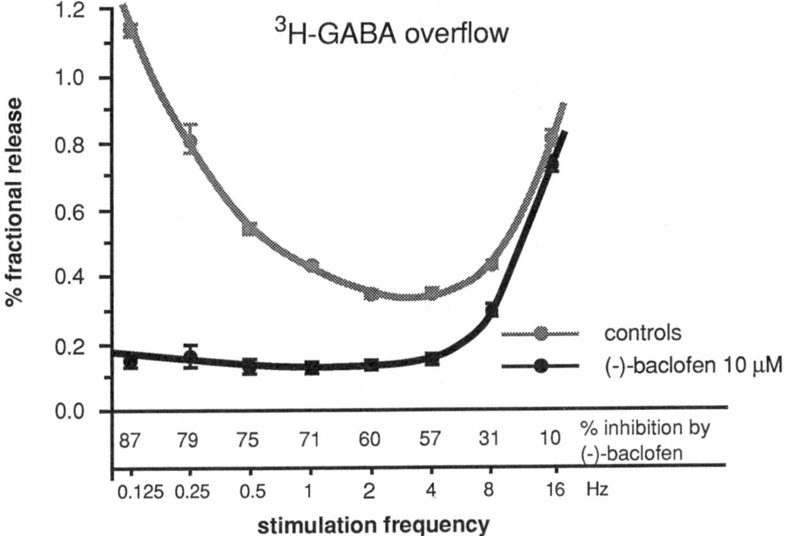

FIGURE 2. Frequency-dependence of the effect of $(-)$-baclofen on electrically evoked [³H]GABA overflow. The data summarize the results from a series of experiments in each of which 2 groups of 4 rat cortical slice preparations preloaded with [³H]GABA were stimulated 3 times by a constant number of 60 pulses. The first stimulation served as a reference for the normalization of the results from the different experiments and was always at 2 Hz. The second and third stimulation was applied at the test frequency. $(-)$-Baclofen (10 μM) was added to the superfusion medium of one group before S_3. Data are means of the stimulated fractional release $\pm$ SEM. In most cases, SEM was smaller than the symbols and therefore omitted. Data are taken from Baumann *et al.*,[37] where also experimental details are available.

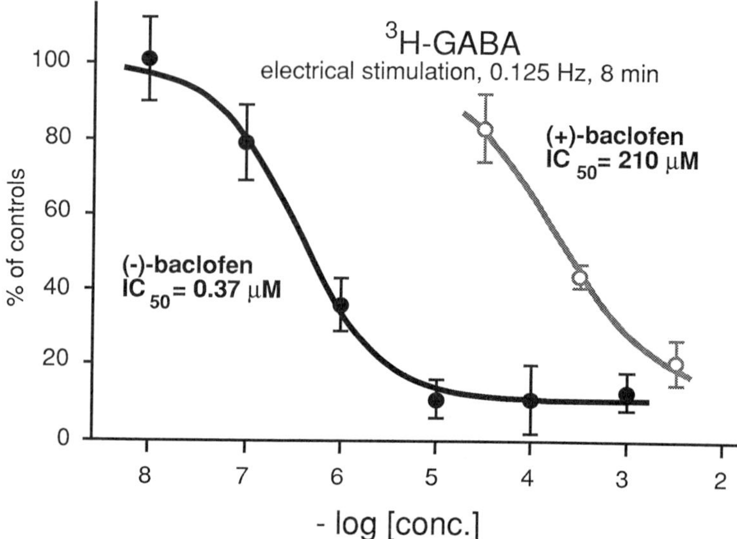

FIGURE 3. Inhibition of the electrically induced release of [³H]GABA from rat cortical slices by (−)- and (+)-baclofen. A separate experiment with its own control was carried out for each concentration ($n = 4$ for drug and control, respectively). The slices were stimulated twice at 0.125 Hz for 8 minutes (60 pulses). Drug was added to the medium of one group of 4 slice preparations after the first stimulation (S_1). Data represent means of the S_2/S_1 ratios in percent of those of the controls and are taken from Baumann et al.[37] The error bars represent 5% confidence limits.

EVIDENCE THAT MUSCIMOL AND BICUCULLINE DO NOT AFFECT GABA RELEASE

Raiteri and his collaborators and our own group have studied muscimol under a variety of conditions, in synaptosomes and slices, stimulating with different concentrations of K^+ or electrically at different frequencies, and have never been able to observe an inhibitory effect of this agent (TABLE 2). Bonanno et al.[28] have used the same conditions as those indicated by Brennan et al.[7] without success. Our own negative results in cortical slices stimulated with 30 mM K^+ were obtained under comparable conditions under which Ennis and Minchin[30] observed a concentration-dependent inhibition by muscimol in the range of 1 nM–1 μM. On the other hand, Neal and Shah were also unable in batch experiments to see an inhibitory effect of 10 μM muscimol on the release of endogenous GABA evoked by 50 mM K^+ from cortical slices of rats pretreated with the GABA transaminase inhibitor γ-vinyl-GABA or of controls.[31]

Moreover, the GABA$_A$ antagonist bicuculline did not affect the release of endogenous GABA elicited from rat cortical slices by electrical stimulation at 4 Hz at a concentration of 10 μM[32] or of [³H]GABA at 5 Hz and 100 μM.[23] Likewise, another GABA$_A$ antagonist, SR 95531, was also inactive at 5 Hz and 100 μM.[23] It should be noted that, under these stimulation conditions, GABA markedly suppressed its own release. Had this been due to an action of GABA$_A$ autoreceptors, a GABA$_A$ antagonist should have increased GABA release.

EVIDENCE THAT GABA$_B$ ANTAGONISTS INCREASE GABA RELEASE

GABA$_B$ antagonists have not been available until 1987, when Kerr *et al.* reported that the phosphono analogue of baclofen, phaclofen, exhibited such properties, although it had a rather low potency.[33] In the following, Bonanno *et al.* showed that the compound was able to antagonize the effects of baclofen on the release of [^{3}H]GABA from cortical synaptosomes at a concentration of 10 μM.[29] Since in superfused synaptosomal preparations, the medium effectively removes released transmitter, they could not demonstrate an enhancing effect of this compound on its own. On the other hand, such an antagonist could increase GABA release in a slice model, provided the frequency is chosen adequately. FIGURE 4 shows the effect of 1mM phaclofen on the release of [^{3}H]GABA in relation to the applied frequency. As expected, the compound increased [^{3}H]GABA release. However, while one would expect an antagonist to raise transmitter release to a maximal level irrespective of the applied frequency, phaclofen just displaced the curve obtained in its absence upwards. Correspondingly, the extent of the enhancement, expressed in percent of the control release at the same frequency, did not increase with increasing frequency, but remained relatively stable, at least up to 16 Hz, where the enhancement was almost lost. For the range of frequency between 1 and 4 Hz, this can be explained by phaclofen's poor potency, i.e., by its partial and progressive displacement from the putative autoreceptor by endogenously released GABA. However, one would have expected that the antagonist was less effective at lower frequencies, where there is little endogenous GABA released to activate the autoreceptor. It cannot be excluded that endogenously released GABA already markedly suppressed its own release at this low frequency. The steepness of the frequency dependence in this range at least does not contradict this idea. Moreover, all

TABLE 2. Inability of Muscimol to Inhibit the Release of Endogenous or [^{3}H]GABA under Various Stimulation Conditions and in Various Preparations[a]

Preparation	Stimulation	Percent Inhibition at Muscimol Concentration (μM)				
		0.01	0.1	1	10	100
Rat cortical slices	0.5 Hz[32]				*−11 ± 4*	
	12 Hz[32]				*−10 ± 7*	
	24 Hz[26]				*1 ± 5*	
	5 Hz[23]			8 ± 4	0 ± 5	10 ± 5
	30 mM K^{+}[26]				*0 ± 12*	
Rat hippocampal slices	0.5 Hz[26]				*9 ± 8*	
Rat striatal slices	0.5 Hz[26]				*13 ± 21*	
Rat cortical synaptosomes	9 mM K^{+}[28]			6 ± 3	1 ± 4	13 ± 5
	15 mM K^{+}[28]			−3 ± 4	−5 ± 4	0 ± 6
	15 mM K^{+}[21]		1 ± 4		2 ± 3	
	25 mM K^{+}[28]			−1 ± 2	−2 ± 4	0 ± 3
	35 mM K^{+}[28]			5 ± 4	−2 ± 2	−4 ± 5
	50 mM K^{+}[28]			1 ± 5	−4 ± 4	
	55 mM K^{+}[28,b]	2 ± 3	3 ± 3	4 ± 3	8 ± 4	
Human cortical synaptosomes	15 mM K^{+}[22]			0 ± 3	2 ± 2	−3 ± 3

[a]Plain figures, [^{3}H]GABA; italic figures, endogenous GABA.
[b]Experiments carried out under the conditions given by Brennan *et al.*[7]

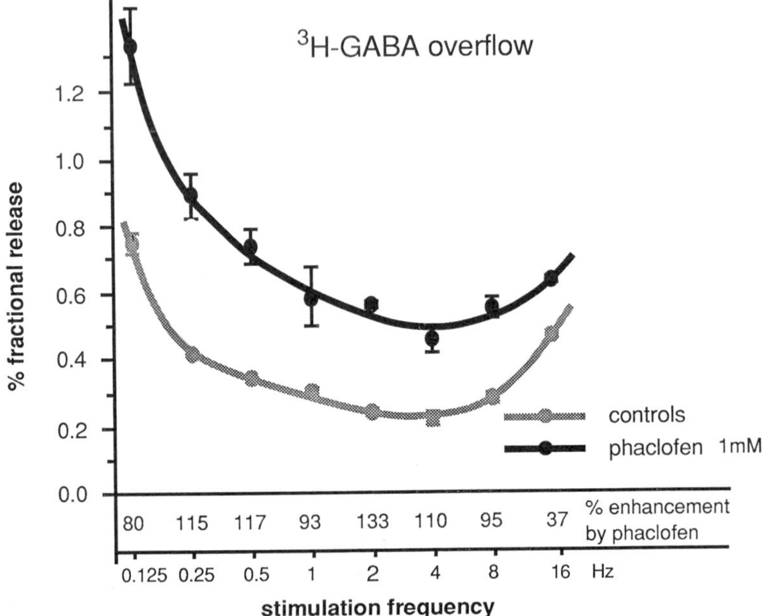

FIGURE 4. Frequency dependence of the effect of phaclofen on [^{3}H]GABA overflow. These experiments were performed analogously to those shown in FIGURE 2, with the exception that the drug used was 1 mM phaclofen instead of ($-$)-baclofen. Data are taken from Baumann *et al.*,[37] where also experimental details are available.

these experiments were performed in the presence of a GABA uptake inhibitor. Not only does this probably shift the range in which autoreceptor activation by endogenously released GABA occurs towards lower frequencies, but it also extends the range of diffusion of the transmitter from the site of release. This may permit cross-talk, i.e., activation of presynaptic autoreceptors by transmitter molecules originating from neighboring terminals, and thus lead to a feedback inhibition of GABA release already at low frequencies. A further possibility can also not be dismissed at present. At least some of the GABA interneurons in our slice preparation are probably intact. They may possess, e.g., somatically or dendritically located $GABA_B$ receptors, through which GABA release may also be modulated. To make this explanation viable, however, one would still have to assume that such receptors are more potently affected by phaclofen that the terminal autoreceptors to explain our results. On the other hand, the postulate of $GABA_B$ receptors with different sensitivity towards phaclofen is not without precedent: the compound antagonized the postsynaptic hyperpolarization of hippocampal pyramidal cells induced by baclofen much more potently than that by GABA, whereas the presynaptic inhibition was unaffected.[34,35] The issue still awaits clarification.

FIGURE 5 shows the concentration-response curve of phaclofen with respect to the release of [^{3}H]GABA at a frequency of 2 Hz. The threshold concentration was about 100 μM, and the control release was about doubled at 3 mM; higher concentrations were not tested, and it is therefore not clear what the maximally possible enhancement is. The extent of the increase caused by 300 μM phaclofen is very similar to that reported very recently by Raiteri *et al.* in cortical slices stimulated at 5 Hz.[23] On the other hand, Neal and Shah observed an increasing effect of the antagonist at 500 μM on the release of endogenous GABA from cortical slices stimulated with 50 mM K[+] only if the animals had been pretreated with the GABA transaminase inhibitor γ-vinyl-GABA.[31]

Due to interference of phaclofen with our assay for endogenous GABA, corresponding investigations could not be made. Instead, we studied CGP 35348 (3-aminopropyldiethoxymethylphosphinic acid), which is markedly more potent as a GABA_B antagonist than phaclofen and seems to be rather specific.[36] This compound increased the release of both endogenous and [3H]GABA very similarly, with a threshold concentration of about 10 μM (FIGURE 5). The curves were displaced to the left with respect to that of phaclofen, indicating that CGP 35348 is about 10–20 times more potent. It may be worth mentioning that this compound also, like phaclofen, increased [3H]GABA release at 0.125 Hz (not shown).

ANTAGONISM OF THE EFFECTS OF BACLOFEN BY GABA_B ANTAGONISTS

The GABA_B antagonist phaclofen was reported to inhibit the release of [3H]GABA from rat cortical synaptosomes evoked by 9 mM K[+] already at a concentration of 10

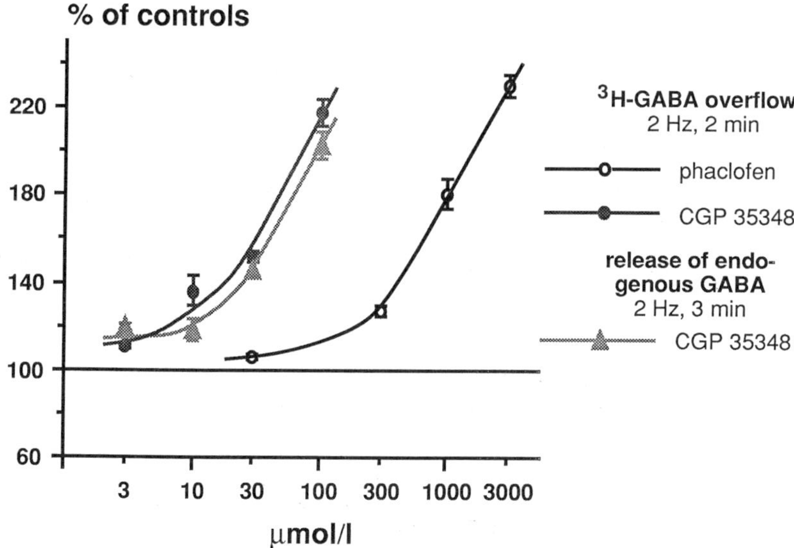

FIGURE 5. Effects of phaclofen and CGP 35348 on the electrically stimulated release of GABA. A separate experiment with its own control was carried out for each concentration of the drugs. Groups of 4 slices were stimulated twice at 2 Hz ([3H]GABA, 2 minutes = 240 pulses; endogenous GABA, 3 minutes = 360 pulses). The drugs were added to the medium of one group after the first stimulation (S_1). Data represent means ± SEM of the S_2/S_1 ratios in percent of those of the controls.

μM, i.e., at much lower concentration than had been found effective in electrophysiological experiments.[28,29]

We have studied the antagonism by 100 μM and 1 mM phaclofen against the inhibitory effect of 1 μM $(-)$-baclofen on the electrically evoked release of [³H]GABA from rat cortical slices at a stimulation frequency of 0.125 Hz, where the effect of the agonist is most marked. In two separate experiments, two groups of slices each were stimulated three times, a first time under control conditions and a second and third time in the presence of $(-)$-baclofen alone. Before the third stimulation, the antagonist was added to the medium of one group. Phaclofen at 100 μM significantly antagonized the effect of 1 μM $(-)$-baclofen by about 30% and at 1 mM by about 80% (TABLE 3). The more potent antagonist, CGP 35348, was tested in a similar experiment and found to markedly overantagonize $(-)$-baclofen at 100 μM. This overantagonism might be related to the increase of [³H]GABA release both this compound and phaclofen caused when given alone at a stimulation frequency of 0.125 Hz (see above).

The pA_2 and pA_{10} values of phaclofen against $(-)$-baclofen were determined at 0.125 and 2 Hz in an additional series of experiments. The IC_{50} values of the agonist at

TABLE 3. Antagonism by Phaclofen and CGP 35348 of the Inhibitory Effect of $(-)$-Baclofen on the Release of [³H]GABA[a]

Stimulation Period	Drugs Added	Percent Fraction Release	Stimulation Period	Drugs Added	Percent Fraction Release
S_1	None	0.704 ± 0.041 (4)	S_1	None	0.624 ± 0.042 (4)
S_2	$(-)$-Baclofen 1 μM	0.287 ± 0.027 (4)	S_2	$(-)$-Baclofen 1 μM	0.238 ± 0.013 (4)
S_3	$(-)$-Baclofen 1 μM	0.309 ± 0.024 (4)	S_3	$(-)$-Baclofen 1 μM + phaclofen 100 μM	0.312 ± 0.019 (4)
S_2/S_1		0.41 ± 0.02 (4)	S_2/S_1		0.38 ± 0.02 (4)
S_3/S_2		1.10 ± 0.10 (4)	S_3/S_2		1.47 ± 0.03^b (3)
S_1	None	0.939 ± 0.044 (4)	S_1	None	0.966 ± 0.053 (3)
S_2	$(-)$-Baclofen 1 μM	0.349 ± 0.023 (4)	S_2	$(-)$-Baclofen 1 μM	0.372 ± 0.024 (3)
S_3	$(-)$-Baclofen 1 μM	0.339 ± 0.008 (4)	S_3	$(-)$-Baclofen 1 μM + phaclofen 1 mM	0.892 ± 0.095 (3)
S_2/S_1		0.37 ± 0.01 (4)	S_2/S_1		0.39 ± 0.01 (4)
S_3/S_2		0.92 ± 0.02 (4)	S_3/S_2		2.29 ± 0.10^a (4)
S_1	None	1.335 ± 0.026 (4)	S_1	None	1.176 ± 0.036 (4)
S_2	$(-)$-Baclofen 1 μM	0.479 ± 0.024 (4)	S_2	$(-)$-Baclofen 1 μM	0.375 ± 0.004 (4)
S_3	$(-)$-Baclofen 1 μM	0.540 ± 0.044 (4)	S_3	$(-)$-Baclofen 1 μM + CGP 35348 100 μM	2.037 ± 0.051 (4)
S_2/S_1		0.36 ± 0.02 (4)	S_2/S_1		0.32 ± 0.01 (4)
S_3/S_2		1.12 ± 0.05 (4)	S_3/S_2		5.41 ± 0.12^a (4)

[a]Two groups of slices preloaded with [³H]GABA were stimulated 3 times for 8 minutes at 0.125 Hz in each experiment. During the first stimulation (S_1), no drug was present; during the second and third (S_2 and S_3), $(-)$-baclofen was present in both groups. The antagonists were added to the medium of one group before S_3. Data are means $\pm$ SEM of the stimulated fraction release.

[b]$p < 0.05$.

[c]$p < 0.01$ vs. controls (Student's t-test).

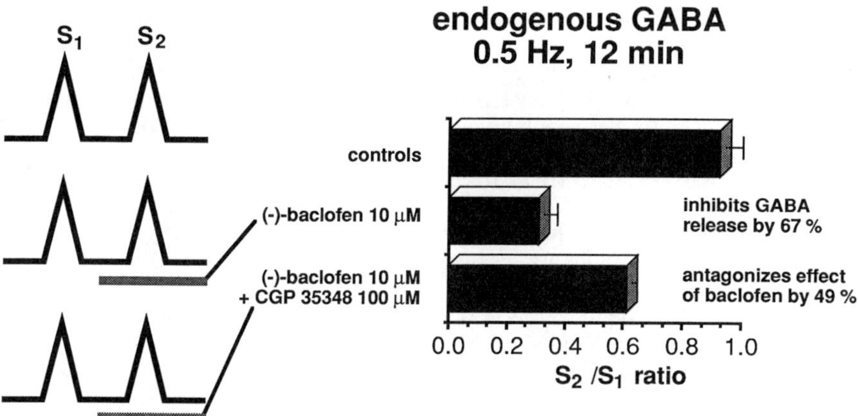

FIGURE 6. Antagonism by CGP 35348 of the inhibitory effect of (−)-baclofen on the release of endogenous GABA. Three groups of 4 slice preparations were stimulated twice at 0.5 Hz for 12 minutes (2 mseconds, 25 mA). Drugs were added after the first stimulation [(−)-baclofen, 10 μM; CGP 35348, 100 μM]. Data are means ± SEM of the S_2/S_1 ratios. Both the effect of baclofen and its antagonism by CGP 35348 were significant ($p < 0.01$; Dunnett's test).

these frequencies were 0.63 ± 0.04 and 4.88 ± 0.45 μM, respectively. The pA_2 and pA_{10} values at 0.125 and 2 Hz were 3.73 and 2.93 (slope = 1.18) and 3.36 and 2.93 (slope 2.32), respectively.[37] These latter results suggest a competitive interaction between phaclofen and baclofen at 0.125 Hz, but not at 2 Hz. This apparent noncompetitivity at 2 Hz should not be overemphasized, however, since endogenously released GABA as a third ligand may have influenced the results, and because the concentration range over which phaclofen could be tested was rather small due to its low potency.

CGP 35348 at 100 μM was also able to antagonize the effect of 10 μM (−)-baclofen on the release of endogenous GABA elicited by electrical stimulation at 0.5 Hz by about 50% (FIGURE 6).

The general impression from the slice experiments with phaclofen is that the compound has a similar potency as has previously been found in electrophysiological experiments,[33–35,38,39] i.e., in the upper micromolar to millimolar range, in contrast to the synaptosome experiments, where it seemed to be at least 10-fold more potent.[28,29] It is conceivable that this apparent discrepancy is due to a poor tissue penetration of the antagonist. On the other hand, the IC_{50} of phaclofen to inhibit baclofen binding in rat brain synaptic membranes is also in the upper micromolar range (118 μM),[40] which is more compatible with the data in slices and does not support the theory of poor tissue penetration.

SUPPORTING EVIDENCE FOR THE EXISTENCE OF GABA$_B$ AUTORECEPTOR FROM ELECTROPHYSIOLOGICAL EXPERIMENTS

Paired-pulse inhibition of hippocampal CA1 pyramidal neurons induced by stimulation of Schaffer collaterals is predominantly mediated by activation of postsynaptic GABA$_A$ receptors, because bicuculline, but not phaclofen, antagonized it at short interpulse intervals.[38] Baclofen reduced paired-pulse inhibition at concentrations that hardly affected the first population spikes, thus rendering a postsynaptic effect on pyramidal cells unlikely. It is more likely that the depression of paired-pulse inhibition was caused by a reduction by baclofen of the release of GABA from interneurons or of their activity.[38] The former possibility was favored by the authors because Harrison *et al.* had shown beforehand that (−)-baclofen reduced the GABA-mediated monosynaptic inhibitory postsynaptic currents in cultured hippocampal neurons from embryonic rats without interfering with postsynaptic membrane conductance or postsynaptic responses to GABA.[41] While phaclofen was unable to enhance paired-pulse inhibition in the above paradigm, probably because of its low potency, CGP 35348 did so. Both GABA$_B$ antagonists, however, attenuated the depressing effect of baclofen.[42]

More evidence for the existence of release-regulating GABA$_B$ autoreceptors comes from experiments in which inhibitory postsynaptic potentials (IPSPs) were assessed by intracellular recording performed on rat and guinea pig neocortical slices after orthodromic stimulation. These IPSPs were reduced if the stimulation frequency was raised from 0.1 to 1 Hz (compare this frequence range to that in FIGURE 1). Nipecotic acid reduced the IPSPs evoked at 0.1, but not at 1 Hz. IPSPs were also attenuated by baclofen. This effect outlasted the postsynaptic effects of the compound, which suggests that the attenuation of the IPSPs was mediated by a reduction of GABA release.[43,44]

GABA$_B$ AUTORECEPTORS IN BRAIN AREAS OTHER THAN THE CORTEX

No other brain area has been nearly as thoroughly investigated as the cortex. Anderson and Mitchell have found baclofen to inhibit the release of [^{3}H]GABA in synaptosomes of the rat median eminence, but not of the neurointermediate lobe of the pituitary,[20] both of which are densely innervated by GABAergic neurons. We have obtained evidence that the agonist decreases the release of endogenous GABA in slices of the rat hippocampus and striatum in a very similar manner as in the cortex.[32] There is, however, considerable confusion with respect to the situation in the substantia nigra. This area is of particular interest not only because it contains one of the highest GABA concentrations in the brain, but also because it receives important GABAergic input from the striatum, globus pallidus, etc. In contrast to most if not all other areas of the brain, the majority of the GABAergic nerve endings in the substantia nigra are those of projection neurons, not of interneurons.

Early on, Arbilla *et al.* reported muscimol and exogenous GABA to inhibit the release of [^{3}H]GABA from nigral slices elicited by 30 mM K$^+$.[3] In contrast, we have been unable to observe such an effect of muscimol on the release of endogenous GABA under otherwise similar conditions. Moreover, muscimol also did not inhibit GABA release evoked electrically at 24 Hz.[27] On the other hand, Floran *et al.* reported that muscimol, but not baclofen, inhibited the release of [^{3}H]GABA elicited by 15 mM K$^+$ from slices of the pars compacta of the substantia nigra, where as the converse was observed in the pars reticulata.[45] In contrast, Giralt *et al.* have studied the release of [^{3}H]GABA evoked by 9 mM K$^+$ from synaptosomes of these two parts of the substantia nigra and found baclofen and GABA (in a phaclofen-sensitive manner), but not muscimol, to inhibit it in both parts.[46] To further complicate the matter, we were unable to find any release at all of endogenous GABA from nigral slices evoked by electrical stimulation at frequences below 12 Hz, and (−)-baclofen was unable to affect that caused by 15 mM K$^+$ in this area.[27] Our inability to find a significant amount of GABA release from nigral slices might have two reasons which could be checked immediately. One is related to the impression obtained before (see above), that [³H]GABA preferentially enters a readily releasable pool; drug effects thereon might be masked by the relatively high basal release in this area when endogenous GABA is measured. On the other hand, due to the high density of GABAergic terminals, the high basal release might markedly activate the autoreceptors and

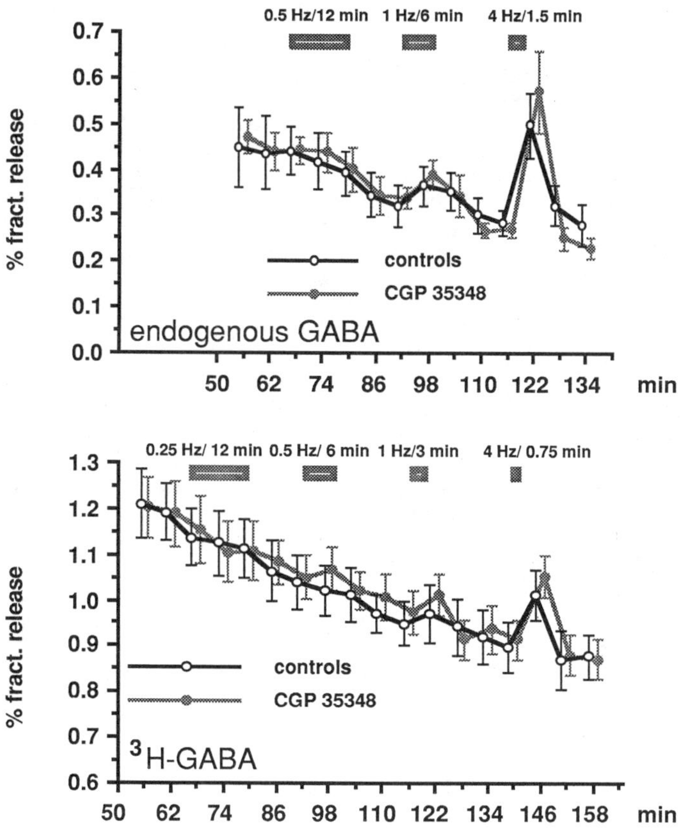

FIGURE 7. Failure to evoke significant release of endogenous or [³H]GABA from nigral slices by electrical stimulation at 0.25–4 Hz, in the absence and presence of CGP 35348. In each of two separate experiments, two groups of 4 slice preparations were stimulated 3 or 4 times at 0.25–4 Hz. One group was exposed to 100 μM CGP 35348 throughout. In both experiments, the medium contained 10 μM SK&F 89976; in that with [³H]GABA, it contained 50 μM aminooxyacetic acid in addition. Data are means ± SEM of the fractional release.

suppress stimulated release almost completely. Such a situation would probably not arise in a synaptosomal preparation. Two similar experiments were made to study these issues, one with [³H]- and the other with endogenous GABA. In each experiment two groups of nigral slices were stimulated 4 and 3 times, respectively, using frequencies between 0.25 and 4 Hz. One group of slices was superfused with control medium, the other with medium containing the $GABA_B$ antagonist, CGP 35348 at 100 μM throughout. The results of these experiments are given in FIGURE 7. The release of endogenous GABA evoked by these frequencies was nil (0.5 Hz) to very low (1 and 4 Hz), very similar to what we reported previously,[27] and this was not changed by the presence of the $GABA_B$ antagonist. The situation was rather similar when [³H]GABA was measured; however, looking at FIGURE 7 closely, one might get the impression that, in the presence of CGP 35348, small peaks emerged at 0.5 and 1 Hz, although statistical significance is not nearly reached. Thus, neither a better ability of [³H]GABA to enter a rapidly releasable pool nor a full activation of an autoreceptor already at low frequencies by a high synaptic concentration of GABA can explain the different results in synaptosomes and slices. It seems possible that the density of GABAergic nerve endings carrying $GABA_B$ autoreceptors is relatively low in this area, and they are better uncovered in a synaptosomal model. Since the majority of GABAergic nerve endings in the substantia nigra belong to projection neurons, one might anticipate that they are not endowed with autoreceptors. Conversely, this might mean that those that are are interneurons.

CONCLUSIONS

Based on the studies of the group of Raiteri on the K^+-stimulated release of [³H]- and endogenous GABA from synaptosomes as well as our investigations on the effects of baclofen and the $GABA_B$ antagonists phaclofen and CGP 35348 on the electrically and K^+-stimulated release of [³H]- and endogenous GABA from brain slices, the existence of $GABA_B$ receptors regulating GABA release seems now fairly probable. Neither group could find an inhibitory effect of the prototypic $GABA_A$ agonist muscimol on GABA release, under a variety of different conditions, in contrast to some earlier and some more recent studies.[30,45] However, some questions remain to be solved, such as that concerning the enhancement of GABA release at low frequencies by $GABA_B$ antagonists. Other issues are the situation in the substantia nigra and the significance of $GABA_B$ autoreceptors in the *in vivo* situation. Especially with respect to the latter question, the availability of a relatively potent antagonist, CGP 35348, may prove beneficial.

REFERENCES

1. MITCHELL, P. R. & J. L. MARTIN. 1978. Is GABA release modulated by presynaptic receptors? Nature **274:** 904–905.
2. KURIYAMA, K., K. KANMORI, J. TAGUCHI & Y. YONEDA. 1984. Stress-induced enhancement of suppression of [³H]GABA release from striatal slices by presynaptic autoreceptors. J.Neurochem. **42:** 943–950.
3. ARBILLA, S., L. KAMAL & S. Z. LANGER. 1979. Presynaptic GABA autoreceptors on GABAergic nerve endings of the rat substantia nigra. Eur. J. Pharmacol. **57:** 211–217.
4. SNODGRASS, S. R. 1978. Use of ³H-muscimol for GABA receptor studies. Nature **273:** 392–394.
5. BRENNAN, M. J. W. & R. C. CANTRILL. 1979. ∂-Aminolaevulinic acid is a potent agonist for GABA autoreceptors. Nature **280:** 514–515.
6. ANDERSON, R. & R. MITCHELL. 1986. Uptake and autoreceptor-controlled release of [³H]GABA by the hypothalamic median eminence and pituitary neurointermediate lobe. Neuroendocrinology **42:** 277–284.

7. BRENNAN, M. J. W., R. C. CANTRILL, M. OLDFIELD & P. KROGSGAARD-LARSEN. 1981. Inhibition of γ-aminobutyric acid release by γ-aminobutyric acid agonist drugs. Pharmacology of the γ-aminobutyric acid autoreceptor. Mol. Pharmacol. **19:** 27–30.
8. LIMBERGER, N., L. SPAETH & K. STARKE. 1986. A search for receptors modulating the release of γ-[^{3}H]aminobutyric acid in rabbit caudate nucleus slices. J. Neurochem. **46:** 1109–1117.
9. BOWERY, N. G. & A. L. HUDSON. 1979. γ-Aminobutyric acid reduces the evoked release of [^{3}H]noradrenaline from sympathetic nerve terminals. Br. J. Pharmacol. **66:** 108P.
10. BOWERY, N. G., A. DOBLE, D. R. HILL, A. L. HUDSON, J. S. SHAW, M. J. TURNBULL & R. WARRINGTON. 1981. Bicuculline-insensitive GABA receptors on peripheral autonomic nerve terminals. Eur. J. Pharmacol. **71:** 53–70.
11. BROWN, D. A. & A. J. HIGGINS. 1979. Presynaptic effects of γ-aminobutyric acid in isolated rat superior cervical ganglia. Br. J. Pharmacol. **66:** 108P–109P.
12. BOWERY, N. G., D. R. HILL, A. L. HUDSON, A. DOBLE, D. N. MIDDLEMISS, J. SHAW & M. TURNBULL. 1980. (–)Baclofen decreases neurotransmitter release in the mammalian CNS by an action at a novel GABA receptor. Nature **283:** 92–94.
13. REIMANN, W., A. ZUMSTEIN & K. STARKE. 1982. γ-Aminobutyric acid can both inhibit and facilitate dopamine release in the caudate nucleus of the rabbit. J. Neurochem. **39:** 961–969.
14. SCHLICKER, E., K. CLASSEN & M. GÖTHERT. 1984. GABA$_B$ receptor–mediated inhibition of serotonin release in the rat brain. Naunyn-Schmiedebergs Arch. Pharmacol. **326:** 99–105.
15. GRAY, J. A. & A. R. GREEN. 1987. GABA$_B$-receptor mediated inhibition of potassium-evoked release of endogenous 5-hydroxytryptamine from mouse frontal cortex. Br. J. Pharmacol. **91:** 517–522.
16. POTASHNER, S. J. 1979. Baclofen: effects on amino acid release and metabolism in slices of guinea-pig cerebral cortex. J. Neurochem. **32:** 103–109.
17. HILL, D. R. & N. G. BOWERY. 1981. ^{3}H-Baclofen and ^{3}H-GABA bind to bicuculline-insensitive GABA$_B$ sites in rat brain. Nature **290:** 149–152.
18. JOHNSTON, G. A. R., M. H. HAILSTONE & C. G. FREEMAN. 1980. Baclofen: stereoselective inhibition of excitant amino acid release. J. Pharm. Pharmacol. **32:** 230–231.
19. COLLINS, G. G. S., J. ANSON & E. P. KELLY. 1982. Baclofen: effects on evoked field potentials and amino acid neurotransmitter release in the rat olfactory cortex slice. Brain Res. **238:** 371–383.
20. ANDERSON, R. A. & R. MITCHELL. 1985. Evidence for GABA$_B$ autoreceptors in median eminence. Eur. J. Pharamcol. **118:** 355–358.
21. PITTALUGA, A., D. ASARO, G. PELLEGRINI & M. RAITERI. 1987. Studies on [^{3}H]GABA and endogenous GABA release in rat cerebral cortex suggest the presence of autoreceptors of GABA$_B$ type. Eur. J. Pharmacol. **144:** 45–52.
22. BONANNO, G., P. CAVAZZANI, G. C. ANDRIOLI, D. ASARO, G. PELLEGRINI & M. RAITERI. 1989. Release-regulating autoreceptors of the GABA$_B$-type in human cerebral cortex. Br. J. Pharmacol. **96:** 341–346.
23. RAITERI, M., G. BONANNO & E. FEDELE. 1989. Release of gamma-[^{3}H]aminobutyric acid (GABA) from electrically stimulated rat cortical slices and its modulation by GABA$_B$ autoreceptors. J. Pharmacol. Exp. Ther. **250:** 648–653.
24. JOHNSTON, G. A. R., A. L. STEPHANSON & B. TWITCHIN. 1976. Uptake and release of nipecotic acid by rat brain slices. J. Neurochem. **26:** 83–87.
25. YUNGER, L. M., P. J. FOWLER, P. ZAREVICS & P. E. SETLER. 1984. Novel inhibitors of γ-aminobutyric acid (GABA) uptake: anticonvulsant actions in rats and mice. J. Pharmacol. Exp. Ther. **228:** 109–115.
26. WALDMEIER, P. C., P. WICKI, J. J. FELDTRAUER & P. A. BAUMANN. 1989. Ca^{2+}-dependent release of endogenous GABA from rat cortical slices from different pools by different stimulation conditions. Naunyn-Schmiedebergs Arch. Pharmacol. **339:** 200–207.
27. WALDMEIER, P. C., P. WICKI, J. J. FELDTRAUER & P. A. BAUMANN. 1989. Release of endogenous GABA from the substantia nigra is not controlled by GABA autoreceptors. Naunyn-Schmiedebergs Arch. Pharmacol. **340:** 372–378.

28. BONANNO, G., G. PELLEGRINI, D. ASARO, G. FONTANA & M. RAITERI. 1989. GABA$_B$ autoreceptors in rat cortex synaptosomes: response under different depolarizing and ionic conditions. Eur. J. Pharmacol. **172:** 41–49.

29. BONANNO, G., G. FONTANA & M. RAITERI. 1989. Phaclofen antagonizes GABA at autoreceptors regulating release in rat cerebral cortex. Eur. J. Pharmacol. **154:** 223–224.

30. ENNIS, C. & M. C. W. MINCHIN. 1988. Modulation of the GABA autoreceptor by benzodiazepine receptor ligands. Neuropharmacology **27:** 1003–1006.

31. NEAL, M. J. & M. A. SHAH. 1989. Baclofen and phaclofen modulate GABA release from slices of rat cerebral cortex and spinal cord but not from retina. Br. J. Pharmacol. **98:** 105–112.

32. WALDMEIER, P. C., P. WICKI, J. J. FELDTRAUER & P. A. BAUMANN. 1988. Potential involvement of a baclofen-sensitive autoreceptor in the modulation of the release of endogenous GABA from rat brain slices in vitro. Naunyn-Schmiedebergs Arch. Pharmacol. **337:** 289–295.

33. KERR, D. J., J. ONG, R. H. PRAGER, B. D. GYNTHER & D. R. CURTIS. 1987. Phaclofen: a peripheral and central baclofen antagonist. Brain Res. **405:** 150–154.

34. DUTAR, P. & R. A. NICOLL. 1988. A physiological role for GABA$_B$ receptors in the central nervous system. Nature. **332:** 156–158.

35. DUTAR, P. & R. A. NICOLL. 1988. Pre- and postsynaptic GABA$_B$ receptors in the hippocampus have different pharmacological properties. Neuron **1:** 585–591.

36. BITTIGER, H., W. FRÖSTL, K. HAUSER, G. KARLSSON, K. KLEBS, H. R. OLPE, M. POZZA, E. RADEKE, M. STEINMANN, H. VAN RIEZEN & A. VASSOUT. Biochemistry, electrophysiology and pharmacology of a new GABA$_B$ antagonist. *In* GABA$_B$ Receptors in Mammalian Functions. Proceedings of the 1st International GABA$_B$ Symposium (Cambridge, UK, September 1989). N. G. Bowery, H. Bittiger & H. R. Olpe, Eds. Wiley & Sons Ltd. Chichester, England. (In press.)

37. BAUMANN, P. A., P. WICKI, C. STIERLIN & P. C. WALDMEIER. 1990. Investigations on GABA$_B$ receptor-mediated autoinhibition of GABA release. Naynyn-Schmiedebergs Arch. Pharmacol. **341:** k88–93.

38. KARLSSON, G. & H. R. OLPE. 1989. Late inhibitory postsynaptic potentials in rat prefrontal cortex may be mediated by GABA$_B$ receptors. Experientia **45:** 157–158.

39. KARLSSON, G., M. POZZA & H. R. OLPE. 1988. Phaclofen: a GABA$_B$ blocker reduces long-duration inhibition in the neocortex. Eur. J. Pharmacol. **148:** 485–486.

40. BOWERY, N. 1989. GABA$_B$ receptors and their significance in mammalian pharmacology. TIPS **10:** 401–407.

41. HARRISON, N. L., C. D. LANGE & J. L. BARKER. 1988. (−)-Baclofen activates presynaptic GABA$_B$ receptors on GABAergic inhibitory neurons from embryonic rat hippocampus. Neurosci. Lett. **85:** 105–109.

42. POZZA, M., G. KARLSSON, F. BRUGGER & H. R. OLPE. Effects of GABA$_A$ and GABA$_B$ blockers on paired-pulse inhibition in the hippocampus CA1 region. In GABA$_B$ Receptors in Mammalian Functions. Proceedings of the 1st International GABA$_B$ Symposium (Cambridge, UK, September 1989). N. G. Bowery, H. Bittiger & H. R. Olpe, Eds. Wiley & Sons Ltd. Chichester, England. (In press.)

43. DEISZ, R. A. & D. A. PRINCE. 1989. Frequency-dependent depression of inhibition in guinea-pig neocortex in vitro by GABA$_B$ receptor feedback on GABA release. J. Physiol. London **412:** 513–541.

44. DEISZ, R. A. & W. ZIEGLGÄNSBERGER. GABA$_B$ receptor-mediated frequency dependence of inhibitory postsynaptic potentials of neocortical neurones. In GABA$_B$ Receptors in Mammalian Functions. Proceedings of the 1st International GABA$_B$ Symposium (Cambridge, UK, September 1989). N. G. Bowery, H. Bittiger & H. R. Olpe, Eds. Wiley & Sons Ltd. Chichester, England. (In press.)

45. FLORAN, B., I. SILVA, C. NAVA & J. ACEVES. 1988. Presynaptic modulation of the release of GABA by GABA$_A$ receptors in pars compacta and by GABA$_B$ receptors in pars reticulata of the rat substantia nigra. Eur. J. Pharmacol. **150:** 277–286.

46. GIRALT, M. T., G. BONANNO & M. RAITERI. 1989. GABA terminal autoreceptors in pars compacta and in pars reticulata of the rat substantia nigra are both GABA$_B$. Eur. J. Pharmacol. **175:** 137–144.
47. WALDMEIER, P. C., P. WICKI, J. J. FELDTRAUER & P. A. BAUMANN. 1988. The measurement of the release of endogenous GABA from rat brain slices by liquid chromatography with electrochemical detection. Naunyn-Schmiedebergs Arch. Pharmacol. **337:** 284–288.
48. YAMAMOTO, T., C. NANJOH & I. KURUMA. 1985. Determination of endogenous GABA released from the cerebral cortex slices of the rat by high-performance liquid chromatography with a series-dual electrochemical detector. Neurochem. Int. **7:** 77–82.

Changes in Cardiovascular Responses of Conscious Rats to Endogenous Opioids following Treatment with Catecholamine-Depleting Agents[a]

WALTER R. DIXON AND ANDI PIANG-LING CHANG

Department of Pharmacology and Toxicology
School of Pharmacy
University of Kansas
Lawrence, Kansas 66045

INTRODUCTION

In many peripheral adrenergic neuronal tissues, activation of prejunctional opioid receptors results in inhibition of nerve stimulation–mediated release of norepinephrine (NE) in the perfused cat spleen,[1,2] superfused rabbit ear artery,[3,4] superfused rabbit ileocolic artery,[5] perfused rabbit heart,[6] perfused rat heart,[7] and superfused guinea pig atria.[8] Moreover, in the isolated perfused rat adrenal gland prelabeled with tritiated NE, Wakade *et al.* found that methionine-enkephalin almost completely inhibited the evoked secretion of tritium but had little effect on the secretion of catecholamines.[9] The authors concluded that chromaffin cells do not possess the NE uptake mechanism and that the uptake of tritiated NE occurs mainly in sympathetic nerve terminals present in the adrenal gland and the surrounding blood vessels.

Dopamine-B-hydroxylase (DBH) and enkephalinlike immunoreactivity have been found in sympathetic ganglion cells.[10] Evidence also supports the coexistence of opiatelike peptides and NE in adrenergic vesicles of the bovine splenic nerve[11] and methionine-enkephalin and epinephrine (Ep) and NE in chromaffin granules of the adrenal medulla of every species studied.[12] Viveros *et al.* have provided direct evidence for the release of enkephalins from peripheral nervous tissue.[13] Using the isolated perfused dog adrenal gland, they found that methionine-enkephalin and leucine-enkephalin were released along with adrenal catecholamines in the same ratio as they are found in chromaffin granules following the injection of acetylcholine. Although many different investigators have found that opioid peptides inhibit nerve-evoked release of neurotransmitter from isolated perfused organs, the cardiovascular responses of the intact animal to the peptides varies markedly depending on the mode of administration and the anesthetic state of the animal.

Szabo *et al.* found that intravenous administration of dynorphin (1-13), leucine-enkephalin, and D-ala^{2}D-leu^5-enkephalin diminished the electrically evoked increase in plasma NE and mean arterial pressure in anesthetized rabbits in which the spinal cord was destroyed.[14] The effects were antagonized by naloxone. In anesthetized rats, leucine-enkephalin given intravenously, intracisternally or intraventricularly produced a pressor response.[15,16] However, intracisternal or intravenous administration of methionine-enkephalin in anesthetized rats produced a depressor response.[15,16] Naloxone blocked the pressor or depressor response to leucine- or methionine-enkephalin.[16] In

[a]This work was supported by the National Institute on Drug Abuse grant number DA03504.

anesthetized rats, dynorphin (1-13) given intravenously, intracerebroventricularly, or intracisternally produced hypotension and bradycardia.[17,18] The hypotension and bradycardia produced by intravenous administration of dynorphin was also observed in pithed rats[17,19] and was not blocked by naloxone.[19]

In awake freely moving rats, intravenous injection of methionine-enkephalin produced an increase in blood pressure and heart rate.[20,21] However, intravenous injection of d-ala^2-met^5-enkephalinamide, which is longer acting than methionine-enkephalin, produced a decrease in blood pressure and heart rate in conscious rats.[22] Intracerebroventricular injection of dynorphin (1-13) in conscious rats produced a pressor response and tachycardia.[23,24] The pressor response and tachycardia produced by dynorphin (1-13) were inhibited by intracerebroventricular injection of naloxone, phentolamine, and prazocin but not by yohimbine.[24] Recently, Thornhill *et al.* found that intravenous injection of dynorphin (1-13) caused a transient increase in mean blood pressure and a decrease in heart rate in conscious, unrestrained rats.[25] Pretreatment with naloxone, yohimbine, and prazocin at a dose of 4.2 mumol/kg blocked the pressor responses to selective doses of dynorphin. Naloxone blocked responses up to 20 nmol/kg of dynorphin; yohimbine up to 60 nmol/kg; and prazocin only the lower doses of dynorphin.

In this study, we have used conscious, unrestrained rats and compared the cardiovascular changes (blood pressure, heart rate, and plasma catecholamines) following intravenous administration of d-ala^2 met^5-enkephalinamide (delta receptor agonist) and dynorphin (1-13) (a kappa receptor agonist) in control, reserpine-treated, and 6-hydroxydopamine-treated rats. Administration of opioid agonists to the intact animal may elicit a number of complex drug responses leading to cardiovascular changes measured. Therefore, we have used reserpine (which depletes catecholamine stores in the central and peripheral nervous system) and 6-hydroxydopamine (which destroys sympathetic nerve endings) to minimize the contribution of neurotransmitter interactions with the opioids in the cardiovascular responses elicited by the opioid agonists and thereby unmask effects of the opioids attributable to direct versus indirect effects resulting from interactions with catecholamines presynaptically or postsynaptically.

METHODS

Male Sprague-Dawley rats weighing between 250 and 350 g were used in this study. Under ether anesthesia, the femoral artery and vein were catheterized with PE-50 tubing so that blood pressure and heart rate could be measured and drugs injected. The PE-50 tubing was guided through the subcutaneous fat on the back of each test animal and brought out through the neck. A heparin solution (285 U/ml) was injected into the tubing to prevent blood clotting. The PE-50 tubing was then connected to a 21-gauge needle linked to a three-way stopcock fastened to a transducer that was in turn connected to a Model 7 Grass polygraph, which was used to record blood pressure and heart rate. The animals were injected with vehicle (pyrogen-free 0.9% normal saline) in the femoral vein to obtain any responses due to vehicle alone. The drug solution was prepared in pyrogen-free glassware. The volume injected was based on body weight (1 ml/kg).

Reserpinized Rats

Rats were given reserpine (Rauserpine, 1 mg/ml, Taipei, China), 1 mg/kg intraperitoneally (ip) every 24 hours for 3 days (a total of 3 injections).

6-Hydroxydopamine (6-OHDA) Treated Rats

Rats were given 6-OHDA intravenously for 5 days starting with 50 mg/kg, followed by 100, 25, 50, and 25 mg/kg as described by Wakade *et al.*[9] Rats were used two days after the last dose of 6-OHDA.

Blood Samples

Blood samples were obtained from the femoral artery in aliquots of 1 ml. Blood removed was replaced by normal saline. Catecholamines were extracted as follows: blood samples were centrifuged at 800 rpm at 4°C for 10 minutes and the plasma separated from the blood cells. An equal volume of 0.4 N $HCLO_4$ was added to the plasma to precipitate protein and then centrifuged at 10,000 rpm at 4°C for 5 minutes. One milliliter of Tris buffer (2 M, 5% EDTA) was then added to each sample along with 50 mg alumina. One nanogram of dihydroxybenzylamine (DHBA) was also added as an internal standard for calculating the % recovery. Each preparation was then shaken for 10–15 minutes and then washed twice with 1 ml deionized water. To each sample, 200 μl 0.1 N $HCLO_4$ were added and then centrifuged at 3000 rpm at 4°C for 5 minutes, filtered, and frozen pending analysis via high pressure liquid chromatography with electrochemical detection (HPLC-EC) according to the procedure of Gaddis and Dixon.[26]

Drugs

The following drugs were obtained from Sigma Chemical Co., St. Louis, Mo.: Naloxone hydrochloride, propranolol hydrochloride, atropine hydrochloride, mecamylamine hydrochloride, 6-hydroxydopamine hydrobromide, dynorphin (1-13) acetate and d-ala^2-met^5-enkephalinamide acetate. Phentolamine mesylate was donated by Ciba-Geigy Corp., Summit, N.J. Diprenorphine hydrochloride was obtained from Research Technology Branch, Division of Preclinical Research, National Institute on Drug Abuse.

Statistics

Changes in blood pressure and heart rate were tested for significance using a one-way analysis of variance (ANOVA) and the Student-Newman-Keuls test for the multiple comparison of means.[27] Plasma catecholamine differences were compared using a Student's t-test.[27]

RESULTS

Effects of Intravenous Injections of Dynorphin (1-13) (Dyn) or d-ala^2-met^5 Enkephalinamide (DAME) on Blood Pressure (BP), Heart Rate (HR), and Plasma Catecholamine Levels in Conscious, Unrestrained Rats

Intravenous injection of Dyn (0.05, 0.1, 0.2, and 0.4 mg/kg) caused a dose-related increase in BP and duration of the pressor response. The HR decreased. Intravenous injection of DAME (0.5, 1, 2, and 4 mg/kg) also caused a dose-related increase in BP and duration of the pressor response. The HR initially fell precipitously followed by a rapid recovery (TABLE 1). There was no change in plasma catecholamine levels.

Effect of Naloxone on BP and HR Response to Intravenous Injections of Dyn and DAME

Naloxone [5 mg/kg, intravenously (iv)] had no effect on BP but produced a decrease in HR. When given 10 minutes prior to DAME, naloxone antagonized the actions of DAME and led to a decrease in the pressor response and bradycardia seen

TABLE 1. Effect of Dynorphin and d-ala^2-met^5-Enkephalinamide (DAME) on Blood Pressure (BP) and Heart Rate (HR) in Conscious, Unrestrained Rats[a]

Drugs	n	Dose (mg/kg)	BP (mmHg)	HR (beats/minute)	DR (minutes)
Saline	4	1 ml/kg	107 ± 2	315 ± 5	—
Dynorphin	4	0.05	122 ± 3[c]	268 ± 22[b]	6.75 ± 1.70
Saline	4	1 ml/kg	103 ± 2	318 ± 3	—
Dynorphin	4	0.1	134 ± 3[c]	220 ± 9[b]	16.50 ± 1.32
Saline	4	1 ml/kg	103 ± 1	316 ± 2	—
Dynorphin	4	0.2	141 ± 2[c]	233 ± 18[b]	18.75 ± 1.38
Saline	4	1 ml/kg	101 ± 1	320 ± 2	—
Dynorphin	4	0.4	153 ± 5[c]	225 ± 12[b]	20.00 ± 0.82
Saline	6	1 ml/kg	108 ± 3	323 ± 6	—
DAME	6	0.5	139 ± 3[c]	165 ± 15[b]	0.98 ± 0.14
Saline	6	1 ml/kg	104 ± 2	323 ± 9	—
DAME	6	1.0	146 ± 5[c]	150 ± 16[b]	2.25 ± 0.42
Saline	6	1 ml/kg	109 ± 2	327 ± 9	—
DAME	6	2.0	141 ± 4	146 ± 14[b]	1.90 ± 0.72
Saline	6	1 ml/kg	105 ± 5	343 ± 8	—
DAME	6	4.0	156 ± 5[c]	135 ± 15[b]	9.17 ± 0.79

[a] n = number of animals tested. BP = mean arterial blood pressure. DR = duration time of blood pressure increase. [b] $p < 0.05$, [c] $p < 0.01$, represents statistical significance compared to saline injection.

with DAME alone (FIGURES 1 and 2). However, naloxone had no effect on the BP increase or bradycardia produced by Dyn.

Effect of Diprenorphine on BP and HR Response to Intravenous Injections of Dyn and DAME

Diprenorphine (5mg/kg, iv) produced a significant but transient increase in BP and decrease in HR. When given 10 minutes prior to DAME (1 mg/kg, iv), diprenorphine antagonized the actions of DAME and led to a decrease in the pressor response and bradycardia seen with DAME alone. However, diprenorphine had no effect on the BP increase or bradycardia produced by Dyn (TABLE 2a).

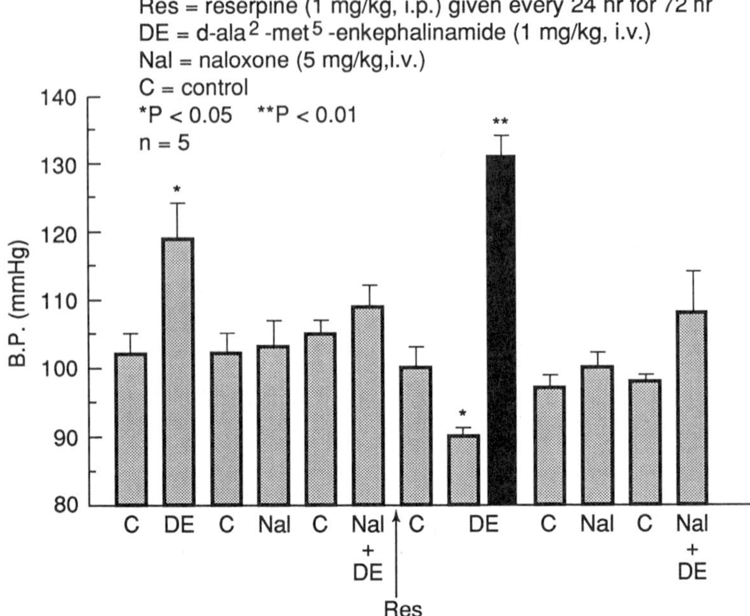

FIGURE 1. Effect of naloxone on the BP response of control and reserpine-treated rats following d-ala^2-met^5-enkephalinamide injection. Naloxone was given 10 minutes prior to d-ala^2-met^5-enkephalinamide. The solid bar represents the increase in BP during the biphasic BP response. *$p < 0.05$, **$p < 0.01$ represents statistical significance compared to control.

Effects of Phentolamine, Propranolol, Atropine, and Mecamylamine on BP and HR Responses to Intravenous Injection of Dyn

Phentolamine (0.8 mg/kg, iv) produced a significant decrease in BP and increase in HR. Injection of phentolamine 10 minutes prior to Dyn (0.1 mg/kg, iv) did not alter the increase in BP and decrease in HR produced by Dyn alone (TABLE 2a). Propranolol (1 mg/kg, iv) and mecamylamine (5 mg/kg, iv) produced significant decreases in BP but did not affect the HR. When given prior to Dyn, propranolol and mecamylamine were also ineffective in antagonizing the BP and HR response to Dyn. Atropine (1.5 mg/kg, iv) did not affect the BP but produced a significant increase in the HR. When given prior to Dyn, atropine did not antagonize the BP and HR response to Dyn (TABLE 2b).

Depletion of Endogenous Catecholamines by Reserpine

Intraperitoneal (ip) injection of reserpine (1 mg/kg) every 24 hours for 3 days resulted in significant depeletion of norepinephrine (NE) in heart (0.022 ± 0.004 μg NE/g versus 1.32 ± 0.12 μg NE/g in controls) and whole brain (0.044 ± 0.005 μg NE/g versus 0.376 ± 0.007 μg NE/g in controls) and NE and epinephrine (Ep) in the adrenal gland (4.20 ± 0.40 μg NE/gland pair versus 8.51 ± 0.79 μg NE/gland pair in controls and 6.41 ± 0.71 μg Ep/gland pair versus 32.10 ± 1.66 μg Ep/gland pair in controls).

BP and HR Changes during Reserpine Treatment

The animals were given 3 doses of reserpine spaced 24 hours apart. The first dose of reserpine (1 mg/kg, ip) produced a maximum decrease in BP (72 ± 4 mmHg) within 6 hours. The BP returned to control levels (102 ± 5 mmHg) 24 hours after the injection. The second and third injections of reserpine (1 mg/kg, ip) did not change the BP (FIGURE 3). In contrast, the HR (340 ± 15 beats/minute) was decreased to a maximum extent (295 ± 13 beats/minute) 24 hours after the first injection of reserpine and remained at this level with a further reduction (263 ± 9 beats/minute) after the second and third injections of reserpine (FIGURE 4).

BP and HR Responses to Intravenous Injection of Dyn or DAME in Rats Chronically Treated with Reserpine

Dyn (0.1 mg/kg, iv) produced a BP increase that was significantly greater than the BP response to Dyn in control animals (FIGURE 5). The HR which decreased in control animals following Dyn injection exhibits a biphasic response, an initial increase followed by a decrease (FIGURE 6). In contrast, the BP response to intravenous injection of DAME was converted from a pressor response to a biphasic BP response, an initial decrease followed by an increase (FIGURE 7). However, the duration of the

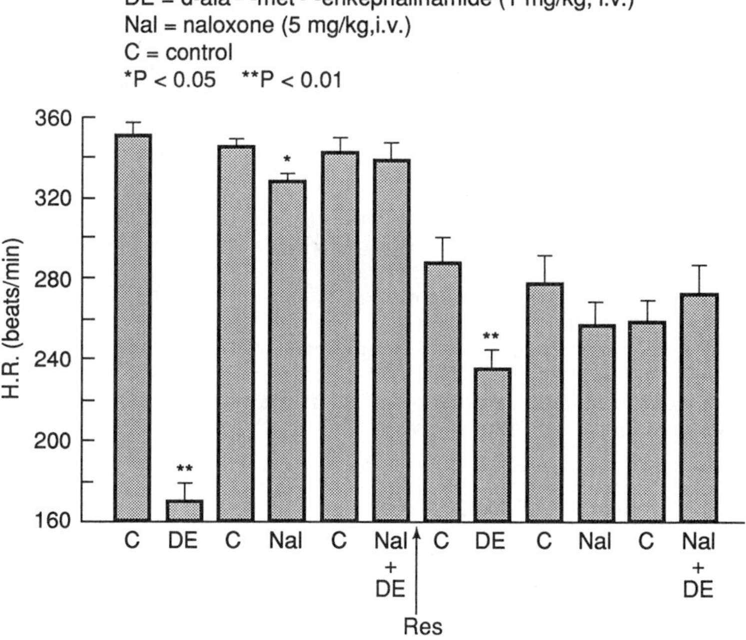

FIGURE 2. Effect of naloxone on the HR response of control and reserpine-treated rats following d-ala^2-met^5-enkephalinamide injection. Naloxone was given 10 minutes prior to d-ala^2-met^5-enkephalinamide. *$p < 0.05$, **$p < 0.01$ represents statistical significance compared to control.

TABLE 2A. Effect of Dynorphin on Blood Pressure (BP) and Heart Rate (HR) Responses in Conscious, Unrestrained Rats Pretreated with Diprenorphine or Phentolamine[a]

Drugs	Dose (mg/kg)	BP (mmHg)	HR (beats/minute)	DR (minutes)
Control	—	103 ± 2	327 ± 10	—
Dynorphin	0.1	138 ± 4^c	279 ± 13^c	18.75 ± 2.75
Control	—	105 ± 2	318 ± 8	—
Diprenorphine	5.0	129 ± 2^c	294 ± 11^b	0.60 ± 0.08
Control	—	104 ± 3	318 ± 6	—
Diprenorphine	5.0			
+ dynorphin	0.1	141 ± 2^c	254 ± 9^c	19.25 ± 1.49
Control	—	102 ± 2	328 ± 5	—
Dynorphin	0.1	141 ± 5^c	298 ± 3^c	15.50 ± 1.65
Control	—	105 ± 2	330 ± 8	—
Phentolamine	0.8	91 ± 3^c	454 ± 6^c	2.70 ± 0.50
Control	—	101 ± 4	365 ± 3	—
Phentolamine	0.8			
+ dynorphin	0.1	141 ± 4^c	303 ± 5^c	13.00 ± 2.00

[a]Number of animals tested = 6. BP = mean arterial blood pressure. DR = duration time of blood pressure increase or decrease. [b]$p < 0.05$, [c]$p < 0.01$, represents statistical significance compared to control by ANOVA.

pressor response was significantly longer than the pressure response in controls. The HR decreased in both control and reserpine-treated rats.

Effect of Naloxone on BP and HR Responses to Intravenous Injections of Dyn or DAME in Rats Chronically Treated with Reserpine

Naloxone (5 mg/kg, iv), given alone, did not change the BP but the HR decreased. Injection of naloxone 10 minutes prior to Dyn had no effect on the pressor response or biphasic HR response produced by Dyn. However, injection of naloxone 10 minutes prior to DAME converted the biphasic BP response seen after DAME injection to a pressor response of reduced amplitude and completely prevented the decrease in HR (FIGURES 1 and 2).

Effect of Atropine on BP and HR Responses to Dyn or DAME in Control Rats

Atropine (1.5 mg/kg, iv) did not change the BP. However, the HR increased from 335 ± 5 beats/minute to 465 ± 9 beats/minute. Injection of atropine 10 minutes prior to DAME led to a potentiation of the BP response (an increase from 101 ± 1 mmHg to 118 ± 2 mmHg after DAME alone compared to an increase from 106 ± 2 mmHg to 151 ± 2 mmHg after DAME injection in rats pretreated with atropine) (FIGURE 8). In contrast, the HR decrease normally seen after DAME injection was blocked (a decrease from 338 ± 5 beats/minute to 149 ± 11 beats/minute after DAME alone compared to comparable levels in rats pretreated with atropine (465 ± 9 beats/minute) and then given DAME (460 ± 4 beats/minute) (FIGURE 9).

BP and HR Responses of 6-Hydroxydopamine (6-OHDA) Treated Rats to Intravenous Injection of Dyn or DAME

Rats were given intraperitoneal injections of 6-OHDA for 5 days to destroy sympathetic nerve terminals. The effectiveness of the treatment was judged by measuring norepinephrine (NE) content of the heart and salivary gland which were depleted over 90 ± 7%. The BP and HR of control rats were 104 ± 2 mmHg and 342 ± 14 beats/minute, respectively. Three days after the last dose of 6-OHDA, the BP (98 ± 3 mmHg) and the HR (323 ± 10 beats/minute) were within control levels. At this time, Dyn (0.1 mg/kg, iv) produced an increase in Bp to 142 ± 3 mmHg and a decrease in HR to 254 ± 8 beats/minute which is comparable to the increase in BP and decrease

TABLE 2B. Effect of Dynorphin on Blood Pressure (BP) and Heart Rate (HR) Responses in Conscious, Unrestrained Rats Pretreated with Propranolol, Atropine, or Mecamylamine[a]

Drugs	Dose (mg/kg)	BP (mmHg)	HR (beats/minute)	DR (minutes)
Control	—	130 ± 2	343 ± 9	—
Dynorphin	0.1	133 ± 2[c]	295 ± 9[c]	14.25 ± 1.10
Control	—	104 ± 2	335 ± 5	—
Propranolol	1.0	95 ± 4[b]	326 ± 6	1.50 ± 0.64
Control	—	101 ± 2	331 ± 7	—
Propranolol	1.0			
+ dynorphin	0.1	133 ± 4[c]	293 ± 9[c]	14.00 ± 0.91
Control	—	103 ± 2	325 ± 10	—
Dynorphin	0.1	139 ± 2[c]	285 ± 9[c]	13.25 ± 0.85
Control	—	102 ± 1	325 ± 6	—
Atropine	1.5	109 ± 1	408 ± 5[c]	1.20 ± 0.27
Control	—	103 ± 2	443 ± 8	—
Atropine	1.5			
+ dynorphin	0.1	140 ± 4[c]	295 ± 9[c]	14.70 ± 0.85
Control	—	103 ± 2	323 ± 6	—
Dynorphin	0.1	136 ± 2[c]	280 ± 7[c]	13.75 ± 1.10
Control	—	100 ± 3	310 ± 8	—
Mecamylamine	5.0	68 ± 6[c]	320 ± 10	2.70 ± 0.50
Control	—	73 ± 4	318 ± 8	—
Mecamylamine	5.0			
+ dynorphin	0.1	142 ± 6[c]	309 ± 6	17.00 ± 2.65

[a]Number of animals tested = 6. BP = mean arterial blood pressure. DR = duration time of blood pressure increase or decrease. [b]$p < 0.05$, [c]$p < 0.01$, represents statistical significance compared to control by ANOVA.

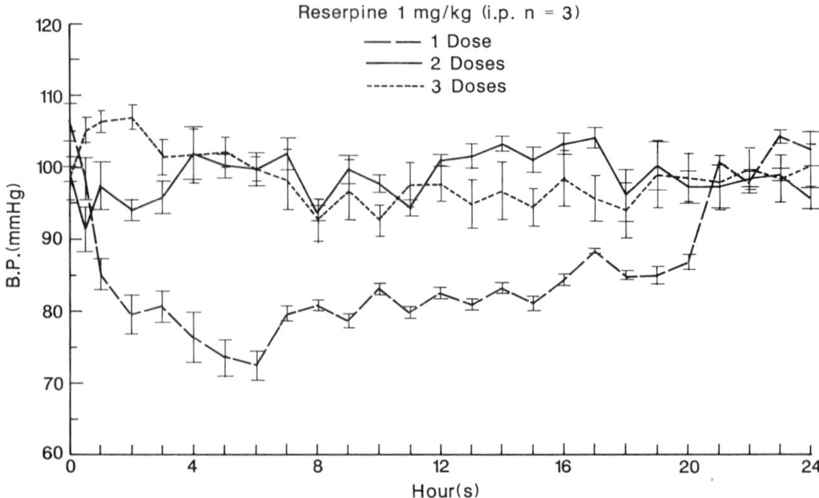

FIGURE 3. Effect of reserpine on BP in conscious unrestrained rats. Reserpine (1 mg/kg, ip) was given every 24 hours for 3 days—a total of 3 injections. The BP was recorded continuously via a cannula in the femoral artery.

in HR produced by Dyn in control rats. In contrast, the BP response to DAME (1 mg/kg, iv) was markedly altered in 6-OHDA treated rats. The pressor response seen in control rats was converted to a biphasic BP response, an initial decrease to 63 ± 3 mmHg followed by an increase to 110 ± 2 mmHg. However, the HR was decreased throughout the biphasic BP response which was similar to the effect of DAME on HR in control rats (TABLE 3).

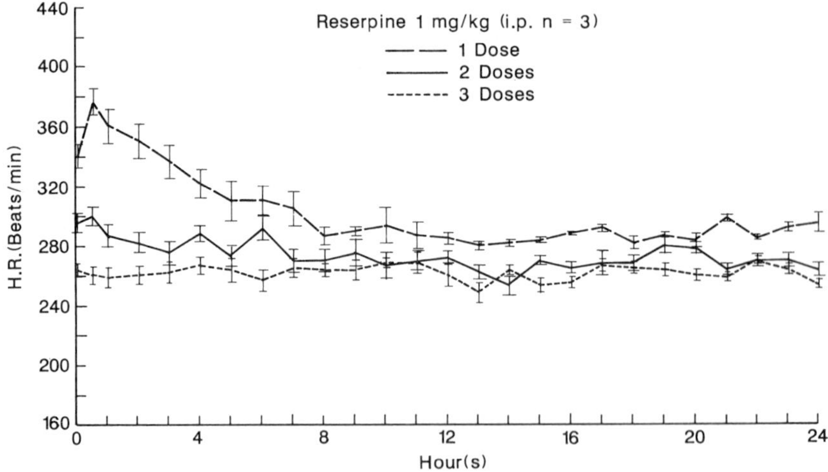

FIGURE 4. Effect of reserpine on HR in conscious unrestrained rats. Reserpine (1 mg/kg, ip) was given every 24 hours for 3 days—a total of 3 injections. The HR was recorded continuously.

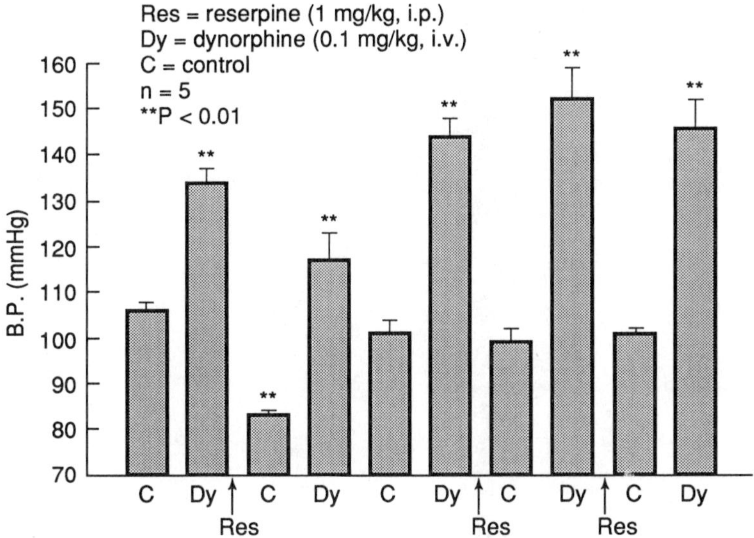

FIGURE 5. BP response to dynorphin during repeated administration of reserpine to conscious unrestrained rats. Reserpine (1 mg/kg, ip) was given every 24 hours for 3 days—a total of 3 injections. **$p < 0.01$ represents statistical significance compared to control.

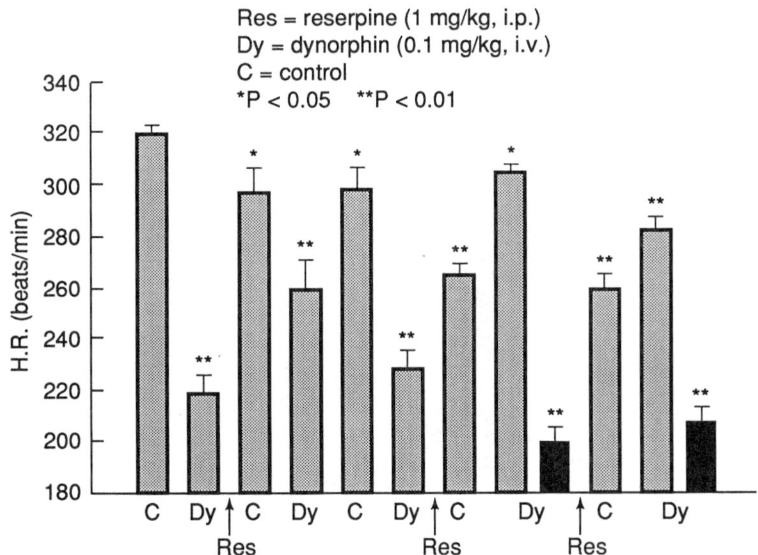

FIGURE 6. HR response to dynorphin during repeated administration of reserpine to conscious unrestrained rats. Reserpine (1 mg/kg, ip) was given every 24 hours for 3 days—a total of 3 injections. *$p < 0.05$, **$p < 0.01$ represents statistical significance compared to control.

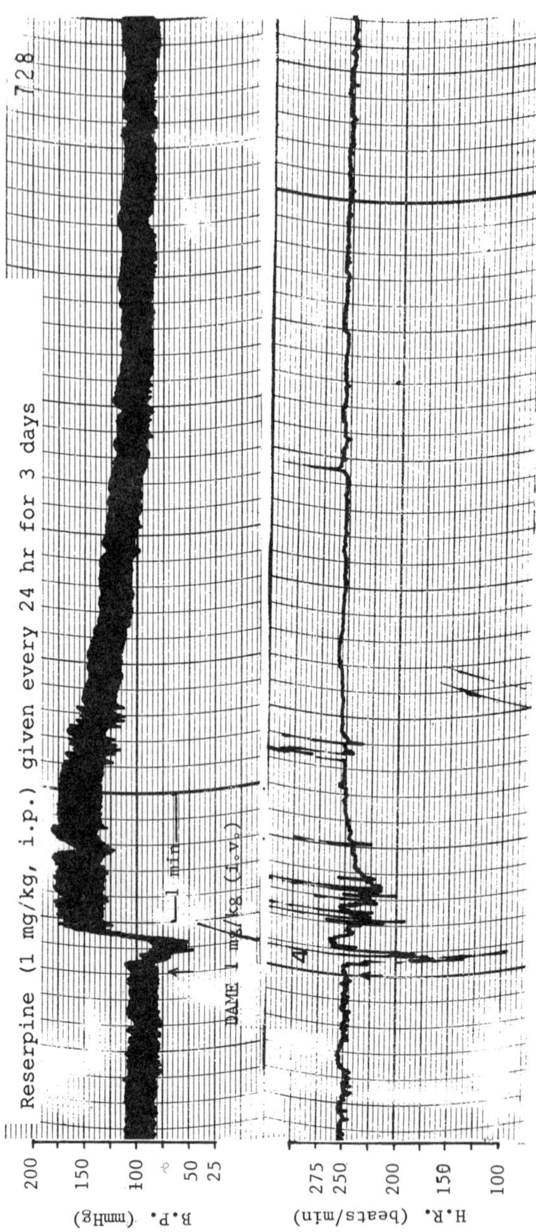

FIGURE 7. Tracing of BP and HR response of reserpine-treated conscious unrestrained rats to intravenous injection of d-ala^2-met^5-enkephalinamide (DAME). Reserpine (1 mg/kg, ip) was given every 24 hours for 3 days—a total of 3 injections.

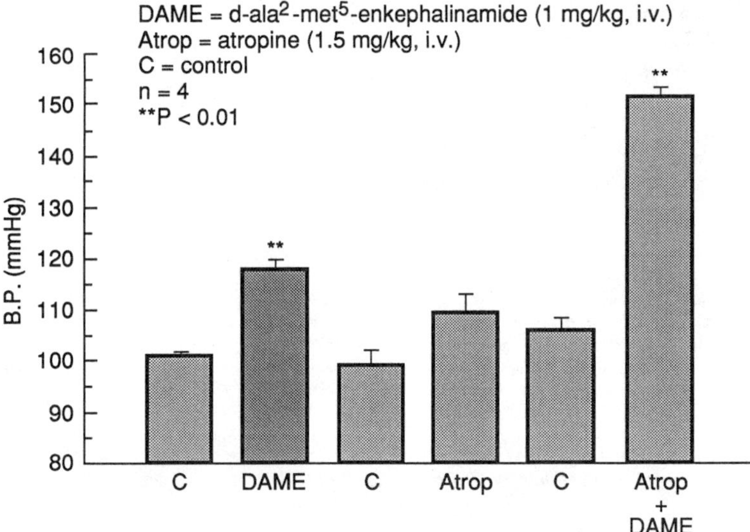

FIGURE 8. Effect of atropine on the BP response of conscious unrestrained rats following d-ala^2-met^5-enkephalinamide injection. Atropine was given 10 minutes prior to d-ala^2-met^5-enkephalinamide. **p < 0.01 represents statistical significance compared to control.

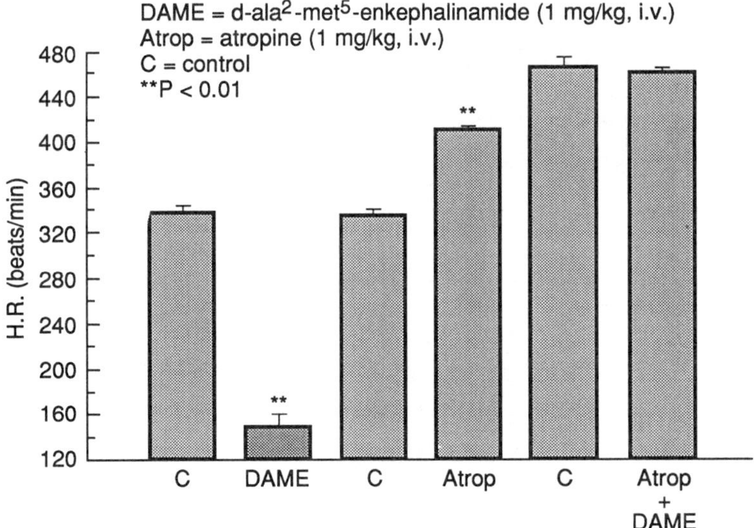

FIGURE 9. Effect of atropine on the HR response of conscious unrestrained rats following d-ala^2-met^5-enkephalinamide injection. Atropine was given 10 minutes prior to d-ala^2-met^5-enkephalinamide. **p < 0.01 represents statistical significance compared to control.

TABLE 3. Effect of d-ala[2]-met[5]-Enkephalinamide (DAME) on Blood Pressure (BP) and Heart Rate (HR) in Conscious, Unrestrained 6-Hydroxydopamine (6-OHDA) Treated Rats[a]

Drugs	Time	BP (mmHg)	HR (beats/minute)	DR (minute)
Control	0	97 ± 3	347 ± 18	—
6-OHDA	192 hours	94 ± 2	320 ± 13	—
DAME	—	63 ± 3[b]	184 ± 20[b]	—
	—	110 ± 2	—	0.82 ± 0.08

[a]Rats were treated with 6-hydroxydopamine (6-OHDA) intravenously for 5 days with a different dose for each day (50, 100, 25, 50, 25 mg/kg). Three days after the last dose of 6-OHDA, the rats were given DAME (1 mg/kg, iv). Number of animals tested = 5. Time represents the time from initial dose of 6-OHDA. BP = mean arterial blood pressure. DR = duration of BP increase or decrease. [b]$p < 0.01$, represents statistical significance compared to control.

Effect of Dyn and DAME on Plasma Catecholamine Levels in Conscious, Unrestrained Control and 6-OHDA-Treated Rats

Plasma norepinephrine (NE), epinephrine (Ep), and dopamine (DA) levels of control rats were 0.202 ± 0.019 ng/ml, 0.180 ± 0.030 ng/ml, and 0.026 ± 0.016 ng/ml. Injection of Dyn (0.1 mg/kg, iv) or DAME (1 mg/kg, iv) did not change plasma catecholamine levels. Following 6-OHDA treatment, plasma catecholamine levels were unchanged from control levels. Injection of Dyn (0.1 mg/kg, iv) did not change plasma catecholamine levels in 6-OHDA-treated rats. DAME (1 mg/kg, iv) did not change plasma NE or DA levels in 6-OHDA-treated rats. However, a significant increase in plasma Ep levels was seen at 45 seconds (0.600 ± 0.107 ng/ml) and 5 minutes (0.584 ± 0.093 ng/ml) after injection of DAME which corresponds to the time of maximum decrease and increase in the biphasic BP response to DAME (TABLE 4).

Effect of Verapamil on BP and HR Response to Dyn and DAME

Verapamil, 2 mg/kg, iv, produced a transient decrease in Bp from 101 ± 2 mmHg to 38 ± 2 mmHg (duration 3.18 ± 0.28 minutes) and an increase in HR from 368 ± 7

TABLE 4. Effect of d-ala[2]-met[5]-Enkephalinamide (DAME) on Plasma Catecholamine Levels in Conscious, Unrestrained 6-Hydroxydopamine (6-OHDA) Treated Rats[a]

Drugs	Time	NE (ng/ml)	Epi (ng/ml)	DA (ng/ml)
Control	0	0.218 ± 0.034	0.200 ± 0.023	0.008 ± 0.004
6-OHDA	191 hours	0.213 ± 0.033	0.201 ± 0.027	0.007 ± 0.003
DAME	192 hours			
	45 seconds later	0.228 ± 0.035	0.600 ± 0.107[c]	0.004 ± 0.004
	5 minutes later	0.278 ± 0.032	0.584 ± 0.093[b]	0.005 ± 0.003

[a]Rats were treated with 6-hydroxydopamine (6-OHDA) intravenously for 5 days with a different dose for each day (50, 100, 25, 50, 25 mg/kg). Three days after the last dose of 6-OHDA, the rats were given DAME (1 mg/kg, iv). Number of animals tested = 5. Time represents the time from initial dose of 6-OHDA. NE = norepinephrine. Epi = epinephrine. DA = dopamine. [b]$p < 0.05$, [c]$p < 0.01$, represents statistical significance compared to control.

TABLE 5. Effect of Dynorphin on Blood Pressure (BP) and Heart Rate (HR) Responses in Conscious, Unrestrained Rats Pretreated with Verapamil[a]

Drugs	Dose (mg/kg)	BP (mmHg)	HR (beats/minute)	DR (minutes)
Control	—	102 ± 1	382 ± 3	—
Dynorphin	0.1	128 ± 2[b]	322 ± 6[b]	12.50 ± 1.80
Control	—	101 ± 2	368 ± 7	—
Verapamil	2	38 ± 2[b]	425 ± 8[b]	3.18 ± 0.28
● 5 minutes later	—	88 ± 1	408 ± 3	—
+ dynorphin	0.1	92 ± 2	348 ± 6	—

[a]Dynorphin (0.1 mg/kg) and verapamil (2 mg/kg) were given intravenously. Number of animals tested = 6. BP = mean arterial blood pressure. DR = duration time of blood pressure increase or decrease. [b]$p < 0.01$, represents statistical significance compared to control by ANOVA.

beats/minute to 425 ± 8 beats/minute. Five minutes later, the BP was 88 ± 1 mmHg and the HR 408 ± 3 beats/minute. At this time, the pressor response normally seen with intravenous injections of Dyn was abolished. The BP response was essentially unchanged (92 ± 2 mmHg versus 88 ± 1 mmHg in controls). However, verapamil did not prevent the decrease in HR. Dyn injection led to a decrease in HR to 348 ± 6 beats/minute compared to a control level of 408 ± 3 beats/minute after verapamil injection (TABLE 5). However, it should be emphasized that verapamil injection led to an increase in HR that persisted up to the time that Dyn was injected. Therefore, Dyn reduced the HR to levels normally found in control rats. In contrast, verapamil had a different effect on the BP and HR response to DAME (1 mg/kg, iv). DAME was able to produce a BP increase from 86 ± 2 mmHg to 105 ± 3 mmHg and an HR decrease from 368 ± 7 beats/minute to 150 ± 9 beats/minute in rats pretreated with verapamil (TABLE 6).

TABLE 6. Effect of d-ala^2-met^5-Enkephalinamide (DAME) on Blood Pressure (BP) and Heart Rate (HR) Responses in Conscious, Unrestrained Rats Pretreated with Verapamil[a]

Drugs	Dose (mg/kg)	BP (mmHg)	HR (beats/minute)	DR (minutes)
Control	—	102 ± 3	351 ± 10	—
DAME	1	137 ± 4[b]	174 ± 13[b]	0.76 ± 0.10
Control	—	98 ± 3	332 ± 6	—
Verapamil	2	46 ± 4[b]	387 ± 16[b]	1.48 ± 0.05
● 5 minutes later	—	86 ± 2	368 ± 7	—
+ DAME	1	105 ± 3[c]	150 ± 9[c]	0.52 ± 0.08

[a]DAME (1 mg/kg) and verapamil (2 mg/kg) were given intravenously. Number of animals tested = 5. BP = mean arterial blood pressure. DR = duration time of blood pressure increase or decrease. [b]$p < 0.01$, represents statistical significance compared to control by ANOVA. [c]$p < 0.01$, represents statistical significance (by ANOVA, one-way analysis) compared to 5 minutes after verapamil treatment.

DISCUSSION

We have compared the cardiovascular responses to intravenous administration of DAME and Dyn in control animals and animals treated with catecholamine-depleting agents. Although both opioids produced a blood pressure increase and a heart rate decrease, Dyn caused comparable changes to DAME in the BP and HR of control rats at 1/10th lower dose. The significance of this difference is not clear but may reside in differences in susceptibility to metabolism within the body. Plasma catecholamines do not change following injection of the peptides suggesting that catecholamine release from the adrenal gland is not involved.

In reserpine-treated rats, the BP response to DAME was markedly changed. The pressor response seen in control rats was converted to a biphasic response; an initial decrease followed by an increase. In addition, the duration of the pressor response was prolonged. Normally, the increase in BP in control rats is of short duration and suggests that an interaction involving catecholamines is involved in termination of DAME's cardiovascular effect. The initial decrease in BP seen after DAME injection in reserpine-treated rats is transitory in nature and may be a result of acetylcholine (ACh) release from vagal nerve endings leading to a decrease in HR and cardiac output unopposed by sympathetic nerve activity. This interpretation is supported by the finding that pretreatment with atropine prevents the bradycardia seen after DAME in control as well as reserpine-treated rats. Naloxone also prevented the bradycardia seen after DAME showing that DAME initially interacts with opiate receptors on vagal nerve endings, thereby increasing ACh release which acts on muscarinic receptors in the heart. Therefore, in reserpine-treated rats, the transitory decrease in BP following DAME may result from increased cholinergic activity unopposed by adrenergic activity.

The BP response to Dyn was potentiated in reserpine-treated rats while the HR was converted to a biphasic response, an initial increase followed by a decrease. Since the BP increase and HR decrease following Dyn injection was not blocked by naloxone, phentolamine, propranolol, atropine, or mecamylamine, it is difficult to propose a receptor-mediated mechanism for the cardiovascular changes. However, verapamil (2 mg/kg, iv) blocked the BP increase caused by Dyn (0.1 mg/kg, iv) but not the HR decrease. This finding suggests a possible postsynaptic site of action.

Some additional information was obtained in 6-hydroxydopamine (6-OHDA) treated rats given DAME or Dyn. Sympathetic nerve terminals are destroyed by 6-OHDA including nerve terminals in the heart.[9] Therefore, the initial transient decrease in the BP following DAME injection in 6-OHDA-treated rats can be explained by an increase in ACh release by DAME which leads to an HR decrease. The cholinergic effect is unopposed by adrenergic activity and leads to a decrease in cardiac output and BP. Atropine and naloxone effectively block the cardiovascular responses to DAME for the same reasons already explained for effects of DAME in reserpine-treated rats. Dame injection in 6-OHDA-treated rats also caused a significant elevation in plasma epinephrine levels without changing plasma norepinephrine and dopamine levels. This finding supports the notion that DAME increases the neuronal activity along the splanchnic nerve innervating the adrenal gland leading to an increase in ACh release from the splanchnic nerve and an increase in epinephrine secretion from the adrenal gland.

The BP and HR changes following Dyn injection in control and 6-OHDA-treated rats were similar and provided no additional information on the mechanisms responsible for the cardiovascular changes.

In summary, the BP increase and HR decrease following intravenous administration of Dyn and DAME have been shown to be affected differently by selected blocking agents and treatments that alter neuronal activity in the conscious, unrestrained animal. These finding support the idea of multiple opioid receptor regulation of cardiovascular function in the intact animal.

REFERENCES

1. GADDIS, R. R. & W. R. DIXON. 1982. Modulation of peripheral adrenergic neurotransmission by methionine-enkephalin. J. Pharmacol. Exp. Ther. **211:** 282–288.

2. GADDIS, R. R. & W. R DIXON. 1982. Presynaptic opiate receptor–mediated inhibition of endogenous norepinephrine and dopamine-beta-hydroxylase release in the cat spleen, independent of the presynaptic alpha-adrenoceptor. J. Pharmacol. Exp. Ther. **223:** 77–83.

3. ILLES, P., N. PFEIFFER, I. VON KUGELGEN & K. STARKE. 1985. Presynaptic opioid receptor subtypes in the rabbit ear artery. J. Pharmacol. Exp. Ther. **232:** 526–533.

4. DIXON, W. R., S. H. NELSON & O. S. STEINSLAND. 1985. Opioid inhibition of adrenergic neurotransmission in the rabbit ear artery. Pharmacologist **27:** 205.

5. VON KUGELGEN, I., P. ILLES, D. WOLF & K. STARKE. 1985. Presynaptic inhibitory opioid delta- and kappa-receptors in a branch of the rabbit ileocolic artery. Eur. J. Pharmacol. **118:** 97–105.

6. STARKE, K., E. SCHOFFEL & P. ILLES. 1985. The sympathetic axons innervating sinus node of the rabbit possess presynaptic opioid kappa but not mu or delta receptors. Arch. Pharmacol **329:** 206–209.

7. SEMAFUKO, W. E. B., D. L. FOLLETT & W. R. DIXON. 1988. Effect of etorphine on adrenergic neurotransmission in the rat and guinea pig heart. Pharmacology **37:** 195–202.

8. FUDER H., M. BUDER, H. D. RIERS & G. ROTHACHER. 1986. On the opioid receptor subtype inhibiting the evoked release of ^{3}H-noradrenaline from guinea-pig atria. Arch. Pharmacol. **332:** 148–155.

9. WAKADE, A. R., R. W. MALHOTRA, T. D. WAKADE & W. R. DIXON. 1986. Simultaneous secretion of catecholamines from the adrenal medulla and of ^{3}H-norepinephrine from sympathetic nerves from a single test preparation by various agents. Neurosceince **18:** 877–888.

10. SCHULTZBERG, M., T. HOKFELT, L. TERENIUS, L. G. ELFVIN, J. M. LUNDBERG, J. BRANDT, R. P. ELDE & M. GOLDSTEIN. 1979. Enkephalin immunoreactive nerve fibers and cell bodies in sympathetic ganglia of the guinea-pig and rat. Neuroscience **4:** 249–270.

11. WILSON, S. P., R. L. KLEIN, K. J. CHANG, M. S. GASPARIS, O. H. VIVEROS & W. H. YANG. 1980. Are opioid peptides co-transmitters in noradrenergic vesicles of sympathetic nerves? Nature London **288:** 707–709.

12. VIVEROS, O. H., E. J. DILIBERTO, E. HAZUM & K. J. CHANG. 1979. Opiate-like materials in the adrenal medulla: evidence for storage and secretion with catecholamines. Mol. Pharmacol. **16:** 1101–1108.

13. VIVEROS, O. H., E. J. DILIBERTO, E. HAZUM & K. J. CHANG. 1980. Adv. Biochem. Psychopharmacol. **22:** 191–204.

14. SZABO, B., L. HEDLER, H. ENSINGER & K. STARKE. 1986. Opiate peptides decrease noradrenaline release and blood pressure in the rabbit at peripheral receptors. Arch. Pharmacol. **332:** 50–56.

15. BOLME, P., K. FUXE & L. F. AGNATI. 1978. Cardiovascular effects of morphine and opioid peptides following intracisternal administration in chloralose- anesthetized rats. Eur. J. Pharmacol. **48:** 319–324.

16. MOORE, R. H. & D. A. DOWLING. 1980. Effects of intravenously administered leu- or met-enkephalin on arterial blood pressure. Regul. Pept. **1:** 77–87.

17. GAUTRET, B. & H. SCHMITT. 1985. Central and peripheral sites for cardiovascular actions of dynorphin (1-13) in rats. Eur. J. Pharmacol. **111:** 263–266.

18. LAURENT, S., & H. SCHMITT. 1983. Central cardiovascular effects of K agonists—

dynorphin (1-13) and ethylketocyclazocine in the anesthetized rat. Eur. J. Pharmacol. **96:** 165–169.

19. EIMERL, J. & G. FEUERSTEIN. 1986. The effect of mu, delta, kappa and epsilon opioid receptor agonists on heart rate and blood pressure of the pithed rat. Neuropeptides **8:** 351–358.

20. SIMON, W., K. SCHAZ, U. GANTEN, G. STOCK, K. H. SCHLOR & D. GANTEN. 1978. Effects of enkephalins on arterial blood pressure are reduced by propranolol. Clin. Sci. Mol. Med. **55:** 237s–241s.

21. THORNHILL, J. A., M. EWEN, A. A. WILFONG, L. GREGOR & W. S. SAUNDERS. 1986. Pressor effects of systemic administration of methionine and leucine enkephalin in the conscious rat. Can. J. Physiol. Pharmacol. **64:** 1353–1360.

22. THORNHILL, J. A. & W. S. SAUNDERS. 1985. Blood pressure responses of conscious rats to intravenous administration of enkephalin derivatives (D-ala^2 methionine and leucine enkephalinamide, and methionine and leucine enkephalinamide. Peptides **6:** 1252–1256.

23. HOLADAY, J. W., J. R. KENNER, C. E. GLATT & J. B. LONG. 1984. Dynorphin: cardiovascular consequences of opioid receptor interactions in normal and endotoxemic rats. Proc. West. Pharmacol. Soc. **27:** 429–433.

24. SAUNDES, W. S. & J. A. THORNHILL. 1987. Pressor, tachycardic and feeding responses in conscious rats following i.c.v. administration of dynorphin. Central blockade by opiate and alpha$_1$-receptor antagonists. Regul. Pept. **19:** 209–220.

25. THORNHILL, J. A., L. GREGOR & W. S. SAUNDERS. 1989. Opiate and alpha-receptor antagonists block the pressor responses of conscious rats given intravenous dynorphin. Peptides **10:** 171–177.

26. GADDIS, R. R. & W. R. DIXON. 1985. Effects of ethylketocyclazocine on adrenergic transmission in the isolated prefused cat spleen. Pharmacology **30:** 205–214.

27. ZAR, J. H. 1974. *In* Biostatistical Analysis. 2nd edit. Prentice Hall. Englewood Cliffs, N.J.

Dopamine Autoreceptors[a]

Biochemical, Pharmacological, and Morphological Studies

M. GOLDSTEIN,[b] K. HARADA,[b] E. MELLER,[c]
M. SCHALLING,[d] AND T. HOKFELT[d]

[b]Neurochemistry Research Laboratories
New York University Medical Center
550 First Avenue
New York, New York 10016

[c]Millhauser Laboratories
New York University Medical Center
550 First Avenue
New York, New York 10016

[d]Department of Histology
Karolinska Institute
Stockholm, Sweden

INTRODUCTION

Pharmacological studies in the early 1970s suggested the existence of striatal presynaptic dopamine (DA) receptors (DA autoreceptors) which regulate the synthesis and release of Dopa.[1,2] The findings that the enzyme tyrosine hydroxylase (TH), which catalyzes the first step in catecholamine biosynthesis, is inhibited in synaptosomal preparations by low concentrations of apomorphine (Apo) and that this inhibition is partially reversed by DA receptor antagonists[3,4] led to the conclusion that regulation of DA synthesis by DA autoreceptors occurs at the stage of tyrosine hydroxylation. Since the enzymatic activity of TH is stimulated by cAMP-dependent phosphorylation of the enzyme,[5] it was of interest to determine whether the regulation of DA synthesis by DA autoreceptors involves the state of phosphorylation-dephosphorylation of the enzyme. Indeed, it was shown that activation of synaptosomal TH by phosphorylation (e.g., in the presence of forskolin) is inhibited by DA agonists and this inhibition is reversed by DA antagonists.[6]

Since DA autoreceptors pharmacologically resemble D_2 DA receptors,[7] we have investigated whether guanine nucleotide–sensitive proteins, "G proteins," are involved in the coupling of this receptor to the effector. In this presentation we will review our data on the effects of DA agonists on autoreceptor-mediated inhibition of synaptosomal TH activity in vitro and on inhibition of Dopa synthesis in vivo. Evidence is presented that a receptor reserve exists for DA agonists at autoreceptors regulating synthesis and release of DA, and that such a reserve also exists at supersensitive but not at normosensitive postsynaptic DA receptors. The selective activity of some DA agonists might be related to the existence of receptor reserve at DA autoreceptors. In

[a]This study was supported by grant MH 43230 from the National Institute of Mental Health and grant NS 23618 from the National Institutes of Health.

addition we investigated the distribution of D_2 DA receptor (D_2-R) mRNA in brain and adrenal glands by *in situ* hybridization histochemistry. The distribution of the D_2 DA receptor gene expression supports the physiological and pharmacological findings on the existence of DA autoreceptors in the substantia nigra and adrenal glands.

INHIBITION OF FORSKOLIN-ACTIVATED SYNAPTOSOMAL TH ACTIVITY BY DA AGONISTS

Forskolin stimulates in a concentration-dependent manner the activity of TH in synaptosomes[6] and slices,[8] and the enzyme activity is inhibited by Apo and by other DA agonists. The results presented in FIGURE 1 show the inhibition of synaptosomal TH activity by Apo and by the selective DA autoreceptor agonist 3-PPP, and the reversal of the inhibition by the selective D_2 DA antagonist ($-$)sulpiride. It is evident from the data in FIGURE 1 that the inhibition of the enzyme activity by low concentrations of Apo is completely reversed while that by higher concentrations is only partially reversed by sulpiride. ($+$)-3-PPP inhibits the enzyme activity at higher concentrations than Apo, and sulpiride reverses the inhibition elicited by lower, and partially that elicited by higher concentrations of the agonists. The results presented in FIGURE 2 show that the DA agonists quinpirole, quinelorane, ($+$)PHNO, and ($+$)-3PPP inhibit synaptosomal TH activity in a concentration-dependent manner (10^{-6}–10^{-4}M). Thus, the measurements of the inhibition of synaptosomal TH provide a means to determine the pharmacological profile of DA agonists.

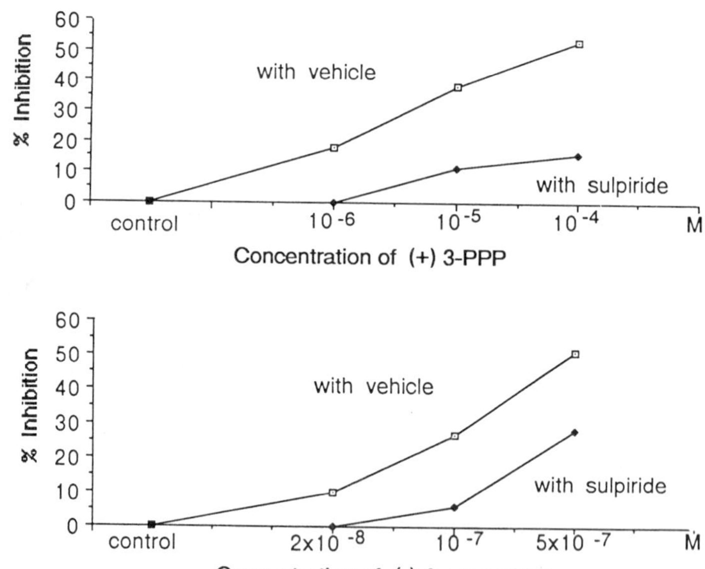

FIGURE 1. Inhibition of striatal synaptosomal TH activity by apomorphine and by ($+$)3-PPP, and the reversal of inhibition by ($-$)sulpiride. [Results are the means from these experiments ± standard error (SE) of 3–7%.]

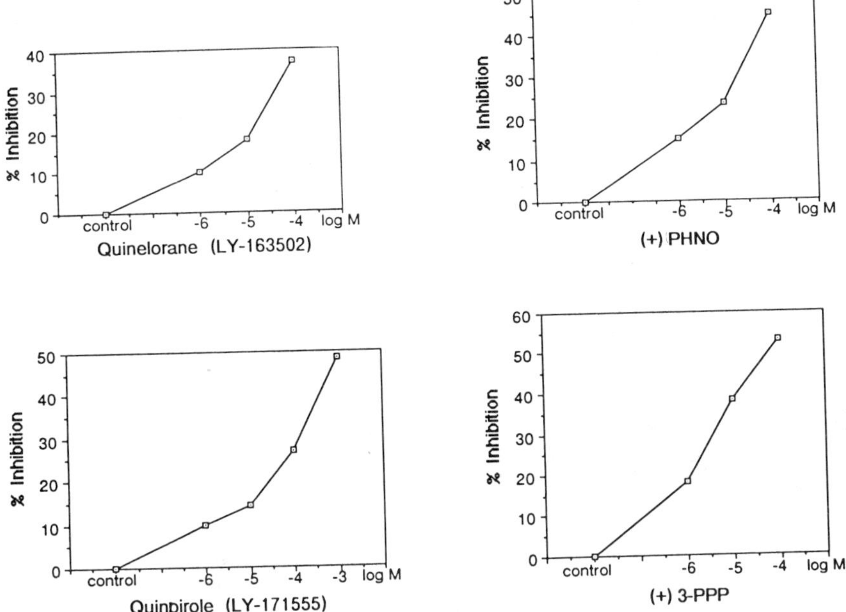

FIGURE 2. Inhibition of striatal synaptosomal TH activity by noncatechol DA agonists. (Results are the means from 2–4 experiments ± SE of 3–10%.)

EFFECT OF PERTUSSIS TOXIN ON INHIBITION OF SYNAPTOSOMAL TH ACTIVITY BY APOMORPHINE

Pertussis toxin (PTX) attenuates the APO-elicited inhibition of forskolin-stimulated TH activity.[9] The inhibition of synaptosomal TH activity by low concentrations of Apo was completely blocked, while that elicited by higher concentrations of Apo was only partially blocked by PTX. Exposure of synaptosomes to PTX abolished the sulpiride-induced attenuation of Apo-induced inhibition of TH activity in the presence and absence of forskolin (TABLE 1). The degree of inactivation of G proteins by PTX was determined by measuring its effect on the back-ADP ribosylation of the G_i and G_o proteins.[9]

RECEPTOR RESERVE AND AUTORECEPTOR SELECTIVITY OF DA AGONISTS

Recently several partial DA agonists with preferential activity at autoreceptors and supersensitive Da receptors were described.[10,11] To explain the autoreceptor selectivity it was postulated that the degree of previous agonist occupancy on the receptor determines the responsiveness of the agonist. The autoreceptors and supersensitive postsynaptic DA receptors have a reduced availability of DA at the receptor sites, and therefore the responsivity of the receptor to the agonist increases.[12] On the other hand, changes in receptor reserve (e.g., changes in agonist-induced receptor-effector coupling

TABLE 1. The Attenuation by $(-)$Sulpiride of $(-)$Apomorphine-Induced Inhibition of Synaptosomal TH Activity in the Presence or Absence of Forskolin[a]

Apomorphine (M)	Sulpiride (M)	Dopa Accumulation with Forskolin[b] (% inhibition) (ng/5 mg original tissue)	Dopa Accumulation without Forskolin (% inhibition) (ng/5 mg original tissue)
0	0	5.360 ± 0.089	4.111 ± 0.054
	10^{-4}	5.357 ± 0.041	3.902 ± 0.074
2×10^{-8}	0	4.621 ± 0.082 (13.8)	3.672 ± 0.049 (10.7)
	10^{-4}	5.442 ± 0.059 (-1.6)	3.746 ± 0.070 (4.0)
10^{-7}	0	3.873 ± 0.063 (27.7)	3.085 ± 0.077 (25.0)
	10^{-4}	5.029 ± 0.053 (6.1)	3.425 ± 0.025 (12.2)
5×10^{-7}	0	2.682 ± 0.044 (50.0)	1.871 ± 0.026 (54.5)
	10^{-4}	3.860 ± 0.054 (27.9)	2.421 ± 0.120 (38.0)

[a]Each value is the mean $\pm$ standard error of 4 experiments.
[b]The concentration of forskolin was 10^{-7} M.

efficiency) could also explain why DA agonists with low intrinsic efficacy are effective at autoreceptors and at supersensitive, but not at normosensitive, postsynaptic DA receptors. The results of our studies show that a large receptor reserve exists for DA agonists at autoreceptors regulating the synthesis[13] and release of the transmitter.[14] In contrast of DA autoreceptors, no such receptor reserve for DA agonists was demonstrated at normosensitive postsynaptic DA autoreceptors regulating striatal cholinergic activity.[15] However, 6 OH-DA-induced denervation of nigrostriatal DA neurons generates a small receptor reserve for DA agonists at supersensitive postsynaptic DA receptors regulating cholinergic activity.[16] Based on these results, we postulate that the preferential activity of some DA agonists [e.g., $(+)$-3-PPP, EMD 23,448] at autoreceptors and at supersensitive postsynaptic receptors is related to the presence of a receptor reserve for DA agonists at these receptors.

THE DISTRIBUTION OF D_2-R mRNA IN RAT BRAIN AND ADRENAL GLANDS

Since ligand-binding studies do not differentiate between pre- and postsynaptic DA receptors, the existence of presynaptic receptors has been questioned by some investigators.[17] The recent cloning of the genes for the D_2 DA receptor, and for the enzyme TH and aromatic amino acid decarboxylase (AADC),[18,19,20] has made it possible to investigate whether the D_2 DA receptor gene is expressed in dopaminergic cells. The results presented in FIGURE 3 show that D_2-R mRNA and AADC mRNA have similar distributions in the mesencephalon of the rat and monkey brain. The D_2-R-labeled cells show the highest density in the substantia nigra pars compacta and the ventral tegmental area, with scattered cells also observed in the zona reticulata.[21] Furthermore, the adrenal medulla, but not the adrenal cortex, expresses D_2-R mRNA (FIGURE 3). These findings provide morphological evidence for the existence of DA autoreceptors which are probably involved in the regulation of synthesis and release of DA in the substantia nigra and adrenal glands.

DISCUSSION

Our recent studies wtih forskolin-stimulated synaptosomal TH[9] further demonstrate that synaptosomal preparations can be used for investigations of the molecular properties of nerve terminal DA autoreceptors. The finding that not only apomorphine (Apo), but also other noncatechol DA agonists (e.g., 3-PPP, quinpirole) inhibit synaptosomal TH and that this inhibition is attenuated by ($-$)sulpiride indicates that

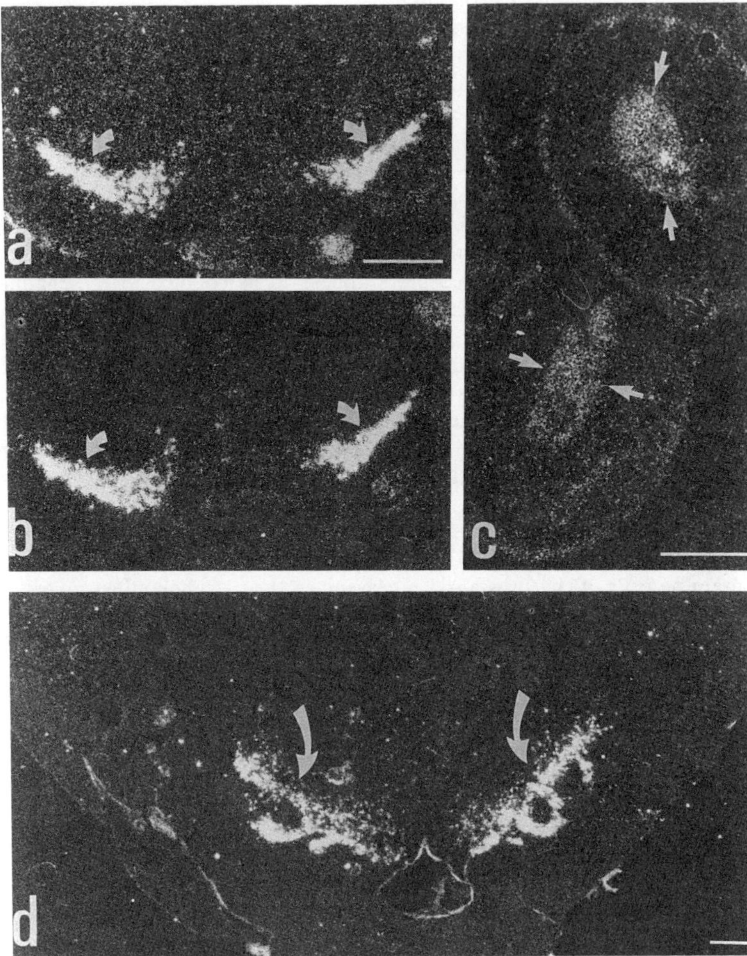

FIGURE 3. Autoradiographs of the ventral mesencephalon of rat (a,b) and monkey (d) and of rat adrenal glands (c) after hybridization with ^{35}S-labeled probe complementary to the D_2 receptor mRNA (a,c,d) or with ^{35}S-labeled probe complementary to AADC mRNA (b). D_2 receptor mRNA is present in DA cells in the zona compacta (curved arrows in a and d) both in rat (a) and monkey (d). Note close overlap with AADC mRNA as a marker for the DA neurons (b). Also the adrenal medulla contains D_2 receptors. Bars indicate 1 mm (a = b).

the inhibitory action is related to stimulation of synaptosomal DA autoreceptors and that these receptors have the pharmacological characteristics of D_2 DA receptors. The attenuation of the Apo-induced inhibition by pertussis toxin (PTX) indicates that DA autoreceptors that regulate the synthesis of DA are coupled to PTX-sensitive proteins (G_i and/or G_o). The inhibition of synaptosomal TH elicited by low concentrations of Apo is reversed by ($-$)sulpiride, but in PTX pretreated synaptosomes the inhibition is not reversed by the D_2 DA antagonist. Thus, PTX pretreatment uncouples the DA autoreceptor from the second messenger system and the DA agonist–induced response is completely dependent on this coupling.

The findings that DA autoreceptors are linked to a PTX-sensitive G protein and that there is a large receptor reserve for DA agonists at autoreceptors regulating their synthesis[13] suggest that these receptors are efficiently coupled to the effector and that the linkage to G proteins might be the limiting step in the agonist-induced response. This idea is completely supported by the finding that inactivation of 40–50% of the synaptosomal G proteins by PTX completely attenuates the DA agonist response. Whether DA autoreceptors linked to the G proteins are negatively coupled to the adenylate cyclase system and/or to other effector systems remains to be elucidated. It is noteworthy that there is also a large receptor reserve for DA agonists regulating the release of DA,[16] although these receptors appear to be coupled to K^+ channels not linked to PTX-sensitive G proteins.[22] It will be of interest to determine whether DA autoreceptors regulating synthesis and those regulating release are differently coupled and whether they also differ structurally.

The observation that there is a receptor reserve at DA autoreceptors and at supersensitive postsynaptic[13] but not at normosensitive postsynaptic receptors[15] could explain the selective action of some DA agonists at DA autoreceptors and at supersensitive receptors, and provides a rationale for designing selective antiparkinsonian and antipsychotic drugs.

It has been suggested that the evidence for the existence of presynaptic autoreceptors is indirect and circumstantial.[17] The presence of D_2 DA receptor mRNA in DA-containing cells of the substantia nigra and in the catecholaminergic cells of the adrenal medulla is in agreement with a large number of pharmacological and physiological studies on DA agonist responses at autoreceptors and provides morphological evidence for the existence of DA autoreceptors.

CONCLUSIONS

The results presented in this study show that the inhibition of synaptosomal TH activity by DA agonists is mediated by DA autoreceptors. Evidence is presented that the DA autoreceptors are linked to PTX-sensitive G proteins and that the receptors are efficiently coupled to the effector. The presence of a receptor reserve for DA agonists at autoreceptors and at supersensitive postsynaptic, but not at normosensitive, receptors might explain the selective activity of some partial DA agonists and provide a rationale for designing selective antiparkinsonian and antipsychotic drugs. The presence of D_2 DA receptor mRNA in DA-containing cells of the substantia nigra and in the catecholaminergic cells of the adrenal glands provides additional evidence for the existence of DA autoreceptors.

REFERENCES

1. EBSTEIN, B., C. ROBERGE, I. TABCHNICK & M. GOLDSTEIN. 1974. J. Pharm. Pharmacol. **26**: 975.
2. CARLSSON, A., W. KEHR, M. LINDQUIST, T. MAGNUSSON & C. W. ATACK. 1972. Pharmacol. Rev. **24**: 371–375.
3. CHRISTIANSEN, J. & R. J. SQUIRES. 1974. J. Pharm. Pharmacol. **26**: 367–369.
4. GOLDSTEIN, M., R. L. BRONAUGH, B. EBSTEIN & C. ROBERGE. 1976. Brain Res. **109**: 563–574.
5. ANAGNOSTE, B., C. SHIRRON, E. FRIEDMAN & M. GOLDSTEIN. 1974. J. Pharmacol. Exp. Ther. **191**: 370–376.
6. STRAIT, K. & R. KUCZENSKI. 1986. Mol. Pharmacol. **29**: 561–569.
7. GOLDSTEIN, M. 1984. Ann. N.Y. Acad. Sci. **340**: 1–7.
8. BOHMAKER, K., T. PUZA, M. GOLDSTEIN & E. MELLER. 1989. J. Pharmacol. Exp. Ther. **248**: 97–103.
9. HARADA, K., E. MELLER & M. GOLDSTEIN. Eur. J. Pharmacol. (In press.)
10. GOLDSTEIN, M., K. FUXE, E. MELLER, C. A. SEYFRIED, L. AGNATI & F. M. MASCAGNI. 1987. J. Neural Transm. **70**: 193–215.
11. HJORTH, S., A. CARLSSON, H. WIKSTROM. P. LINDBERG, D. SANCHEZ, U. HACKSELL, L. E. ARVIDSSON, U. SVENSSON & J. L. G. NILSSON. 1981. Life Sci. **28**: 1225–1238.
12. CLARK, D., S. HJORTH & A. CARLSSON. 1985. J. Neural Transm. **62**: 171.
13. MELLER, E., K. BOHMAKER, Y. NAMBA, A. J. FRIEDHOFF & M. GOLDSTEIN. 1987. Mol. Pharmacol. **31**: 592–597.
14. YOKOO, H., M. GOLDSTEIN & E. MELLER. 1988. Eur. J. Pharmacol. **155**: 323–327.
15. MELLER, E., A. ENZ & M. GOLDSTEIN. 1988. Eur. J. Pharamcol. **155**: 151–154.
16. ENZ, A., M. GOLDSTEIN & E. MELLER. 1990. Mol. Pharmacol. (In press.)
17. LADURON, P. M. 1984. Trends Pharmacol. Sci. **5**: 459–461.
18. BUNZOW, J. R., H. H. M. VAN TOL, D. K. GRANDY, P. ALBERT, J. SALON, M. C. CHRISTIE, C. A. MACHIDA, K. A. NEVE & O. CIVELLI. 1988. Nature **336**: 783–787.
19. GRIMA, B., A. LAMOUROUX, F. BLANOT, N. FAUCON BIGUET & J. MALLET. 1985 Proc. Nat. Acad. Sci. USA **82**: 617–621.
20. TANAKA, T., H. YOSHIYUKI, M. TAKETOSHI, I. IMAMURA, M. ANDO-YAMAMOTO, K. KANGAWA, M. HISAYUKI, M. KURODA & W. HIROSHI. 1989. Proc. Nat. Acad. Sci. USA **86**: 8142–8146.
21. SCHALLING, M., A. DAGERLIND, M. GOLDSTEIN, M. EHRLICH, P. GREENGARD & T. HOKFELT. Eur. J. Pharmacol. (In Press.)
22. BOWYER, J. F. & N. WEINER. 1989. J. Pharmacol. Exp. Ther. **248**: 514–519.

Transmitter Release from Sympathetic Nerve Terminals on an Impulse-by-Impulse Basis and Presynaptic Receptors

JAMES A. BROCK[a] AND THOMAS C. CUNNANE

University Department of Pharmacology
South Parks Road
Oxford OX1 3QT, United Kingdom

INTRODUCTION

Studies of the effects of α-adrenoceptor agonists and antagonists on electrically evoked transmitter release from postganglionic sympathetic nerve terminals suggest that release is modulated by prejunctional α-adrenoceptors.[1] In most studies the effects of drugs on transmitter overflow from electrically field stimulated tissues have been investigated. Such studies provide good evidence for the existence of presynaptic receptors but provide little direct information about the mechanism of inhibition of transmitter release. In order to understand the fundamental mechanisms underlying the regulation of the transmitter release process by presynaptic α-adrenoceptors, it is necessary to establish the characteristic features of transmitter release on an impulse-by-impulse basis at the level of the individual varicosity.

Burnstock and Holman were the first to use electrophysiological techniques to investigate chemical transmission at the sympathetic neuroeffector junction.[2,3] In the absence of stimulation, spontaneous excitatory junction potentials (SEJPs) were recorded from individual smooth muscle cells of the guinea pig vas deferens using intracellular microelectrodes, suggesting that spontaneous transmitter release from sympathetic nerve endings was packeted and presumably quantal. The spontaneous and evoked end plate potentials recorded intracellularly at the skeletal neuromuscular junction have closely similar time courses,[4] but the time course of the electrically evoked excitatory junction potential (EJP) lasts several times longer than that of the SEJP.[3] The explanation for this lies in the fact that smooth muscle cells are electrically coupled to one another and receive a multiple innervation.[5] For this reason direct comparisons between spontaneous and evoked junction potentials could not readily be made. As a result much of our understanding of the process of evoked transmitter release from postganglionic sympathetic nerve terminals comes from biochemical studies, which revealed a disparity between the amount of transmitter released from the "average" varicosity per stimulus and that estimated to be contained in a single transmitter storage vesicle;[6,7] a varicosity secreting approximately 1% of the transmitter content of a vesicle per stimulus. Two hypotheses were put forward to explain the transmitter release process at the level of the individual varicosity: either the vesicle is not the basic unit of transmitter release or individual varicosities intermittently secrete the whole transmitter content of a single vesicle.

[a]J.B. is a Squibb Research Fellow.

NOVEL ELECTROPHYSIOLOGICAL TECHNIQUES

Two electrophysiological methods have been used to investigate the transmitter release process at the level of the individual release site which is presumed to be the varicosity. In the first the rising phases of the intracellularly recorded EJPs in the guinea pig and mouse vas deferens were electronically differentiated.[8,9] The first time differential of the EJP revealed occasional peaks in the rate of depolarization (*discrete events*) of individual EJPs which occurred at one or a few fixed latencies after the stimulus. The important observation was made that at any one latency discrete events occurred intermittently. Discrete events occurring at a single latency measure indirectly the release of transmitter from one or a few varicosities located close to the intracellular recording electrode, therefore, transmitter release from any one varicosity occurs intermittently. The differentiated SEJPs and the electrically evoked discrete events had closely similar amplitudes and time courses suggesting that both spontaneous and evoked transmitter release was quantal. Furthermore, Cunnane and Stjärne concluded from an analysis of the discrete events resulting from the secretory activity of individual varicosities that under normal conditions only single quanta are released.[9] Taken together these studies indicate that action potential evoked transmitter release from individual varicosities is intermittent and monoquantal. Similar findings have been reported for the transmitter release mechanism in the sympathetic nerves innervating the guinea pig mesenteric arterioles.[10] However, both these studies gave no indication as to the cause of intermittence.

The second method is a focal extracellular recording technique which allows simultaneous recording of both the nerve terminal impulse (NTI) and transmitter release from single postganglionic sympathetic nerve fibres innervating the guinea pig and mouse vas deferens[11,12] and blood vessels.[12–14] Briefly, a small glass electrode (tip diameter approximately 50 μm) is applied to the muscle surface with slight suction and the signals recorded through an a.c. amplifier. In the absence of stimulation, spontaneous negative-going potentials are recorded (FIGURE 1A) which have similar characteristics to the intracellularly recorded SEJPs.[11] There is reason to believe that these extracellular records measure faithfully the time course of the conductance change underlying the intracellular potential change, therefore these events have been termed spontaneous excitatory junction currents (SEJCs). Following low-frequency (0.5–2 Hz) electrical stimulation, stimulus-locked excitatory junction currents (EJCs) are evoked intermittently, whose amplitude and time course are similar to those of individual SEJCs. EJCs are always preceded by a nerve impulse, which normally displays no conduction failures (FIGURE 1B). For a number of reasons we believe that the nerve impulse preceding the EJC is likely to be the extracellular equivalent of the nerve terminal action potential.[11,15] This extracellular method therefore confirms the previous observation that transmitter release from individual varicosities is intermittent and monoquantal but in addition establishes that intermittence results from a low probability of depolarization-secretion coupling in the invaded varicosities.

On all occasions that postganglionic fibers have been stimulated propagation failure does not occur. However, in the guinea pig vas deferens when the hypogastric nerve trunk is stimulated close to the prostatic end of the tissue, recordings have occasionally been made from postganglionic fibers stimulated preganglionically. On most occasions the nerve impulses recorded from these fibers display no propagation failures. However, some preganglionically stimulated fibers displayed frequency dependent conduction failure; fewer failures being recorded at higher frequencies of stimulation. The nerve impulses recorded from these 'intermittent' fibers were abolished

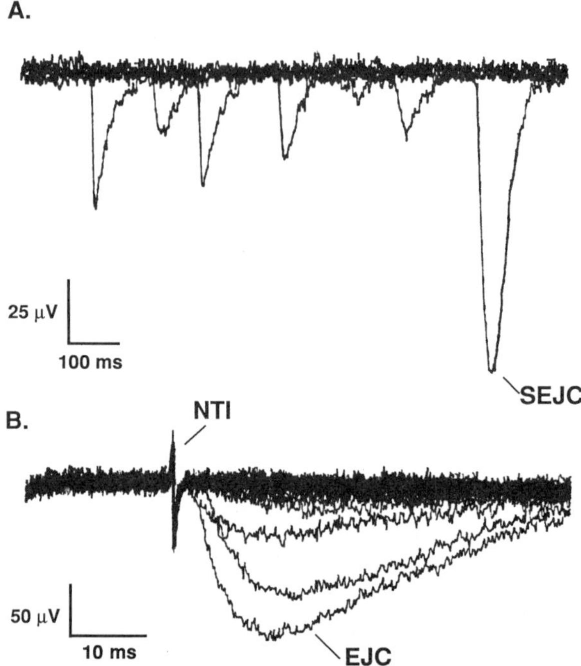

FIGURE 1. Electrical activity recorded extracellularly with a suction electrode from the surface of the guinea-pig vas deferens. **A:** Spontaneous excitatory junction currents (SEJCs). **B:** Simultaneous recording of the nerve terminal impulse (NTI) and evoked excitatory junction currents (EJCs) (train of 25 stimuli at 1 Hz). Note that although the NTI always arrives, transmitter release is intermittent.

reversibly by hexamethonium (30–100 μM) and the removal of Ca^{2+} from the bathing medium. Thus it seems likely that in these cases intermittent conduction occurred as a result of failure of chemical transmission across the hypogastric ganglia, facilitation of acetylcholine release at higher frequencies of stimulation overcoming the block. Similar findings were reported by Cunnane and Stjärne for impulses recorded in preterminal fibers innervating the guinea pig vas deferens. However, in this study the sensitivity of impulse conduction to hexamethonium was not assessed. Although the possibility exists that failure of transmission at the hypogastric ganglia is an artifact of the *in vitro* methodology, the integrating role of autonomic ganglia in regulating activity in sympathetically innervated organs may be functionally important.

PHARMACOLOGY OF THE POSTJUNCTIONAL ELECTRICAL RESPONSE

SEJPs, EJPs, SEJCs, and EJCs recorded from rodent vasa deferentia and various vascular tissues are unaffected by α-adrenoceptor antagonists[11,13,14,17] but are abolished by the P_{2x} purinoceptor desensitizing agent α,β-methylene ATP (1–10 μM)[11,13,14,18] and by pretreatment of tissues with 6-hydroxydopamine,[11,19,20] indicating that they result from the release of ATP from sympathetic nerve terminals. It is likely that noradrena-line (NA) and ATP are costored in the same vesicle and coreleased, and it is therefore reasonable to assume that postjunctional potentials can be used to measure the packeted release of NA. In this article when we refer to transmitter release, we mean ATP release as measured by the postjunctional response of individual smooth muscle cells.

EFFECTS OF α-ADRENOCEPTOR AGONISTS AND ANTAGONISTS

α_2-Adrenoceptor agonists reduce the amplitude of EJPs in the guinea pig and mouse vas deferens, an effect reversed by selective α_2-adrenoceptor antagonists.[21,22] The addition of an α_2-adrenoceptor antagonist alone increases the amplitude of EJPs.[21-23] The potentiating effects of α-adrenoceptor antagonists are not seen in response to the first pulse in a train of stimuli but are clearly apparent by the fifth (see FIGURE 5A), suggesting that transmitter release is regulated by endogenous noradren-aline (NA), and that the inhibition is only seen when sufficient NA accumulates in the extracellular space. These findings suggest that α-adrenoceptors are functionally operative in the guinea pig and mouse vas deferens.

In sympathetically innervated tissues drugs usually have to be applied to the whole nerve terminal arborization. One advantage of the extracellular recording technique is that drugs can be applied focally by internal perfusion of the suction electrode. In this way the application of drugs can be limited to the varicosities enclosed within the recording electrode. Using this method it is clear that clonidine (0.1 μM) reduces and

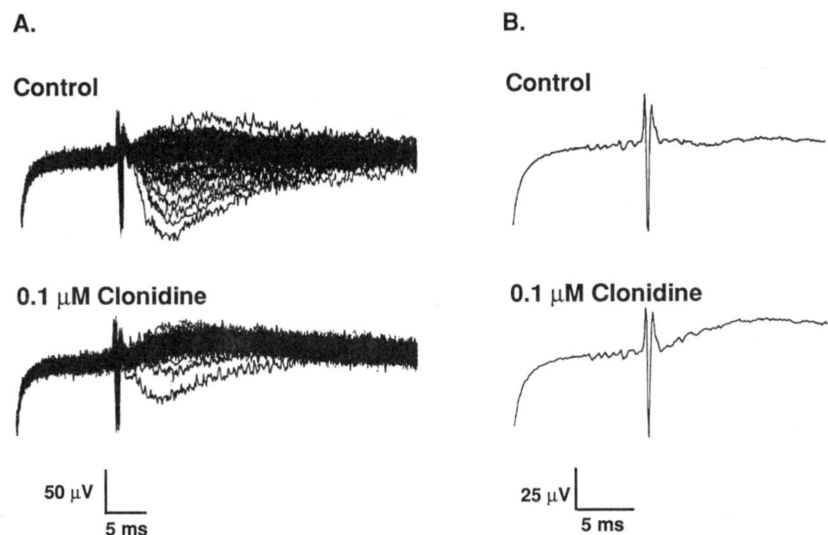

FIGURE 2. Effects of local application of clonidine on evoked electrical activity in the guinea pig vas deferens. **A:** NTIs and EJCs evoked by trains of 100 stimuli at 1 Hz before and 10 minutes after the addition of clonidine (0.1 μM). **B:** Averages of 100 traces recorded at the same times as in **A.** Clonidine inhibited the occurrence of EJCs without altering the configuration of the NTI.

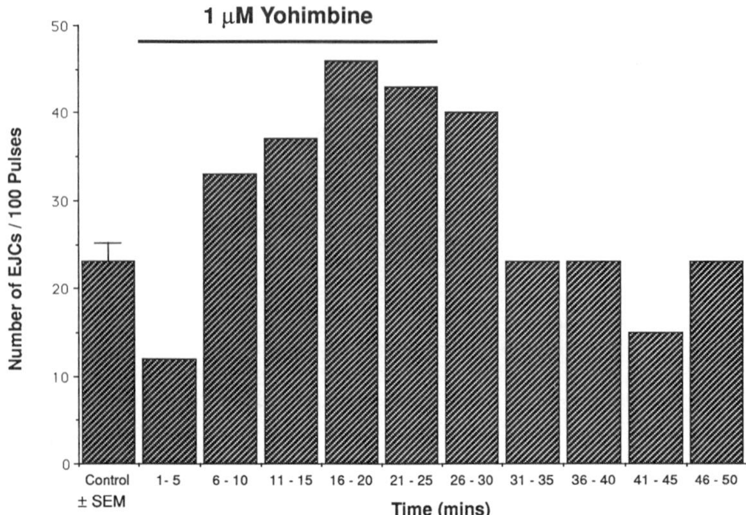

FIGURE 3. Effects of locally applied yohimbine on transmitter release from the guinea pig vas deferens as measured by the EJC. The histogram shows the total number of EJCs recorded during 5 minute periods (20 pulses, 1 Hz every minute) before, during, and after the application of yohimbine (1 μM). The control represents the mean number of EJCs recorded during four 5-minute periods of stimulation [± standard error of the mean (SEM)]. Yohimbine reversibly increased the probability of occurrence of EJCs.

yohimbine (1 μM) increases the number of EJCs recorded from the guinea pig vas deferens during trains of stimuli at 1 Hz without any apparent alteration in the NTI (FIGURES 2A and B, 3, and 4A). Neither of these agents significantly alters the size or the frequency of occurrence of SEJCs. These results suggest that activation of presynaptic α-adrenoceptors interferes with depolarization-secretion coupling directly and does not involve inhibition of action potential propagation in the secretory terminals. Local application of a solution containing both tetraethylammonium (TEA) (2 mM) and 4-aminopyridine (4-AP) (0.2) mM) produced a clear alteration in the shape of the NTI (FIGURE 4B). The effects of yohimbine are not therefore due to K^+ channel block.

AUTOREGULATION

The results with α_2-adrenoceptor antagonists in a wide variety of tissues clearly support the hypothesis that transmitter release is regulated by endogenous NA. The possibility remains, however, that the effects of α-adrenoceptor antagonists are not solely due to blockade of α-adrenoceptors located on the nerve terminals.[24] For this reason the effects of yohimbine on transmitter release in vasa deferentia removed from guinea pigs pretreated with reserpine to deplete the NA stores have been investigated. Yohimbine (1 μM) was without effect on the amplitude of EJPs in reserpinized tissues (FIGURE 5). It is interesting that in reserpinized tissues and in yohimbine-treated control tissues that the development of full facilitation of EJPs requires more stimuli, suggesting that activation of prejunctional α-adrenoceptors receptors normally limits the magnitude of facilitation.

LOCAL REGULATION?

It is commonly assumed that transmitter released from one varicosity acts locally to inhibit subsequent transmitter release from the same varicosity. However, the observation that release is highly intermittent means that transmitter is unlikely to accumulate in the vicinity of a varicosity. Furthermore, it is difficult to envisage how, on an impulse-by-impulse basis, transmitter can feed back locally if release from a particular varicosity is unlikely to occur again within the next 100 or so stimuli. Previous studies have failed to demonstrate that endogenous transmitter can regulate its own release locally in the absence of prejunctional α-adrenoceptor antagonists.[25] Using extracellular recording techniques we have asked the question, Does the occurrence of an EJC alter the probability of observing subsequent EJCs?, by analyzing the stochastic features of the transmitter release process using the correlogram method.[26] The occurrence of EJCs recorded during long trains of stimuli (500–2000 pulses) delivered at various frequencies (0.5–4 Hz) was analyzed in the guinea pig vas deferens. FIGURE 6 shows a correlogram for a train of 2000 pulses

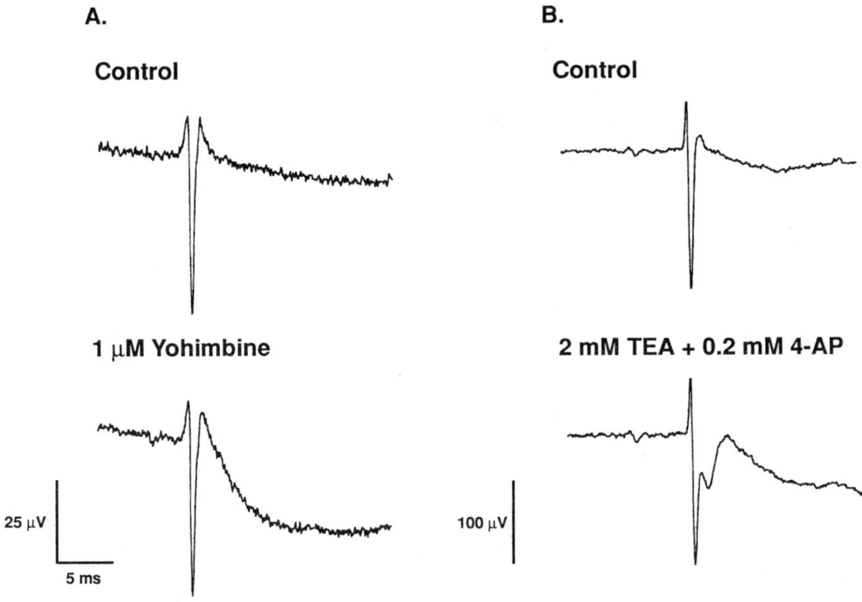

FIGURE 4. Effects of local application of (**A**) yohimbine (1 μM) and (**B**) K$^+$ channel blockers (2 mM TEA + 0.2 mM 4-AP) on the NTI and transmitter release in the guinea pig vas deferens. Each trace represents the average of 20 pulses delivered at 1 Hz before and 10 minutes after application of the drugs. Yohimbine increased transmitter release without altering the NTI. The local application of K$^+$ channel blockers had a relatively small facilitatory effect on transmitter release but markedly altered the configuration of the NTI.

  ANNALS NEW YORK ACADEMY OF SCIENCES

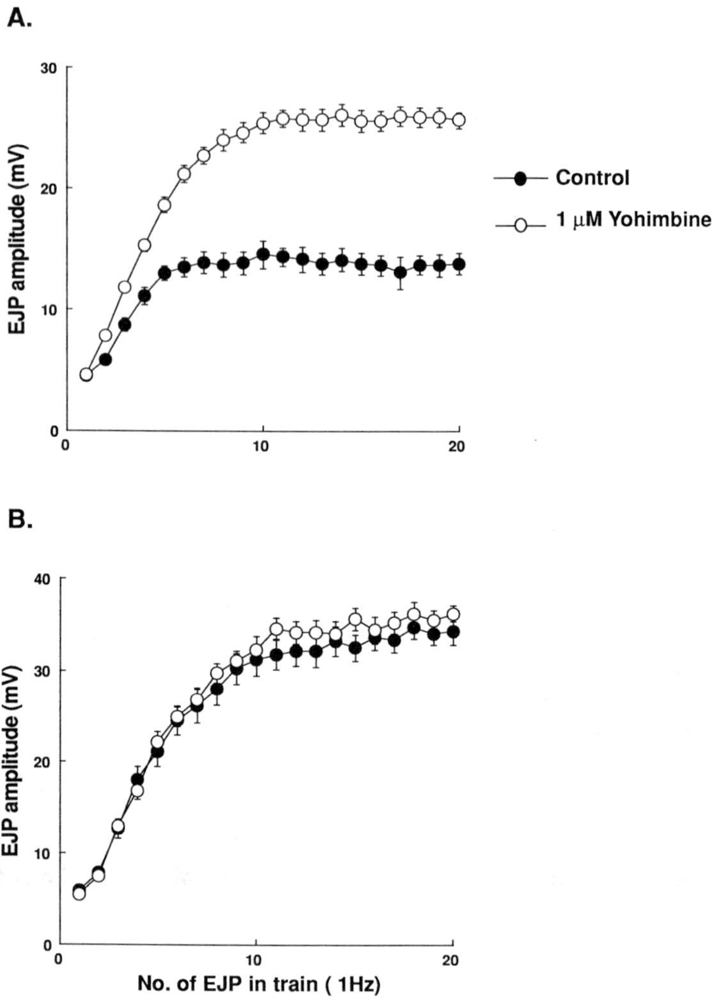

FIGURE 5. Effects of yohimbine (1 μM) on excitatory junction potential (EJP) amplitude in (**A**) control and (**B**) reserpinized guinea-pig vas deferens. In control tissues the EJPs facilitated over the first 5 pulses of a train of stimuli, whereas in reserpinized tissues facilitation continued for up to 10 to 15 pulses. Yohimbine markedly increased the amplitude of the fully facilitated EJP in control tissues but was without effect in reserpinized tissues. Interestingly, yohimbine prolonged the time course of facilitation in control tissues suggesting that activation of prejunctional α-adrenoceptor limits the magnitude of facilitation. The data presented are the mean amplitude $\pm$ SEM of 8 control cells and 8 cells in the presence of yohimbine in the same tissue (from Reference 32).

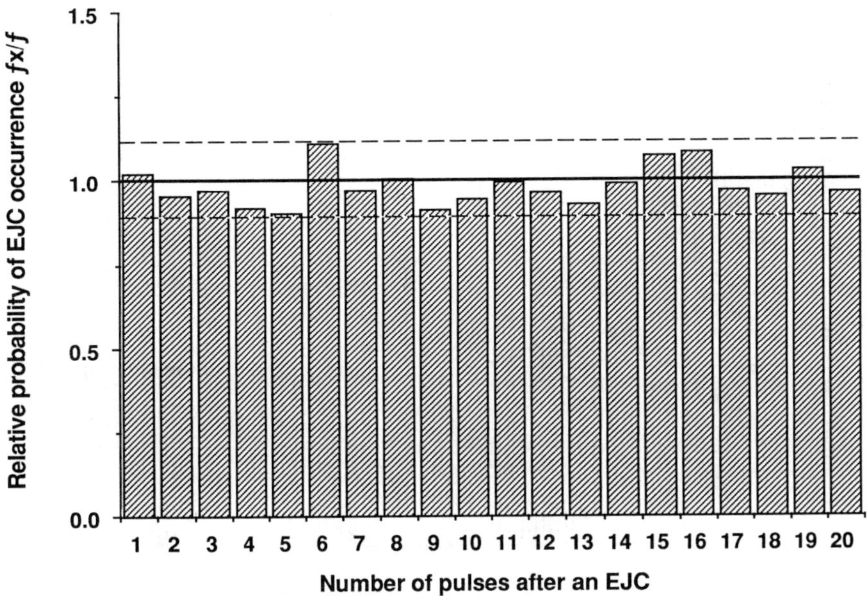

FIGURE 6. Correlogram plotted for EJC interval data obtained from the guinea pig vas deferens during a train of 2000 stimuli delivered at 2 Hz in a single experiment. The correlogram was constructed by calculating the number of EJCs (fx) occurring 1, 2, 3, etc. pulses following each EJC during the train of stimuli. The expected number of EJCs in each interval class (f) was determined by averaging the number of EJCs (fx) occurring 21 to 25 pulses following each EJC, during which period it was assumed that EJCs occurred independently. The probability of EJCs occurring 1, 2, 3, etc. pulses following an EJC relative to that expected if EJCs occurred independently of one another was determined by dividing fx by f. The continuous line is that predicted if EJCs occur independently , and the dashed lines represent the 95% confidence limits of this prediction. The occurrence of an EJC does not inhibit the occurrence of subsequent EJCs on an impulse-by-impulse basis.

delivered at 2 Hz. In all the experiments no evidence for a decreased probability of occurrence of EJCs immediately following an EJC was found. Indeed, the occurrence of an EJC had no obvious effect on the probability of observing subsequent EJCs.

Broadly similar results were found when the activity of individual varicosities was analyzed using discrete events,[9] but if anything it appeared that following release there was a transient increase in the probability of subsequent secretion from the same varicosity. Essentially similar findings have been reported for release sites at the skeletal neuromuscular junction.[27] Interestingly, the occurrence of SEJCs in the guinea pig vas deferens when analyzed by the correlogram method does not appear to be random; the occurrence of an SEJC transiently increasing the probability of observing subsequent SEJCs (FIGURE 7). This so called clustering has been reported for SEJPs in guinea pig submucous arterioles.[10] Indeed, studies at both vertebrate and invertebrate neuromuscular junctions suggest that spontaneous quantal release is not a random process, and that events are clustered.[28–30] Thus, a transient increase in the probability of subsequent secretion from a particular release site following its activation may be a characteristic feature of both spontaneous and electrically evoked transmitter release in many different nerve terminals.

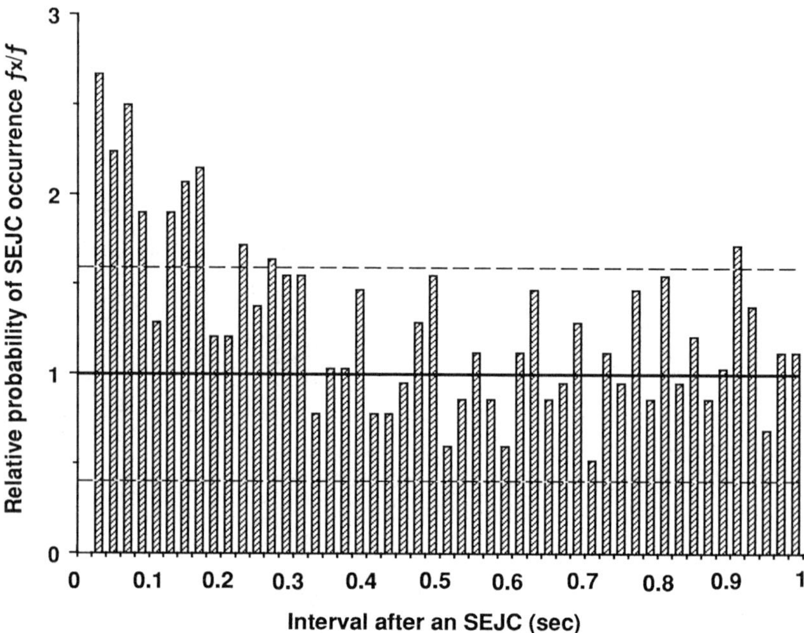

FIGURE 7. Correlogram plotted for SEJC interval data obtained from the guinea pig vas deferens. The correlogram was constructed by calculating the number of SEJCs (fx) occurring in successive 20-msecond bin intervals for a period of 25 seconds following each SECJ recorded during a 10 minute period. The expected number of SEJCs in each interval class (f) was determined by averaging the number of SEJCs (fx) occurring 20 to 25 seconds following each SEJC, during which period it was assumed that SEJCs occurred independently. The probability of SEJC occurrence relative to the expected probability of SEJC occurrence was determined by dividing fx by f. The continuous line is that predicted if SEJCs occur independently and the dashed lines represent the 95% confidence limits of this prediction. Note that within the first 250 mseconds after an SEJC the probability of observing subsequent SEJCs is significantly greater than the random prediction. The 0–20 msecond bin has been left blank owing to the difficulty of identifying with any degree of certainty an individual event within this period.

The question is raised, therefore, whether endogenous NA is able to inhibit transmitter release locally when it is released from nerve terminals by the indirectly acting sympathomimetic amine tyramine. Local application of tyramine (10–100 μM) powerfully inhibited electrically evoked transmitter (FIGURE 8) in the guinea pig vas deferens, an effect reversed by yohimbine (1 μM). In tissues from reserpinized guinea pigs, tyramine had little or no inhibitory effect but clonidine (0.1 μM) still powerfully inhibited electrically evoked transmitter release showing that the presynaptic α-adrenoceptors were functionally intact. The simplest explanation for these findings is that NA released from sympathetic nerve terminals by tyramine stimulates presynaptic α-adrenoceptors and inhibits electrically evoked transmitter release.

DISCUSSION

Electrophysiological studies of transmitter release from postganglionic sympathetic nerve terminals indicate that secretion from individual release sites, presumed to be varicosities, is intermittent and monoquantal. Focal extacellular recording of electrical activity at the sympathetic neuroeffector junction has demonstrated that intermittence results from a low probability of secretion in the invaded varicosities and is not due to intermittent failure of the action potential to propagate in the secretory terminals. The intermittent character of the transmitter release process is in accord with biochemical studies of the transmitter release process, which indicated that individual varicosities secrete on average only about $1/100^{th}$ of the NA content of a single storage vesicle per stimulus. However, recent studies indicate that the transmitter responsible for the generation of postjunctional electrical activity in many sympathetically innervated tissues, including the rodent vas deferens, is not NA but ATP. Evidence suggests that NA and ATP are coreleased from the same sympathetic nerve terminal and it is therefore reasonable to assume that the EJC is an indirect measure of the quantal release of NA.[11] The effects of α_2-adrenoceptor agonists and antagonists on EJPs and EJCs in various tissues support this view, since such studies indicate that the release of ATP, like that of NA, is regulated by activation of prejunctional α_2-

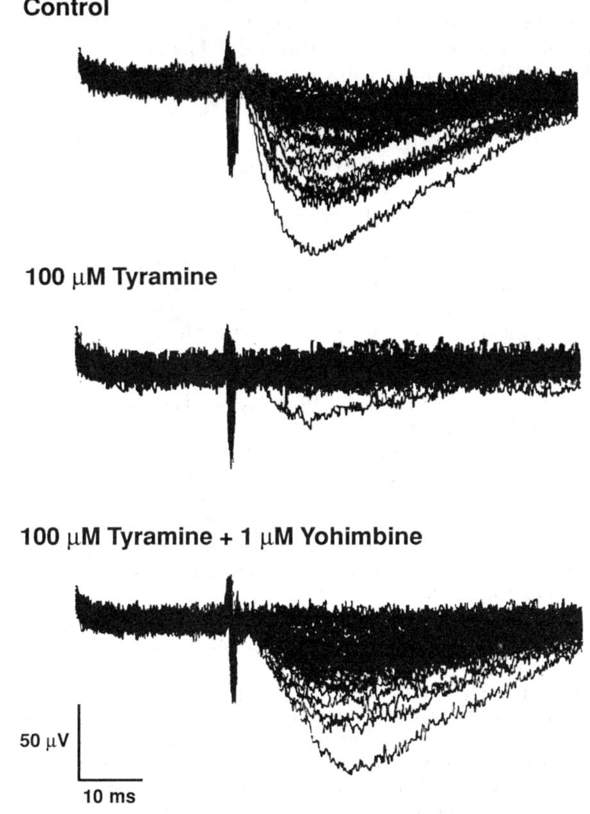

FIGURE 8. Inhibition of transmitter release by locally applied tyramine (100 μM) and its reversal by yohimbine (1 μM) in the guinea pig vas deferens. NTIs and EJCs evoked by trains of 50 stimuli delivered at 1 Hz before, 10 minutes after the addition of tyramine, and a further 10 minutes after the addition of yohimbine. Tyramine inhibited transmitter release without altering the configuration of the NTI.

adrenoceptors. Interestingly, local application of clonidine or yohimbine at concentrations sufficient to clearly inhibit or potentiate transmitter release, respectively, caused no apparent alteration in the NTI. This finding suggests that activation of prejunctional α_2-adrenoceptors inhibits depolarization-secretion coupling in the invaded varicosity and does not involve effects on the propagating nerve impulse.

The observation that yohimbine is without effect on the amplitude of EJPs in reserpinized tissues provides strong supporting evidence for the idea that transmitter release is regulated by endogenous NA.[32] To date, however, in the absence of drugs, there is no evidence to support the view that transmitter release is regulated locally; i.e., transmitter secreted from an individual varicosities inhibits subsequent release from the same varicosity. Indeed, it seems that release from an identified varicosity transiently increases the probability of subsequent secretion from the same varicosity rather than the expected inhibition. Despite this finding, endogenous NA released from the sympathetic nerve terminals by focal application of tyramine can readily inhibit electrically evoked transmitter release.

In conclusion our results support the idea that transmitter release from sympathetic nerve terminals is regulated by prejunctional α_2-adrenoceptors. The evidence to date does not support the view that transmitter feeds back locally to inhibit subsequent secretion from its own site of release. Instead, it seems that NA acts at sites remote from its own site of release (*lateral inhibition*).[31,32] The reasons why action-potential-evoked NA release fails to inhibit transmitter release locally while tyramine-evoked NA release does remain to be established.

REFERENCES

1. STARKE, K. 1987. Presynaptic α-adrenoceptors. Rev. Physiol Biochem. Pharmacol. **107:** 73–146.
2. BURNSTOCK, G. & M. E. HOLMAN. 1961. The transmission of excitation from autonomic nerve to smooth muscle. J. Physiol. London **155:** 115–133.
3. BURNSTOCK, G. & M. E. HOLMAN. 1962. Spontaneous potentials at the sympathetic nerve endings in smooth muscle. J. Physiol. London **160:** 446–460.
4. FATT, P & B. KATZ. 1952. Spontaneous subthreshold activity at motor nerve endings. J. Physiol. London **117:** 109–128.
5. HIRST, G. D. S. & F. R. EDWARDS. 1989. Sympathetic neuroeffector transmission in arteries and aterioles. Physiol. Rev. **69** 546–604.
6. FOLKOW, B., J. HÄGGENDAL & B. LISANDER. 1967. Extent of release and elimination of noradrenaline at peripheral adrenergic nerve terminals. Acta Physiol. Scand. Suppl. 307.
7. BEVAN, J. A., G. B. CHESHER & C. SU. 1969. Release of adrenergic transmitter from terminal plexus in artery. Agents Actions. **1:** 20–26.
8. BLAKELEY, A. G. H. & T. C. CUNNANE. 1979. The packeted release of transmitter from the sympathetic nerves of the guinea pig vas deferens: an electrophysiological study. J. Physiol. London **296:** 85–96.
9. CUNNANE, T. C. & L. STJÄRNE. 1984. Transmitter secretion from individual varicosities of guinea-pig and mouse vas deferens: highly intermittent and monoquantal. Neuroscience. **13:** 1–20.
10. HIRST, G. D. S. & T. O. NEILD. 1980. Some properties of spontaneous excitatory junction potentials recorded from arterioles of guinea-pigs. J. Physiol. London **303:** 43–60.
11. BROCK, J. A. & T. C. CUNNANE. 1988. Electrical activity at the sympathetic neuroeffector junction in the guinea-pig vas deferens. J. Physiol. London **399:** 607–632.
12. ÅSTRAND, P., J. A. BROCK, & T. C. CUNNANE. 1988. Time course of transmitter action at the sympathetic neuroeffector junction in rodent vascular and non-vascular smooth muscle. J. Physiol. London **401:** 632-657.
13. ÅSTRAND, P. & L. STJÄRNE. 1988. On the secretory activity of single varicosities in the sympathetic nerves innervating the rat tail artery. J. Physiol. London **409:** 207–220.

14. ÅSTRAND, P. & L. STJÄRNE. 1989. ATP as a sympathetic co-transmitter in rat vasomotor nerves—further evidence that individual release sites respond to nerve impulses by intermittent release of single quanta. Acta Physiol. Scand. **136:** 355–365.

15. BROCK, J. A. & T. C. CUNNANE. 1987. The relationship between the nerve action potential and transmitter release from sympathetic nerve terminals. Nature **326:** 605–607.

16. CUNNANE, T. C. & L. STJÄRNE. 1984. Frequency dependent intermittency and ionic basis of impulse conduction in postganglionic fibres of guinea-pig vas deferens. Neuroscience **11:** 211–229.

17. SNEDDON P. & D. P. WESTFALL. 1984. Pharmacological evidence that adenosine triphosphate and noradrenaline are co-transmitters in the guinea-pig vas deferens. J. Physiol. London **347:** 561–580.

18. SNEDDON, P. & G. BURNSTOCK. 1984. Inhibition of excitatory junction potentials in the guinea-pig vas deferens by a α, β-methylene ATP: further evidence for ATP and noradrenaline as co-transmitters. Eur. J. Pharmacol. **100:** 85–90.

19. ALLCORN, R. J., T. C. CUNNANE & K. KIRKPATRICK. 1986. Action of α, β-methylene ATP and 6-hydroxydopamine on sympathetic neurotransmission in the vas deferens of the guinea-pig, rat and mouse: support for co-transmission. Br. J. Pharmacol. **89:** 647–659.

20. CUNNANE, T. C. & R. J. EVANS. 1990. Nature and localization of the transmitter mediating EJPs in the rabbit jejunal artery. J. Auton. Pharmacol. **10:** 14–15.

21. BLAKELEY, A. G. H., T. C. CUNNANE & S. A PETERSEN. 1981. An electropharmacological analysis of the effects of some drugs on neuromuscular transmission in the vas deferens of the guinea-pig. J. Auton. Pharmacol. **1:** 367–375.

22. ILLES, P. & K. STARKE. 1983. An electrophysiological study of presynaptic α-adrenoceptors in the vas deferens of the mouse. Br. J. Pharmacol. **78:** 365–373.

23. BLAKELEY, A. G. H., T. C. CUNNANE., T. MASKELL., A. MATHIE & S. A. PETERSEN. 1984. α-Adrenoceptors and facilitation at a sympathetic neuroeffector junction. J. Auton. Pharmacol. **4:** 53–58.

24. KALSNER, S. & M. QUILLAN. 1984. A hypothesis to explain the presynaptic effects of adrenoceptor antagonists. Br. J. Pharmacol. **82:** 515–522.

25. BLAKELEY, A. G. H., T. C. CUNNANE & S. A. PETERSEN. 1982. Local regulation of transmitter release from rodent sympathetic nerve terminals? J. Physiol. London 325: 93–109.

26. ROTSHENKER, S. & R. RAHAMIMOFF. 1970. Neuromuscular synapse: stochastic properties of spontaneous release of transmitter. Science. **170:** 648–649.

27. ROBITAILLE, R. & J. P. TREMBLAY. 1987. Non-uniform release at the frog neuromuscular junction: evidence of morphologic and physiological plasticity. Brain Res. Rev. **12:** 95–116.

28. USHERWOOD, P. R. N. 1972. Transmitter release from insect excitatory motor nerve terminals. J. Physiol. London **227:** 527–551.

29. COHEN, I., H. KITA. & W. VAN DER KLOOT. 1974. The intervals between miniature end-plate potentials in the frog are unlikely to be independently or exponentially distributed. J. Physiol. London **239:** 327–339.

30. BENNETT. M. R. & A. G. PETTIGREW. 1975. The formation of synapses in amphibian striated muscle during development. J. Physiol. London **252:** 203–239.

31. STJÄRNE, L. 1981. On sites and mechanisms of presynaptic control of noradrenaline secretion. *In* Chemical Neurotransmission 75 Years. L. Stjärne, P. Hedqvist, P. Lagercrantz & Å. Wennmalm, Eds: 257–272 Academic Press. London, New York, Sydney & San Francisco.

32. BROCK, J. A., T. C. CUNNANE, K. STARKE & C. WARDELL. α_2-Adrenoceptor mediated autoinhibition of sympathetic transmitter release in guinea pig vas deferens studied by intracellular and focal extracellular recording of junction potentials and currents. Naunyn Schmiedebergs Arch. Pharmacol. (In press.)

The Mechanism of Inhibition of ^{3}H-Norepinephrine Release by Norepinephrine in Cultured Sympathetic Neurons

SANJIV V. BHAVE, DENNIS A. PRZYWARA,
ANJALI S. BHAVE, TARUNA D. WAKADE,
AND ARUN R. WAKADE[a]

Department of Pharmacology
Wayne State University
School of Medicine
540 East Canfield
Detroit, Michigan 48201

Modulation of release of sympathetic transmitter by alpha$_2$-receptor agonists and antagonists has been demonstrated in peripheral organs for over 20 years. In addition to adrenergic receptors a number of other types of receptors are also known to modulate stimulus-evoked release of norepinephrine (NE) from postganglionic sympathetic neurons.[1,2] Whether alpha$_2$-receptors play a physiological role in the control of sympathetic transmitter release has been repeatedly challenged by Kalsner.[3] While this controversy remains unresolved, several investigators have focused their attention on the mechanism of presynaptic regulation because several endogenous substances such as NE, acetylcholine (ACh), adenosine (Adn), dopamine (DA), etc., significantly modulate the release of NE.[4–7] For example, it was suggested that alpha$_2$-agonists could depress invasion of action potentials into the varicosities and reduce the probability of the release of NE during intermittent stimulation.[8] A reduction in neuronal cAMP by exogenous NE was considered to be another means of inhibiting NE release by alpha$_2$ agonists.[9] Inhibition of Ca^{++} influx either through voltage-gated Ca^{++} channels or secondary to enhanced efflux of K$^+$ by NE was suggested by others.[10–12]

We studied the inhibitory effects of ACh, Adn, NE, and epinephrine (EPI) on ^{3}H-NE release in the guinea pig heart in the presence of K$^+$ channel blocker, tetraethylammonium (TEA), to know if these agents modulate ^{3}H-NE release by altering K$^+$ efflux.[13,14] Observations and conclusions of these experiments are summarized in FIGURE 1. ACh, Adn, NE, and EPI lost their inhibitory effects on ^{3}H-NE release induced by stimulation in the presence of TEA. However, lowering of Ca^{++} from 2.5 to 0.25 mM (in the presence of 20 mM TEA) restored the inhibitory action of NE and EPI but not of ACh and Adn. This observation led us to propose that alpha$_2$ agonists directly inhibit Ca^{++} entry without the involvement of K$^+$ efflux and thereby inhibit the stimulus-evoked release of ^{3}H-NE. In contrast, ACh and Adn act on some other sites (i.e., K$^+$ channels) and in turn reduce Ca^{++} entry to inhibit the release. Unfortunately we could not provide more direct support for this proposal because measurements of ionic conductances and Ca^{++} concentrations in the sympathetic neurons of the guinea pig heart are technically impossible.

---

[a] Author to whom correspondence should be addressed.

Availability of pure neuronal cultures has eliminated this problem. Sympathetic neurons of different species and age can be grown in primary culture and used to study the noradrenergic mechanisms. Cultured sympathetic neurons take up and retain ^{3}H-NE and the newly stored ^{3}H-NE can be released by depolarizing stimuli in a Ca^{++}-dependent manner.[15,16] These facts tempted us to reinvestigate the question of mechanism of alpha-adrenergic agents in regulating the release of ^{3}H-NE in cultured sympathetic neurons. In this communication we present preliminary data derived from biophysical and fluorescent-probe techniques to support our original proposal that NE inhibits its own release by restricting the entry of Ca^{++} in sympathetic neurons.

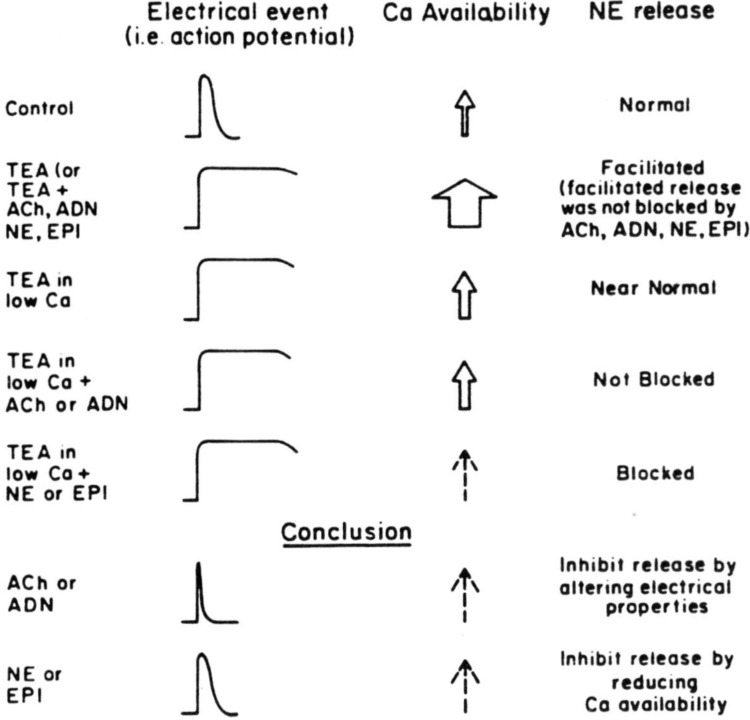

FIGURE 1. Hypothetical model used to describe the site and mechanism of action of various presynaptic agents on sympathetic neurotransmission. ADN, adenosine; EPI, epinephrine; ACh, acetylcholine. (Reproduced from Wakade and Wakade[14] with permission from Pergamon Press Ltd.)

We have established the existence of adrenergic receptors in cultured sympathetic neurons of the chick embryo.[17] The existence was assessed by measuring changes in cAMP content because of the known coupling between alpha-adrenergic receptors and the adenylate cyclase-cAMP system. As shown in TABLE 1, alpha$_2$ agonists by themselves did not alter cAMP content of sympathetic neurons but that enhanced by forskolin was greatly facilitated by NE, clonidine, and DA. The facilitatory effect was mainly due to activation of alpha$_2$-receptors because phenylephrine did not facilitate cAMP formation and phentolamine and yohimbine blocked the effects of alpha$_2$

TABLE 1. Effects of Adrenergic Agents on cAMP Content of Embryonic Chick Sympathetic Neurons[a]

Test Agent (μM)	cAMP Content (pmol/mg protein)	Percent of Control
1. None (Control)	21.5 ± 2.9	100
2. Norepinephrine (30)	22.5 ± 2.5^b	105
3. Clonidine (10)	18.2 ± 3.5^b	85
4. Phenylephrine (10)	19.5 ± 2.2^b	90
5. Forskolin (10)	148.0 ± 21.7^c	688
6. + Norepinephrine	431.8 ± 18.5^d	2008
7. + Clonidine	344.3 ± 20.8^d	1600
8. + Phenylephrine	153.7 ± 13.5^e	715
9. Yohimbine + forskolin + norepinephrine	245.7 ± 45.5^f	1142
10. Yohimbine + forskolin + clonidine	180.0 ± 22.5^g	837
11. Prazosin + forskolin + clonidine	358.7 ± 27.5^h	1668

[a]Sympathetic neurons derived from lumbosacral paravertebral ganglia of 10-day-old chick embryos were prepared for culture as described before.[20,21] Briefly, sympathetic ganglia were digested with trypsin (0.1%) for 30 minutes, washed with excess of phosphate-buffered saline, and then dissociated by trituration in F-14 medium containing 10% heat-inactivated horse serum. Dissociated cells were plated in a plastic dish and kept in a CO_2 incubator for 90 minutes. Unattached neuronal cells were removed, centrifuged, and the pellet (cells) resuspended in F-14 medium and counted. Neurons were cultured in a chemically defined medium, F-14, in the presence of nerve growth factor (40 ng/ml) and insulin and transferrin (1 μg/ml of each). Sympathetic neurons (75,000 cells/dish) cultured for 2 days were washed with Krebs solution and treated with various agents as shown. To prevent degradation of cAMP, 10 μM 3-isobutyl-1-methylxanthine was present in all the dishes. Neurons were pretreated with adrenergic agents for 15 minutes and then forskolin was added for 15 minutes. In other cases, neurons were treated with adrenergic blocking agents for 15 minutes and then agonist plus forskolin was added for 15 minutes. Adrenergic agonist or antagonist was added for 30 minutes. Concentrations of agonists and antagonists used in combination experiments were the same as those when used singly. At the end of the treatment, cultures were extracted in 6% trichloroacetic acid and the extract processed for the analysis of cAMP. The results described in this table are taken from Wakade *et al.*[17] Each value is a mean $\pm$ standard error of the mean of 5 experiments.

[b]Statistically not significant compared to control.

[c]$p < 0.001$ compared to control.

[d]$p < 0.001$ compared to 5.

[e]$p < 0.2$ compared to 5.

[f]$p < 0.001$ compared to 6.

[g]$p < 0.001$ compared to 7.

[h]$p < 0.5$ compared to 7.

agonists. We do not know the mechanism of action of enhancing cAMP by alpha$_2$ agonists but such a phenomenon has been observed in other systems[18,19] and appears to be mediated by alpha$_2$-receptors. We have not yet identified alpha$_2$-receptors by the above or other criteria in cultured sympathetic neurons of the rat.

Recently we have introduced a method to evoke the release of ^{3}H-NE from cultured sympathetic neurons by intermittent electrical stimulation.[16] This technique was used here to evoke release of ^{3}H-NE from cultured sympathetic neurons of the rat superior cervical ganglia. As shown in FIGURE 2, field stimulation of ^{3}H-NE-loaded sympathetic neurons significantly increased the release of tritium over the spontaneous release. After omission of $CaCl_2$ from the medium the stimulus-evoked release of

^{3}H-NE was abolished. The release was restored when Ca^{++} was reintroduced. Although not shown the stimulus-evoked release was susceptible to Ca^{++} channel blockers such as cadmium and lanthanum. These data establish the need for Ca^{++} in the release of NE from the rat cultured sympathetic neurons and justify the use of this model to study presynaptic mechanisms.

Effects of NE on the stimulus-evoked release of ^{3}H-NE are shown in FIGURE 3. 30 μM NE significantly reduced the tritium release induced by electrical stimulation. The inhibitory effect was reversible. A more detailed investigation is in progress to see if other alpha agonists inhibit the release and alpha$_2$ antagonists reverse the inhibitory effects of the agonists.

Effects of NE on Ca^{++} movements were determined by taking two different approaches. Patch-clamp technique was used to measure whole cell Ca^{++} current from the cell body region of sympathetic neurons and Indo-1 probe was used to measure intracellular Ca^{++} in the cell body and the growth cone of sympathetic neurons.

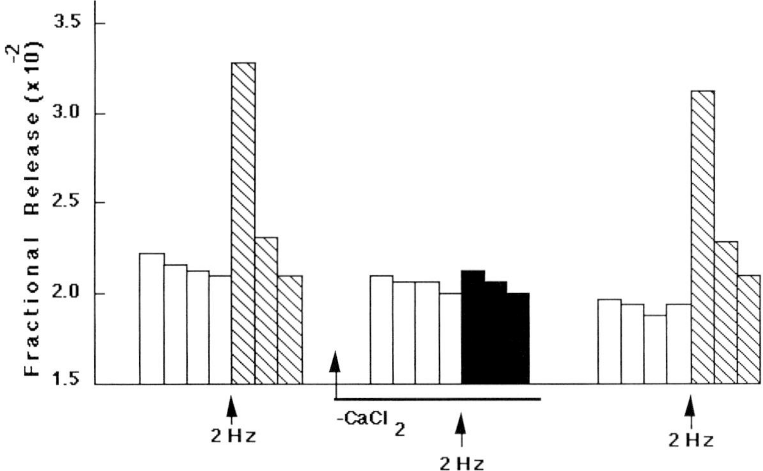

FIGURE 2. Effects of Ca^{++} omission on the stimulation-evoked release of ^{3}H-NE from sympathetic neurons. Sympathetic neurons were incubated with ^{3}H-norepinephrine (4 μCi/ml; 43.9 Ci/mmol, New England Nuclear Co.) for 60 minutes followed by three washes with Krebs solution over a 60 minute period to remove loosely bound ^{3}H-NE. Once the spontaneous outflow of tritium was steady(open columns), neurons were stimulated by delivering 900 pulses at 2 Hz from a Grass stimulator (hatched and solid columns). Neurons were washed with CaCl$_2$-free Krebs solution for 15 minutes, and the above protocol was repeated. Finally, the neurons were returned to Ca^{++}-containing medium for 15 minutes and the release was measured before and after stimulation. Radioactivity determined in 2-minute samples is shown by each column. Each column represents a mean of 5 observations. Method: Sympathetic neurons derived from superior cervical ganglia of neonatal rats were cultured as described before[22] with some modifications. Briefly, superior cervical ganglia obtained from neonatal rats, between 0 to 48 hours after birth, were used. Sympathetic ganglia were digested first with collagenase (0.1%) for 45 minutes and then with trypsin (0.1%) for another 45 minutes. After the washout of trypsin with excess of phosphate-buffered saline, ganglia were dissociated mechanically by trituration in F-14 medium. The number of cells in the neuronal suspension thus obtained was counted using a hemocytometer. Seventy-five thousand neurons were plated in 35 mm dishes in presence of nerve growth factor (100 ng/ml), and insulin and transferrin (2 μg/ml of each). Cultured neurons were used after 2 to 3 days to study the stimulation-induced outflow of tritium.

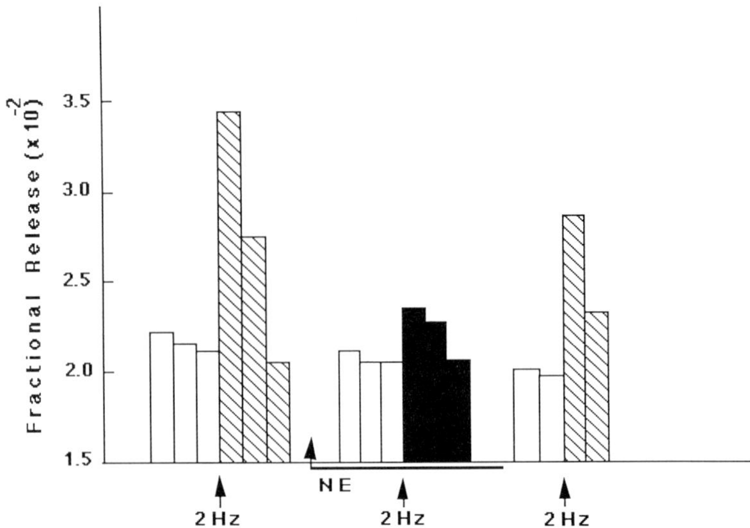

FIGURE 3. Effect of NE on the stimulation-evoked release of ^{3}H-NE from sympathetic neurons. The experimental protocol was identical to that described in FIGURE 2 with the following exceptions. Krebs solution used to wash out the excess of ^{3}H-NE contained 0.2 μM desipramine. The drug was included to prevent exchange of ^{3}H-NE with NE. Thirty micromolar NE was added for 15 minutes prior to collection of samples and the recovery was estimated 15 minutes after washout of NE. Each bar represents a mean of 5 observations.

Cesium and TEA were included to block outward K$^+$ currents and tetrodotoxin (TTX) to block inward current through Na$^+$ channels. As shown in FIGURE 4a, stepping up the holding potential from -70 mV to 0 mV for 200 mseconds produced a robust Ca^{++} current that had a sharp rising phase and a slow noninactivating phase. Application of 30 μM NE did not produce a significant decrease in Ca^{++} current. The trace in FIGURE 4a is a representative of 5 experiments.

The effects of NE on the outward K$^+$ currents in the cell body of sympathetic neurons were also examined by the patch-clamp technique. Cadmium (30 μM) and TTX were added to the bath solution to block inward Ca^{++} and Na$^+$ currents, respectively. As shown in FIGURE 4b, voltage-elicited outward K$^+$ current was unaffected by 30 μM NE. Similar observations were made in two other cell bodies. Although not shown, TEA and 4-aminopyridine caused over 80% reduction in K$^+$ current of sympathetic neurons that were nonresponsive to NE. Thus, patch-clamp analysis provided no evidence of inhibition of Ca^{++} current and enhancement of K$^+$ current in the cell bodies of sympathetic neurons by NE.

It has been known for a long time that release of transmitter occurs mainly from the terminal regions of neurons. Therefore it is likely that modulation of transmitter release occurs at the nerve terminal region rather than at the cell body region. For this reason we examined the effects of NE on Ca^{++} movements in the nerve ending or the growth cone as well as the cell body of cultured sympathetic neurons by using a laser photometer. The results are summarized in FIGURE 5. Field stimulation (1 Hz, 10 seconds) caused a significant increase in Ca^{++} concentration of the cell body. The effect of stimulation on Ca^{++} rise was unaffected by NE. In sharp contrast, the increase in Ca^{++} concentration observed in the growth cone after field stimulation was completely blocked by 30 μM NE. The inhibitory effect was fully reversible 10 minutes after washout of NE. Similar observations were made in five other preparations.

Unlike paravertebral sympathetic neurons of the chick embryo,[17] cultured sympathetic neurons of the rat superior cervical ganglia exhibited modulation of ^{3}H-NE release by exogenous NE. Whether the rat neurons (obtained from 1- to 2-day neonates) are developmentally more mature than the chick neurons (obtained from 10-day-old embryos) to account for the difference in their response to NE is not clear. Whatever the case may be, expression of alpha$_2$ modulation of ^{3}H-NE release by a pure population of mammalian sympathetic neurons provides a valuable model to study the molecular mechanism of transmitter release.

We used this model to investigate the mechanism of alpha$_2$-receptor-mediated modulation of ^{3}H-NE release and came up with two clear findings. One, NE did not enhance either outward K^+ current or depress inward Ca^{++} current in the cell bodies of sympathetic neurons. Two, NE was effective in blocking inward Ca^{++} current only in the growth cones of sympathetic neurons.

We believe these observations offer new insights not only in understanding the mechanism of modulation of ^{3}H-NE release but improving our knowledge about the behavior of different regions of sympathetic neurons growing in culture. Since NE

FIGURE 4. Effects of NE on the Ca^{++} and K^+ currents of the sympathetic neurons. (**a**) Ca^{++} currents. Peak voltage-elicited Ca^{++} currents were recorded from the same neuron during control and the indicated NE treatment. Methods: Coverslips with attached sympathetic neurons were placed in a 1.5-ml superfusion chamber mounted on the stage of an inverted microscope. The bathing solution contained (in mM) 120 NaCl, 4.7 KCl, 1 MgCl$_2$, 5 CaCl$_2$, 10 HEPES, 10 glucose, 20 TEA-Cl, and 200 nM tetrodotoxin, pH 7.3 with NaOH. Electrodes of 0.8 to 3 megohm resistance contained (in mM) 100 CsCl, 5 MgCl$_2$, 10 EGTA, 2 Mg-ATP, 40 HEPES, pH 7.3 with CsOH. Cells were voltage clamped at 22°C using the patch-clamp technique.[23] Cells without extensive neurites were chosen to insure adequate voltage control. Cells exhibiting signs of inadequate space clamp (notches or latency at the leading edge of the current trace or long-lasting tails after return to holding potential) were not used. Cells were tested for current rundown before exposure to NE and only those showing stable currents were used. NE was added by superfusing 5 ml of desired solution through the bath. Inward currents were elicited when cells held at -70 mV were depolarized to 0 mV for 200 mseconds. The signal from the patch clamp amplifier (List EPC7) was filtered at 3 KHz through an 8 pole Bessel filter (Frequency Devices 902) and digitally stored and analyzed using a personal computer (TL-1-125 kHz DMA interface and pCLAMP software, Axon instruments). (**b**) K^+ currents. Effect of 30 μM NE on K^+ current in rat sympathetic neuron held at -70 mV and depolarized to 0 mV for 200 mseconds. Solutions used for recording K^+ currents were external (in mM) 120 NaCl, 5 KCl, 1 MgCl$_2$, 2.5 CaCl$_2$, 10 HEPES, 10 glucose, 60 μM CdCl$_2$, and 200 nM tetrodotoxin, pH 7.3 with NaOH; internal (in mM) 125 KCl, 5 MgCl$_2$, 10 EGTA, 2 Mg-ATP, 10 HEPES, pH 7.3 with KOH. For other details of cell culture see FIGURE 2.

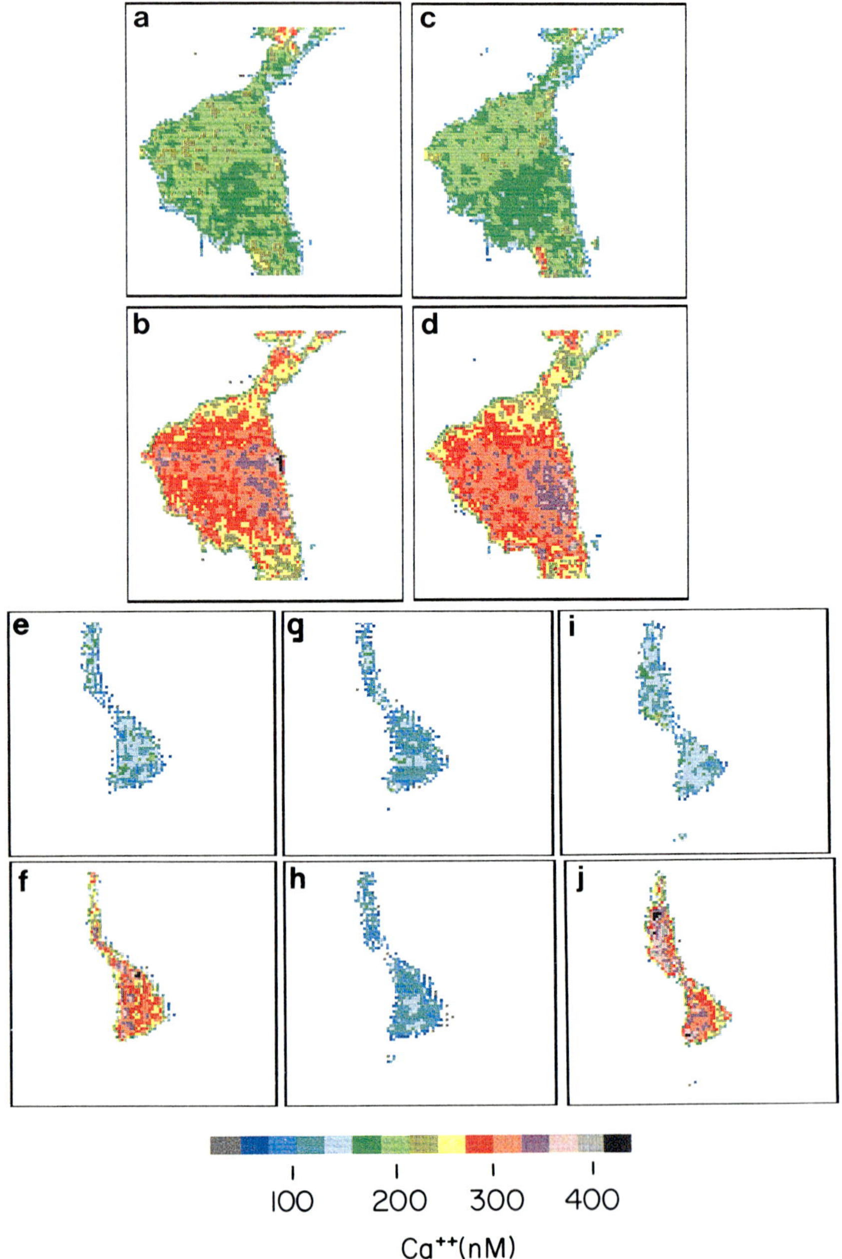

a
c
b
d
e
g
i
f
h
j
100   200   300   400
Ca⁺⁺(nM)

depressed Ca^{++} influx in the growth cone region but not in the cell body region and since NE inhibited the stimulation-evoked release of ^{3}H-NE we have to deduce that the main sites of transmitter release are extrasomatic. This is certainly not a surprise because nerve terminals of sympathetic neurons innervating the effector organs have been known for many years to release NE. For this reason sympathetic neuroeffector organs, rather than sympathetic ganglia, were used extensively to study the release mechanism. This perspective was somewhat understated when sympathetic neurons were placed in culture and used to study the release of ^{3}H-NE. It was largely assumed that the release occur from the entire sympathetic neuron. It is technically easy to measure ionic conductances and Ca^{++} transport properties of the cell bodies compared to that of the growth cones of sympathetic neurons. As a consequence, information derived from the cell body is extrapolated to the growth cone region to explain pharmacological actions on ^{3}H-NE release. Our results clearly demonstrate that such an approach would be misleading. It seems that one has to examine ionic transport or other properties of sympathetic neurons in the growth cone region to draw any correlation with the transmitter release mechanism.

We do not know the reason for ineffectiveness of NE in affecting Ca^{++} transport in the cell body. Although the present study offers a clear demonstration of inhibitory effects of NE on Ca^{++} influx in the growth cone, we do not know whether this is a direct effect on Ca^{++} channels or the reduction in Ca^{++} entry is a secondary consequence of enhanced K^+ efflux in the growth cone. Clearly additional experiments are necessary to fully understand the molecular mechanisms of modulation of transmitter release in sympathetic neurons. The present study has at least provided an important guideline to search for such a mechanism by focusing attention on the region that is primarily involved in the release of NE.

FIGURE 5. Effects of NE on Ca^{++} concentration in the cell body and the growth cone of sympathetic neurons. Images were recorded from sympathetic neural cell body at rest (a and c in absence or presence of 30 μM NE, respectively) and during stimulation, 1 Hz 5 sec (b and d in absence or presence of NE, respectively). Similar protocol was followed for the growth cone at rest (e and g in absence or presence of NE, and i after the washout of NE) and during stimulation, 1 Hz 5 seconds (f and h in absence or presence of NE, and j after the washout of NE). Methods: Sympathetic neurons, cultured on glass coverslips, were loaded with 0.25 μM indo-1-acetoxymethyl ester (Molecular Probes) for 60 minutes and washed 2 times with Krebs solution at 37°C. The coverslip was placed in a Leiden chamber, which was then securely anchored on the stage of the microscope of ACAS laser photometer (Meridian Instruments). Neurons were illuminated by passing a laser beam between 340 to 360 nm, and the light emitted from the cell body or the nerve terminal was recorded in the form of fluorescence or images at 405 nm (Ca^{++}-bound Indo-1) and 485 nm (Ca^{++}-free Indo-1) wavelengths. Neutral density filter was used to reduce photobleaching and obtain reproducible responses from the same sites. Computer programs were used to take the ratios of fluorescence values (405/485) and convert into Ca^{++} concentration from the Ca^{++} calibration curve. Images recorded at each wavelength were also ratioed to quantitate distribution of Ca^{++} within the cell. A small platinum wire electrode was placed in the Leiden chamber and connected to the Grass stimulator for the purpose of electrical stimulation of neurons. Supramaximal parameters were used at 1 Hz for 5 seconds (120 mA strength and 1.0 mseconds duration). All the recording and optical parameters were constant for the cell body and the growth cone except for the scan strength and step size. Scan strength for the cell body was between 10 to 20% and for the growth cone 40 to 60%. Step size, which determines the distance between two scanning points, along x and y axes, for the cell body and growth cone was 0.75 and 0.25 μm, respectively. Decrease in step size provides an enlarged image at the same optical setting. Magnification 400×. Reduced to 65%.

REFERENCES

1. STARKE, K. 1977. Rev. Physiol. Biochem. Pharmacol. **77:** 1–124.
2. ILLES, P. 1986. Neuroscience **17:** 909–928.
3. KALSNER, S. 1985. Biochem. Pharmacol. **34:** 4085–4097.
4. KIRPEKAR, S. M., R. F. FURCHGOTT, A. R. WAKADE & J. C. PRAT. 1973. J. Pharmacol. Exp. Ther. **187:** 529–537.
5. LOFFELHOLZ, K. & E. MUSCHOLL. 1969. Arch. Exp. Pathol. Pharmacol. **267:** 181–184.
6. WAKADE, A. R. & T. D. WAKADE. 1978. J. Physiol. London **282:** 35–49.
7. DUBOCOVICH, M. L. & S. Z. LANGER. 1980. J. Pharmacol. Exp. Ther. **212:** 144–152.
8. STJÄRNE, L. 1989. Rev. Physiol. Biochem. Pharmacol. **112:** 1–137.
9. MULDER, A. H. & A. N. M. SCHOFFELMEER. 1985. Adv. Cyclic Nucleotide Protein Phosphorylation Res. **19:** 273–286.
10. HORN, J. P. & D. A. MCAFEE. 1980. J. Physiol. London **301:** 191–204.
11. MORITA, K. & R. A. NORTH. 1981. Br. J. Pharmacol. **74:** 419–428.
12. LIPSCOMBE, D., S. KONGSAMUT & R. W. TSIEN. 1989. Nature **340:** 639–642.
13. WAKADE, A. R. & T. D. WAKADE. 1982. Neuroscience **7:** 2267–2276.
14. WAKADE, A. R. & T. D. WAKADE. 1983. Neuroscience **9:** 673–677.
15. WAKADE, A. R. & T. D. WAKADE. 1982. J. Pharmacol. Exp. Ther. **223:** 125–129.
16. WAKADE, A. R. & T. D. WAKADE. 1988. Neuroscience **27:** 1007–1019.
17. WAKADE, A. R., T. D. WAKADE, S. V. BHAVE & R. K. MALHOTRA. 1988. Neuroscience **27:** 1021–1028.
18. JONES, J. B., M. L. TOEWS, J. T. TURNER & D. B. BYLUND. 1987. Proc. Nat. Acad. Sci. USA **84:** 1294–1298.
19. SABOL, S. L. & M. NIRENBERG. 1979. J. Biol. Chem. **254:** 1913–1920.
20. EDGAR, D., Y. A. BARDE & H. THOENEN. 1981. Nature **289:** 294–295.
21. WAKADE, A. R., D. EDGAR & H. THOENEN. 1982. Exp. Cell Res. **140:** 71–78.
22. HEFTI, F., H. GNAHN, M. E. SCHWAB & H. THOENEN. 1982. J. Neurosci. **2:** 1554–1566.
23. HAMIL, O. P., A. MARTY, E. NEHER, B. SAKMANN & F. J. SIGWORTH. 1981. Pflugers Arch. **391:** 85–100.

Receptor Interactions at Noradrenergic Neurones[a]

P. ILLES, H.-D. WEBER, J. NEUBURGER, B. BUCHER,[b]
J. T. REGENOLD, AND W. NÖRENBERG

Department of Pharmacology
University of Freiburg
Hermann-Herder-Strasse 5
D-7800 Freiburg, Federal Republic of Germany

[b]*Laboratory of Cellular and Molecular Pharmacology CNRS*
UA 600, B.P. 10
F-67400 Illkirch, France

INTRODUCTION

Classical neurotransmitters were shown to coexist with peptide and nonpeptide cotransmitters both in the peripheral and central nervous system.[1,2] The present overview will deal with two cotransmitters of noradrenaline, namely, with the opioid peptides and neuropeptide Y (NPY). Both noradrenaline and its cotransmitters may activate pre- and postsynaptic (or somatodendritic) receptors. Noradrenaline acts at adrenoceptors, which may be classified as α_1 and α_2 types.[3] Three major groups of opioid peptides have been described; they stimulate the μ- (β-endorphin), δ- (enkephalins), and κ- (dynorphins) receptors.[4] NPY activates Y_1- and Y_2-receptors.[5]

Recently, it was reported that the concomitant application of catecholamines and their peptide cotransmitters produces selective changes in the number and affinity of the respective binding sites.[6,7] Our aim was to demonstrate a similar interaction in electrophysiological experiments. Moreover, we clarified the signal transduction mechanisms triggered by receptor activation and tried to elucidate whether the interaction takes place at the level of these mechanisms or at a step prior to that.

SIGNAL TRANSDUCTION MECHANISMS

Two major second messenger pathways have been described. Firstly, adenylate cyclase hydrolyzes ATP to cAMP, and secondly, phospholipase A hydrolyzes the membrane phospholipid phosphatidylinositol 4,5-bisphosphate to inositol 1,4,5-trisphosphate and diacylglycerol. cAMP and diacylglycerol activate two protein kinases, namely, protein kinase A and C, respectively, which may phosphorylate ionic channels thereby affecting their conductance.[8,9] Adenylate cyclase may be coupled either positively (G_s) or negatively (G_i) to presynaptic receptors via GTP-binding proteins.[10] It is assumed that a negative coupling of phospholipase C to receptors occurs also via a G protein, termed G_p.[11] G_i (and G_o) are ADP ribosylated by pertussis toxin, whereas G_s is ADP ribosylated by cholera toxin. The intracellular application of nonhydrolyzable analogues of guanosine diphosphate (GDP) or triphosphate (GTP), namely GDP-β-S or GTP-γ-S, also interferes with the coupling of receptors to G

[a]This work was supported by the Deutsche Forschungsgemeinschaft (SFB 325).

proteins. GDP-β-S blocks the agonist effects, while GTP-γ-S converts the normally reversible responses to agonists into irreversible ones.

Presynaptic Receptors

Presynaptic receptors may be divided into two classes. The activation of α_2-adrenoceptors, opioid, and NPY receptors depresses nerve stimulation–induced noradrenaline release, whereas the activation of β_2-adrenoceptors enhances it. In general, transmitter release depends on the level of free axoplasmic calcium, which triggers exocytosis.[12] Thus, the ultimate reason for presynaptic inhibition is a reduction of Ca^{2+} influx into the axoplasm via voltage-sensitive Ca^{2+} channels.[13] By contrast, presynaptic facilitatory receptors increase Ca^{2+} entry into the nerve terminals.[13]

There is general agreement that the β_2-adrenoceptor-mediated potentiation of noradrenaline release is mainly due to an enhanced intraneuronal level of cAMP.[14] This effect involves the activation of G_s, which seems to increase indirectly via the cAMP-dependent protein kinase A the opening of Ca^{2+} channels; a minor direct effect on the channel cannot be excluded either (see FIGURE 5).

The situation is much less clear with inhibitory α_2-adrenoceptors. Based on experiments performed in brain slices, some authors suggest that α_2-adrenoceptor stimulation depresses the release of noradrenaline subsequent to a decrease in cAMP concentration.[15,16] However, recent data disagree with this conclusion.[17] It is more likely that an inhibitory G protein, which can be inactivated both by pertussis toxin[17,18] and by N-ethylmaleimide,[19] directly increases a potassium conductance.[13] N-Ethylmaleimide, in concentrations up to 50 μmol/l, rather selectively reacts with cystein residues of inhibitory G proteins, and only in higher concentrations blocks the catalytic subunit of adenylate cyclase[20] or the binding of radioligands to the α_2-adrenoceptor.[21] The involvement of G_p with the subsequent activation of protein kinase C by diacylglycerol was excluded.[22] α_2-Adrenoceptors and opioid receptors of the μ, δ,[23] and κ type[24] may be coupled to a common inhibitory G protein.

It was unequivocally shown that, in postganglionic sympathetic neurones, presynaptic α_2-adrenoceptors do not inhibit transmitter release via a decreased level of cAMP.[25,26] We were also able to confirm this finding by using the sympathetically innervated rat isolated tail artery. The preparation was incubated in [^{3}H]noradrenaline in order to load the transmitter pools of perivascular nerve terminals with the labeled amine. Thereafter, electrical stimulation (24 pulses at 0.4 Hz, 0.3 msecond, 200 mA) led to [^{3}H]noradrenaline release. Both the membrane-permeating analogue of cAMP, 8-Br-cAMP (100 μmol/l), and forskolin (3 μmol/l), which is thought to act directly on the catalytic subunit of adenylate cyclase,[27] concentration dependently increased the electrically induced release of [^{3}H]noradrenaline. B-HT 933 (30 μmol/l), a highly selective α_2-adrenoceptor agonist,[28] was inhibitory (FIGURE 1). The effect of B-HT 933 30 μmol/l did not change in the presence of forskolin 3 μmol/l, and slightly decreased in the presence of 8-Br-cAMP 100 μmol/l (FIGURE 1A). Since 8-Br-cAMP 100 μmol/l facilitated the release to a greater extent than forskolin 3 μmol/l (see legend to FIGURE 1), we performed experiments in which, in spite of the application of 8-Br-cAMP 100 μmol/l, the release of [^{3}H]noradrenaline was unaltered with respect to the predrug situation. Thus, control stimulation was performed with 24 pulses at 0.4 Hz, but in the presence of 8-Br-cAMP 100 μmol/l, stimulation was with only 12 pulses at 0.2 Hz. Under these conditions, 8-Br-cAMP 100 μmol/l failed to alter the effect of B-HT 933 30 μmol/l (FIGURE 1B).

The next question was whether in the terminals of perivascular nerves, α_2-adrenoceptors are linked to an inhibitory G protein. In fact, pretreatment of the tail arteries with N-ethylmaleimide 3 μmol/l for 60 minutes, and a subsequent washout for

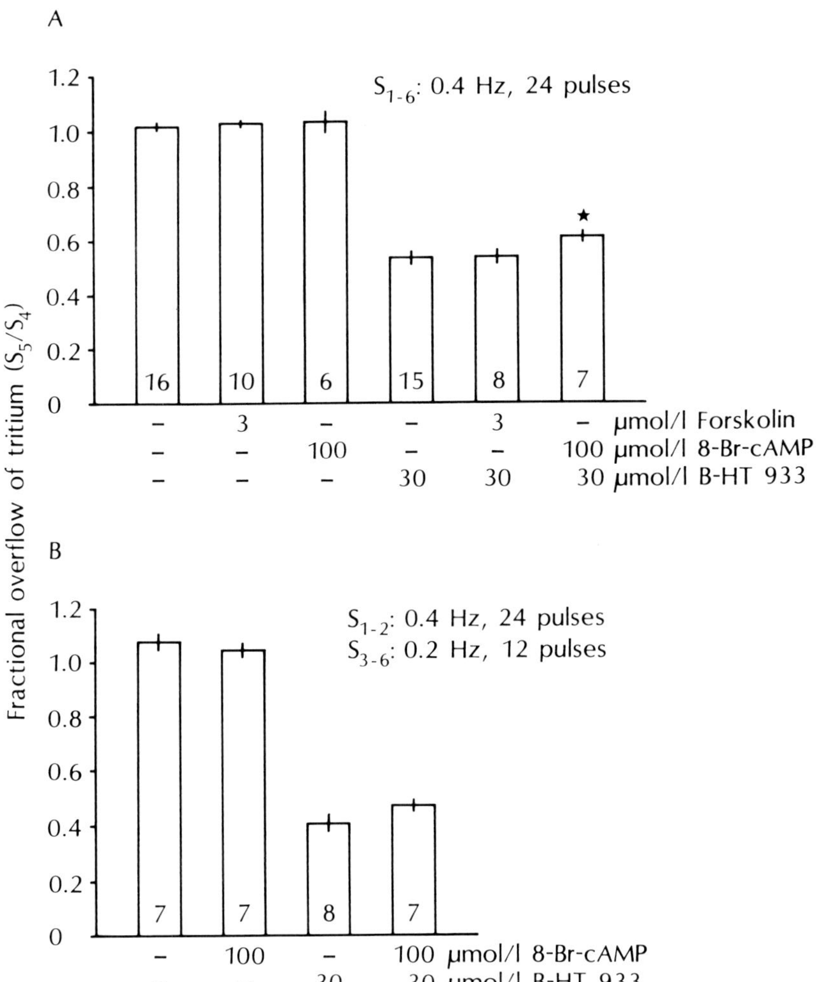

FIGURE 1. Effects of forskolin and 8-Br-cAMP on the overflow of tritium from rat tail arteries preincubated with [^{3}H]noradrenaline, as well as interaction with B-HT 933. Preparation and perfusion/superfusion was as previously described.[66] Six stimulation periods (S_1–S_6) were applied every 16 minutes either with 24 pulses at 0.4 Hz or with 12 pulses at 0.2 Hz (0.3 msecond, 200 mA in both cases) as indicated. Forskolin 3 μmol/l or 8-Br-cAMP 100 μmol/l was added 10 minutes before S_3 and maintained throughout the whole experiment. B-HT 933 30 μmol/l was added 10 minutes before S_5 and was also present throughout. (**A**) When the stimulation conditions were identical during S_1–S_6, forskolin 3 μmol/l did not change the inhibitory effect of B-HT 933 30 μmol/l, and 8-Br-cAMP slightly reduced it. Forskolin 3 μmol/l and 8-Br-cAMP 100 μmol/l increased the overflow of tritium by about 40% and 90%, respectively. (**B**) When the number of pulses and the frequency of stimulation was decreased after S_2, 8-Br-cAMP 100 μmol/l failed to change the inhibitory effect of B-HT 933 30 μmol/l. 8-Br-cAMP 100 μmol/l did not alter the overflow under these conditions. The number of experiments is indicated in the columns. *$p <$ 0.05; significant difference from the effect of B-HT 933 30 μmol/l in the absence of forskolin 3 μmol/l and 8-Br-cAMP 100 μmol/l.

another 30 minutes, markedly facilitated the electrically (0.4 Hz, 24 pulses, 0.3 mseconds, 200 mA) evoked release of [³H]noradrenaline (FIGURE 2). BH-T 933 10 μmol/l produced a larger percent inhibition without (FIGURE 2A), than after a preincubation with N-ethylmaleimide 3 μmol/l (Figure 2B). When the current strength was decreased from 200 mA to 90 mA, the release decreased, but the percent inhibition by B-HT 933 10 μmol/l was almost unchanged. Thus, at a low current strength N-ethylmaleimide 3 μmol/l failed to increase the release; under these conditions B-HT 933 10 μmol/l had no effect at all.

In accordance with our results, both N-ethylmaleimide (mouse vas deferens)[29] and pertussis toxin (rat vas deferens;[30] guinea pig mesenteric artery[31]) prevented the effect of the α_2 agonist clonidine in sympathetically innervated isolated smooth muscle preparations. By contrast, in pithed rats, the presynaptic cardioinhibitory action of α_2 agonists was not altered by pertussis toxin pretreatment.[32,33] Similar negative results were reported also for the mouse isolated atrium.[34] Pertussis toxin may be ineffective because of its slow penetration through the neuronal membrane by a temperature-dependent process. Moreover, the biological activity of the toxin is highly variable.

Since our results indicated that α_2 agonists act via an inhibitory G protein, but without the involvement of protein kinase A, the possible role of protein kinase C was investigated. Phorbol 12-myristate 13-acetate (0.1, 1 μmol/l) concentration and time dependently increased [³H]noradrenaline release in the rat tail artery, while phorbol 13-acetate (1 μmol/l) was inactive (FIGURE 3A). Phorbol 13-acetate is structurally related to phorbol 12-myristate 13-acetate, but it does not affect protein kinase C.[35] Although phorbol 12-myristate 13-acetate 1 μmol/l almost doubled the release, it did not alter appreciably the effect of B-HT 933 10 μmol/l; the percentual inhibition by B-HT 933 10 μmol/l was the same both in the absence and presence of the phorbol ester (FIGURE 3B). Thus, we suggest that in the terminals of perivascular nerves α_2-adrenoceptors are coupled to G_i or G_o, and the G protein directly inhibits the opening of voltage-dependent Ca^{2+} channels (see FIGURE 5).

Presynaptic α_2-adrenoceptors and opioid receptors of central noradrenergic neurones are probably linked to a K^+ channel.[13,36] A potassium permeability increase may enhance the intermittency of action potential propagation to the varicosity, or may shorten the duration of the action potential. In consequence, less Ca^{2+} will enter the axoplasm and less transmitter will be released.

Somatodendritic Receptors

Noradrenaline inhibited the Ca^{2+}-dependent shoulder of action potentials in postganglionic sympathetic neurones.[37] The hyperpolarizing after potential was also decreased, indicating a direct effect of noradrenaline on voltage-sensitive Ca^{2+} conductance. In fact, voltage clamp measurements revealed a reduction of the inward Ca^{2+} current (I_{ca}) by noradrenaline, probably by reducing the number of available calcium channels.[38] Patch clamp experiments showed that α_2-adrenoceptor activation by noradrenaline depressed I_{ca}, and a diffusible second-messenger was not involved in this effect.[39] With GTP-γ-S in the electrode, noradrenaline caused a persistent inhibition of the Ca^{2+} current, indicating receptor coupling to a G protein.[40]

The nucleus locus ceruleus (LC) is situated in the pons and consists of a compact group of noradrenergic cell bodies, which project into various areas of the central nervous system.[41] In LC neurones, stimulation of somatic (and/or dendritic) α_2-adrenoceptors[42,43] and opioid μ-receptors[44,45] increases the same potassium conductance[46] and, thereby, leads to hyperpolarization and inhibition of spontaneous firing. The 2coupling between these receptors and the inward rectifying K^+ channels[47] involves a pertussis toxin–sensitive G protein.[48,49]

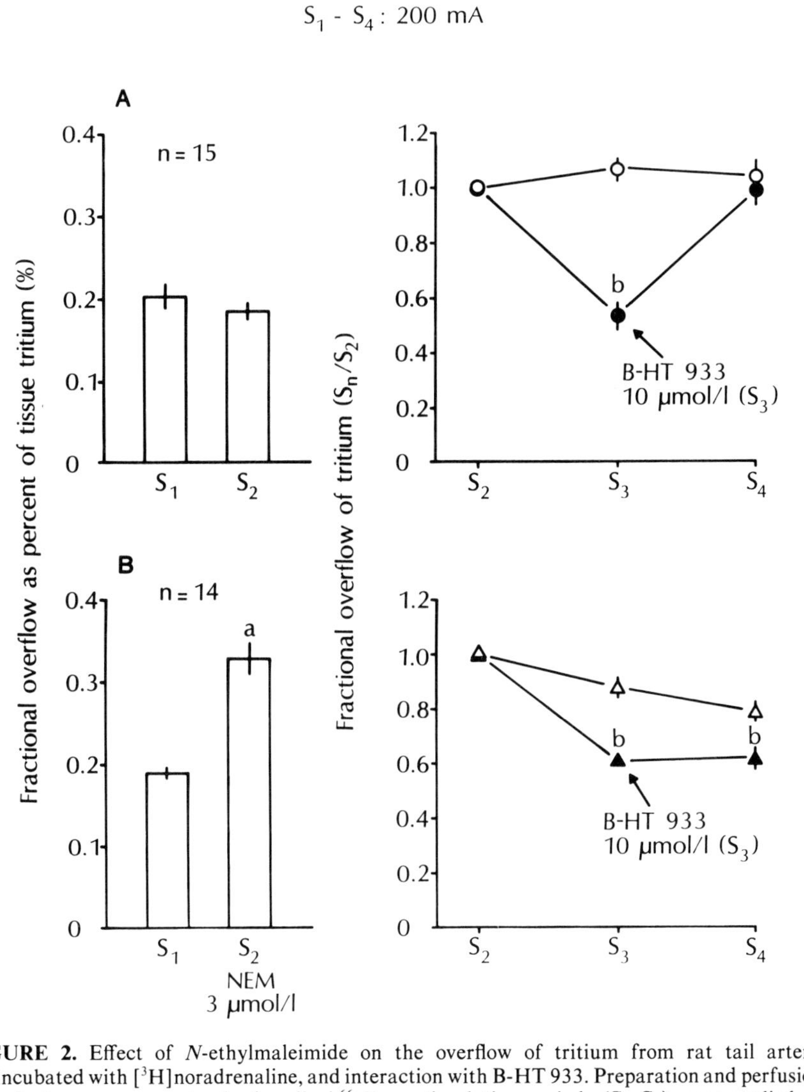

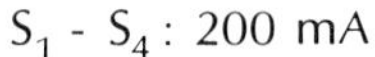

FIGURE 2. Effect of *N*-ethylmaleimide on the overflow of tritium from rat tail arteries preincubated with [³H]noradrenaline, and interaction with B-HT 933. Preparation and perfusion/superfusion was as previously described.[66] Four stimulation periods (S_1–S_4) were applied (24 pulses at 0.4 Hz, 0.3 msecond, 200 mA). The interval between S_1 and S_2 was 103 minutes; then further stimulation periods followed every 27 minutes. *N*-Ethylmaleimide 3 μmol/l was added 12 minutes after S_1 for 60 minutes and was subsequently washed out for 30 minutes. B-HT 933 10 μmol/l was added 21 minutes before S_3 and was present for another 5 minutes. (**A**) Constancy of tritium overflow in the absence of drugs (open circles; $n = 8$) and inhibition by B-HT 933 10 μmol/l (filled circles; $n = 7$). (**B**) Increase of tritium overflow by *N*-ethylmaleimide 3 μmol/l (open triangles; $n = 7$) and prevention of the inhibitory effect of B-HT 933 10 μmol/l by this treatment (filled triangles; $n = 7$). [a]$p < 0.01$; significant difference from the effect of S_1. [b]$p < 0.05$–0.01; significant differences from the respective ratios in the absence of B-HT 933 10 μmol/l.

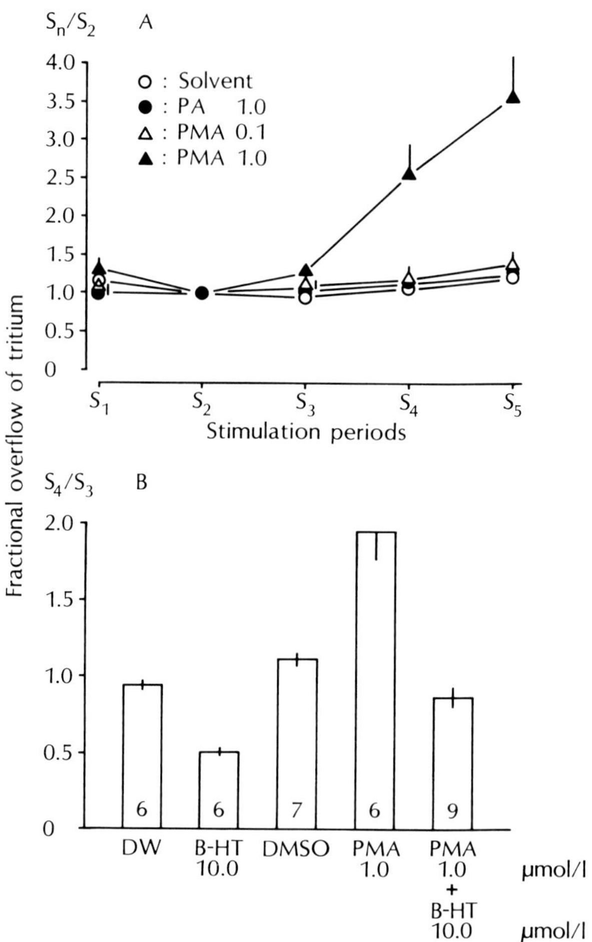

FIGURE 3. Effects of phorbol 12-myristate 13-acetate (PMA) and phorbol 13-acetate (PA) on the overflow of tritium from rat tail arteries preincubated with [^{3}H]noradrenaline, as well as interaction with B-HT 933. Preparation and perfusion/superfusion were as previously described.[66] Five stimulation periods (S_1–S_5) were applied every 14 minutes (24 pulses at 0.4 Hz, 0.3 mseconds, 200 mA). PMA (0.1, 1 µmol/l) or PA 1 µmol/l was added 8 minutes before S_3 and maintained throughout the whole experiment. B-HT 933 10 µmol/l was added 4 minutes before S_4 and was also present throughout. (**A**) PMA 1 µmol/l (filled triangles; $n = 6$), but not PMA 0.1 µmol/l (open triangles; $n = 6$) or PA 1 µmol/l (filled circles; $n = 7$), enhanced the overflow of tritium, when compared with the effect of the solvent (open circles; $n = 7$). (**B**) B-HT 933 10 µmol/l produced the same percent inhibition in the presence and absence of PMA 1 µmol/l. PMA and PA were dissolved in dimethyl sulphoxide (DMSO). B-HT 933 (B-HT) was dissolved in distilled water (DW). The number of experiments is indicated in the columns.

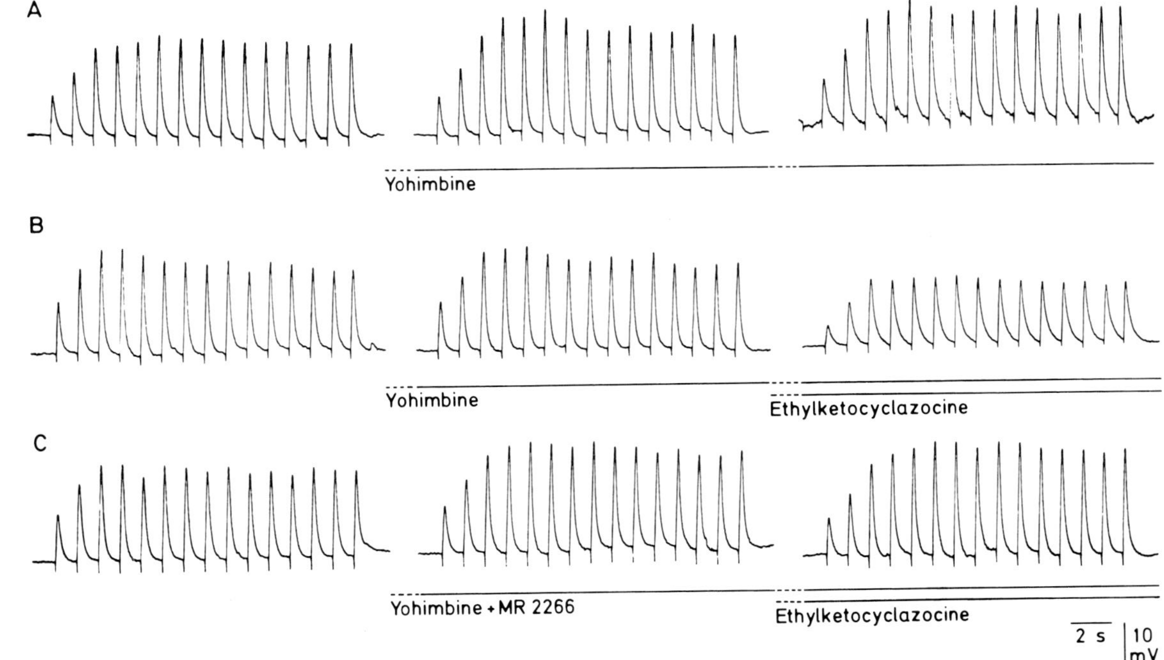

FIGURE 4. Effect of ethylketocyclazocine on excitatory junction potentials (EJPs) of rabbit jejunal arteries, in the presence of yohimbine. Interaction of ethylketocyclazocine with MR 2266. Preparation and experimental conditions were as previously described.[58] Single smooth muscle cells were impaled with glass microelectrodes filled with KCl 2.5 mol/l. Five stimulation periods (S_1–S_5) were applied every 3 minutes (15 pulses at 1 Hz, 1 msecond, 3–8 V). The voltage was adjusted so that an EJP of about 10 mV amplitude was evoked by a single pulse. Responses to S_1, S_3, and S_5 are shown in each panel. Yohimbine 0.3 μmol/l, with (**C**) or without MR 2266 0.1 μmol/l (**A, B**), was added immediately after S_1 and maintained throughout the whole experiment. Ethylketocyclazocine 0.1 μmol/l was added immediately after S_3 and was also present throughout (**B, C**). Representative tracings from 7 (**A, B**) and 5 (**C**) arteries.

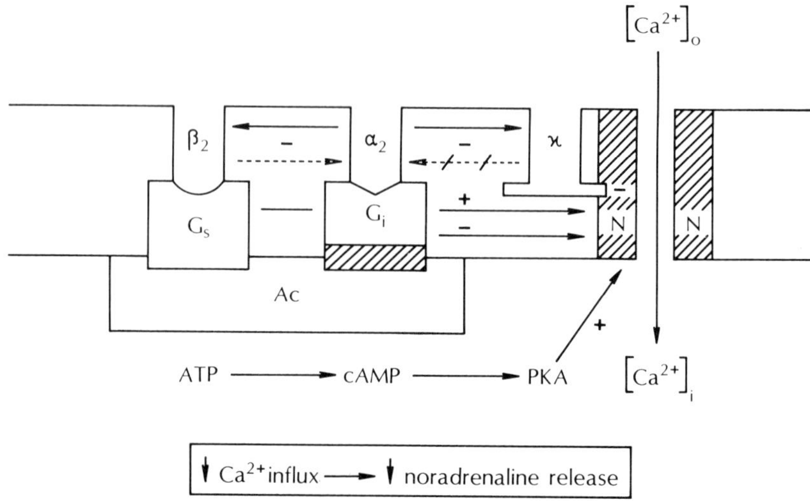

FIGURE 5. The transduction mechanisms of α_2- and β_2-adrenoceptors, as well as opioid κ-receptors in perivascular nerve terminals. Negative interactions between the receptors are indicated by arrows. α_2, α_2-adrenoceptor; β_2, β_2-adrenoceptor; κ, opioid κ-receptor; G_s, stimulatory G protein; G_i, inhibitory G protein; Ac, adenylcyclase; PKA, protein kinase A; N, N-type Ca^{2+} channel; $[Ca^{2+}]_o$, extracellular Ca^{2+} concentration; $[Ca^{2+}]_i$, intracellular Ca^{2+} concentration. The positive and negative signs designate facilitation or inhibition of the Ca^{2+} channel opening, respectively. Downward arrows show that the release of noradrenaline is decreased, when less Ca^{2+} enters the nerve terminals. For further details see the text.

We found that NPY 0.1 μmol/l reduced the discharge of action potentials in LC neurones. When impalement of the cells was with microelectrodes containing GTP-γ-S 2 mmol/l, the spontaneous firing stopped and the cells were hyperpolarized even in the absence of agonists. Thus, it was impossible to investigate the alteration of the firing rate by NPY, although noradrenaline became gradually less effective on repeated application and eventually completely lost its activity. However, when the electrodes contained GDP-β-S 2 mmol/l, the firing did not stop. Under these conditions, the effect of noradrenaline also decreased on repeated application and, at the same time, NPY 0.1 μmol/l became inactive.

The input resistance of LC neurones was reduced in the presence of NPY 0.1 μmol/l. The ionic species involved may be potassium and not chloride, since the microelectrodes were filled with KCl, and under these conditions an increased permeability of the membrane to Cl^- depolarizes rather than hyperpolarizes LC neurones.[50] As suggested by our experiments in which recording was with GDP-β-S containing electrodes, K^+ channels may be opened via the activation of the same type of G protein as in the case of opioid μ-receptors and α_2-adrenoceptors (see FIGURE 9). This possibility is strengthened by the finding that in rats, hypotension induced by the intracisternal injection of NPY and clonidine was reduced by pretreatment of the animals with pertussis toxin.[51] The cardiovascular effects of both agonists were suggested to be due, at least partly, to the inhibition of catecholamine neurones in the medulla oblongata.[52]

RECEPTOR INTERACTIONS

Presynaptic Receptors

Presynaptic α_2- and β_2-adrenoceptors (rabbit ear artery;[53] rabbit pulmonary artery;[54] see FIGURE 5) as well as α_2-adrenoceptors and muscarinic acetylcholine receptors (rabbit ear artery)[55] of perivascular nerves were shown to interact with each other. In the above blood vessels, the release of [³H]noradrenaline was facilitated by β_2 agonists only after α_2-adrenoceptor blockade. Moreover, the inhibitory effect of

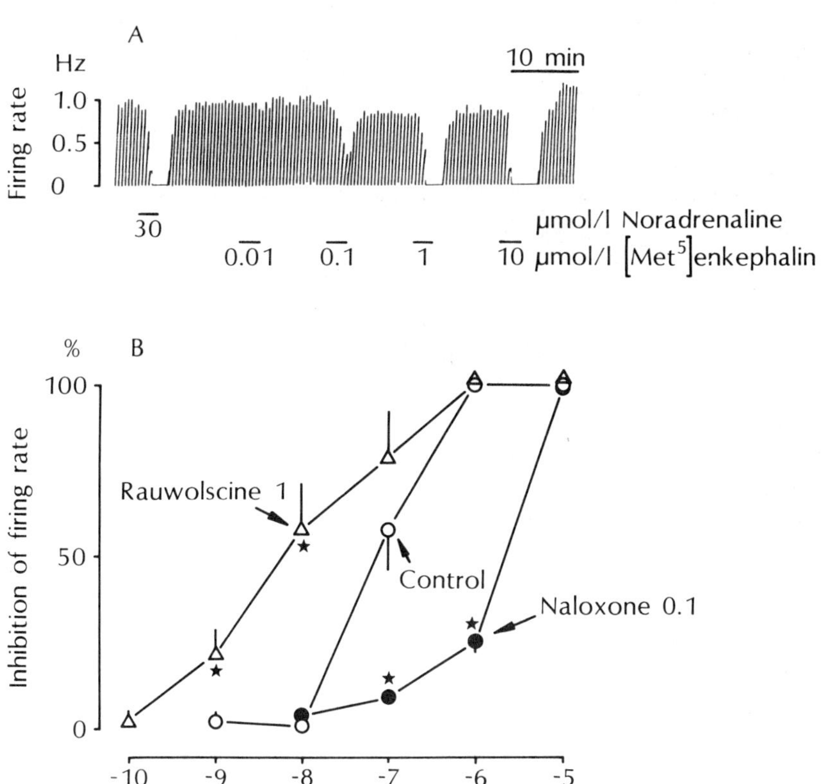

FIGURE 6. Effect of [Met⁵]enkephalin on the firing rate of locus ceruleus (LC) neurones, and interaction with naloxone and rauwolscine. Pontine slices of the rat brain were prepared as previously described.[67] The frequency of spontaneous action potentials was recorded extracellularly as consecutive 30 second samples; glass microelectrodes filled with NaCl 4 mol/l were used. [Met5]enkephalin (0.01–10 μmol/l) was added for 3 minutes every 13 minutes. Naloxone 0.1 μmol/l and rauwolscine 1 μmol/l were present in the bath from at least 15 minutes before addition of the first agonist concentration until the end of the experiment. (**A**) Representative tracing. Noradrenaline 30 μmol/l was added in order to identify the neurones. (**B**) Concentration-response curves of [Met⁵]enkephalin were determined in the absence (open circles) or in the presence of naloxone 0.1 μmol/l (filled circles) or rauwolscine 1 μmol/l (open triangles). Means ± standard error (SE) from 6 slices each. $*p < 0.05$–0.01; significant differences from the effect of the same agonist concentration without antagonists.

acetylcholine on [³H]noradrenaline release was markedy enhanced by α_2 antagonists. We found that in rabbit jejunal arteries opioid δ, but not μ or κ agonists depressed the amplitude of excitatory junction potentials (EJPs),[56] which are a measure of transmitter release.[57] However, after the application of the α_2-adrenoceptor antagonist yohimbine (0.3 μmol/l), the opioid κ-agonist ethylketocyclazocine (0.1 μmol/l) became active (FIGURE 4);[58] The inhibitory effect of ethylketocyclazocine was abolished by the preferential κ antagonist MR 2266 (0.1 μmol/l). Thus, it was concluded that previously silent κ-receptors became operative only when presynaptic α_2-adrenoceptors had been blocked (see FIGURE 5).

In brain slices of rats and rabbits, the inhibition of [³H]noradrenaline release by opioids was reduced in the presence of the α_2 agonist clonidine[59,60] and, conversely, the action of clonidine was diminished in the presence of opioid agonists.[61] Moreover, α_2 antagonists such as yohimbine potentiated the opioid agonist effects.[59,60,62,63] Hence, opioid receptors and α_2-adrenoceptors were suggested to interact with each other in an inhibitory manner.

NPY depressed the release of [³H]noradrenaline from brain slices of rats.[64] Moreover, NPY potentiated the effect of α_2-adrenoceptor agonists, whereas α_2 antago-

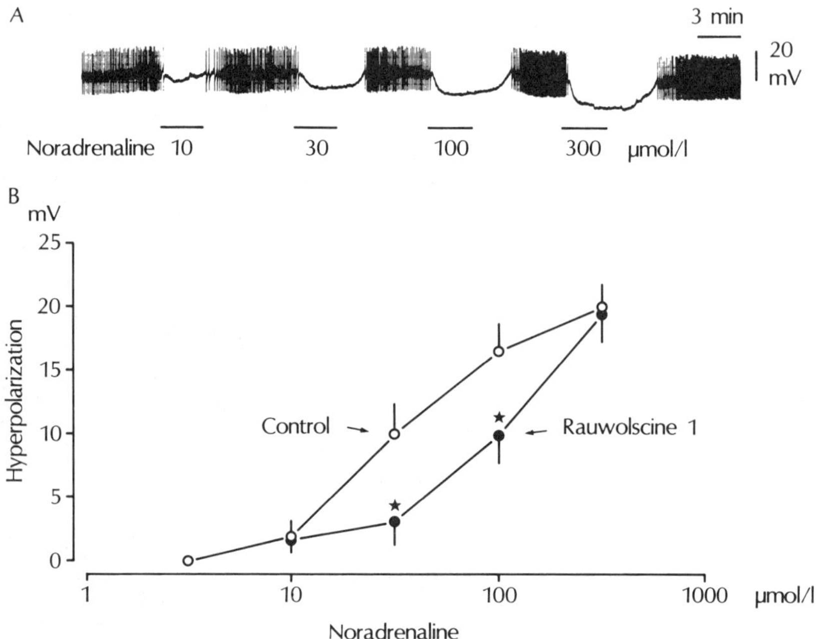

FIGURE 7. Effect of noradrenaline on the membrane potential of LC neurones, and interaction with rauwolscine. Pontine slices of the rat brain were prepared as previously described,[67] and LC cells were impaled with glass microelectrodes filled with KCl 2 mol/l. Noradrenaline (10–300 μmol/l) was added for 3 minutes as indicated by the horizontal bars. Rauwolscine 1 μmol/l was present in the bath from at least 15 minutes before addition of the first agonist concentration until the end of the experiment. (**A**) Representative tracing. (**B**) Concentration-response curves of noradrenaline were determined in the absence (open circles) or in the presence of rauwolscine 1 μmol/l (filled circles). Means ± SE from 6 (open circles) and 7 (filled circles) slices. *$p < 0.05$; significant differences from the effect of the same agonist concentration without antagonists.

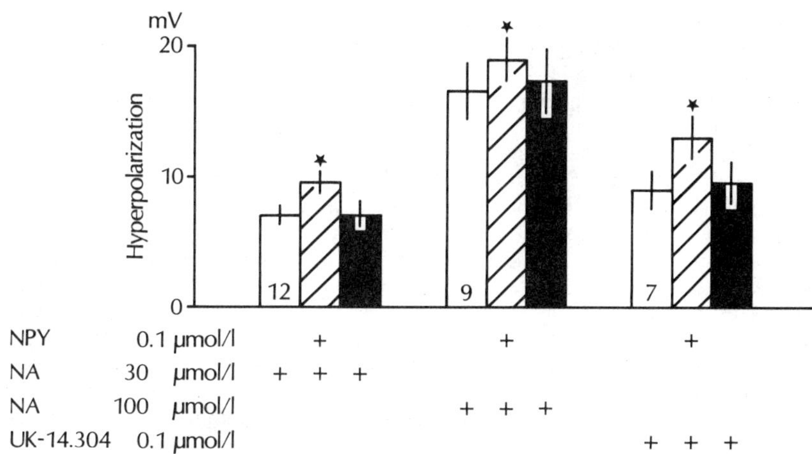

FIGURE 8. Potentiation of the hyperpolarizing effect of noradrenaline and UK 14.304 by neuropeptide Y (NPY) in LC neurones. Pontine slices of the rat brain were prepared as previously described,[67] and LC cells were impaled with glass microelectrodes filled with KCl 2 mol/l. Noradrenaline (30, 100 μmol/l) and UK 14.304 (0.1 μmol/l) were added for 3 minutes, both before (white bars) and 3 minutes after the addition of NPY 0.1 μmol/l (hatched bars). Then NPY 0.1 μmol/l was washed out, and at least 7 minutes were allowed to elapse before the effects of noradrenaline and UK 14.304 (black bars) were again determined. Means ± SE are shown; the number of experiments in each series is presented in the first column. *$p < 0.01$; significant differences from the effect of the same concentration of the α_2-adrenoceptor agonist before the application of NPY 0.1 μmol/l.

nists diminished the effect of NPY. Thus, a facilitatory interaction between α_2- and NPY-receptors appeared to be present.

The site of interaction may be located in the neuronal membrane between the receptors themselves, or in the signal transduction system.[4,58,60] It was suggested that the availability of a common G protein for coupling with the activated opioid receptor may depend on the extent of α_2-adrenoceptor stimulation.[24] A similar conclusion was reached from another series of experiments. It has been shown that α_2-adrenoceptors must be activated by endogenous noradrenaline before α_2 antagonists are able to potentiate the effects of opioid agonists; a mere occupation of the α_2-receptors by antagonists is not sufficient.[63]

Somatodendritic Receptors

Noradrenaline inhibited the spontaneous firing of rat LC neurones in a slice preparation; we found that the effect of noradrenaline was prevented by α_2- but not α_1-adrenoceptor antagonists. [Met⁵]Enkephalin (0.01–10 μmol/l) also depressed the firing (FIGURE 6A). The opioid antagonist naloxone (0.1 μmol/l) shifted the concentration-response curve of [Met⁵]enkephalin to the right, while the α_2 antagonist rauwolscine (1 μmol/l) displaced it to the left (FIGURE 6B). Thus, the blockade of α_2-adrenoceptors potentiated the opioid μ-receptor-mediated inhibition, although rauwolscine alone did not increase the firing rate. This indicates that a tonic control of neuronal activity by endogenously released noradrenaline did not operate under *in vitro* conditions.

FIGURE 7A shows the concentration-dependent hyperpolarization of LC neurones by noradrenaline. The antagonism of noradrenaline by rauwolscine 1 μmol/l proves the presence of α_2-adrenoceptors (FIGURE 7B). NPY 0.1 μmol/l reduced the discharge of action potentials in a manner partly prevented by rauwolscine 1 μmol/l. Moreover, NPY 0.1 μmol/l potentiated the effect of both 30 and 100 μmol/l noradrenaline; this potentiation was completely reversible on washout (FIGURE 8). The action of the selective α_2-adrenoceptor agonist UK 14,304[65] 0.1 μmol/l was also enhanced by NPY 0.1 μmol/l. Unexpectedly the hyperpolarization by opioid agonists, such as [Met5]enkephalin and [D-Ala2, D-Leu5]enkephalin, was not increased by NPY.

In contrast to presynaptic μ-receptors,[63] α_2 antagonists potentiated the opioid agonist effects at somatic μ-receptors even when somatic α_2-adrenoceptors were not activated by endogenous noradrenaline. Under the same conditions, α_2 antagonists counteracted NPY. In LC neurones, the stimulation of α_2-, μ-, and NPY-receptors probably leads to the activation of the same second messenger systems (G proteins), and the effector mechanisms (inward rectifying K^+ channels) are also identical. Since a mere occupation of α_2-adrenoceptors by α_2 antagonists enhanced the potency of [Met5]enkephalin and [D-Ala2, D-Leu5]enkephalin, but decreased the potency of NPY, it is most likely that the site of interaction is located between the respective receptors themselves, and not in the common signal transduction system (see FIGURE 9). Thus, the mechanism of interaction appears to be different from that operating between presynaptic receptors of postganglionic sympathetic neurones.

Locus coeruleus neurones

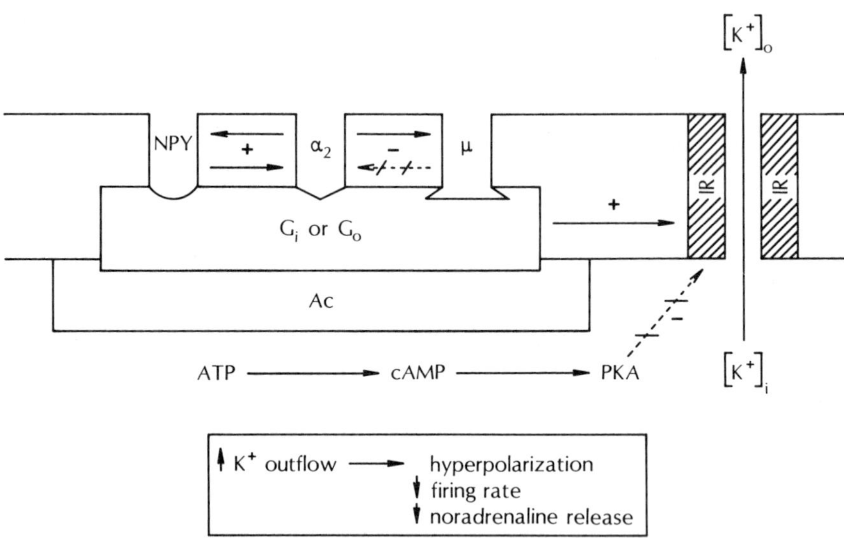

FIGURE 9. The transduction mechanisms of α_2-adrenoceptors as well as of opioid μ- and NPY-receptors in the cell somata of LC neurones. Positive and negative interactions between the receptors are indicated by arrows. α_2, α_2-adrenoceptor; μ, opioid μ-receptor; NPY, NPY-receptor; G_s, stimulatory G protein; G_i or G_o, inhibitory G proteins; Ac, adenylcyclase; PKA, protein kinase A; IR, inward rectifying K^+ channel; $[K^+]_o$, extracellular K^+ concentration; $[K^+]_i$, intracellular K^+ concentration. The positive and negative signs designate facilitation or inhibition of the K^+ channel opening, respectively. Upward and downward arrows show that an enhanced K^+ outflow hyperpolarizes the membrane; this leads to a decreased firing rate and eventually to a decrease of transmitter release from the nerve terminals. For further details see the text.

REFERENCES

1. BURNSTOCK, G. 1986. Acta Physiol. Scand. **126:** 67–91.
2. BARTFAI, T., K. IVERFELDT, G. FISONE & P. SERFÖZÖ. 1988. Ann. Rev. Pharmacol. Toxicol. **28:** 285–310.
3. STARKE, K. 1977. Rev. Physiol. Biochem. Pharmacol. **77:** 1–124.
4. ILLES, P. 1989. Rev. Physiol. Biochem. Pharmacol. **112:** 139–233.
5. WAHLESTEDT, C., L. EDVINSSON, E. EKBLAD & R. HAKANSON. 1987. *In* Neuronal Messengers in Vascular Function. A. Nobin, C. Owman & B. Arneklo-Nobin, Eds.: 231–242. Elsevier Science Publishers. Amsterdam, The Netherlands.
6. FUXE, K., L. F. AGNATI, F. BENFENATI, A. HÄRFSTRAND, G. VON EULER, E. M. PICH & F. FONTANA. 1987. *In* Neuronal Messengers in Vascular Function. A. Nobin, C. Owman & B. Arneklo-Nobin, Eds.: 305–325. Elsevier Science Publishers, Amsterdam, The Netherlands.
7. FUXE, K., L. F. AGNATI, M. ZOLI, A. CINTRA, A. HÄRFSTRAND, G. VON EULER, R. GRIMALDI, M. KALIA & P. ENEROTH. 1988. *In* Regulatory Roles of Opioid Peptides. P. Illes & C. Farsang, Eds.: 33–68. VCH Publishers, Weinheim, FRG.
8. ERULKAR, S. D. 1983. Rev. Physiol. Biochem. Pharmacol. **98:** 64–175.
9. KACZMAREK, L. K. 1987. Trends Neurosci. **10:** 30–34.
10. ROSENTHAL, W., J. HESCHELER, W. TRAUTWEIN & G. SCHULTZ. 1988. FASEB J. **2:** 2784–2790.
11. LINDEN, J. & T. M. DELAHUNTY. 1989. Trends Pharmacol. Sci. **10:** 114–120.
12. BLAUSTEIN, M. P. 1979. *In* The Release of Catecholamines from Adrenergic Neurones. D. M. Paton, Ed.: 39–58. Pergamon Press. Oxford, England
13. ILLES, P. 1986. Neuroscience **17:** 909–928.
14. MAJEWSKI, H. 1983. J. Auton. Pharmacol. **3:** 47–60.
15. WEMER, J., A. N. M. SCHOFFELMEER & A. H. MULDER. 1982. J. Neurochem. **39:** 349–345.
16. SCHOFFELMEER, A. N. M. & A. H. MULDER. 1983. Naunyn-Schmiedebergs Arch. Pharmacol. **323:** 188–192.
17. FREDHOLM, B. B. & E. LINDGREN. 1987. Acta Physiol. Scand. **130:** 95–105.
18. ALLGAIER, C., T. J. FEUERSTEIN, R. JACKISCH & G. HERTTING. 1985. Naunyn-Schmiedebergs Arch. Pharmacol. **331:** 235–239.
19. ALLGAIER, C., T. J. FEUERSTEIN & G. HERTTING. 1986. Naunyn-Schmiedebergs Arch. Pharmacol. **333:** 104–109.
20. UKENA, D., E. POESCHLA, E. HÖTTEMANN & U. SCHWABE. 1984. Naunyn-Schmiedebergs Arch. Pharmacol. **327:** 247–253.
21. KITAMURA, Y. & Y. NOMURA. 1987. J. Neurochem. **49:** 1894–1901.
22. ALLGAIER, C., G. HERTTING, H. Y. HUANG & R. JACKISCH. 1987. Br. J. Pharmacol. **92:** 161–172.
23. WERLING, L. L., P. N. MCMAHON & B. M. COX. 1989. Naunyn-Schmiedebergs Arch. Pharmacol. **339:** 509–513.
24. ALLGAIER, C., B. DASCHMANN, J. SIEVERLING & G. HERTTING. 1989. J. Neurochem. **53:** 1629–1635.
25. ALBERTS, P., V. R. ÖGREN & A. I. SELLSTRÖM. 1985. Naunyn-Schmiedebergs Arch. Pharmacol. **330:** 114–120.
26. JOHNSTON, H., H. MAJEWSKI & I. F. MUSGRAVE. 1987. Br. J. Pharmacol. **91:** 773–781.
27. SEAMON, K. B. & J. W. DALY. 1983. Trends Pharmacol. Sci. **4:** 120–123.
28. KOBINGER, W. & L. PICHLER. 1981. Eur. J. Pharmacol. **73:** 313–321.
29. KASCHUBE, M. 1989. Naunyn-Schmiedebergs Arch. Pharmacol. **339:** R35.
30. LAI, R.-T., Y. WATANABE & H. YOSHIDA. 1983. Eur. J. Pharmacol. **90:** 453–456.
31. NOZAKI, M. & N. SPERELAKIS. 1989. Am. J. Physiol. **256:** H455–H459.
32. DOCHERTY, J. R. 1988. J. Auton. Pharmacol. **8:** 197–201.
33. NICHOLS, A. J., E. D. MOTLEY & R. R. RUFFOLO. 1988. Eur. J. Pharmacol. **145:** 345–349.
34. MUSGRAVE, I., P. MARLEY & H. MAJEWSKI. 1987. Naunyn-Schmiedebergs Arch. Pharmacol. **336:** 280–286.
35. CASTAGNA, M., Y. TAKAI, K. KAIBUCHI, K. SANO, U. KIKKAWA & Y. NISHIZUKA. 1982. J. Biol. Chem. **257:** 7847–7851.
36. SCHOFFELMEER, A. N. M. & A. H. MULDER. 1984. Eur. J. Pharmacol. **105:** 129–135.
37. HORN, J. P. & D. A. MCAFEE. 1980. J. Physiol. **301:** 191–204.

38. GALVAN, M. & P. R. ADAMS. 1982. Brain Res. **244:** 135–144.
39. LIPSCOMBE, D., S. KONGSAMUT & R. W. TSIEN. 1989. Nature **340:** 639–642.
40. SONG, S.-Y., K. SAITO, K. NOGUCHI & S. KONISHI. 1989. Brain Res. **494:** 383–386.
41. FOOTE, S. L., F. E. BLOOM & G. ASTON-JONES. 1983. Physiological Rev. **63:** 844–915.
42. AGHAJANIAN, G. K. & C. P. VANDERMAELEN. 1982. Science **215:** 1394–1396.
43. WILLIAMS, J. T., G. HENDERSON & R. A. NORTH. 1985. Neuroscience **14:** 95–101.
44. WILLIAMS, J. T., T. M. EGAN & R. A. NORTH. 1982. Nature **229:** 74–77.
45. WILLIAMS, J. T. & R. A. NORTH. 1984. Mol. Pharmacol. **26:** 489–497.
46. NORTH, R. A. & J. T. WILLIAMS. 1985. J. Physiol. **364:** 265–280.
47. WILLIAMS, J. T., R. A. NORTH & T. TOKIMASA. 1988. J. Neurosci. **8:** 4299–4306.
48. AGHAJANIAN, G. K. & Y. Y. WANG. 1986. Brain Res. **371:** 390–394.
49. NORTH, R. A., J. T. WILLIAMS, A. SURPRENANT & M. J. CHRISTIE. 1987. Proc. Nat. Acad. Sci. USA **84:** 5487–5491.
50. CHERUBINI, E., R. A. NORTH & J. T. WILLIAMS. 1988. J. Physiol. **406:** 431–442.
51. FUXE, K., G. VON EULER, I. VAN DER PLOEG, B. B. FREDHOLM & L. F. AGNATI. 1989. Neurosci. Lett. **101:** 337–341.
52. WARD-ROUTLEDGE, G. & C. A. MARSDEN. 1988. Trends Pharmacol. Sci. **9:** 209–214.
53. MAJEWSKI, H. & M. J. RAND. 1981. Eur. J. Pharmacol. **69:** 493–498.
54. JOHNSTON, H. & H. MAJEWSKI. 1986. Br. J. Pharmacol. **87:** 553–562.
55. LOIACONO, R. E., M. J. RAND & D. F. STORY. 1985. Br. J. Pharmacol. **84:** 697–705.
56. ILLES, P., D. RAMME & K. STARKE. 1986. J. Physiol. **379:** 217–228.
57. ILLES, P. 1983. Trends Pharmacol. Sci. **4:** 335–338.
58. RAMME, D., P. ILLES, L. SPÄTH & K. STARKE. 1986. Naunyn-Schmiedebergs Arch. Pharmacol. **334:** 48–55.
59. LIMBERGER, N., L. SPÄTH, T. HÖLTING & K. STARKE. 1986. Naunyn-Schmiedebergs Arch. Pharmacol. **334:** 166–171.
60. SCHOFFELMEER, A. N. M., J. PUTTERS & A. H. MULDER. 1986. Naunyn-Schmiedebergs Arch. Pharmacol. **333:** 377–380.
61. LIMBERGER, N., L. SPÄTH & K. STARKE. 1988. Naunyn-Schmiedebergs Arch. Pharmacol. **338:** 53–61.
62. JACKISCH, R., M. GEPPERT & P. ILLES. 1986. J. Neurochem. **46:** 1802–1810.
63. LIMBERGER, N., E. A. SINGER & K. STARKE. 1988. Naunyn-Schmiedebergs Arch. Pharmacol. **338:** 62–67.
64. YOKOO, H., D. H. SCHLESINGER & M. GOLDSTEIN. 1987. Eur. J. Pharmacol. **143:** 283–286.
65. TIMMERMANS, P. B. M. W. M. & P. A. VAN ZWIETEN. 1982. J. Med. Chem. **25:** 1389–1401.
66. ILLES, P., R. BETTERMANN, I. BROD & B. BUCHER. 1987. Naunyn-Schmiedebergs Arch. Pharmacol. **335:** 420–427.
67. REGENOLD, J. T., H. L. HAAS & P. ILLES. 1988. Neurosci. Lett. **92:** 347–350.

Pulse-to-Pulse Modulation of Transmitter Release in the Central Nervous System

Basic and Pharmacological Aspects

M. CEJNA, E. AGNETER, H. DROBNY,
B. VALENTA, AND E. A. SINGER[a]

Institute of Pharmacology
University of Vienna
Währinger Strasse 13a
A-1090 Vienna, Austria

INTRODUCTION

Electrical field stimulation of brain slices previously incubated with ^{3}H-labeled transmitters is a widely employed method to study transmitter release. Of particular interest has been the study of presynaptic autoreceptors, the activation of which by released transmitter causes inhibition of action potential–evoked transmitter release.[1] Since the introduction of a method to elicit transmitter release with a single pulse of electrical stimulation,[2] a condition that allows the examination of release in the absence or near absence of presynaptic inhibition, detailed analyses regarding this mechanism have been carried out. Thus, investigations on autoinhibition using one-pulse stimulation have been performed on release of ^{3}H-noradrenaline (^{3}H-NA) from slices of rat brain cortex,[3] as well as of release of ^{3}H-NA, ^{3}H-dopamine (^{3}H-DA), and ^{3}H-acetylcholine from slices of rabbit brain cortex or striatum.[4]

The present study, in its first part, expands these investigations by examining release of ^{3}H-NA and ^{3}H-DA evoked by single pulses as well as by four pulses delivered at frequencies from 0.0125 Hz to 100 Hz, i.e., pulse-to-pulse intervals of 80 seconds to 10 milliseconds. It was the aim of the experiments to fully explore the properties of negative feedback mediated through presynaptic alpha$_2$ and dopamine receptors in the rat brain. The second part of the study deals with the problem of pharmacological analyses of the release-modulating presynaptic receptor under ongoing autoinhibition of release. The determination of the dissociation constant (K_A) of the alpha$_2$-adrenoceptor agonist 5-bromo-6-(2-imidazolin-2-ylamino)quinoxaline (UK-14,304)[5] was chosen as example. Concentration-response curves for the agonist were generated using two types of electrical stimulation, i.e., without and with activation of autoinhibition, in brain slices from control rats and in slices of rats treated with the irreversible alpha-2 adrenoceptor antagonist N-ethoxycarbonyl-2-ethoxy-1,2-dihydroquinoline (EEDQ).[6] Classical Furchgott analysis[7] was used to determine K_A. The results bear out the expectation that released transmitter in the biophase, by acting as endogenous agonist, will lead to an underestimation of the dissociation constant of the exogenous agonist.

---

[a]Author to whom correspondence should be addressed.

METHODS

Experiments on Stimulation-Evoked Release of ^3H-NA and ^3H-DA

Preparation of Tissue, Incubation, and Superfusion

Male Sprague-Dawley rats (200–250 g, Forschungsanstalt für Versuchstierzucht, Himberg, Austria) were decapitated, and the brain removed rapidly. Slices of parietooccipital cortex and corpus striatum were prepared (McIlwain tissue chopper) for experiments on the release of noradrenaline and dopamine, respectively. Then the tissue samples were incubated for 30 minutes at 30°C in 2 ml of medium containing 0.125 μM ^3H-noradrenaline (^3H-NA; specific activity, 42.1 Ci/mmol) or in 3 ml of medium containing 0.04 μM ^3H-dopamine (^3H-DA; specific activity, 42.1 Ci/mmol), respectively (for composition of medium for uptake and superfusion see Valenta *et al.*).[3] After incubation, the tissues were transferred to superfusion chambers in which they were positioned between two platinum wire electrodes 5 mm apart (one slice per chamber, 12 chambers in parallel; 30°C, 0.7 ml/minute). After a 45-minute washout period the experiments were started with the collection of 5-minute fractions. The superfusion, but not the incubation medium used for labeling with ^3H-NA or ^3H-DA, also contained the uptake inhibitors desipramine (1 μM) or nomifensine (3 μM), respectively. At the end of the experiments the slices were homogenized in 2% (v/v) perchloric acid by sonication, and the radioactivity in slices and superfusate fractions was determined by liquid scintillation counting.

Stimulation and Drug Administration

The slices were stimulated electrically 15 minutes (S_1) and 65 minutes (S_2) after the start of the collection of 5-minute fractions with single pulses, or trains of 4 or 36 pulses delivered at frequencies of 0.0125–100 Hz, as indicated in figures and tables (monophasic rectangular pulses, 2 mseconds, 16 V/cm, 24 mA). Drugs were added 20 minutes before S_2 for the rest of the experiment.

Calculations

Outflow of tritium per fraction was expressed as percentage of radioactivity in the slice at the onset of the respective 5-minute collection period. Stimulation-evoked overflow was calculated as the difference between total outflow during and after stimulation, and estimated basal outflow, which was assumed to decline linearly; it was expressed as the percentage of total amount of tritium in the slice at the onset of stimulation ($S_1\%$, $S_2\%$). Drug effects were expressed as the ratio between the overflow evoked by S_2 and the overflow evoked by S_1 (S_2/S_1). Effects of drugs on basal outflow of tritium were estimated by calculating the ratio between the value of fractional outflow during the 5-minute collection period preceding S_2 and the 5-minute collection period preceding S_1 (L_2/L_1).

Experiments on ^3H-NA Release following Partial Receptor Inactivation

The agonist-receptor dissociation constant of the alpha-2 adrenoceptor agonist UK-14,304 was determined using the method of partial receptor inactivation according to Furchgott.[7] For this purpose, rats were treated with the irreversible alpha-adrenoceptor antagonist *N*-ethoxycarbonyl-2-ethoxy-1,2-dihydroquinoline[6] (EEDQ). The animals received 0.8 mg/kg EEDQ intraperitoneally (ip), and were killed 18 hours later. Control rats received an ip injection of saline (1 ml/kg).

The incubation and superfusion protocol was the same as described above for ^{3}H-NA with the following changes: superfusion was performed at a temperature of 37°C with the collection of 4-minute fractions. Four test periods of electrical stimulation (S_1–S_4) were applied after 8, 36, 64, and 92 minutes, respectively, after the start of the collection of 4-minute fractions. The test periods consisted either of 4 pulses at 100 Hz or 36 pulses at 3 Hz. The procedure for the determination of agonist affinity was as follows: six slices prepared from the brain of an EEDQ-treated animal and six slices from a control animal were used in a single experiment. Four slices of each group were used for the generation of cumulative concentration-response curves for UK-14,304. The drug was added at increasing concentrations 20 minutes before S_2, S_3, and S_4, respectively (six concentrations, each concentration tested at two slices). The remaining two slices of each group served as controls, i.e., they did not receive the agonist (further details, see Limberger *et al.*).[8] Drug effects on stimulated overflow were expressed as the ratio between the overflow evoked by S_2, S_3, or S_4 and the overflow evoked by S_1 (S_2/S_1, S_3/S_1, S_4/S_1; also termed S_x/S_1). Each concentration-response curve was generated from pooled data of at least three superfusion experiments.

Calculation of Agonist Affinity

The method published by Furchgott[7] allows a determination of the activation constant for an agonist (K_A) at steady-state conditions and the fraction of receptors not inactivated (q) by comparing agonist concentration-response curves before and after partial irreversible receptor inactivation. The following equation is used:

$$\frac{1}{A} = \frac{1}{q} \cdot \frac{1}{A'} + \frac{1-q}{q \cdot K_A}$$

where A is the concentrations of agonist needed to give a specific percentage of the maximal response, and A' the concentrations of agonist needed to give the same percentage of the control maximal response after receptor inactivation.

To obtain values for equieffective concentrations of agonist A and A' the best fit concentration-response curves were generated using the ALLFIT computer program of De Lean *et al.*[9] A' values were taken from the concentration response curve after EEDQ treatment (in most cases between 40 and 95% of effect, as outlined by Kenakin,[10] and corresponding A values were calculated from the control curve. The reciprocals of these values, $1/A$ and $1/A'$, were then plotted and analyzed by least-squares linear regression yielding a slope of $1/q$ and a y-intercept of $(1 - q)/(q \cdot K_A)$.[7] From these values, the fraction of receptors remaining active, $q = 1/\text{slope}$, and the equilibrium activation constant, $K_A = (\text{slope-1})/\text{y-intercept}$, were calculated.

Statistics

All data are given as means ± standard errors of the mean (SEM); n = number of observations; one observation = one slice. Student's unpaired t-test was used for determination of statistical significances.

Drugs

Desipramine HCl from Ciba-Geigy (Basel, Switzerland); nomifensine hydrogen-maleate from Hoechst AG (Frankfurt, FRG); idazoxan from Reckitt & Colman (Hull, United Kingdom); sulpiride racemate from Bender (Vienna, Austria); levo-(ring-2,5,6-^{3}H)-norepinephrine (specific activity 42.1 Ci/mmol) and 3,4-(ring-2,5,6-^{3}H)-dihydroxyphenylethylamine hydrochloride from NEN (Dreieich, FRG).

N-Ethoxycarbonyl-2-ethoxy-1,2-dihydroquinoline (EEDQ) was obtained from Research Biochemicals Incorporated (ANAWA Trading, Wangen, Switzerland); 5-bromo-6-(2-imidazolin-2-ylamino)quinoxaline (UK-14,304) from Pfizer Inc. (New York, N.Y.); EEDQ was dissolved in absolute ethanol and sequentially diluted with propylene glycol and distilled water (0.2:0.25:0.55; v/v/v) immediately before injection.

RESULTS

Experiments on Autoinhibition of Transmitter Release

The basal outflow of ^{3}H-NA and ^{3}H-DA during the 5-minute collection period immediately before S_1 amounted to 0.64 ± 0.02% ($n = 114$) and 0.76 ± 0.02% ($n = 199$), respectively. These values corresponded to 3287 ± 133 dpm of tritium in ^{3}H-NA experiments and 3253 ± 86 dpm in ^{3}H-DA experiments. Neither the alpha$_2$-adrenoceptor antagonist idazoxan (^{3}H-NA experiments) nor the dopamine D$_2$-receptor antagonist sulpiride (^{3}H-DA experiments) affected basal outflow of radioactivity at the concentration of 1 μM (L_2/L_1; control ^{3}H-NA, 0.80 ± 0.01, $n = 57$; idazoxan, 0.83 ± 0.01, $n = 57$; control ^{3}H-DA, 0.96 ± 0.01, $n = 126$; sulpiride, 0.99 ± 0.01, $n = 73$).

The percentages of release of ^{3}H-NA and ^{3}H-DA evoked by a single pulse in the absence and presence of idazoxan (1 μM) and sulpiride (1 μM), respectively, are shown in TABLE 1. Both antagonists slightly increased the overflow of tritium by about 16%.

The frequency dependence of the effect of idazoxan (1 μM) on overflow of ^{3}H-NA elicited by 4 pulses is shown in FIGURE 1. As judged by the S_2/S_1 ratios, the facilitatory effect of the antagonist followed a biphasic curve reaching a peak at 1 Hz (FIGURE 1a). There was no significant facilitation at frequencies of stimulation above 10 Hz. The percentages of tissue radioactivity released (S_1%, S_2%) are shown in FIGURE 1b. S_1% decreased with increasing frequencies, reached a minimum at 1 Hz, and sharply increased above 10 Hz, amounting to about 3.6× the release evoked by single pulses. The drop in S_2/S_1 ratios at frequencies above 10 Hz appears to be a consequence of the increase in S_1%, since S_2% remained the same in the range of frequencies from 0.033 to 100 Hz. At the very low frequency of 0.0125 Hz (pulse-to-pulse interval of 80 seconds), the highest percentages of radioactivity were released. The values obtained in the absence of idazoxan reached 90% of the theoretically possible fourfold of the percentage released by a single pulse (1.30% and 0.36%, respectively; FIGURE 1b, TABLE 1).

A similar analysis was performed regarding the effects of sulpiride on the evoked overflow of ^{3}H-DA from striatum slices (FIGURE 2). Again, a biphasic course of S_2/S_1 ratios reaching peak values at 1–3 Hz was observed (FIGURE 2a). There was no significant facilitatory effect of sulpiride at the frequencies above 3.3 Hz. Analysis of released percentages in the absence (S_1%) and presence (S_2%) of sulpiride (FIGURE 2b) revealed that, unlike the values in ^{3}H-NA experiments, S_1% and S_2% decreased from 0.0125–1 Hz in an almost parallel fashion. S_2% did not decline any further at the high frequencies of 10, 33.3, and 100 Hz, and S_1% slightly increased, thus causing a corresponding drop in S_2/S_1 ratios. The increase in release at higher frequencies was very small as compared to the increase seen in ^{3}H-NA experiments, amounting to only the 1.6-fold of the value for a single pulse. In contrast, release evoked by 4 pulses at 0.0125 Hz (4p/0.0125 Hz) in the absence of sulpiride reached almost 80% of the possible fourfold of the percentage released by a single pulse (FIGURE 2b and TABLE 1). However, when stimulation with one pulse and with 4p/0.0125 Hz was applied to the same slice, an exactly fourfold difference in percentages of release was observed (0.19 ± 0.04% and 0.76 ± 0.13%, respectively; $n = 12$).

Experiments on ^{3}H-NA Release following Partial Receptor Inactivation

The agonist-receptor dissociation constant of the alpha$_2$-adrenoceptor agonist UK-14,304[5] was determined according to the method of Furchgott[7] using partial

TABLE 1. Effect of Drugs on Overflow of Tritium Evoked by a Single Pulse[a]

	Stimulation-Evoked Overflow (mean ± SEM)		
^{3}H-NA			
No antagonist, S_1	0.36 ± 0.01%	1066 ± 52 dpm	(12)
S_2	0.33 ± 0.02%	923 ± 54 dpm	(6)
S_2/S_1	0.93 ± 0.04	—	(6)
Idazoxan 1 μM, S_2	0.39 ± 0.01%	1009 ± 86 dpm	(6)
S_2/S_1	1.08 ± 0.05[b]	—	(6)
^{3}H-DA			
No antagonist, S_1	0.21 ± 0.02%	514 ± 49 dpm	(10)
S_2	0.16 ± 0.02%	362 ± 48 dpm	(5)
S_2/S_1	0.81 ± 0.04	—	(5)
Sulpiride 1 μM, S_2	0.21 ± 0.04%	393 ± 40 dpm	(5)
S_2/S_1	0.94 ± 0.04[b]	—	(5)

[a]Stimulation-evoked overflow of tritium from cortex slices labeled with ^{3}H-NA or from striatum slices labeled with ^{3}H-DA. Electrical stimulation of slices was performed 15 minutes (S_1) and 65 minutes (S_2) after the start of sample collection with a single pulse. Idazoxan or sulpiride was added 20 minutes before S_2. The amount of radioactivity released is shown as percentage of tissue tritium (S_1%, S_2%) and as dpm released above baseline efflux. The effects of idazoxan and sulpiride are also shown as S_2/S_1 ratios (see Methods). All values represent means ± SEM of (n) observations.
[b]$p < 0.05$ vs. S_2/S_1 of antagonist-free condition.

receptor inactivation. UK-14,304-mediated inhibition of stimulation-evoked release of ^{3}H-NA was determined 18 hours post-EEDQ (0.8 mg/kg ip). Two different types of electrical stimulation, 4 pulses at 100Hz (4p/100Hz) and 36 pulses at 3 Hz (36p/3Hz), were used for the generation of concentration-response curves for UK-14,304. In control slices, the percentages of tissue tritium released during the agonist-free reference stimulations (S_1%) amounted to 1.21 ± 0.03% ($n = 18$) and 1.53 ± 0.06% ($n = 21$) at 4p/100Hz and 36p/3Hz, respectively. The corresponding values in EEDQ-treated animals were 1.33 ± 0.06% ($n = 18$) and 4.88 ± 0.17% ($n = 18$, $p < 0.001$ vs control). Thus, release of tritium evoked by 4p/100Hz was not increased in EEDQ-treated tissues, whereas a marked enhancement of release evoked by 36p/3Hz was observed. The concentration-response curves for UK-14,304 in control and EEDQ-treated tissues are shown in FIGURE 3a. UK-14,304 decreased stimulation-evoked release of tritium more potently in experiments with 4p/100Hz than with

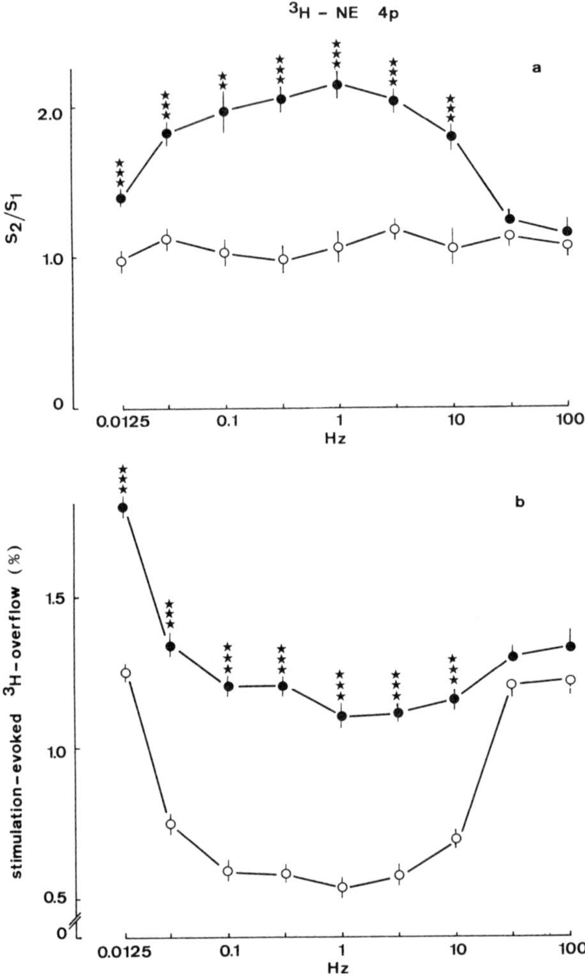

FIGURE 1. Frequency dependence of the effects of idazoxan on stimulation-evoked overflow of tritium from superfused cortex slices labeled with ^{3}H-NA. Electrical stimulation of slices was performed 15 minutes (S_1) and 65 minutes (S_2) after the start of sample collection with 4 pulses at frequencies of 0.-0125–100 Hz. (**a**) Results are expressed as S_2/S_1 ratios (see Methods) obtained from drug-free control experiments (open circles) or experiments in which idazoxan (1 μM) (filled circles) was added 20 minutes before S_2. (**b**) Same experiments as in a; the symbols represent means ± SEM of percentages of tissue tritium released by electrical stimulation in the absence (open circles, $S_1\%$) or presence (filled circles, $S_2\%$) of idazoxan (1 μM). Numbers of observations are 10–22 for S_1-, and 5–10 for S_2-derived data. ***$p < 0.001$ vs. idazoxan-free condition.

36p/3Hz. The IC_{50} values amounted to 8 nM and 16 nM, and the S_x/S_1 ratios at maximal inhibition to 0.08 and 0.23, respectively. At 4p/100Hz EEDQ shifted the IC_{50} of UK-14,304 ninefold to the right (control = 8 nM, EEDQ = 72nM), and maximal response decreased to an S_x/S_1 ratio of 0.42. The EEDQ-induced shift at 36p/3Hz was sevenfold (control $=$ 16 nM, EEDQ $=$ 112 nM), and maximal response decreased to an S_x/S_1 ratio of 0.54. Equieffective concentrations were calculated for each pair of concentration-response curves and used to construct double reciprocal plots according to Furchgott (see Methods), from which the fraction of receptors remaining active, $q = 1/\text{slope}$, and the agonist dissociation constant, $K_A = (\text{slope-1})/\text{y-intercept}$, were

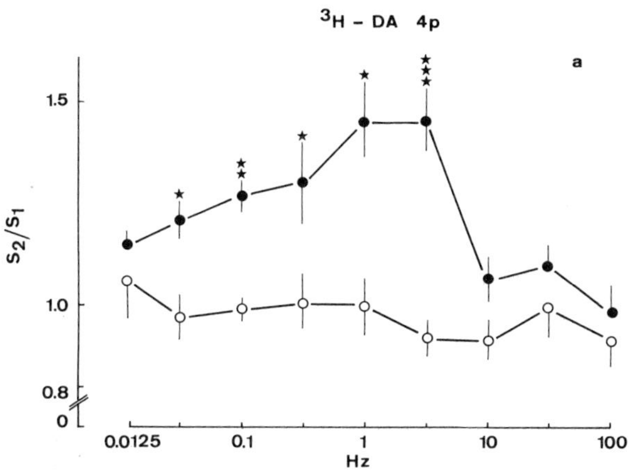

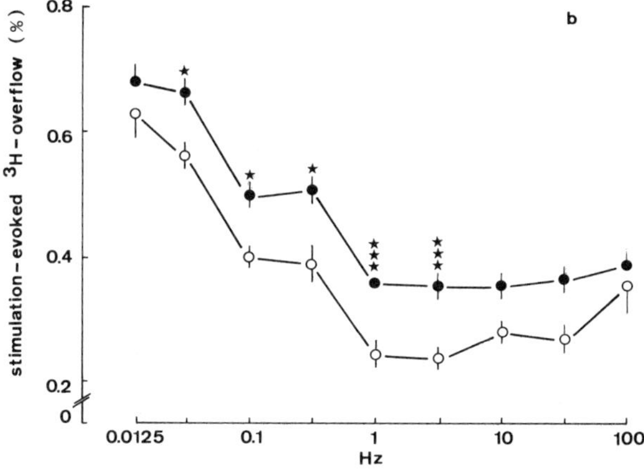

FIGURE 2. Frequency dependence of the effects of sulpiride on stimulation-evoked overflow of tritium from superfused striatum slices labeled with ^{3}H-DA. Electrical stimulation of slices was performed 15 minutes (S_1) and 65 minutes (S_2) after the start of sample collection with 4 pulses at frequencies of 0.0125–100 Hz. (a) Results are expressed as S_2/S_1 ratios (see Methods) obtained from drug-free control experiments (open circles) or experiments in which sulpiride (1 μM) (filled circles) was added 20 minutes before S_2. (b) Same experiments as in a; the symbols represent means $\pm$ SEM of percentage of tissue tritium released by electrical stimulation in the absence (open circles, $S_1\%$) or presence (filled circles, $S_2\%$) of sulpiride (1 μM). Numbers of observations are 6–14 for S_1- and S_2-derived data. *$p < 0.05$, **$p < 0.01$, ***$p < 0.001$ vs. sulpiride-free condition.

calculated. The results are shown in FIGURE 3b. The plot derived from experiments with 4p/100Hz yielded a value for q of 0.08, or 8%, and a value for K_A of 136 nM. The corresponding values obtained from experiments with 36p/3Hz were $q = 0.09$, or 9%, and $K_A = 217$ nM. Thus it appears, that stimulation of slices with 36p/3Hz led to an about 1.6-fold increase in estimated K_A.

DISCUSSION

It has been shown repeatedly that electrical stimulation of superfused brain slices labeled with ^{3}H-NA or ^{3}H-DA results in overflow of radioactivity which, in the

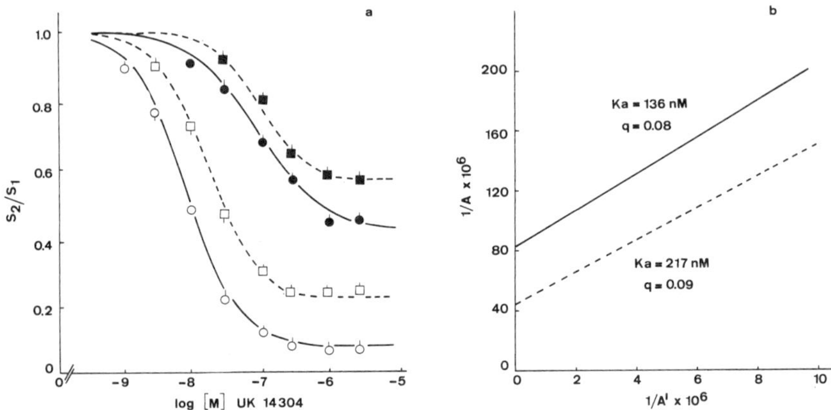

FIGURE 3. (a) Concentration-response relationships of the effects of UK-14,304 on stimulation-evoked overflow of tritium from superfused rat brain cortex slices labeled with ^{3}H-NA. Animals were pretreated with saline (open symbols) or with 0.8 mg/kg of EEDQ ip (filled symbols) and killed 18 hours later. Electrical stimulation of slices was performed at 8, 36, 64, and 92 minutes (S_1–S_4) after the start of the collection of 4 minute fractions. The test periods consisted either of 4 pulses delivered at 100 Hz (circles; solid lines) or 36 pulses delivered at 3 Hz (squares; dashed lines). Effects of UK-14,304 are expressed as S_x/S_1 ratios (see Methods). For amounts of radioactivity released during the agonist-free reference stimulations S_1 see text. Symbols represent means ± SEM of 6–8 observations. (b) Double reciprocal plot of the equieffective concentrations of UK-14,304 for inhibiting ^{3}H-NA release (experiments of 3a). Equieffective concentrations of agonist A (control curves) and A' (curves after EEDQ treatment) were calculated (see Methods), and the reciprocals $(1/_A)$, $1/_{A'}$) plotted. The fraction of receptors remaining active, $q = 1/\text{slope}$, and the agonist-receptor dissociation constant, $K_A = (\text{slope-1})/\text{y-intercept}$, were calculated from the unweighted linear regression lines (solid line, stimulation with 4 pulses/100 Hz; dashed line, stimulation with 36 pulses/3 Hz).

presence of a reuptake inhibitor, consists mainly of unmetabolized ^{3}H-monoamines and reflects release of endogenous transmitter.[11,12] The stimulation-evoked overflow of radioactivity in the present experiments will, therefore, also be referred to as release of ^{3}H-NA and ^{3}H-DA, respectively. Furthermore, the experiments on pulse-to-pulse modulation of ^{3}H-DA release were performed at 30°C for technical reasons (lower and more stable spontaneous efflux of tritium and higher evoked overflow). To achieve optimal comparability of results, the release of ^{3}H-NA in this part of the study was also carried out at 30°C.

A single pulse induced overflow of a reliably measurable amount of tritium from slices labeled with ^{3}H-NA or ^{3}H-DA. According to the theory of presynaptic autoinhibition, receptor antagonists should not increase the release of transmitter evoked by a single pulse. This prediction can be considered as fulfilled in the present experiments. The alpha$_2$-adrenoceptor antagonist idazoxan and the dopamine D$_2$-receptor antagonist sulpiride caused only a slight facilitation of one-pulse-induced overflow of radioactivity, most likely due to the presence of some tonic autoinhibition caused by spontaneous release of transmitter. A similar observation with ^{3}H-DA release from rabbit brain slices has been published.[4] By contrast, marked and frequency-dependent autoinhibition was observed when trains of four pulses were employed for electrical stimulation. In the following, some differences as well as analogies between the release of ^{3}H-NA and ^{3}H-DA are pointed out.

First, autoinhibition of ^{3}H-NA release was abolished at frequencies of more than 10 Hz (pulse-to-pulse intervals of less than 100 mseconds), i.e., idazoxan did not increase S$_2$/S$_1$ ratios above control values, and, in the absence of idazoxan, 4p/100Hz or 4p/33.3Hz caused overflow of radioactivity close to four times the amount of tritium released by a single pulse (1.29% and 0.36%, respectively; FIGURE 1b and TABLE 1). The result indicates that the pulse-to-pulse intervals were too short for development of negative feedback[3] and corroborates findings of an earlier study[13] as well as similar findings in cortex slices of the rabbit.[4] However, release of ^{3}H-DA did not show such properties. Although sulpiride did not increase S$_2$/S$_1$ ratios above control at frequencies of more than 3.3 Hz (pulse-to-pulse intervals of less than 300 mseconds), stimulation with 4p/100Hz in the absence of sulpiride caused only a 1.6-fold increase in electrically evoked release of ^{3}H-DA. A very similar observation has been made by Mayer *et al.* in the rabbit.[4] The interpretation of this finding has been that a fall in the release of ^{3}H-DA occurs as a consequence of factors other than presynaptic inhibition, e.g., failure of the dopaminergic nerve terminals to generate action potentials at such high frequencies or slow recruitment of vesicles for exocytotic release. A further point to be discussed in this context is the claim based on experiments on rat[14] and rabbit brain slices[15] that ^{3}H-DA release is virtually independent of the frequency of stimulation. This is in contrast to the present study in which a marked frequency dependence of ^{3}H-DA release, as would be predicted by the theory of presynaptic inhibition, was observed in the range of frequencies from 0.0125–0.3 Hz. In the above-mentioned experiments on ^{3}H-DA release in the rat and rabbit, however, the lowest frequencies tested were 0.25 Hz (15 pulses) and 0.1 Hz (10 pulses), respectively. Thus, an important part of the frequency ^{3}H-overflow relationship was not examined (most likely due to technical limitations).

Second, idazoxan increased electrically evoked release of ^{3}H-NA to rather constant values (S$_2$% of 1.1–1.3%, FIGURE 1b; exception 0.0125 Hz, discussed below) which were virtually independent of the frequency of stimulation, whereas sulpiride increased release of ^{3}H-DA to values spanning a factor of two (S$_2$% of 0.34–0.69%), declining from 0.0125 to 1 Hz). The decline in percentages of ^{3}H-DA released by four pulses in the presence of sulpiride may be, though not likely at this high concentration of sulpiride, a consequence of released dopamine effectively competing with the antagonist for the presynaptic receptor. Alternative explanations would be concomitant release of other release-inhibiting transmitters from neighboring synapses or changes in release due to factors not involved in presynaptic inhibition (see above).

Third, an analogy between release of ^{3}H-NA and ^{3}H-DA was observed when electrical stimulation was carried out with 4 pulses delivered at 0.0125 Hz, i.e., at a pulse-to-pulse interval of 80 seconds. This interval was obviously long enough to disrupt the negative feedback loop activated by released transmitter and, hence, led to stimulation-evoked release which amounted to almost exactly 4 times the release of 4 single stimuli (TABLE 1, FIGURES 1b and 2b). The increase in the release of ^{3}H-NA and ^{3}H-DA observed in the presence of idazoxan and sulpiride, respectively, is, as in the case of single pulses, most likely due to blockade of tonic inhibition caused by spontaneously released transmitter. When antagonists were present, the percentages of tritium overflow evoked by 4p/0.0125Hz again amounted to 4 times the percentages released by a single pulse (TABLE 1, FIGURES 1b and 2b). The very long time necessary for interruption of pulse-to-pulse modulation of transmitter release in the absence of antagonists is most likely a consequence of the inhibition of reuptake, favoring a slow elimination of released transmitters from the synaptic cleft. It follows from this thought that the data obtained in the present experiments do not bear immediate relevance for the physiological situation in which transmitter uptake is intact. One may expect, however, that no basic difference exists between the two situations and that the operative range of autoinhibition *in vivo* is shifted to higher frequencies than observed *in vitro*.

In the second part of the study the dissociation constant of the alpha$_2$-adrenoceptor agonist UK-14,304 at the inhibitory presynaptic alpha$_2$-adrenoceptor was determined after partial receptor inactivation according to the method of Furchgott.[7] Stimulation of slices was carried out with 4p/100Hz or 36p/3Hz. Since release of ^{3}H-NA evoked by 4p/100Hz is not, or only slightly, increased by reversible[13,8] or irreversible alpha$_2$-adrenoceptor antagonists (present study), and, therefore, shows similar characteristics as release evoked by a single pulse, it was termed pseudo-one-pulse (POP) stimulation.[2] In contrast to POP stimulation, which causes release free of autoinhibition, stimulation with 36p/3Hz strongly activates negative feedback. Thus, ^{3}H overflow induced by 36p/3Hz was only slightly higher than overflow elicited by POP stimulation in control slices (1.53% vs. 1.21%), but was strongly increased after receptor inactivation with EEDQ. The consequences of the presence of endogenous NA in the biophase of the presynaptic alpha$_2$-adrenoceptor on pharmacological analyses at this receptor are clearly demonstrated in the present experiments. The concentration-response curve for UK-14,304 at 4p/100Hz was shifted to the left and reached a higher maximal effect as compared to the curve obtained at 36p/3Hz. Furthermore, the agonist-receptor dissociation constant (K_A), estimated from the double-reciprocal plots obtained from the pairs of concentration-response relationships before and after partial receptor inactivation, was distinctly lower at 4p/100Hz than at 36p/3Hz. Moreover, preliminary experiments in which ^{3}H-NA release was evoked with 72 pulses at 3 Hz suggest that a further shift in the K_A value for UK 14,304 occurs at this experimental condition, yielding values of 400–500 nM. Such autoinhibition-dependent distortion of drug constants has also been described for competitive antagonists at the presynaptic release-inhibiting alpha$_2$-adrenoceptor in the cortex of the rat.[8] The K_A for UK-14,304 of 136 nM determined with POP stimulation in the present experiments was about 10 times lower than the K_A published by Adler *et al.* in the same tissue (1410 nM) using potassium for stimulation of ^{3}H-NA release.[6] This discrepancy may be explained in part by the fact that EEDQ facilitated potassium-evoked overflow of tritium to some degree in these experiments indicating operative autoinhibition during transmitter release. However, another factor may be the use of potassium stimulation. It is well known that agonists at alpha$_2$-adrenoceptors exhibit less potency when transmitter release is evoked by elevated concentrations of potassium. For instance, the IC_{30} of clonidine is about 30 nM on potassium-depolarized synaptosomes of rat cerebral cortex[16] as opposed to an IC_{50} of 4–5 nM on POP-stimulated slices[13] of the same tissue. Likewise, the IC_{50} for UK-14,304 determined by Adler *et al.* amounted to 24 nM,[6] whereas in the present experiments it was 8 nM at 4p/100Hz and 16 nM at 36p/3Hz.

In conclusion, the present experiments demonstrate pulse-to-pulse modulation of stimulation-evoked release of ^{3}H-NA and ^{3}H-DA *in vitro*. They also demonstrate that the endogenous transmitter complicates the determination of agonist-receptor dissociation constants at the presynaptic release-inhibiting autoreceptor.

ACKNOWLEDGMENT

We thank E. Kotai for preparing the graphs.

REFERENCES

1. STARKE, K., M. GÖTHERT & H. KILBINGER. 1989. Modulation of neurotransmitter release by presynaptic autoreceptors. Physiol. Rev. **69:** 864–989.

2. SINGER, E. A. 1988. Transmitter release from brain slices elicited by single pulses: a powerful method to study presynaptic mechanisms. Trends Pharmacol. Sci. **9:** 274–276.

3. VALENTA, B., H. DROBNY & E. A. SINGER. 1988. Presynaptic autoinhibition of central noradrenaline release in vitro: operational characteristics and effects of drugs acting at alpha-2 adrenoceptors in the presence of uptake inhibition. J. Pharmacol. Exp. Ther. **245:** 944–949.

4. MAYER, A., N. LIMBERGER & K. STARKE. 1988. Transmitter release patterns of noradrenergic, dopaminergic and cholinergic axons in rabbit brain slices during short pulse trains, and the operation of presynaptic autoreceptors. Naunyn-Schmiedebergs Arch. Pharmacol. **338:** 632–643.

5. CAMBRIDGE, D. 1981. UK-14,304, a potent and selective alpha-2 agonist for the characterization of alpha-adrenoceptor subtypes. Eur. J. Pharmacol. **72:** 413–415.

6. ADLER, C. H., E. MELLER & M. GOLDSTEIN. 1987. Receptor reserve at the alpha-2 adrenergic receptor in the rat cerebral cortex. J. Pharmacol. Exp. Ther. **240:** 508–515.

7. FURCHGOTT, R. F. 1966. The use of beta-haloalkylamines in the differentiation of receptors and in the determination of dissociation constants of receptor-agonist complexes. Adv. Drug Res. **3:** 21–55.

8. LIMBERGER, N., A. MAYER, G. ZIER, B. VALENTA, K. STARKE & E. A. SINGER. 1989. Estimation of pA_2 values at presynaptic alpha-2 autoreceptors in rabbit and rat brain cortex in the absence of autoinhibition. Naunyn-Schmiedebergs Arch. Pharmacol. **340:** 639–647.

9. DE LEAN, A., P. MUNSON & D. RODBARD. 1978. Simultaneous analysis of families of sigmoidal curves: application to bioassay, radioligand assay, and physiological dose-response curves. Am. J. Physiol. **235:** E97–E102.

10. KENAKIN, T. P. 1987. Pharmacologic analysis of drug-receptor interaction. Raven Press. New York, NY.

11. TAUBE, H. D., K. STARKE & E. BOROWSKI. 1977. Presynaptic receptor systems on the noradrenergic neurones of rat brain. Naunyn-Schmiedebergs Arch. Pharmacol. **299:** 123–141.

12. ZUMSTEIN, A., W. KARDUCK & K. STARKE. 1981. Pathways of dopamine metabolites in the rabbit caudate nucleus in vitro. Naunyn-Schmiedebergs Arch. Pharmacol. **316:** 205–217.

13. ZIER, G., H. DROBNY, B. VALENTA & E. A. SINGER. 1988. Evidence against a functional link between noradrenaline uptake mechanisms and presynaptic alpha-2 adrenoceptors. Naunyn-Schmiedebergs Arch. Pharmacol. **337:** 118–121.

14. DWOSKIN, L. P. & N. R. ZAHNISER. 1986. Robust modulation of ^{3}H-dopamine release from rat striatal slices by D-2 dopamine receptors. J. Pharmacol. Exp. Ther. **239:** 442–453.

15. CUBEDDU, L. X. & I. S. HOFFMANN. 1982. Operational characteristics of the inhibitory feedback mechanism for regulation of dopamine release via presynaptic receptors. J. Pharmacol. Exp. Ther. **223:** 497–501.

16. MAURA, G., A. GEMIGNANI & M. RAITERI. 1985. Alpha-2 adrenoceptors in rat hypothalamus and cerebral cortex: functional evidence for pharmacologically distinct subpopulations. Eur. J. Pharmacol. **116:** 335–339.

Prostaglandins and the Release of the Adrenergic Transmitter

KAFAIT U. MALIK AND ELMIR SEHIC

Department of Pharmacology
College of Medicine
The University of Tennessee, Memphis
874 Union Avenue
Memphis, Tennessee 38163

Prostaglandins (PGs) are derived from arachidonic acid, which is esterified mainly at the Sn-2 position of phospholipids.[1,2] Arachidonic acid is deacylated from tissue lipids in response to various neurohumoral stimuli, including activation of the sympathetic nervous system or administration of catecholamines.[3-5] Several mechanisms for the release of arachidonic acid from tissue lipids have been proposed (FIGURE 1). It may be released from tissue lipids by the activation of phospholipase A_2[6] or phospholipase C, which promotes breakdown of phosphoinositides and generates diacylglycerol and inositol phosphates.[7] Diacylglycerol can be phosphorylated by diacylglycerol kinase to phosphatidic acid which in turn can serve as a substrate for phosphatidic acid–specific phospholipase A_2 and release arachidonic acid[8] or be metabolized by diacylglycerol lipase.[9] Triglycerides may also release arachidonic acid in some tissues such as renal medullary interstitial cells.[10] The relative contribution of each of these pathways, in different tissues, is not known. A hormone or a neurotransmitter may cause activation of more than one lipase (phospholipases A_2 and C), but selectively utilize one pathway to release arachidonic acid for PG synthesis.[11]

Arachidonic acid released from tissue lipids can be metabolized by cyclooxygenase, lipoxygenase, or cytochrome P-450 NADPH-dependent epoxygenase pathways[2,12-14] (FIGURE 1). The contribution of each of these pathways may vary in different tissues and cell types. Products of arachidonic acid formed through cyclooxygenase, lipoxygenase, or cytochrome P-450 expoxygenase pathways have profound biological actions.[1,2,12,15] The finding that products of arachidonic acid derived via cyclooxygenase particularly PGE_2 inhibit release of adrenergic transmitter and the associated response to sympathetic nerve stimulation together with the demonstration that (a) stimulation of sympathetic nerves promotes PG synthesis and (b) inhibition of cyclooxygenase enhances release of norepinephrine (NE) and the response of effector organs to adrenergic nerve stimulation or exogenous (NE) in several tissues has led to the proposal that PGs function as physiological modulators of adrenergic transmission.[3-5] We will briefly review the effect of sympathetic nerve stimulation and catecholamines on PG synthesis and then discuss the mechanism of action of PGs on release of the adrenergic transmitter.

EFFECT OF SYMPATHETIC NERVE STIMULATION AND CATECHOLAMINES ON THE RELEASE AND METABOLISM OF ARACHIDONIC ACID

Although in peripheral tissues innervated by sympathetic fibers, exogenous arachidonic acid is metabolized to a variable degree via cyclooxygenase, lipoxygenase, and cytochrome P-450 epoxygenase pathways, the full profile of the products of endogenously released arachidonic acid in response to neurohormonal stimuli in various tissues has not yet been established. In most tissues, stimulation of adrenergic nerves or administration of NE promotes synthesis of the products of arachidonic acid derived primarily via the cyclooxygenase pathway, mainly PGE_2 and PGI_2 and to a lesser extent, $PGF_{2\alpha}$ and other products[3,5,16–22] (FIGURE 2). Moreover, the principal arachidonic acid metabolite synthesized in the kidney and spleen in response to stimulation of adrenergic nerves or catecholamines is PGE_2 accompanied by smaller amounts of PGI_2

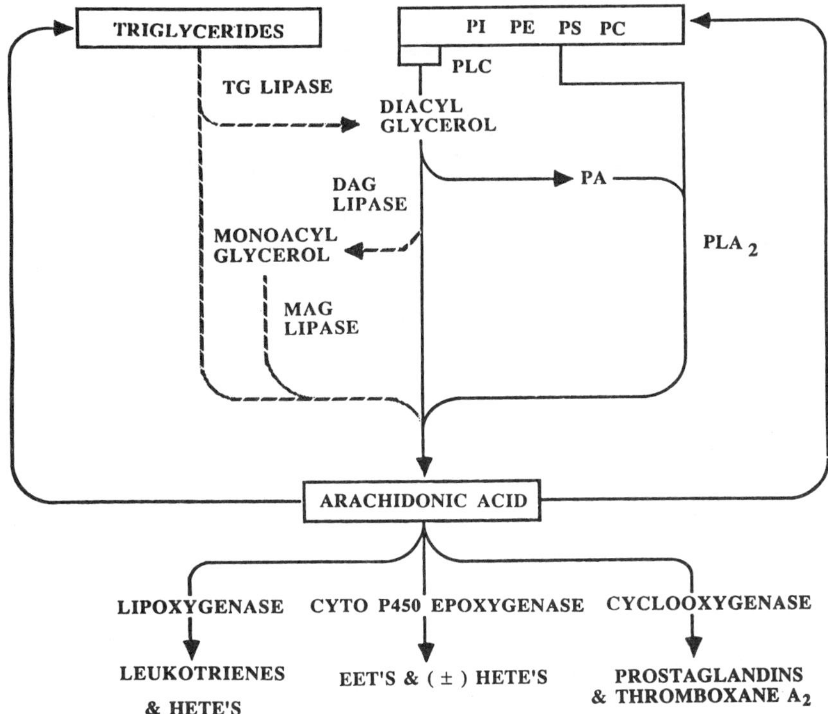

FIGURE 1. Possible pathways for the deesterification of arachidonic acid from tissue lipids and its metabolism. PI = phosphatidylinositol; PE = phosphatdylethanolamine; PS = phosphatidylserine; PC = phosphatidylcholine; PLC = phospholipase C; PLA_2 = phospholipase A_2; DAG = diacylglycerol; MAG = monoacylglycerol; HETE'S = hydroxyeicosatetraenoic acids; EET'S = epoxyeicosatrienoic acids.

and $PGF_{2\alpha}$.[16–19] However, in the heart and blood vessels, PGI_2 is the principal product of arachidonic acid released in response to adrenergic nerve stimulation or NE.[20–22] Whether these differences in the profile of the principal arachidonic acid metabolites formed in response to adrenergic stimuli are due to (a) differences in the activity of enzymes involved in the synthesis of various prostanoids; (b) effects of adrenergic transmitter and cotransmitters on the activity of enzymes involved in the generation of arachidonic acid metabolites; or (c) differences in the presence of various cofactors,

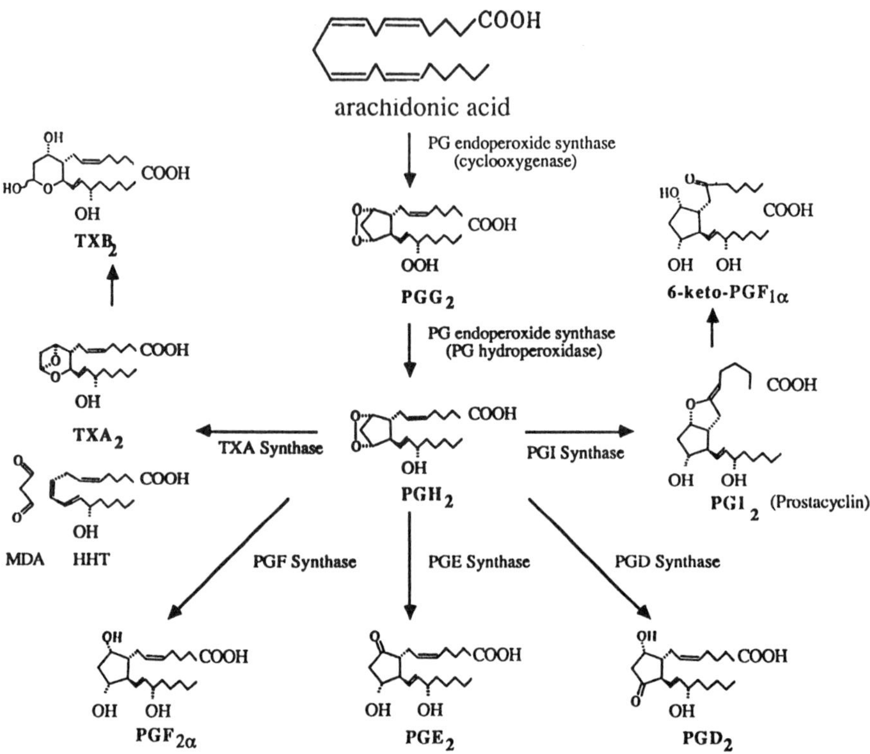

FIGURE 2. Products of arachidonic acid generated via cyclooxygenase pathway. PG = prostaglandin; HHT = 12(s)-hydroxy-4,8,10-heptadecaenoic acid; MDA = malondialdehyde.

inhibitors, or activators of enzymes involved in arachidonic acid metabolism remains to be determined.

Although most mammalian cells, including neuronal cells,[23] are capable of metabolizing arachidonic acid, the major site for PG synthesis at the adrenergic neuroeffector junction in peripheral tissues appears to be the postjunctional effector cells. Supporting this view, it has been demonstrated that adrenergic denervation did not alter PG efflux produced by direct electrical stimulation in the vas deferens.[24] Moreover, PG synthesis produced by either adrenergic nerve stimulation or NE in various tissues was abolished by postjunctional adrenergic receptor blockade.[3,21,25,26] However, studies by Petkov and Radomirov[27] and Trachte[28] in vas deferens and by Greenberg[29] in portal vein indicate that adrenergic nerves release PGs. The mechanism and sequence of events involved in PG synthesis in response to adrenergic transmitter including the type of adrenergic receptors have been recently reviewed.[26]

EFFECTS OF PGs ON ADRENERGIC NEUROEFFECTOR JUNCTION

Products of arachidonic acid derived via cyclooxygenase influence both the release of adrenergic transmitter and the actions of NE at the postjunctional sites. The principal products of arachidonic acid, PGE_2 and PGI_2, synthesized in response to sympathetic nerve stimulation or NE, inhibit depolarization-induced release of NE and/or the action of this neurotransmitter at postjunctional sites in various tissues, including vas deferens, heart, kidney, spleen, and numerous blood vessels *in vitro.*[3–5,30–36] PGE$_2$ and PGI$_2$ also inhibit the effect of sympathetic nerve stimulation and exogenous NE and release of the adrenergic transmitter *in vivo.*[37–41] For example in the heart of pentobarbital anesthetized dogs treated with indomethacin to inhibit production of endogenous PGs, intracoronary infusion of PGE$_2$ or PGI$_2$ attenuated the effect of cardiac sympathetic nerve stimulation and exogenous NE to increase the myocardial contractility as indicated by a decrease in dP/dt_{max} (FIGURES 3 and 4).[37] Infusion of PGE$_2$ and PGI$_2$ in animals pretreated with indomethacin also reduced the coronary sinus output of NE elicited by sympathetic nerve stimulation (FIGURE 5).[37] Although in most tissues PGE$_2$ and PGI$_2$ exert an inhibitory effect at the pre- and postjunctional sites, viz., reduce adrenergic transmitter release and the action of NE at the effector cell, these agents in some tissues exert a differential effect at the pre- and postjunctional sites.[33–36,42] For example in the isolated rat kidney perfused with Tyrode's solution at 5 ml/minute and prelabeled with [³H]NE, stimulation of periarterial nerves at 2 Hz for 1 min (40 V, 1 msecond duration) increased the overflow of tritium that consisted primarily of intact [³H]NE. Renal nerve stimulation (RNS) at 10-minute intervals for two periods increased the overflow of tritium and raised perfusion pressure. The increase in tritium overflow, but not the associated rise in perfusion pressure, during the second period (P$_2$) was less than in the first period (P$_1$). However, the ratio of tritium overflow elicited by RNS between P$_2$ and P$_1$ remained constant in different preparations. Infusion of PGE$_2$ (5–25 ng/ml) during P$_2$ produced a concentration-dependent rise in perfusion pressure and reduced RNS-induced tritium overflow, as indicated by a decrease in the ratio of P$_2$/P$_1$ compared to the corresponding ratio

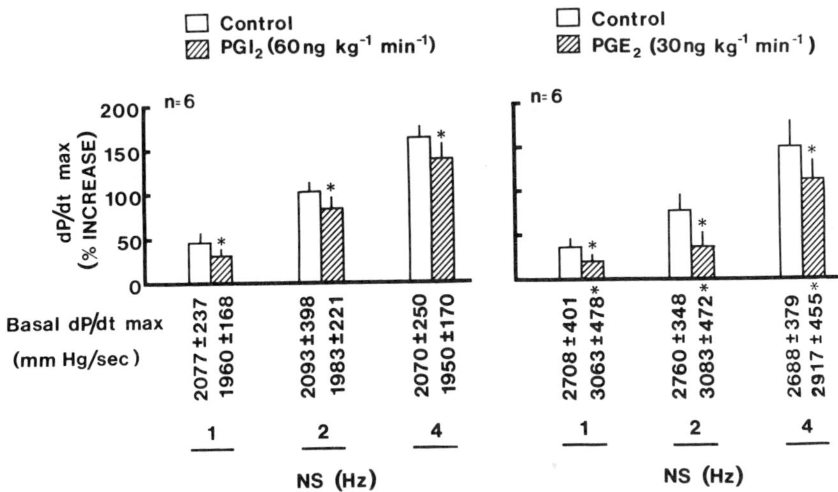

FIGURE 3. Effect of intracoronary PGI$_2$ (left panel) and PGE$_2$ (right panel) infusion on basal and the increase in left ventricular dP/dt_{max} produced by cardiac sympathetic nerve stimulation (NS) at 1–4 Hz in animals pretreated with indomethacin (10 mg/kg). The number of dogs in each group is indicated by *n*. The increase in left ventricular dP/dt_{max} caused by NS above basal values before (open columns) and during PGI$_2$ or PGE$_2$ infusion (hatched columns) was significant ($p < 0.05$). Asterisks denote value different from the corresponding value obtained before PGI$_2$ or PGE$_2$ infusion.

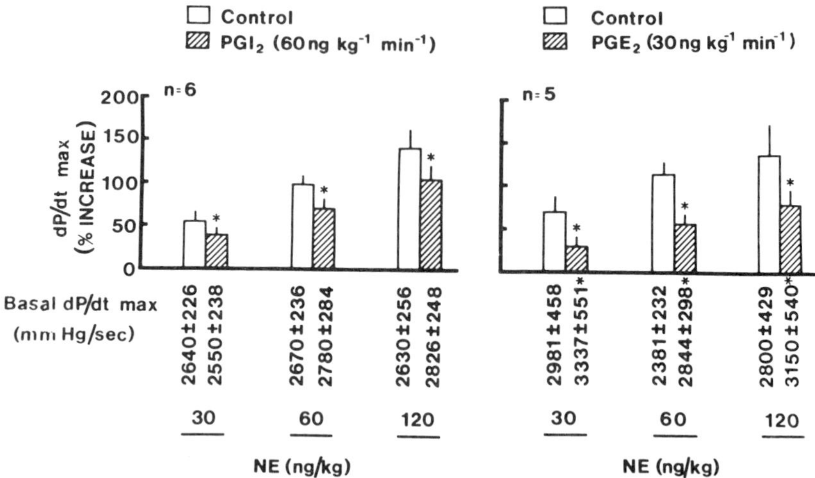

FIGURE 4. Effect of PGI_2 (left panel) and PGE_2 (right panel) on basal and on the increase in left ventricular dP/dt_{max} produced by intracoronary injections of norepinephrine (NE) at 30–120 ng/kg in animals pretreated with indomethacin (10 mg/kg). The number of dogs in each group is indicated by *n*. Augmentation in left ventricular dP/dt_{max} caused by NE above basal values before (open columns) and during (hatched columns) PGI_2 or PGE_2 infusion was significant ($p < 0.05$). Asterisks denote value different from the corresponding value obtained before PGI_2 or PGE_2 infusion.

obtained in the vehicle-treated group (FIGURE 6). In contrast to its inhibitory effect at prejunctional sites, PGE_2 potentiated the rise in perfusion pressure produced by RNS. Since PGE_2 also increased the rise in perfusion pressure caused by exogenous NE, the potentiation of RNS-induced vasoconstriction produced by this prostanoid was due to increased vascular reactivity to the neurotransmitter released from nerve fibers. Similar effects were obtained with higher concentrations of PGI_2 (25–100 ng/ml) and $PGF_{2\alpha}$ (25–100 ng/ml); PGD_2 (100–500 ng/ml), which produced renal vasoconstriction and enhanced the vasoconstrictor response to RNS, did not alter tritium overflow.

In contrast, in the rabbit kidney, PGE_2, which also inhibited NE release elicited by RNS, did not enhance, but rather attenuated the associated renal vasoconstrictor response and produced renal vasodilation.[42] PGI_2 also attenuated the renal vasoconstrictor response to RNS in the rabbit kidney.[34] However, this species difference in the actions of PGE_2 and PGI_2 obtained *in vitro* was not observed *in vivo*. For example, in the rat kidney perfused with blood from the carotid artery, PGE_2 decreased the effect of sympathetic nerve stimulation or exogenous NE in increasing renal vascular resistance (FIGURE 7).[41] PGI_2 produced effects similar to those of PGE_2.[41] Opposite effects of PGE_2 on adrenergically induced vasoconstriction *in vitro* and *in vivo* have also been reported in rat mesenteric vessels.[43,44] Our recent studies indicate that this difference in the action of PGE_2 *in vitro* and *in vivo* is independent of the basal vascular tone (unpublished work). Whether these opposite effects of PGE_2 in the renal and mesenteric vasculature of the rat *in vitro* and *in vivo* could be due to its transformation into some other active metabolite or to the presence *in vivo* of a permissive substance required for PGE_2 to exert its inhibitory effects at the postjunctional sites remains to be determined.

The minor products of arachidonic acid generated via cyclooxygenase, $PGF_{2\alpha}$, PGD_2, and TxA_2, or its analogues or hydrolysis product, exert variable effects at pre- and postjunctional sites in various tissues.[45–50] In most tissues, these agents potentiate the effect of sympathetic nerve stimulation and inhibit $(PGF_{2\alpha})$[36] or enhance $(PGF_{2\alpha}$, PGD_2, TxA_2 analogues or TxB_2) NE release.[46–48,50] However, inhibitors of cyclooxygenase in most tissues do not inhibit but rather enhance release of adrenergic transmitter and/or the response of effector organs to sympathetic nerve stimulation and exogenous NE.[3–5,30,37,41] FIGURE 8 illustrates the effect of indomethacin to increase the rise in the

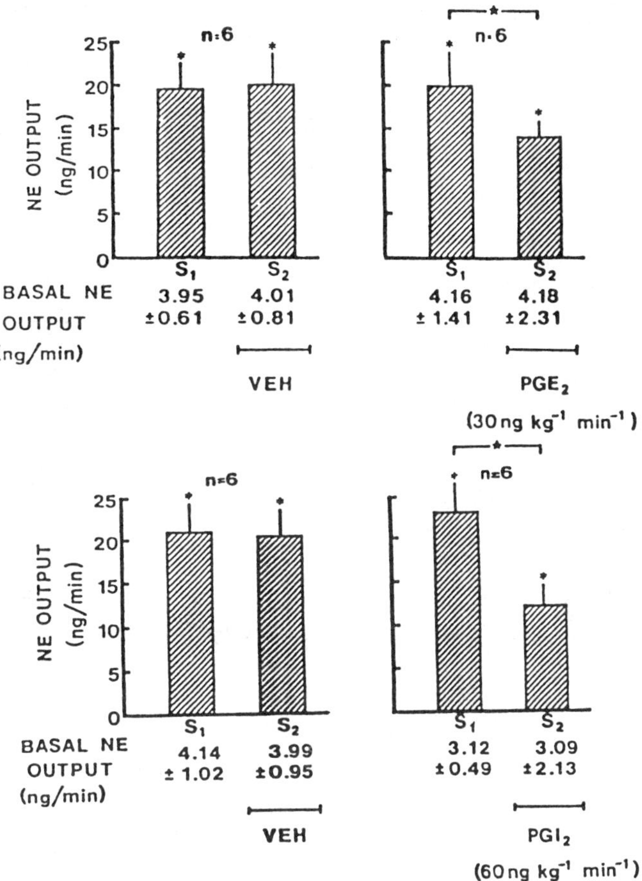

FIGURE 5. Effect of PGE_2 (30 ng kg^{-1} minutes^{-1}) or vehicle (VEH, 30 μl kg^{-1} minutes^{-1} ethanol) (upper panel) and PGI_2 (60 ng kg^{-1} minutes^{-1}) or vehicle (0.6 μl kg^{-1} minutes^{-1} Tris buffer pH 10.5) (lower panel) on the basal and on the increase in coronary sinus output of NE elicited by cardiac sympathetic nerve stimulation (S) in animals pretreated with indomethacin (10 mg/kg). The left cardiac sympathetic nerves were stimulated (2 Hz, supramaximal voltage) for 1.5 minutes at 15 minute intervals (S_1 and S_2). The number of dogs in each group is indicated by n. Asterisks denote values different from basal ($p < 0.05$). Stars denote values different from corresponding value obtained during the first collection period (S_1) ($p < 0.05$).

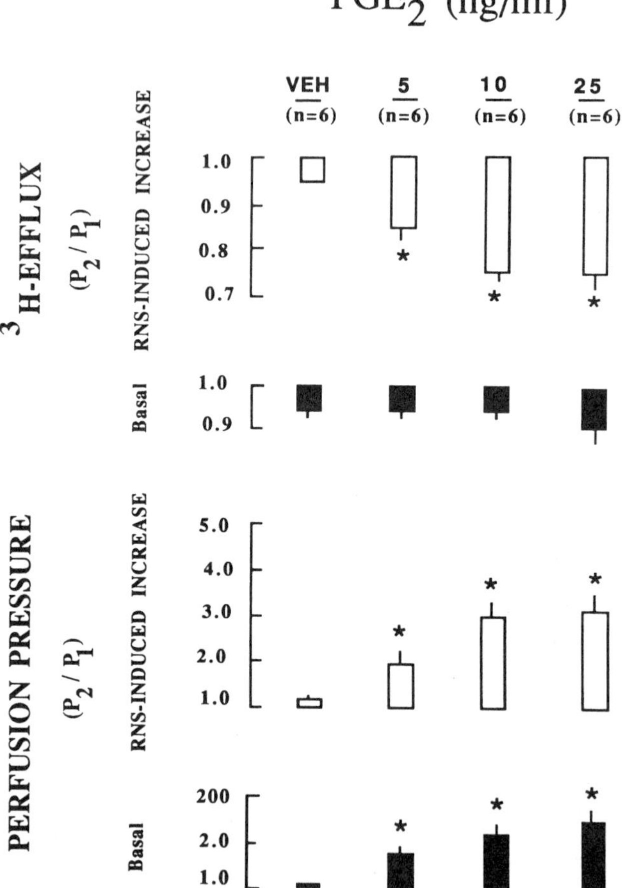

FIGURE 6. Effect of PGE$_2$ and its vehicle (VEH) on the basal and on the increase in tritium efflux (upper two panels) and perfusion pressure elicited by renal nerve stimulation (RNS, 2 Hz) in the isolated rat kidney perfused with Tyrode's solution at 5 ml/minute and prelabeled with ^{3}H-norepinephrine. P$_1$ and P$_2$ represent the first and second periods of collection of the perfusate, respectively. In the first period (P$_1$), the kidney was perfused with normal Tyrodes solution, whereas in the second period (P$_2$), the solution contained different concentrations of PGE$_2$. During P$_1$, the average value of basal tritium efflux and perfusion pressure in various experiments were 4.3 ± 5 dpm and 50 ± 4 mmHg, respectively, and the RNS-induced increase in tritium efflux and perfusion pressure were 20 ± 2 dpm and 35 ± 3 mmHg ($n = 6$) respectively. Asterisks denote value different from the corresponding value obtained in the presence of vehicle.

myocardial contractility and the coronary sinus output of NE elicited by sympathetic nerve stimulation in the heart of the pentobarbital-anesthetized dog.[37] Therefore, it appears that the effects of prostanoids that inhibit, rather than of those that may facilitate (PGF$_{2\alpha}$, PGD$_2$, TxA$_2$), adrenergic transmission predominate at the adrenergic neuroeffector junction. Moreover, these findings support the proposition that

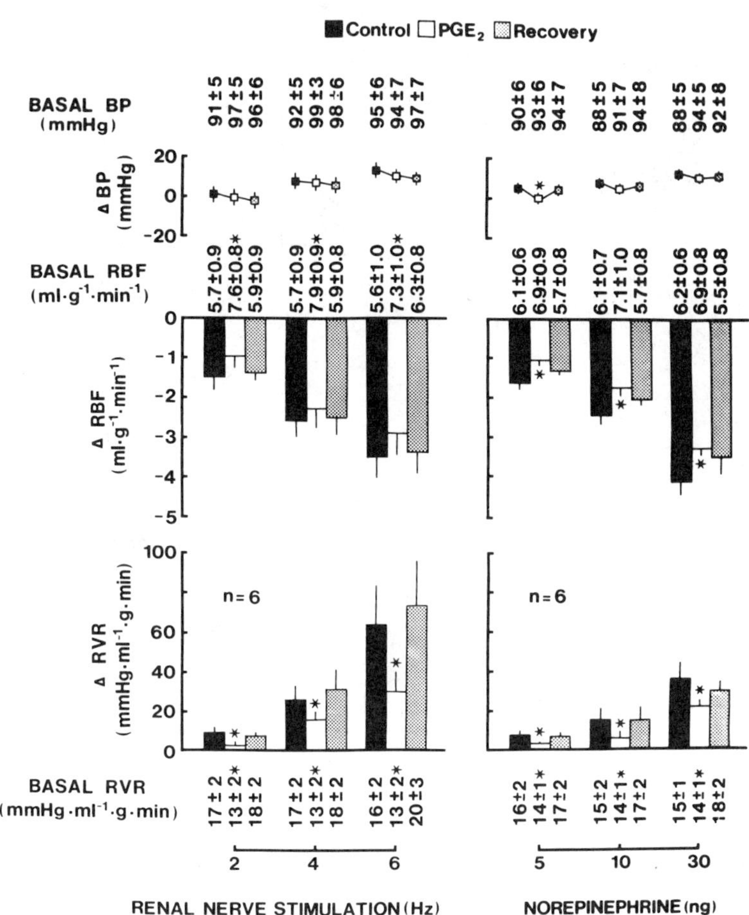

FIGURE 7. Effect on intrarenal arterial infusion of prostaglandin (PGE$_2$) (0.4 μg kg^{-1} minute^{-1}) on arterial blood pressure (BP), renal blood flow (RBF), renal vascular resistance (RVR), and decrease in RBF and rise in RVR elicited by renal nerve stimulation (left panel) and by injected norepinephrine (right panel). Renal nerves were stimulated at 2–6 Hz, and norepinephrine was injected at 5–30 ng into the renal arterial circuit. Values of BP, RBF, RVR, Δ BP, ΔRBF, and ΔRVR during infusion of PGE$_2$ were compared with corresponding values obtained during 0.15 M NaCl infusion before administration of PG. Basal values of BP, RBF, and RVR were obtained immediately before stimulation or administration of norepinephrine (control). ΔBP, ΔRBF, and ΔRVR were calculated by subtracting the values of BP, RBF, and RVR from the values of basal BP, RBF, and RVR, respectively, measured just before nerve stimulation or norepinephrine injections and obtained during nerve stimulation or norepinephrine injection. n denotes number of rats and bars in columns are ± standard error (SE). Asterisks denote value during PGI$_2$ infusion different from that obtained before infusion of prostaglandin (control) ($p < 0.05$).

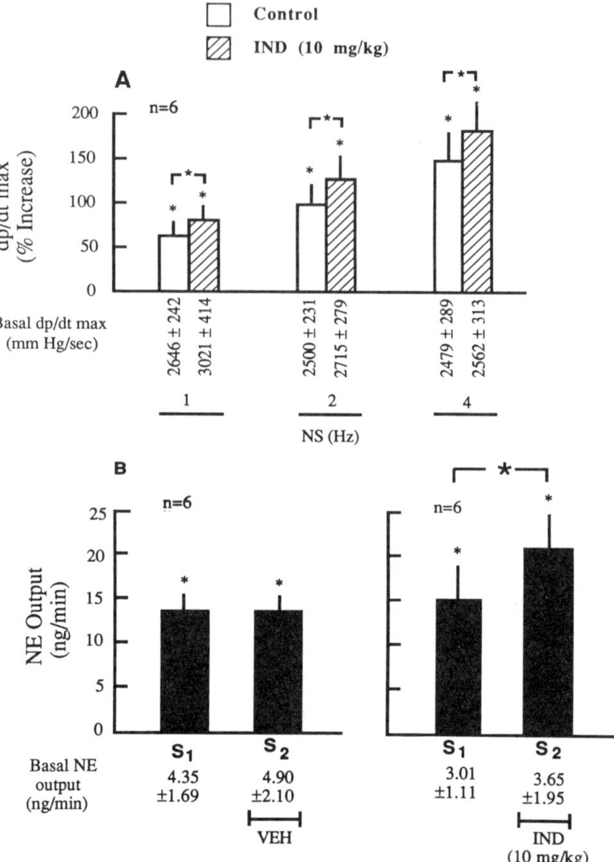

FIGURE 8. Effect of indomethacin (IND) [10 mg/kg intravenously (iv)] or its vehicle (VEH) (0.5 ml ethanol iv in 25 ml of saline) on basal and on the increase in ventricular dP/dt_{max} (upper panel, A) and on the coronary sinus output of norepinephrine (NE) (lower panel, B) elicited by left cardiac sympathetic nerve stimulation (NS) at 1–4 Hz for 20 seconds at 4 minute intervals (A) or at 2 Hz (S_1 and S_2) for 1.5 minutes at 15 minute intervals (B) using supramaximal voltage and 1.5 mseconds duration. The number of dogs in each group is indicated by n. Asterisks denote value different from corresponding value obtained in the control group (A) or from that obtained during the first collection period (B).

prostaglandins particulary PGE_2 and/or PGI_2 act as inhibitory modulators of adrenergic transmission.[3–5]

MECHANISM OF THE INHIBITORY EFFECT OF PGs ON RELEASE OF THE ADRENERGIC TRANSMITTER

The inhibitory effect of PGs on the release of adrenergic transmitter has been shown to operate independently of alpha$_2$-mediated feedback restriction of NE release.[51]

It is well established that NE released from sympathetic fibers by depolarization in response to action potential or by agents such as KCl is absolutely dependent upon extracellular Ca^{++}.[52,53] Depolarization of adrenergic nerve terminals by electrical stimulation causes opening of fast sodium channels and increased sodium entry. K^+ produces depolarization of the nerve terminal by reducing the concentration gradient between the intra- and extracellular spaces.[54] The depolarization caused by these stimuli is associated with an increased influx of Ca^{++} which in turn promotes release of the adrenergic transmitter from the nerve fibers.[52,53,55] The demonstration that PGE_2 inhibited depolarization-induced Ca^{++}-dependent, but not tyramine-induced Ca^{++}-independent, release of NE has led to the proposition that PGE_2 inhibits the release of adrenergic transmitter by restricting the availability of Ca^{++} to this process.[56] Supporting this view are the findings that increasing the extracellular Ca^{++} concentration counteracted the inhibitory effect of PGE_2 on NE release and restored it to normal levels in the cat spleen.[56] Similarly, in the guinea pig vas deferens, prolonged pulse duration or addition of tetraethylammonium or rubidium, interventions known to prolong duration of action potential and pobably increase Ca^{++} influx, also minimized the inhibitory effect of PGE_2 on release of NE elicited by transmural stimulation.[57] Moreover, the inhibitory effect of PGE_2 on NE release in this preparation was found to be inversely related to extracellular Ca^{++} concentration.[58–60] A kinetic analysis of the inhibitory effects of PGE_2 revealed that PGE_2 reduced the apparent sensitivity of the release process for Ca^{++}. These observations and the findings that PG in the vas deferens may also originate from nerve fibers has led Stjärne to suggest that PGE_2 may operate in a positive feedback loop enhancing the initial reduction in available Ca^{++}.[60,61] Taylor and Einhorn reported that in the mouse vas deferens, low concentrations of PGE_2 that did not alter resting membrane potential inhibited the excitatory postjunctional potential and contraction elicited by transmural stimulation; these effects were antagonized by elevated external Ca^{++}.[62]

The mechanism by which Ca^{++} enters the nerve terminal and PGs inhibit its availability to the process of secretion has not yet been established. It has been postulated that Na^+ entering the nerve fiber during depolarization by nerve action potentials may regulate Ca^{++} metabolism by influencing its influx into nerve fibers and extrusion from axoplasm via carrier-mediated Na^+-Ca^{++} exchange.[63,64] Ca^{++} influx through Na^+-Ca^{++} exchange has been shown to occur in squid axon and synaptosomes.[63,64] However, in our recent study in the isolated rat kidney perfused with Tyrode's solution and prelabeled with $[^3H]NE$, amiloride and Li^+, which are known to inhibit Na^+-Ca^{++} exchange,[65,66] did not reduce but rather enhanced tritium overflow elicited by RNS, suggesting that Na^+-Ca^{++} exchange is not involved in the depolarization-induced overflow of tritium.[67] The findings that in the rat kidney (a) ω-conotoxin (ω-CgTx), a neurotoxin known to block "L" and "N" but not "T"-type Ca^{++} channels in chicken sensory neurons,[68,69] inhibited tritium overflow elicited by RNS and the associated vasoconstrictor response but not the vasoconstrictor effect of exogenous NE; and (b) nifedipine, a dihydropyridine Ca^{++} channel antagonist that inhibits L- but not N- and T-type Ca^{++} channels,[70] enhanced RNS-induced tritium overflow suggest that Ca^{++} enters the nerve fibers via Ca^{++} channels ("N-type") that are distinct from those present in the vascular or cardiac muscle.[71] Since PGs inhibit release of NE as well as the postjunctional actions of the adrenergic neurotransmitter in several tissues, it is possible that PGE_2 inhibits influx of Ca^{++} through neuronal ("N-type") as well as vascular ("L-type") Ca^{++} channels. To test this view, we investigated the effect of PGE_2 on the release of adrenergic transmitter elicited by experimental interventions that increase influx of Ca^{++} through ω-CgTx ("N-type") and nifedipine sensitive ("L-type") Ca^{++} channels. We have reported that infusion of ouabain (10^{-4} M) in the presence of low extracellular K^+ (0.54 mM) which inhibits Na^+-K^+-ATPase increased

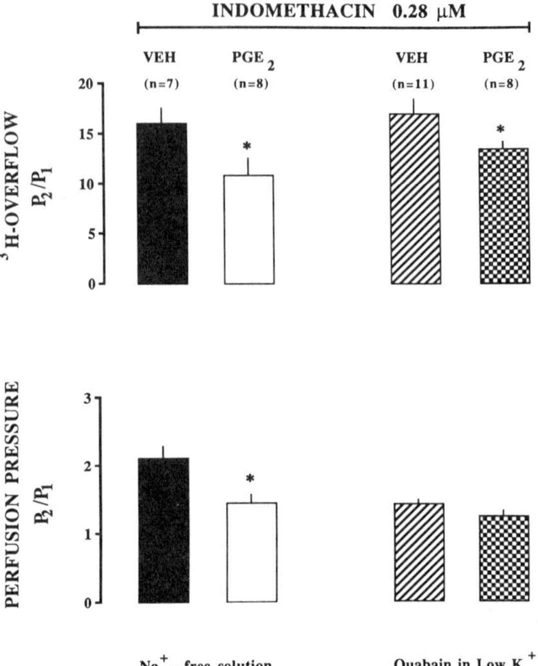

FIGURE 9. Effect of PGE_2 (84 nM) and its vehicle (VEH) on the increase in tritium overflow (upper panel) and perfusion pressure (lower panel) elicited by external Na^+ depletion (left panel) or by ouabain 10^{-4} M in low K^+ 0.40 mM (right panel) in the isolated rat kidney prelabeled with [³H]NE and perfused with Tyrode's solution containing indomethacin (0.28 μM). Infusion of PGE_2 or its vehicle was initiated 5 minutes before and continued during external Na^+ depletion or administration of ouabain in low K^+. P_1 represent the period of collection of perfusate in the presence of normal Tyrode's solution containing indomethacin and PGE_2 or its vehicle. P_2 represents the period 25 minutes after commencing the infusion of PGE_2 or its (VEH) in Na^+-free Tyrode's solution or with Tyrode's solution containing low K^+ and ouabain. Data are presented as ratios of the values of tritium efflux and perfusion pressure obtained during Na^+ depletion or infusion of ouabain in low K^+ in the presence of PGE_2 or its vehicle (P_2) to the corresponding values obtained immediately before Na^+ depletion or administration of ouabain in low K^+ (P_1). Bars in columns represent $\pm$ standard error of the mean (SEM). Asterisks denote values significantly different from the corresponding values in the vehicle group within the same period ($p < 0.05$). Values during the first period (P_1) of tritium efflux (dpmx10³) before: Na^+ depletion, VEH 4.9 $\pm$ 6, PGE_2 4.7 $\pm$ 5; ouabain in low K^+, VEH 4.7 $\pm$ 7, PGE_2 4.4 $\pm$ 7; perfusion pressure (mmHg) before Na^+-free depletion, VEH 77 $\pm$ 5, PGE_2 114 $\pm$ 7; ouabain in low K^+, VEH 61 $\pm$ 2, PGE_2 79 $\pm$ 5.

the output of tritium, which, like that caused by RNS, was enhanced by amiloride and inhibited by the Na^+ channel blocker tetrodotoxin and by extracellular Ca^{++} depletion and ω-CgTx, but not by nifedipine, in the rat kidney.[72] On the other hand, extracellular Na^+ depletion, an intervention also known to inhibit Na^+-K^+-ATPase, enhanced tritium efflux, which was inhibited by omission of Ca^{++} from the perfusion medium or by nifedipine, but not by ω-CgTx, tetrodotoxin, or amiloride.[71] Infusion of PGE_2 (30 ng/ml) inhibited the overflow of tritium elicited by oubain in low K^+ or by Na^+ depletion (FIGURE 9). Since the efflux of tritium elicited by oubain in low K^+ exhibited the characteristics of the physiological release process activated by RNS, it appears that PGE_2 inhibits the availability of Ca^{++} to the secretion process via ω-CgTx Ca^{++} channels. Inasmuch as PGE_2 also reduced the tritium efflux elicited by Na^+ depletion, it appears that PGE_2 also inhibits Ca^{++} availability via dihydropyridine-sensitive Ca^{++} channels that become operative in the absence of extracellular Na^+ in the rat kidney. However, the possibility that PGE_2 reduces release of the adrenergic transmitter by increasing Ca^{++} efflux from the nerve terminal cannot be excluded.

SUMMARY

Prostaglandins (PG) are synthesized from arachidonic acid, which is deesterified from tissue lipids in response to various stimuli including adrenergic transmitter, consequent to activation of one or more lipase(s). The profile of arachidonic acid metabolites generated in response to sympathetic nerve stimulation or administration of norepinephrine (NE) may vary in different tissues. For example, in the kidney and spleen, PGE_2 is the major and PGI_2 and $PGF_{2\alpha}$ the minor products; whereas in the heart and blood vessels, PGI_2 is the principal product of arachidonic acid generated in response to sympathetic nerve stimulation. PGE_2 and PGI_2 inhibit release of NE and/or the postjunctional actions of this neurotransmitter in several tissues. These observations and the findings that inhibitors of cyclooxygenase enhance NE release and the response of effector organs to nerve stimulation suggest that PGs act as physiological modulators of adrenergic transmission. The mechanism by which PGs modulate release of the adrenergic transmitter has not yet been established. NE appears to be released from sympathetic fibers during depolarization by influx of Na^+, which is associated with entry of Ca^{++} through ω-conotoxin-sensitive Ca^{++} channels that are distinct from those in the vascular smooth muscle, which are sensitive to nifedipine. Ouabain in low external K^+ activates the former, whereas external Na^+ depletion activates the latter type of Ca^{++} channels in the nerve fiber and promotes release of NE. PGs (PGE_2) may inhibit release of NE from nerve fibers by interfering with the availability of Ca^{++} through these Ca^{++} channels or promoting efflux of Ca^{++} from the nerve terminal.

REFERENCES

1. McGIFF, J. C. 1981. Prostraglandins, prostacyclin and thromboxanes. Annu. Rev. Pharmacol. Toxicol. **21:** 479–509.
2. NEEDLEMAN, P., J. TURK, B. A. JACKSCHIK, A. R. MORRISON & J. B. LEFKOWITH. 1986. Arachidonic acid metabolism. Annu. Rev. Biochem. **55:** 69–102.
3. HEDQVIST, P. 1977. Basic mechanisms of prostaglandin action on autonomic neurotransmission. Annu. Rev. Pharmacol. Toxicol. **17:** 259–279.
4. WESTFALL, T. C. 1977. Local regulation of adrenergic neurotransmission. Physiol. Rev. **57:** 659–728.
5. MALIK, K. U. 1978. Prostaglandin modulation of adrenergic nervous system. Fed. Proc. **37:** 203–207.
6. KUNZE, H. & W. VOGT. 1971. Significance of phospholipase A for prostaglandin formation. Ann. N.Y. Acad. Sci. **180:** 123–125.
7. IRVINE, R. F. 1982. How is the level of free arachidonic acid controlled in mammalian cells? Biochem. J. **204:** 3–16.

8. BILLAH, M. M., E. G. LAPETINA & P. CUATRECASAS. 1981. Phospholipase A_2 activity specific for phosphatidic acid. A possible mechanism for the production of arachidonic acid in platelets. J. Biol. Chem. **256:** 5399–5403.

9. PRESCOTT, S. M. & P. W. MAJERUS. 1983. Characterization of 1,2-diacyglycerol hydrolysis in human platelets. J. Biol. Chem. **258:** 764–769.

10. BOJESEN, I. 1974. Quanitive and qualitive analyses of isolated lipid droplets from interstitial cells in renal papillae from various species. Lipids **9:** 835–843.

11. SLIVKA, S. & P. A. INSEL. 1987. α_1-Adrenergic receptor-mediated phosphoinositide hydrolysis and prostaglandin E_2 formation in Madin-Darby canine kidney cells. J. Biol. Chem. **262:** 4200–4207.

12. SAMUELSSON, B. 1983. Leukotrienes: mediators of immediate hypersensitivity reactions and inflammation. Science **220:** 568–575.

13. CAPDEVILA, J., L. J. MARNETT, N. CHACOS, R. A. PROUGH & R. W. ESTABROOK. 1982. Cytochrome P-450 dependent oxygenation of arachidonic acid to hydroxyeicosatetraenoic acids. Proc. Nat. Acad. Sci. USA **79:** 767–770.

14. FERRERI, N. R., M. SCHWARTZMAN, N. G. IBRAHAM, P. N. CHANDER & J. C. McGIFF. 1984. Arachidonic acid metabolism in a cell suspension isolated from rabbit renal outer medulla. J. Pharmacol. Exp. Ther. **231:** 441–448.

15. MONCADA, S. & J. R. VANE. 1978. Pharmacology and endogenous roles of prostaglandin endoperoxides, thromboxane A_2 and prostacyclin. Pharmacol. Rev. **30:** 293–331.

16. DAVIES, B. N., E. W. HORTON & P. G. WITHRINGTON. 1968. The occurrence of prostaglandin E_2 in splenic venous blood of the dog following splenic nerve stimulation. Br. J. Pharmacol. Chemother. **32:** 127–135.

17. DAVIS, H. A. & E. W. HORTON. 1972. Output of prostaglandins from the rabbit kidney, its increase on renal nerve stimulation and its inhibition by indomethacin. Br. J. Pharmacol. **46:** 658–675.

18. COOPER, C. L. & K. U. MALIK. 1985. Prostaglandin synthesis and renal vasoconstriction elicited by adrenergic stimuli are linked to activation of alpha-1 adrenergic receptors in the isolated rat kidney. J. Pharmacol. Exp. Ther. **223:** 24–31.

19. HIDAKA, T. & K. U. MALIK. 1980. Release of prostaglandins evoked by neurohormonal stimuli in the isolated spleen of rabbit. Eur. J. Pharmacol. **62:** 97–105.

20. WENNMALM, A. 1978. Prostaglandin-mediated inhibition of norepinephrine release. III. Separation of prostaglandins released from stimulated hearts and analysis of their neurosecretion inhibitory capacity. Prostaglandins **15:** 113–121.

21. SHAFFER, J. E. & K. U. MALIK. 1982. Enhancement of prostaglandin output during activation of beta-1 adrenoceptors in the isolated rabbit heart. J. Pharmacol. Exp. Ther. **223:** 729–735.

22. PIPILI, E. & N. L. POYSER. 1981. Effects of nerve stimulation and of administration of noradrenaline or potassium chloride upon the release of prostaglandin I_2, E_2 and $F_{2\alpha}$ from the perfused mesenteric arterial bed of the rabbit. Br. J. Pharmacol. **72:** 89–93.

23. BIRKLE, D. L. & E. F. ELLIS. 1983. Conversion of arachidonic acid to cyclooxygenase and lipoxygenase products, and incorporation into phospholipids in the mouse neuroblastoma clone, Neuro-2A. Neurochem. Res. **8:** 319–332.

24. HEDQVIST, P. 1973. *In* Autonomic Neurotransmission in the Prostaglandins. P. W. Ramwell, Ed.: 101–131. Plenum Press. New York, N.Y.

25. NEEDLEMAN, P., J. DOUGLAS, JR., B. JAKSCHIK, P. B. STOECKLEIN & E. M. JOHNSON, JR. 1974. Release of renal prostaglandin by catecholamines: relationship to renal endocrine function. J. Pharmacol. Exp. Ther. **188:** 453–460.

26. MALIK, K. U. 1988. Interaction of arachidonic acid metabolites and adrenergic nervous system. Am. J. Med. Sci. **31:** 280–286.

27. PETKOV, V. & R. RADOMIROV. 1980. On the origin of prostaglandin and its role in sympathetic nerve transmission in vas deferens. Gen. Pharmacol. **11:** 275–282.

28. TRACHTE, G. 1988. Adrenergic nerves mediate angiotensin-induced prostaglandin release in the rabbit vas deferens. J. Pharmacol. Exp. Ther. **246:** 211–217.

29. GREENBERG, R. 1978. The neuronal origin of prostaglandin released from the rabbit portal vein in response to electrical stimulation. Br. J. Pharmacol. **63:** 79–85.

30. KHAN, M. T. & K. U. MALIK. 1983. Modulation of prostaglandins of the release of [^{3}H] noradrenaline evoked by potassium and nerve stimulation in the isolated rat heart. Eur. J. Pharmacol. **78:** 213–218.

31. MARQUES, M., C. A., E. M. P. LADOSKY & M. C. FONTELES. 1983. The presnaptic effects of aspirin and prostaglandin E_2 in the isolated cat heart. Brazilian J. Med. Biol. Res. **16:** 133–140.

32. WENNMALM, M., G. A. FITZGERALD & A. WENNMALM. 1987. Prostacyclin as neuromodulator in the sympathetically stimulated rabbit heart. Prostaglandins **33:** 675–691.

33. MALIK, K. U. 1978. Inhibitory effect of prostaglandin (PGE_2) on the release of norepinephrine evoked by renal nerve stimulation in the rat kidney. *In* Abstracts. 7th International Congress of Pharmacology, P-807.

34. HEDQVIST, P. 1979. Actions of prostacyclin (PGI_2) on adrenergic neuroeffector transmission in the rabbit kidney. Prostaglandin **17:** 249–258.

35. RUMP, L. C. & P. SCHOLLMEYER. 1989. Effect of endogenous and synthetic prostanoids, the thromboxane A_2 receptor agonist U-46619 and arachidonic acid on [^{3}H]-noradrenaline release and vascular tone in rat isolated kidney. Br. J. Pharmacol. **97:** 819–828.

36. FRAME, M. H. 1976. A comparison of the effects of prostaglandins A_2, E_2, $F_{2\alpha}$ on the sympathetic neuroeffector system of the isolated rabbit kidney. *In* Advances in Prostaglandin and Thromboxane Research. B. Samuelson and R. Paoletti, Eds. **1:** 369–373. Raven Press. New York, N.Y.

37. LANIER, S. M. & K. U. MALIK. 1985. Inhibition by prostaglandins of adrenergic transmission in the left ventricular myocardium of anesthetized dogs. J. Cardiovasc. Pharmacol. **7:** 653–659.

38. CHAPNICK, B. M., P. W. PAUSTIAN, E. KLAINER, P. D. JOINER, A. HYMAN & P. J. KADOWITZ. 1976. Influence of prostaglandins E, A and F on vasoconstrictor responses to norepinephrine, renal nerve stimulation and angiotensin in the feline kidney. J. Pharmacol. Exp. Ther. **196:** 44–52.

39. LONIGRO, A. J., N. A. TERRAGNO, K. U. MALIK & J. C. McGIFF. 1973. Differential inhibition by prostaglandins of the renal actions of pressor stimuli. Prostaglandins **3:** 595–606.

40. SUSIC, H. & K. U. MALIK. 1981. Prostacyclin and prostaglandin E_2 effects on adrenergic transmission in the kidney of anesthetized dog. J. Pharmacol. Exp. Ther. **218:** 588–592.

41. INOKUCHI, K. & K. U. MALIK. 1984. Attenuation by prostaglandins of adrenergically-induced renal vasoconstriction in anesthetized rats. Am. J. Physiol. **246:** R228–R235.

42. MALIK, K. U. & J. C. McGIFF. 1975. Modulation by prostaglandins of adrenergic transmission in the isolated perfused rabbit and rat kidney. Circ. Res. **36:** 599–609.

43. MALIK, K. U., P. RYAN, & J. C. McGIFF. 1976. Modification by prostaglandins E_1 and E_2, indomethacin, and arachidonic acid of the vasoconstrictor responses of the isolated perfused rabbit and rat mesenteric arteries to adrenergic stimuli. Circ. Res. **39:** 163–168.

44. JACKSON, E. K. & W. B. CAMPBELL. 1980. The in situ blood perfused rat mesentery; a model for assessing modulation of adrenergic neurotransmission. Eur. J. Pharmacol. **66:** 217–224.

45. HEMKER, D. P. & J. W. AIKEN. 1981. Actions of indomethacin and prostaglandin E_2 and D_2 on nerve transmission in the nicitating membrane of the cat. Prostaglandins **22:** 599–611.

46. NAKAJIMA, M. & N. TODA. 1984. Neuroeffector actions of prostaglandin D_2 on isolated dog mesenteric arteries. Prostaglandins **27:** 407–419.

47. TRACHTE, G. J. 1986. Thromboxane agonist (U 46619) potentiates norepinephrine efflux from adrenergic nerves. J. Pharmacol. Exp. Ther. **237:** 473–477.

48. NAKAJIMA, M. & N. TODA. 1986. Prejunctional and postjunctional actions of prostaglandins $F_{2\alpha}$ and I_2 and carbocyclic thromboxane A_2 in isolated dog mesenteric arteries. Eur. J. Pharmacol. **120:** 309–318.

49. TRACHTE, G. J. & E. STEIN. 1988. Platelet-generated thromboxane A_2 enhances norepinephrine release from adrenergic nerves. J. Pharmacol. Exp. Ther. **247:** 1139–1145.

50. OKAMURA, T., M. NAKAJIMA & N. TODA. 1988. Neuroeffector actions of thromboxane B_2 in dog isolated mesenteric arteries. Br. J. Pharmacol. **93:** 367–374.

51. STJÄRNE, L. 1973. Prostaglandin-versus α-adrenoceptor-mediated control of sympathetic

neurotransmitter secretion in guinea-pig isolated vas deferens. Eur. J. Pharmacol. **22:** 233–238.

52. HUKOVIC, S. & E. MUSCHOLL. 1962. Die noradrenalin-abgabe aus dem isolierten kaninchen-herzen bei sympathischer nervenreizung und ihre Pharmacologische beeinflussung. Naunyn-Schmiedebergs Arch. Exp. Pathol. Pharmacol. **244:** 81–96.

53. KIRPEKAR, S. M. & A. R. WAKADE. 1968. Release of catecholamines from the cat spleen by potassium. J. Physiol. London **94:** 595–608.

54. SHANES, A. M. 1958. Electrochemical aspects of physiological and pharmacological action in excitable cells. Pharmacol. Rev. **10:** 59–164.

55. BLAUSTEIN, M. P. 1975. Effect of potassium, veratridine and scorpion venom on calcium accumulation and transmitter release by nerve terminals *in vitro.* J. Physiol. London **247:** 617–655.

56. HEDQVIST, P. 1970. Studies on the effect of prostaglandin E_1 and E_2 on the sympathetic neuromuscular transmission in some animal tissues. Acta Physiol. Scand. **79**(Suppl. 345): 1–40.

57. HEDQVIST, P. 1976. Further evidence that prostaglandins inhibit release of noradrenaline from adrenergic nerve terminals by restriction of availability of calcium. Br. J. Pharmacol. **58:** 599–603.

58. HEDQVIST, P. 1974. Interaction between prostaglandins and calcium ions on noradrenaline release from the stimulated guinea pig vas deferens. Acta Physiol. Scand. **90:** 153–157.

59. STJÄRNE, L. 1973. Inhibitory effect of prostaglandin E_2 on noradrenaline secretion from sympathetic nerves as a function of external calcium. Prostaglandins **3:** 105–109.

60. STJÄRNE, L. 1973. Kinetics of secretion of sympathetic neurotransmitter as a function of external calcium: mechanism of inhibitory effect of prostaglandin E. Acta Physiol. Scand. **87:** 428–430.

61. STJÄRNE, L. 1972. Prostaglandin E restricting noradrenaline secretion—Neural in origin? Acta Physiol. Scand. **86:** 574–576.

62. TAYLOR, G. S. & V. F. EINHORN. 1972. The effect of prostaglandins on junctional potentials in the mouse vas deferens. Eur. J. Pharmacol. **20:** 40–45.

63. BAKER, P. F. 1972. Transport and metabolism of calcium ions in nerve. Prog. Biophys. Mol. Biol. **24:** 177–223.

64. BLAUSTEIN, M. P. & C. J. OBORN. 1975. The influence of sodium on Ca^{++} fluxes in pinched-off nerve terminals *in vitro.* J. Physiol. London **247:** 657–686.

65. SCHELLENBERG, G. D., L. ANDERSON & P. D. SWANSON. 1983. Inhibition of Na^+-Ca^{++} exchange in rat brain by amiloride. Mol. Pharmacol. **24:** 251–258.

66. KINSELLA, J. L. & P. S. ARONSON. 1981. Interaction of NH^{+4} and Li^+ with the renal microvillus membrane Na^+-H^+ exchange. Am. J. Physiol. **241:** C220–C226.

67. MOHY EL-DIN, M. M. & K. U. MALIK. 1987. Mechanism of norepinephrine release elicited by renal nerve stimulation, veratridine and potassium chloride in the isolated rat kidney. J. Pharmacol. Exp. Ther. **243:** 76–85.

68. MCCLESKEY, E. W., A. P. FOX, D. FELDMAN, B. M. OLIVERA, R. W. TSIEN & D. YOSHIKAMI. 1986. The peptide toxin ω-CgTx blocks particular types of neuronal Ca channels. (Abstr.) Biophys. J. **49:** 431a.

69. HIRNING, L. D., A. P. FOX, E. W. MCCLESKEY, R. J. MILLER, B. M. OLIVER, S. A. THAYER & R. W. TSIEN. 1986. Dominant role of N-type calcium channels in K-evoked release of norepinephrine from rat sympathetic neurons. (Abstr.) 16th Ann. Meet. Neurosci. **12:** 28.

70. NOWYCKY, M. C., A. P. FOX & R. W. TSIEN. 1985. Three types of neuronal calcium channel with different calcium agonist sensitivity. Nature **316:** 440–443.

71. MOHY EL-DIN, M. M. & K. U. MALIK. 1988. Differential effect of ω-conotoxin on release of the adrenergic transmitter and the vasoconstrictor response to noradrenaline in the rat isolated kidney. Br. J. Pharmacol. **94:** 355–362.

72. MOHY EL-DIN, M. M. & K. U. MALIK. 1988. Mechanism of norepinephrine release elicited by Na^+-K^+ adenosine triphosphate inhibition in the isolated rat kidney: involvement of voltage-dependent Ca^{++} channels. J. Pharmacol. Exp. Ther. **245:** 436–443.

On the Role of Adenylate Cyclase in Presynaptic Modulation of Neurotransmitter Release Mediated by Monoamine and Opioid Receptors in the Brain

ARIE H. MULDER, ANTON N. M. SCHOFFELMEER,
AND JOHANNES C. STOOF[a]

Department of Pharmacology
[a]Department of Neurology
Free University Medical Faculty
Van der Boechorststraat 7
1081 BT Amsterdam, the Netherlands

How is activation of presynaptic receptors translated into a change in neurotransmitter release from nerve terminals? Due to the complexity of the phenomenon of depolarization-secretion coupling, as briefly outlined below, this question will be difficult to answer unambiguously. It is well known that the depolarization following a sudden increase in Na^+ permeability of the nerve terminal (or varicosity) membrane upon arrival of an action potential initiates opening of voltage-sensitive calcium channels, resulting in an influx of Ca^{2+}. The increase in the intracellular Ca^{2+} concentration is thought to be the critical stimulus inducing exocytotic neurotransmitter release, involving fusion of the vesicular with the neuronal membrane.[1,2] Subsequent repolarization of the nerve terminal membrane is effected by a decrease in Na^+ and a simultaneous increase in K^+ permeability and an active transport of Na^+ and K^+, involving $[Na^+, K^+]$ATPase, to restore the ionic gradients over the nerve terminal membrane.[3] In addition, the Ca^{2+} that enters the nerve terminal during depolarization must eventually be extruded and/or buffered. Extrusion of Ca^{2+} is thought to be mediated by an Na^+-Ca^{2+} exchange and an ATP-dependent Ca^{2+} transport mechanism,[4,5] whereas mitochondria, smooth endoplasmic reticulum, and intracellular calcium-binding proteins are important sequestration sites involved in Ca^{2+} buffering in nerve terminals.[4,6]

The mechanisms transforming the influx of Ca^{2+} resulting from depolarization of the nerve terminals into neurotransmitter release still are themes of intensive research. Neurochemical studies suggest that protein phosphorylation reactions may be of paramount importance for the process of chemical neurotransmission. In the brain a number of protein kinases and a large variety of substrate proteins have been detected, which might play diverse roles in pre- as well as postsynaptic processes.[7-9] It has been shown, for instance, that depolarization of brain slices or synaptosomes by high K^+ concentrations or the alkaloid veratrine results in enhanced phosphorylation of certain nerve terminal proteins that appear to be associated with synaptic vesicles and might be involved in the movement of the vesicles to the sites of release and/or the exocytotic process.[10-13] Furthermore, calmodulin, a ubiquitous high-affinity Ca^{2+}-binding protein, found in high concentrations in the brain, is likely to play a pivotal role in the sequence of events, including protein phosphorylation, leading to neurotransmitter release.[6,14] In addition to calcium- and calmodulin-dependent protein kinases, cyclic nucleotide–dependent kinases have been found in the brain.[8,9] Cyclic AMP–dependent protein kinase is thought to mediate phosphorylation of a variety of neuronal proteins, some of which may be intimately involved in the process of neurotransmitter secretion, such as the synaptic vesicle protein synapsin I.[8,15] It has also been suggested[16,17] that cyclic AMP–dependent protein kinase may mediate phosphorylation of calcium channels in the nerve terminal membrane, thereby promoting the influx of Ca^{2+}. Obviously, the view that cyclic AMP–dependent phosphorylation might play a physiological role in the presynaptic processes connected with neurotransmission necessarily implies the presence of an adenylate cyclase system in the nerve terminal membrane.

This paper surveys neurochemical evidence pro or contra (a) the possible involvement of presynaptic adenylate cyclase systems in the regulation of neurotransmitter release from noradrenergic, serotonergic, dopaminergic, and cholinergic neurons in the brain, and (b) the possibility that inhibition of transmitter release through activation of (some) presynaptic receptors may, at least in part, be due to an inhibition of adenylate cyclase activity. Other and perhaps more common primary effects of presynaptic receptor activation (for recent reviews see References 18 and 19), such as alterations in the permeability of the nerve terminal membrane for certain ions, particularly K^+, Na^+, and Ca^{2+}, that are involved in the process of depolarization and the ensuing neurotransmitter secretion, will only be mentioned occasionally, whenever appropriate. All of the studies reviewed here have been carried out using *in vitro* methods, measuring the depolarization-induced overflow of (radiolabeled) neurotransmitters from brain slices or synaptosomes.

NORADRENALINE RELEASE AND PRESYNAPTIC α_2-ADRENOCEPTORS AND μ-OPIOID RECEPTORS

Thus far, the strongest evidence for a presynaptic adenylate cyclase system involved in the regulation of neurotransmitter release concerns noradrenergic neurons. Several studies have demonstrated enhancing effects of cyclic AMP analogues and phosphodiesterase (PDE) inhibitors on the electrically evoked release of radiolabeled noradrenaline (NA) from peripheral tissues,[20–24] suggesting that in postganglionic sympathetic neurons cyclic AMP may play a facilitatory role in the depolarization-secretion coupling process. More recently we[25–28] and others[29] have presented evidence indicating that the same holds true for noradrenergic neurons in the brain. Thus, cyclic AMP analogues (dibutyryl-cAMP, 8-bromo-cAMP) and various PDE inhibitors (e.g., IBMX, ZK62771) as well as the diterpene forskolin, a potent activator of adenylate cyclase,[30] were shown to enhance the electrically evoked release of [³H]NA from rat or rabbit cortex slices[27–29] (FIGURE 1). Further data suggest that depolarization as such may result in an activation of adenylate cyclase.[31] Thus, although [³H]NA release from cortical slices induced by the calcium ionophore A23187 was increased by forskolin and cyclic AMP analogues, it remained unchanged in the presence of PDE inhibitors. Therefore, exposure of brain slices to the calcium ionophore, which (in contrast to electrical stimulation) does not involve membrane depolarization, appears not to result in enhanced endogenous cyclic AMP levels in noradrenergic nerve terminals.[31] A presynaptic localization of the adenylate cyclase system involved is strongly supported by further experiments showing that [³H]NA release from cortex synaptosomes following depolarization by the alkaloid veratrine was enhanced by forskolin, as well as by 8-bromo-cyclic AMP and ZK62771[31] (FIGURE 2).

Very recently it was reported that forskolin and IBMX enhanced electrically evoked [³H]NA release from slices of rat hypothalamus,[32] indicating that also in this brain region an adenylate cyclase system is present in the noradrenergic nerve terminals. Surprisingly, in another recent study, using similar methods but examining [³H]NA release from slices of rat hippocampus, no evidence was found for the involvement of a presynaptic adenylate cyclase in the regulation of NA release.[33] Although the possibility of regional differences may be invoked, it remains difficult to understand why the enzyme would play a regulatory role in the secretory mechanisms in some noradrenergic nerve terminals, but not in others. We have explored several possible mechanisms, including a role of adenylate cyclase, that might be involved in the inhibition of [³H]NA release from rat brain cortical slices by activation of presynaptic α_2-adrenoceptors or μ-opioid receptors.[25,26,28,31,34,35] Strong evidence for the localization of these receptors on noradrenergic nerve terminals has resulted from

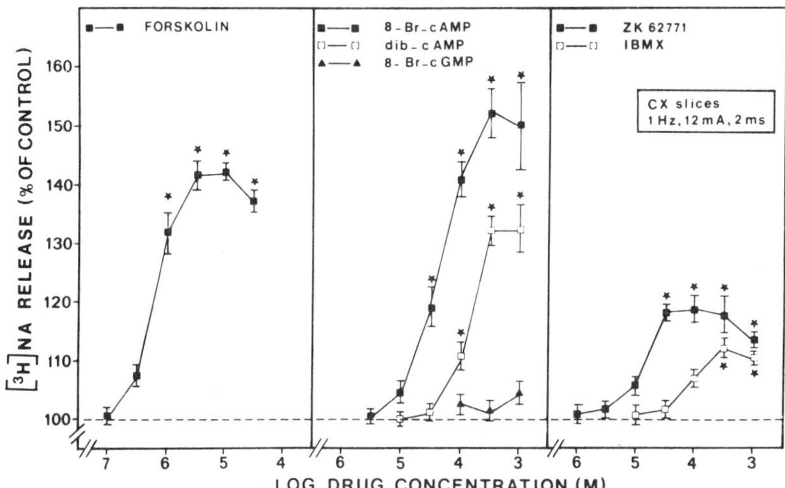

FIGURE 1. Effect of increasing intracellular cyclic AMP levels on the electrically evoked release of [³H]NA from rat brain cortex slices. Slices, prelabeled with [³H]NA, were superfused with physiological medium and, after 50 minutes of superfusion, stimulated electrically (1 Hz, 12 mA, 2 msecond pulses) for 5 minutes. Forskolin, cyclic AMP (GMP) analogues, or PDE inhibitors were added to the superfusion medium 20 minutes before stimulation. Control release averaged 4.3% of total tissue [³H]NA content in excess of spontaneous efflux. Data represent means ± standard errors of the mean (SEM) of 8–12 observations; *significantly different from control ($p < 0.01$; one-way ANOVA). (Modified from Reference 27.)

experiments with synaptosomes.[36–39] Early studies had suggested that activation of these receptors might reduce the availability of Ca^{2+} for the depolarization-secretion process by primarily inhibiting voltage-sensitive calcium channels.[37,40,41] However, another possible mechanism of the inhibitory effect of presynaptic receptor activation on NA release might be a decrease in the *utilization* of intracellular Ca^{2+} by the depolarization-secretion coupling process, perhaps in addition to an inhibition of Ca^{2+} *influx*. This possibility was examined by studying the effects of 4-aminopyridine and an increase in pulse duration on the presynaptic modulation of the electrically evoked release of [³H]NA from rat brain cortex slices mediated by α_2-adrenoceptors and μ-opioid receptors.[35] At the concentration used (0.1 mM) 4-aminopyridine is thought to increase the influx of Ca^{2+} not only by blocking K^+ channels, which prolongs the

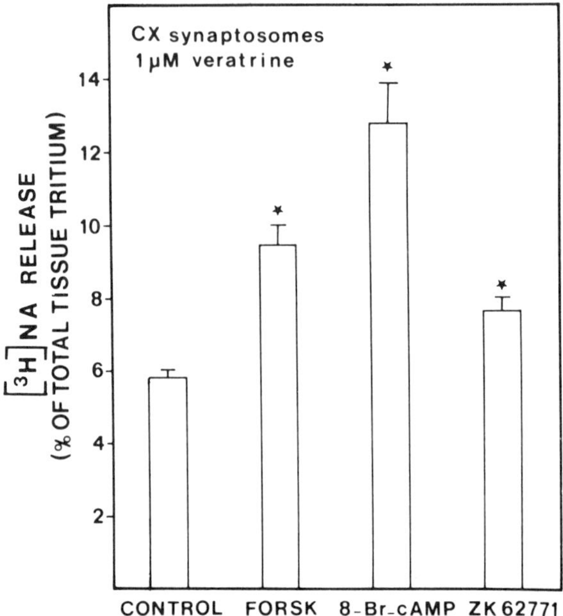

FIGURE 2. Effect of increasing intracellular cyclic AMP levels on veratrine-induced [³H]NA release from rat brain cortex synaptosomes. Synaptosomes prelabeled with [³H]NA were superfused with Ca^{2+}-free medium and, after 50 minutes of superfusion, exposed to 1 μM veratrine for 5 minutes in the presence of 1.2 mM Ca^{2+} to induce neurotransmitter release. Drugs were added to the medium 20 minutes before stimulation. Data are means ± SEM of 16 observations; *significantly different from control ($p < 0.01$; one-way ANOVA). (Modified from Reference 27.)

depolarization time, but also and perhaps primarily by a direct stimulatory action on the calcium channels.[42] In contrast, prolongation of the pulse duration is likely to increase the intracellular Ca^{2+} concentration by keeping calcium channels open for a longer period of time. In both cases (with [³H]NA release approximately doubled) presynaptic α_2-adrenoceptor efficacy appeared to be strongly reduced, whereas presynaptic μ-opioid receptor efficacy, like that of the potent calcium channel blocker Cd^{2+}, was reduced by 4-aminopyridine, but not by prolonging the pulse duration.[35] Therefore, the inhibitory effect of μ-opioid receptor activation appears to resemble that of Cd^{2+}, suggesting that activation of these receptors reduces NA release by decreasing the availability of free Ca^{2+} for the secretion process, subsequent to inhibition of voltage-sensitive calcium channels and/or changes in the intracellular Ca^{2+}-buffering process as suggested by others.[43,44] In contrast, activation of presynaptic α_2-adrenoceptors presumably diminishes the utilization of Ca^{2+} by the secretory mechanism, e.g., the phosphorylation of relevant intracellular proteins by Ca^{2+}/calmodulin- and/or cyclic AMP–dependent kinases.

This conclusion, implying that activation of presynaptic α_2-adrenoceptors does not (or not only) inhibit NA release primarily by reducing the Ca^{2+} influx through voltage-sensitive calcium-channels, is further supported by the finding that presynaptic α_2-adrenoceptor efficacy was not changed when [³H]NA release from brain cortex slices was induced by veratrine *in the absence of extracellular* Ca^{2+}[26] (FIGURE 3). It has been shown that in this experimental model, which precludes the possibility of a depolarization-induced Ca^{2+} influx, veratrine-induced [^{3}H]NA release is triggered by Ca^{2+} mobilized from intracellular stores, such as mitochondria, subsequent to the strong increase in the intracellular Na^+ concentration caused by the alkaloid.[45] In a recent study we found that not only activation of α_2-adrenoceptors, but also that of μ-opioid receptors, inhibited the release of [^{3}H]NA induced by veratrine in the absence of extracellular Ca^{2+}.[46] This observation is compatible with the hypothesis that activation of presynaptic μ-opioid receptors, perhaps in addition to inhibiting voltage-sensitive calcium channels (see above), interferes with the extrusion and/or buffering of intracellular Ca^{2+}.

The difference in the transduction mechanisms connected with presynaptic α_2-adrenoceptors and μ-opioid receptors also appears from studies examining the possible involvement of adenylate cyclase. In view of the postulated facilitatory role of adenylate cyclase in the regulation of NA release (see above), activation of presynaptic receptors might cause inhibition of release by reducing the activity of this enzyme.

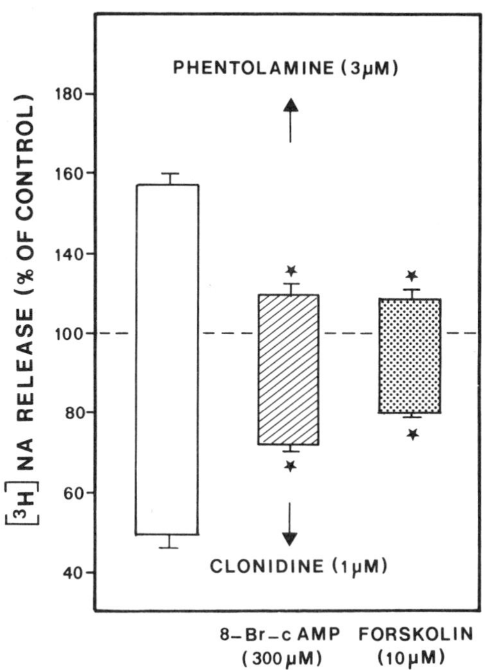

FIGURE 3. Presynaptic α_2-adrenergic modulation of veratrine-induced [^{3}H]NA release from rat cortical slices in the absence of extracellular Ca^{2+} and effects of 8-bromo-cAMP and forskolin thereon. Slices previously labeled with [^{3}H]NA were superfused with Ca^{2+}-free EGTA (1 mM) containing medium and exposed to 3 μM veratrine for 10 minutes to induce [^{3}H]NA release (see Reference 45) either in the absence (open column) or presence of 8-bromo-cAMP or forskolin. Drugs were added to the medium 20 minutes before stimulation. Average control [^{3}H]NA release induced by 3 μM veratrine: 12.7% (drug free), 18.0% (8-Br-cAMP), and 19.8% (forskolin). Data are means ± SEM of 8–12 observations; *significantly different modulation ($p < 0.01$) from that found in the absence of 8-bromo-cAMP or forskolin. (Modified from Reference 26.)

With regard to presynaptic α_2-adrenoceptors, preliminary support for this hypothesis was obtained by showing that the inhibitory effect of α_2 agonists as well as the enhancing effect of the antagonist phentolamine on depolarization-induced [³H]NA release and on release induced by veratrine in the absence of extracellular Ca^{2+} was reduced upon exposure of brain cortex slices to 8-bromo-cyclic AMP or forskolin[25,26,28] (FIGURES 3 and 4). In a later study, examining [³H]NA release from cortex slices induced by the calcium ionophore A23187, firm evidence for a role of adenylate cyclase in presynaptic α_2-adrenoceptor efficacy was obtained.[31] Normally, A23187-induced [³H]NA release is not inhibited by activation of α_2-adrenoceptors, which is considered to be due to the fact that voltage-sensitive calcium-channels are not involved.[37,40] Yet an additional reason might be that A23187, in contrast to depolarization by eletrical stimulation, does not activate the adenylate cyclase system thought to be present in noradrenergic nerve terminals. Indeed, although clonidine did not affect A23187-induced [³H]NA release, it completely blocked the release-facilitating effect of activation of adenylate cyclase by forskolin. Similarly, phentolamine only enhanced A23187-induced release if the slices were simultaneously exposed to forskolin[31] (FIGURE 5). Neither clonidine nor phentolamine changed A23187-induced [³H]NA

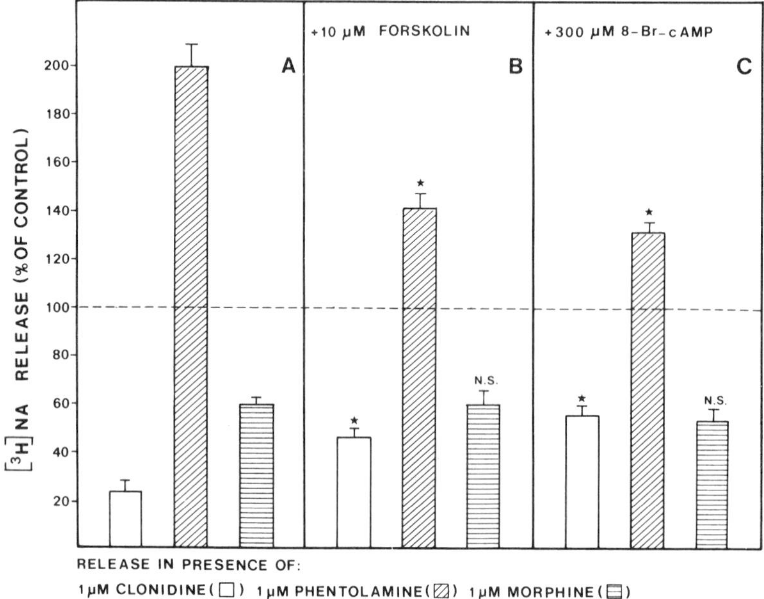

FIGURE 4. Presynaptic modulation of electrically evoked [³H]NA release from rat brain cortex slices mediated by α_2-adrenoreceptors or μ-opioid receptors (A) and effects thereon of forskolin (B) and 8-bromo cyclic AMP (C). Slices previously labeled with [³H]NA were superfused and, after 50 minutes of superfusion, exposed to electrical stimulation (1 Hz, 12 mA, 2 mseconds; 1.2 mM Ca^{2+}) for 5 minutes. Drugs were present in the medium from 20 minutes before depolarization. In the absence of drugs, [³H]NA release in excess of spontaneous efflux averaged 4.8%, and forskolin and 8-bromo cyclic AMP enhanced release to 7.8% and 7.2%, respectively, of total tissue [³H]NA content. Data represent means ± SEM of five experiments carried out in quadruplicate; *significantly different modulation ($p < 0.01$) from that found in the absence of forskolin or 8-bromo cyclic AMP; NS no significant difference. (Modified from Reference 28.)

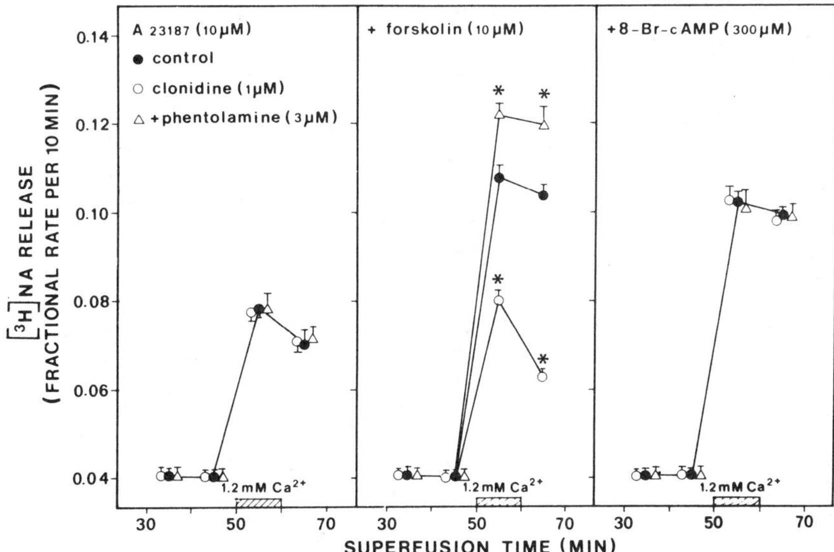

FIGURE 5. Presynaptic α_2-adrenoceptor-mediated modulation of [³H]NA release from rat brain cortex slices induced by the calcium ionophore A23187 only found in the presence of forskolin. Slices, prelabeled with [³H]NA, were superfused with Ca²⁺-free medium. Drugs, including 10 µM A23187, were added to the superfusion medium 30 minutes after the start of superfusion. After 50 minutes of superfusion the slices were exposed to 1.2 mM Ca² for 10 minutes to induce [³H]NA release. Data represent means ± SEM of 12 observations; *release significantly different ($p < 0.01$) from respective control. (Reproduced from Reference 31 with permission.)

release in the presence of 8-bromo-cyclic AMP, indicating that activation of presynaptic α_2-adrenoceptors does not affect a step in the secretory process beyond the generation of cyclic AMP, but results in an inhibition of adenylate cyclase activity. Compatible with this hypothesis are recent findings of Ramdine *et al.*, who showed that forskolin counteracted the inhibitory effect of α_2-adrenoceptor activation on the electrically evoked release of [³H]NA (a shift to the right of the agonist concentration-effect curve) from rat hypothalamic slices.[32]

An involvement of adenylate cyclase in the mechanism by which activation of presynaptic μ-opioid receptors reduces NA release in the brain appears to be very unlikely. Thus, the inhibitory effect of presynaptic μ-opioid receptor activation on electrically evoked [³H]NA release from cortex slices remained unaltered in the presence of 8-bromo-cyclic AMP or forskolin[28,31] (FIGURE 4). Moreover, whereas in the presence of forskolin A23187-induced [³H]NA release became liable to inhibition via activation of α_2-adrenoceptors, activation of μ-opioid receptors remained ineffective.[31]

SEROTONIN RELEASE AND PRESYNAPTIC SEROTONIN RECEPTORS AND α_2-ADRENOCEPTORS

Evidence for the involvement of a presynaptic adenylate cyclase system in the regulation of neurotransmitter release from serotonergic nerve terminals in the brain has recently been reported by a number of groups.[32,47,48] We[47] and others[48] have shown that forskolin and 8-bromo-cyclic AMP enhanced the electrically evoked release of serotonin ([3H]5-HT) from rat neocortical slices. The PDE-inhibitor ZK62771 was found to be ineffective by itself, but enhanced the increasing effect of forskolin on [3H]5-HT release. Very recently similar findings were reported with regard to the electrically evoked release of [3H]5-HT from rat hypothalamic slices.[32] Schlicker et al. also demonstrated that forskolin and 8-bromo-cyclic AMP enhanced the K^+-induced release of [3H]5-HT from cortical synaptosomes,[48] in support of the view that the adenylate cyclase involved is localized in the serotonergic nerve terminals. It should be noted, however, that the release-enhancing effects shown in that study were much smaller than those found in our study (see above) on synaptosomal [3H]NA release induced by veratrine.[27]

The few studies examining a possible role of adenylate cyclase in presynaptic modulation of 5-HT release pertain to the inhibitory 5-HT autoreceptors[48] and α_2-adrenoceptors.[32,48] The view that these receptors are indeed localized on serotonergic nerve terminals is strongly supported by data obtained from studies using synaptosomes.[38,49,50] Interestingly, analogous to our findings with regard to the presynaptic modulation of NA release through μ-opioid receptors and α_2-adrenoceptors,[28,31] Schlicker et al. presented evidence indicating that, in contrast to the presynaptic α_2-adrenoceptors, the 5-HT autoreceptors involved in inhibition of [3H]5-HT release from rat cortical slices are not linked to adenylate cyclase.[48] With regard to presynaptic α_2-adrenoceptors it thus appears that those localized on serotonergic as well as those on noradrenergic nerve terminals may be coupled to adenylate cyclase. Recent data of Ramdine et al. also suggest the involvement of adenylate cyclase in the α_2-adrenoceptor-mediated inhibition of NA as well as 5-HT release from rat hypothalamic slices.[32] However, in their study the inhibitory effect of α_2-adrenoceptor activation on [3H]5-HT release, in contrast to that on [3H]NA release, was not decreased by forskolin alone, but only if the slices were exposed to both forskolin and the PDE inhibitor IBMX.[32] Therefore, Ramdine et al. doubt a direct coupling of presynaptic α_2-adrenoceptors to adenylate cyclase and point out the complexity of the regulation of neurotransmitter release, which may involve several presynaptic receptor–mediated mechanisms. We would like to emphasize here, subscribing to the views expressed in a recent authoritative review by Starke et al.,[19] that a direct coupling of a presynaptic receptor to any transducing mechanism, whether a second messenger system or an ion channel, is difficult to prove, particularly in the brain. Furthermore, although we believe that the evidence for an involvement of an inhibition of adenylate cyclase in the transduction of presynaptic α_2-adrenoceptor activation is quite convincing, particularly in the case of α_2 autoreceptors in rat cerebral cortex, this by no means excludes the simultaneous involvement of other and presumably even mutually interacting transduction mechanisms.

DOPAMINE AND ACETYLCHOLINE RELEASE AND D-2 DOPAMINE, κ-, AND δ-OPIOID RECEPTORS

The electrically evoked release of dopamine ([3H]DA) from rat striatal slices was found to be increased by forskolin, 8-bromo cyclic AMP, and ZK62771,[47] suggesting that also in dopaminergic neurons a presynaptic adenylate cyclase may be involved in the regulation of neurotransmitter release. Very recently, Bowyer and Weiner reported that also the (Ca^{2+}-induced) release of [3H]DA from striatal synaptosomes was increased by forskolin and the PDE inhibitor IBMX.[51]

Although it is well known that D-2 dopamine receptors in the mammalian striatum can be coupled to adenylate cyclase in an inhibitory way,[52,53] the D-2 autoreceptors mediating presynaptic inhibition of DA release apparently use another transduction mechanism. This conclusion is mainly based on the following findings. Memo *et al.* reported that D-2 receptor activation inhibited the K$^+$-induced release of endogenous DA from striatal slices obtained from rats treated with kainic acid to the same extent as that from untreated controls.[54] However, the D-2 receptors mediating inhibition of adenylate cyclase activity could not be demonstrated anymore in striatal tissue from kainic acid–treated rats, suggesting that these receptors are located postsynaptically and that the presynaptic D-2 receptor–mediated inhibition of DA release does not involve a reduction in the activity of a presynaptic adenylate cyclase.[54] Data from a recent study using striatal synaptosomes[51] support this conclusion and suggest that, rather than adenylate cyclase, K$^+$ channels are involved in the presynaptic inhibition of DA release produced by activation of D-2 receptors. Similarly, the inhibitory effect of D-2 receptor activation on the depolarization-induced release of radiolabeled acetylcho-

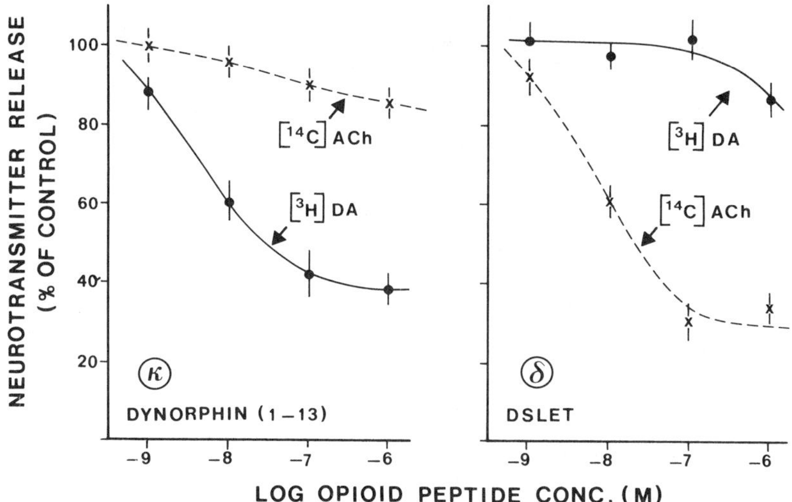

FIGURE 6. Differential inhibitory effects of the κ-selective opioid peptide dynorphin (1–13) and the δ-selective peptide [D-Ser^2]Leu-enkephalin-Thr (DSLET) on the electrically evoked release of [^{3}H]DA and [^{14}C]ACh from rat striatal slices. (Modified from Reference 60.)

line (ACh) release[55] appears not to involve adenylate cyclase. First of all, to our knowledge there is no evidence in support of the possible involvement of this enzyme in the regulation of ACh release in the brain. Secondly, neither forskolin, nor 8-bromo-cyclic AMP nor IBMX affected the inhibitory effect of D-2 receptor activation on electrically evoked or K$^+$-induced [^{3}H]ACh release from striatal slices.[55] In fact, recent studies in our laboratory indicate that presynaptic inhibition of striatal ACh release mediated by D-2 dopamine receptors as well as muscarinic autoreceptors involves the opening of K$^+$ channels as the major transduction mechanism.[56,57]

Depending on their receptor selectivity, opioid drugs may differentially inhibit the depolarization-induced release of radiolabeled DA and ACh from rat striatal slices. Thus, DA release appears to be inhibited by activation of κ-, but not δ- or μ-, opioid receptors, whereas ACh release is inhibited by opioid drugs only if they activate δ-opioid receptors[58–60] (FIGURE 6). We are currently investigating the transduction mechanisms involved. Preliminary findings with forskolin suggest that a reduction in adenylate cyclase may be involved in the inhibitory effect of κ-opioid receptor activation on DA release, but not in the δ-receptor-mediated inhibition of ACh release.

REFERENCES

1. LLINÁS, R. & J. E. HEUSER. 1977. Depolarization-release coupling systems in neurons. Neurosci. Res. Progr. Bull. **15:** 557–687.
2. BLAUSTEIN, M. P. 1979. The role of calcium in catecholamine release from adrenergic nerve terminals. *In* The Release of Catecholamines from Adrenergic Neurons. D. M. Paton, Ed.: 39–58. Pergamon Press. Oxford, England.
3. ALMERS, W. 1978. Gating currents and charge movements in excitable membranes. Rev. Physiol. Biochem. **82:** 96–190.
4. BLAUSTEIN, M. P., R. W. RATZLAFF & E. S. SCHWEITZER. 1980. Control of intracellular calcium in presynaptic nerve terminals. Fed. Proc. **39:** 2790–2795.
5. COUTINHO, O. P., A. P. CARVALHO L & C. A. M. CARVALHO. 1983. Effect of monovalent cations on Na^+/Ca^{2+} exchange and ATP-dependent Ca^{2+} transport in synaptic plasma membranes. J. Neurochem. **41:** 670–676.
6. CHEUNG, W. Y. 1980. Calmodulin plays a pivotal role in cellular regulation. Science **207:** 19–27.
7. GREENGARD, P. 1978. Phosphorylated proteins as physiological effectors. Science **199:** 146–152.
8. NESTLER, E. J. & P. GREENGARD. 1983. Protein phosphorylation in the brain. Nature **305:** 583–588.
9. NAIRN, A. C., H. C. HEMMINGS & P. GREENGARD. 1985. Protein kinases in the brain. Annu. Rev. Biochem. **54:** 931–976.
10. KRUEGER, B. K., J. FORN & P. GREENGARD. 1977. Deploarization-induced phosphorylation of specific proteins, mediated by calcium ion influx, in rat brain synaptosomes. J. Biol. Chem. **252:** 2764–2773.
11. FORN, J. & P. GREENGARD. 1978. Depolarizing agents and cyclic nucleotides regulate the phosphorylation of specific neuronal proteins in rat cerebral cortex slices. Proc. Nat. Acad. Sci. USA **75:** 5195–5199.
12. HUTTNER, W. B. & P. GREENGARD. 1979. Multiple phosphorylation sites in protein I and their differential regulation by cyclic AMP and calcium. Proc. Nat. Acad. Sci. USA **76:** 5204–5406.
13. BURKE, B. E. & R. J. DELORENZO. 1982. Ca^{2+} and calmodulin-regulated endogenous tubulin kinase activity in presynaptic nerve terminal preparations. Brain Res. **236:** 393–415.
14. DELORENZO, R. J. 1981. Calcium, calmodulin and synaptic function: modulation of neurotransmitter release, nerve terminal protein phosphorylation and synaptic vesicle morphology by calcium and calmodulin. *In* Regulatory Mechanisms of Synaptic Transmission. R. Tapia & C. W. Cotman, Eds.: 205–239. Plenum Press. New York, N.Y.
15. DUNKLEY, P. R., C. M. BAKER & P. J. ROBINSON. 1986. Depolarization-dependent protein phosphorylation in rat cortical synaptosomes: characterization of active protein kinases by phosphopeptide analysis of substrates. J. Neurochem. **46:** 1692–1703.
16. SULAKHE, P. V. & P. J. ST. LOUIS. 1980. Passive and active calcium fluxes across plasma membranes. Prog. Biophys. Mol. Biol. **35:** 135–195.
17. REUTER, H. 1983. Calcium channel modulation by neurotransmitters, enzymes and drugs. Nature **301:** 569–574.
18. ILLES, P. 1986. Mechanisms of receptor-mediated modulation of transmitter release in noradrenergic, cholinergic and sensory neurons. Neuroscience **17:** 909–928.
19. STARKE, K., M. GÖTHERT & H. KILBINGER. 1989. Modulation of neurotransmitter release by presynaptic autoreceptors. Physiol. Rev. **69:** 864–989.
20. CUBEDDU, L. X., E. BARNES & N. WEINER. 1975. Release of norepinephrine and dopamine-beta-hydroxylase by nerve stimulation. IV. An evaluation of a role for cyclic adenosine monophosphate. J. Pharmacol. Exp. Ther. **193:** 105–127.

21. PELAYO, F., M. DUBOCOVICH & S. Z. LANGER. 1978. Possible role of cyclic nucleotides in regulation of noradrenaline release from rat pineal through presynaptic adrenoceptors. Nature **274:** 76–78.

22. STJÄRNE, L., T. BARTFAI & P. ALBERTS. 1979. The influence of 8-Br-3′5′-cyclic nucleotide analogs and of inhibitors of 3′5′-cyclic nucleotide phosphodiesterase on noradrenaline secretion and neuromuscular transmission in guinea-pig vas deferens. Naunyn-Schmiedebergs Arch. Pharmacol. **308:** 99–105.

23. WEINER, N. 1979. Multiple factors regulating the release of norepinephrine consequent to nerve stimulation. Fed. Proc. **38:** 2193–2202.

24. ALBERTS, P., V. R. ÖGREN & A. I. SELLSTRÖM. 1985. Role of adenosine 3′5′-cyclic monophosphate in adrenoceptor-mediated control of ^{3}H-noradrenaline secretion in guinea-pig ileum myenteric nerve terminals. Naunyn Schmiedebergs Arch. Pharmacol. **330:** 114–120.

25. WEMER, J., A. N. M. SCHOFFELMEER & A. H. MULDER. 1982. Effects of cyclic AMP analogues and phosphodiesterase inhibitors on K^+-induced ^{3}H-noradrenaline release from rat brain slices and on its presynaptic α-adrenergic modulation. J. Neurochem. **39:** 349–356.

26. SCHOFFELMEER, A. N. M. & A. H. MULDER. 1983. ^{3}H-Noradrenaline release from rat neocortical slices in the absence of extracellular Ca^{2+} and its presynaptic alpha2-adrenergic modulation. A study on the possible role of cyclic AMP. Naunyn Schmiedebergs Arch. Pharmacol. **323:** 188–192.

27. SCHOFFELMEER, A. N. M., F. HOGENBOOM & A. H. MULDER. 1985. Evidence for a presynaptic adenylate cyclase system facilitating [^{3}H] norepinephrine release from rat brain neocortex slices and synaptosomes. J. Neurosci. **5:** 2685–2689.

28. MULDER, A. H. & A. N. M. SCHOFFELMEER. 1985. Catecholamine and opioid receptors, presynaptic inhibition of CNS neurotransmitter release, and adenylate cyclase. Adv. Cyclic Nucleotide Protein Phosphorylation Res. **19:** 273–286.

29. MARKSTEIN, R., K. DIGGES, N. R. MARSHALL & K. STARKE. 1984. Forskolin and the release of noradrenaline in cerebrocortical slices. Naunyn-Schmiedebergs Arch. Pharmacol. **325:** 17–24.

30. SEAMON, K. B., W. PADGETT & J. W. DALY. 1981. Forskolin: a unique diterpene activator of adenylate cyclase in membranes and in intact cells. Proc. Nat. Acad. Sci. USA **78:** 3363–3367.

31. SCHOFFELMEER, A. N. M., E. A. WIERENGA & A. H. MULDER. 1986. Role of adenylate cyclase in presynaptic α2-adrenoceptor and μ-opioid receptor–mediated inhibition of [^{3}H] noradrenaline release from rat brain cortex slices. J. Neurochem. **46:** 1711–1717.

32. RAMDINE R., A.-M. GALZIN & S. Z. LANGER. 1989. Involvement of adenylate cylase and protein kinase C in the α_2-adrenoceptor-mediated inhibition of noradrenaline and 5-hydroxytryptamine release in rat hypothalamic slices. Naunyn Schmiedebergs Arch. Pharmacol. **340:** 386–395.

33. FREDHOLM, B. B. & E. LINDGREN. 1988. Protein kinase C activation increases noradrenaline release from the rat hippocampus and modifies the inhibitory effect of α_2-adrenoceptor and adenosine A_1-receptor agonists. Naunyn Schmiedebergs Arch. Pharmacol. **337:** 477–483.

34. SCHOFFELMEER, A. N. M. & A. H. MULDER. 1984. Presynaptic opiate receptor- and α_2-adrenoceptor-mediated inhibition of noradrenaline release in the rat brain: Role of hyperpolarization? Eur. J. Pharmacol. **105:** 129–135.

35. SCHOFFELMEER, A. N. M. & A. H. MULDER. 1983. Differential control of Ca^{2+}-dependent ^{3}H-noradrenaline release from rat brain slices through presynaptic opiate receptors and α-adrenoceptors. Eur. J. Pharmacol. **87:** 449–458.

36. DE LANGEN, C. D. J., F. HOGENBOOM & A. H. MULDER. 1979. Presynaptic noradrenergic alpha-receptors and modulation of ^{3}H-noradrenaline release from rat brain synaptosomes. Eur. J. Pharmacol. **60:** 79–89.

37. DE LANGEN, C. D. J. & A. H. MULDER. 1980. On the role of calcium ions in the presynaptic alpha-receptor mediated inhibition of ^{3}H-noradrenaline release from rat brain cortex synaptosomes. Brain. Res. **185:** 399–408.

38. RAITERI, M., G. MAURA & P. VERSACE. 1983. Functional evidence for two stereochemically

different α_2-adrenoceptors regulating central norepinephrine and serotonin release. J. Pharmacol. Exp. Ther. **224:** 679–684.

39. MULDER, A. H., F. HOGENBOOM, G. WARDEH & A. N. M. SCHOFFELMEER. 1987. Morphine and enkephalins potently inhibit ^{3}H-noradrenaline release from rat brain synaptosomes: further evidence for a presynaptic localization of μ-opioid receptors. J. Neurochem. **48:** 1043–1047.

40. GÖTHERT, M., I. M. POHL & E. WEHKING. 1979. Effects of presynaptic modulators on Ca^{2+}-induced noradrenaline release from central noradrenergic neurons. Naunyn Schmiedebergs Arch. Pharmacol. **307:** 21–27.

41. STARKE, K. 1987. Presynaptic α-autoreceptors. Rev. Physiol. Biochem. Pharmacol. **107:** 73–146.

42. ROGAWSKI, M. A. & J. L. BARKER. 1983. Effects of 4-aminopyridine on calcium action potentials and calcium current under voltage clamp in spinal neurons. Brain Res. **280:** 180–185.

43. YAMAMOTO, H., R. A. HARRIS, H. H. LOH & E. L. WAY. 1978. Effects of acute and chronic morphine treatments on calcium localization and binding in brain. J. Pharmacol. Exp. Ther. **205:** 255–264.

44. WEST, R. E. & R. J. MILLER. 1983. Opiates, second messengers and cell response. Br. Med. Bull. **39:** 53–58.

45. SCHOFFELMEER, A. N. M. & A. H. MULDER. 1983. ^{3}H-Noradrenaline release from brain slices induced by an increase in the intracellular sodium concentration: role of intracellular calcium stores. J. Neurochem. **40:** 615–621.

46. SCHOFFELMEER, A. N. M., F. HOGENBOOM & A. H. MULDER. 1988. Sodium dependent ^{3}H-noradrenaline release from rat neocortical slices in the absence of extracellular calcium: presynaptic modulation by μ-opioid receptor and adenylate cyclase activation. Naunyn Schmiedebergs Arch. Pharmacol. **338:** 548–552.

47. SCHOFFELMEER, A. N. M., G. WARDEH & A. H. MULDER. 1985. Cyclic AMP facilitates the electrically evoked release of radiolabelled noradrenaline, dopamine and 5-hydroxytryptamine from rat brain slices. Naunyn Schmiedebergs Arch. Pharmacol. **330:** 74–76.

48. SCHLICKER, E., K. FINK, K. CLASSEN & M. GÖTHERT. 1987. Facilitation of serotonin (5-HT) release in the rat brain cortex by cAMP and probable inhibition of adenylate cyclase in 5-HT nerve terminals by presynaptic α_2-adrenoceptors. Naunyn Schmiedebergs Arch. Pharmacol. **336:** 251–256.

49. MARTIN, L. L. & SANDERS-BUSCH. 1982. Comparison of the pharmacological characteristics of 5 HT_1 and 5 HT_2 binding sites with those of serotonin autoreceptors which modulate serotonin release. Naunyn Schmiedebergs Arch. Pharmacol. **321:** 165–170.

50. MAURA, G., E. ROCCATAGLIATA & M. RAITERI. 1986. Serotonin autoreceptor in rat hippocampus: pharmacological characterization as a subtype of the 5-HT_1 receptor. Naunyn Schmiedebergs Arch. Pharmacol. **334:** 323–326.

51. BOWYER, J. F. & N. WEINER. 1989. K^+-channel and adenylate cyclase involvement in regulation of Ca^{2+}-evoked release of $[^3H]$dopamine from synaptosomes. J. Pharmacol. Exp. Ther. **248:** 514–520.

52. STOOF, J. C. & J. W. KEBABIAN. 1981. Opposing roles for D-1 and D-2 dopamine receptors in efflux of cyclic AMP from rat neostriatum. Nature **294:** 366–368.

53. ONALI, P. L., M. C. OLIANAS & G. L. GESSA. 1985. Characterization of dopamine receptors mediating inhibition of adenylate cyclase activity in rat striatum. Mol. Pharmacol. **28:** 138–145.

54. MEMO, M., C. MISSALE, M. O. CARRUBA & P. F. SPANO. 1986. D-2 dopamine receptors associated with inhibition of dopamine release from rat neostriatum are independent of cyclic AMP. Neurosci. Lett. **71:** 192–196.

55. STOOF, J. C. & J. W. KEBABIAN. 1982. Independent in vitro regulation by the D-2 dopamine receptor of dopamine stimulated cyclic AMP efflux and K^+-stimulated release of acetylcholine from rat neostriatum. Brain Res. **250:** 263–270.

56. DRUKARCH, B., E. SCHEPENS, A. N. M. SCHOFFELMEER & J. C. STOOF. 1989. Stimulation of D-2 dopamine receptors decreases the evoked in vitro release of $[^3H]$-acetylcholine from rat neostriatum: role of K^+ and Ca^{2+}. J. Neurochem. **52:** 1680–1685.

57. DRUKARCH, B., E. SCHEPENS & J. C. STOOF. Muscarinic receptor activation attenuates D-2

receptor mediated inhibition of acetylcholine release in rat striatum: indications for a common signal transduction pathway. Neuroscience. (In press.)

58. MULDER, A. H., G. WARDEH, F. HOGENBOOM & A. L. FRANKHUYZEN. 1984. Kappa- and delta-opioid receptor agonists differentially inhibit striatal dopamine and acetylcholine release. Nature **308:** 278–280.
59. SCHOFFELMEER, A. N. M., K. C. RICE, A. E. JACOBSON, J. G. VAN GELDEREN, F. HOGENBOOM, M. H. HEIJNA & A. H. MULDER. 1988. μ-, δ-, and κ-opioid receptor mediated inhibition of neurotransmitter release and adenylate cyclase activity in rat brain slices: studies with fentanyl isothioyanate. Eur. J. Pharmacol. **154:** 169–178.
60. MULDER, A. H., G. WARDEH, F. HOGENBOOM & A. L. FRANKHUYZEN. 1989. Selectivity of various opioid peptides towards delta-, kappa- and mu-opioid receptors mediating presynaptic inhibition of neurotransmitter release in the brain. Neuropeptides **14:** 99–104.

Presynaptic Receptors and Modulation of Noradrenaline and ATP Secretion from Sympathetic Nerve Varicosities[a]

LENNART STJÄRNE, JIAN-XIN BAO,
FRANCOIS G. GONON,[b]
CLAIRE MERMET,[b] MUSSIE MSGHINA,
EIVOR STJÄRNE, AND PER ÅSTRAND

Department of Physiology
Karolinska Institute
S-10401 Stockholm, Sweden

INTRODUCTION

This paper describes results obtained in our laboratory in studies of the prejunctional control of the secretion of noradrenaline (NA) and a cotransmitter, presumed to be adenosine 5′-triphosphate (ATP), evoked by stimulating at low frequency (0.1–2 Hz) the sympathetic nerves in two model tissues, the mouse vas deferens [1] and rat tail artery.[2] No attempt is made to refer to all original papers in the voluminous literature; references are given mainly to recent review articles.[3-5] The main objectives were (1) to discuss whether modulation of NA secretion parallels that of ATP,[4] (2) to express presynaptic control[3] in terms of P, the probability of monoquantal release from the average varicosity,[4] (3) to clarify the relative importance of α_2-adrenoceptor-mediated autoinhibition *versus* other mechanisms responsible for the low P,[4,6] and (4) to show that prejunctional control of P is exerted, at least in part, in a biophase "upstream" of the varicosity.[5,7]

METHODS

Preparations and Experimental Setup

In the experiments described here we used male C57 mice (15–25 g) or Sprague-Dawley rats (150–300 g). The methods employed have been explained in more detail elsewhere.[2,5,7-12] Briefly, the animals were stunned and bled and the vasa deferentia in the mice or the proximal region of the central tail artery in the rats dissected out and mounted in 2–3 ml Perspex chambers perfused at 36°C at 1 ml minute^{-1} with modified Tyrode's solution (Ca^{2+} 1.3 mmol l^{-1}) with or without Tris buffer and gassed either with O_2 or a mixture of 93.5% O_2/6.5% CO_2; pH was 7.3–7.4. The same setup was used

[a]The research in the paper was supported by the Swedish Medical Research Council (projects B90-14X-03027-21A), and Karolinska Institutets Fonder. C. Mermet is a visiting scientist supported by a Fellowship from the Swedish Medical Research Council and by INSERM, and F.G. Gonon's traveling expenses were supported by the Wenner Gren Foundation and by the European Science Foundation (twinning grant no. 8912).
[b]INSERM U171, CNRS UA, Ste Eugénie Hospital Pav. 4H, 69230 St Genis Laval, France.

for ATP and NA release; a different chamber was used to record the neurogenic contraction.

Method to Study the Secretion of ATP

Extracellular recording of the excitatory junction current (EJC) was employed to monitor on an impulse-by-impulse basis the release of quanta of ATP;[1] the approach has been described in detail elsewhere[7,13,14] (FIGURES 1 and 2).

Previous papers from our laboratory have shown (1) that ATP release from varicosities of sympathetic nerves in mouse vas deferens and rat tail artery, just as in

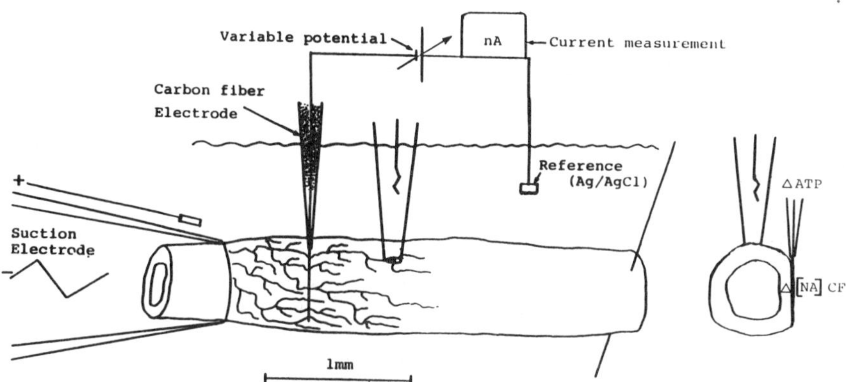

FIGURE 1. Diagram showing experimental setup. It was used to monitor (1) in rat tail artery, by an electrochemical method, $\Delta[NA]_{CF}$, i.e, the nerve stimulation–induced increase in the concentration of endogenous NA at the probe, a carbon fiber (CF) electrode,[11,12] or (2) in rat tail artery and mouse vas deferens, by extracellular recording of the excitatory junction current (EJC),[10] ΔATP, i.e., the quantal release of ATP.[13] A suction electrode into which the proximal end of the artery or the prostatic end of the vas deferens was drawn was used to stimulate electrically the sympathetic postganglionic nerves. The bath was perfused with modified Tyrode's solution (Ca^{2+} 1.3 mmol l^{-1}); for electrochemistry the medium was phosphate buffer solution (pH 7.4) containing prazosin (0.1 μmol l^{-1}) to prevent neurogenic contraction.[12] Indicated are also the electrochemically treated CF electrode[11,12] (diameter, 12 μm), held vertically with the active part apposed to the arterial wall and used to measure $[NA]_{CF}$, and the extracellular electrode, a glass micropipette (internal tip diameter, 40–100 μm) filled with a medium that during recording could be changed by internal perfusion of the pipette.

guinea pig vas deferens,[6,10] is monoquantal and highly intermittent (P 0–0.03)[2,4], (2) that the low P is not due to failure of invasion,[2,4] and (3) that tetrodotoxin (TTX) in the recording electrode blocks the nerve impulse–induced (but not the spontaneous) quantal release.[2,4,14]

As a working hypothesis we have assumed (1) that regenerative Na^+ channels in the varicosities ensure that each impulse invades all regions of the terminals,[4] (2) that the release mechanisms in the varicosity ignore in 99% of the cases the Ca^{2+} current caused by the invading all-or-none action potential,[2,4,5] (3) that nerve impulses at low frequency release single quanta, i.e., the transmitter content of a small dense cored

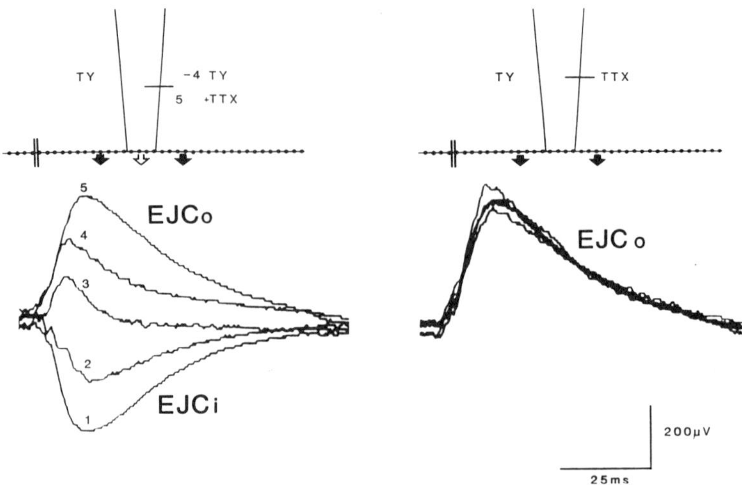

FIGURE 2. Extracellular recording[10] in mouse vas deferens. Shown are a varicose nerve terminal running on the surface of the muscle fiber, the stimulating electrode (vertical double bar), and the recording electrode (internal diameter 100 μm). **Left:** Originally when the electrode was filled with Tyrode's solution the response to nerve stimulation (0.5 Hz) varied from "failure" to a pure, negative-going EJC [here termed EJCi,[13] as it reflects ATP release from sites inside the electrode[10] (open arrow; see trace 1 which is average of 6 sweeps)] to mixed, positive-negative-going deflections (traces 2, 3 and 4, single sweeps). Addition of tetrodotoxin (TTX, 1 μmol l^{-1}) totally abolished the EJCi; all stimuli now caused a positive-going EJC [here termed EJCo,[13] as it reflects ATP release from sites outside the electrode[10] (filled arrows; see trace 5 which is average of 6 sweeps)] of relatively constant size. **Right:** Same experiment, example of the pure, nonintermittent, nonfluctuating EJCo response to six consecutive pulses when TTX had abolished the EJCi. Note that TTX in the electrode prevented release from sites in the patch i.e., that the nerve impulse could not by passive depolarization release ATP even from varicosities 5–10 μm beyond the site of block.[2,4,10,14]

vesicle (SDV),[4] and (4) that the parameter controlled by prejunctional receptors is P, the probability of monoquantal release from the varicosity.[4,5]

We have therefore used, tentatively, the extracellularly recorded[10] EJCo or EJCi, i.e., the excitatory junction current caused by release of ATP from sites outside (EJCo) or inside (EJCi) the recording electrode[13] (FIGURE 2), to monitor on an impulse-by-impulse basis the average probability (P) of monoquantal release in the examined group of varicosities (FIGURE 3).

Estimation of the Secretion of NA

At present, methods are lacking that resolve directly on an impulse-by-impulse basis the secretion of endogenous NA. To study NA release *per se* we have therefore used the conventional tracer method,[8] prelabeling the neuronal stores by preincubation with ^{3}H-NA (0.7 μmol l^{-1}) at 37°C for 30 minutes and then measuring the field stimulation–induced fractional rise in the overflow of ^{3}H. In these experiments, which were performed in rat tail artery, the nerves were stimulated with 150 shocks at 20 Hz, delivered either in one sequence or as 30 five shock trains separated by 10 second intervals.[9]

Electrochemical Determination of the Extracellular Endogenous NA

An electrochemical method[11,12] was employed to study on line in rat tail artery $\Delta[\text{NA}]_{\text{CF}}$, i.e., the nerve stimulation-induced increase in the extracellular concentration of endogenous NA (FIGURE 3). The differential pulse oxidation current caused by a pulsed potential (at $+0.1$ V, where the oxidation current of NA is maximal) applied to the carbon fiber (CF) electrode against a reference electrode was recorded by differential pulse amperometry; the current was shown to be proportional to the NA concentration at the carbon fiber.

When the carbon fiber electrode was applied to the rat tail artery as shown in FIGURE 1, no oxidation current at $+0.1$ V was detectable in the resting state, but nerve stimulation with 100 pulses (0.3 msecond, 0.2–0.4 mA) at 1–2 Hz induced a current similar to that caused by 10–50 nmol l^{-1} NA *in vitro*. The current grew throughout the

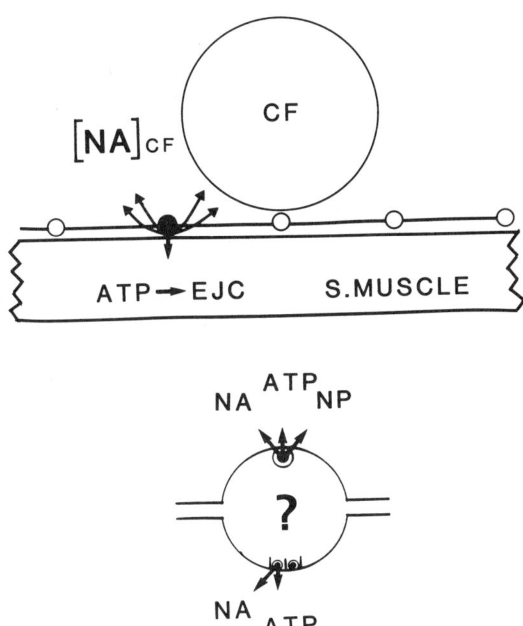

FIGURE 3. Diagram showing a secretory response of a varicose nerve fiber running on the surface of muscle, and the carbon fiber (CF) electrode. Open rings are inactive varicosities, filled ring a varicosity that releases a single quantum of ATP (causing an EJC) and NA. The CF does not detect per se release but the increase in the concentration of NA caused by release, i.e., $\Delta[\text{NA}]_{\text{CF}}$. Note the size relations between the CF (diameter, 12 μm), the nerve fiber (varicosity diameter, 0.5–1.5 μm; interval between varicosities, 4 μm)[4] and the much larger extracellular electrode used to detect the EJC (internal diameter, 40–100 μm; not shown). **Lower panel:** At low frequency, nerve impulses are believed to release preferentially the NA and ATP contents of a small, dense-cored vesicle (SDV) from a preferred release site;[4,6] nerve impulses in a burst at high frequency may in addition release the contents, NA, ATP, and often a neuropeptide (NP), from other sites in the varicosity.[4] The average varicosity in sympathetic nerves in rat tail artery or mouse vas deferens probably contains about 500 SDVs and 25 LDVs.[4] The question mark indicates uncertainty that the SDV quantum is fixed in size and composition, i.e., that it always contains the same amount and proportions of NA and ATP[4] (see also FIGURE 8).

stimulation period (see FIGURES 6–8), at a rate proportional to the stimulation frequency (not shown). The response was abolished by removal of Ca^{2+} from the medium, or by addition of the unselective Ca^{2+} channel–blocking agent Cd^{2+} (10 μmol l^{-1}) but not by the L-type Ca^{2+} channel blocker nifedipine (10 μmol l^{-1}); it was also blocked by TTX (0.1 μmol l^{-1}) or guanethidine (10 μmol l^{-1}). The position of the maximum (at +0.1 V) and the pharmacological criteria show that the 'current response' was caused by oxidation of released NA.[12]

The method provides an extremely useful tool to monitor on line the kinetics of the nerve stimulation–induced changes in the level of endogenous NA at the surface of the carbon fiber[11,12] (referred to as $\Delta[NA]_{CF}$). It is important to keep in mind, however, that $\Delta[NA]_{CF}$ does not reflect directly release *per se* (namely, of NA), but the balance between release and removal of NA, by reuptake and diffusion (FIGURE 3). This is in contrast to the EJC which is a direct measure of release *per se* (namely, of ATP).

Usefulness and Limitations of the NA-Mediated Neurogenic Contraction

A 3 mm ring segment from the proximal regions of rat tail artery[9] was mounted between two parallel L-shaped stainless steel holders (resting tension, 5 mN) in a 2 ml Perspex chamber perfused with Tyrode's solution (1.3 mmol l^{-1} Ca^{2+}) at 37°C and 1 ml minute^{-1}. The contraction evoked by field stimulation at 1–2 Hz, via two parallel platinum electrodes along the preparation, was recorded isometrically. The neurogenic, i.e., TTX-sensitive, contraction in this tissue is mediated by NA via postjunctional α_1- and to a lesser extent α_2-adrenoceptors, and in part by ATP via postjunctional P_{2X}-purinoceptors; it is modulated by NA via prejunctional α_2-adrenoceptors.[9]

The postjunctional α_2-adrenoceptors make the neurogenic contraction unsuitable as a tool to analyze prejunctional α_2-adrenoceptor-mediated modulation of transmitter release. However, we have used the neurogenic contraction in the presence of α,β-methylene ATP (mATP, 10 μmol l^{-1}), which could be blocked by a combination of prazosin and yohimbine (0.1 μmol l^{-1}), and hence was caused exclusively by NA,[9] to find out if this contractile response is "synaptic," i.e., due to NA release, or "nonsynaptic," i.e., a function of the extrajunctional concentration of NA.

RESULTS AND DISCUSSION

Occurrence of Prejunctional Control of NA and ATP Secretion

Regulation of ATP Secretion by Presynaptic Receptors

We have considered, but finally rejected, the possibility that ATP released as a cotransmitter depresses via prefunctional P_{2X}-purinoceptors the secretory mechanisms of the sympathetic nerves in rat tail artery.[9] Thus, we concentrate here on control via prejunctional α_2-adrenoceptors.

Effects of Exogenous α_2-Adrenoceptor Agonist on ATP Release. In preliminary experiments we examined the sensitivity of ATP release (as reflected in the EJCo) to clonidine, an α_2-adrenoceptor agonist. This agent depressed dose dependently the EJCo; the effect was inversely related to the stimulation frequency (at 0.1–2 Hz), as well as to the number of pulses (at 2 Hz). At 0.1 Hz the IC_{50} for clonidine was about 0.05 μmol l^{-1}; at 2 Hz at least 20-fold higher. The effect of clonidine could be prevented or reversed by the α_2-adrenoceptor blocking agent yohimbine and, hence, was exerted via prejunctional α_2-adrenoceptors. Examples of these results, expressed in P, are given in FIGURE 4. It can be seen (1) that P in controls during a train of 100 pulses at 2 Hz showed initial facilitation followed by defacilitation, (2) that clonidine depressed profoundly (by 70%) the "unfacilitated" P (the EJCo at 0.1 Hz), (3) that the facilitating effect of a train of pulses at 2 Hz overcame in part the inhibitory effect of clonidine (steady-state inhibition 30%), and (4) that yohimbine reversed the effects of clonidine, showing that the effects were exerted via α_2-adrenoceptors.[3]

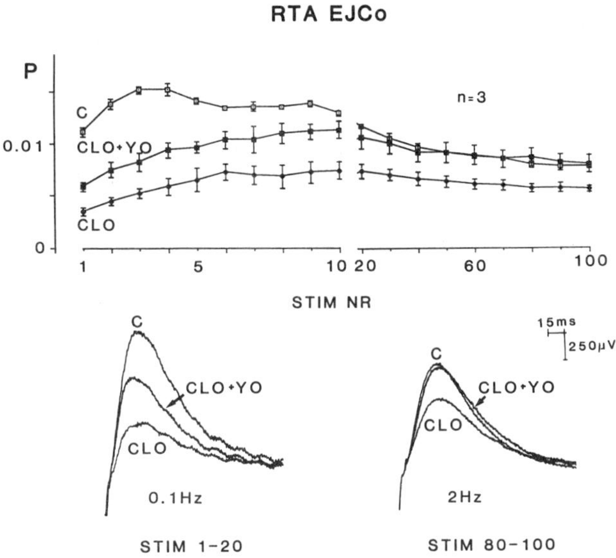

FIGURE 4. Effects of clonidine and yohimbine (1 μmol l^{-1}) on the EJCo response in rat tail artery to nerve stimulation with 1–100 pulses (0.3 mseconds, 0.2–0.4 mV) at 0.1 or 2 Hz. **Upper panel:** Effects of clonidine and yohimbine expressed in P (the probability of monoquantal release in the average varicosity in the small group examined) as a function of the number of pulses applied at 2 Hz (each point is average of values from 3 different preparations). Ordinate: presumed P levels, the first EJCo at 2 Hz in the control is taken to correspond to $P = 0.01$ (see the text). The inhibitory effect of clonidine and the reversal by yohimbine were significant in all points ($p < 0.05$). **Lower panel:** Typical EJCo recordings (averages of 20 consecutive sweeps) from one of the three experiments in the upper panel.

Effects of Endogenous NA on ATP Release. Yohimbine (1 μmol l^{-1}) did not affect the EJCo caused by nerve impulses at 0.1 Hz or by the first few pulses at 2 Hz, but enhanced by 20–50% the EJCo caused by later pulses in a train at 2 Hz (not shown). Yohimbine also enhanced the EJCo when desipramine (DMI, 1 μmol l^{-1}) was present to block reuptake of NA and thereby enhance the biophase concentration of NA. Examples are given in FIGURE 5, which shows (1) that DMI seemed to depress the EJCo (i.e., P) by 10–20% (effect not statistically significant), and (2) that subsequent addition of yohimbine, which had no effect on EJCo at 0.1 Hz (not shown) or in response to the first few pulses at 2 Hz, enhanced the EJCo amplitude (i.e., P) significantly (by about 50%) after the first (2–) 5 pulses at 2 Hz.

Discussion. Similar effects of clonidine and yohimbine on the EJCo were observed in the mouse vas deferens.[14] Thus, yohimbine increased NA release by blocking prejunctional α_2-adrenoceptors, not by a "release-enhancing" effect of its own. The results imply therefore that endogenous NA when accumulating extraneuronally restricts via

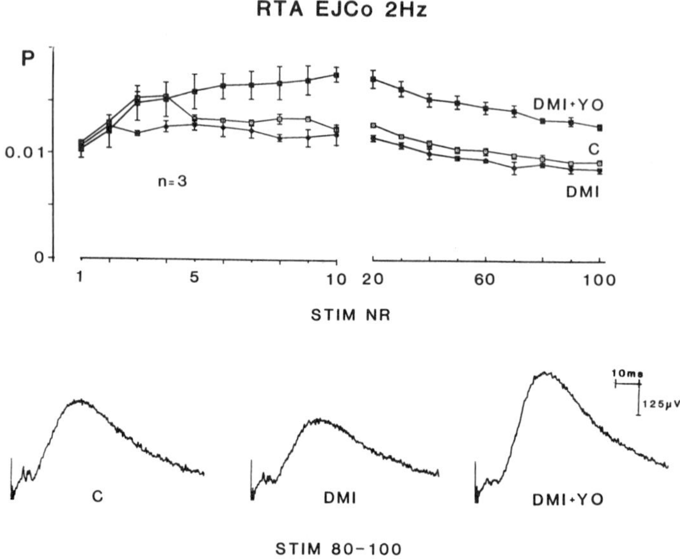

FIGURE 5. Effects of block of reuptake of NA by desipramine (DMI, 1 μmol l^{-1}) and of further addition of yohimbine (Yo, 1 μmol l^{-1}) to block autoinhibition mediated by endogenous NA via prejunctional α_2-adrenoceptors, on the EJCo response of rat tail artery to nerve stimulation with 100 pulses (0.3 msecond 0.2–0.4 mA) at 0.1 or 2 Hz. **Upper panel:** Effects of DMI and yohimbine on P (cf. FIGURE 4; each point is average of values from 3 different preparations). The inhibitory effect of DMI was not statistically significant ($p > 0.05$); the enhancing effect of yohimbine was significant from pulse 6 ($p < 0.05$). **Lower panel:** Typical EJCo recordings (averages of 20 consecutive sweeps) from one of the three experiments in the upper panel.

prejunctional α_2-adrenoceptors the secretion of ATP from the sympathetic nerves in these tissues.[3,4]

Conclusions. We conclude tentatively (for rat tail artery and mouse vas deferens under the experimental conditions) (1) that the release of ATP could be inhibited by α_2-adrenoceptor agonists and that their efficiency was inversely related to the length and frequency of stimulus trains, (2) that the resting level of endogenous NA extracellularly was too low to activate the prejunctional α_2-adrenoceptors, which therefore did not prevent stray nerve impulses from releasing ATP, (3) that trains of more than 5 impulses at low frequency increased the extracellular NA level sufficiently to activate the receptors and depress the release of ATP (this may explain the "defacilitation" of the EJCo in the absence but not in the presence of yohimbine, FIGURES 4 and 5), (4) that autoinhibition of ATP release via α_2-adrenoceptors operated within a narrow range (P = "0"–0.015), and thus, was not a major cause of the low P in these varicosities.[2,4]

Regulation of NA Secretion by Presynaptic Receptors

These experiments in rat tail artery were performed under conditions similar to those in the previous section where ATP release was examined.

Autoinhibition of NA Secretion, Evidence by Tracer Method. The fractional secretion of ^{3}H-NA evoked by field stimulation at 20 Hz with a continuous train of 150 shocks was depressed by clonidine (1 μmol l^{-1}) and significantly (by about 80%) enhanced by yohimbine (1 μmol l^{-1}). However, yohimbine had no effect on the secretory response to 30 trains of five pulses (separated by 10 second intervals) at the same frequency (data not shown). Thus, the release caused by long but not short trains seems to be restricted by endogenous NA via prejunctional α_2-adrenoceptors.[9]

Prejunctional Control of the Extracellular Level of Endogenous NA. As mentioned under Methods, field stimulation of the sympathetic nerves in rat tail artery with trains of 100 pulses at 2 Hz caused a signal, which we have termed $\Delta[NA]_{CF}$, i.e., an increase in the concentration of endogenous NA at the carbon fiber electrode.[12] The responses to stimulation once every 5 minutes showed little or no "fatigue" for at least 2 hours. The effects of some drugs are summarized in FIGURE 6. Clonidine (1 μmol l^{-1}) depressed, and yohimbine (1 μmol l^{-1}) enhanced, $\Delta[NA]_{CF}$ (FIGURE 6A). DMI (1 μmol l^{-1}) or cocaine (10 μmol l^{-1}), added to block reuptake of NA, increased $\Delta[NA]_{CF}$ to a variable extent in different experiments, for DMI to 340 ± 74%, and for cocaine to 165 ± 39% [means ± standard errors of the mean (SEM), 4 experiments], of the $\Delta[NA]_{CF}$ in the same experiment before addition of drug.[12] Subsequent addition of yohimbine (1 μmol l^{-1}) approximately doubled $\Delta[NA]_{CF}$ (FIGURE 6B).

Discussion. The following *caveats* must be kept in mind when using $\Delta[NA]_{CF}$ to study NA secretion: (1) that $\Delta[NA]_{CF}$ reflects the net difference between release and removal of endogenous NA, not NA release *per se*[11,12] (FIGURE 2); tentative comparison with EJCo results (FIGURE 5) suggests that $\Delta[NA]_{CF}$ may deviate markedly from NA release, especially when reuptake of NA is prominent (FIGURE 7), and (2) that conclusive proof is lacking for the assumption that ATP and NA are released in parallel, underlying the application of EJCo principles to interpret $\Delta[NA]_{CF}$ data (FIGURE 8E).[4]

Conclusions. Based on comparison between the EJCo and $\Delta[NA]_{CF}$ recorded under similar conditions (FIGURE 7), we propose as an interpretation of the results (1) that exogenous (clonidine) and endogenous (NA) α_2-agonists restricted under appropriate conditions the nerve impulse–induced release of endogenous NA and ATP to about the same extent, and (2) that α_2-adrenoceptor-mediated autoinhibition did not contribute more than marginally to the basically low release probability (P) in these nerves.[4]

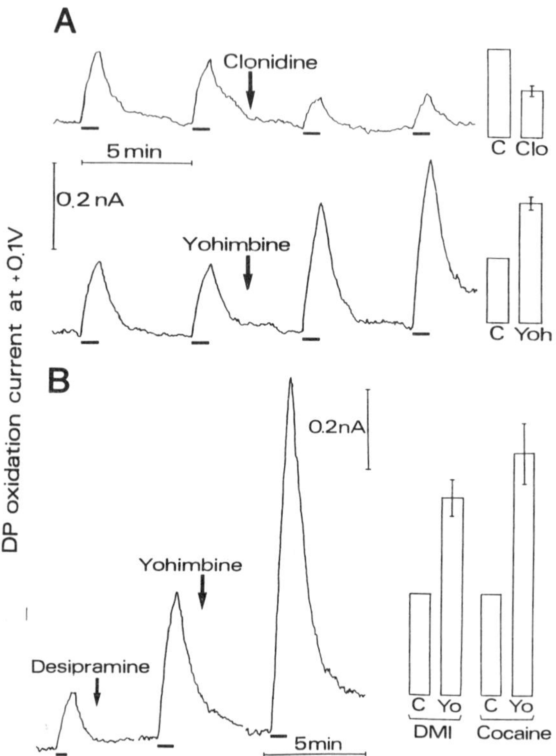

FIGURE 6. (A) Effects of clonidine (Clo) and yohimbine (Yoh) on $\Delta[\text{NA}]_{\text{CF}}$, i.e., the evoked increase in the local concentration of endogenous NA at the carbon fiber (CF) electrode. Four successive periods of electrical stimulation with 100 pulses at 2 Hz (S_1–S_4), with 5 minute intervals, induced a $\Delta[\text{NA}]_{\text{CF}}$, i.e., increase in the DP oxidation current at $+0.1$ V, due to oxidation of the released NA (see the text). When added to the perfusion medium (after S_2), clonidine (1 μmol l^{-1}) reduced and yohimbine (1 μmol l^{-1}) enhanced $\Delta[\text{NA}]_{\text{CF}}$. The figure shows two typical recordings. The results of 4 identical experiments for each drug are summarized in the right part of the figure. For each experiment the control $\Delta[\text{NA}]_{\text{CF}}$ value (C) was the mean of the values in S_1 and S_2. The mean $\Delta[\text{NA}]_{\text{CF}}$ values in S_3 and S_4 in the presence of Clo and Yoh were 55 ± 6% and 181 ± 11% (mean ± SEM, 4 experiments), respectively, of that in the controls. (B) Effects of the inhibitors of neuronal reuptake of NA, desipramine (DMI) and cocaine, and of subsequent addition of yohimbine (Yo, 1 μmol l^{-1}), on $\Delta[\text{NA}]_{\text{CF}}$. Shown are the results of eight successive periods of electrical stimulation (S_1–S_8) with 100 pulses at 2 Hz, applied with 5 minute intervals. DMI (1 μmol l^{-1}) or cocaine (10 μmol l^{-1}) was added in the perfusion medium after S_2 and yohimbine after S_5. The figure shows S_2, S_5, and S_8 in a typical experiment. The right part of the figure summarizes the effect of yohimbine on $\Delta[\text{NA}]_{\text{CF}}$, expressed in percent of the $\Delta[\text{NA}]_{\text{CF}}$ (means of S_4 and S_5) in the presence of DMI or cocaine. When present together with the uptake blockers, yohimbine increased $\Delta[\text{NA}]_{\text{CF}}$ (mean value of S_7 and S_8) to 196 ± 10% and 239 ± 28% (mean ± SEM, $n = 4$), respectively, of that with DMI or cocaine alone.

Modes and Mechanisms of Regulation of P

Which Factors Are the Main Determinants of P?

Modes and mechanisms of prejunctional control[3,4] are not the main issues of this paper, but are commented on to clarify our views concerning the scope of α_2-adrenoceptor-mediated control of NA and ATP secretion.[4]

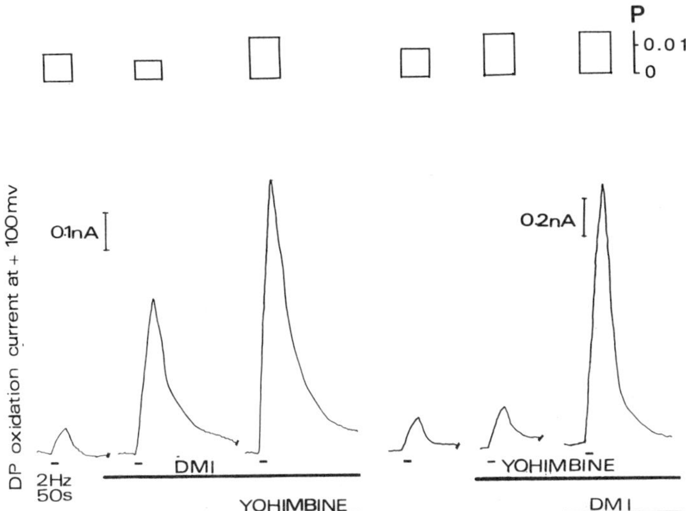

FIGURE 7. Speculative attempt to express the effects of desipramine (DMI, 1 μmol l^{-1}) and yohimbine (1 μmol l^{-1}) on $\Delta[\text{NA}]_{\text{CF}}$ in rat tail artery caused by field stimulation with 100 pulses at 2 Hz, in terms of P, the probability of nerve impulse-induced release of (the NA and ATP in) an SDV quantum from the average varicosity. Experimental data (lower panels) from two preparations of rat tail artery in which NA reuptake was unusually prominent and DMI consequently had large effects on $\Delta[\text{NA}]_{\text{CF}}$. In controls P is assumed to be 0.01; the effects of DMI and yohimbine on P (upper panels) are derived from effects on EJCo under similar conditions in other experiments (FIGURE 5). We propose the following interpretation. **Left panel:** DMI increased $\Delta[\text{NA}]_{\text{CF}}$ by fivefold, but release was actually depressed by 20% (reducing P from 0.01 to 0.008). Subsequent addition of yohimbine doubled both $\Delta[\text{NA}]_{\text{CF}}$ and release (increasing P from 0.008 to 0.016). **Right panel:** Yohimbine alone increased by 50% both $\Delta[\text{NA}]_{\text{CF}}$ and release (enhancing P from 0.01 to 0.015). Further addition of DMI had no effect on release (P remained at 0.015); the sixfold increase in $\Delta[\text{NA}]_{\text{CF}}$ was due to block or reuptake of NA. Note that, in spite of the large differences in $\Delta[\text{NA}]_{\text{CF}}$, P would be the same for stimulus 3 in the left, and stimulus 2 in the right panel.

Block of α_2-Autoinhibition Increases P to "Physiological Ceiling"? The modestly (less than twofold) increased P (from 0.01 to 0.015) in the presence of yohimbine (FIGURES 5, 6, and 7), i.e., when the secretory mechanisms were liberated from α_2-adrenoceptor-mediated autoinhibition, may represent the "physiological ceiling" of P.[4]

Block of K$^+$ Channels Increases P to "Pharmacological Ceiling"? K$^+$ channel blockers, such as tetraethylammonium (TEA) and 4-aminopyridine (4AP) are the

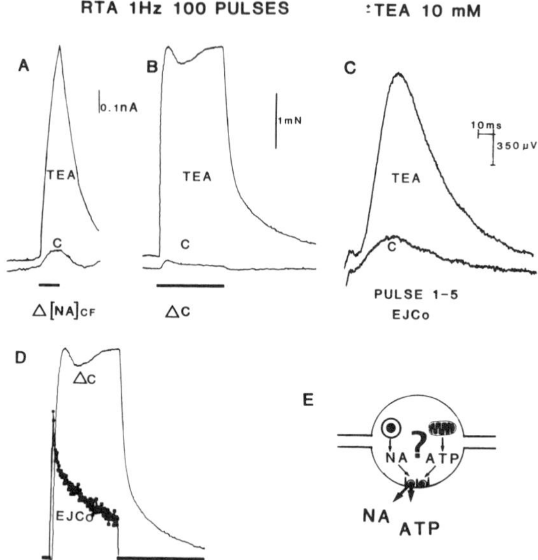

FIGURE 8. Effects in rat tail artery (RTA) of field stimulation with 100 pulses at 1 Hz before and after adding the K^+ channel blocking agent tetraethylammonium (TEA, 10 mmol l^{-1}). The contractile response (ΔC) in (B) and (D) was recorded in the presence of α,β-methylene ATP (mATP, 10 μmol l^{-1}), added to eliminate the ATP-mediated component. The neurogenic contraction could now be blocked completely by prazosin and yohimbine (0.1 μmol l^{-1}) and hence was mediated exclusively by NA. The horizontal bars in (A) and (B) represent the duration of the stimulation with 100 pulses at 1 Hz. Shown are typical effects of TEA on $\Delta[NA]_{CF}$ (A) and the NA-mediated neurogenic contraction (B, D), the averaged EJCo response to the first five shocks in the train (C), and the EJCo response to each shock in the 100 pulse train (D). The results imply that TEA (probably by block of neuronal K^+ channels) increased dramatically the secretion of both ATP and NA. The EJCo curve in (D) reflects directly the kinetics of secretion of ATP (and, possibly, of NA if released in parallel). The $\Delta[NA]_{CF}$ curve in (A) does not directly reflect the kinetics of NA release; observations during the experiment in combination with close inspection of the $\Delta[NA]_{CF}$ curve in (A) indicate that $\Delta[NA]_{CF}$ in the presence of TEA was maximal from the first pulse and that per-pulse increments became progressively smaller throughout the stimulus train. The NA-mediated neurogenic contraction differed in time course from that of NA release by being maintained throughout the stimulation period, and from that of $\Delta[NA]_{CF}$ by having a lag phase (the upstroke of the large contractile response began only after the first few pulses). (E) is a reminder that NA is synthesized in LDVs and ATP in mitochondria and taken up from the cytosol into SDVs by separate carriers. Differences in turnover of NA and ATP in SDVs may cause the quantum to vary in composition, especially under conditions when P is high.[4]

most potent known amplifiers of the nerve impulse–induced release of transmitters in several systems. The effect is generally thought to be due to broadening of the action potential and increased Ca^{2+} influx into the nerve terminals.[15]

In rat tail artery TEA (10 mmol l^{-1}) enhanced much more strongly than yohimbine (FIGURES 5, 6, and 7) the secretion of NA (as reflected in the $\Delta[NA]_{CF}$ and in the NA-mediated component of the neurogenic contraction) as well as ATP (as reflected in the EJCo), evoked by nerve stimulation with 100 pulses at 1 Hz (FIGURE 8). Higher concentrations of TEA (20–30 mmol l^{-1}), alone or in combination with 4AP (1 mmol

l^{-1}), cause even larger effects, increasing in the guinea pig or mouse vas deferens or rat tail artery by up to 100-fold the secretion of ^{3}H-NA[4] and/or of ATP (as reflected in EJCo)[14] evoked by stimulation at low frequency (0.1 Hz or less), without changing their resting release.[4,5] Furthermore, the presence of TEA and 4AP within the extracellular recording electrode increased in the guinea pig or mouse vas deferens the amplitude and duration of the nerve impulse–induced EJCi[14] without changing the frequency, amplitude, or time course of spontaneous EJCi.[5] Under all conditions the large nerve impulse–induced secretion in the presence of TEA and 4AP was fully Ca^{2+} dependent and TTX sensitive.[4,5]

In conclusion, the available evidence indicates that block of neuronal K^+ channels may increase P under some conditions by up to 100-fold, from 0.01 towards 1.0, the theoretical maximum if release is always monoquantal.[4] The low P may thus be due mainly to a high, "unregulated," K^+ efflux.[4]

Site(s) of Regulation of P

We have used extracellular recording to find out at what sites e.g., K^+ channel blockers or prejunctional agonists modulate P, by studying how the EJCi was affected by pharmacological agents added inside or outside the recording electrode[5,7,14] (Fig. 9).

K^+ Channel Block "Upstream" of the Varicosity Enhanced P. The presence of TEA (20 mmol l^{-1}) and 4AP (1 mmol l^{-1}) within the recording electrode did not change the frequency, amplitude or time course of the spontaneous EJCI[5] but (1) caused the nerve terminal spike (NTS) to acquire a late negative component (LNC), carried at least in part by Ca^{2+} and therefore proposed to represent the increased action potential–induced Ca^{2+} influx into the nerve terminals caused by K^+ channel block,[5,16] and (2) caused the EJCi to become nonintermittent and greatly increased in amplitude and duration[5,7] (FIGURE 9). These effects appear plausible; they are expected if K^+ channel block increased the EJCi (i.e., ATP release) by increasing the action potential–induced Ca^{2+} influx into the varicosities.[15]

However, to our surprise, TEA and 4AP amplified the EJCi, i.e., ATP release inside the patch, equally strongly when added outside the recording electrode, in this case without causing the NTS to acquire an LNC[5,17] (not shown). The development of the amplifying effect was independent of nerve stimulation; the EJCi response to the first stimulus, 10 minutes after adding TEA and 4AP to the outside medium, was amplified equally strongly in preparations rested during this period as in those continuously stimulated at 0.1 Hz. Did TEA and 4AP exert this effect directly on or in the varicosities in the patch, either (1) by leaking across the rim into the pipette, or (2) by being carried intraaxonally into the terminals within the patch, blocking from the inside K^+ channels in the varicosities? Against (1) speaks the finding that mATP when present in the external medium at 10 μmol l^{-1} blocked the EJCo without affecting the EJCi, but when present within the pipette profoundly depressed or abolished the EJCi. Against (2) speaks the fact that the large EJCi was not in this case preceded by an LNC.[5] Thus, apparently TEA and 4AP amplified the EJCi, i.e., ATP release from varicosities inside the patch, by an action exerted "upstream" of them.[5]

In conclusion, block of action potential–induced K^+ efflux and increase in Ca^{2+} influx into the varicosity may not be the explanation of the TEA- and 4AP-induced enhancement of P, as currently thought;[15] the mechanism may be block of the resting K^+ efflux (and consequent increase in Ca^{2+} influx) "upstream" of the varicosity[5,17] (FIGURE 10).

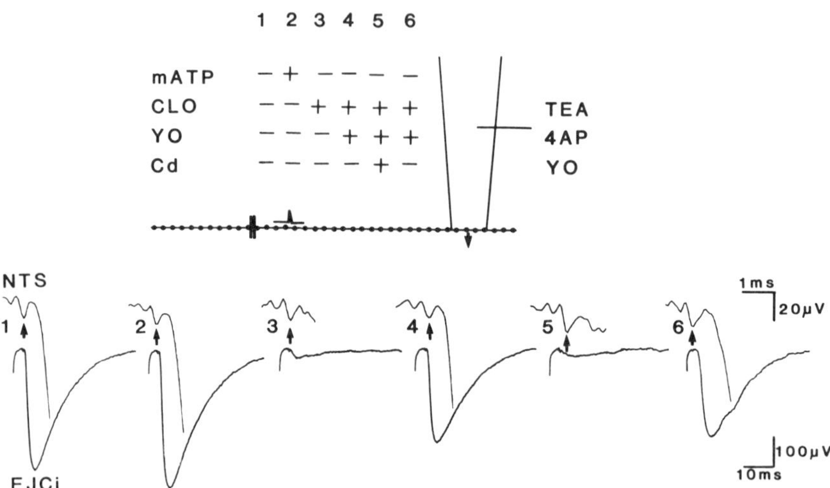

FIGURE 9. "Upstream" inhibition of EJCi in mouse vas deferens.[4,5,7] Continuous nerve stimulation (0.3 msecond, 0.4 mA) at 0.1 Hz. The modified Tyrode's solution (Ca^{2+} 1.3 mmol l^{-1}), buffered at pH 7.4 by 20 mmol l^{-1} Tris and gassed with O_2, contained the dihydropyridine L-type Ca^{2+} channel blocker nifedipine (10 μmol l^{-1}). **Upper panel:** A varicose nerve terminal running on the surface of the smooth muscle, the stimulating electrode (vertical bars) and the recording electrode (internal tip diameter, 100 μm) filled with Tyrode's solution containing tetraethylammonium (TEA, 20 mmol l^{-1}), 4-aminopyridine (4AP, 1 mmol l^{-1}), and yohimbine (YO, 1 μmol l^{-1}). Additions to the external medium during periods 1–6 (cf. lower panel). **Lower panel:** The paired traces 1–6 show averaged records (20 continuous sweeps), the upper trace in each pair is an amplification for better visualization of the nerve terminal spike (NTS) with its late negative component[5,16] (LNC, arrows). Addition of TEA and 4AP had caused the NTS to acquire an LNC, and the EJCi to become nonintermittent, very large, and prolonged in duration when compared with the EJCi before addition of these agents (not shown). (**1**) The NTS, LNC, and EJCi when no drug was present in the external medium. (**2**) Recording 20 minutes after adding α,β-methylene ATP (mATP, 10 μmol l^{-1}) outside the pipette. When present within the pipette at this concentration mATP abolished the EJCi, but here had no effect, showing that agents in the outside medium did not leak appreciably into the pipette. (**3**) Ten minutes after addition to the outside, clonidine (CLO, 1 μmol l^{-1}), in spite of the presence of yohimbine in the pipette, had almost abolished the EJCi, i.e., ATP release from sites inside the patch, without affecting the LNC. (**4**) Addition of yohimbine (YO, 10 μmol l^{-1}) to the external medium reversed within 20 minutes the effects of clonidine. (**5**) Ten minutes after addition to the outside, cadmium (Cd^{2+}, 10 μmol l^{-1}) had almost abolished the EJCi without affecting the LNC. (**6**) The effects of Cd^{2+} were reversed by wash for 30 minutes.

P Control via α_2-Adrenoceptors "Upstream" of the Varicosity. As first shown in the guinea pig vas deferens,[18] addition of clonidine outside an extracellular recording electrode containing yohimbine inhibited the EJCi without blocking the conduction of nerve impulses to the release sites. Two alternative explanations were considered, namely, that clonidine (1) blocked a Ca^{2+} component of the action potential, or (2) acted via an intraaxonally transported second messenger.[18]

We have confirmed and extended these findings, showing in mouse vas deferens[7] and rat tail artery,[5] both when the Tyrode's solution in the recording electrode was drug free[7] and when it contained TEA and 4AP (which caused an LNC and greatly amplified the EJCi) as well as yohimbine (added to "protect" the prejunctional

α_2-adrenoceptors within the patch),[5,7] (1) that addition of α_2 agonists (clonidine or xylazine) or neuropeptide Y to the outside medium inhibited the EJCi,[4,5,7] (2) that the effects of clonidine and xylazine (but not of neuropeptide Y) could be reversed by adding yohimbine to the external medium, (3) that the agonists inhibited the EJCi without affecting the LNC, which may reflect the action potential–induced Ca^{2+} influx into the varicosity,[5,16] and (4) that all effects of prejunctional agonists could be mimicked by adding Cd^{2+} (10 μmol l^{-1}) to the outside medium[5] (FIGURE 9). Our conclusions and working hypothesis for these modes and mechanisms of prejunctional control are summarized in FIGURE 10.

Discussion. The present results are in good agreement with two concepts proposed in earlier work from this laboratory, namely, (1) that inhibition of sympathetic transmitter secretion by prejunctional agonists may not be exerted (exclusively) by inhibiting Ca^{2+} influx *per se,* but at least in part by reducing the availability of a permissive factor, "CaX" formed by binding of Ca^{2+} to a hypothetical substance "X" in the varicosity[19] (FIGURE 10), and (2) that modulation of transmitter secretion (by prejunctional agonists or by changes in the Ca^{2+} concentration of the medium) may be exerted, at least in part, (in present terms) "upstream" of the varicosity.[20] More direct

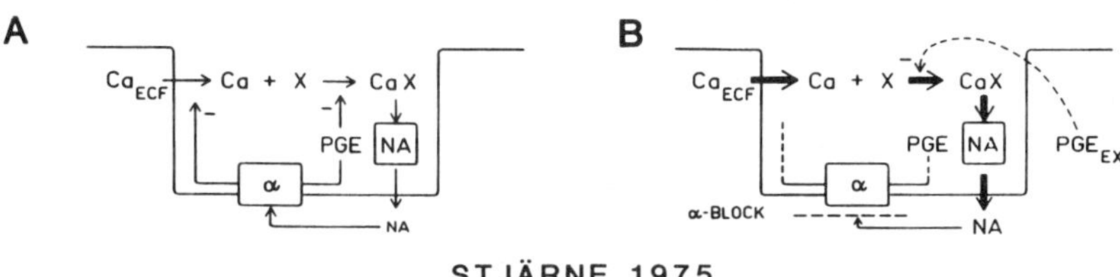

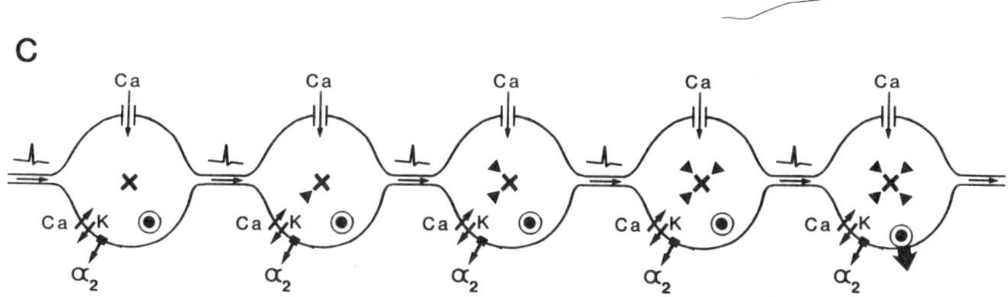

FIGURE 10. "Old"[19] and "new"[5] models from our laboratory, showing possible mechanisms of prejunctional modulation of sympathetic transmitter secretion. (**A, B**) In a conventional "old" model,[19] the action potential opens Ca^{2+} (Ca) channels, Ca combines with a hypothetical compound X, and the concentration of CaX determines the amount of transmitter (in present terms, the number of quanta) released. The prejunctional agonists NA and prostaglandin E_2 (PGE_2) inhibit transmitter secretion either by depressing Ca influx, or the availability of X, or the formation of CaX. (**C**) In a speculative "new" string model,[5] (1) all varicosities are invaded by the all-or-none action potential, (2) the resulting Ca current is necessary but not sufficient to trigger release, (3) P is determined by the availability of a fully activated (still hypothetical), intraaxonally transported permissive factor (X) which allows a vesicle to move to the preferred release site, (4) the activation and/or transport of X is restricted by the resting K^+ efflux and promoted by the resting Ca^{2+} influx upstream of the varicosity, and (5) prejunctional agonists inhibit transmitter secretion by preventing the activation and/or transport of X to (and within) the varicosity, and thereby the mobilization of a vesicle to the preferred release site.

experimental evidence by extracellular recording suggests that the permissive factor may be carried by fast intraaxonal transport, and that K^+ or Ca^{2+} channel blockers, as well as prejunctional agonists, modulate P by altering the activation and/or transport of the permissive factor[5] (FIGURE 10).

The nearest analogies in the literature seem to be the findings in squid giant synapse (1) that injection of dephospho-synapsin I or Ca^{2+}/calmodulin protein kinase II into the presynaptic terminals depressed or enhanced the depolarization-induced transmitter release without changing the Ca^{2+} current,[21] and (2) that transmitter vesicles were carried to the release sites by an active transport mechanism which required the presence of Ca^{2+} in the external medium.[22]

Finally, it should be pointed out (1) that presynaptic agonists may even in synaptosomes restrict transmitter secretion by preventing vesicles from moving to the preferred release site, rather than by restricting the depolarization-induced Ca^{2+} influx, as currently thought,[3] and (2) that our hypothesis that presynaptic agonists exert this effect by restricting the activation and/or intraaxonal transport of a permissive factor does not exclude that they enhance K^+ efflux and/or depress Ca^{2+} influx by mechanisms that do not involve soluble second messengers.[23]

SUMMARY AND CONCLUSIONS

Our results in the model tissues examined show (1) that α_2 agonist(s) depressed the secretion of NA and ATP caused by nerve stimulation at low frequency, (2) that the secretion of both NA and ATP was moderately autoinhibited, under conditions when endogenous NA was shown to accumulate extracellularly, (3) that a K^+ channel blocking agent increased much more strongly than α_2-adrenoceptors block the secretion of both NA and ATP, and also amplified enormously the NA-mediated neurogenic contraction, (4) that, therefore, a high K^+ efflux is likely to be much more important than α_2-adrenoceptor-mediated autoinhibition for maintaining a low release probability in sympathetic nerve varicosities, and (5) that the α_2-adrenoceptor agonist, clonidine, or the Ca^{2+} channel blocking agent, Cd^{2+}, inhibited transmitter secretion, at least in part, via targets "upstream" of the varicosity.

REFERENCES

1. STJÄRNE, L. & P. ÅSTRAND. 1984. Neuroscience **13:** 21–28.
2. ÅSTRAND, P. & L. STJÄRNE. 1989. J. Physiol. London **409:** 207–220.
3. STARKE, K., M. GÖTHERT & H. KILBINGER. 1989. Physiol. Rev. **69:** 864–989.
4. STJÄRNE, L. 1989. Rev. Physiol. Biochem. Pharmacol. **112:** 1–138.
5. STJÄRNE, L., M. MSGHINA & E. STJÄRNE. Neuroscience. (In press.)
6. CUNNANE, T. C. & L. STJÄRNE. 1984. Neuroscience **13:** 1–20.
7. STÄRNE, L. & E. STJÄRNE. 1988. Acta Physiol. Scand. **133:** 433–434.
8. ALBERTS, P., T. BARTFAI & L. STJÄRNE. 1981. J. Physiol. London **312:** 297–334.
9. BAO, J. X., I. E. ERIKSSON & L. STJÄRNE. Acta Physiol. Scand. (In press.)
10. BROCK, J. A. & T. C. CUNNANE. 1988. J. Physiol. London **399:** 607–632.
11. MERMET, C. & F. G. GONON. 1986. Neuroscience **19:** 829–838.
12. MERMET, C., F. G. GONON & L. STJÄRNE. Acta Physiol. Scand. (In press.)
13. STJÄRNE, L. & E. STJÄRNE. 1989. Acta Physiol. Scand. **135:** 217–226.
14. STJÄRNE, L. & E. STJÄRNE. 1989. Acta Physiol. Scand. **135:** 227–239.
15. KOKETSU, K. 1958. Am. J. Physiol. **193:** 213–218.
16. ÅSTRAND, P. & L. STJÄRNE. 1989. *In* Extracellular Recording of Electric Activity at the Sympathetic Neuroeffector Junction in Blood Vessels. P. Åstrand. Thesis. Karolinska Institute. Stockholm, Sweden.

17. STJÄRNE, L., M. MSGHINA & E. STJÄRNE. 1989. Acta Physiol. Scand. **137:** 151–152.
18. BROCK, J. A. & T. C. CUNNANE. 1988. In Vascular Neuroeffector Mechanisms. J. A. Bevan, H. Majewski, R. A. Maxwell & D. F. Story, Eds.: 161–168. IRL Press. Oxford, England.
19. STJÄRNE, L. 1975. In Biogenic amine Receptors. L. L. Iversen, S. D. Iversen & S. H. Snyder, Eds. Handbook of Psychopharmacology **6:** 179–233. Plenum Press. New York, N.Y.
20. STJÄRNE, L. 1978. Neuroscience **3:** 1147–1155.
21. LLINÁS, R., T. L. McGUINNESS, C. S. LEONARD, C. S. SUGIMORI & P. GREENGARD. 1985. Proc. Nat. Acad. Sci. USA **82:** 3035–3039.
22. LLINÁS, R., M. SUGIMORI, J. W. LIN, P. L. LEOPOLD & S. T. BRADY. 1989. Proc. Nat. Acad. Sci. USA **86:** 5656–5660.
23. LIPSCOMBE, D., S. KONGSAMUTH & R. W. TSIEN. 1989. Nature **340:** 639–642.

Second Messengers Are Involved in Facilitatory but Not Inhibitory Receptor Actions at Sympathetic Nerve Endings

H. MAJEWSKI, M. COSTA, S. FOUCART, T. V. MURPHY,
AND I. F. MUSGRAVE

Department of Pharmacology
University of Melbourne
Parkville, Victoria, Australia, 3052

INTRODUCTION

In general, receptors coupled to adenylate cyclase and phospholipase C play a pivotal role in the regulation of intracellular concentrations of cyclic AMP and inositol trisphosphate (IP_3)/diacylglycerol respectively (see reviews in References 1 and 2). Protein kinases controlled by these second messengers such as protein kinase A (cyclic AMP) and protein kinase C (diacylglycerol) are now regarded as important modulators of cellular activity (see review in Reference 3). A convenient hypothesis is that prejunctional receptors on nerve endings and specifically on sympathetic nerve endings may exert some of their actions through these second messenger pathways.

SECOND MESSENGERS IN SYMPATHETIC NERVES

Several second messenger systems are operative in sympathetic nerve endings and may modulate transmitter release. First, there is good evidence that activation of the adenylate cyclase/cyclic AMP system enhances action-potential-evoked norepinephrine release. Cell permeable analogues of cyclic AMP, the adenylate cyclase activator forskolin, and phosphodiesterase inhibitors which prevent the breakdown of cyclic AMP all enhance norepinephrine release from sympathetic nerve endings (see Reference 4).

The evidence for the involvement of the guanylate cyclase/cyclic GMP system is less clear-cut with cell-permeable analogues of cyclic GMP having either a small facilitatory effect on norepinephrine release[5,6] or no effect at all.[7,8] The difficulty in interpretation of these results is that analogues of cyclic GMP can in some cases increase cyclic AMP levels.[6] Furthermore, relatively selective inhibitors of cyclic GMP phosphodiesterase such as M&B 22948 had no effect on norepinephrine release in mouse atria when used in selective concentrations.[5]

With regard to the phospholipase C–IP_3/diacylglycerol system, there is little evidence to implicate IP_3 in the modulation of norepinephrine release, although such a role may be suggested due to the ability of IP_3 to increase intracellular calcium levels.[9] However, it has been suggested that the calcium stores in sympathetic nerves are small[10] and therefore IP_3 may not play a role in transmitter release (see Reference 11). On the other hand, derivatives of diacylglycerol and diacylglycerol mimetics such as phorbol esters have been shown to enhance norepinephrine release in a number of sympathetically innervated tissues (see Reference 12). This is likely due to the activation of protein kinase C since the effect can be blocked by protein kinase C inhibitors (see Reference 12).

The focus of our research has been on the adenylate cyclase/cyclic AMP system as well as the phospholipase C/diacylglycerol system. The reason for considering these two systems in presynaptic receptor operation is not only because of their effects on norepinephrine release but also because in other cell types these two systems are often linked to membrane receptors (see References 13 and 14).

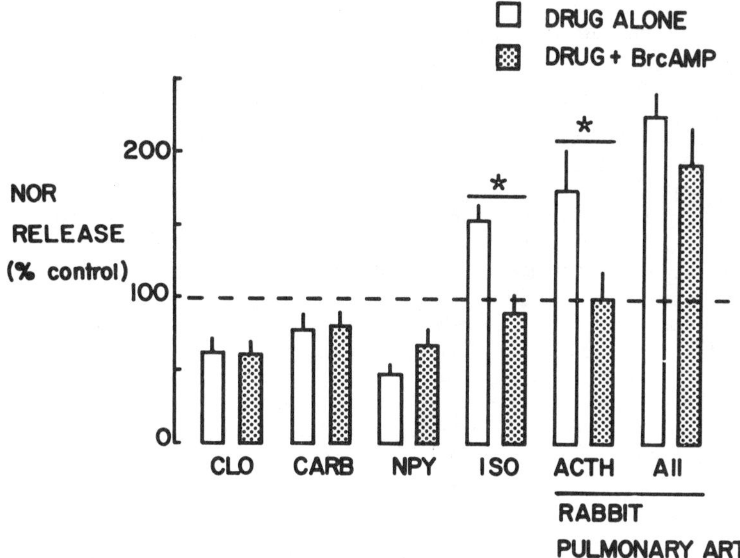

FIGURE 1. The effect of saturating concentrations of 8-bromo-cyclic AMP (BrcAMP) on prejunctional modulation of norepinephrine release. The tissues (mouse atria and rabbit pulmonary artery) were incubated with [³H]norepinephrine and the release of radioactivity in response to electrical field stimulation (2 or 5 Hz) was taken as an index of norepinephrine release. In some cases the tissues were incubated with 8-bromo-cyclic AMP (90 or 270 μM) and in some cases (NPY and carbachol) also IBMX (3-isobutyl-1-methyl xanthine 100 μM). 8-Bromo-cyclic AMP by itself or in combination with IBMX increased norepinephrine release in the control situation. The following drugs were used: clonidine (CLO, 0.03 μM), carbachol (CARB, 1.0 μM), neuropeptide Y (NPY, 0.3 μM) by themselves inhibited norepinephrine release, and isoproterenol (ISO, 0.1 μM) enhanced norepinephrine release in mouse atria. $ACTH_{1-24}$ (0.1 μM) and angiotensin II (AII, 0.01 μM) were used in rabbit pulmonary artery in the presence of 1 μM phentolamine and both enhanced norepinephrine release; *represents significant difference from experiments with 8-bromo-cyclic AMP (two way analysis of variance). (The data are taken from the following papers: References 5, 15, 37, and 38.)

ADENYLATE CYCLASE

In view of the facilitatory effect of cyclic AMP on norepinephrine release, it is a reasonable starting hypothesis that facilitatory prejunctional receptors increase cyclic AMP levels and inhibitory prejunctional receptors decrease cyclic AMP levels to effect changes in transmitter release. FIGURE 1 illustrates the operation of several inhibitory presynaptic receptor systems in mouse atria where the tissue had been incubated with [^{3}H]norepinephrine and the electrical stimulation–induced outflow of radioactivity measured. In these studies some of the tissues were incubated with high concentrations of the cell-permeable cyclic AMP analogue 8-bromo-cyclic AMP such that the concentration used was supramaximal for enhancing norepinephrine release. Our hypothesis was that this would result in the elevation of total neuronal cyclic AMP levels (cyclic AMP + 8-bromo-cyclic AMP) such that changes in endogenous cyclic AMP production would not appreciably alter total neuronal cyclic AMP levels, and would not affect norepinephrine release. In some studies, the phosphodiesterase inhibitor IBMX was also present for the same reason. It can be seen from FIGURE 1 that the inhibitory effects on norepinephrine release of clonidine (α-adrenoceptors), neuropeptide Y (NPY receptors), and carbachol (muscarinic receptors) in mouse atria were not affected by the 8-bromo-cyclic AMP treatment which makes it unlikely that these systems are operating through an adenylate cyclase pathway. On the other hand, the enhancement of norepinephrine release by isoproterenol (β-adrenoceptors) was totally abolished by 8-bromo-cyclic AMP treatment (FIGURE 1). In rabbit pulmonary artery, the facilitatory effect of ACTH$_{1-24}$ and isoproterenol on norepinephrine release, but not that of angiotensin II, was inhibited in the presence of the cyclic AMP analogue (Reference 15, Figure 1). These results with ACTH and β-adrenoceptors are consistent with the linking of these receptors to activation of adenylate cyclase. Similar conclusions have been reached in other studies in other tissues where phosphodiesterase inhibitors were found to potentiate the facilitatory effects of isoproterenol and ACTH$_{1-24}$.[5,17] Furthermore, this association is similar to that found with β-adrenoceptors and ACTH receptors in other cell types (see References 2 and 18).

The above conclusion is persuasive for the positive results (β-adrenoceptors, ACTH). However, in those experiments where no evidence of an interaction with the cyclic AMP analogue was observed (α-adrenoceptors, NPY receptors, muscarinic receptors), there must be some reservations since it may be that even with maximal exogenous cyclic AMP effects, a reduction in endogenous cyclic AMP production could still possibly decrease norepinephrine release. Therefore the interaction of these inhibitory prejunctional mechanisms with cyclic AMP pathways was studied using an alternative approach. Firstly, the effects of cyclic AMP are thought to be mediated through a specific cyclic AMP–dependent protein kinase (protein kinase A) and in the presence of inhibitors of protein kinase A (K252 and staurosporine), the inhibitory effects of neuropeptide Y[19] and carbachol[16] on norepinephrine release were still observed (see also FIGURE 2). This further supports the contention that neuropeptide Y and muscarinic receptors are not linked to adenylate cyclase. Another line of argument can be used for prejunctional α-adrenoceptors. If α-adrenoceptor inhibition of norepinephrine release is due to inhibition of adenylate cyclase, then disruption of α-adrenoceptor automodulation by α-adrenoceptor-blocking drugs, such as phentolamine, should enhance norepinephrine release because of increased cyclic AMP levels. However, phentolamine was still effective in enhancing norepinephrine release in the presence of a maximally effective concentration of 8-bromo-cyclic AMP even though under the same conditions the facilitatory effect of isoproterenol was abolished.[5] This again suggests that cyclic AMP is not involved in α-adrenoceptor modulation of norepinephrine release.

PROTEIN KINASE C

The role of protein kinase C in receptor modulation of norepinephrine release was also studied in atria whose transmitter stores were loaded with [^{3}H]norepinephrine.

The inhibitory effects of clonidine (α-adrenoceptors), neuropeptide Y, and carbachol (muscarinic receptors) were not affected by protein kinase C inhibitors which is indicative that these receptors are not linked to protein kinase C (FIGURE 2). However, when it came to the study of facilitatory processes, difficulties with the effects of protein kinase C inhibitors became evident. Firstly, polymyxin B, at concentrations that by themselves had no effect on transmitter release, inhibited not only the release-enhancing effects of angiotensin II,[20] but also the release-enhancing effects of isoproterenol, idazoxan, high-frequency stimulation, and tetraethylammonium.[12] The effect of idazoxan is thought to be due to the blockade of prejunctional α-autoreceptors, thus disinhibiting norepinephrine release. Therefore, if the inhibitory effect of polymyxin B

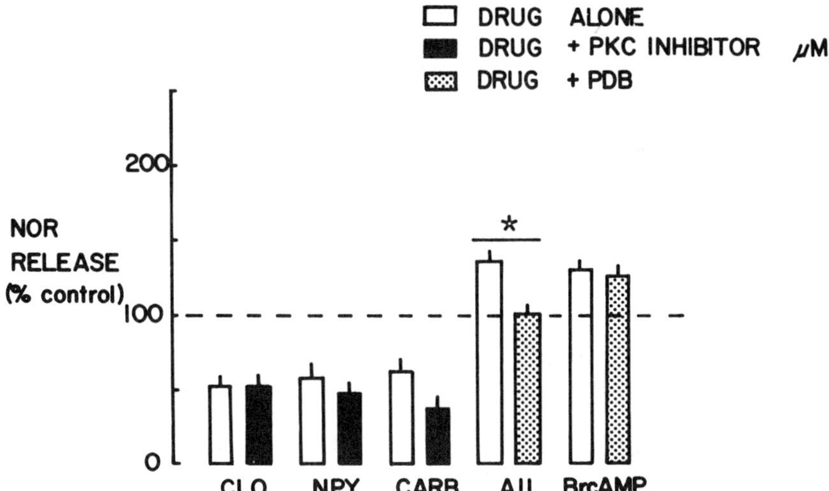

FIGURE 2. The effect of protein kinase C inhibitors and saturating concentrations of phorbol dibutyrate (PDB, 0.1 μM) on prejunctional modulation of norepinephrine release. The tissues (mouse atria and rat atria) were incubated with [³H]norepinephrine and the release of radioactivity in response to electrical field stimulation (1 or 5 Hz) was taken as an index of norepinephrine release. The following drugs were used: clonidine (CLO, 0.03 μM), neuropeptide Y (NPY, 0.3 μM), angiotensin II (AII, 0.01 μM), 8-bromo-cyclic AMP (BrcAMP, 90 μM) in mouse atria and carbachol (CARB, 1.0 μM) in rat atria. The following protein kinase C inhibitors were used: with clonidine, polymyxin B (70 μM); with NPY, K252 (1.0 μM); and with carbachol, staurosporine (0.1 μM); *represents significant difference from protein kinase C inhibitor or PDB experiments (two-way analysis of variance). (These data are taken from References 12, 16, 19, and 20.)

were due to disruption of this α-adrenoceptor system, then the inhibitory effect of the α-adrenoceptor agonist clonidine on norepinephrine release should also have been blocked, but this was not observed.[19] It seems likely that the inhibition of the various facilitatory processes (some not receptor mediated) by protein kinase C inhibitors is a general phenomenon, operative when release is elevated beyond a critical level. This also applies to inhibitors other than polymyxin B since another protein kinase C inhibitor, K-252, reduced the facilitatory effects of angiotensin II on norepinephrine release but it also reduced the facilitatory effects of the potassium channel–blocking drug tetraethylammonium.[12] In view of these effects it is probably unsuitable to examine the effects of protein kinase C inhibitors with drugs that enhance transmitter release since one cannot be sure as to the correct interpretation of the data.

Another way of looking at the involvement of protein kinase C is to maximally activate protein kinase C with phorbol esters. In an early study the protein kinase C activator phorbol myristate acetate was used in a maximally effective concentration for enhancing norepinephrine release.[20] In the presence of this agent angiotensin II still enhanced norepinephrine release and it was concluded that the effect of angiotensin II was independent of protein kinase C.[20] The assumption was that phorbol myristate acetate maximally activated protein kinase C. However, in a later study it was found that another phorbol ester, phorbol dibutyrate enhanced norepinephrine release to a greater extent than phorbol myristate acetate.[21] Therefore it was possible that there was a pool of protein kinase C activated by phorbol dibutyrate but not by phorbol myristate acetate. This has functional consequences since angiotensin II facilitation of norepinephrine release was not observed in the presence of phorbol dibutyrate (FIGURE 2), although the facilitatory effect of 8-bromo-cyclic AMP was unaltered by phorbol dibutyrate, showing that it was not a nonspecific effect. Thus, it is likely that prejunctional angiotensin II receptors act through protein kinase C to enhance norepinephrine release. The linking of angiotensin II receptors to protein kinase C activation has been observed in nonneuronal tissues.[22]

G PROTEINS

Receptors on cell membranes are often associated with a family of trimeric GTP-binding proteins, termed G proteins, which link the receptors to second messenger generating systems, enzymes, and ion channels. Various tools have been employed to assess the role of G proteins in receptor function in neuronal systems. One of these is pertussis toxin, which ADP ribosylates the α subunit of some G proteins to cause their inactivation.[23] One G protein that is inactivated by pertussis toxin is the one that links receptors to inhibition of adenylate cyclase (see below).

Pertussis toxin treatment leads to the loss of α_2-adrenoceptor inhibition of adenylate cyclase activity in both membranes and intact cells in a number of tissues[24] as well as a loss of inhibition of adenylate cyclase through neuropeptide Y[25] and muscarinic receptors.[26] FIGURE 3 shows the effects of pertussis toxin on modulation of norepinephrine release from mouse atria. The inhibitory effects of clonidine (α_2-adrenoceptor) and neuropeptide Y on norepinephrine release in mouse atria were unaffected by pertussis toxin treatment, and the inhibitory effect of carbachol (muscarinic receptor) was only slightly attenuated (FIGURE 3). In rat atria the inhibitory effect of carbachol was not affected by pertussis toxin.[16] These results suggest that a pertussis toxin susceptible G protein was not involved in any of the prejunctional inhibitory mechanisms tested.

There is considerable controversy as to whether prejunctional α_2-adrenoceptors are coupled to a pertussis toxin susceptible G protein. Clonidine inhibition of contractile responses to electrical stimulation of the rat vas deferens was inhibited by pertussis toxin, but the contractile response to norepinephrine was unaffected.[27] However, in another study in rat vas deferens was inhibitory effect of the α_2-adrenoceptor agonist xylazine on contractile responses to nerve stimulation was not affected by pertussis toxin.[28] Similarly, in mouse vas deferens clonidine inhibition of the contractile response to electrical stimulation was not altered by pertussis toxin treatment.[29] Furthermore, in the pithed rat the inhibition of responses to cardioaccelerator nerve stimulation was not affected by pertussis toxin.[28,30]

Although the majority of these results suggest that a pertussis toxin susceptible G protein is not involved in α_2-adrenoceptor inhibition of norepinephrine release, some caution must be used in the interpretation of these largely negative results. There is the question of whether the pertussis toxin treatment was effective. In many of the above studies, some postjunctional event was shown to be blocked by the pertussis toxin treatment (constrictor responses to xylazine in the vas deferens;[28] α_2-adrenoceptor vasoconstriction in the pithed rat,[28,30] and muscarinic inhibition of atrial rate.[31] However, it is not clear if pertussis toxin has a prejunctional effect. In a recent study, although the inhibitory effect of clonidine on norepinephrine release in rat kidney was not abolished by pertussis toxin (FIGURE 4), the facilitatory effect of angiotensin II was reduced and the inhibitory effect of the α_1-adrenoceptor agonist methoxamine was abolished (see FIGURE 4), suggesting indeed that pertussis toxin was affecting neuronal

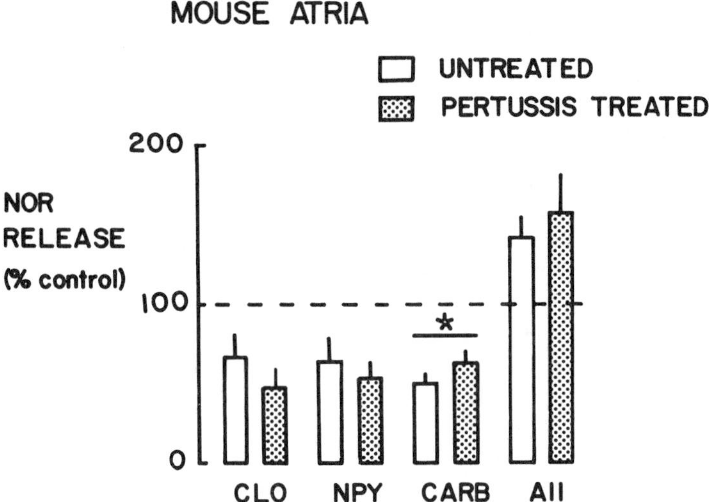

FIGURE 3. Effect of pertussis toxin on modulation of norepinephrine release in mouse atria incubated with [³H]norepinephrine and subjected to field stimulation (5 Hz). Mice were treated with pertussis toxin (1 or 1.5 μg/mouse) 4 days prior to the experiment. The following drugs were used: clonidine (CLO, 0.03 μM), neuropeptide Y (NPY, 0.3 μM), carbachol in the presence of phentolamine (1 μM) (CARB, 1.0 μM) inhibited norepinephrine release and angiotensin II (AII, 0.01 μM) enhanced norepinephrine release; *represents significant difference from pertussis toxin experiments (two-way analysis of variance). (Data are taken from References 12, 31, and 37 and Costa & Majewski, unpublished.)

G proteins. Central noradrenergic neurones appear different to sympathetic nerves since pertussis toxin reduces α_2-adrenoceptor inhibition of norepinephrine release in rabbit hippocampal slices.[32]

Another approach is to use N-ethylmaleimide which is reported to alkylate G proteins (see Reference 33) and α_2-adrenoceptor inhibition of norepinephrine release was reduced by this agent in rat tail artery.[34] However this is not inconsistent with the lack of effect of pertussis toxin since N-ethylmaleimide has a wider spectrum of activity than pertussis toxin. Indeed, in mouse atria N-ethylmaleimide reduced the inhibitory effect of clonidine on norepinephrine release (Murphy, Foucart, and Majewski, unpublished) whereas pertussis toxin did not (FIGURE 3). Similarly, although pertussis toxin did not reduce the inhibitory effects of neuropeptide Y on norepinephrine release in mouse atria (FIGURE 4), N-ethylmaleimide did reduce the inhibitory effect of NPY (Foucart, Murphy, and Majewski, unpublished).

Pertussis toxin had no effect on the facilitation of norepinephrine release by angiotensin II in mouse atria (FIGURE 3) but attenuated the facilitation of norepinephrine release by angiotensin II in slices of rat kidney cortex (FIGURE 4). Therefore it may be that prejunctional angiotensin II receptors are linked to different G proteins in different tissues. Interestingly, N-ethylmaleimide treatment decreased the facilitatory effect of angiotensin II in mouse atria (Foucart and Majewski, unpublished observations).

CONCLUSIONS

In the systems that we have studied to date, a clear pattern emerges. The first point is that we were able to establish clear signal transduction processes for the facilitatory prejunctional receptors: adenylate cyclase in the case of prejunctional β-adrenoceptors and prejunctional ACTH receptors, and protein kinase C in the case of prejunctional angiotensin II receptors. However, as far as the inhibitory prejunctional receptors are concerned it is unlikely that either protein kinase C or adenylate cyclase pathways are involved. Furthermore it is unlikely that the guanylate cyclase system is operative in view of the inconsistent effects of cyclic GMP analogues on norepinephrine release (see above). Therefore another avenue must be sought to explain prejunctional inhibition of

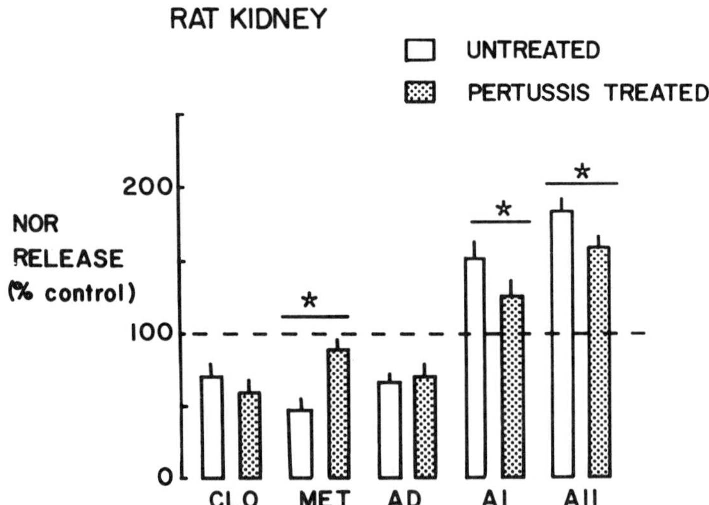

FIGURE 4. Effect of pertussis toxin on modulation of norepinephrine release in rat kidney slices incubated with [³H]norepinephrine and subjected to field stimulation (1 Hz or 5 Hz in the case of angiotensin). Rats were treated with pertussis toxin (8 μg/rat) 4 days prior to the experiment. Clonidine (CLO, 0.1 μM), methoxamine (MET, 10 μM) and adenosine (AD, 10 μM) by themselves inhibited norepinephrine release, angiotensin I (AI, 0.1 μM) and angiotensin II (AII, 0.1 μM) increased norepinephrine release; $n = 4$–8 for each group; *represents significant difference from pertussis toxin experiments (two-way analysis of variance). (Data taken from Murphy & Majewski, unpublished.)

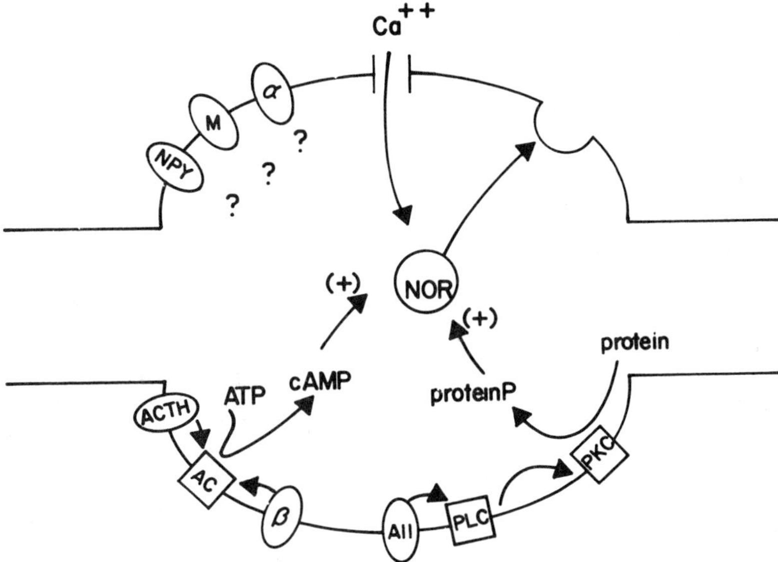

FIGURE 5. Possible signal transduction processes for prejunctional receptors at sympathetic nerve endings. Inhibitory neuropeptide Y (NPY), muscarinic (M), and α-adrenoceptors may be linked directly to a calcium channel; whereas facilitatory β-adrenoceptors and ACTH receptors are coupled to adenylate cyclase (AC) to stimulate cyclic AMP production. Angiotensin II receptors (AII) may be linked to phospholipase C (PLC) which in turn, through protein kinase C (PKC) activation, leads to the phosphorylation of proteins involved in the release process.

norepinephrine release. Recent studies have indicated that there is coupling of prejunctional α-adrenoceptors to calcium channels in frog sympathetic neurones and that a G protein is directly involved without an intervening second messenger.[35] There is also substantial indirect evidence that α-adrenoceptor inhibition of norepinephrine release is due to decreased calcium entry (see Reference 36). The data presented in the present paper tend to support the contention that no second messenger pathway is involved. Furthermore, our preliminary data with N-ethylmaleimide suggest also that a G protein is involved (Murphy, Foucart, and Majewski unpublished observations) even if it is not a pertussis toxin susceptible G protein.[31] Therefore a working hypothesis for inhibitory prejunctional receptors is that they are linked via a G protein to a calcium channel without intervening second messengers. This scheme also seems applicable to the other inhibitory receptors that we have studied: neuropeptide Y receptors and muscarinic receptors. The possible coupling of prejunctional receptors to signal transduction pathways is summarized in FIGURE 5.

REFERENCES

1. NISHIZUKA, Y. 1986. Studies and perspectives of protein kinase C. Science **233:** 305–312.
2. LEVITSKI, A. 1986. β-Adrenergic receptors and their mode of coupling to adenylate cyclase. Physiol. Rev. **66:** 819–854.
3. BROWNING, M. D., R. HUGANIR & P. GREENGARD. 1985. Protein phosphorylation and neuronal function. J. Neurochem. **45:** 11–23.

4. MAJEWSKI, H., E. J. N. ISHAC & I. F. MUSGRAVE. 1988. Intraneuronal mechanisms in the modulation of noradrenaline release through prejunctional receptors. *In* Vascular Neuroeffector Mechanisms. J. A. Bevan, H. Majewski, R. Maxwell, D. F. Story, Eds.: 243–250. IRL Press. Oxford, England.

5. JOHNSTON, H., H. MAJEWSKI & I. F. MUSGRAVE. 1987. Involvement of cyclic nucleotides in prejunctional modulation of noradrenaline release in mouse atria. Br. J. Pharmacol. **91:** 773–781.

6. CUBEDDU, L., X. E. BARNES & N. WEINER. 1975. Release of norepinephrine and dopamine-β-hydroxylase by nerve stimulation. IV. An evaluation of a role for cyclic adenosine monophosphate. J. Pharmacol. Exp. Ther. **193:** 105–127.

7. STJÄRNE, L., T. BARTFAI & P. ALBERTS. 1979. The influence of 8-Br 3'5'-cyclic nucleotide analogs and inhibitors of 3',5'-cyclic nucleotide phosphodiesterase, on noradrenaline secretion and neuromuscular transmission in the guinea-pig vas deferens. Naunyn-Schmiedebergs Arch. Pharmacol. **308:** 99–105.

8. ALBERTS, P., V. R. ÖGREN, & Ä. I. SELLSTRÖM. 1985. Role of adenosine 3',5'-cyclic monophosphate in adrenoceptor-mediated control of ^{3}H-noradrenaline secretion in guinea-pig myenteric nerve terminals. Naunyn-Schmiedebergs Arch. Pharmacol. **330:** 114–120.

9. STREB, H., R. F. IRVINE, M. J. BERRIDGE & I. SCHULZ. 1983. Release of Ca^{2+} from a nonmitochondrial intracellular store in pancreatic acinar cells by inositol 1,4,5-triphosphate. Nature **306:** 67–69.

10. LIPSCOMBE, D., D. V. MADISON, M. POENIE, H. REUTER, R. Y. TSIEN & R. W. TSIEN. 1988. Spatial distribution of calcium channels and cytosolic calcium transients in growth cones and cell bodies of sympathetic neurons. Proc. Nat. Acad. Sic. USA **85:** 2398–2402.

11. BHAVE, S. V., R. K. MALHOTRA, T. D. WAKADE & A. R. WAKADE. 1988. Formation of inositol triphosphate by muscarinic agents does not stimulate transmitter release in cultured sympathetic neurons. Neurosci. Lett. **90:** 234–238.

12. MUSGRAVE, I. F. & H. MAJEWSKI. 1989. Effect of phorbol esters and polymyxin B on modulation of noradrenaline release in mouse atria. Naunyn Schmiedebergs Arch. Pharmacol. **339:** 48–53.

13. JAKOBS, K. H., K. AKTORIES & G. SCHULTZ. 1981. Inhibition of adenylate cyclase by hormones and neurotransmitters. Adv. Cyclic Nucleotide Res. **14:** 183–186.

14. BERRIDGE, M. J. 1987. Inositol triphosphate and diacylglycerol: two interacting second messengers. Annu. Rev. Biochem. **56:** 158–193.

15. COSTA, M. & H. MAJEWSKI. 1988. Facilitation of noradrenaline release from sympathetic nerves through activation of ACTH receptors, β-adrenoceptors and angiotensin II receptors. Br. J Pharmacol. **95:** 993–1001.

16. COSTA, M. & H. MAJEWSKI. 1990. Further evidence that muscarinic inhibition of noradrenaline release does not involve cyclic AMP. Clin. Exp. Pharmacol. Physiol. (Suppl. 16): 193.

17. GÖTHERT, M. & F. HENTRICH. 1984. Role of cAMP for regulation of impulse-evoked noradrenaline release from the rabbit pulmonary artery and its possible relationship to presynaptic ACTH receptors. Naunyn-Schmiedebergs Arch. Pharmacol. **328:** 127–134.

18. GRAHAME-SMITH, D. G., R. W. BUTCHER, R. L. NEY & E. W. SUTHERLAND. 1967. Adenosine 3',5'-monophosphate as the intracellular mediator of the action of adrenocorticotrophic hormone on the adrenal cortex. J. Biol. Chem. **242**(23): 5535–5541.

19. FOUCART, S., I. F. MUSGRAVE & H. MAJEWSKI. 1990. Inhibition of noradrenaline release by neuropeptide Y does not involve protein kinase C in mouse atria. Neuropeptides **15:** 179–185.

20. MUSGRAVE, I. F. & H. MAJEWSKI. 1989. Effect of phorbol ester and pertussis toxin on the enhancement of noradrenaline release by angiotensin II in mouse atria. Br. J. Pharmacol. **96:** 609–616.

21. MUSGRAVE, I. F., S. FOUCART & H. MAJEWSKI. Evidence that angiotensin II enhances noradrenaline release in mouse atria by activating protein kinase C. J. Neurochem. (Submitted.)

22. PFEILSCHIFTER, J., A. KURTZ & C. BAUER. 1986. Role of phospholipase C and protein kinase C in vasoconstrictor-induced prostaglandin synthesis in cultured rat renal mesangial cells. Biochem. J. **234:** 125–130.

23. UI, M. 1986. Pertussis toxin as a probe of receptor coupling to inositol lipid metabolism. *In* Phosphoinositides and Receptor Mechanisms. J. W. Putney, Ed.: 163–195. A.R. Liss. New York, N.Y.

24. DOLPHIN, A. C. 1987. Nucleotide binding proteins in signal transduction and disease. Trends Pharmacol. Sci. **10**: 53–57.

25. KASSIS, S., M. OLASMAA, L. TERENIUS & P. H. FISHMAN. 1987. Neuropeptide Y inhibits cardiac adenylate cyclase through a pertussis toxin–sensitive G protein. J. Biol. Chem. **262**: 3429–3431.

26. HAZEKI, O. & M. UI. 1981. Modification by islet-activating protein of receptor-mediated regulation of cyclic AMP accumulation in isolated heart cells. J. Biol. Chem. **256**: 2856–2862.

27. LAI, R.-T., Y. WATANABE & H. YOSHIDA. 1983. Effect of islet-activating protein (IAP) on contractile responses of rat vas deferens: evidence for participation of Ni (inhibitory GTP binding regulating protein) in the α_2-adrenoceptor-mediated response. Eur. J. Pharmacol. **90**: 453–456.

28. DOCHERTY, J. R. 1988. Pertussis toxin and pre-junctional α_2-adrenoceptors in rat heart and vas deferens. J. Auton. Pharmacol. **8**: 197–201.

29. LUX, B. & R. SCHULTZ. 1986. Effect of cholera toxin and pertussis toxin on opioid tolerance and dependence in the guinea-pig myenteric plexus. J. Pharmacol. Exp. Ther. **237**: 995–1000.

30. NICHOLS, A. J., E. D. MOTLEY & R. R. RUFFOLO. 1988. Differential effect of pertussis toxin on pre- and postjunctional α_2-adrenoceptors in the cardiovascular system of the pithed rat. Eur. J. Pharmacol. **145**: 345–349.

31. MUSGRAVE, I. F., P. MARLEY & H. MAJEWSKI. 1987. Pertussis toxin does not attenuate α_2-adrenoceptor mediated inhibition of noradrenaline release in mouse atria. Naunyn-Schmiedebergs Arch. Pharmacol. **336**: 280–286.

32. ALLGAIER, C., T. J. FEUERSTEIN, R. JACKISCH & G. HERTTING. 1985. Islet-activating protein (pertussis toxin) diminishes α_2-adrenoceptor mediated effects on noradrenaline release. Naunyn-Schmiedebergs Arch. Pharmacol. **331**: 235–239.

33. HERTTING, G. & C. ALLGAIER. 1988. Participation of protein kinase C and regulatory G proteins in modulation of the evoked noradrenaline release in brain. Cell. Mol. Neurobiol. **8**: 105–114.

34. WEBER, H. D. 1989. Presynaptic α_2-adrenoceptors at the terminals of perivascular sympathetic nerves are coupled to an N-ethylmaleimide-sensitive G-protein. Naunyn-Schmiedebergs Arch. Pharmacol. **339**: R35.

35. LIPSCOMBE, D., S. KOMGSAMUT & R. W. TSIEN. 1989. α-Adrenergic inhibition of sympathetic neurotransmitter release mediated by modulation of N-type calcium-channel gating. Nature **340**: 639–642.

36. STARKE, K. 1987. Presynaptic α-autoreceptors. Rev. Physiol. Biochem. Pharmacol. **107**: 73–146.

37. FOUCART, S. & H. MAJEWSKI. 1989. Inhibition of noradrenaline release by neuropeptide Y in mouse atria does not involve inhibition of adenylate cyclase or a pertussis toxin–susceptible G protein. Naunyn-Schmiedebergs Arch. Pharmacol. **340**: 658–665.

38. COSTA, M. & H. MAJEWSKI. 1989. Inhibition of noradrenaline release by carbachol does not involve cyclic AMP. Proc. Aust. Physiol. Pharmacol. Soc. **20**(1): 41P.

Role of G Proteins, Cyclic AMP, and Ion Channels in the Inhibition of Transmitter Release by Adenosine[a]

BERTIL B. FREDHOLM, MARIANNE DUNÉR-ENGSTRÖM,
JOHAN FASTBOM, PING-SHENG HU, AND
INGEBORG VAN DER PLOEG

Department of Pharmacology
Karolinska Institute
Box 60400
S-104 01 Stockholm, Sweden

INTRODUCTION

That adenosine and adenine nucleotides are able to decrease the release of neurotransmitters was first shown with electrophysiological techniques by Ginsborg and Hirst in 1972.[1] Soon thereafter it was directly demonstrated that the release of noradrenaline and acetylcholine evoked by electrical stimulation could be reduced by these compounds.[2–4] Since that time it has become clear that adenosine is able to inhibit the release of a variety of neurotransmitters both centrally and peripherally[5–7] However, the magnitude of the adenosine effect may vary. For example, in peripheral noradrenergic neurons, adenosine seems to inhibit noradrenaline release in most species except the cat,[6,8] and in the central nervous system there are marked differences in the magnitude of the presynaptic inhibitory effect of adenosine in striatal and hippocampal neurons even though serotonin release was examined in both instances.[9]

At approximately the same time it was shown that nerve stimulation can also enhance the release of adenosine.[10,11] Much of this release could be blocked by blocking the actions of the neurotransmitter substance released from the neuron. Thus, it was postulated that adenosine could act as a "transsynaptic modulator" or "a retrograde transmitter"[12,13] Later work has also emphasized the release of adenosine triphosphate (ATP) either as a transmitter or as a cotransmitter with classical transmitter substances.[14] Thus, there are several possible sources of the adenosine that is released following nerve stimulation and that has a potential role as a feedback modulator of transmitter release (see FIGURE 1).

As already mentioned, in several peripheral systems the release of adenosine can be blocked by receptor agents, and mimicked by adding the transmitter substance itself.[5] More recently we have carried out experiments in the central nervous system using hippocampus slices, the same preparation that we have used to study transmitter release. It is clear that also in the central nervous system the characteristics of electrically evoked purine release are different from the release of classical neurotransmitters.[15] Quite recently it was shown that the adenosine that is released from hippocampus slices by electrical stimulation does not derive from extracellular adenine nucleotides, but is formed intracellularly.[16] Thus, the biological roles of

[a]These studies were supported by the Swedish Medical Research Council (proj no 2553), Ostermans foundation, 75 years fund against stroke, Astra Alab, and Karolinska Institute.

adenosine in neurotransmission appear to be fundamentally different from the biological roles of ATP.

PHYSIOLOGICAL ROLE OF ADENOSINE AS A PRESYNAPTIC MODULATOR

There are several lines of evidence that indicate that adenosine may play a physiologically important role in regulating neurotransmitter release. First, the adeno-

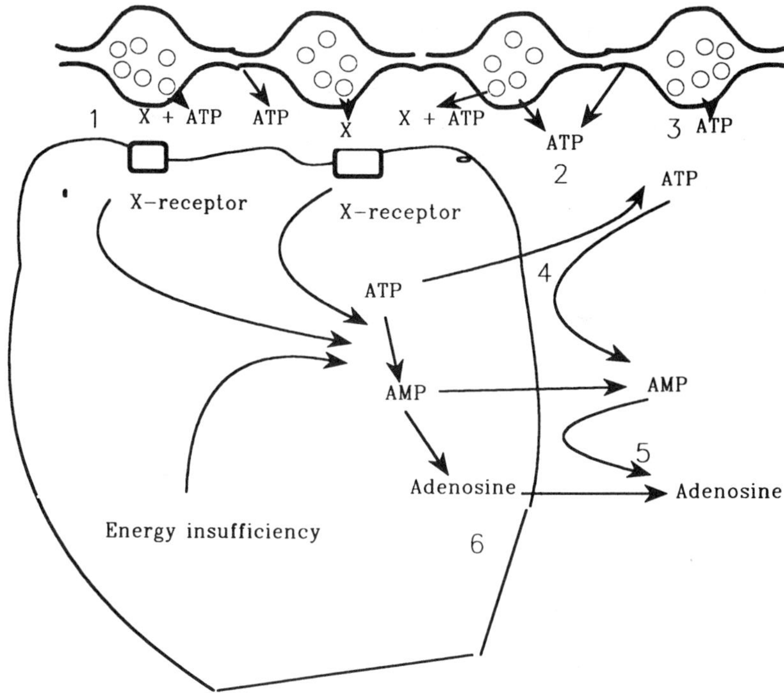

FIGURE 1. Schematic illustration of several possible sources of adenosine released following nerve activation. Six different possibilities are indicated. 1–5 all represent different ways in which adenosine would derive from an adenine nucleotide released to the outside of the cell. (1) ATP released as a co-transmitter; (2) ATP released in a nonexocytotic fashion from the nerve; (3) ATP released as a transmitter in its own right; (4) ATP released from an effector cell;[13] (5) AMP or ADP rather than ATP is released; (6) intracellular formation of adenosine.

sine receptor antagonist theophylline as well as related xanthine adenosine antagonists are able to increase the release of neurotransmitters both centrally and peripherally. While these effects may be indirect and primarily due to the central nervous system excitation that the compound produces, it cannot be excluded that the effect is due to the antagonism of the endogenous adenosine acting as a brake on the transmitter release. Secondly, it has been shown that adenosine acts to inhibit transmitter release in concentrations starting at submicromolar concentrations, reaching a maximal effect at less than 100 μM.[5,6] It now seems clear that the endogenous concentrations of adenosine in tissues and body fluids range between a few times 10^{-8} M up to perhaps 100 μM.[7] Thus, there is good agreement between the concentrations required to produce the effect under the in vitro conditions and the concentrations that are actually achieved in vivo. Thirdly, it has been shown that chronic treatment with adenosine receptor antagonists including theophylline and caffine leads to an up regulation of adenosine receptors. This up regulation in adenosine receptors is in all probability due to the antagonism of tonically active endogenous adenosine. Thus, the up regulation must be viewed as evidence for tonic occupation of these receptors. The increase in the number of receptors is associated with an increase in the potency of adenosine to antagonize the release of noradrenaline from cortical noradrenergic neurons. Together, these three lines of evidence provide good evidence that adenosine is one of the factors that do control transmitter release via presynaptic receptors under physiological conditions.

The importance of adenosine is likely to increase under conditions of hypoxia, hypoglycemia, or ischemia when the background level of adenosine is increased. For example the presynaptic inhibitory effect of adenosine is likely to be quite important under conditions of ischemia in brain or in the heart. Removal of the effect of adenosine in limiting the release of glutamate may be an important reason why acute theophylline or caffeine can aggravate ischemic brain damage.[17,18]

Similarly, the amount of adenosine released by nerve activity increases markedly if the nerve stimulation is prolonged. The release of adenosine is delayed and little or no release occurs for the first minute or minutes. This means that the role of adenosine to limit transmitter release is much less following short impulse trains than following prolonged periods of activation. This also makes teleological sense, because it is only prolonged and/or high frequency nerve activation that has the potential of causing major tissue damage.

MEDIATION VIA G PROTEIN–COUPLED A$_1$ RECEPTORS

Since the pioneering work of Van Calker *et al.*[19] and of Londos *et al.*,[20] it is clear that adenosine can exert its biological effects via two types of receptors, A$_1$ and A$_2$. Even though there have been several attempts to introduce further classes of adenosine receptors, the opinion of most workers in the field is that there is at present no compelling evidence to go beyond these two major receptor classes. Given the precedent of several other receptor systems it does seem likely, however, that these are in fact families of adenosine receptors.

The A$_1$-receptors were originally defined on the basis of decreases in cyclic AMP and the A$_2$-receptors with increases in this second messenger. It has become clear that also effects that are not directly dependent on changes in cyclic adenosine monophosphate (cAMP) may be produced by adenosine acting on receptors very similar to A$_1$ and A$_2$. Therefore, mechanism of action should not be used as a criterion for subdividing receptor subtypes.[21]

There is good evidence that the receptors mediating inhibition of transmitter release are of the A$_1$ subtype.[6,7,22] The adenosine A$_1$ receptors can be studied by binding techniques because both potent agonists and antagonists are available.[23–25] Using these ligands in binding assays it has been conclusively shown that the A$_1$-receptors are associated with G proteins.[26,27] Even though there may be differences in the degree of coupling between the A$_1$-receptors and the guanosine triphosphate (GTP) binding proteins, possibly reflecting differences in the type of the G protein that the receptor associates with, there is no evidence for any binding sites that are not associated with G proteins and that do not show the expected affinity shift upon addition of guanine nucleotide. Thus, the adenosine A_1-receptor is likely to fall into the growing class of receptors that show characteristics of being associated with classical G proteins including having seven membrane-spanning regions with a critically important cytoplasmic loop between the fifth and the sixth membrane-spanning domain.

ROLE OF G PROTEINS

With these results as a background it seemed reasonable to try to find more direct proof that G proteins are involved in the adenosine receptor–mediated inhibition of transmitter release. We tried to use pertussis toxin in rat hippocampus slices, but found that even with prolonged periods of incubation we could not find any adequate degree of adenosine diphosphate (ADP) ribosylation. Next we used N-ethylmaleimide which has been shown to produce significant inactivation of the G proteins associated with adenosine receptors.[28,29] Incubating the slices for a brief period of time with a carefully titrated concentration of N-ethylmaleimide, it was possible to block approximately 60 to 80% of the G proteins, as assessed by inhibition of back [^{32}P]-ADP ribosylation caused by pertussis toxin, without any effect on adenosine receptor, G_S-mediated effects, adenylate cyclase, or evoked transmitter release (FIGURE 2).This caused a virtually complete inhibition of the effect of adenosine analogues to reduce noradrenaline release from hippocampus slices[30] and a marked decrease in the potency of phenyl-isopropyladenosine to block acetylcholine release (FIGURE 2). In parallel the adenosine receptor–mediated inhibition of adenylate cyclase was reduced, but there was no effect on the bonding of adenosine antagonists to its receptors.[29] These results are clearly compatible with an involvement of an N-ethylmaleimide-sensitive G protein in the effects of adenosine.

Next we used in vivo pertussis toxin treatment and found that the administration of pertussis toxin close to the hippocampus 60 hours before experiment caused 40 to 75% inhibition of the ADP ribosylation induced by added exogenous pertussis toxin in vitro.[31] This suggests that 40 to 75% of the G proteins in the hippocampus had been ADP ribosylated by the pertussis toxin. Following this treatment there was a complete blockade of several known G protein–linked effects. For example, the adenosine-mediated decrease in cAMP accumulation was blocked, so was the adenosine effect on hippocampal CA1 neurons that is mediated by an effect on K channels.[31] By contrast, and in the same slices, several parameters measuring the presynaptic inhibitory effect of adenosine analogues were unaffected. For example the inhibition of field excitatory postsynaptic potentials (EPSPs), which is largely due to an inhibition of the release of glutamate from excitatory afferents, was virtually unaltered by pertussis toxin pretreatment. Similarly, the presynaptic inhibition of acetylcholine and noradrenaline release were unaffected by such pretreatment (FIGURE 3).These results could indicate that the G proteins on nerve endings are difficult to ADP ribosylate by exogenous pertussis toxin. The results could also mean that the G proteins that link adenosine A_1-receptors to inhibition of transmitter release are not pertussis toxin sensitive. They would therefore be different from those G proteins that mediate A_1-receptor linked inhibition of adenylate cyclase as well as the adenosine receptor associated G protein that mediates an enhancement of a K channel. These two alternatives are diagrammatically illustrated in FIGURE 4.

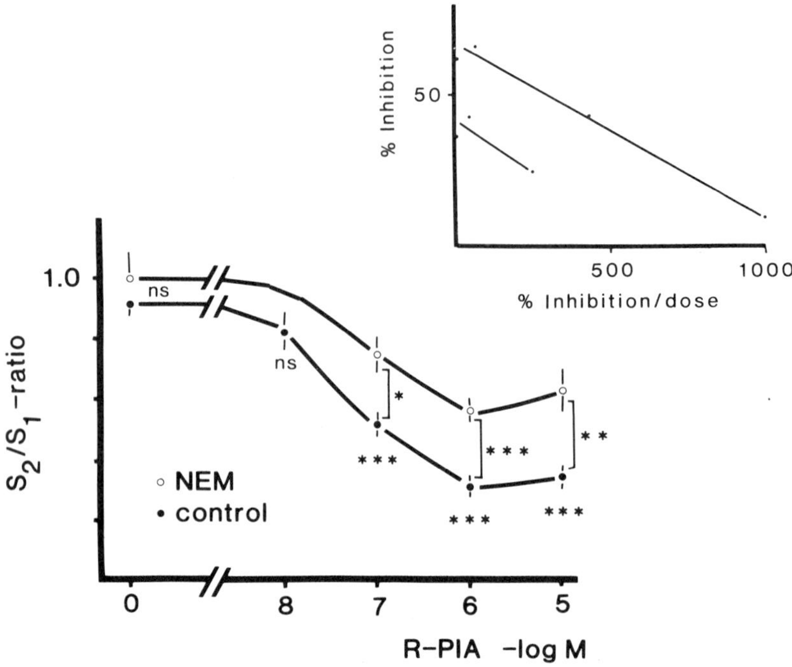

FIGURE 2. The effect of NEM treatment to block G proteins on the release of acetylcholine (ACh) from hippocampus and the presynaptic inhibitory effect of the adenosine analogue R-PIA. **FIGURE 2A** (above). The concentration-dependent inhibition of [^{3}H]-acetylcholine release by R-PIA in control slices and slices treated with NEM (reprinted from Dunér-Engström and Fredholm. 1988. Acta Physiol. Scand. **134:** 119–126, by permission).

ROLE OF CYCLIC AMP

Since it was originally demonstrated that adenosine receptors are linked to cAMP formation, causing inhibition (A_1) or stimulation (A_2),[19,20] it was reasonable to examine whether cAMP could mediate the effects of adenosine on neurotransmission. The above-mentioned results with pertussis toxin are circumstantial evidence that an inhibition of cAMP formation is not an obligatory link. It has also been shown that the stimulation of cAMP formation by combination of forskolin and a phosphodiesterase inhibitor that raises cAMP at least 10-fold in brain does not alter the presynaptic inhibitory effect of adenosine analogues.[30] Similarly the inhibition of cAMP formation by means of drugs that directly effect the adenylatecyclase, including 2'5'-dideoxy-adenosine, do not modify transmitter release. Together, these findings provide good evidence that changes in cAMP cannot be the only, or even the major, mechanism by which adenosine analogues decrease transmitter release. However, inhibition of cAMP formation may play a role in regulating transmitter release under such circumstances when cAMP elevation is important for the level of transmitter release that actually occurs.

INVOLVEMENT OF POTASSIUM CHANNELS

There is very good evidence that the adenosine A_1-like receptors can open K channels.[32] Such adenosine receptor–mediated increase in K conductance is dependent on a G protein. It is likely that this type of mechanism underlies the hyperpolarization of pyramidal CA1 neurons and the inhibition of neuronal excitation that occurs at the postjunctional level.[33] It has been argued that this type of action could well explain the presynaptic inhibitory effects of adenosine as well as other presynaptically active substances.[34,35] We have therefore used inhibitors of K channels as a tool to assess the importance of such channels in the mechanismal action of presynaptic inhibitors. The results using 4-aminopyridine (4-AP) as the inhibitor are summarized in FIGURE 5. 4-AP is able to increase the overflow of neurotransmitter in a concentration-dependent manner. At a concentration of 30 to 100 μM the effect is close to maximal and the evoked release of both acetylcholine and noradrenaline is increased at least threefold. Under these circumstances the presynaptic inhibitory effect of an α_2-adrenoceptor ligand acting on autoreceptors on noradrenergic neurons is virtually abolished.[36]

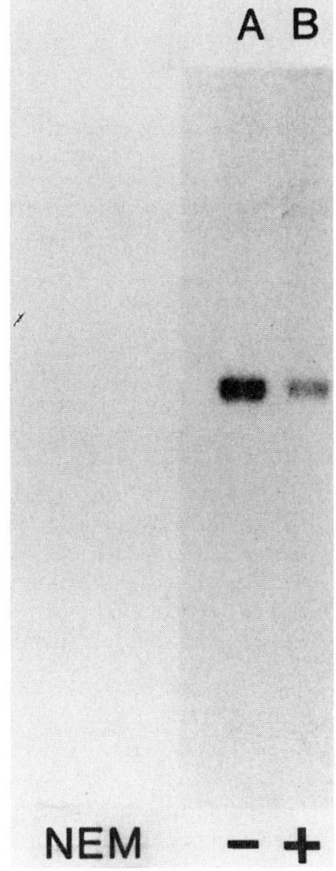

FIGURE 2B. $[^{32}P]$-ADP ribosylation of membranes from rat hippocampal slices pretreated with NEM (10 minutes, 100 mM) toxin added *in vitro*.

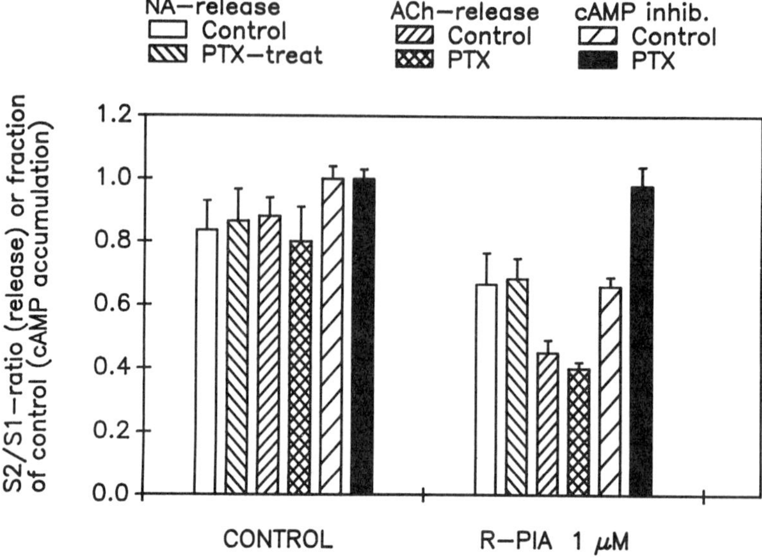

FIGURE 3. The effect of pertussis toxin treatment in vivo 60 hours prior to sacrifice and preparation of slices on biochemical parameters. The effect of R-PIA to block nonadrenalin (NA) release and ACh release was unaltered by PTX pretreatment, whereas the adenylatecyclase inhibition was completely antagonized. The PTX treatment caused a 35 to 72% inhibition of back-ADP ribosylation in these experiments. (Data redrawn with permission from Fredholm *et al.* 1989 Eur. J. Pharmacol. Mol. Pharmacol Sect. **172:** 249–262.)

Similarly, the ability of carbachol to inhibit the release of acetylcholine by a presynaptic mechanism is also essentially gone.[37] By contrast, the ability of the adenosine analogue R-PIA to block the release of both acetylcholine and noradrenaline is virtually unaffected.[36,37] The same was true for ω-conotoxin (ω-cgTOX). Since ω-cgTOX is known to act on n-type Ca channels,[38] these results are compatible with the opinion that the adenosine analogue does not utilize to any major extent the 4-AP-sensitive K channel to bring about an additional transmitter release but it may use an N-type Ca channel. Another important consequence of these findings is that two presynaptic ligands, in this case a ligand acting at an autoreceptor and an adenosine receptor, may employ different signal transduction mechanisms in order to bring about the same net effect, namely, an inhibition of neurotransmitter release.

THE INVOLVEMENT OF CALCIUM CHANNELS

Evoked neurotransmitter release is consistently reduced by eliminating extracellular Ca. Thus, Ca entry is an obligatory part of the transmission process. Addition of cadmium ions, which nonspecifically block Ca entry, causes a concentration-dependent decrease in the evoked release of both noradrenaline and acetylcholine from hippocampal slices.[36] By contrast, the dihydropyridine nifedipine was virtually without effect and the dihydropyridine L-type Ca channel agonist BayK 8644 was also without effect (see FIGURE 6) Thus, there is no good evidence that L-type Ca channels are directly involved in the presynaptic actions of adenosine analogues. There are, however, a number of complicating aspects. Thus, high concentrations of dihydropyridines have been shown to interact directly with adenosine receptors.[39] Another complication is that under some circumstances BayK 8644 can inhibit the presynaptic effects of R-PIA in the rat hippocampus.[40] The reason is probably that by blocking other types of Ca channels by R-PIA, the relative importance of L-type channels increases and therefore stimulation of such L-type channels may become functionally important.

Whereas inhibitors and activators of L-type Ca channels were ineffective, ω-conotoxin, which acts on N-type Ca-channels,[38] was quite active. In fact, in several instances the pharmacology of ω-conotoxin and of adenosine analogues was quite similar. This is at least compatible with the opinion that one important effect of adenosine could be to inhibit N-type Ca channels.[7]

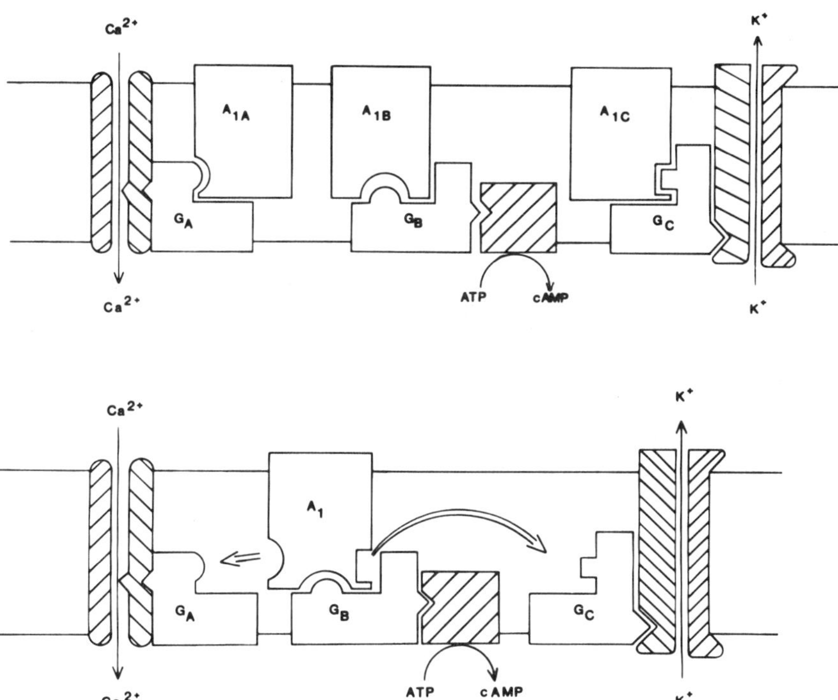

FIGURE 4. Two alternative explanations of why adenosine could act on several different types of effector mechanisms. **Upper panel:** There are several subtypes of adenosine receptors—each of which is linked to a different effector system. **Lower panel:** There is only one class of A_1-receptor but it can, via different G proteins, influence several different types of effector systems (Redrawn with permission from Fredholm and Dunwiddie. 1988. Trends Pharmacol. Sci. **9:** 130–134.) Note that β,γ-subunits liberated from G proteins associated with K^+ or Ca^{2+} channels may be responsible for inhibition of adenylate cyclase, and that a distinct G protein is not required.

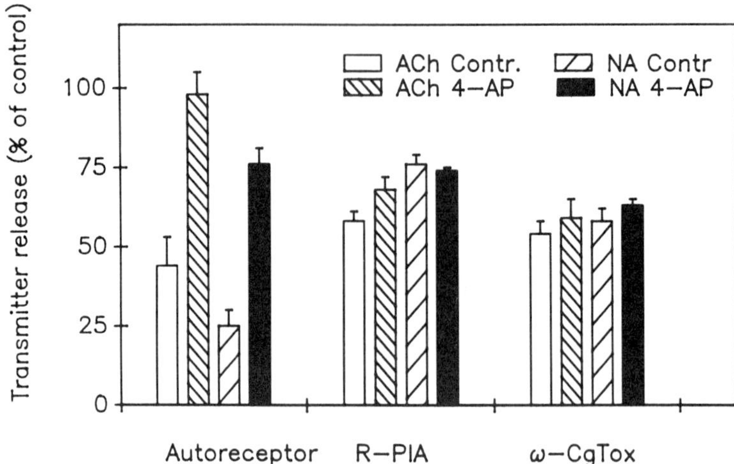

FIGURE 5. The effect of 4-aminopyridine (4-AP; 100 mM) on the presynaptic inhibition of noradrenaline (NA) and acetylcholine (ACh) release by autoreceptor stimulation and adenosine analogue and by an inhibitor of N-type Ca^{2+} channels. (Redrawn with permission from Hu and Fredholm. 1989. Acta Physiol. Scand. **136:** 347–353.)

ROLE OF PROTEIN KINASE C

It has been suggested that the effects of presynaptic agonists on Ca channels are indirect and due to the stimulation of protein kinase C.[41] Thus some neuronal noradrenaline effects can be mimicked by protein kinase C stimulation[41] and abolished by selective inhibition of protein kinase C.[42] However, we found no direct evidence that protein kinase C is involved in the regulation of acetylcholine release from rat hippocampal slices. The results summarized in FIGURE 7A show that activation of protein kinase C enhances the release of acetylcholine. Therefore, activation of protein kinase C is unlikely to be the mechanism by which adenosine inhibits such release.

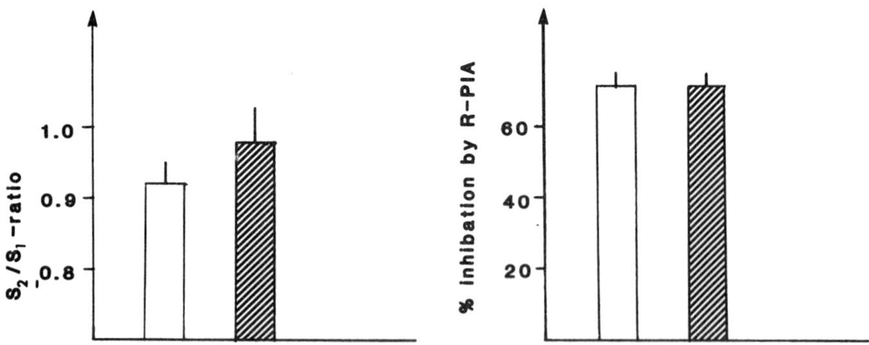

FIGURE 6. Lack of effect of BayK 8644 (1 mM; shaded bar) on the release of ACh induced by field stimulation (0.3 Hz) or on the presynaptic inhibitory action of R-PIA.

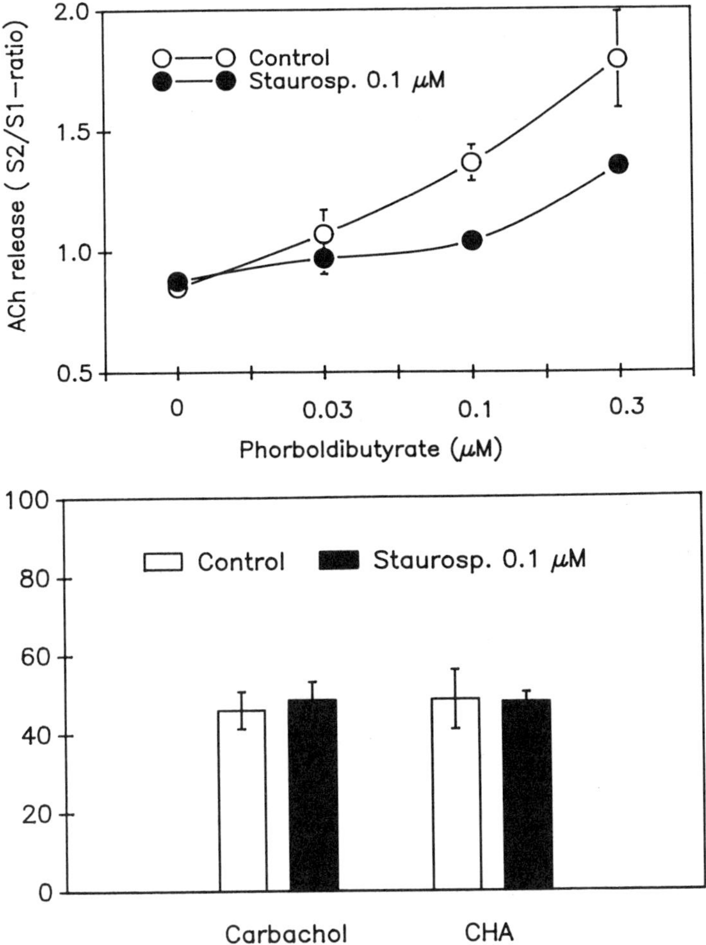

FIGURE 7. Evidence against an involvement of protein kinase C in the presynaptic effect of adenosine on acetylcholine release from rat hippocampal slices. (**A**) Upper panel shows the effect of increasing doses of phorbol-12,13-dibutyrate on the release of ACh and the inhibitory effect of staurosporine thereon. (**B**) Lower panel shows that staurosporine does not influence the presynaptic effects of either the adenosine analogue cyclohexyladenosine (CHA) or carbachol.

Furthermore, staurosporine in a concentration that blocks the effects of phorbolester was completely unable to block any effect of presynaptic agonists either carbachol activating the autoreceptors or cyclohexyladenosine activating adenosine receptors (FIGURE 7B).

The situation is somewhat more complicated in the case of noradrenaline release. There we showed that phorbolesters were unable to influence the presynaptic effects of adenosine analogues but that activation of protein kinase C with phorbolesters could in fact reduce the effects of α_2-adrenoceptor agonists.[43] In the case of noradrenaline there was also a marked increase in the evoked release. This might be related to the recent finding that an antibody to the protein kinase C substrate GAP-43 blocks the efflux of noradrenaline from synaptosomes permeabilized to Ca.[44]

CONCLUSIONS

The results briefly summarized here indicate that adenosine acting on receptors that have the pharmacological properties of A_1-receptors can decrease the release of several neurotransmitters. These receptors are linked to G proteins, but it is probable that not all the receptors are linked to the same type of G proteins.

These A_1 receptors can influence the level of cyclic AMP, but this does not seem to be a major factor in causing the inhibition of transmitter release. The presynaptic effects are similarly not due to changes in PKC or to interactions with a 4-AP sensitive K^+ channel, even though other presynaptic effectors are. It seems reasonable to assume that one important mechanism is A_1-receptor-mediated inhibition of a voltage-dependent Ca^{2+} channel. Since dihydropyridines, acting on L-type channels, are largely ineffective but ω-conotoxin and Cd^{2+} are effective, it is assumed that these channels are of the N-type.

Some very general conclusions can also be drawn:

1. Adenosine has the potential of inhibiting transmitter by several different mechanisms, including decreased cAMP, increased K^+ entry, decreased Ca^{2+} channel activity, and perhaps changes in intracellular Ca^{2+} actions. Since these possibilities exist, it seems probable that they are, in fact, also sometimes used. Not all of them in all instances, but perhaps each one of them in at least some situations.
2. The mechanisms underlying presynaptic control seem to vary depending upon several factors including type of transmitter, location, and mode of evoking transmitter release. Furthermore, different presynaptic agonists modulating the release of transmitter from one and the same type of nerve may do so via different mechanisms.

ACKNOWLEDGMENTS

The technical assistance of Mr. Sten Ågren and Mrs. Karin Lindström is gratefully acknowledged. We want to thank Dr. T. V. Dunwiddie for important suggestions and for major contributrions to several of the reported studies.

REFERENCES

1. GINSBORG, B. L. & G. D. S. HIRST. 1972. The effect of adenosine on the release of transmitter from the phrenic nerve of the rat. J. Physiol. Lond. **224:** 629–645.
2. FREDHOLM, B. B. 1974. Vascular and metabolic effects of theophylline, dibuturyl cyclic AMP, and dibuturyl cyclic GMP in canine subcutaneous adipose tissue in situ. Acta Physiol. Scand. **90:** 226–236.
3. HEDQVIST, P. & B. B. FREDHOLM. 1976. Effects of adenosine on adrenergic neurotransmission; prejunctional inhibition and postjunctional enhancement. Naunyn Schmiedebergs Arch. Pharmacol. **293:** 217–223.
4. VIZI, E. S. AND J. KNOLL. 1976. The inhibitory effect of adenosine and related nucleotides on the release of acetylcholine. Neuroscience **1:** 391–398.
5. FREDHOLM, B. B. & P. HEDQVIST. 1980. Modulation of neurotransmission by purine nucleotides and nucleosides. Biochem. Pharmacol. **29:** 1635–1643.

6. FREDHOLM, B. B., L. GUSTAFSSON, P. HEDQVIST & A. SOLLEVI. 1983. Adenosine in the regulation of neurotransmitter release in the peripheral nervous system. *In* Regulatory Function of Adenosine, R. M Berne, T. W. Rall & R. Rubio, Eds.: 479–495, Martinus Nijhoff Publishers. The Hague, Boston & London.

7. FREDHOLM, B. B. & T. V. DUNWIDDIE. 1988. How does adenosine inhibit transmitter release? Trends Pharmacol. Sci. **9:** 130–134.

8. LUCHELLI-FORTIS, M. A., B. B. FREDHOLM, & S. Z. LANGER. 1981. Evidence against the presence of presynaptic inhibitory adenosine receptors in the cat nictitating membrane. J. Pharmacol. Exp. Ther. **219:** 235–242.

9. FEUERSTEIN, T. J., L. I. BÄR & C. H. LÜCKING. 1988. Activation of A_1 adenosine receptors decreases the release of serotonin in the rabbit hippocampus, but not in the caudate necleus. Naunyn Schmiedebergs Arch. Pharmacol. **338:** 664–670.

10. MCILWAIN, H. 1972. Regulatory significance of the release and actin of adenine derivatives in cerebral systems. Biochem. Soc. Symp. **36:** 69–85.

11. FREDHOLM, B. B. 1976. Release of adenosine-like material from isolated perfused dog adipose tissue following sympathetic nerve stimulation and its inhibition by adrenergic alpha-receptor blockade. Acta Physiol. Scand. **96:** 122–130.

12. FREDHOLM, B. B. & P. HEDQVIST. 1979. Adenosine—a transsynaptic modulator of norepinephrine release? *In* Catecholamines: Basic and Clinical Frontiers. E. Usdin, I. J. Kopin & J. Barchas, Eds. **2:** 1146–1148. Pergamon Press. New York, N.Y.

13. ISRAEL, M., B. LESBATS, F. M. MEUNIER & J. STINNAKRE. 1976. Postsynaptic release of adenosine triphosphate induced by single impulse transmitter action. Proc. R. Soc. London Ser. B **193:** 461–468.

14. WHITE, T. D. 1988. Role of adenine compounds in autonomic neurotransmission. Pharmacol. Ther. **38:** 129–168.

15. JONZON, B. & B. B. FREDHOLM. 1985. Release of purines, noradrenaline and GABA from rat hippocampal slices by field stimulation. J. Neurochem. **44:** 217–224.

16. LLOYD, H. G. E. & B. B. FREDHOLM. Sources of adenosine released from hippocampal slices following electrical and hypoxic/hypoglycemic stimulation. *In* Biological Actions of Extracellular ATP. Ann. N.Y. Acad. Sci. **603.** (In press.)

17. RUDOLPHI, K. A., M. KEIL & H. J. HINZE. 1987. Effect of theophylline on ischaemically induced hippocampal damage in Mongolian gerbils: a behavioural and histopathological study. J. Cereb. Blood Flow Metab. **7:** 74–81.

18. RUDOLPHI, K. A., M. KEIL, J. FASTBOM & B. B. FREDHOLM. 1989. Ischaemic damage in gerbil hippocampus is reduced following upregulation of adenosine (A_1) receptor by caffeine treatment. Neurosci. Lett. **103:** 275–280.

19. VAN CALKER, D., M. MULLER & B. HAMPRECHT. 1979. Adenosine regulates via two different types of receptors, the accumulation of cyclic AMP in cultured brain cells. J. Neurochem. **33:** 999–1005.

20. LONDOS, C., D. M. F. COOPER & J. WOLFF. 1980. Subclasses of external adenosine receptors. Proc. Nat. Acad. Sci. USA. **77:** 2551–2554.

21. FREDHOLM, B. B. 1982. Adenosine receptors. Med. Biol. **60:** 289–293.

22. DUNWIDDIE, T. V. & B. B. FREDHOLM. 1989. Adenosine A_1-receptors inhibit adenylate cyclase activity, inhibit neurotransmitter release, and hyperpolarize pyramidal neurons in rat hippocampus. J. Pharmacol. Exp. Ther. **249:** 31–37.

23. BRUNS, R. F., J. W. DALY & S. H. SNYDER. 1980. Adenosine receptors in brain membranes: binding of N6-cyclohexyl[3H]adenosine and 1,3-diethyl-8-[3H] phenylxanthine. Proc Nat. Acad. Sci. USA **77:** 5547–5551.

24. SCHWABE, U. & T. TROST. 1980. Characterization of adenosine receptors in rab brain by [^{3}H]-N6-phenylisopropyladenosine. Naunyn Schmiedebergs Arch. Pharmacol. **313:** 179–187.

25. BRUNS, R. F., R. E. DAVIS, F. W. NINTEMAN, B. P. H. POSCHEL, J. N. WILEY & T. G. HEFFNER. 1988. Adenosine agonists as pharmacological tools. *In* Adenosine and Adenine Nucleotides. D. M. Paton, Ed.: 39–49. Taylor & Francis. London England.

26. GOODMAN, R. R., M. J. COOPER, M. GAVISH & S. H. SNYDER. 1982. Guanine nucleotide and cation regulation of the binding of [3H]cyclohexyladenosine and [3H]diethyl-

phenylxanthine to adenosine A1 receptors in brain membranes. Mol. Pharmacol. **21:** 329–335.

27. LOHSE, M. J., V. LENSCHOW & U. SCHWABE. 1984. Two affinity states of Ri adenosine receptors in brain membranes. Analysis of gaunine nucleotides and temperature effects on radioligand binding. Mol. Pharmacol. **26:** 1–9.

28. UKENA, D., E. POESCULA, E. HÜTTEMANN & U. SCHWABE. 1984. Effects of N-ethylmaleimide on adenosine receptors of rat fat cells and human platelets. Naunyn Schmiedebergs Arch. Pharmacol. **327:** 247–253.

29. FREDHOLM, B. B., E. LINDGREN & K. LINDSTRÖM. 1985. Treatment with N-ethylmaleimide selectively reduces adenosine receptor mediated decreases in cyclic AMP accumulation in rat hippocampal slices. Br. J. Pharmacol. **86:** 509–513.

30. FREDHOLM, B. B. & E. LINDGREN. 1987. Effects of N-ethylmaleimide and forskolin on noradrenaline release from rat hippocampal slices. Evidence that prejunctional adenosine and α-receptors are linked to N-proteins, but not to adenylate cyclase. Acta Physiol. Scand. **130:** 95–105.

31. FREDHOLM, B. B., W. PROCTOR. I. VAN DER PLOEG & T. V. DUNWIDDIE. 1989. In vivo pertussis toxin treatment attenuates some, but not all, adenosine A_1 effects in slices of the rat hippocampus. Eur. J. Pharmacol. Mol. Pharmacol. **172:** 249–262.

32. TRUSSELL, L. O. & M. B. JACKSON. 1989. Potassium currents coupled to an A_1-receptor by a G-protein cultured central neurons. *In* Adenosine Receptors in the Nervous System. J. A. Ribeiro, Ed.: 177–185. Taylor and Francis. London, England.

33. GREENE, R. W. & H. L. HAAS. 1985. Adenosine action on CA1 pyramidal neurons in rat hippocampal slices. J. Physiol. London **366:** 119–127.

34. STARKE, K. 1987. Presynaptic α-autoreceptors. Rev. Physiol. Biochem. Pharmacol. **107:** 73–146.

35. STJÄRNE, L. 1989. Basic mechanisms and local modulation of nerve impulse–induced secretion of neurotransmitters from individual sympathetic nerve varicosities. Rev. Physiol. Biochem. Pharmacol. **112:** 1–137.

36. HU, P.-S. & B. B. FREDHOLM. 1989. 4-Aminopyridine blocks the inhibition of [³H]-noradrenaline release from rat hippocampus caused by an α_2-adrenoceptor agonist, but not that caused by an adenosine analogue or ω-conotoxin. Acta Physiol. Scand. **136:** 347–353.

37. FREDHOLM, B. B. 1990. Differential sensitivity to blockade by 4-aminopyridine of presynaptic receptors regulating [³H]-acetylcholine release from rat hippocampus. J. Neurochem. **54:** 1386–1390.

38. MILLER, R. J. 1987. Multiple calcium channels and neuronal function. Science **235:** 46–52.

39. HU, P.-S., E. LINDGREN, K. A. JACOBSON & B. B. FREDHOLM. 1987. Interaction of dihydropyridine calcium channel agonists and antagonists with adenosine receptors. Pharmacol Toxicol. **61:** 121–125.

40. FREDHOLM, B. B., P.-S. HU & E. LINDGREN. 1986. The dihydropyridine calcium-channel agonist BayK 8644 inhibits the presynaptic effects of R-phenylisopropyl adenosine in the rat hippocampus. Acta Physiol. Scand. **128:** 659–660.

41. RANE, S. G. & K. DUNLAP. 1986. Kinase C activator, 1,2-oleoylacetyl-glycerol attenuates voltage-dependent calcium current in sensory neurons. Proc. Nat. Acad. Sci. USA **83:** 184–188.

42. RANE, S. G., M. P. WALSH, J. R. MCDONALD & K. DUNLAP. 1989. Specific inhibitors of protein kinase C block transmitter-induced modulation of sensory neuron calcium current. Neuron **3:** 239–245.

43. FREDHOLM, B. B. & E. LINDGREN. 1988. Protein kinase C activation increases noradrenaline release from rat hippocampus and modifies the inhibitory effect of α_2-adrenoceptor and A_1-receptor agonists. Naunyn Schmiedebergs Arch. Pharmacol. **337:** 477–483.

44. DEKKER, L. V., P. N. E. DE GRAAN, A. B. OESTREICHER, D. H. G. VERSTEEG & W. H. GISPEN. 1989. Inhibition of noradrenaline release by antibodies to B-50 (GAP-43). Nature **342:** 74–76.

Regulatory Proteins in Presynaptic Function

GEORG HERTTING, SIEGFRIED WURSTER, AND
CLEMENS ALLGAIER

Institute of Pharmacology
University of Freiburg
D-7800 Freiburg, Federal Republic of Germany

INTRODUCTION

Neurotransmitter release evoked by electrical field stimulation closely resembles action potential–induced transmitter release in that it is tetrodotoxin sensitive and Ca^{2+} dependent. The present study deals exclusively with receptor-mediated modulation of stimulation-induced noradrenaline release from slices or synaptosomes of rabbit hippocampus or rat brain cortex. The cell bodies of these noradrenergic neurons are located in the locus ceruleus; thus, modulation of transmitter release occurs either at the level of axons or at nerve terminals of which the latter is most probable. Noradrenergic nerve terminals of the rabbit hippocampus are endowed with α_2-autoreceptors[1]—activation of which decreases the evoked release in a concentration-dependent manner of the transmitter in the synaptic cleft—and with κ-opioid[2,3] and A_1-adenosine receptors[4] which both influence transmitter release similar to that of the α_2-autoreceptors. In addition, noradrenergic nerve terminals of rat brain cortex possess inhibitory prostaglandin E (PGE) receptors (FIGURE 1).[5–9]

The transduction of extracellular signals to specific intracellular events often involves the participation of regulatory GTP-binding proteins (G proteins) coupled to enzymes or ionic channels.[10–12] G proteins are heterotrimers sharing common β- and γ-subunits; their α-subunits which bind GTP and possess GTPase activity differ in the peptide sequences. The α-subunits of some G proteins, e.g., G_i and G_o, can be specifically ADP ribosylated by Pertussis toxin (PTX), thereby inhibiting signal transduction.[13,14] The α-subunits of PTX-sensitive G proteins can also be inactivated by N-ethylmaleimide (NEM) through the alkylation of the SH group found at or near the site that is ADP ribosylated by PTX.[15–19]

The intention of this paper is to discuss (a) whether the presynaptic α_2-autoreceptors and heteroreceptors (κ-opioid, A_1-adenosine, PGE), which both inhibit evoked noradrenaline release, are linked to PTX-sensitive GTP-binding proteins, and (b) whether these receptors operate via a common signal transduction pathway.

INVOLVEMENT OF PTX-SENSITIVE G PROTEINS IN RECEPTOR-MEDIATED MODULATION OF NORADRENALINE RELEASE: INTERACTION OF SIGNAL TRANSDUCTION

When membranes of synaptosomes, prepared from hippocampal tissue, were subjected to PTX-catalyzed ^{32}P-ADP ribosylation, the autoradiographs of sodium dodecyl sulfate (SDS) gels revealed three bands of polypeptides with molecular weights (FIGURE 2) corresponding to the α-subunits of G_o (39 kDa) and the G_i proteins

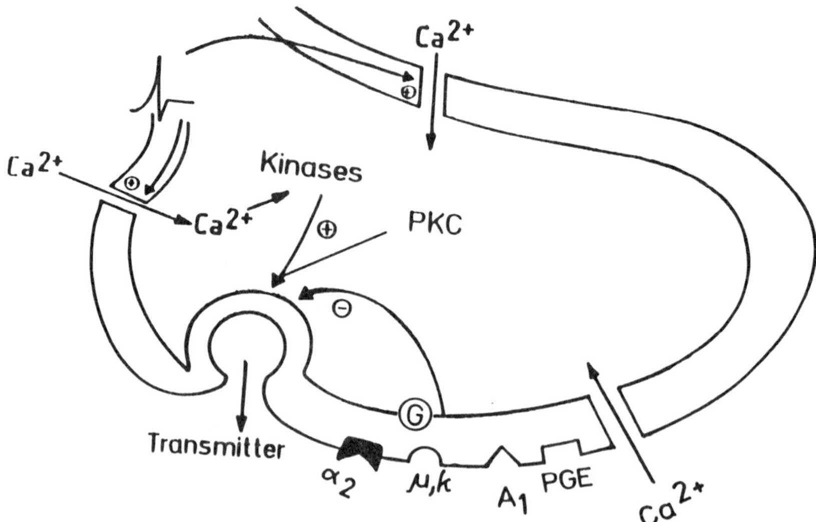

FIGURE 1. Scheme of the primary mechanisms presumed to be involved in noradrenaline release. Depolarization of the nerve terminal membrane triggers off an influx of Ca^{2+} through voltage-dependent Ca^{2+} channels and a subsequent transmitter release. The specific intracellular events following Ca^{2+} influx are still under investigation. The involvement of protein kinases, such as the cAMP-dependent kinase and the protein kinase C (PKC), has been demonstrated previously. The noradrenergic nerve terminals are endowed with certain receptors, activation of which inhibits depolarization-induced release of noradrenaline (for detail, see text).

(40, 41 kDa).[20,21] The labeling of the 39 kDa polypeptide was most prominent. Pretreatment of the slices with NEM (30 minutes) reduced ADP ribosylation of each polypeptide in a concentration-dependent manner. At an NEM concentration of 30 μM, the ^{32}P labeling was decreased by about 60%. A concentration of 30 μM was chosen in experiments on brain slices where NEM was used instead of PTX in order to avoid larger increases of basal outflow seen at higher NEM concentrations. Recently, a fourth PTX-sensitive G protein has been found in bovine brain.[22]

Presynaptic α₂-Autoreceptors

Indirect evidence for the participation of PTX-sensitive G proteins in the modulation of noradrenaline release by α_2-adrenoceptors was first provided by experiments performed on guinea pig vas deferens.[23] PTX pretreatment diminished the inhibitory effect of the α_2-adrenoceptor agonist clonidine on the contraction of electrically stimulated vas deferens, whereas PTX did not affect the contractions induced by exogenous noradrenaline. In contrast, some experiments on peripheral tissues, e.g., heart, blood vessels, or rat vas deferens, failed to show any effects of PTX treatment on the control of noradrenaline release by presynaptic α_2-adrenoceptors.[24–26] However, PTX in these experiments was effective in abolishing events mediated by postsynaptic α_2-adrenoceptors and muscarine receptors. On the other hand, NEM experiments on rat tail artery and mouse vas deferens demonstrated an attenuation of α_2-adrenoceptor agonist–induced inhibition of [^{3}H]noradrenaline release by the SH reagent.[27,28] These discrepancies may be due to NEM's higher membrane permeability and to the fact that PTX's special binding site needed to penetrate membranes may not be present at some sympathetic nerve terminals.

In the following experiments performed on brain slices, the tissue prelabeled with [3H]noradrenaline was continuously superfused with physiological medium in the presence of a reuptake inhibitor. The electrically evoked release of [3H]noradrenaline represents the difference between total and basal radioactivity in the superfusate. The release of [3H]noradrenaline was expressed as a percentage of the radioactivity in the slice at the onset of each respective stimulation period (% S_1; %S_2).[29]

In PTX experiments, the brain slices were pretreated with PTX in sterile medium gassed with 95% O_2/5% CO_2 at 37°C before ^{3}H labeling. PTX increased the [3H]noradrenaline release evoked by 360 pulses/3 Hz in a time- and concentration-dependent manner;[18,19,30,31] at 8 μg/ml and an incubation period of 18 hours, an increase of approximately threefold was seen (FIGURE 3). In order to investigate the influence of the α_2-adrenoceptor antagonist yohimbine and the α_2-adrenoceptor agonist clonidine on [3H]noradrenaline release after treatment with PTX, the slices were stimulated twice (S_1, S_2). The drugs were added to the medium 15 minutes before the second stimulation. Their effects were evaluated by the ratio of [3H]noradrenaline release evoked by S_2 and by S_1 (S_2/S_1). The ratio S_2/S_1 of drug-free controls was about 1.0. In control slices, the increase of the evoked transmitter release by yohimbine revealed a pronounced autoinhibition (FIGURE 4).[18,19,30,31] When evoked [3H]noradrenaline release was enhanced by pretreatment with PTX (8 μg/ml; 18 hours), the facilitatory effect of yohimbine was markedly reduced and the inhibition of release caused by clonidine was almost abolished (FIGURE 4).[18,19,30,31] Also, in rat brain cortex, PTX suppressed the clonidine-induced inhibition of [3H]noradrenaline release caused by K^+-depolarization.[32] These PTX-results on the α_2-receptor-mediated release modulation are comparable to those obtained using NEM.[33,34] From receptor protec-

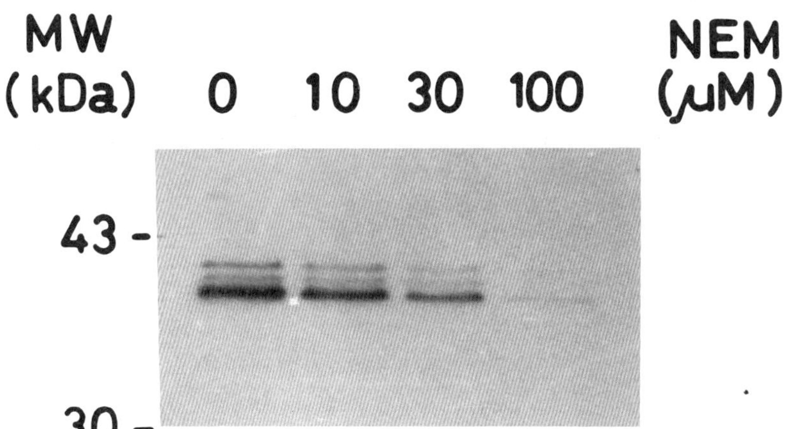

FIGURE 2. Influence of NEM on the PTX-catalyzed [32P]ADP ribosylation of synaptosomal membrane proteins of rabbit hippocampus. Synaptosomes from hippocampal tissue, incubated for 30 minutes in normal or NEM-containing medium, were prepared by Percoll gradient centrifugation.[60] Gradient fractions three and four were washed in a Krebs-Ringer solution and resuspended in 50 mM Tris-HCl, 1 mM Na$_2$EDTA, pH 8.0; 60 μg of protein were subjected to PTX-induced [32P]ADP ribosylation, sodium dodecyl sulfate (SDS)–polyacrylamide gel electrophoresis and autoradiography.[61]

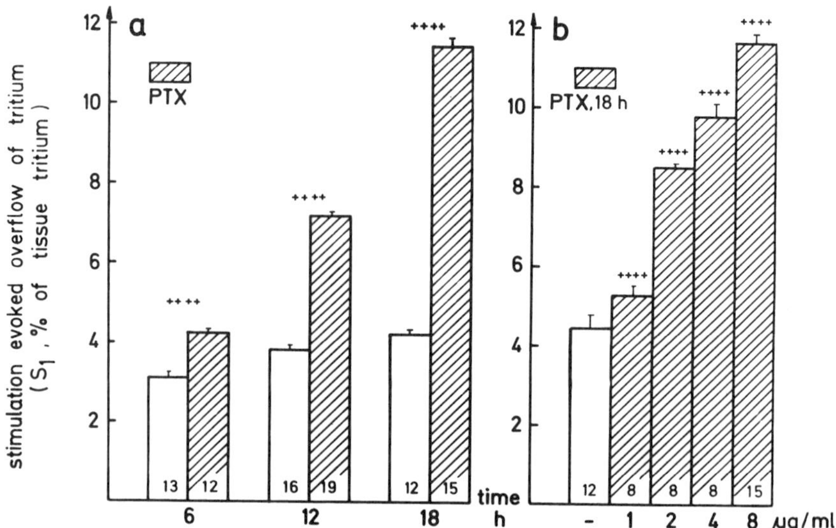

FIGURE 3. Time (a; PTX 8 μg/ml) and concentration (b; 18 hours) dependence of the effect of PTX on electrically evoked [^{3}H]noradrenaline release from slices of rabbit hippocampus. Slices were preincubated at 37°C in gassed medium (95% O_2/5% CO_2) containing PTX (hatched columns) or free of PTX (open columns), followed by labeling with [^{3}H]noradrenaline (0.1 μM; 30 minutes).[31] After rinsing, the slices were superfused (flow rate, 1 ml/minute); electrically evoked stimulation (360 rectangular pulses of 2 mseconds duration at 3 Hz, 24 mA, 5 V/cm) occurred after 60 minutes of superfusion. Reuptake of noradrenaline was inhibited by 30 μM cocaine. The stimulation-evoked overflows (expressed as a percentage of the radioactivity in the tissue at the onset of stimulation) ± standard error of the mean (SEM) are given. Numbers of experiments are found in the columns. Significant differences between control and PTX-treated slices: ****$p < 0.0001$.

tion experiments, it can be excluded that NEM directly acted on the active center of the α_2-receptors.[33] These aforementioned results indicate that PTX-sensitive G proteins are involved in the α_2-autoreceptor-mediated inhibition of evoked noradrenaline release in brain tissue.

As evident from FIGURE 5, the impairment of autoinhibition caused by PTX was more pronounced than by NEM and the effects of PTX and NEM on evoked [^{3}H]noradrenaline release were not additive. Due to the nonadditive results of NEM and PTX, it can be assumed that both drugs affect the same mechanism. In conclusion, the release data, as well as the ribosylation experiments, prove NEM as a valuable tool for exploring the role of G proteins in neurotransmitter release.

FIGURE 6 shows the effects of yohimbine and/or NEM on [^{3}H] noradrenaline release. When [^{3}H]noradrenaline release was evoked with 360 pulses/3 Hz, autoinhibition was operating; whereas, [^{3}H]noradrenaline release evoked with 4 pulses/100 Hz gives no autoinhibition due to insufficient time for the released noradrenaline to activate the α_2-autoreceptors.[35–37] In the presence of autoinhibition, NEM increased the evoked release twofold and yohimbine by about fivefold. The effects of both drugs were not additive. When stimulation was carried out with 4 pulses/100 Hz, NEM and yohimbine only negligibly increased transmitter release. These results indicate that both drugs influence the α_2-autoreceptor-mediated release inhibition either by blocking the receptor (yohimbine) or by alkylating a regulatory G protein (NEM).

Presynaptic Heteroreceptors

Opioid Receptors

Noradrenergic nerve terminals of rabbit hippocampus possess inhibitory opioid receptors which have been characterized as belonging to the κ subtype.[2,3] The preferential κ-opioid receptor agonist ethylketocyclazocine (EKC) diminished the evoked [^{3}H]noradrenaline release in a concentration-dependent manner (FIGURE 7A). When autoinhibition was prevented by yohimbine, the EKC-induced inhibition of [^{3}H]noradrenaline release was more pronounced, indicating that both the α_2- and the κ-opioid receptors share a common postreceptor signal transduction pathway.[31] A similar interaction between α_2- and μ-opioid receptors has also been demonstrated on rabbit brain cortex.[38]

Pretreatment of rabbit hippocampal tissue with PTX attenuated markedly the effect of EKC in the presence or absence of autoinhibition (FIGURE 8).[31] Similar effects on the κ-receptor-mediated inhibition of [^{3}H]noradrenaline release were obtained with NEM,[31] suggesting that this event is also mediated by G protein. The functional interaction of α_2- and κ-receptors as well as the coupling of both receptors to PTX-sensitive G proteins would be compatible with the idea that these receptors may compete for a common pool of G proteins. However, differences in the coupling of

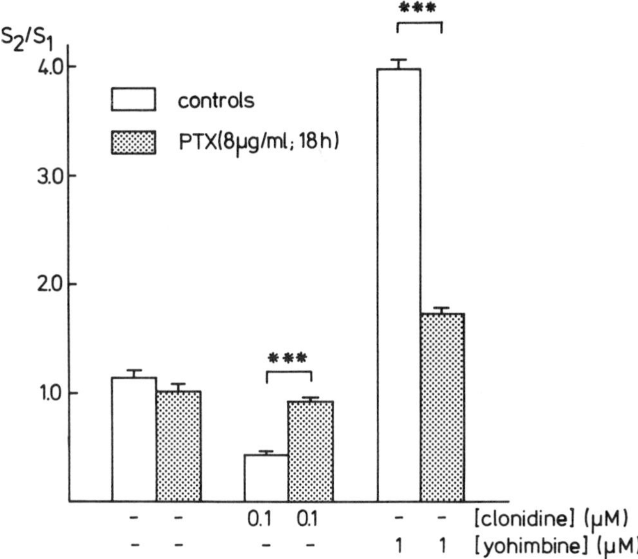

FIGURE 4. PTX impairment of α_2-adrenoceptor-mediated modulation of [^{3}H]noradrenaline release from rabbit hippocampal slices. Slices were preincubated (PTX; 8 μg/ml; 18 hours) and labeled with [^{3}H]noradrenaline as described in the legend of FIGURE 3. The tissue was electrically stimulated twice (360 pulses/3 Hz) during superfusion with medium containing 30 μM cocaine. Clonidine (0.1 μM) or yohimbine (1 μM) was added to the superfusion medium 15 minutes prior to S_2. S_1 $\pm$ SEM (in percent of the radioactivity in the tissue at the onset of stimulation) of controls was 4.42 $\pm$ 0.39% ($n = 8$) and of PTX-treated slices it was 10.79 $\pm$ 0.39 ($n = 8$). Means of S_2/S_1 $\pm$ SEM of 4–8 experiments per group are given. Significant differences between control and PTX-treated slices: ***$p < 0.001$.

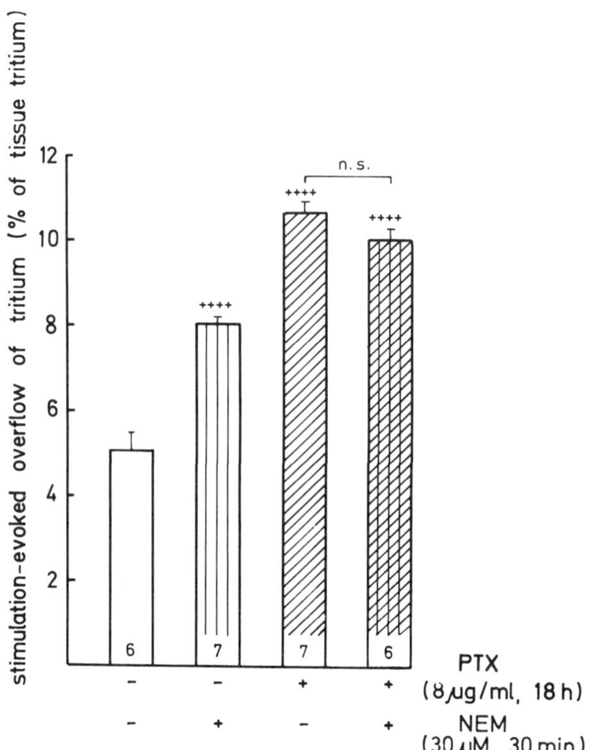

FIGURE 5. Effects of NEM and/or PTX on electrically evoked [³H]noradrenaline release from slices of rabbit hippocampus. An 18 hour incubation in control medium or medium containing PTX was followed by a 30 minute incubation in the absence or presence of NEM. After labeling with [³H]noradrenaline, slices were superfused with medium (30 μM cocaine) and stimulated electrically with 360 pulses/3 Hz after 60 minutes of superfusion. The stimulation-evoked overflows ± SEM are given. Numbers of experiments are given in the columns; significant differences between control slices and slices treated with PTX and/or NEM are given.

presynaptic κ-receptors to PTX-sensitive G proteins may vary according to species. It has been shown in a recent paper that, in guinea pig hippocampus, PTX inhibited the effects of μ- and δ-receptor agonists, but not that of κ agonists on K⁺-evoked [³H]noradrenaline release.[39] Similarly, the κ-receptor-mediated inhibition of [³H]dopamine release was not changed by PTX in rat striatum and cortex.

Adenosine Receptors

In rabbit hippocampus, endogenous adenosine permanently inhibited noradrenaline release via A_1-adenosine receptors from 10 to 15%.[4] (−)Phenylisopropyladenosine [(−)PIA], a preferential A_1 agonist, inhibited the evoked [³H]noradrenaline release in a concentration-dependent manner (FIGURE 7B). The A_1-adenosine receptor–mediated inhibition also depended on the extent of autoinhibition induced by released noradrenaline (FIGURE 7B). This indicates a similar interaction between the A_1- and

α_2-receptors as obtained for the κ- and α_2-receptors. NEM pretreatment significantly decreased the inhibitory effect of $(-)$PIA (FIGURE 9).[29] Thus, it can be concluded that A_1-adenosine receptors, located at noradrenergic nerve terminals, employ a common pathway involving PTX-sensitive G proteins similar to that of the α_2- and κ-receptors. Participation of G proteins in the adenosine receptor–mediated modulation of glutamate release from cultured cerebellar rat neurons and noradrenaline release from rat hippocampus has been additionally demonstrated.[34,40–42] Further experiments in which PTX was injected into rat hippocampus in order to obtain high local concentrations have shown prevention of both the inhibitory pre- and postsynaptic effects of adenosine on synaptic transmission and the impairment of the PTX-catalyzed $[^{32}P]$ADP ribosylation of G proteins.[43] In contrast, in experiments where only lower tissue concentrations of PTX were achieved by intracerebroventricular administration, solely the postsynaptic effects of adenosine were inhibited.[44] Apparently, higher local concentrations of PTX are necessary to exhibit effects on the presynaptic than on the postsynaptic site of the synapse. This may explain some failures to show any presynaptic effects of PTX on peripheral structures.[24–26]

The K^+-induced $[^3H]$noradrenaline release from rabbit synaptosomes was used to obtain functional data confirming a presynaptic localization of PTX-sensitive G

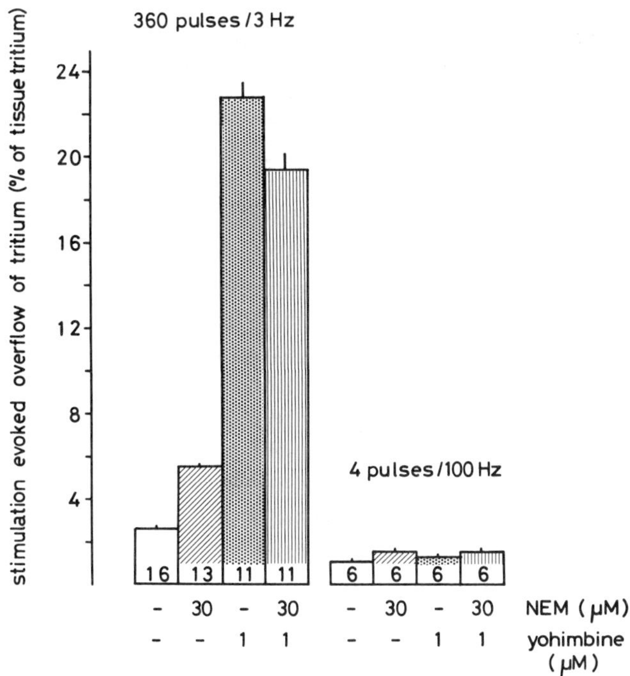

FIGURE 6. Effects of NEM and/or yohimbine on electrically evoked $[^3H]$noradrenaline release from rabbit hippocampal slices. Following a 30 minute preincubation in medium with or without NEM, slices were labeled with $[^3H]$noradrenaline, superfused in the presence of cocaine, and stimulated electrically, as indicated. In some experiments yohimbine (1 μM) was present throughout. The stimulation-evoked overflows ± SEM are given. Numbers of experiments are found in the columns.

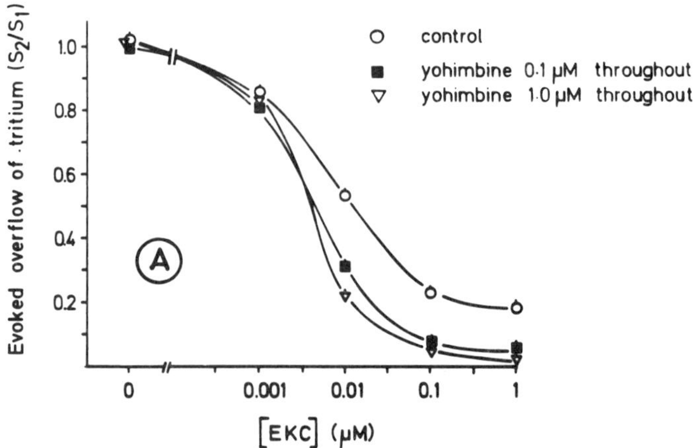

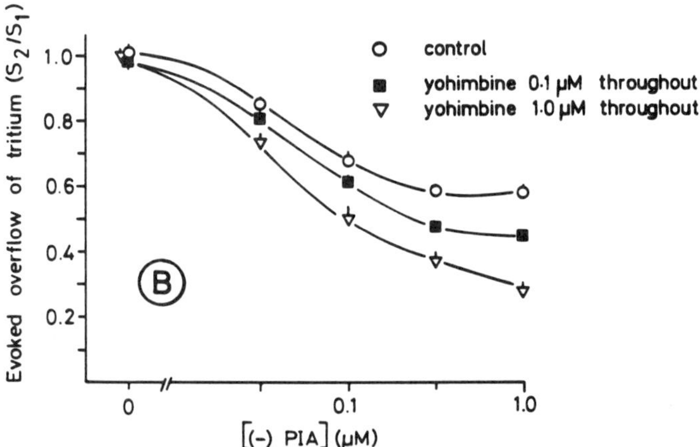

FIGURE 7. Influence of yohimbine on effects of EKC (A) and $(-)$ PIA (B) on electrically evoked [³H]noradrenaline release from rabbit hippocampal slices. Tissue slices, prelabeled with [³H]noradrenaline, were superfused and stimulated twice $(S_1; S_2)$ with 360 pulses/3 Hz. EKC or $(-)$PIA was added to the superfusion medium 15 minutes before second stimulation. Yohimbine $(0.1, 1.0\ \mu M)$ was present throughout superfusion. $S_1 \pm$ SEM of controls (open circles) was $3.54 \pm 0.54\%$ ($n = 30$) (A) and 3.84 ± 0.07 ($n = 44$) (B). In the presence of 0.1 μM yohimbine, S_1 was $11.79 \pm 0.41\%$ ($n = 38$) (A) and 12.92 ± 0.57 (B). In the presence of 1 μM yohimbine, S_1 was $21.54 \pm 0.46\%$ ($n = 25$) (A) or 21.39 ± 0.40 ($n = 19$) (B). Means of $S_2/S_1 \pm$ SEM of 4–10 experiments/group are given.

proteins. [³H]Noradrenaline release, free of autoinhibition,[45] was accomplished by exposing a thin layer of synaptosomes to 15 mM K^+/75 μM Ca^{2+}.[46] Under these conditions, clonidine, EKC, and $(-)$PIA inhibited depolarization-induced [³H]noradrenaline release in a concentration-dependent manner. NEM treatment of the synaptosomes ($2\ \mu M$, 15 minutes) reduced the subsequent PTX-catalyzed [³²P]ADP ribosylation of the three PTX substrates (FIGURE 2) by about 40% (data not shown) and also markedly diminished the inhibitory effects of all three agonists without affecting their EC_{50} values(FIGURE 10). These results provide direct evidence for the presynaptic localization of G proteins involved in receptor-mediated modulation of stimulation-induced [^{3}H]noradrenaline release.[47] Release-modulating muscarine receptors, κ-opioid receptors, and A_1-adenosine receptors, all of which are located at cholinergic nerve endings in rabbit hippocampus, appear to be similarly coupled to PTX-sensitive G proteins.[19,48]

PGE Receptors

Release-inhibiting prostaglandin E receptors have been found on noradrenergic nerve terminals of rat brain.[5-9] The stimulation parameters of the rat tissue experiments were 36 pulses/3 Hz. Under these conditions autoinhibition was fully developed; therefore, α_2-adrenoceptor blockade with yohimbine increased the evoked transmitter release by about 220% (legend to FIGURE 11). Similar to A_1- or κ-receptors, the inhibitory effect of PGE_2 on [^{3}H]noradrenaline release was markedly increased in the absence of autoinhibition (36 pulses/3 Hz and yohimbine or 4 pulses/100 Hz) (FIGURE 11).[49] These data suggest that PGE receptors share a common pathway with α_2-receptors, as formerly shown for A_1- and κ-opioid receptors.

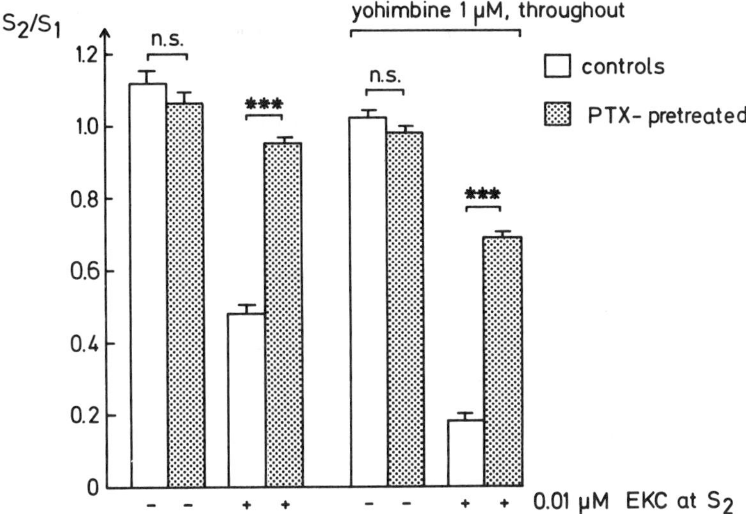

FIGURE 8. Effects of PTX on κ-opioid receptor–coupled inhibition of [^{3}H]noradrenaline release. Slices were firstly preincubated in normal medium (open columns) or in medium containing PTX (8 μg/ml, 18 hours; dotted columns), then labeled with [^{3}H]noradrenaline, superfused, and stimulated twice electrically (360 pulses/3 Hz). EKC was added 15 minutes before S_2. In some experiments, 1 μM yohimbine was present throughout. S_1 $\pm$ SEM of PTX-untreated slices was 4.10 $\pm$ 0.14% ($n = 18$), and of PTX-treated slices it was 8.98 $\pm$ 0.24% ($n = 15$). In the presence of yohimbine, S_1 $\pm$ SEM of PTX-untreated slices was 23.89 $\pm$ 0.87% ($n = 11$) and 20.00 $\pm$ 0.44% ($n = 15$). Means of S_2/S_1 $\pm$ SEM of 6–10 experiments per group are depicted. Significant differences between control and PTX-treated slices: ***$p < 0.001$.

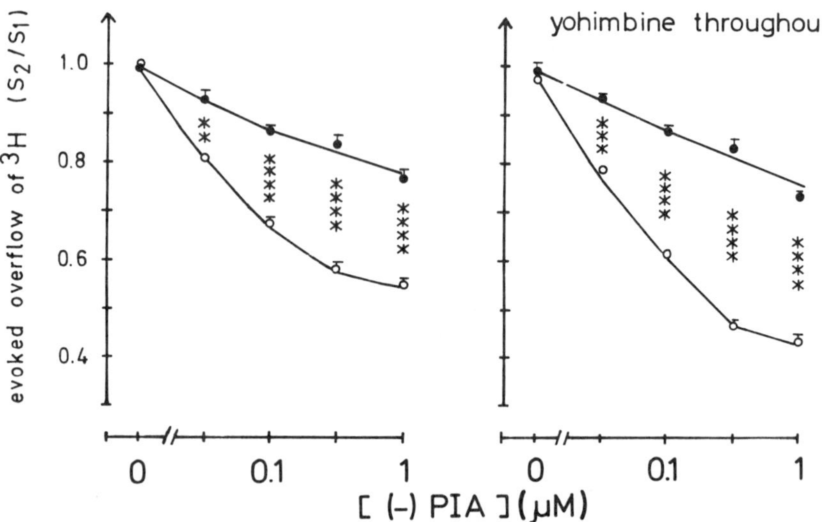

FIGURE 9. Influence of NEM on the effect of ($-$) PIA on electrically evoked [^{3}H]noradrenaline release from rabbit hippocampal slices. Following a 30 minute preincubation in NEM-free or NEM-containing (30 μM) medium, slices were loaded with [^{3}H]noradrenaline (0.1 μM). Subsequently the slices were superfused in the presence of cocaine and stimulated with 360 pulses/3 Hz. ($-$) PIA was added to the medium 15 minutes before second stimulation. In some experiments 0.1 μM yohimbine was present throughout (right panel). $S_1 \pm$ SEM was 3.84 $\pm$ 0.07% ($n = 44$) for the control slices (open circles) and 6.02 $\pm$ 0.13% for the NEM-treated slices (filled circles). When yohimbine was present, S_1 was 12.92 $\pm$ 0.57 ($n = 23$) in control slices and 12.10 $\pm$ 0.33 ($n = 27$) in NEM-treated slices. Numbers of experiments, 3–8. Significant differences between untreated and NEM-treated slices: **$p < 0.01$; ***$p < 0.001$; ****$p < 0.0001$.

FIGURE 12 shows the influence of NEM pretreatment on the inhibition of noradrenaline release in absence of autoinhibition (4 pulses/100 Hz) by clonidine and PGE$_2$. Under these experimental conditions, clonidine (0.1 μM) and PGE$_2$ (10 μM) similarly inhibited [^{3}H]noradrenaline release by about 80%. NEM pretreatment attenuated markedly the inhibition of [^{3}H]noradrenaline release caused by clonidine, but did not affect the inhibition of release caused by PGE$_2$ (FIGURE 12). Thus in rat brain cortex, the involvement of a PTX- and NEM-sensitive G protein occurs in signal transduction of α_2-receptors which is consistent with the rabbit experiments. However, a participation of a PTX-sensitive G protein on the PGE$_2$ response can be excluded. Since the effect of PGE$_2$ in rat brain (FIGURE 11) is similarly impaired by endogenous noradrenaline as the effects of EKC and ($-$)PIA in rabbit brain tissue (FIGURE 7), a common pathway for the α_2-autoreceptors and the PGE heteroreceptors is suggested. Since NEM did not affect the PGE response, it appears that the interaction between the α_2-receptors and the PGE receptors does not occur at the level of a G protein but at some subsequent step of the signal transduction mechanism, possibly at common ion channels. This may also be the case for the interaction between the α_2-autoreceptors and the κ-opioid and A$_1$-adenosine receptors in rabbit hippocampus.

At present, the postreceptor mechanisms participating on the modulation of noradrenaline release by presynaptic receptors are uncertain.[50–52] Discussions include enzymes, such as the cAMP-dependent protein kinase and protein kinase C, as well as a direct coupling to various ion channels.[53,54] Results from rat brain tissue concerning an interaction between protein kinase C activation and receptor-mediated inhibition of release are contradictory.[55-57] In rabbit hippocampus, however, the phorbol ester–induced enhancement of noradrenaline release was fully additive to the facilitatory effect of the α_2-adrenoceptor antagonist yohimbine.[58,59] The interaction of cyclic nucleotides and presynaptic receptors will be dealt with in detail in this volume by several authors. Some recent data suggest a coupling of presynaptic α_2-receptors via a G protein with N-type voltage-sensitive Ca^{2+} channels without affecting adenylate cyclase.[54]

SUMMARY AND CONCLUSION

Activation of α_2-adrenoceptors, opioid, A_1-adenosine, and PGE receptors inhibited the stimulation-induced [^{3}H]noradrenaline release in brain tissue in a concentration-

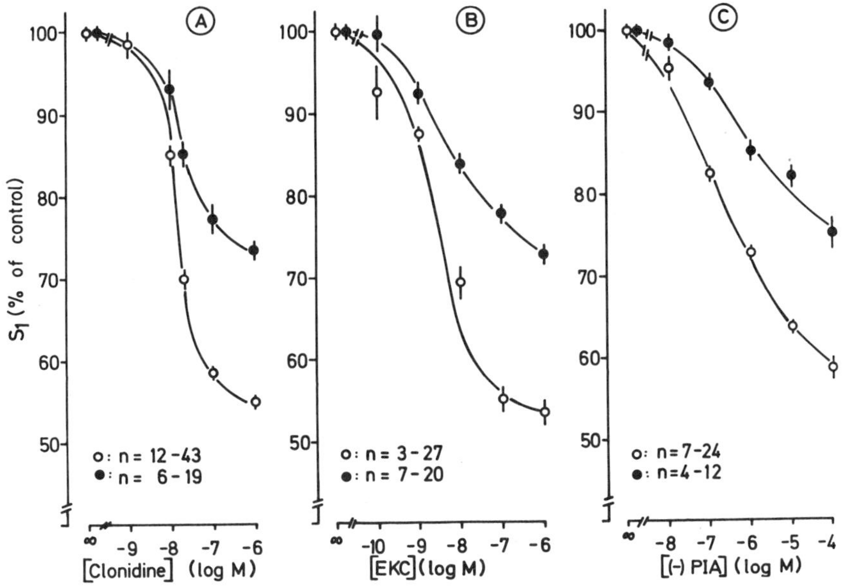

FIGURE 10. Agonist effects on potassium-evoked [^{3}H]noradrenaline release from rabbit hippocampal synaptosomes treated with NEM. P_2 pellet from rabbit hippocampus was resuspended and incubated for 10 minutes in Ca^{2+}-free Krebs-Ringer solution containing 0.1 μM [^{3}H]noradrenaline. Synaptosomes were continuously superfused at a rate of 0.8 ml/minute and treated with 2 μM NEM for 15 minutes during superfusion. Depolarization-induced release was accomplished after 60 minutes by exposing the synaptosomes to 15 mM K^+/75 μM Ca^{2+} for three minutes. Clonidine (A), EKC (B), or (−)PIA (C) was added 10 minutes before depolarization. Radioactivity remaining in the synaptosomes at the end of the experiment was extracted with 1% Triton TX-100. Agonists effects were expressed as a percentage of evoked release of NEM-free or NEM-treated controls. The evoked release was not influenced by NEM and amounted to about 6.5% of tissue tritium. Significant differences between NEM-treated (filled circles) and untreated (open circles) groups: not significant for 10^{-10} M EKC and 10^{-8} M (−)PIA; $p < 0.01$ for 10^{-8} M clonidine and 10^{-9} M EKC; $p < 0.001$ for all other agonist concentrations.

dependent manner. Under experimental conditions (360 pulses/3 Hz) where the released noradrenaline activated the presynaptic α_2-autoreceptors, the effects of the heteroreceptor (κ-opioid, A_1-adenosine, PGE) agonists were decreased. By avoiding autoinhibition by either blockade of the α_2-autoreceptors with yohimbine or stimulating the tissue with four pulses/100 Hz, the heteroreceptor-mediated inhibition of [^{3}H]noradrenaline release was markedly increased. The dependence of the heteroreceptor-mediated inhibition of evoked noradrenaline release on the extent of α_2-autoreceptor activation suggests a common postreceptor signal transduction pathway.

PTX-catalyzed [^{32}P]ADP ribosylation of synaptosomal membrane proteins revealed three bands of polypeptides with molecular weights corresponding to the α subunits of G_o (39,000) and the G_i proteins (40,000, 41,000). Pretreatment with NEM reduced the PTX-induced ^{32}P labeling by alkylating the α subunits at or near the site that is ADP ribosylated by PTX in a concentration-dependent manner. K^+-evoked release of [^{3}H]noradrenaline from synaptosomes indicated the presynaptic localization of the PTX-sensitive G proteins coupled to α_2-, κ-, and A_1-receptors of noradrenergic nerve terminals.

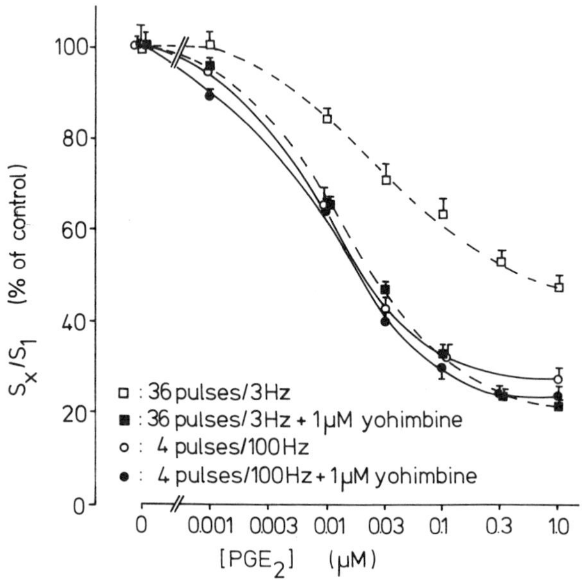

FIGURE 11. Effects of PGE$_2$ on electrically evoked [^{3}H]noradrenaline release from rat brain cortex slices. After ^{3}H labeling, the slices were superfused with medium containing 1 µM oxaprotiline and stimulated electrically (S_1, S_2, S_3, S_4) using either 36 pulses/3 Hz (dashed lines) or 4 pulses/100 Hz (solid lines). PGE$_2$ was present in increasing concentrations from 16 minutes before stimulation until 16 minutes after the onset of S_2, S_3, S_4, respectively. In some experiments, 1 µM yohimbine was present throughout (filled symbols). When stimulation was performed with 36 pulses/3 Hz S_1 ± SEM was 1.00 ± 0.55% ($n = 24$); in the presence of yohimbine it was 3.16 ± 0.11 ($n = 12$). When stimulation was carried out with 4 pulses/100 Hz S_1 ± SEM was 1.025 ± 0.08% ($n = 35$); in the presence of yohimbine it was 1.30 ± 0.07% ($n = 20$). Means of S_x/S_1 (S_2/S_1, S_3/S_1, S_4/S_1) ± SEM of 6–12 observations are given. The ratios obtained in the presence of PGE$_2$ were expressed as a percentage of the mean ratios of the respective drug-free controls. Control S_x/S_1 ratios did not significantly differ from each other. They were about 1.10.

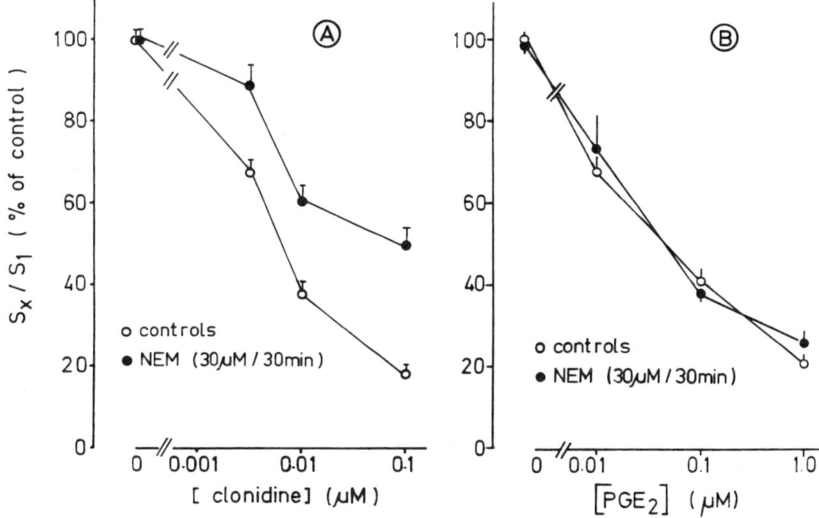

FIGURE 12. Influence of NEM on clonidine (A) and PGE_2 (B) induced inhibition of electrically evoked [^{3}H]noradrenaline release from rat brain cortex slices. Following a 30 minute preincubation in normal medium or medium containing 30 μM NEM, slices were loaded with [^{3}H]noradrenaline and superfused in the presence of 1 μM oxaprotiline. Stimulation was carried out with 4 pulses/100 Hz (S_1, S_2, S_3, S_4). Clonidine and PGE_2 were present in increasing concentrations from 16 minutes before stimulation until 16 minutes after the onset of S_2, S_3, S_4, respectively. S_1 ± SEM of NEM-untreated slices wat 1.11 ± 0.55% ($n = 18$), S_1 ± SEM of NEM-treated slices was 1.81 ± 0.09 ($n = 18$). Means of S_x/S_1 (S_2/S_1, S_3/S_1, S_4/S_1) ± SEM of 6–10 observations are given. The ratios obtained in the presence of the agonists were expressed as a percentage of the mean ratios of the respective agonist-free controls. Control S_x/S_1 ratios did not significantly differ from each other.

Electrically evoked [^{3}H]noradrenaline release was only increased by PTX or NEM in a time- and concentration-dependent manner when autoinhibition was present. The α_2-, opioid, and A_1-adenosine receptor–mediated inhibition of [^{3}H]noradrenaline release was impaired similarly by PTX or NEM treatment. In contrast, the inhibitory effect of PGE_2 remained unaffected. These results indicate that presynaptic α_2-, opioid, and A_1-receptors but not PGE receptors of noradrenergic nerve terminals are linked to PTX-sensitive G proteins. The interaction between the α_2-autoreceptors and the PGE receptors therefore does not occur at the level of a common pool of G proteins but at some subsequent step of the signal transduction mechanism.

REFERENCES

1. JACKISCH, R., E. WERLE & G. HERTTING. 1984. Identification of mechanisms involved in the modulation of release of noradrenaline in the hippocampus of the rabbit in vitro. Neuropharmacology **23:** 1363–1371.
2. JACKISCH, R., M. GEPPERT & P. ILLES. 1986. Characterization of opioid receptors modulating noradrenaline release in the hippocampus of the rabbit. J. Neurochem. **46:** 1802–1810.
3. JACKISCH, R., M. GEPPERT, A. LUPP, H. Y. HUANG & P. ILLES. 1988. Types of opioid

receptors modulating neurotransmitter release in discrete brain regions. In Regulatory Roles of Opioid Peptides. P. Illes & C. Farsang, Eds.: 240–258. VCH Weinheim.

4. JACKISCH, R., R. FEHR & G. HERTTING. 1985. Adenosine: an endogenous modulator of hippocampal noradrenaline release. Neuropharmacology **24:** 499–507.

5. BERGSTROM, S., L. O. FARNEBO & K. FUXE. 1973. Effect of prostaglandin E_2 on central and peripheral catecholamine neurons. Eur. J. Pharmacol. **21:** 362–368.

6. TAUBE, H. D., K. STARKE & E. BOROWSKI. 1977. Presynaptic receptor systems on the noradrenergic neurons of rat brain. Naunyn Schmiedebergs Arch. Pharmacol. **299:** 123–141.

7. HILLIER, K. & W. W. TEMPLETON. 1980. Regulation of noradrenaline overflow in rat cerebral cortex by prostaglandin E_2. Br. J. Pharmacol. **70:** 469–473.

8. REIMANN, W., H. B. STEINHAUER, L. HEDLER, K. STARKE & G. HERTTING. 1981. Effect of prostaglandins D_2, E_2 and $F_{2\alpha}$ on catecholamine release from slices of rat and rabbit brain. Eur. J. Pharmacol. **69:** 421–427.

9. SCHLICKER, E., K. FINK & M. GÖTHERT. 1987. Influence of eicosanoids on serotonin release in the rat brain: inhibition by prostaglandins E_1 and E_2. Naunyn Schmiedebergs Arch. Pharmacol. **335:** 646–651.

10. BIRNBAUMER, L., J. CODINA, R. MATTERA, A. YATANI, N. SCHERER, M.-J. TORO & A. M. BROWN. 1987. Signal transduction by G proteins. Kidney Int. 32(Suppl. 23): 14–37.

11. GILMAN, A. G. 1987. G proteins: transducers of receptor-generated signals. Annu. Rev. Biochem. **56:** 615–649.

12. ROSS, E. M. 1989. Signal sorting and amplification through G protein–coupled receptors. Neuron **3:** 141–152.

13. STERNWEIS, P. C. & J. D. ROBISHAW. 1984. Isolation of two proteins with high affinity for guanine nucleotides from membranes of bovine brain. J. Biol. Chem. **259:** 13806–13813.

14. UI, M. 1984. Islet-activating protein, pertussis toxin: a probe for functions of the inhibitory guanine nucleotide regulatory component of adenylate cyclase. Trends Pharmacol. Sci. **5:** 277–279.

15. JAKOBS, K. H., K. AKTORIES & G. SCHULTZ 1984. Mechanisms and components involved in adenylate cyclase inhibition by hormones. Adv. Cyclic Nucleotide Res. **17:** 135–143.

16. ASANO, T. & N. OGASAWARA. 1986. Uncoupling of y-aminobutyric acid B receptors from GTP-binding proteins by N-ethylmaleimide: effect of N-ethylmaleimide on purified GTP-binding proteins. Mol. Pharmacol. **29:** 244–249.

17. NORDSTEDT, C. & B. B. FREDHOLM. 1987. Phorbol-12,13-dibutyrate enhances the cyclic AMP accumulation in rat hippocampal slices induced by adenosine analogues. Naunyn Schmiedebergs Arch. Pharmacol. **335:** 136–142.

18. HERTTING, G. & C. ALLGAIER. 1988. Participation of protein kinase C and regulatory G proteins in the modulation of the evoked noradrenaline release in brain. Cell. Mol. Neurobiol. **8:** 105–114.

19. HERTTING, G., S. WURSTER, P. GEBICKE-HÄRTER & C. ALLGAIER. 1988. Participation of regulatory G-proteins and protein kinase C in the modulation of transmitter release in hippocampus. In Modulation of Synaptic Transmission and Plasticity in Nervous System. G. Hertting & H.-C. Spatz, Eds. NATO ASI Series **H19:** 147–164. Springer-Verlag. Berlin-Heidelberg.

20. KATADA, T., M. OINUMA, K. KUSAKABE & M. UI. 1987. A new GTP-binding protein in brain tissues serving as the specific substrate of islet-activating protein, pertussis toxin. FEBS Lett. **213:** 353–358.

21. BRABET, P., A. DUMUIS, M. SEBBEN, C. PANTALONI, J. BOCKAERT & V. HOMBURGER. 1988. Immunocytochemical localization of the guanine nucleotide–binding protein G_o in primary cultures of neuronal and glial cells. J. Neurosci. **8:** 701–708.

22. GOLDSMITH, P., P. S. BACKLUND, K. ROSSITER, A. CARTER, G. MILLIGAN, C. G. UNSON & A. SPIEGEL. 1988. Purification of heterotrimeric GTP-binding proteins from brain: identification of a novel form of G_o. Biochemistry **27:** 7085–7090.

23. LAI, R.-T., Y. WATANABE & H. YOSHIDA. 1983. Effect of islet-activating protein (IAP) on contractile responses of rat vas deferens: evidence for participation of N_i (inhibitory GTP binding regulating protein) in the α_2-adrenoceptor-mediated response. Eur. J. Pharmacol. **90:** 453–456.

24. MUSGRAVE, I., P. MARLEY & H. MAJEWSKI. 1987. Pertussis toxin does not attenuate α_2-adrenoceptor mediated inhibition of noradrenaline release in mouse atria. Naunyn Schmiedebergs Arch. Pharmacol. **336:** 280–286.

25. DOCHERTY, J. R. 1988. Pertussis toxin and pre-junctional α_2-adrenoceptors in rat heart and vas deferens. J. Auton. Pharmacol. **8:** 197–201.

26. NICHOLS, A. J., E. D. MOTLEY & R. R. RUFFOLO. 1988. Differential effective of pertussis toxin on pre- and postjunctional α_2-adrenoceptors in the cardiovascular system of the pithed rat. Eur. J. Pharmacol. **145:** 345–349.

27. KASCHUBE, M. 1989. Effect of *N*-ethylmaleimide on presynaptic inhibition in the field-stimulated mouse vas deferens. Naunyn Schmiedebergs Arch. Pharmacol. **339**(Suppl.): 137.

28. WEBER, H.-D. 1989. Presynaptic α_2-adrenoceptors at the terminals of perivascular sympathetic nerves are coupled to an *N*-ethylmaleimide-sensitive G-protein. Naunyn Schmiedebergs Arch. Pharmacol. **339**(Suppl.): 138.

29. ALLGAIER, C., G. HERTTING & O. V. KÜGELGEN. 1987. The adenosine receptor–mediated inhibition of noradrenaline release possibly involves an N-protein and is increased by α_2-autoreceptor blockade. Br. J. Pharmacol. **90:** 403–412.

30. ALLGAIER, C., T. J. FEUERSTEIN, R. JACKISCH & G. HERTTING. 1985. Islet-activating protein (pertussis toxin) diminishes α_2-adrenoceptor mediated effects on noradrenaline release. Naunyn Schmiedebergs Arch. Pharmacol. **331:** 235–239.

31. ALLGAIER, C., B. DASCHMANN, J. SIEVERLING & G. HERTTING. 1989. Presynaptic k-opioid receptors on noradrenergic nerve terminals couple to G proteins and interact with the α_2-adrenoceptors. J. Neurochem. **53:** 1629–1635.

32. KAWATA, K. & Y. NOMURA. 1987. Suppressing effect of pertussis toxin on clonidine-induced inhibition of noradrenaline release from cerebral cortical slices of rats. Neurosci. Res. **4:** 236–240.

33. ALLGAIER, C., T. J. FEUERSTEIN & G. HERTTING. 1986. N-Ethylmaleimide (NEM) diminishes α_2-adrenoceptor mediated effects on noradrenaline release. Naunyn Schmiederbergs Arch. Pharmacol. **333:** 104–109.

34. FREDHOLM, B. B. & E. LINDGREN. 1987. Effects of *N*-ethylmaleimide and forskolin on noradrenaline release from rat hippocampal slices. Evidence that prejunctional adenosine and α-receptors are linked to N-proteins but not to adenylate cyclase. Acta Physiol. Scand. **130:** 95–105.

35. SINGER, E. A. 1988. Transmitter release from brain slices elicited by single pulses: a powerful method to study presynaptic mechanisms. Trends Pharmacol. Sci. **9:** 274–276.

36. VALENTA, B., H. DROBNY & E. A. SINGER. 1988. Presynaptic autoinhibition of central noradrenaline release in vitro: operational characteristics and effects of drugs acting at alpha-2 adrenoceptors in the presence of uptake inhibition. J. Pharmacol. Exp. Ther. **245:** 944–949.

37. ZIER, G., H. DROBNY, B. VALENTA & E. A. SINGER. 1988. Evidence against a functional link between noradrenaline uptake mechanisms and presynaptic alpha-2 adrenoceptors Naunyn Schmiedebergs Arch. Pharmacol. **337:** 118–121.

38. LIMBERGER, N., L. SPÄTH & K. STARKE. 1988. Presynaptic α_2-adrenoceptor, opioid k-receptor and adenosine A_1-receptor interactions on noradrenaline release in rabbit cortex. Naunyn Schmiedebergs Arch. Pharmacol. **338:** 53–61.

39. WERLING, L. L., P. N. MCMAHON & B. M. COX. 1989. Effects of pertussis toxin on opioid regulation of catecholamine release from rat and guinea pig brain slices. Naunyn Schmiedebergs Arch. Pharmacol. **339:** 509–513.

40. DOLPHIN, A. C. & S. A. PRESTWICH. 1985. Pertussis toxin reverses adenosine inhibition of neuronal glutamate release. Nature **316:** 148–150.

41. FREDHOLM, B. B. & E. LINDGREN. 1986. Possible involvement of the N_1-protein in the prejunctional inhibitory effect of a stable adenosine analogue (R-PIA) on noradrenaline release in the rat hippocampus. Acta Physiol. Scand. **126:** 307–309.

42. FREDHOLM, B. B. & T. V. DUNWIDDIE. 1988. How does adenosine inhibit transmitter release? Trends Pharmacol. Sci. **9:** 130–134.

43. STRATTON, K. R., A. J. COLE, J. PRITCHETT, C. U. ECCLES, P. F. WORLEY & J. M.

BARABAN. 1989. Intrahippocampal injection of pertussis toxin blocks adenosine suppression of synaptic responses. Brain Res. **494:** 359–364.

44. DUNWIDDIE, T. V., W. R. PROCTER & B. B. FREDHOLM. 1988. Pertussis toxin pretreatment blocks postsynaptic but not presynaptic effects of adenosine in the rat hippocampus. Soc. Neurosci. **14**(Suppl): 113.

45. RAITERI, M., G. BONNANO, M. MARCHI & G. MAURA. 1984. Is there a functional linkage between neurotransmitter uptake mechanisms and presynaptic receptors? J. Pharmacol. Exp. Ther. **231:** 671–677.

46. DE LANGEN, C. D. J. & A. H. MULDER. 1980. On the role of calcium ions in the presynaptic alpha-receptor mediated inhibition of [^{3}H]noradrenaline release from rat brain cortex synaptosomes. Brain Res. **185:** 399–408.

47. WURSTER, S. 1989. Coupling of N-ethylmaleimide-sensitive G-proteins to pharmacologically different presynaptic receptors on noradrenergic terminals. Naunyn Schmiedebergs Arch. Pharmacol. **339**(Suppl.): 355.

48. ALLGAIER, C. 1989. Presynaptic M_2-autoreceptors in hippocampus couple to N-ethylmaleimide-sensitive G proteins. Naunyn Schmiedebergs Arch. Pharmacol. **339**(Suppl.): 334.

49. ALLGAIER, C., T. JÄGER & G. HERTTING. 1989. Endogenous noradrenaline impairs the prostaglandin-induced inhibition of noradrenaline release. Naunyn Schmiedebergs Arch. Pharmacol. **340:** 472–474.

50. MULDER, A. H., A. L. FRANKHUYZEN, J. C. STOOF, J. WEMER & A. N. M. SCHOFFELMEER. 1984. Catecholamine receptors, opiate receptors, and presynaptic modulation of neurotransmitter release in the brain. In Catecholamines: Neuropharmacology and Central Nervous System—Theoretical Aspects. E. Usclin, R. Carlsson, A. Dahlström & I. Engel, Eds.: 47–58. Alan R. Liss, Inc. New York, N.Y.

51. STARKE, K. 1987 Presynaptic α-autoreceptors. Rev. Physiol. Biochem. Pharmacol. **107:** 73–146.

52. STARKE, K., M. GÖTHERT & H. KILBINGER. 1989. Modulation of neurotransmitter release by presynaptic autoreceptors. Physiol. Rev. **69:** 864–989.

53. HESCHELER, J., W. ROSENTHAL, W. TRAUTWEIN & G. SCHULTZ. 1987. The GTP-binding protein, G_o, regulates neuronal calcium channels. Nature **325:** 445–447.

54. LIPSCOMBE, D., S. KONGSAMUT & R. W. TSIEN. 1989. α-Adrenergic inhibition of sympathetic neurotransmitter release mediated by modulation of N-type calcium-channel gating. Nature **340:** 639–642.

55. VERSTEEG, D. H. G. & W. J. FLORIJN. 1987. Phorbol-12,13-dibutyrate enhances electrically stimulated neuromessenger release from rat dorsal hippocampal slices in vitro. Life Sci. **40:** 1237–1243.

56. FREDHOLM, B. B. & E. LINDGREN. 1988. Protein kinase C activation increases noradrenaline relase from the rat hippocampus and modifies the inhibitory effect of α_2-adrenoceptor and adenosine A_1-receptor agonists. Naunyn Schmiedebergs Arch. Pharmacol. **337:** 477–483.

57. RAMDINE, R., A.-M. GALZIN & S. Z. LANGER. 1989. Involvement of adenylate cyclase and protein kinase C in the α_2-adrenoceptor-mediated inhibition of noradrenaline and 5-hydroxytryptamine release in rat hypothalamic slices. Naunyn Schmiedebergs Arch. Pharmacol. **340:** 386–395.

58. ALLGAIER, C., O. V. KÜGELGEN & G. HERTTING. 1986. Enhancement of noradrenaline release by 12-O-tetradecanoyl-phorbol-13-acetate, an activator of protein kinase C. Eur. J. Pharmacol. **129:** 389–392.

59. ALLGAIER, C., G. HERTTING, H. HUANG & R. JACKISCH. 1987. Protein kinase C activation and α_2-autoreceptor-modulated release of noradrenaline. Br. J. Pharmacol. **92:** 161–172.

60. DUNKLEY, P. R., P. E. JARVIE, J. W. HEATH, G. J. KIDD & J. A. P. ROSTAS. 1986. A rapid method for isolation of synaptosomes on percoll gradients. Brain Res. **372:** 115–129.

61. GEBICKE-HAERTER, P. J., A. SEREGI, S. WURSTER, A. SCHOBERT, C. ALLGAIER & G. HERTTING. 1988. Multiple pertussis toxin substrates as candidates for regulatory G proteins of adenylate cyclase coupled to the somatostatin receptor in primary rat astrocytes. Neurochem. Res. **13:** 997–1001.

Presynaptic Modulation of Sympathetic Neurotransmitter Release by Modulators of Cyclic 3′,5′-Guanosine Monophosphate in Canine Vascular Smooth Muscle [a]

STAN S. GREENBERG, FRIEDRICH P. J. DIECKE,
F. A. CURRO,[b] KEITH PEEVY,[c] AND
TOSHIOKO P. TANAKA

Department of Physiology and
[b]St. Josephs Hospital
UMDNJ–New Jersey Medical School
Newark, New Jersey 07103

[c]Departments of Pediatrics and Pharmacology
College of Medicine
University of South Alabama
Mobile, Alabama 36688

INTRODUCTION

Extracellular calcium ion is required for neurotransmitter release consequent to electrical and receptor-mediated stimulation of sympathetic and other neurons. A large number of substances are known to modify the quanta of norepinephrine (NE) released per nerve impulse. It has been proposed that cyclic 3′5′-adenosine monophosphate (cAMP) may enhance, and endogenous cyclic 3′,5′-guanosine monophosphate (cGMP) may inhibit, neurotransmitter release. However, no consistent relationship between the effects of the several modulators of neurally mediated NE release and their effects on adenylate and guanylate cyclase is as yet apparent. This research has been hampered by the relative absence of techniques that allow the measurement of cyclic nucleotide levels in the presynaptic sympathetic nerve terminal after exposure to the putative modulators of release and consequent to nerve stimulation. Although intraneuronal cGMP or cAMP still has not been measured in intact peripheral sympathetic nerve terminals, new information on the effects of modulators of cAMP and cGMP on sympathetic neuroeffector transmission has emerged through the use of pharmacologic modulators of the cyclic nucleotides system. This review evaluates the role of cAMP and cGMP in sympathetic neurotransmitter release to vascular smooth muscle (VSM) and from adrenal chromaffin cells. The concept that emerges is that cGMP, in addition to its role as a directly acting smooth muscle relaxant, may decrease smooth muscle tone by acting as a physiologic feedback inhibitory modulator of the release of the sympathetic neurotransmitters NE and epinephrine (EPI).

The release of epinephrine and norepinephrine from adrenal medullary cells by depolarization and acetylcholine (ACh), and from sympathetic nerve terminals innervating peripheral vascular smooth muscle during sympathetic nerve stimulation (SNS),

[a]Supported in part by a grant from the New Jersey Affiliate of the American Heart Association and by UMDNJ–New Jersey Medical School.

respectively, is associated with activation of a pertussis toxin–sensitive G protein, phospholipases A_2 and C, protein kinase C (PKC), and inositol trisphosphate ($INSP_3$) and increased calcium entry into the sympathetic neuron. Cyclic 3'5'-guanosine monophosphate and cyclic 3'5'-adenosine monophosphate also increase in rat superior cervical ganglion,[1] rat PC12 adrenal chromaffin cells, *in culture,*[2] and perfused bovine adrenomedullary cells.[3–5] However, the function of these cyclic nucleotides in postganglionic neurotransmitter release from sympathetic neurons is uncertain. Guanosine triphosphate, the precursor for cGMP, facilitates exocytosis of histamine from neutrophils,[6] probably via stimulation of G proteins. Polymyxin B, an inhibitor of protein kinase C, inhibits NE release from sympathetic nerves caused by phorbol ester and SNS.[7–10] Activation of PKC by diacylglycerol, especially in the presence of low concentrations of intracellular calcium ion concentrations (Ca_i^{+2}), facilitates exocytosis of NE.[11] Elevated levels of cAMP within the sympathetic neuron may facilitate NE release.[12–19] less is certain, however, about the role of cGMP in SNS.

MODULATION OF cGMP AND NEUROTRANSMISSION

Both direct and indirect evidence support a role for cGMP in SNS to vascular smooth muscle. The release of ACh from brain slices is inhibited by 8-bromo-cGMP.[20] Compounds that increase cGMP inhibit the contractions of porcine and guinea pig mesenteric arteries to SNS, in part by inhibiting release of NE from sympathetic nerves[21–25] in concentrations that increase cGMP within smooth muscle.[26–29] Kalix showed that cGMP elevated extracellular potassium ion and electrical stimulation increase cGMP in rat superior cervical ganglion.[1] Nicorandil increases cGMP in porcine and guinea pig mesenteric artery and inhibits the contractile responses of these tissues to sympathetic nerve stimulation.[21] The presence of a functional endothelium, and presumably endothelium-derived relaxing factor (EDRF), which increases cGMP in blood vessels,[28,30] inhibited the responses of arteries and veins to sympathetic nerve stimulation[31,32] as well as the efflux of 2-[^{14}C]-NE.[22–24,31] Finally, Drewett *et al.* showed that atrial natriuretic hormone inhibited the release of radiolabeled norepinephrine from sympathetic nerves innervating rabbit vas deferens and PC12 cells.[3,25]

Indirect evidence for a role of cyclic GMP in neurotransmission comes from the data with capsaicin and bradykinin and sensory neurons. Capsaicin and bradykinin both increase cGMP in rat sensory neurons in a calcium dependent manner.[33,34] It has been suggested that calcium is the trigger that activates guanylate cyclase indirectly through activation of phospholipases A and C.[35,36] Capsaicin has been shown to inhibit sympathetic neurotransmission and release of NE from sympathetic nerves innervating rabbit ear artery and in guinea pig vas deferens.[37,38] In the latter, an intermediate released by capsaicin, either calcium-gene related peptide or substance P, has been invoked as the actual inhibitor of SNS. Thus, intraneuronal cGMP may inhibit the release of NE from sympathetic nerves innervating smooth muscle.

To determine whether modulation of cGMP affected, in a consistent manner, NE release from sympathetic nerve endings innervating VSM we measured the efflux of radiolabeled norepinephrine during modulation of three of the pathways involved in the synthesis and degradation of cGMP (FIGURE 1). Activation of soluble guanylate cyclase with nitroprusside (SNP) or SIN-1 and inhibition of type II cGMP phosphodiesterase with M&B-22948 and verofyllin, interventions that increase cGMP in blood vessels, inhibited the efflux of 2^{14}C-NE from sympathetic nerves innervating canine pulmonary and mesenteric arteries and veins (FIGURE 2).[22–24] Moreover, bradykinin also inhibited the efflux of NE from the nerve terminals of canine pulmonary and mesenteric arteries (FIGURE 3). These effects are due to cyclic GMP and not to direct effects of the compounds on other second messengers within the sympathetic nerve

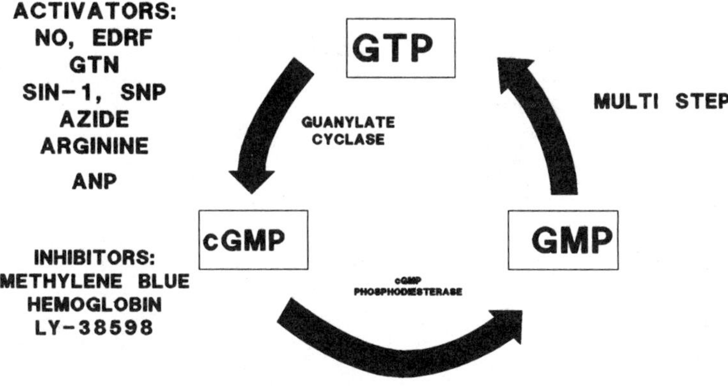

FIGURE 1. Schema of cyclic GMP synthesis and degradation and the site of action of the compounds that modulate the enzyme synthesis of cyclic GMP.

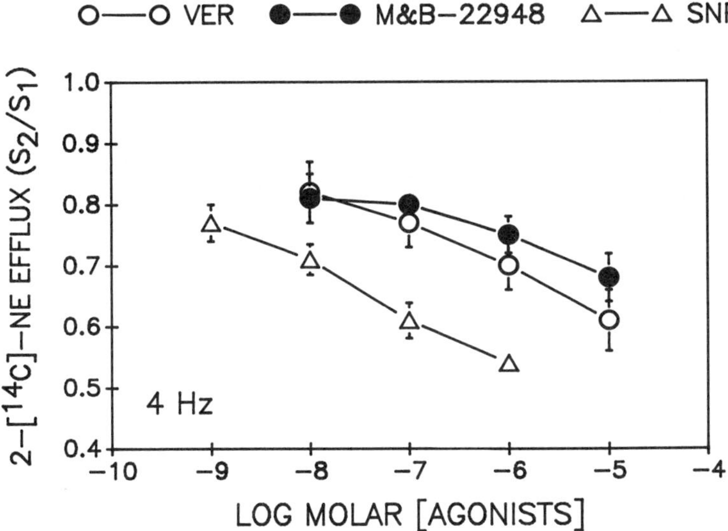

FIGURE 2. Inhibition by sodium nitroprusside (SNP), verofyllin, and M&B-22948 of the efflux of 2-[^{14}C]-norepinephrine from sympathetic nerve endings innervating canine mesenteric artery and pulmonary vein. The ordinate is the efflux of 2-[^{14}C]-NE expressed as the fractional efflux of the vehicle or drug-treated stimulus (S$_2$) and the control efflux (S$_1$). The abscissa is the concentration of drug. Vertical lines are standard errors of the mean (SEM). An asterisk denotes the responses differ ($p < 0.05$) from vehicle.

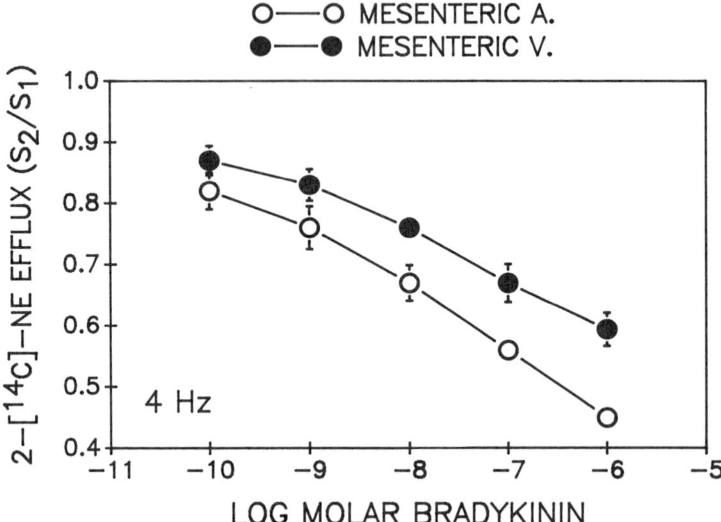

FIGURE 3. Inhibition by bradykinin of the efflux of 2-[^{14}C]-norepinephrine from sympathetic nerve endings innervating canine mesenteric artery and pulmonary vein. The ordinate is the efflux of 2-[^{14}C]-NE expressed as the fractional efflux of the vehicle or drug-treated stimulus (S$_2$) and the control efflux (S$_1$). The abscissa is the concentration of drug. Vertical lines are SEM. The responses differ from control which was 0.87 ± 0.04.

terminals. The inhibitory effect of SNP, nitroglycerin (GTN), bradykinin, M&B-22948, and verofyllin were enhanced by inhibition of cGMP phosphodiesterase (PDE) with M&B-22948 and inhibited by blockade of guanylate cyclase with methylene blue (FIGURE 4). Moreover, cGMP inhibition of SNS is frequency dependent, with greater inhibition of SNS observed at low, physiologic frequencies of SNS (0.5–8 Hz) and less inhibition at pathophysiologic frequencies of SNS (16 and 32 Hz).[22-24]

BOTH SOLUBLE AND PARTICULATE GUANYLATE CYCLASE ARE THE SOURCE OF cGMP

Both soluble and particulate guanylate cyclase appear to be the enzymes that produce the cGMP involved in modulating NE release from sympathetic, as well as purinergic nerve endings in smooth muscle. Atriopeptin III (ANPIII) decreases the excitability of the inferior mesenteric ganglion of the guinea pig, as determined by intracellular electrode recordings.[39,40] Although cGMP and release of NE were not measured, Cheung *et al.*[39] measured contractions to nerve stimulation and the excitatory postsynaptic potentials of guinea pig saphenous artery during SNS and administration of exogenous NE in the presence of ANPIII and 0.3 μM SNP. ANPIII inhibited the responses to SNS whereas SNP did not. Both ANPIII and SNP inhibited the postsynaptic potentials to exogenous NE. McKay and Cheung concluded that ANPIII inhibits NE release via a presynaptic mechanism of action.[40] However, ANPIII attenuates the effector response to neurally released NE. Therefore, Drewett *et al.* extended these observations by showing that ANPII inhibited the release of NE and purine nucleotides from sympathetic nerves innervating rabbit carotid artery, vas deferens, and rat PC-12 cells, in culture.[2,25] Moreover, ANPIII and SNP increase cGMP and decrease release of radiolabeled NE from cultured P-12 cells and adrenal medulla. Thus, particulate guanylate cyclase may play a role in the generation of cGMP and the modulation of sympathetic neurotransmitter release to vascular and nonvascular smooth muscle.

In an attempt to ascertain that the response to ANP was mediated by cGMP, we investigated its effects on the efflux of NE from canine mesenteric artery in the presence and absence of modulators of the cGMP system. [8-33]*h*ANP inhibited the efflux of NE (FIGURE 5). The effect was enhanced by M&B-22948 but not affected by methylene blue or hemoglobin.[23] Thus, we also concluded that ANP inhibits the release of NE from adrenergic nerve terminals by a mechanism that involves particular guanylate cyclase. Similar conclusions were reached in the ganglion-blocked anesthetized dog.[41] Although questions have been raised concerning the relationship between cGMP and the pharmacologic actions of ANP, ANP increases cGMP within astroglial cells from rat brain[42] and membrane-bound guanylate cyclase is an ANP receptor, with increased guanylate cyclase activity linked to ANF binding.[43] Two ANP-binding sites of 66,000 and 130,000 daltons exist in many preparations including bovine adrenal chromaffin cells. The 130,000-dalton ANP-binding sites bound to a guanosine ticphosphate (GTP)-agarose affinity column, and the specific activity of guanylate cyclase was increased by 90-fold in this fraction.[44] Thus, the increase in cyclic GMP accumulation and particulate guanylate cyclase activity by ANP may not correlate with the affinity and number of ANP-binding sites because of the existence of two distinct populations of ANP receptors.

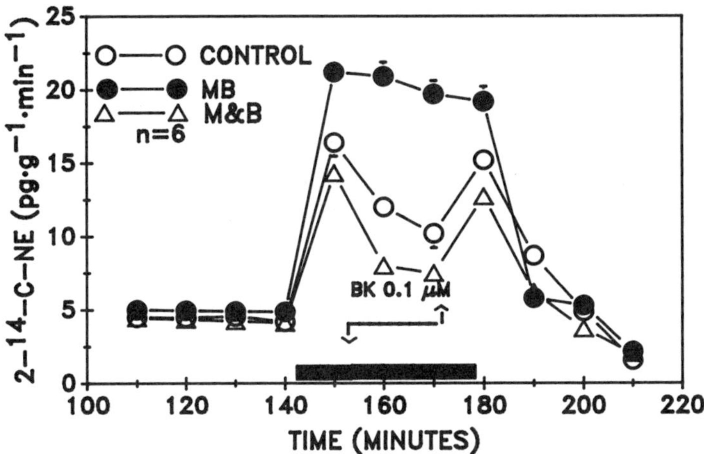

FIGURE 4. Effect of M&B-22948 (500 μM) and methylene blue (500 μM) on the inhibitory effect of bradykinin on release of NE from sympathetic nerves at 4 Hz transmural nerve stimulation. The ordinate is the actual efflux of radiolabeled NE, corrected for the metabolites, in pg/minute per g of tissue, wet weight. The abscissa is the time in minutes. Bradykinin was added at arrows. M&B-22948 (10 μM) and methylene blue (0.5 mM) were present throughout the experiment. Vertical lines are SEM. The inhibitory effects of bradykinin are enhanced by M&B-22948 and inhibited by methylene blue ($p < 0.05$).

cGMP-DEPENDENT REGULATION OF NE RELEASE BY ENDOTHELIUM AND NERVE TERMINAL

Vascular endothelial cells (EC) play an active role in the regulation of VSM tone by releasing factors that promote VSM relaxation and efflux of cGMP.[22–24,31,45–49] Endothelium-derived relaxing factor (EDRF)[46] is one of the most potent vasodilator substances released from EC by circulating and neurogenically released humoral substances such as serotonin (5-HT), histamine, thrombin, arachidonic acid (AA), bradykinin, catecholamines, substance P, adenosine triphosphate (ATP), and calcitonin-gene-related peptide. EDRF is released from EC of arteries and veins of every species tested, including man. EDRF was initially thought to be a hydroperoxide or free

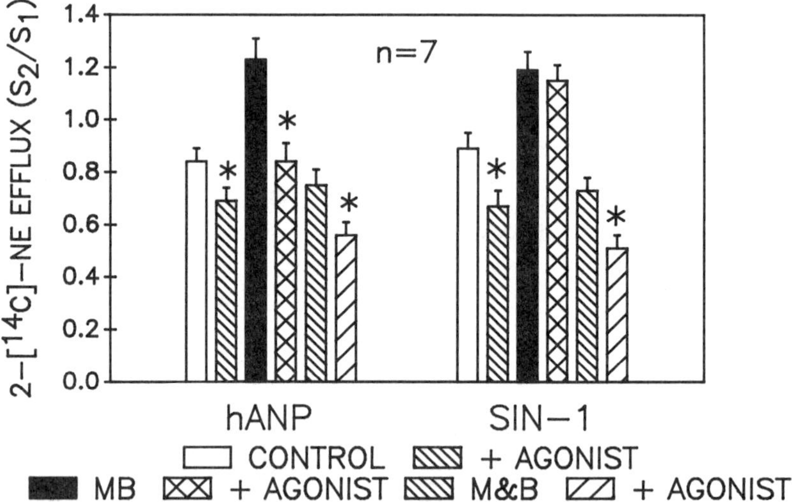

FIGURE 5. Effect of M&B-22948 (500 μM) and methylene blue (500 μM) on the inhibitory effect of [8-33]hANP on the efflux of 2-[^{14}C]-norepinephrine from sympathetic nerve endings innervating canine mesenteric artery. The ordinate is the efflux of 2-[^{14}C]-NE expressed as the fractional efflux of the vehicle or drug-treated stimulus (S$_2$) and the control efflux (S$_1$) at 4Hz. On the abscissa are the different drugs. Vertical lines are SEM. An asterisk denotes the responses differ ($p < 0.05$) from vehicle.

radical, derived from fatty acids, which stimulated guanylate cyclase within the VSM. The resultant increase of cGMP within VSM is believed to mediate the relaxant effects of EDRF.[48–51]

The chemical nature of EDRF is unknown, but EDRF is neither a prostanoid nor a fatty acid derived from the lipoxygenase pathway. Both EDRF and nitric oxide (NO free radical) are inactivated by hemoglobin; this action is reversed by carbon monoxide. Superoxide dismutase prolongs the half-life of EDRF and NO from 6 seconds to 30–60 seconds acidification inhibits the relaxation produced by nitrite. Hemoglobin inactivated and SOD potentiates the relaxation responses to EDRF and NO. This suggested that EDRF was a free radical similar to, or identical to, nitric oxide.[52,53] Palmer et al., using chemiluminescence under conditions that could generate NO, demonstrated NO in extracts of EDRF.[54,55] These investigators claim that EDRF is or generates NO, which is the active moiety of EDRF (for references see References 45–51).

EDRF and Modulation of SNS

The endothelium, via tonic release of EDRF(s) (NO, hyperpolarizing factor, nitrocysteine,[56–62] may modulate NE release from sympathetic nerves during SNS. Endothelium rubbing enhances the contractions of VSM to NE and other agonists, including SNS[22–24,31,32,62–67] Some investigators do not find enhanced responses to SNS following endothelium rubbing.[68–70] Enhancement of the responses to both NE and SNS was found to be similar forcing the conclusion that the endothelium, through tonic release of EDRF, desensitized the VSM to vasoconstrictor stimuli rather than inhibiting NE.[62,63] However, these investigators did not measure the NE release from sympathetic nerves. Moreover, Meehan *et al.* did not find any enhancement of the contractions to SNS.[70] However, careful inspection of the data suggests that the contractions to SNS were enhanced in the endothelium-rubbed arteries.[69,70]

The discrepancy among these studies may reside in the process of sympathetic nerve stimulation itself. Electrical stimulation (9 V, 1–2 mseconds, 4 Hz) of bovine and rat arteries inhibits endothelium-dependent relaxation. However, relaxation responses persist in vessels where an antioxidant is included in the electrically stimulated buffer. Electron microscopy reveals endothelial damage in tissues exposed to electrical stimulation without antioxidant protection.[71–73] These results suggested that SNS itself damages the endothelium in isolated VSM. Thus, SNS, by destruction of the endothelium, can interfere with the study of endothelium-mediated effects of SNS. With the use of a more physiologic oxygen concentration in physiologic salt solution (PSS), 20% O_2 rather than 95% O_2 during nerve stimulation, and superfusion of the blood vessel with an antioxidant present, the accumulation of oxygen-derived and SNS-derived free radicals should be minimized.

We tested this postulate in isolated canine pulmonary arteries and veins using the technique of superfusion and measurement of the efflux of radiolabeled NE during transmural nerve stimulation at 1, 2, 4, 8, 16, and 32 Hz for 10 minutes. In pulmonary artery and vein with intact endothelium, the contractile responses to low frequency SNS were decreased when compared to endothelium-rubbed blood vessels. Electrical stimulation of arteries and veins with intact endothelium for 10 minutes released less 2-[^{14}C]-NE than rubbed blood vessels, especially at low frequencies of SNS (1, 2, and 4 Hz). Using the technique of bioassay (FIGURE 6). In addition, EDRF from porcine thoracic aorta inhibited the efflux of 2-[^{14}C]-NE from the pulmonary artery and vein (FIGURE 7). Enhancement of NE efflux from sympathetic nerve endings to VSM following endothelium rubbing does not result from physical trauma to the blood vessel or removal of a barrier to diffusion or a site of metabolism of NE.[22,31,73] Endothelium rubbing increased the fractional release of NE, but not its metabolites, during SNS. The effects of rubbing persist after blockade of NE metabolism by catechol-*O*-methyl transferase and monoamine oxidase. Removal of a barrier to diffusion of 2-[^{14}C]-NE from the synaptic cleft to the effluent would produce an equivalent percent increase of the efflux of 2-[^{14}C]-NE at all frequencies of SNS. Instead, enhancement of NE efflux was greater at the lower frequencies of SNS than at the higher frequencies. Finally, effluent from endothelium competent blood vessels inhibited NE efflux from pulmonary arteries and veins during SNS. These findings suggest that the endothelium, possibly via EDRF, may modulate NE release from sympathetic nerves innervating VSM. However, the possibility must also be considered that the enhanced NE efflux

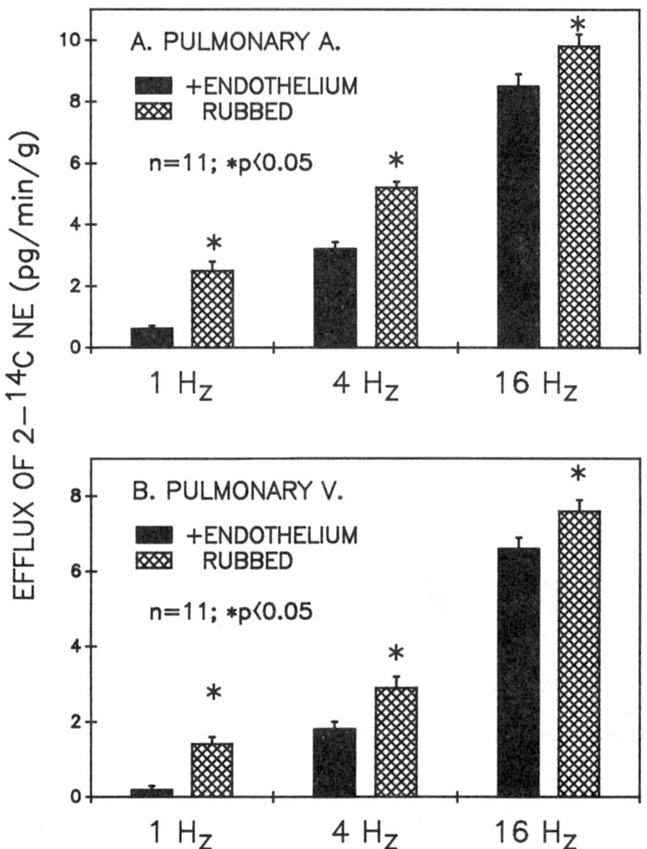

FIGURE 6. Efflux of 2-[^{14}C]-NE before and during transmural nerve stimulation (4 Hz, 9–10 V, 2 mseconds duration and delay) in endothelium-competent and endothelium-rubbed canine pulmonary arteries (top panel) and veins (bottom panel). The ordinate is the efflux rate of 2-[^{14}C]-NE (expressed as pg/g blood vessel per minute). The abscissa is the frequency of stimulation (Hz). Vertical lines are the SEM. Stimulation was for 10 minute periods followed by a 10 minute rest period. An asterisk denotes responses differ from endothelium-competent tissues ($p < 0.05$). (Reproduced from Reference 22, with permission.)

observed following endothelium rubbing was due, in part, to removal of the inhibitory effect of the EC-derived peptide endothelin.[74]

Basal and 5-HT-induced effluents containing EDRF inhibit the release of NE from sympathetic nerve endings innervating canine pulmonary arteries and veins.[22–24,31] If cGMP is involved in the release of NE from sympathetic nerves, then inhibition of soluble and/or particulate guanylate cyclase should enhance the overflow of radiolabeled or endogenous NE during SNS. Methylene blue inhibits guanylate cyclase (soluble) and variably enhances the release of NE from sympathetic nerves innervating blood vessels during SNS, since it enhances the responses to nerve stimulation and the overflow of NE during SNS.[31] Enhancement of contractions may result from inhibition of the vascular smooth muscle depressant effects of cGMP on the effector responses to NE released by SNS.

Cohen and Weisbrodt measured Ne release and concluded that methylene blue did not prevent endothelium-dependent inhibition of NE overflow in rabbit carotid arteries.[31] Although based on a limited number of experiments (2) with a large variability, their data showed that methylene blue may have stimulated the overflow of radiolabeled NE during SNS. Linnick and Lee also reported that hemoglobin, which inhibits guanylate cyclase and EDRF(s), enhanced the neurogenically mediated responses and neurotransmitter release during SNS to isolated cerebral arteries of the cat.[75] Finally, Cohen *et al.*[76] and Greenberg *et al.*[77] demonstrated that nitric oxide (NO) and sodium nitroprusside (SNP) inhibited the efflux of NE from the sympathetic nerves innervating rabbit carotid and canine pulmonary artery, respectively. Using measurement of radioactivity during efflux and measurement of endogenous NE, Cohen concluded that the effect of NO was mediated by oxidation of NE by the NO. In contrast, using measurement of 2-[^{14}C]-NE in the presence of ascorbate we concluded that the effect of SNP was mediated by stimulation of cGMP. Inhibition of guanylate cyclase with methylene blue inhibited and verofyllin enhanced the effect of EDRF on the efflux of NE from sympathetic nerves innervating in the blood vessels (FIGURE 8). These findings support the conclusion that the endothelium can inhibit release of NE from sympathetic nerves innervating canine pulmonary artery and vein by a mechanism involving cGMP. Thus, cGMP and the endothelium, in part through EDRF, can act as an endogenous inhibitor of sympathetic neurotransmitter release.

Sympathetic nerves maintain vascular tone in animals and man by tonic and phasic release of NE from these nerves. The presence of a presynaptic alpha-adrenoceptor-mediated feedback inhibitory mechanism inhibits the quanta of NE released during SNS.[18] Chlonidine-induced feedback inhibition of NE release from central sympathetic neurons has been associated with a decrease in cAMP. Two cGMP-sensitive PDE exist in adrenal chromaffin cells.[78] One hydrolyzes cGMP with hyperbolic kinetics but cAMP with sigmoidal kinetics. cGMP stimulates the hydrolysis of cAMP

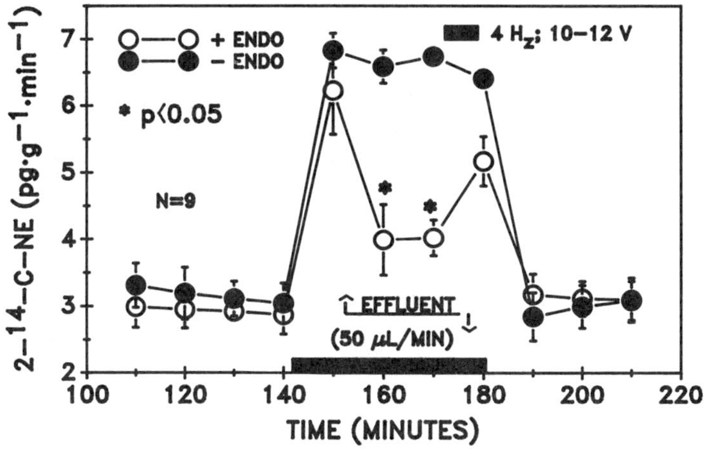

FIGURE 7. Effect of effluent from endothelium-competent and rubbed porcine thoracic aortae on the efflux of 2-[^{14}C]-NE before and during transmural nerve stimulation (4 Hz, 9–10 V, 2 mseconds duration and delay) in endothelium-rubbed canine pulmonary artery. The ordinate is the efflux rate of 2-[^{14}C]-NE (expressed as pg/g blood vessel per minute). The abscissa is the condition of the effluent. Vertical lines are the SEM from 6 experiments. (Reproduced from Reference 22, with permission.)

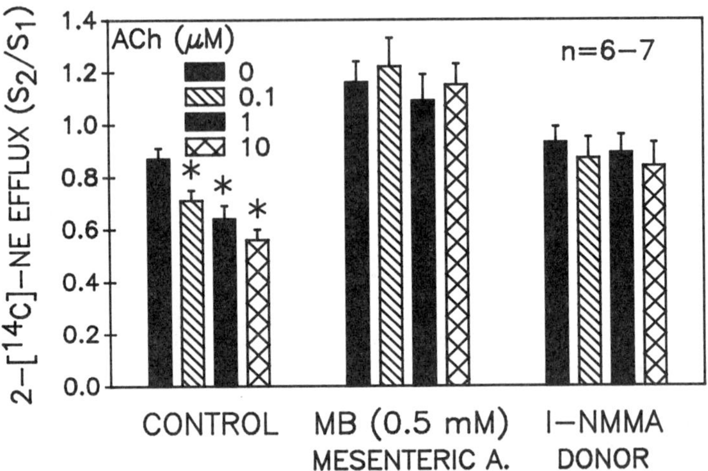

FIGURE 8. Effect of methylene blue (0.5 mM and 1-*N*-monomethylarginine (10 μM) on the inhibitory effect of acetylcholine-stimulated effluent from endothelium-competent canine mesenteric arteries on the efflux of 2-[^{14}C]-NE (4 Hz, 9–10 V, 2 mseconds duration and delay) in endothelium-rubbed canine mesenteric artery. The ordinate is the efflux rate of 2-[^{14}C]-NE (expressed as the fractional release). The abscissa is the treatment groups. Methylene blue was given to the endothelium-rubbed mesenteric artery. 1-NMMA was given to the donor. Atropine was superfused over the rubbed mesenteric artery. Vertical lines are the SEM.

and vice versa. Thus, if endogenous levels of intraneuronal cAMP maximally facilitate release of NE by enhancing the breakdown of cGMP, then cGMP may inhibit NE release by enhancing hydrolysis of cAMP. Thus, the decrease in cAMP produced by clonidine may inhibit NE release by a cGMP-dependent mechanism.

Destruction of the endothelium and removal of EDRF enhance the intrinsic contractile response of VSM to clonidine.[62,63] In contrast, endothelium rubbing decreases the inhibitory effect of clonidine on low frequency SNS to canine mesenteric arteries (FIGURE 9A). Inactivation of guanylate cyclase with methylene blue also inhibits the feedback inhibition of NE efflux by clonidine (FIGURE 9B). Finally, in the presence of phenoxybenzamine, clonidine-induced feedback inhibition of NE release is attenuated less by inactivation of guanylate cyclase (FIGURE 9C). Thus, clonidine, by a direct effect on presynaptic alpha$_2$-receptors, and indirectly by releasing EDRF from the endothelium, may act to inhibit the release of NE from sympathetic nerve terminals of blood vessels, by a cGMP-dependent mechanism. Alternatively, endogenous cGMP may interact with or facilitate the inhibitory mechanism utilized by clonidine to inhibit NE efflux from sympathetic nerves (see A Working Hypothesis on the Mechanism of Inhibition of SNS by Cyclic GMP).

CYCLIC GMP INHIBITS ONLY CALCIUM-DEPENDENT
RELEASE OF NOREPINEPHRINE

Extracellular calcium ion is required for neurotransmitter release consequent to electrical and receptor-mediated stimulation of sympathetic nerves. However, tyramine releases NE by a calcium-independent mechanism.[18] During generation of the action potential towards the terminal boutons of the sympathetic nerve endings innervating VSM, there is a graded depolarization of the nerve terminals associated with an increase in calcium conductance, a decrease in potassium conductance, and release of NE from within the nerve terminals. In contrast, with maintained depolarization, such as is produced by elevations in extracellular potassium ion or inhibition of the membrane sodium-potassium-activated, magnesium-dependent adenosine triphosphatase (ATPase), potassium conductance is inactivated and sodium and calcium accumulation occur with maintained NE efflux as a result of an inability to "turn off" the release process that occurs during the normal depolarization and repolarization of the nerve terminal (for references see Reference 18). Application of nitroprusside to canine mesenteric arteries during tyramine-induced efflux of NE failed to inhibit the efflux of NE from sympathetic nerve endings innervating this VSM (FIGURE 10). This finding supports a conclusion that nitro-compound mediated inhibition of NE efflux (*vide infra* cyclic GMP dependent) is mediated through an effect on calcium ion.

DEPOLARIZATION AND K-CHANNEL INACTIVATION PREVENT THE ACTIONS OF cGMP

Graded depolarization produced by increasing concentrations of KCl increased the efflux of NE from sympathetic nerves innervating the mesenteric and pulmonary arteries. Although SNP inhibited the increase in NE efflux at low concentrations of KCl, high concentrations abolished the inhibitory effect of SNP (FIGURE 10). Similarly, in canine pulmonary artery treated with ouabain, the inhibitory effect of SNP was no longer evident (FIGURE 10). Finally, depolarization induced by inactivation of

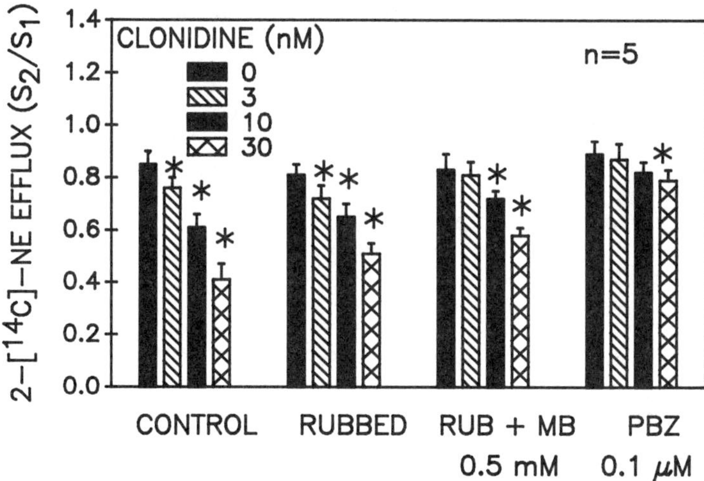

FIGURE 9. Effect of clonidine (100 nM) on the efflux of 2-[^{14}C]-norepinephrine from sympathetic nerve endings innervating canine pulmonary artery. The ordinate is the efflux of 2-[^{14}C]-NE expressed as fractional efflux of the vehicle or drug-treated stimulus (S$_2$) and the control efflux (S$_1$). The abscissa is the treatment group. Vertical lines are SEM. An asterisk denotes the responses differ ($p < 0.05$) from vehicle. MB = methylene blue (0.5 mM). PBZ = phenoxybenzamine (0.1 μM).

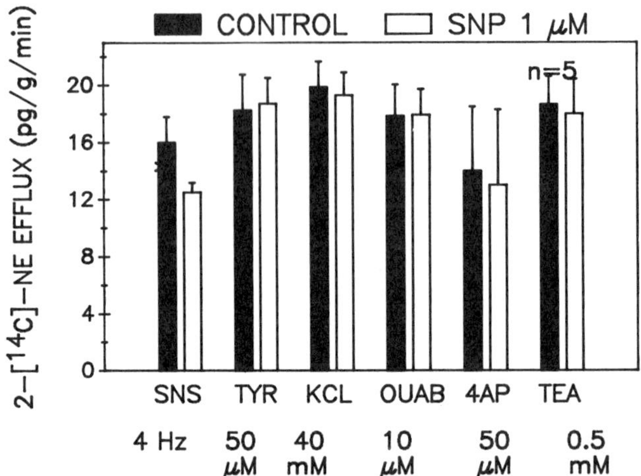

FIGURE 10. Effect of nitroprusside (SNP) on the efflux of 2-[^{14}C]-norepinephrine from sympathetic nerve endings innervating canine mesenteric artery. The ordinate is the efflux of 2-[^{14}C]-NE expressed as pg of NE/g blood vessel per minute before and after administration of SNP. The abscissa is the agonist used to evoke release of NE. Vertical lines are the SEM. The efflux before and after SNP does not differ during any treatment. TYR = tryamine; OUAB = ouabain; 4AP = 4-aminopyridine; TEA = tetraethylammonium bromide.

the potassium current by tetraethylammonium and 4-aminopyridine also attenuated the inhibitory effect of SNP and SIN-1 on NE efflux from sympathetic nerve terminals innervating canine pulmonary artery (FIGURE 10).

In canine femoral and mesenteric arteries[79–81] and rat thoracic aorta[82] it was shown that inhibitors of the Na$^+$, K$^+$-pump (ouabain and elevated potassium ion) depress responsiveness of the VSM to EDRF released in response to acetylcholine. This was not mediated by a specific effect of ouabain on guanylate cyclase since the relaxation responses to isoproterenol were similarly inhibited in bioassay tissues treated with ouabain. Thus, the antagonistic effect of ouabain, KCl, TEA and 4-aminopyridine on cGMP-mediated inhibition of NE efflux is probably mediated by membrane-potential-dependent inactivation of cellular process affected by cGMP and probably not due to inactivation of guanylate cyclase.

SECOND MESSENGERS AND NEURONAL CYCLIC GMP

Cyclic GMP may decrease sympathetic neurotransmitter release by decreasing the influx of extracellular Ca^{+2} as a result of decreasing G protein activation of second messengers, stimulating calcium sequestration within, or efflux from, the cell or inhibiting the release of Ca$_i^{+2}$ from the sarcoplasmic reticulum, respectively. Intraneuronal cGMP may hasten repolarization by increasing potassium conductance. cGMP may also decrease the hydrolysis of phosphoinositides to INSP$_3$.[1,18,29,82,83] Suppression of intraneuronal levels of INSP$_3$ can decrease Ca$_i^{+2}$. Moreover, TEA and 4-aminopyridine increase efflux of NE as a consequence of inactivation of potassium conductance in the nerve terminal. This also depolarizes the nerve terminal membrane. Membrane depolarization is associated with enhanced phosphatidylinositol turnover in intact nerve terminals.[83] Cyclic GMP has been shown to inhibit the generation of inositol trisphosphate in smooth muscle.[82,83] The combined effect of decreased Ca_i^{+2} and inhibited phosphoinositide hydrolysis (decreased $INSP_3$ and diacylglycerol) can decrease activation of protein kinase C and inhibit exocytosis of NE.[7-9,17,18,84-86]

Phorbol myristoyl acetate (PMA) stimulates and polymyxin B inhibits protein kinase C and the efflux of NE from sympathetic nerve terminals.[7-9,17,18] A three hour exposure to PMA enhanced the efflux of NE from sympathetic nerves innervating pulmonary artery, but did not affect SNP-induced inhibition of NE efflux. Phorbol dibutyrate was inactive (FIGURE 11). However, after a 24 hour preincubation with PMA, which down-regulates protein kinase C, and after a 1 hour incubation with polymyxin B, which inhibits protein kinase C, the inhibitory effects of SNP were no longer evident (FIGURE 11). Although polymyxin B has several actions and is not a clean drug, its inhibitory effect on SNP-mediated inhibition of sympathetic neurotransmission combined with the inhibition produced by down regulation of protein kinase C supports the conclusion that protein kinase C is involved in the inhibitory effect of cGMP on NE efflux from sympathetic nerves innervating blood vessel.

ROLE OF CYCLIC AMP IN PERIPHERAL ADRENERGIC NEUROTRANSMISSION

A discussion of cyclic nucleotide regulation of sympathetic neurotransmission would not be complete without a discussion of cAMP. Cyclic AMP modulates cholinergic neurotransmitter release in brain and skeletal muscle by activation of

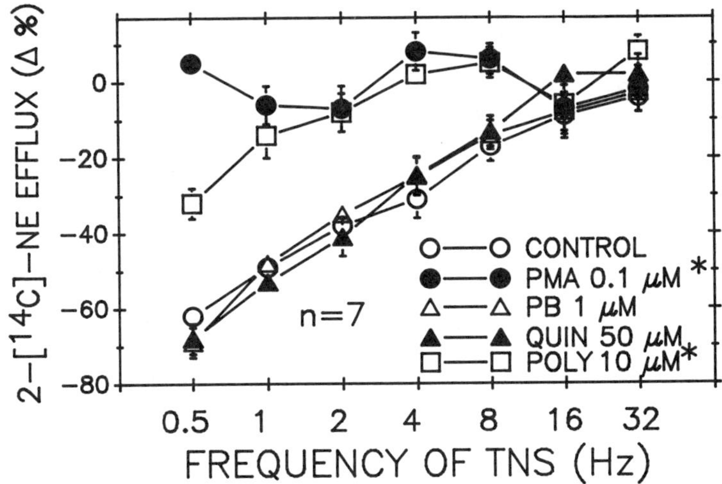

FIGURE 11. Effect of 24 hour preincubation with phorbol myristoyl acetate (PMA) (0.1 μM and phorbol 4-butyrate (PB) (1 μM) and a 1 hr incubation with polymyxin B (10 μM) and quinacrine (50 μM) on SNP-mediated inhibition of the efflux of 2-[14C]-norepinephrine from sympathetic nerve endings innervating canine pulmonary artery. The ordinate is the percent inhibition of efflux of 2-[14C]-NE. The abscissa is the concentration of SNP. Vertical lines are the SEM. An asterisk denotes the responses differ ($p < 0.05$) from vehicle.

protein kinases within the nerve terminal. The activated protein kinase may phosphorylate both the vesicles containing neurotransmitter and sites on the inner portion of the nerve terminal membrane, providing anionic sites to which calcium ions appear to bind.[85,87] Intraneuronal calcium ion serves as a cation bridge between the anionic sites of the phosphorylated vesicles containing neurotransmitter and those of the inner surface of the nerve terminal membrane. This is believed to increase contact between these surfaces allowing for membrane fusion and exocytosis of the contents of the vesicles (for references see References 85, 87, and 88).

The effects of cAMP on peripheral sympathetic neurotransmission are less well defined. cAMP phosphodiesterase inhibitors (PDEI) and membrane permeable analogues of cAMP facilitate calcium-dependent corelease of NE and dopamine B-hydroxylase from vas deferens evoked by high frequency sympathetic nerve stimulation. Calcium-independent release of NE by tyramine is not affected by these agents.[11,18] However, the release of NE by low frequency (5 Hz) SNS (the upper limit of physiological frequencies for NE release) was not affected (or was marginally enhanced) by theophyllin and dibutyryl cAMP.[14,17,89] Sympathetic neurotransmission to rabbit[11,15,90] and human pulmonary arteries[90] was also affected in a variable manner by manipulation of the cAMP system. Forskolin (0.01 mM) increases the overflow of ^{3}H-NE and dopamine efflux from isolated synaptosomes by 40% in concentrations that increase the cAMP in smooth muscles and nerve by more than 5000% from basal values.[87,91] Although the level to which cAMP may increase in nerve terminals of peripheral nerves is unknown, the physiologic significance of a 40% increase in NE release with these concentrations of forskolin must be questioned. Moreover, cAMP-dependent increases of neurally evoked release of ^{3}H-NE only occur during combined administration of an activator of adenylate cyclase and a PDEI. Dibutyryl cAMP and rolipram (a neuronal specific cAMP PDEI)[92,93] increase the overflow of ^{3}H-NE, do not affect the postsynaptic responses to exogenous NE, yet fail to enhance the responses to SNS. Consistent enhancement of the contractions to SNS and the overflow of NE from sympathetic nerves of rabbit pulmonary artery is obtained with exceedingly high (0.33 mM) concentrations of 8-bromo cAMP.[11,12,15,89] Finally, compounds such as prostaglandin I_2 (PGI$_2$) and iloprost, a stable analogue, which raise cAMP within nerve and muscle, do not increase or minimally increase the efflux of NE from sympathetic nerves.[22–24] These data suggest that nonspecific increases in cAMP, such as those produced by exceedingly high concentrations of forskolin and prostanoids, cAMP PDEI, and cAMP analogues, may facilitate the release of NE from sympathetic nerves *in vitro,* but that the physiologic and pathophysiologic importance of cAMP-mediated facilitation of NE release is questionable.

This conclusion can be challenged since stimulation of presynaptic β-adrenoceptor autoreceptors consistently increases intraneuronal cAMP and the efflux of ^{3}H-NE from sympathetic nerves.[90,93,94] β-Adrenoceptor-mediated stimulation of adenylate cyclase may increase cAMP within a specific compartment of the sympathetic neuron linked to exocytosis of NE. However, using human saphenous vein, no evidence for β-adrenoceptor mediated facilitation of NE release appeared to exist.[94] Moreover, in isolated sympathetic neurons in culture[86,95] both α-adrenoceptors and dopamine receptors were coupled to adenylate cyclase but not to the release of NE. Thus, the role of cAMP as a modulator of NE release is unclear.

A WORKING HYPOTHESIS ON THE MECHANISM OF INHIBITION OF SNS BY CYCLIC GMP

Depolarization of sympathetic nerves is associated with activation of phospholipase and protein kinase C, elevated levels of inositol trisphosphate (INSP$_3$), and increased calcium ion within the sympathetic neuron. Calcium ion activates phospholipases A_2 and C and lipoxygenase[96] which results in arachidonic acid and hydroperoxides that can stimulate guanylate cyclase activity.[97] This can result in an increase of cGMP and a decrease in free GTP levels within the sympathetic neuron with resultant inhibition of NE release (FIGURE 12). Guanosine triphosphate, the precursor for cGMP, facilitates exocytosis from neutrophils[6] and may facilitate NE release by acting on G proteins linked to activation of phospholipase C. NE also releases EDRF which increases cGMP efflux from smooth muscle. This may contribute to inhibition of NE release from sympathetic nerve endings. cGMP-induced suppression of intraneuronal levels of inositol trisphosphate may also decrease release of intracellular calcium ion.

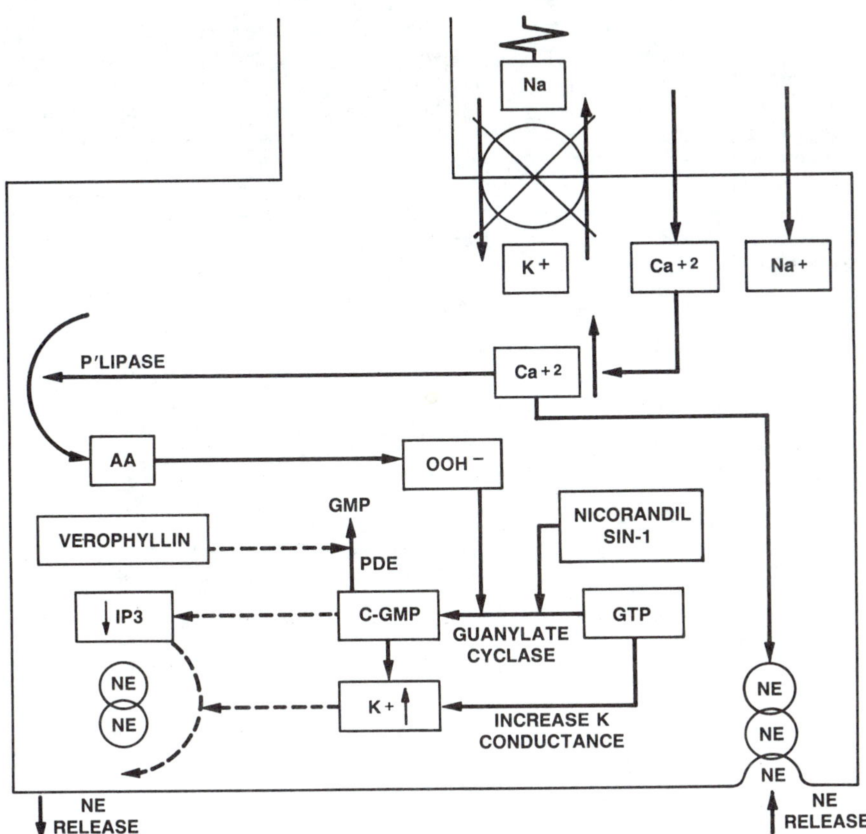

FIGURE 12. Proposed mechanism by which cyclic GMP inhibits NE efflux from nerve endings innervating smooth muscle. Increased Ca_i^{+2} activates phospholipase A_2 and C and lipoxygenase, the end products of which may stimulate guanylate cyclase, thereby elevating cGMP and decreasing GTP. Decreased levels of GTP or increased levels of GMP may decrease Ca_i^{+2} by decreasing G protein stimulation, stimulating calcium sequestration within, or efflux from, the cell or inhibiting the release of Ca_i^{+2} from the sarcoplasmic reticulum, respectively. Intraneuronal cGMP may hasten repolarization by increasing potassium conductance. cGMP may also decrease the hydrolysis of phosphoinositides to $INSP_3$ and Ca_i^{+2}. The combined effect of decreased Ca_i^{+2} and inhibited phosphoinositide hydrolysis (decreased $INSP_3$ and diacylglycerol) can decrease activation of protein kinase C and inhibit exocytosis of NE.

Polymyxin B (amphotericin B) an inhibitor of protein kinase C inhibits NE release from sympathetic nerves caused by phorbol ester and electrical stimulation.[8] Down regulating protein kinase C with prolonged incubation with phorbol esters inhibits cGMP-dependent inhibition of neurotransmitter release. cGMP may act at neuronal membranes to increase calcium-activated potassium conductance and enhance membrane repolarization. Inhibition of potassium conductance and membrane depolarization by KCl, ouabain, TEA, and 4-aminopyridine inhibits SNP-induced decreases in the release of NE from sympathetic nerves innervating VSM. Maintaining membrane depolarization increases the turnover of phosphoinositides in neurons. The activation of the enzyme may be beyond the capacity of cGMP to produce significant inhibition consistent with reductions in the efflux of NE. Decreased levels of GTP may also inhibit calcium influx directly by interfering with protein kinase C activation or by decreasing the activity of G proteins involved in neurosecretion. Both soluble and particulate guanylate cyclase must be involved because atrial natriuretic factor inhibits sympathetic neurotransmitter release. The system is probably functional under nonperturbed, *in vitro* conditions because 1-arginine, a modulator of guanylate cyclase and a precursor to EDRF, inhibits the release of NE from sympathetic nerves innervating rabbit myocardium[98] and methylene blue facilitates the release of NE in endothelium-rubbed mesenteric arteries. Finally if endogenous levels of intraneuronal cAMP maximally facilitates release of NE then cGMP may inhibit NE release by enhancing hydrolysis of cAMP. Further studies are required to clarify the role of cGMP in peripheral sympathetic neurotransmitter release.

REFERENCES

1. KALIX, P. 1976. J. Neurochem. **27:** 1563–1564.
2. DREWETT, J. G., R. J. ZIEGLER, G. R. MARCHAND & G. J. TRACHTE. 1989. J. Pharmacol. Exp. Ther. **250:** 428–432.
3. YANAGIHARA. N., M. ISOSAKI, T. OHUCHI & M. OKA. 1979. FEBS. Lett. **105:** 296–298.
4. SCHNEIDER, A. S., H. T. CLINE & S. LEMAIRE. 1979. Life Sci. **24:** 1389–1394.
5. SITARAMAYYA, A., J. A. CAMPBELL & F. L. SIEGEL. 1978. J. Neurochem. **30:** 1281–1285.
6. BARROWMAN, M. M., S. COCKROFT & B. D. GOMPERTS. 1986. Nature **319:** 504–507.
7. ITO, S., M. NEGISHI, H. HAYASHI & O. HAYAISHI. 1989. Ann. N.Y. Acad. Sci. **559:** 453–454.
8. ALLGAIRE, C. & G. HERTTING. 1986. Naunyn Schmiedebergs Arch. Pharmacol. **334:** 218–221.
9. WAKADE, A. R., R. K. MALBOTRA & T. D. WAKADE. 1985. Naunyn Schmiedebergs Arch. Pharmacol. **331:** 122–124.
10. ALLGAIER, C., G. HERTTING, H. Y. HUANG & R. JACKISCH. 1987. Br. J. Pharmacol. **92:** 161–172.
11. WINKLER, H. 1988. *In* Handbook of Pharmacology. Catecholamines I.: 43–188. Springer Verlag. Berlin, FRG.
12. TREIMAN, M., W. WEBER & M. GRATZL. 1983. J. Neurochem. **40:** 661–669.
13. CUBEDDU, L., E. BARNES & N. WEINER. 1975. J. Pharmacol. Exp. Ther. **193:** 105–112.
14. FREDHOLM, B. B., K. BRODIN & K. STRANDBERG. 1979. Acta Pharmacol. Toxicol. **45:** 336–344.
15. GOTHERT, M. 1984. Blood Vessels **21:** 117–125.
16. GOTHERT, M. & F. HENTRICH. 1984. Naunyn Schmiedebergs Arch. Pharmacol. **328:** 127–134.
17. STJARNE, L. 1976. Neuroscience **1:** 19–22.
18. TRENDELENBURG, U. & N. WEINER. 1988. Handbook of Experimental Pharmacology. Catecholamines I and II. Springer Verlag. Berlin, FRG.
19. WOTTEN, G. F., N. B. THOA, I. J. KOPIN & J. AXELROD. 1973. Mol. Pharmacol. **9:** 178–193.
20. NORDSTROM, O. & T. BARTFAI. 1981. Brain Res. **213:** 467–471.
21. ITOH, T., K. FURUKAWA, T. INOUE, *et al.* 1981. J. Pharmacol. Exp. Ther. **218:** 260–270.

22. GREENBERG, S., K. PEEVEY, T. P. TANAKA & F. P. J. DIECKE. 1989. Eur. J. Pharmacol. **162:** 67–80.

23. GREENBERG, S. S., F. P. J. DIECKE, K. PEEVY & T. P. TANAKA. 1990. Am. J. Hypertens. **3:** 263–274.

24. GREENBERG, S., F. P. J. DIECKE & K. PEEVY. Eur. J. Pharmacol. (In press.)

25. DREWETT, J. G., G. J. TRACHTE & G. R. MARCHAND. 1989. J. Pharmacol. Exp. Ther. **248:** 135–141.

26. COCKS, T. M. & J. A. ANGUS. 1983. Nature **305:** 627–630.

27. ENDOH, M. & N. TAIRA. 1983. Naunyn Schmiedebergs Arch. Pharmacol. **322:** 319–321.

28. FURCHGOTT, R. F. & D. JOTHIANANDAN. 1983. (Abstr.) Fed. Proc. **42:** 619.

29. MURAD, F. 1986. J. Clin. Invest. **78:** 1–5.

30. RAPOPORT, R. M., M. B. DRAZNIN & F. MURAD. 1983. Nature **306:** 174–176.

31. COHEN, R. A. & H. WEISBRODT. 1988. Am. J. Physiol. **255:** H871–H879.

32. TESFAMARIAM, B., R. M. WEISBROD & R. A. COHEN. 1987. Am. J. Physiol. **253:** H792–H798.

33. WOOD, J. N., P. R. COOTE, A. MINHAS, I. MULLANEY, M. MCNEILL & G. M. BURGESS. 1989. J. Neurochem. **53:** 1203–1211.

34. BURGESS, G. M., I. MULLANEY, M. MCNEILL, P. R. COOTE, A. MINHAS & J. N. WOOD. 1989. J. Neurochem. **53:** 1212–1218.

35. PERNEY, T. M., L. D. HIRNING, S. E. LEEMAN & R. J. MILLER. 1986. Proc. Nat. Acad. Sci. **83:** 6656–6659.

36. VILLARROEL, A., N. V. MARRION, H. LOPEZ & P. R. ADAMS. 1989. FEBS Lett. **255:** 42–46.

37. WATANABE, T., T. KAWADA, M. YAMAMOTO & K. IWAI. 1987. Biochem. Biophys. Res. Commun. **142:** 259–264.

38. ELLIS, J. L. & G. BURNSTOCK. 1989. Br. J. Pharmacol. **98:** 707–713.

39. CHEUNG, D. W. 1988. Biochem. Biophys. Res. Commun. **154:** 411–415.

40. MacKAY, M. J. & D. W. CHEUNG. 1987. Can. J. Physiol. Pharmacol. **65:** 1988–1990.

41. BERGEY, J. L., D. R. MUCH, V. KRATUNIS & M. ASAAD. 1989. Eur. J. Pharmacol. **159:** 103–112.

42. FRIEDL, A. C. HARMENING, F. SCHMALZ, *et al.* 1989. J. Neurochem. **52:** 589–597.

43. CHINKERS, M., D. L. GARBERS, M. S. CHANG, *et al.* 1989. Nature **338:** 78–83.

44. LEITMAN, D. C., J. W. ANDRESEN, R. M. CATALANO, S. A. WALDMAN, J. J. TUAN & F. MURAD. 1988. J. Biol. Chem. **263:** 3720–3728.

45. FURCHGOTT, R. F. 1984. Annu. Rev. Pharmacol. Toxicol. **24:** 175–197.

46. FURCHGOTT, R. F. & J. V. ZAWADZKI. 1980. Nature **288:** 373–376.

47. FURCHGOTT, R. F., J. V. ZAWADZKI & P. D. CHERRY. 1985. *In* Mechanisms of Vasodilatation. P. M. Vanhoutte & I. Leusen, Eds.: 34–43. S. Karger. Basel, Switzerland.

48. FURCHGOTT, R. F. & P. M. VANHOUTTE. 1989. FASEB J. **3:** 2007–2018.

49. VANHOUTTE, P. M. 1988. The Endothelium in Cardiovascular Function. Humana Press. Clifton, N.J.

50. GREENBERG, S. S., F. P. J. DIECKE & T. P. TANAKA. 1986. Drug Dev. Res. **10:** 269–309.

51. GREENBERG, S. & F. P. J. DIECKE. 1988. Drug Dev. Res. **12:** 131–150.

52. MARTIN, W., J. A. SMITH & D. G. WHITE. 1986. Br. J. Pharmacol. **89:** 563–571.

53. MARTIN, W., G. M. VILLANI, D. JOTHIANANDAN & R. F. FURCHGOTT. 1985. J. Pharmacol. Exp. Ther. **233:** 679–685.

54. PALMER, R. F., R. RIENZE & S. MONCADA. 1987. Nature **339:** 635–637.

55. PALMER, R. F., R. RIENZE & S. MONCADA. 1987. Br. J. Pharmacol. **92:** 365–374.

56. GREENBERG, S., G. M. RUBANYI & D. E. WILCOX. Circulation Res. (Accepted.)

57. GUERRERA, R. & D. MYERS. 1990. *In* Proceedings of the First International Conference on Endothelium-Derived Relaxing and Contracting Factors. Elsevier Press. New York, NY.

58. FELETOU, M. & P. M. VANHOUTTE. 1988. Br. J. Pharmacol. **93:** 515–524.

59. HOLZMANN, S. 1982. J. Cyclic Nucleotide Res. **8:** 409–419.

60. IGNARRO, L. J., G. M. BUGA, K. S. WOOD, *et al.* 1987. Proc. Nat. Acad. Sci. **84:** 9265–9269.

61. PALMER, R. M. J., D. A. FERRIGE & S. MONCADA. 1987. Nature London **327:** 524–526.

62. ESTRADA, C. C., C. GONZALEZ, S. LLUCH, *et al.* 1988. (Abstr.) FASEB J. **2:** 3925.

63. GONZALEZ, R. R. & S. P. DUCKLES. 1988. (Abstr.) FASEB J. **2:** 5179.

64. HYNES, M. R., H. DANG & S. P. DUCKLES. 1988. Life Sci. **42:** 357–365.

65. KOMORI, K., G. CHEN & H. SUZUKI. 1989. Pflugers Arch. **413:** 359–364.

66. KOTECHA, N. & T. O. NEILD. 1985. Neurosci. Lett. (Suppl.) **19:** S78–S80.
67. SOARES-DASILVA, P. & J. S. GILLESPIE. 1987. Blood Vessels **24:** 287–288.
68. LOIACONO, R. E. & D. F. STORY. 1984. J. Pharm. Pharmacol. **36:** 262–264.
69. RORIE, T. & G. M. TYCE. 1985. Am. J. Physiol. **248:** H193–199.
70. MEEHAN, A. G., D. F. STORY & I. C. MEDGETT. 1987. Eur. J. Pharmacol. **141:** 339–344.
71. GREENBERG, B., K. RHODEN & P. J. BARNES. 1986. J. Moll. Cell. Cardiol. **18:** 975–981.
72. GREENBERG, B., K. RHODEN & P. J. BARNES. 1986. Am. J. Physiol. Heart Circ. Physiol. **257:** H241–H248.
73. LAMB, F. S., C. M. KING, K. HARRELL, *et al.* 1987. Am. J. Physiol. **252:** H1041–H1046.
74. WIKLUND, N. P., A. ÖHLEN & B. CEDERQVIST. 1988. Acta Physiol. Scand. **134:** 311–312.
75. LINNIK, M. D. & T. J. F. LEE. 1989. J. Cereb. Blood Flow Metab. **9:** 219–225.
76. COHEN, R. A., B. TESFAMARIAM & R. M. WEISBRODT. 1990. *In* Proceedings of the First International Conference on Endothelium-Derived Relaxing and Contracting Factors. Elsevier Press. New York, N.Y.
77. GREENBERG, S., G. M. RUBANYI & D. E. WILCOX. 1990. *In* Proceedings of the First International Conference on Endothelium-Derived Relaxing and Contracting Factors. Elsevier Press. New York, N.Y.
78. SABATINE, J. M. & C. J. COFFEE. 1986. Arch. Biochem. Biophys. **249:** 95–105.
79. RUBANYI, G. M. & P. M. VANHOUTTE. 1988. Eur. J. Pharmacol. **145:** 351–354.
80. HOEFFNER, U. & P. M. VANHOUTTE. 1987. Fed. Proc. **46:** 644.
81. RAPOPORT, R. M., K. SCHWARTZ & F. MURAD. 1985. Eur. J. Pharmacol. **110:** 203–208.
82. AHLNER, J., K. L. AXELSSON, J. O. G. KARLSSON & R. G. G. ANDERSSON. 1988. Life Sci. **43:** 1241–1248.
83. AUDIGIER, S. M., J. K. WANG & P. GREENGARD. 1988. Proc. Nat. Acad. Sci. USA **85:** 2859–2863.
84. QUENZER, L. F. & R. L. VOLLE. 1983. J Auton. Nerv. Syst. **8:** 161–164.
85. GREENGARD, P. & M. D. BROWNING. 1988. Adv. Second Messenger Phosphoprotein Res. **21:** 133–146.
86. WAKADE, A. R., T. D. WAKADE, S. V. BHAVE & R. K. MALHOTRA. 1988. Neuroscience **27:** 1021–1028.
87. RAHAMIMOFF, R. & B. KATZ. 1986. Calcium, Neuronal Function and Transmitter Release. Martinus Nijhoff Publishing. Boston, Mass.
88. BEVAN, J. A., R. D. BEVAN & S. P. DUCKLES. 1983. *In* Handbook of Physiology. Section II. Vascular Smooth Muscle: 516–566. American Physiological Society Williams and Wilkins. Bethesda, Md.
89. HEDQUIST, P., B. B. FREDHOLM & S. OHLUNDH. 1978. Circulation Res. **43:** 592–598.
90. HENTRICH, F., M. GOTHERT & D. GRESCHUCHNA. 1985. Naunyn Schmiedebergs Arch. Pharmacol. **330:** 245–247.
91. MULLER, M. J. & H. P. BAER. 1983. Naunyn Schmiedebergs Arch. Pharmacol. **322:** 78–82.
92. WACHTEL, H. 1983. J. Neural Transm. **56:** 139–152.
93. JOHNSTON, H. & H. MAJEWSKI. 1986. Br. J. Pharmacol. **87:** 553–562.
94. MOLDERINGS, G., J. LIKMUGU, H. R. ZERKOWSKI & M. GOTHERT. 1988 Naunyn Schmiedebergs Arch. Pharmacol. **337:** 408–414.
95. WAKADE A. R., T. D. WAKADE & R. K. MALHOTRA. 1988. J. Neurochem. **51:** 820–829.
96. SNIDER, R. M., M. MCKINNEY, C. FORRAY & E. RICHELSON. 1984. Proc. Nat. Acad. Sci. USA **81:** 3905–3910.
97. HIDAKA, H. & T. ASANO. 1977. Proc. Nat. Acad. Sci. USA **74:** 3657–3661.
98. KARWATOWSKA-PROKOPCZUK, E. & A. WENNMALM. 1989. Acta Physiol. Scand. **135:** 419–420.

Autoreceptor Regulation of Dopamine Synthesis

MARINA E. WOLF[a] AND ROBERT H. ROTH[b]

[a]*Department of Psychiatry*
Wayne State University School of Medicine
Lafayette Clinic
951 East Lafayette
Detroit, Michigan 48207

[b]*Departments of Pharmacology and Psychiatry*
Yale University School of Medicine
. 333 Cedar Street
New Haven, Connecticut 06510

The first evidence for the modulation of dopamine (DA) synthesis by presynaptic dopamine receptors was presented in 1972 by Carlsson and coworkers, who observed that when impulse flow in nigrostriatal DA neurons is interrupted by axotomy the rate of tyrosine hydroxylation in striatal terminals is increased.[1] This was interpreted as suggestive of the existence of a presynaptic site at which released DA can act to exert feedback inhibition on DA synthesis. At the time, the concept of autoregulation in the central nervous system (CNS) was novel. Since then, it has become clear that autoreceptors provide a means for many types of neurons to regulate cellular functions such as neurotransmitter release, synthesis, and impulse flow. Although the popularity of DA autoreceptors has waxed and waned over the years, autoreceptor reviews have been plentiful (see Reference 2 for a recent example). This paper will therefore be quite selective in its focus. First, it will concentrate on the autoreceptor regulation of DA synthesis, since release and impulse flow are being discussed by other authors in the volume. Second, the emphasis will be on new developments in this field.

TYPES OF DOPAMINE AUTORECEPTORS

Autoreceptors can exist on most portions of DA cells, including the soma, dendrites, and nerve terminals. Simulation of DA autoreceptors in the somatodendritic region slows the firing rate of DA neurons, while stimulation of autoreceptors located on DA nerve terminals results in an inhibition of DA snythesis and release.[2] Somatodendritic autoreceptors may also regulate DA synthesis.[3] Thus, somatodendritic and nerve termnal autoreceptors work in concert to exert feedback regulatory effects on dopaminergic transmission. For the purposes of discussion, three types of autoreceptors will be defined according to their functional effect: impulse-modulating autoreceptors, release-modulating autoreceptors, and synthesis-modulating autoreceptors. It is not clear whether the same receptor protein actually performs all three functions (see below). In general, however, all DA autoreceptors can be classified as D2 DA receptors. Their pharamacological properties are similar to those of postsynaptic D2 receptors (see below), but they are 5–10 times more sensitive to the effects of DA and apomorphine than postsynaptic DA receptors in behavioral, biochemical and electrophysiological models (for example, see Reference 4).

323

MODEL SYSTEMS FOR STUDYING SYNTHESIS-MODULATING AUTORECEPTORS

In order to study the regulation of tyrosine hydroxylation by nerve terminal autoreceptors, it is necessary to remove all other mechanisms by which DA agonists can influence the rate of tyrosine hydroxylase (TH) activity. The major concern stems

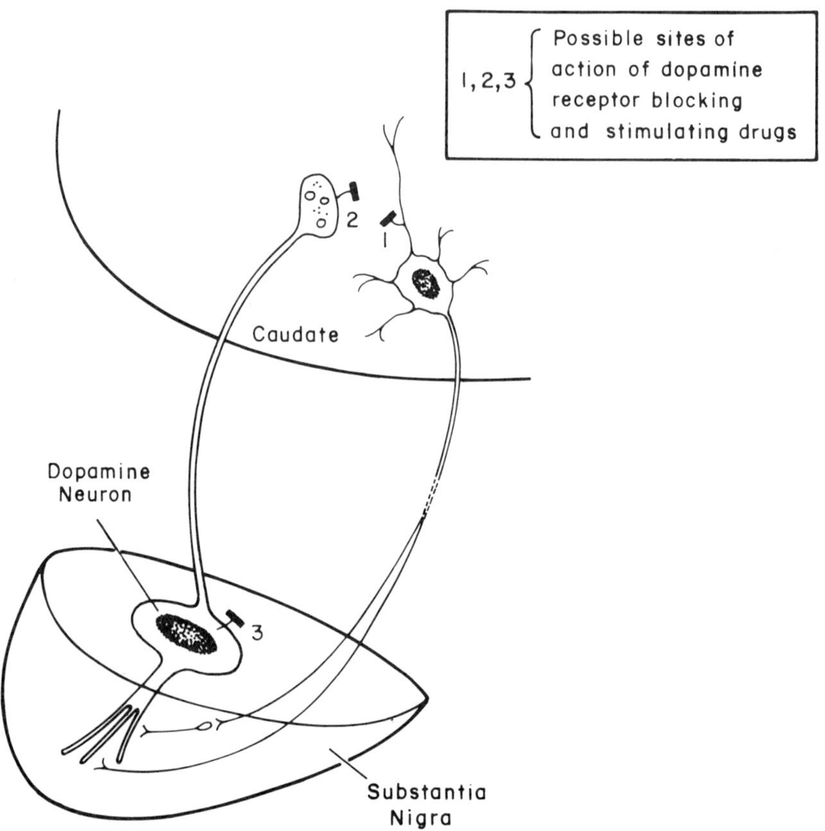

FIGURE 1. Sites at which dopaminergic drugs can act to alter DA synthesis. Lesions of nigrostriatal fibers mechanically or pharmacologically (GBL) eliminate the influence of drug interactions with postsynaptic DA receptors (1) and somatodendritic autoreceptors (3). Similarly, during electrical stimulation of the nigrostriatal pathway, drugs will not be able to have a further effect on dopaminergic impulse flow and thus will alter DA synthesis only by direct presynaptic action (2). (Reprinted from Reference 150 with permission.)

from the fact that TH activity is coupled to the rate of impulse flow; thus, it is necessary to eliminate the possibility that DA agonists are acting either at somatodendritic autoreceptors or at postsynaptic DA receptors to slow DA cell firing either directly or via neuronal feedback pathways. This can be accomplished using several experimental models (FIGURE 1).

GBL or Axotomy

DA neurons are similar to other monoaminergic neurons in that they respond to increases in impulse flow by increasing transmitter synthesis. However, if impulse flow is interrupted in DA neurons, either mechanically or pharamacologically by administration of gamma-butyrolactone (GBL), the neurons respond in a unique fashion by increasing the rate of DA synthesis and increasing DA levels in the terminals.[5-7] The increase in DA levels appears to reflect the fact that inhibiting impulse flow in DA neurons also eliminates impulse-dependent DA release. *In vivo* dialysis studies have now provided direct evidence for a sharp drop in extracellular DA levels after GBL administration.[8] This decrease in synaptic DA levels leads to a disinhibition of those events normally under the tonic control of nerve terminal synthesis-modulating autoreceptors, resulting in an increased rate of tyrosine hydroxylation. DA agonists reverse the GBL-induced increase in both DA levels and the rate of tyrosine hydroxylation. These effects can be attributed entirely to stimulation of nerve terminal autoreceptors, since effects on cell body autoreceptors and postsynaptic receptors become moot under conditions of zero impulse flow. In GBL and other *in vivo* experiments, the rate of DA synthesis can be determined by administering a decarboxylase inhibitor (NSD-1015) to block the conversion of dihydroxyphenylalanine (DOPA) to DA. DOPA accumulation can then be used as an index of *in vivo* tyrosine hydroxylation.

Electrical Stimulation

The influence of drug action at somatodendritic and postsynaptic DA receptors can also be eliminated by stimulating nigrostriatal axons (in the medial forebrain bundle) at supramaximal frequencies, thereby "setting" the firing rate of these neurons at a constant level and overriding drug effects at somatodendritic autoreceptors or postsynaptic feedback pathways. The ability of DA agonists to reverse the stimulation-induced increase in DA synthesis and DA metabolite levels is used as an index of autoreceptor activation.[9,10]

In vitro Studies

It is possible to selectively study drug effects at the nerve terminal by using brain slices or synaptosomal preparations obtained from dopaminergic projection areas. The advantages and pitfalls of this approach have been reviewed elsewhere.[2]

In vivo Dialysis

It was shown recently that dialysate levels of DOPA after local infusion of NSD-1015 can be used to monitor *in vivo* striatal tyrosine hydroxylation in conscious rats.[11] In excellent agreement with the results of postmortem studies, DOPA formation was decreased by apomorphine administration and increased by haloperidol or GBL. This technique should enable autoreceptor function to be studied in discrete brain regions in intact animals.

IS IT NECESSARY TO POSTULATE THE EXISTENCE OF AUTORECEPTORS THAT MODULATE DA SYNTHESIS DIRECTLY?

This question arises because of the ability of DA and other catechols to exert direct inhibitory effects on TH activity. Thus, DA agonists could decrease TH activity through two mechanisms in addition to an action at synthesis-modulating autoreceptors: (1) they could be taken up into the nerve terminal and inhibit TH directly, and (2) they could act at release-modulating autoreceptors to decrease DA release, increase the size of intraneuronal DA pools, and thereby increase the level of feedback inhibition to which TH is subjected. Do these possibilities obviate the need for autoreceptors that modulate synthesis directly?

The answer is clearly no. With respect to the first alternative, inhibition of tyrosine hydroxylation can be produced by DA agonists that do not enter DA nerve terminals by either active uptake or diffusion and that do not inhibit TH directly.[12] With respect to the second alternative, a number of arguments can be made. First, DA agonists can inhibit TH activity under experimental conditions in which the level of impulse flow and DA release vary widely. This is best illustrated by the effectiveness of DA agonists in all of the models discussed above: (a) the GBL model, in which impulse flow and DA release are eliminated, (b) the electrical stimulation model, in which impulse flow and DA release are supramaximal, (c) *in vitro* preparations, in which the level of DA release depends on experimental conditions, and (d) *in vivo* dialysis studies, which presumably reflect normal rates of impulse flow and release in intact animals. Second, studies of mesoprefrontal DA neurons, which possess release-modulating autoreceptors but not synthesis-modulating autoreceptors, suggest that autoregulation of DA release and synthesis can occur independently of each other (see below). Most convincingly, recent studies have shown that DA autoreceptor stimulation decreases the phosphorylation of TH in striatal slices.[13] As will be discussed below, this provides a mechanism for the autoregulation of DA synthesis that is clearly unrelated to manipulations of DA release and intraneuronal DA levels.

REGIONAL DISTRIBUTION OF SYNTHESIS-MODULATING AUTORECEPTORS

A schematic representation of the efferent projections of midbrain dopamine cell groups is shown in FIGURE 2. The nerve terminals of all midbrain DA neurons examined to date have been found to possess release-modulating DA autoreceptors[14–20] (see TABLE 1), although the density of these receptors may differ in some projection areas.[21] In contrast, while the majority of ventral tegmental DA neurons appear to possess snythesis-modulating autoreceptors, those that project to the prefrontal and cingulate cortices appear either to have a greatly diminished number of these receptors or to lack them entirely.[22–24] More recently, a lack of synthesis-modulating autoreceptors was reported in the amygdaloid nuclei, the anterior amygdaloid area, septal nuclei, and subdivisions of the interstitial (bed) nucleus of the stria terminalis. However, their presence in other mesolimbic regions (nucleus accumbens and olfactory tubercles) was confirmed.[25] Heterogeneity with respect to autoreceptor function is also apparent among hypothalamic DA neurons. Incertohypothalamic DA neurons and tuberohypophyseal DA neurons terminating in the intermediate lobe possess synthesis-modulating autoreceptors, while tuberohypophyseal DA neurons terminating in the neural lobe do not.[26–28] Tuberoinfundibular DA neurons also lack snythesis-modulating DA autoreceptors.[26,29,30] The regional distribution of synthesis-modulating autorecep-

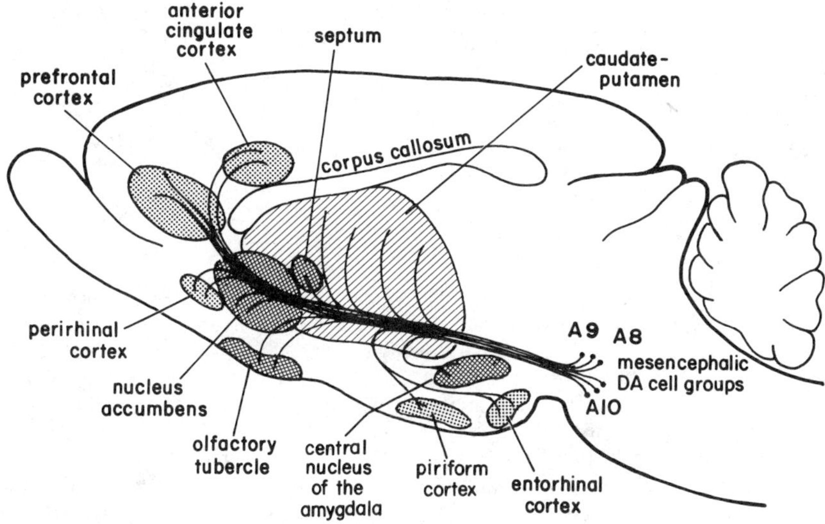

FIGURE 2. Schematic diagram illustrating the anatomy of nigrostriatal, mesolimbic, and mesocortical dopaminergic neuronal systems in rat brain. (Reprinted from Reference 2 with permission.)

tors is summarized in TABLES 2 and 3. It should be noted that end-product inhibition of TH activity by intraneuronal DA appears to be a regulatory property common to all DA neurons.[25,26,31]

The absence of synthesis-modulating autoreceptors appears in part to be responsible for some of the unique characteristics of mesoprefrontal and mesocingulate DA neurons when compared with midbrain DA neurons which possess autoreceptors (see References 2 and 32 for reviews). For example, the turnover rate of DA is much faster in the prefrontal and cingulate cortices than in the projection fields of other midbrain DA neurons which possess autoreceptors.[23,31,33] In addition, synthesis-modulating autoreceptors appear to play a prominent role in determining the response of DA neurons to acute and chronic treatment with DA agonists and antagonists. DA neurons that lack autoreceptors show little response (mesoprefrontal and mesocingulate systems) or no response (tuberoinfundibular system) to acute DA antagonist or agonist challenge. In contrast, DA systems that possess autoreceptors (nigrostriatal, mesolimbic, mesopyriform, and tuberohypophyseal systems) exhibit a dramatic response to the administration of DA agonists and antagonists.

It seems likely that nerve terminal synthesis-modulating autoreceptors may also be involved in the dvelopment of biochemical tolerance to antipsychotic drugs after

TABLE 1. All DA Nerve Terminals Possess Release-Modulating Autoreceptors[a]

1. Striatum
2. Nucleus acumbens
3. Prefrontal cortex
4. Cingulate cortex
5. Entorhinal cortex

[a]References 14, 15, and 17–20.

TABLE 2. DA Neurons Possessing Synthesis-Modulating Autoreceptors[a]

1. Nigrostriatal
2. Mesolimbic—nucleus accumbens
—olfactory tubercles
3. Mesocortical—pyriform cortex
4. Tuberohypophyseal—intermediate lobe
5. Incertohypothalamic

[a]References 6, 7, 22, 24–28, 30, 31, and 45.

long-term administration. A number of laboratories have demonstrated that DA neurons projecting to the prefrontal and cingulate cortices are relatively resistant to the development of biochemical tolerance following the chronic administration of therapeutically revelant doses of antipsychotic drugs.[34–38] This is in contrast to the nigrostriatal, mesolimbic, and mesopyriform DA neurons, which readily develop tolerance during chronic antipsychotic drug administration. The time course for the development of tolerance in these systems parallels the development of autoreceptor supersensitivity.[39,40] These studies in rat have been extended to nonhuman primates, where tolerance development is observed in the caudate, putamen, and olfactory tubercle, but not in the frontal and cingulate cortices.[41,42]

MECHANISMS FOR REGULATING SYNTHESIS IN MESOPREFRONTAL DA NEURONS

Studies using the GBL model have suggested that mesoprefrontal DA neurons lack synthesis-modulating autoreceptors (above). In these studies, rats were administered both GBL and the decarboxylase inhibitor NSD-1015, and DOPA formation was used as an index of tyrosine hydroxylation. However, in experiments in which DOPA formation was measured in the *absence* of GBL, autoreceptor-selective doses of apomorphine were found to produce an inhibition of DOPA accumulation in prefrontal cortex.[30,43,44] The ability of apomorphine to inhibit prefrontal DOPA formation in the absence but not in the presence of GBL is shown in FIGURE 3.

Recent work in our laboratory suggests that these apparently contradictory findings can be resolved by proposing that mesoprefrontal DA neurons lack functional synthesis-modulating autoreceptors but possess release-modulating autoreceptors that are indirectly responsible for the effects of agonists on synthesis. The basis of this hypothesis is that the ability of DA agonists to inhibit prefrontal DOPA formation depends on their ability to modulate intraneuronal DA levels. In GBL experiments, impulse flow and impulse-dependent DA release are prevented. In contrast, in experiments in which NSD-1015 is administered alone, impulse-dependent release is still

TABLE 3. DA Neurons Lacking Synthesis-Modulating Autoreceptors[a]

1. Mesolimbic—amygdaloid nuclei
2. Mesocortical—prefrontal cortex
—cingulate cortex
3. Tuberohypophyseal—neural lobe
4. Tuberoinfundibular

[a]References 6, 7, 22, 24–28, 30, 31, and 45.

occurring during the period of time (generally 30 minutes) between administration of NSD and sacrifice. We have found that, in rapidly firing mesoprefrontal (and mesocingulate) DA neurons, treatment with NSD-1015 results in a dramatic (70%) depletion of DA within 30 minutes (TABLE 4). This presumably reflects the fact that impulse-dependent DA release is continuing under conditions in which DA stores cannot be replenished, since the conversion of DOPA to DA is blocked. In contrast, NSD-1015 produces little change in DA levels in striatum over this time period (TABLE 4). This may be because nigrostriatal cells fire more slowly and would therefore be expected to release transmitter more gradually. The inhibitory effects of DA agonists on DOPA synthesis in the prefrontal and cingulate cortices appear to be related to the ability of these agents to retard the disapperance of DA induced by treatment with NSD-1015. Thus the inhibitory effects of DA agonists on DOPA formation could be

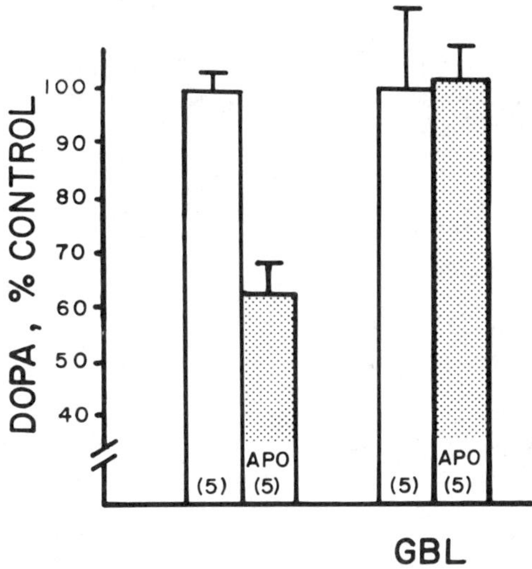

FIGURE 3. GBL pretreatment interferes with the ability of apomorphine [50 μg/kg, subcutaneously (sc)] to inhibit prefrontal DOPA formation, suggesting that this effect is dependent on the maintenance of dopaminergic impulse flow and impulse-dependent DA release (data taken from Reference 31).

related to end-product-inhibitory effects on TH exerted by the elevated DA (relative to NSD-1015 controls) in these nerve terminals.[31]

The mechanism by which DA agonists decrease the rate of DA loss from prefrontal cortex after NSD-1015 may involve activation of release-modulating autoreceptors. Consistent with this possibility, several laboratories have demonstrated the presence of release-modulating autoreceptors on mesoprefrontal nerve terminals.[16,17,19] Under conditions in which effects of DA agonists on prefrontal DA release are prevented, agonists are no longer able to inhibit prefrontal DOPA formation. For example, the apomorphine-induced inhibition of prefrontal, but not striatal, DOPA synthesis is blocked by either cessation of impulse-dependent DA release (GBL pretreatment) or by depletion of releasable DA (by reserpine pretreatment).[31] Additional support for this hypothesis is provided by experiments in which DOPA formation was measured in

TABLE 4. Prefrontal DOPA Accumulation Is Inversely Dependent on DA Concentration[a]

	DOPA (% NSD Control)	DA (% Untreated Control)[d]
Striatum		
NSD	100 ± 6^{b}	92 ± 4
NSD/apomorphine (100 μg/kg)	65 ± 6^{b}	101 ± 5
NSD/apomorphine (2 mg/kg)	37 ± 3^{b}	107 ± 5
NSD/reserpine	318 ± 6^{c}	4 ± 1
NSD/reserpine/apo	53 ± 1^{c}	10 ± 1
Prefrontal cortex		
NSD	100 ± 4^{b}	33 ± 5
NSD/apomorphine (100 μg/kg)	71 ± 7^{b}	79 ± 3
NSD/apomorphine (2 mg/kg)	62 ± 7^{b}	88 ± 4
NSD/reserpine	152 ± 3^{c}	5 ± 1
NSD/res/apo (1 mg/kg)	147 ± 2^{c}	9 ± 1

[a]Data taken from Reference 31.

[b]Control value for striatum = 2.11 ± 0.12 μg DOPA/g per 30 minutes; control value for prefrontal cortex = 156 ± 5 ng DOPA/g per 30 minutes ($n = 6$ each).

[c]Control value for striatum = 1.93 ± 0.07 μg DOPA/g per 30 minutes ($n = 4$); control value for prefrontal cortex = 154 ± 11 ng DOPA/g per 30 minutes ($n = 5$).

[d]Control value for striatum = 17.2 ± 1.1 μg DA/g; control value for prefrontal cortex = 104 ± 8 ng DA/g ($n = 4$ each).

brain slices. Unlike the *in vivo* situation, decarboxylase inhibition does not result in rapid depletion of DA levels in prefrontal cortical slices, presumably because impulse-dependent release is attenuated in this preparation. Under these conditions, DOPA formation in striatal slices was modulated by DA agonists and antagonists, whereas prefrontal DOPA formation was not.[45]

These studies indicate that mesoprefrontal DA nerve terminals, although lacking synthesis-modulating autoreceptors, do contain a release-modulating mechanism. In contrast, striatal DA nerve terminals appear to possess autoreceptors capable of modulating release as well as directly modulating the rate of tyrosine hydroxylation. The question of whether this occurs through one receptive site coupled to both release and synthesis, or through two types of autoreceptors, will be discussed below.

MECHANISM OF ACTION OF SYNTHESIS-MODULATING AUTORECEPTORS

Depolarization of striatal slices or synaptosomes, as well as increases in the firing rate of nigrostriatal DA neurons *in vivo*, results in an increase in DA synthesis which is accompanied by alterations in the kinetic constants of TH prepared from the treated tissue (primarily a decrease in the K_m for cofactor).[46–57] Stimulation of nerve terminal DA autoreceptors in striatal slices and synaptosomes has been reported to reverse the kinetic activation of TH produced by depolarization or treatment with the adenylate cyclase activator forskolin.[58–60] However, DA agonists were unable to reverse the kinetic activation elicited by dibutyryl cyclic AMP. The ability of DA agonists to overcome the effect of forskolin but not dibutyryl cyclic AMP suggests that autoreceptor inhibition of TH activity may involve a decrease in adenylate cyclase activity. This is consistent with the observation that D2 DA receptors in striatum are negatively coupled to adenylate cyclase.[61,62] Furthermore, while forskolin increases TH activity in both the striatum and the prefrontal cortex, DA agonists reverse the effect of forskolin only in the striatum (FIGURE 4) (Knorr and Roth, unpublished findings). This suggests that the reversal is mediated by synthesis-modulating autoreceptors, which are absent in prefrontal cortex. It should be noted, however, that not all evidence is consistent with the notion that synthesis-modulating autoreceptors regulate TH activity by decreasing adenylate cyclase activity. For example, while synthesis-modulating autoreceptors are functional in the nucleus accumbens, D2 DA receptors in this region do not appear to be coupled in an inhibitory manner to adenylate cyclase.[20,63,64]

The phosphorylation of TH is consistently associated with increases in enzyme activity.[65] Several kinases are capable of phosphorylating and activating TH, including Ca^{2+}-phospholipid-dependent protein kinase (C kinase),[66] Ca^{2+}-calmodulin-dependent protein kinase II (CaM kinase),[67-69] and cyclic-AMP-dependent protein kinase.[66,70-72] Work in peripheral catecholaminergic tissues has demonstrated that depolarization and other physiological stimuli increase the phosphorylation of TH. Trypic peptide analysis of TH phosphorylated *in situ* indicates that this phosphorylation occurs at multiple sites, and that specific physiological stimuli and kinases alter phosphorylation at these sites differentially.[73-77] Importantly, similar results were obtained recently in CNS nerve terminals; in these studies, K^+ depolarization of striatal synaptosomes was found to increase phosphate incorporation into two phosphopeptides obtained after limited tryptic digestion of ^{32}P-labeled TH.[78]

Based on the importance of phosphorylation as a regulatory mechanism for TH, it has long been speculated that autoreceptor function and the regulation of TH phosphorylation are mechanistically linked. Direct evidence to support this hypothesis was provided recently by studies showing that incubation of striatal slices with DA agonists decreased phosphate incorporation into TH under both basal and K^+-stimulated conditions (FIGURE 5).[13] Changes observed in the phosphorylation of TH were mirrored by changes in the rate of tyrosine hydroxylation, suggesting that a

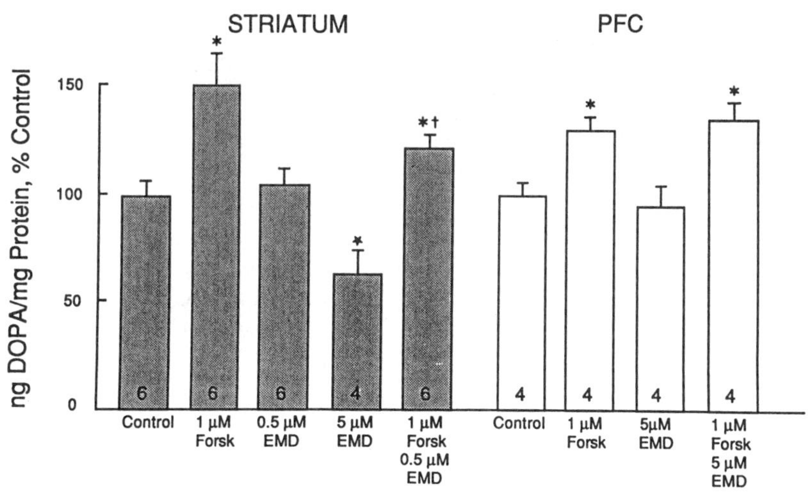

FIGURE 4. Forskolin increases DOPA formation in both striatal and prefrontal cortical brain slices. EMD 23 448 reverses this effect in striatal slices only. (Knorr and Roth, unpublished results.)

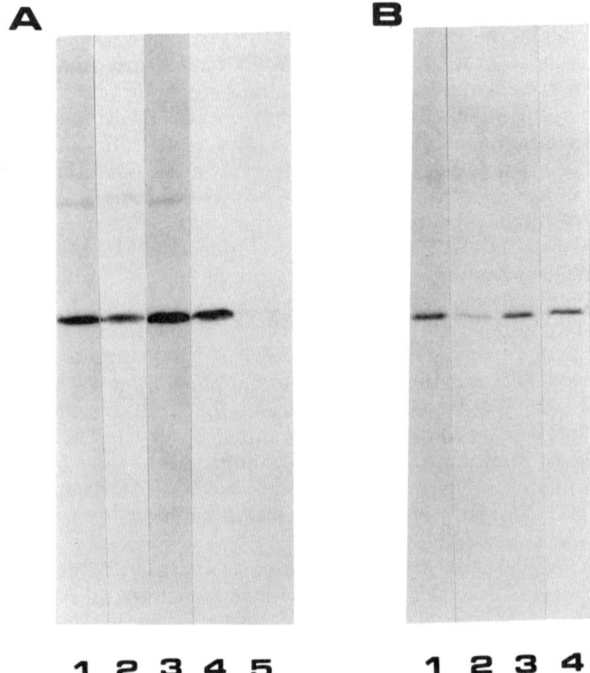

FIGURE 5. Stimulation of DA autoreceptors decreases the phosphorylation of TH. Striatal slices were incubated with ^{32}P-labeled inorganic phosphate, and ^{32}P incorporation into TH was assessed as described in the legend to TABLE 5. **Part A:** Pergolide decreases the optical density of the TH monomer (M_r = 60,000) under both control and K$^+$-stimulated conditions. Control (lane 1), 10 μM pergolide (lane 2), 55 mM K$^+$ (lane 3), and pergolide and K$^+$ (lane 4). **Part B:** The D2 selective antagonist eticlopride blocks the decrease in optical density produced by pergolide. Control (lane 1), pergolide (lane 2), 3 μM eticlopride (lane 3), and pergolide and eticlopride (lane 4). (Data taken from Reference 13.)

reduction in TH phosphorylation is the mechanism by which autoreceptors decrease the rate of DA synthesis (TABLE 5).

Which kinase(s) is affected by autoreceptor stimulation? This question is difficult to answer given current uncertainty about the identity of the kinases involved in the physiological activation of the enzyme. Depolarization-induced increases in TH activity are thought to be mediated by Ca^{2+}-dependent rather than cyclic-AMP-dependent kinases.[47–49,79–83] In fact, recent work from our laboratory suggests that both C kinase and CaM kinase may be involved in the depolarization-induced increase in DA synthesis in striatal slices. TH appears to be kinetically activated by a Ca^{2+}-dependent, calmodulin-independent mechanism—possibly C kinase. However, an additional Ca^{2+}-calmodulin-dependent event is required in order for this kinetic activation to be manifested as an increase in DA synthesis in the intact slice.[49] These data do not imply that autoreceptors must regulate the activity of Ca^{2+}-dependent kinases in order to reverse the kinetic effects of depolarization since the level of TH activation appears to reflect the summation of the activity of multiple kinases, including cylic-AMP-dependent protein kinase. In fact, as discussed above, most evidence is consistent with the idea that autoreceptors regulate TH activity by decreasing cyclic AMP levels. Stimulation of D2 DA receptors fails to decrease intracellular calcium levels in striatal synaptosomes[84] or affect phosphoinositide hydrolysis in striatal slices,[85] arguing against a direct linkage between striatal synthesis-modulating autoreceptors and C kinase or CaM kinase.

PROTEIN CARBOXYLMETHYLATION

Protein-*O*-carboxylmethyltransferase (PCM) catalyzes the *S*-adenosylmethionine-dependent methylation of carboxyl groups on substrate methyl acceptor proteins (MAPs). This enzyme has been implicated in stimulus-secretion coupling in a number of biological systems[86] and is present in synaptosomal preparations,[87] consistent with a possible role in modulating presynaptic function. One functional role of PCM in brain is to regulate the activity of Ca^{2+}-calmodulin-dependent enzymes.[88-94] Since Ca^{2+} plays a pivotal role in the regulation of neurotransmitter synthesis and release, it seemed reasonable to hypothesize that PCM might play a role in autoreceptor function.

Our laboratory became interested in PCM after discovering that stimulation of DA autoreceptors increased [^{3}H]MAP formation in striatal synaptosomes and slices (FIGURE 6).[95,96] Based on these studies, we hypothesized that autoreceptor occupation was coupled to inhibition of DA synthesis and release through a mechanism involving stimulation of PCM activity. Anatomical support for this hypothesis was provided by studies demonstrating that PCM exhibits a unique neuron-specific distribution in rat brain,[97] with relatively high levels of both enzyme activity and immunoreactivity in catecholamine-containing cell body regions and projection areas.[98] In addition, TH and PCM were found to be colocalized in substantia nigra cells and processes.[99]

In order to test the possible involvement of PCM in DA autoreceptor function, we attempted to correlate changes in DA synthesis, DA release, and [^{3}H]MAP formation in striatal slices preincubated with [^{3}H]methionine. For studies of DA synthesis, DOPA accumulation after addition of NSD-1015 was used as an index of tyrosine hydroxylation in the slice. Concentrations of the methyltransferase inhibitor *S*-adenosylhomocysteine (AdoHcy) which reduced [^{3}H]MAP formation to 10% of

TABLE 5. DOPA Formation Correlates with ^{32}Phosphate Incorporation into Tyrosine Hydroxylase[a]

	TH Activity (ng DOPA/mg Protein per 30 Minutes)	^{32}P Incorporation into TH (Optical Density)
Control	100 ± 5	100 ± 2
Pergolide (1 μM)	64 ± 2	58 ± 10
30 mM K$^+$	143 ± 8	178 ± 6
Pergolide/30 mM K$^+$	100 ± 3	79 ± 12

[a]Data taken from Reference 13. DOPA formation was measured in striatal slices after NSD-1015 administration. In parallel experiments, slices were incubated with ^{32}P-labeled inorganic phosphate for 45 minutes and then subjected to the same incubation conditions (e.g., pergolide) used in DOPA formation experiments. ^{32}P incorporation into TH was assessed by immunoprecipitation of samples with rabbit anti-TH serum and subsequent autoradiographic analysis of sodium dodecyl sulfate polyacrylamide gels. Scanning densitometry was used to compare the optical density of the TH monomer (M_r = 60,000) in different lanes. All values are expressed as percent control.

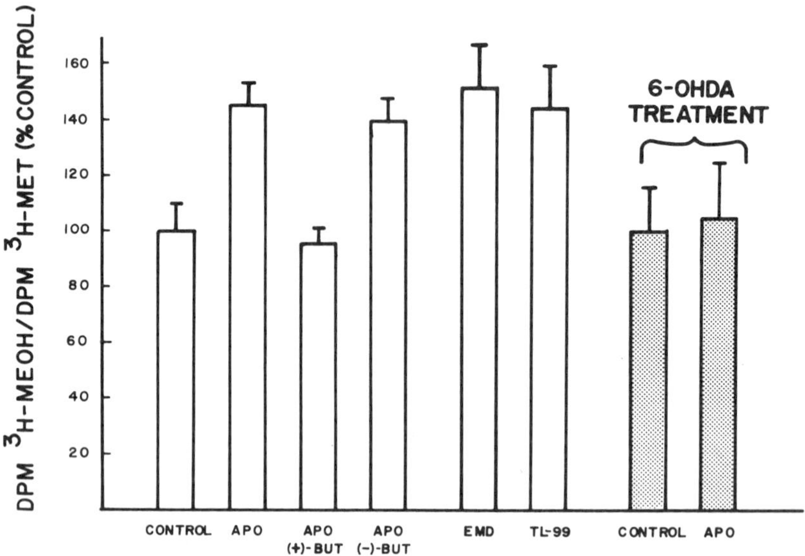

FIGURE 6. DA agonists (apomorphine, EMD 23 448, and TL-99) stimulate PCM in striatal slices. This stimulation is blocked stereospecifically by the DA antagonist (+)-butaclamol and is not observed in slices prepared from rats with 6-hydroxydopamine (6-OHDA) lesions of the medial forebrain bundle, a treatment that destroys DA nerve terminals and their associated autoreceptors in the striatum. This suggests that stimulation of PCM by DA agonists is mediated via nerve terminal autoreceptors.

control values had no effect on basal or K^+-stimulated DOPA formation, or on the ability of DA agonists to inhibit DOPA formation.[100] Similarly, Saller and Salama reported that methyltransferase inhibition did not alter the ability of DA agonists to inhibit synaptosomal DA synthesis.[101] These results suggest that autoreceptor modulation of DA synthesis does not involve a PCM-dependent step.

For studies of DA release, striatal slices were stimulated twice in each experiment (S_1 and S_2) by superfusing for 30 seconds with medium containing 30 mM K^+. The D2 antagonist sulpiride was found to enhance K^+-evoked overflow of radioactivity from striatal slices when present for 10 minutes prior to S_2. This presumably reflects antagonism by sulpiride of ongoing stimulation of autoreceptors by released DA. We reasoned that if PCM is important in coupling DA autoreceptor occupation to inhibition of DA release, then inhibition of PCM should also result in enhancement of K^+-evoked overflow of radioactivity. PCM inhibitors would be predicted to interrupt autoreceptor-mediated modulation of DA release at the signal transduction step, while DA antagonists would act at an earlier step in the same pathway, i.e., occupation of the autoreceptor by a DA agonist. In fact, AdoHcy was found to enhance S_2/S_1 when present in the superfusion medium for 20 minutes prior to S_2 (FIGURE 7).[100]

However, other *in vitro* findings were not consistent with the involvement of PCM in the autoreceptor regulation of DA release. While methyltransferase inhibition did attenuate the ability of very high doses of apomorphine to decrease DA release from striatal slices, it did not alter the ability of lower concentrations of apomorphine to do so.[100,101] These results suggest that autoreceptor regulation of DA release does not involve a PCM-dependent step. However, they do suggest that PCM plays some role in the regulation of DA release or turnover.

Additional support for this idea was provided by *in vivo* studies of the role of PCM in dopaminergic function. For example, pretreatment of rats and mice with D,L-homocysteine (which elevates brain AdoHcy levels, resulting in an inhibition of methyltransferase reactions) blocked the ability of apomorphine to decrease striatal DOPAC levels.[102] We found that pretreatment with D,L-homocysteine thiolactone [HcyTL; 400 mg/kg, intraperitoneally (ip)] increased DOPAC levels in the projection areas of nigrostriatal, mesolimbic, and mesocortical DA neurons.[103] The increase in DOPAC did not reflect an increase in DA synthesis since HcyTL pretreatment had no effect on the rate of *in vivo* DOPA formation after NSD-1015 administration. In order to rule out the involvement of synthesis-modulating autoreceptors in HcyTL effects on DOPAC levels, we examined the effect of HcyTL pretreatment in the GBL model. HcyTL had no effect on the ability of apomorphine to decrease DA synthesis in the GBL model in three brain regions known to possess synthesis-modulating autoreceptors: the striatum, olfactory tubercles, and pyiriform cortex (FIGURE 8).[103] These findings are consistent with previous studies in striatal slices showing that the autoreceptor-mediated regulation of DA synthesis does not require a methyltransferase-dependent step. HcyTL also had no effect on DOPA formation in the absence of GBL in regions possessing synthesis-modulating autoreceptors (not shown).

In contrast, HcyTL pretreatment attenuated the ability of apomorphine to decrease DOPA formation in cortical regions lacking synthesis-modulating autoreceptors (FIGURE 9). Apomorphine-induced inhibition of DOPA formation in these regions has

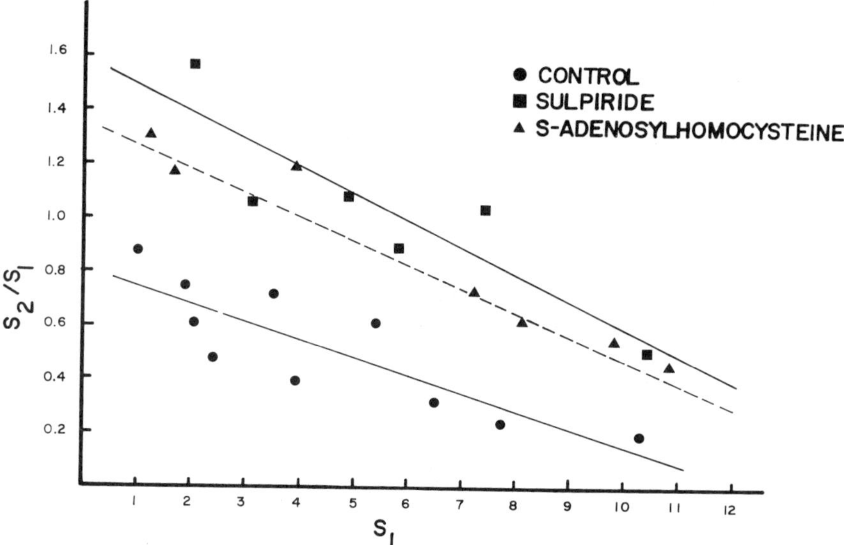

FIGURE 7. Effect of methyltransferase inhibition on the autoreceptor regulation of [³H]DA release from striatal slices. In control experiments (circles), the magnitude of S_2/S_1 was found to depend on the size of the first stimulation (S_1). As the size of S_1 increases, the ratio S_2/S_1 decreases. This relationship might be due to partial depletion of a releasable pool during S_1. However, for all values of S_1, both sulpiride and *S*-adenosylhomocysteine increased S_2/S_1 significantly above control levels when added prior to S_2.

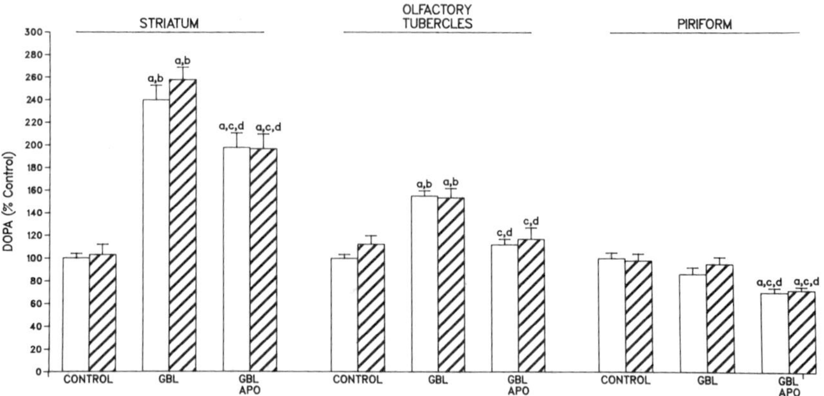

FIGURE 8. Pretreatment with D,L-homocysteine thiolactone (HcyTL; 400 mg/kg, ip) has no effect on synthesis-modulating autoreceptor function in the striatum, olfactory tubercles, and pyriform cortex as assessed using the GBL model. Apomorphine (30 μg/kg, sc) inhibits DOPA formation to the same extent in control rats (open bars) and HcyTL pretreated rats (hatched bars). Statistical analysis: a = different from NSD only, b = different from HcyTL/NSD, c = different from GBL/NSD, d = different from HcyTL/GBL/NSD ($p < 0.05$).

been shown to depend on activation of release-modulating autoreceptors and subsequent decreases in DA release (see above). This finding therefore supports the idea that PCM is involved in the regulation of cortical DA release. However, HcyTL pretreatment did not attentuate the ability of apomorphine to retard the decline in cortical DA levels normally associated with NSD-1015 administration (data not shown), suggesting that HcyTL is not exerting its effect on synthesis by interfering with the function of cortical release-modulating autoreceptors.[103]

These data do not enable clear-cut conclusions to be drawn about the role of PCM in DA neurons. However, it appears that PCM can affect functions regulated by autoreceptors (e.g., DA turnover) even though a direct linkage between PCM activity and autoreceptor function does not exist.

ARE ALL DA AUTORECEPTORS IDENTICAL AND DO THEY DIFFER FROM POSTSYNAPTIC RECEPTORS?

In previous sections of this chapter, autoreceptors have been defined functionally in terms of the events they regulate, e.g., synthesis-modulating, release-modulating, and impulse-modulating autoreceptors. It is not yet possible to determine whether these functions are mediated by distinct autoreceptor proteins. There is no pharmacological basis for distinguishing between the three types of autoreceptors. Both synthesis-modulating and impulse-modulating autoreceptors appear to be coupled to a pertussis-toxin-sensitive G protein,[104,105] although this may not be the case for release-modulating autoreceptors.[106] The existence of DA neurons that possess autoreceptors that regulate DA release but not DA synthesis suggests that the autoregulation of DA synthesis and release can proceed independently. However, this does not necessarily imply that modulation of DA release involves a distinct receptor. For example, it is possible that all DA cells possess a common receptor protein responsible for the initial DA recognition event which ultimately results in regulation of all three functions.

However, mesoprefrontal DA neurons may lack some coupling molecule necessary for translating receptor occupation into inhibition of DA synthesis. Studies of many neurotransmitter receptors indicate that it may not be uncommon for receptors with similar pharmacology to behave differently when placed in cells with different characteristics.

At present, it is uncertain whether there are significant pharmacological differences between DA autoreceptors and postsynaptic D2 DA receptors. A number of compounds have been characterized in recent years that exert selective effects on DA autoreceptors in biochemical, behavioral, and electrophysiological studies (see References 107 and 108 for review). It has been suggested more recently, however, that the apparent autoreceptor selectivity of these compounds is not related to pharmacological differences between autoreceptors and postsynaptic receptors but, instead, reflects the fact that these compounds are partial agonists. Meller and colleagues have demonstrated that their partial agonist character, in combination with the presence of a larger receptor reserve at presynaptic receptors as compared to postsynaptic receptors, appears to account for their autoreceptor selectivity.[109]

It should be noted, however, that the existence of DA antagonists which show autoreceptor selectivity is not entirely consistent with this model.[101] Similarly, this model does not account for some of the odd pharmacological characteristics of drugs such as (−)-3-(3-hydroxyphenyl-*N*-*n*-propylpiperidine (3-PPP). For example, (−)-3-PPP exhibits 65% intrinsic activity (relative to apomorphine) at dopamine autoreceptors controlling DA synthesis in the GBL and 4-hour reserpine models. In other words, it produces only 65% of the reversal of dopamine synthesis obtained with a full agonist (100% intrinsic activity) such as apomorphine and will therefore partially reverse the maximal inhibitory actions of the full DA agonist (i.e., act as an antagonist). However, (−)-3-PPP exhibits 90–95% intrinsic activity in animals with an 18-hour reserpine

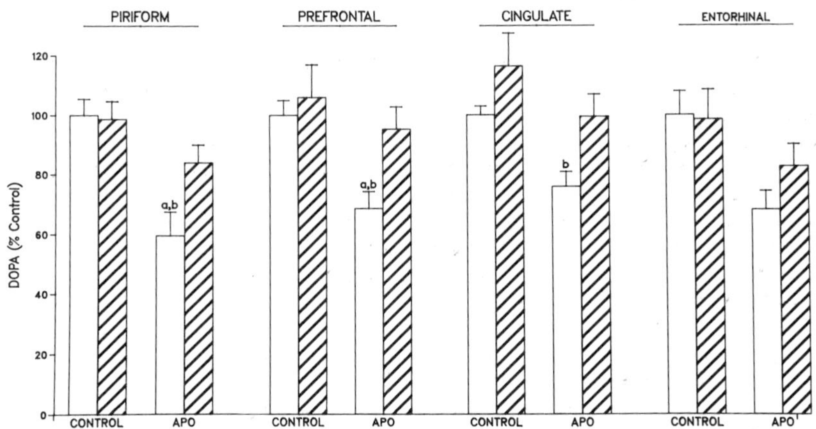

FIGURE 9. In regions lacking synthesis-modulating autoreceptors (prefrontal, cingulate, and entorhinal cortices), DA agonists fail to alter DOPA formation in the presence of GBL. However, they can inhibit DOPA formation in these regions in the absence of GBL by acting at release-modulating autoreceptors.[31] Pretreatment with D,L-homocysteine thiolactone (HcyTL; 400 mg/kg, ip) blocks the ability of apomorphine (30 µg/kg, sc) to decrease cortical DOPA formation, suggesting a role for PCM in the regulation of cortical DA turnover (control rats = open bars, HcyTL rats = hatched bars). Statistical analysis: a = different from NSD, b = different from HcyTL/NSD ($p < 0.05$).

pretreatment.[110–113] Partial agonist theory would not predict a difference in intrinsic activity of (−)-3-PPP in the 4- and 18-hour reserpine models, since in neither case is DA present to interact with the drug at dopamine autoreceptors. Instead, it is interesting to consider these anomalies in light of the proposal by Carlsson that the intrinsic activity of these agents is dependent on the adaptational state of the relevant DA receptors, which, in turn, is related to the degree of previous agonist occupancy of these receptors.[114]

Finally, it should be noted that the existence of drugs that are, in practice, selective for autoreceptors does not necessarily imply that autoreceptors differ from postsynaptic receptors. For example, many compounds that interact with DA receptors also exert effects at other neurotransmitter receptors; the pharmacological profile of certain DA agonists may be such that their effects on other neurotransmitter systems mask their effects at postsynaptic D2 DA receptors. It is especially important to consider this possibility since some of the most compelling evidence for the existence of autoreceptor-selective compounds is based on behavioral studies, which would be most susceptible to this problem. Should this explanation prove correct, however, it would in no way diminish the potential clinical importance of compounds that demonstrate functional autoreceptor selectivity.[115–119]

In addition to pharmacological considerations, it may be possible to distinguish between types of autoreceptors and between autoreceptors and postsynaptic DA receptors based on differences in signal transduction mechanisms. D2 DA receptors in

TABLE 6. Signal Transduction Mechanisms Associated with D2 DA Receptors

1. Inhibition of adenylate cyclase[61]
2. Inhibition of Ca^{2+} entry through voltage-sensitive Ca^{2+} channels[84,120–127]
3. Enhancement of K^+ conductances[128–132]
4. Modulation of phosphoinositide metabolism[133]

brain and pituitary have been shown to utilize mutiple signal transduction mechanisms (TABLE 6) and D2 DA receptors on different cell types appear to differ in their ability to interact with these signal transduction systems. For example, the pituitary D2 DA receptor has long been considered a model for the CNS D2 DA receptor. However, stimulation of D2 DA receptors on lactotrophs inhibits Ca^{2+} entry through voltage-sensitive Ca^{2+} channels whereas stimulation of D2 DA receptors on striatal nerve terminals has no effect on intracellular calcium levels when studied under identical experimental conditions.[84] Differences also exist between D2 DA receptors in different brain regions. For example, D2 DA receptors in the nucleus accumbens do not inhibit adenylate cyclase activity,[63,64] while the striatum appears to contain both D2 DA receptors that inhibit adenylate cyclase and D2 DA receptors that act independently of adenylate cyclase.[20,134–136] The striatal D2 DA receptors that inhibit adenylate cyclase appear to be located on intrinsic neurons while those not coupled to adenylate cyclase are located on nerve terminals.[62,136] The D2 DA receptors not coupled to adenylate cyclase activity are those that mediate inhibition of DA and acetylcholine release in the striatum and nucleus accumbens.[20,136,137] These release-modulating D2 DA receptors may work by increasing a K^+ conductance.[138] Indeed, if any signal transduction proves universal to all D2 DA receptors, it may be the ability to open K^+ channels and thereby hyperpolarize target cells. This has been observed for somatodendritic DA autoreceptors,[129,130] postsynaptic D2 DA receptors on striatal neurons,[131] and pituitary lactotrophs.[132] Terminal excitability studies suggest that stimulation of nerve terminal autoreceptors may also exert a hyperpolarizing effect.[139,140] However, there are exceptions. For example, stimulation of D2 DA receptors on the terminals of hippocampal-accumbens neurons increases their terminal excitability, suggesting that D2 DA receptors exert a depolarizing action.[141] It appears that heterogeneity is the most consistent feature of D2 DA receptor function.

Autoreceptors also differ from postsynaptic D2 DA receptors with respect to their interaction with D1 DA receptors. It is now well established that D1 and D2 DA act synergistically in many behavioral and electrophysiological models. Stimulation of D1 receptors appears to play an enabling role, i.e., D1 receptor occupation is necessary for the expression of the functional effects of postsynaptic D2 DA receptor stimulation (see Reference 142 for review). Interestingly, this does not appear to be the case for either synthesis-modulating or impulse-modulating autoreceptors.[143]

Ultimately, it should be possible to use molecular cloning techniques to identify subtypes of the D2 DA receptor and to determine whether autoreceptors differ from each other or from postsynaptic D2 DA receptors. The D2 DA receptor has now been cloned and it appears that there are at least 2 genes that produce D2 DA receptor subtypes.[144,145] For one of these genes,[144] additional diversity is generated by alternatively spliced mRNAs with tissue specific expression.[146–149] It will be very exciting to learn whether D2 DA receptors coded by different mRNAs correspond to subtypes of D2 DA receptors defined on the basis of anatomy and function.

ACKNOWLEDGMENTS

We thank Drs. Susan R. Sesack, Francis J. White, Amy M. Knorr, and Ray Salah for helpful discussions.

REFERENCES

1. KEHR, W., A. CARLSSON, M. LINDQVIST, T. MAGNUSSON & C. ATACK. 1972. J. Pharm. Pharmacol. **24:** 744–746.
2. WOLF, M. E. & R. H. ROTH. 1987. *In* Receptor Biochemistry Methodology: Dopamine Receptors. I. Creese & C. M. Fraser, Eds.: 45–96. Alan R. Liss, Inc. New York, N.Y.
3. ARGIOLAS, A., M. R. MELIS, F. FADDA & G. L. GESSA. 1982. Brain Res. **234:** 177–181.
4. SKIRBOLL, L. R., A. A. GRACE & B. S. BUNNEY. 1979. Science **206:** 80–82.
5. WALTERS, J. R., R. H. ROTH & G. K. AGHAJANIAN. 1973. J. Pharmacol. Exp. Ther. **186:** 630–639.
6. WALTERS, J. R. & R. H. ROTH. 1974. J. Pharmacol. Exp. Ther. **191:** 82–91.
7. WALTERS, J. R. & R. H. ROTH. 1976. Naunyn Schmiedebergs Arch. Pharmacol. **296:** 5–14.
8. IMPERATO, A. & G. DI CHIARA. 1984. J. Neurosci. **4:** 966–977.
9. MURRIN, L. C. & R. H. ROTH. 1976. Mol. Pharmacol. **12:** 463–475.
10. MURRIN, L. C. & R. H. ROTH. 1987. Neuropharmacology **26:** 591–595.
11. WESTERINK, B. H. C., J. B. DE VRIES & R. DURAN. 1990. J. Neurochem. **54:** 381–387.
12. WAGGONER, W. G., J. MCDERMED & H. J. LEIGHTON. 1980. Mol. Pharmacol. **18:** 91–99.
13. SALAH, R. S., D. M. KUHN & M. P. GALLOWAY. 1989. J. Neurochem. **52:** 1517–1522.
14. PLANTJÉ, J. F., F. A. DIJCKS, P. F. H. M. VERHEIJDEN & J. C. STOOF. 1985. Eur. J. Pharmacol. **114:** 401–402.
15. PLANTJÉ, J. F., H. W. M. STEINBUSCH, J. SCHIPPER, F. A. DIJCKS, P. F. H. M. VERHEIJDEN & J. C. STOOF. 1987. Neuroscience **20:** 157–168.
16. PLANTJÉ, J. F., J. SCHIPPER, P. F. H. M. VERHEIJDEN & J. C. STOOF. 1987. Brain Res. **413:** 205–212.
17. TALMACIU, R. K., I. S. HOFFMANN & L. X. CUBEDDU. 1986. J. Neurochem. **47:** 865–870.
18. ALTAR, C. A., W. C. BOYAR, E. OEI & P. L. WOOD. 1987. J. Pharmacol. Exp. Ther. **242:** 115–120.

19. WOLF, M. E. & R. H. ROTH. 1987. Neuropharmacology. **26:** 1053–1059.
20. STOOF, J. C., P. F. H. M. VERHEIJDEN & J. E. LEYSEN. 1987. Brain Res. **423:** 364–368.
21. HOFFMAN, I. S., R. K. TALMACIU, C. P. FERRO & L. X. CUBEDDU. 1988. J. Pharmacol. Exp. Ther. **245:** 761–772.
22. BANNON, M. J., R. L. MICHAUD & R. H. ROTH. 1981. Mol. Pharmacol. **19:** 270–275.
23. BANNON, M. J., M. E. WOLF & R. H. ROTH. 1983. Eur. J. Pharmacol. **91:** 119–125.
24. CHIODO, L. A., M. J. BANNON, A. A. GRACE, R. H. ROTH & B. S. BUNNEY. 1984. Neuroscience **12:** 1–16.
25. KILTS, C. D., C. M. ANDERSON, T. D. ELY & J. K. NISHITA. 1987. J. Neurosci. **7:** 3961–3975.
26. DEMAREST, K. T. & K. E. MOORE. 1979. J. Neural Transm. **46:** 263–277.
27. LOOKINGLAND, K. J. & K. E. MOORE. 1984. Brain Res. **304:** 329–338.
28. LOOKINGLAND, K. J., J. M. FARAH, K. L. LOVELL & K. E. MOORE. 1985. Neuroendocrinol. **40:** 145–151.
29. MOORE, K. E., G. D. RIEGLE & K. T. DEMAREST. 1985. *In* Catecholamines as Hormone Regulators. N. Ben-Jonathan, J. M. Bahr & R. I. Weiner, Eds. **18:** 31–49. Raven Press. New York, N.Y.
30. ANDÉN, N.-E., M. GRABOWSKA-ANDÉN & B. LILJENBERG. 1983. J. Neural Transm. **57:** 129–137.
31. GALLOWAY, M. P., M. E. WOLF & R. H. ROTH. 1986. J. Pharmacol. Exp. Ther. **236:** 689–698.
32. ROTH, R. H. & S.-Y. TAM. 1987. *In* Amino Acids in Health and Disease: New Perspectives. S. Kaufman, Ed.: 159–178. Alan R. Liss, Inc. New York, N.Y.
33. BANNON, M. J., E. B. BUNNEY & R. H. ROTH. 1981. Brain Res. **218:** 376–382.
34. SCATTON, B. 1977. Eur. J. Pharmacol. **46:** 363–369.
35. JULOU, L., B. SCATTON & J. GLOWINSKI. 1977. Adv. Biochem. Psychopharmacol. **16:** 617–624.
36. BANNON, M. J., J. F. REINHARD, E. B. BUNNEY, JR. & R. H. ROTH. 1982. Nature **296:** 444–446.
37. GALLOWAY, M. P. & R. H. ROTH. 1983. Soc. Neurosci. Abstr. **9:** 1003.
38. MATSUMOTO, T., H. UCHIMURA, M. HIRANO, J. S. KIM, H. YOKOO, M. SHIMOMURA, T. NAKAHARA, K. INOUE & K. OOMAGARI. 1983. Eur. J. Pharmacol. **89:** 27–33.
39. NOWYCKY, M. & R. H. ROTH. 1977. Naunyn Schmiedebergs Arch. Pharmacol. **300:** 247–254.
40. NOWYCKY, M. & R. H. ROTH. 1978. Prog. Neuropsychopharmacol. **2:** 139–158.
41. BACAPOULOS, N. G., G. BUSTOS, D. E. REDMOND, J. BAULU & R. H. ROTH. 1978. Brain Res. **157:** 396–401.
42. ROTH, R. H., N. G. BACAPOULOS, G. BUSTOS & D. E. REDMOND. 1980. Adv. Biochem. Psychopharmacol. **24:** 513–520.
43. ANDÉN, N.-E., H. NILSSON, E. ROS & U. THORNSTRÖM. 1983. Acta Pharmacol. Toxicol. **52:** 51–56.
44. FADDA, F., G. L. GESSA, M. MARCOU, E. MOSCA & Z. ROSSETTI. 1984. Brain Res. **293:** 67–72.
45. WOLF, M. E., M. P. GALLOWAY & R. H. ROTH. 1986. J. Pharmacol. Exp. Ther. **236:** 699–707.
46. BUSTOS, G., R. H. ROTH & V. H. MORGENROTH III. 1976. Biochem. Pharmacol. **25:** 2493–2497.
47. EL MESTIKAWAY, S., J. GLOWINSKI & M. HAMON. 1983. Nature **302:** 830–832.
48. EL MESTIKAWAY, S., H. GOZLAN, J. GLOWINSKI & M. HAMON. 1985. J. Neurochem. **45:** 173–184.
49. ROSIN, D. L., A. M. KNORR, M. E. WOLF & R. H. ROTH. 1986. Soc. Neurosci. Abstr. **12:** 133.
50. MURRIN, L. C., V. H. MORGENROTH III & R. H. ROTH. 1976. Mol. Pharmacol. **12:** 1070–1081.
51. ZIVKOVIC, B., A. GUIDOTTI & E. COSTA. 1974. Mol. Pharmacol. **10:** 727–735.
52. ZIVKOVIC, B., A. GUIDOTTI, A. REVUELTA & E. COSTA. 1975. J. Pharmacol. Exp. Ther. **194:** 37–46.

53. ROTH, R. H., V. H. MORGENROTH III & L. C. MURRIN. 1975. *In* Antipsychotic Drugs. Pharmacodynamics & Pharmacokinetics. G. Sedvall, Ed.: 133–145. Pergamon Press. New York, N.Y.
54. LERNER, P., P. NOSE, E. K. GORDON & W. LOVENBERG. 1977. Science **197:** 181–183.
55. KAPATOS, G. & M. J. ZIGMOND. 1979. J. Pharmacol. Exp. Ther. **208:** 468–475.
56. PRADHAN, S., L. ALPHS & W. LOVENBERG. 1981. Neuropharmacology **20:** 149–154.
57. LAZAR, M. A., I. N. MEFFORD & J. D. BARCHAS. 1982. Biochem. Pharmacol. **31:** 2599–2607.
58. E. MESTIKAWAY, S. & M. HAMON. 1986. J. Neurochem. **47:** 1425–1433.
59. E. MESTIKAWAY, S., J. GLOWINSKI & M. HAMON. 1986. J. Neurochem. **46:** 12–22.
60. STRAIT, K. A. & R. KUCZENSKI. 1986. Mol. Pharmacol. **29:** 561–569.
61. STOOF, J. C. & J. W. KEBABIAN. 1981. Nature **294:** 366–368.
62. ONALI, P., M. C. OLIANAS & G. L. GESSA. 1985. Mol. Pharmacol. **28:** 138–145.
63. STOOF, J. C. & P. F. H. M. VERHEIJDEN. 1986. Eur. J. Pharmacol. **129:** 205–206.
64. KELLY, E. & S. R. NAHORSKI. 1987. Naunyn Schmiedebergs Arch. Pharmacol. **335:** 508–512.
65. GOLDSTEIN, M. & L. A. GREENE. 1987. *In* Psychopharmacology: the Third Generation of Progress. H. Y. Meltzer, Ed.: 75–80. Raven Press. New York, N.Y.
66. ALBERT, K. A., E. HELMER-MATYJEK, A. C. NAIRN, T. H. MULLER, J. W. HAYCOCK, L. A. GREENE, M. GOLDSTEIN & P. GREENGARD. 1984. Proc. Nat. Acad. Sci. USA **81:** 7713–7717.
67. YAMAUCHI, T. & H. FUJISAWA. 1981. Biochem. Biophys. Res. Commun. **100:** 807–813.
68. YAMAUCHI, T., H. NAKATA & H. FUJISAWA. 1981. J. Biol. Chem. **256:** 5405–5409.
69. VULLIET, P. R., J. R. WOODGETT & P. COHEN. 1984. J. Biol. Chem. **259:** 13680–13683.
70. YAMAUCHI, T. & H. FUJISAWA. 1979. J. Biol. Chem. **254:** 503–507.
71. VULLIET, P. R., T. A. LANGAN & N. WEINER. 1980. Proc. Nat. Acad. Sci. USA **77:** 92–96.
72. EDELMAN, A. M., J. D. RAESE, M. A. LAZAR & J. D. BARCHAS. 1981. J. Pharmacol. Exp. Ther. **216:**647–653.
73. McTIGUE, M., J. CREMINS & S. HALEGOUA. 1985. J. Biol. Chem. **260:** 9047–9056.
74. TACHIKAWA, E., A. W. TANK, N. YANAGIHARA, W. MOSIMANN & N. WEINER. 1986. Mol. Pharmacol. **30:** 476–485.
75. YANAGIHARA, N., A. W. TANK, T. A. LANGAN & N. WEINER. 1986. J. Neurochem. **46:** 562–568.
76. CAHILL, A. L. & R. L. PERLMAN. 1987. Biochim. Biophys. Acta **930:** 454–462.
77. CAHILL, A. L. & R. L. PERLMAN. 1987. Soc. Neurosci, Astr. **13:** 806.
78. HAYCOCK, J. W. 1987. Brain Res. Bull. **19:** 619–622.
79. ANAGOSTE, B., C. SHIRRON, E. FRIEDMAN & M. GOLDSTEIN. 1974. J. Pharmacol. Exp. Ther. **191:** 370–376.
80. GOLDSTEIN, M., R. L. BRONAUGH, B. EBSTEIN & C. ROBERGE. 1976. Brain Res. **109:** 563–574.
81. SIMON, J. R. & R. H. ROTH. 1979. Mol. Pharmacol. **16:** 224–233.
82. BUSTOS, G. & R. H. ROTH. 1979. Biochem. Pharmacol. **28:** 3026–3028.
83. KNORR, A. M., M. E. WOLF & R. H. ROTH. 1986. Biochem. Pharmacol. **35:** 1929–1932.
84. WOLF, M. E. & G. KAPATOS. 1989. Synapse **4:** 353–370.
85. KELLY, E., I. BATTY & S. R. NAHORSKI. 1988. J. Neurochem. **51:** 918–924.
86. DILIBERTO, E. J., JR. 1982. *In* Cellular Regulation of Secretion and Release. M. P. Conn, Ed.: 147–192. Academic Press. New York, N.Y.
87. DILIBERTO, E. J., JR. & J. AXELROD. 1976. J. Neurochem. **26:** 1159–1165.
88. GAGNON, C., S. KELLY, V. MANGANIELLO, M. VAUGHN, C. ODYA, W. STRITTMATTER, A. HOFFMAN & F. HIRATA. 1981. Nature **291:** 515–516.
89. GAGNON, C. 1983. Can. J. Biochem. Cell Biol. **61:** 921–926.
90. BILLINGSLEY, M. L., P. A. VELLETRI, R. H. ROTH & R. J. DELORENZO. 1983. J. Biol. Chem. **258:** 5332–5357.
91. BILLINGSLEY, M. L., D. KUHN, P. A. VELLETRI, R. KINCAID & W. LOVENBERG. 1984. J. Biol. Chem. **259:** 6630–6635.
92. BILLINGSLEY, M. L., P. A. VELLETRI, W. LOVENBERG, D. KUHN, J. R. GOLDENRING, R. J. DELORENZO. 1985. J. Neurochem. **44:** 1442–1450.

93. BILLNGSLEY, M. L., R. L. KINCAID & W. LOVENBERG. 1985. Proc. Nat. Acad. Sci. **82:** 5612–5616.
94. BILLINGSLEY, M. L., C. D. BALABAN, D. M. KUHN & R. L. KINCAID. 1986. *In* Biological Methylation and Drug Design. R. T. Borchardt, C. R. Creveling & P. M. Ueland, Eds.: 25–41. Humana Press. Clifton, N.J.
95. BILLINGSLEY, M. L. & R. H. ROTH. 1982. J. Pharmacol. Exp. Ther. **223:** 681–688.
96. WOLF, M. E. & R. H. ROTH. 1985. J. Neurochem. **44:** 291–298.
97. BILLINGSLEY, M. L., S. KIM & D. M. KUHN. 1985. Neuroscience **15:** 159–171.
98. BILLINGSLEY, M. L. & C. D. BALABAN. 1985. Brain Res. **358:** 96–103.
99. BILLINGSLEY, M. L., C. D. BALABAN, U. BERRESHEIM & D. M. KUHN. 1986. Neurochem. Int. **8:** 255–265.
100. WOLF, M. E. & R. H. ROTH. 1984. Soc. Neurosci. Abstr. **10:** 878.
101. SALLER, C. F. & A. I. SALAMA. 1985. Eur. J. Pharmacol. **111:** 17–22.
102. SALLER, C. F. & A. I. SALAMA. 1985. Brain Res. **360:** 407–408.
103. WOLF, M. E. & R. H. ROTH. 1986. Soc. Neurosci. Abstr. **12:** 602.
104. INNIS, R. B. & G. K. AGHAJANIAN. 1987. Brain Res. **411:** 139–143.
105. BEAN, A. J., P. D. SHEPARD, B. S. BUNNEY, E. J. NESTLER & R. H. ROTH. 1988. Mol. Pharmacol. **34:** 715–718.
106. BOWYER, J. F. & N. WEINER. 1989. J. Pharmacol. Exp. Ther. **248:** 514–520.
107. CLARK, D., S. HJORTH & A. CARLSSON. 1985. J. Neural Transm. **62:** 1–52.
108. CLARK, D., S. HJORTH & A. CARLSSON. 1985. J. Neural Transm. **62:** 171–207.
109. MELLER, E., K. BOHMAKER, Y. NAMBA, A. J. FRIEDHOFF & M. GOLDSTEIN. 1987. Mol. Pharmacol. **31:** 592–598.
110. SVENSSON, K., S. HJORTH, D. CLARK, A. CARLSSON, H. WIKSTRÖM, D. SANCHEZ, A. M. JOHANSSON, L.-E. ARVIDSSON, U. HACKSELL & J. L. G. NILSSON. 1986. J. Neural Transm. **65:** 1–27.
111. HJORTH, S., A. CARLSSON, D. CLARK, K. SVENSSON, H. WIKSTRÖM, D. SANCHEZ, P. LINDBERG, U. HACKSELL, L.-E. ARVIDSSON, A. JOHANSSON & J. L. G. NILSSON. 1983. Psychopharmacology **81:** 89–99.
112. HJORTH, S., A. CARLSSON, D. CLARK & K. SVENSSON. 1984. Soc. Neurosci. Abstr. **10:** 239.
113. CLARK, D., S. HJORTH & A. CARLSSON. 1984. Eur. J. Pharmacol. **106:** 185–189.
114. CARLSSON, A. 1983. J. Neural Transm. **57:** 309–315.
115. MELTZER, H. 1980. Schizophrenia Bull. **6:** 456–475.
116. HÄGGSTRÖM, J.-E., L. M. GUNNE, A. CARLSSON & H. WIKSTRÖM. 1983. J. Neural Transm. **58:** 135–142.
117. CORSINI, G. U., U. BONUCCELLI, E. RAINER & M. DEL ZOMPO. 1985. J. Neural Transm. **64:** 105–111.
118. HINZEN, D., O. HORNYKIEWICZ, W. KOBINGER, L. PICHLER, C. PIFL & G. SCHINGNITZ. 1986. Eur. J. Pharmacol. **131:** 75–86.
119. KOVACIC, B., P. LE WITT & D. CLARK. 1988. J. Neural Transm. **74:** 97–107.
120. SCHOFIELD, J. G. 1983. FEBS Lett. **159:** 79–82.
121. MEMO, M., L. CASTELLETTI, C. MISSALE, A. VALERIO, M. CARRUBA & P. F. SPANO. 1986. J. Neurochem. **47:** 1689–1695.
122. SCHREY, M. P., J. H. CLARK & S. FRANKS. 1986. J. Endocrinol. **108:** 423–429.
123. LAFOND, J. & R. COLLU. 1986. Endocrinology **119:** 2012–2017.
124. FUJIWARA, H., N. KATO, H. SHUNTOH & C. TANAKA. 1987. Bri. J. Pharmacol. **91:** 287–297.
125. MALGAROLI, A., L. VALLAR, F. R. ELAHI, T. POZZAN, A. SPADA & J. MELDOLESI. 1987. J. Biol. Chem. **262:** 13920–13927.
126. LOGIN, I. S., A. M. JUDD & R. M. MACLEOD. 1988. Biochem. Biophys. Res. Commun. **151:** 913–918.
127. VALLAR, L., L. M. VICENTINI & J. MELDOLESI. 1988. J. Biol. Chem. **263:** 10127–10134.
128. MEMO, M., L. CASTELLETTI, C. MISSALE & P. F. SPANO. 1987. Eur. J. Pharmacol. **139:** 361–362.
129. LACEY, M. G., N. B. MERCURI & R. A. NORTH. 1987. J. Physiol. **392:** 397–416.
130. LACEY, M. G., N. B. MERCURI & R. A. NORTH. 1988. J. Physiol. **401:** 437–453.

131. FREEDMAN, J. E. & F. F. WEIGHT. 1988. Proc. Nat. Acad. Sci. USA **85:** 3618–3622.
132. CASTELLETTI, L., M. MEMO, C. MISSALE, P. F. SPANO & A. VALERIO. 1989. J. Physiol. **410:** 251–265.
133. JARVIS, W. D., A. M. JUDD & R. M. MACLEOD. 1988. Endocrinology **123:** 2793–2799.
134. EUVRARD, C., J. PREMONT, C. OBERLANDER, J. R. BOISSIER & J. BOCKAERT. Naunyn Schmiedebergs Arch. Pharmacol. **309:** 241–245.
135. STOOF, J. C. & J. W. KEBABIAN. 1982. Brain Res. **250:** 263–270.
136. MEMO, M., C. MISSALE, M. O. CARRUBA & P. F. SPANO. 1986. Neurosci. Lett. **71:** 192–196.
137. HERDON, H. 1988. Eur. J. Pharmacol. **154:** 115–116.
138. DRUKARCH, B., E. SCHEPENS, A. N. M. SCHOFFELMEER & J. C. STOOF. 1989. J. Neurochem. **52:** 1680–1685.
139. TEPPER, J. M., P. M. GROVES & S. J. YOUNG. 1985. Trends Pharmacol. Sci. **6:** 251–256.
140. GARIANO, R. F., S. F. SAWYER, J. M. TEPPER, S. J. YOUNG & P. M. GROVES. 1989. Brain Res. Bull. **22:** 517–523.
141. YANG, C. R. & G. J. MOGENSON. 1986. J. Neurosci. **6:** 2470–2478.
142. CLARK, D. & F. J. WHITE. Synapse **1:** 347–388.
143. WACHTEL, S. R., X.-T. HU, M. P. GALLOWAY & F. J. WHITE. 1989. Synapse **4:** 327–346.
144. BUNZOW, J. R., H. H. M. VAN TOL, D. K. GRANDY, P. ALBERT, J. SALON, M. CHRISTIE, C. A. MACHIDA, K. A. NEVE & O. CIVELLI. 1988. Nature **336:** 783–787.
145. TODD, R. D., T. S. KHURANA, P. SAJOVIC, K. R. STONE & K. L. O'MALLEY. 1989. Proc. Nat. Acad. Sci. USA **86:** 10134–10138.
146. DAL TOSO, R., B. SOMMER, M. EWERT, A. HERB, D. B. PRITCHETT, A. BACH, B. D. SHIVERS & P. H. SEEBURG. 1989. Embo J. **8:** 4025–4034.
147. GIROS, B., P. SOKOLOFF, M.-P. MARTRES, J.-F. RIOU, L. J. EMORINE & J.-C. SCHWARTZ. 1989. Nature **342:** 923–926.
148. MONSMA, F. J., JR., L. D. MCVITTIE, C. R. GERFEN, L. C. MAHAN & D. R. SIBLEY. 1989. Nature **342:** 926–929.
149. O'MALLEY, K. L., K. J. MACK, K.-Y. GANDELMAN & R. D. TODD. 1990. Biochemistry **29:** 1367–1371.
150. BANNON, M. J. & R. H. ROTH. 1983. Pharmacol. Rev. **35:** 53–68.

Synaptic and Nonsynaptic Cross Talk between Neurons

Role of Presynaptic Alpha$_2$-Receptors in Mental Disorders

E. S. VIZI

Institute of Experimental Medicine
Hungarian Academy of Sciences
Post Office Box 67
H-1450 Budapest, Hungary

Within a given neuron, signals are transmitted by waves of electrical activity produced by local changes in the permeability of the membrane to different ions. The electrical signals reaching the axon terminal result in depolarizing it and cause the secretion of chemical substances, provided Ca_o is available. The substances released by depolarization cross the synaptic cleft and act on the postsynaptic membrane equipped with receptors. In the past few years, however, neurochemical and pharmacological[1-4] and morphological[5-7] evidence has been found that transmitters can be released from nonsynaptic sites, for diffusion to target cells distant from the release sites to modulate the release of another transmitter. In addition, it was suggested that there is a nonsynaptic interaction between neurons at the presynaptic level.[2-4] It became clear that chemical neurotransmission may be a far more complicated event than previously assumed. Firstly, the release can be presynaptically modulated through auto-, homo-, and heteroreceptors (see S. Kalsner's paper, this volume). Secondly, transmitters/ modulators can be released from sites other than axon terminals and from varicosities situated synaptically or nonsynaptically. Thirdly, regional disparities in the relationships of the inputs to different regions of the brain make possible the regional nonsynaptic/synaptic control of chemical neurotransmission in the neuronal network.

Autoreceptors located on a nerve receive signals from the transmitter released from the same neuron: a transmitter/modulator released from the axon terminal inhibits its own release evoked by subsequent stimuli. *Homoreceptors* receive signals from the adjacant neuron whose machinery is the same as that located on axon terminals. *Heteroreceptors* are located on axon terminals that do not manufacture substances able to exert an effect on them, and these receptors receive signals from other axon terminals through their chemical substances. Thus, autoreceptors tell a nerve what it said, while homoreceptors and heteroreceptors tell a nerve what it heard. While the functional significance of autoreceptors is to keep the output constant, homo- and heteroreceptors are involved in cross talk between neurons.

During recent years, considerable evidence has been provided in favor of presynaptic modulation of transmitter release. Besides negative feedback modulation, which is mediated via autoreceptors located on axon terminals where the transmitters are released, there is another form of modulation when the modulator is released and acts on other axon terminals reducing or inhibiting the release of chemicals from those neurons. This would be an interneuronal communication between neurons at the presynaptic level. This "cross talk" could be made through synapses, but also without making synaptic contact: transmitters/modulators may be released from both synaptic and nonsynaptic sites for diffusion to target cells more distant than those observed in conventional synaptic transmission. A variety of combinations of neurons can act upon each other without making synaptic contact.

ROLE OF PRESYNAPTIC ALPHA$_2$-HETERORECEPTORS IN MODULATION OF CHEMICAL NEUROTRANSMISSION

It has been observed that the stimulation of sympathetic nerve in the gut resulted in an inhibition of acetylcholine (ACh) release,[8,14] an effect that was mediated via presynaptic alpha-receptors without there being synaptic contact between the two neurons. This finding indicated that there is a functional interaction between the two neurons: they can talk to each other.

Similar interaction was observed in the central nervous system (CNS). It has been shown that norepinephrine (NE) reduces the release of ACh from isolated cortical slices induced either by ouabain[9] or by electrical stimulation.[10] The spontaneous and evoked release of acetylcholine was higher in those slices where noradrenergic input was impaired in some way,[2] either by phentolamine, which blocks the action of NE on alpha$_2$-adrenoceptors, or by 6-OHDA pretreatment, or by an ipsilateral locus ceruleus lesion. Following a locus ceruleus lesion the spontaneous and evoked release of ACh from slices dissected from the ipsilateral side was higher in comparison with those on the contralateral side. Norepinephrine significantly reduced the resting release of acetylcholine only in those cases where noradrenergic control has been previously removed.[2] It is suggested that the release of ACh from cholinergic neurons of the cerebral cortex is continuously controlled by NE released from nerves arising from the locus ceruleus (FIGURE 1). The removal of this inhibitory system results in an increase of ACh release. Similar action was observed on isolated human cortical slices;[11] NE inhibited the release of ACh and its effect is mediated via alpha$_2$-adrenoceptors, since phentolamine prevented the effect of noradrenaline (NA). In addition, a widespread neurochemical effect of NA was observed on axon terminals in the rat frontal cortex.[12] Since there is no morphological contact, no axoaxonic synapse between noradrenergic and cholinergic axon terminals,[6,13] and the naked varicosity is able to release NE, which might be capable of exerting a widespread influence on vast neuronal ensembles, a nonsynaptic communication exists between the two neurons.

Since intraventricular administration of noradrenaline or locus ceruleus stimulation has been shown to increase gamma-aminobutyric acid (GABA) release and to decrease ACh release from the cerebral cortex,[14] it seems plausible that the effects on sleep-arousal mechanisms, due to manipulations affecting noradrenergic mechanisms in the locus ceruleus, may be mediated through changes in other transmitters released at the cortical level.

Sedation, sleep, and characteristic behavioral signs of depression can be observed in chicks and rodents treated with clonidine or xylazine (for review, see Reference 15). These effects are apparently related to the agonist action of clonidine and xylazine on central alpha$_2$-adrenoceptors. Activation of central alpha$_2$- and alpha$_1$-adrenoceptors might lead to opposite effects. Thus, selective agonists for alpha$_1$-adrenoceptors (phenylephrine, methoxamine, cirazoline) have no sedative effect in chicks; they produce excitation. Evidence has been obtained that local application of methoxamine on rat cortical neurons enhances the neural firing.[16] Furthermore, stimulation of alpha$_1$-adrenoceptors by cirazoline induces arousal and mediates behavioral excitation in depression of the central nervous system. It was also shown that central alpha$_1$-adrenoceptors have a controlling role in the mechanism of cataplexy.

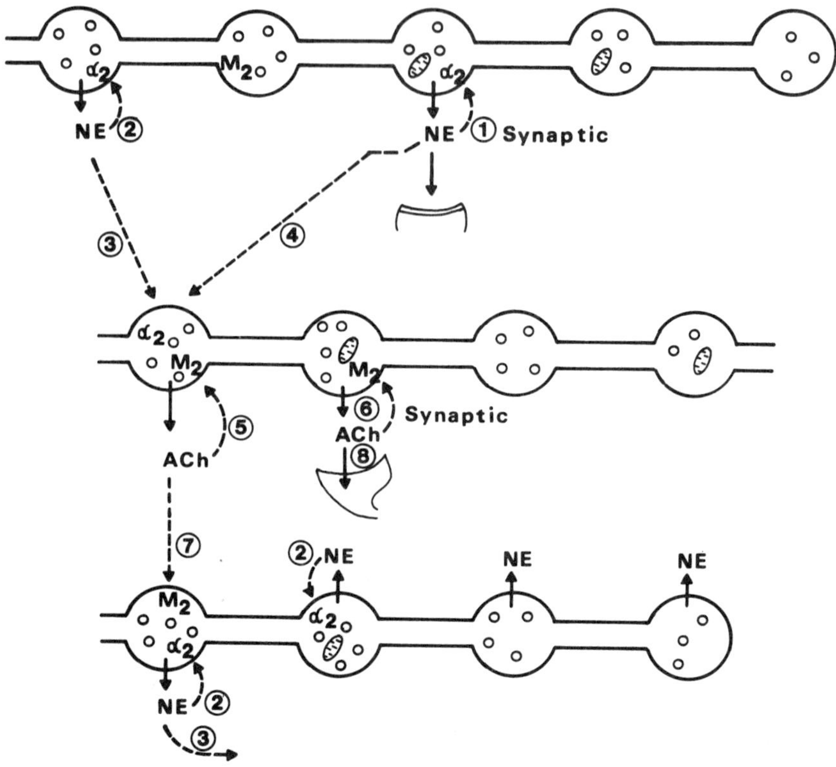

FIGURE 1. Scheme of interactions between neurons with different transmitter machinery. Upper neuron is noradrenergic. (1) Norepinephrine (NE) released from a synaptic axon terminal into the synaptic gap inhibits its own release via stimulation of alpha$_2$-*autoreceptors* (negative feedback). (2) Norepinephrine released from a nonsynaptic axon terminal inhibits its own release. Negative feedback modulation (nonsynaptic condition). (3) Norepinephrine reaches its target cell, a cholinergic varicose axon terminal which is equipped with an alpha$_2$-*heteroreceptor,* and exerts an inhibitory effect on acetylcholine release. If NE reaches its target cell, another noradrenergic axon terminal, it inhibits the release of NE through presynaptic alpha$_2$-homoreceptors (lower neuron). (4) Norepinephrine released into the synaptic gap acts nonsynaptically and stimulates alpha$_2$-adrenoceptors located on the varicose cholinergic axon terminal and inhibits ACh release. (5) Acetylcholine released from a nonsynaptic axon terminal inhibits its own release through muscarinic (M$_2$) receptors. (6) Acetylcholine released from a synaptic axon terminal inhibits its own release. (7) ACh released reaches its target cell, a noradrenergic neuron, and stimulates a muscarinic (M$_2$) receptor located on the noradrenergic axon terminal and inhibits the release of NA. (8) Acetylcholine exerts an effect on the postsynaptic site.

In our experiments alpha$_2$-adrenoceptor agonists (clinidine, xylazine) produced depression (loss of the righting reflex). An alpha$_1$-adrenoceptor antagonist (Prazosin) potentiated this effect (TABLE 1), indicating that while alpha$_2$-adrenoceptors are involved in depression, stimulation of alpha$_1$-adrenoceptors may be responsible for the excitation.

The noradrenergic afferents of the rat cerebral cortex derive exclusively from a relatively small number of varicose neurons. Morphologically, two types of noradrenergic axon terminals have been shown as far as their relationship with other neurons is concerned: there are varicosities that do not make synaptic contact,[17] and that exhibit synaptic specialization.[18] Noradrenergic varicosities engaged in synaptic contact and apposed to dendritic branches or spines might be strategically located to exert a postsynaptic influence on axodendritic communication by other systems of extinsic or intrinsic origin and the release of NE from these varicosities through alpha$_2$-autoreceptors is modulated. These receptors are possibly different from those localized nonsynaptically and easily accessible for exogenous alpha$_2$-adrenoceptor agonists.

NONSYNAPTIC RELEASE OF SIGNAL TRANSMITTER

It has been known for a long time that postganglionic neurons innervating vas deferens,[19] blood vessels, and smooth muscle terminate in a network of varicose areas (*boutons en passant*) that lack synaptic contact with the target cell. The transmitter is released from structurally unspecialized areas along the varicose fibers and must often travel a considerable distance before interacting with its receptors (FIGURE 1). Descarries and coworkers pointed out that nonsynaptic varicosities in the CNS appear to have all the apparatus normally associated with synaptic release.[6] Subsequently, ultrastructural examination of noradrenergic varicosities in several tissues confirmed that both large and small vesicles could undergo exocytosis in the absence of structurally specialized active zones.[20] In addition morphological evidence was provided for exocytosis in nonsynaptic release site in rat median evidence and mesencephalic central grey substance.[21] It is conceivable that a similar exocytosis occurs in monoaminergic neurons in the gut[22,23] and in the brain[6] where active synaptic zones are lacking.

NONSYNAPTIC RECEPTOR LOCALIZATION

There is an interesting difference in localization and sensitivity of receptors on effector cells where the transmitter (e.g., ACh, NE, dopamine) or neuropeptide is

TABLE 1. Role of Alpha$_2$- and Alpha$_1$-Adrenoceptors in Clonidine-Induced Depression[a]

	Clonidine Depression (minutes)
Clonidine (0.7 mg/kg, iv)	32.6 ± 2.5 ($n = 8$)
Clonidine + prazosin (10 mg/kg, ip)	57.2 ± 5.1^b ($n = 10$)
Xylazine (2 mg/kg, iv)	26.7 ± 1.0 ($n = 8$)
Xylazine + prazosin (10 mg/kg, ip)	52.2 ± 1.9^b ($n = 10$)
Clonidine (0.7 mg/kg, iv)	22.3 ± 1.6 ($n = 8$)
Clonidine + CH-38083 (0.5 mg/kg, ip)	2.4 ± 0.4^b ($n = 10$)
Clonidine (0.7 mg/kg, iv)	24.5 ± 1.6 ($n = 8$)
Clonidine + CH-38083 (0.5 mg/kg, ip)	
Prazosin (10 mg/kg, ip)	18.6 ± 3.4 ($n = 8$)

[a]Data taken from Reference 15. The loss of righting reflex in chicks was measured. Clonidine and xylazine, alpha$_2$-adrenoceptor agonists, were injected intravenously (iv). CH-38083, a selective alpha$_2$-adrenoceptor antagonist[44] and prazosin, a selective alpha$_1$-adrenoceptor antagonist, were injected intraperitoneally (ip) 10 minutes prior to clonidine or xylazine administration. Note that prazosin potentiated and CH-38083 prevented the depression produced by clonidine. n, number of experiments. Mean $\pm$ standard error of the mean.
[b]Significant difference, $p < 0.01$.

released into a large extraneuronal space. In the case of synaptic transmission there is a small area of the effector cell where the receptors are concentrated. The neuromuscular junction is an example of a perfect correspondence between release sites of acetylcholine and the location of postsynaptic nicotinic receptors. However, when the transmitter release site and the target cells are widely separated from each other (100–3000 nm) there is no specific subsynaptic arrangement, and the recognition sites, receptors, are unevenly distributed along the surface of the effector cell (presynaptic axon terminals, etc.). This morphological arrangement fits any diffusion type of transmission where the advantage of quantal release cannot be utilized. There seems to be no evidence that nonquantal release can occur solely at diffuse synapses. In addition, it could be argued that the receptors distant from the synaptic sites do not represent active receptors capable of opening ion channels. However it is also known that the sensitivity of extrasynaptic receptors is higher to chemical signals[4] than that of those located in the synaptic gap. In addition, it was shown that the release site/receptor matches are exceptions rather than rules.[24] It was reported that mismatches are more numerous, e.g., in the peptide systems.

CLINICAL IMPLICATIONS

It seems plausible that the effects on depression-manic state, due to manipulations affecting noradrenergic mechanisms in the locus ceruleus, may be mediated through changes in release of other transmitters at the cortical level. Several reports have suggested that chronic but not acute treatment with certain tricyclic antidepressants decreases the functional sensitivity of $alpha_2$-adrenoceptors in brain.[25,26] It has been shown that clonidine inhibition of the acoustic startle reflex in the rat, a behavioral measure of $alpha_2$-adrenoceptor sensitivity following desipramine administration, was also attenuated.[26] Other $alpha_2$-adrenoceptors than autoreceptors are also down regulated by chronic inhibition of NE uptake by tricyclic antidepressant treatment.[27] Although many studies indicate that a down regulation of central beta-adrenoceptor sensitivity accompanies chronic antidepressant administration, the consequences of such treatment on $alpha_2$-adrenoceptor function are more equivocal (for reviews see References 28 and 29).

Endogenous depression appears to be related to a dysfunction of the $alpha_2$-adrenoceptor (for reviews see References 30 and 31). Thus, recent biochemical and functional studies have demonstrated that depressed patients have an increased density and sensitivity of platelet $alpha_2$-adrenoceptors and that these receptor abnormalities are confined to the high affinity state ($alpha_{2H}$) of the receptor that recognizes preferentially agonists.[32–34]

Several lines of evidence suggest disturbances in catecholaminergic neurotransmission in mania.[35] In contrast to depression, mania may be characterized by increased noradrenergic and/or dopaminergic neurotransmission. Elevation in noradrenergic metabolism is supported by the higher excretion of urinary 3-methoxy-4-hydroxyphenylglycol (MHPG) in manic as compared to depressive episodes and the notable increase in cerebrospinal fluid (CSF) NE itself. Moreover, indirect pharmacological data suggest that agents that decrease noradrenergic function, such as reserpine, might be associated with an increased incidence of depression and therapeutic effects in mania. Conversely, most tricyclic and monoamine oxidase inhibitor antidepressants potentiate the noradrenergic system and may potentiate manic shifts. Moreover, the alpha-adrenergic agonist clonidine, which decreases firing of the noradrenergic cells of the locus ceruleus by acting preferentially at presynaptic autoreceptors,[36] has antimanic properties.[37]

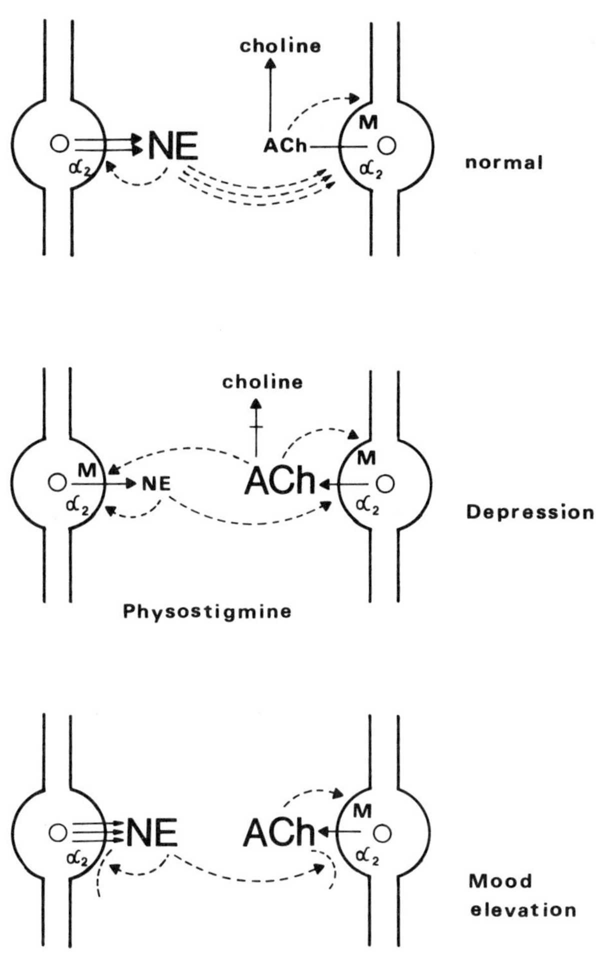

FIGURE 2. Scheme of cross talk between noradrenergic and cholinergic neurons in the cerebral cortex. Imbalance between noradrenergic and cholinergic systems: model of depression/mood elevation. Under physiological conditions the cholinergic transmission is under the tonic control of norepinephrine (NE) released from noradrenergic axon terminals. Physostigmine, an anticholinesterase, prevents the hydrolysis of ACh released from cholinergic axon terminals and multiplies the concentration of ACh and thereby it is able to exert an inhibitory effect on NE release. The effect of ACh is mediated via muscarinic (M_2) receptors. If the release of NE is reduced, the typical behavioral change is depression. However, if the negative feedback effect of NE on NE release is prevented by alpha $_2$-adrenoceptor blockade, the release of NE is enhanced and the possible effect of NE on alpha $_1$-adrenoceptors is multiplied. The effect of NE released in higher concentrations on cortical alpha $_1$-adrenoceptors may be partly responsible for elevation of mood, or inhibition of sleeping behavior.[15]

FIGURE 2 shows a scheme of depression/mood elevation balance as a consequence of cholinergic/noradrenergic interaction.

DISCUSSION

The nonconventional, nonsynaptic release of a substance and the nonsynaptic localization of receptors do not necessarily mean that there is an interaction between the release site and the effector cell. The substance released should be able to diffuse the distance and reach the receptor in a concentration high enough to stimulate it (FIGURE 1). The location of both synaptic and nonsynaptic release sites even within one terminal appeared to be very common in the CNS of different annelids,[38] and was recently also found in the CNS of the invertebrate pond snail *Lymnaea stagnalis*.[21] It seems very likely that even peptides can be used for long-distance signal transmission. An important implication of this communication system is that the release may not be the exclusive property of the axon terminal,[39] and it has been suggested[40–43] that in addition to the classical synaptic form of interaction there is a nonsynaptic form of communication between neurons when the transmitter or modulator released from axon terminals may affect remote target cells.

REFERENCES

1. VIZI, E. S. 1979. Presynaptic modulation of neurochemical transmission. Progr. Neurobiol. **12:** 181–290.
2. VIZI, E. S. 1980. Non-synaptic modulation of transmitter release: pharmacological implication. TIPS: 172–175.
3. VIZI, E. S. 1982. Non-synaptic intercellular communication: presynaptic inhibition. Acta. Biol. Acad. Sci. Hung. **33:** 331–351.
4. VIZI, E. S. 1984. Synaptic Interactions between Neurons: Modulation of Neurochemical Transmission. Pharmacological and Clinical Aspects. Wiley. Chichester, England.
5. DESCARRIES, L., A. BEAUDET & K. C. WATKINS. 1975. Serotonin nerve terminals in adult rat neocortex. Brain Res. **100:** 563–588.
6. DESCARRIES, L., K. C. WATKINS & Y. LAPIERRY. 1977. Noradrenergic axon terminals in the cerebral cortex of the rat. III. Topometric ultrastructural analyses. Brain Res. **133:** 197–222.
7. BEAUDET, A. & L. DESCARRIES. 1978. The monoamine innervation of rat cerebral cortex synaptic and nonsynaptic axon terminals. Neuroscience **3:** 851–860.
8. VIZI, E. S. & J. KNOLL. 1971. The effects of sympathetic nerve stimulation and guanethidine on parasympathetic acetylcholine release. J. Pharm. Pharmacol. **23:** 916–929.
9. VIZI, E. S. 1972. Stimulation by inhibition of (Na+, K+, Mg2+) activated ATPase, of acetylcholine release in cortical slices from rat brain. J. Physiol. London **226:** 95–117.
10. BEANI, L., C. BIANCHI, A. GRACOMELLI & F. TAMBERI. 1978. Noradrenaline inhibition of acetylcholine release from guinea-pig brain. Eur. J. Pharmacol. **48:** 179–183.
11. VIZI, E. S. & E. PASZTOR. 1981. Release of acetylcholine from isolated human cortical slices: inhibitory effect of norepinephrine and phenytoin. Exp. Neurol. **73:** 144–153.
12. MOBLEY, D. & P. GREENGARD. 1985. Evidence for widespread effects of noradrenaline on axon terminals in the rat frontal cortex. Proc. Nat. Acad. Sci. USA **82:** 945–947.
13. AUDET, M. A., G. DOUCET, S. OLESKEVICH & L. DESCARRIES. 1988. Quantified regional and laminar distribution of the noradrenaline innervation in the anterior half of the adult rat cerebral cortex. J. Chem. Neuroanat. **274:** 307–318.
14. MORONI, F., S. TANGANELLI, T. ANTONELLI, V. CARLA, C. BIANCHI & L. BEANI. 1983. Modulation of cortical acetylcholine and aminobutyric acid release in freely moving guinea pigs: effects of choline and other adrenergic drugs. J. Pharmacol. Exp. Ther. **227:** 435–440.

15. HARSING, L. G., JR., J. KAPOCSI & E. S. VIZI. 1989. Possible role of alpha-2 and alpha-1 adrenoceptors in the experimentally-induced depression of the central nervous system. Pharmacol. Biochem. Behav. **32:** 927–932.

16. BRADSHOW, C. M., R. Y. K. PUN, N. T. STATER & E. SZABADI. 1981. Comparison of the effect of methoxamine with those of noradrenaline and phenylephrine on single cerebral cortical neurones. Br. J. Pharmacol. **73:** 47–54.

17. DESCARRIES, L., B. LEMAY, G. DOUCET & B. BERGER. 1987. Regional and laminar density of the dopamine innervation in adult rat cerebral cortex. Neuroscience **21:** 807–824.

18. PARNAVELAS, J. G. & G. C. PAPADOPOULOS. 1989. The monoaminergic innervation of the cerebral cortex is not diffuse and non specific. TINS **12:** 315–319.

19. MERRILLEES, N. C. R., B. BURNSTOCK & M. E. HOLLMAN. 1963. Correlation of fine structure and physiology of the innervation of smooth muscle in the guinea pig vas deferens. J. Cell. Biol. **19:** 529–550.

20. THURESON-KLEIN, A. 1983. Exocytosis from large and small dense cored vesicles in noradrenergic nerve terminals. Neuroscience **10:** 245–252.

21. BUMA, P. 1988/89. Synaptic and nonsynaptic release of neuromediators in the central nervous system. Acta Morphol. Neerl. Scand. **26:** 81–113.

22. GORDON-WEEKS, P. R. 1981. Are there noradrenergic synapses in Auerbach's plexus? Scand. J. Gastroenterol. **16:** 11–182.

23. GORDON-WEEKS, P. R. 1982. Noradrenergic and non-noradrenergic nerves containing small granular vesicles in Auerbach's plexus of the guinea pig. Neuroscience. **11:** 2925–2936.

24. HERKENHAM, M. 1987. Mismatches between neurotransmitters and receptors. Neuroscience **23:** 1–38.

25. SPYRAKI, C. & H. C. FIBIGER. 1980. Functional evidence for subsensitivity of noradrenergic alpha$_2$-receptors after chronic desipramine treatment. Life Sci. **27:** 1863–1867.

26. MCMILLEN, B. A., W. WARNACK, D. C. GERMAN & P. A. SHORE. 1980. Effects of chronic desipramine treatment on rat brain noradrenergic responses to alpha-adrenergic drugs. Eur. J. Pharmacol. **61:** 239–246.

27. BILL, D. J., I. E. HUGHES & R. J. STEPHENS. 1989. The effects of acute and chronic desipramine on the thermogenic and hypoactivity responses to alpha$_2$-agonists in reserpinized and normal mice. Br. J. Pharmacol. **96:** 144–152.

28. CHARNEY, D. S., D. B. MENKES & G. R. HENINGER. 1981. Receptor sensitivity and the mechanism of action of antidepressant treatment. Arch. Gen. Psychiatry **38:** 1160–1180.

29. SUGRUE, M. F. 1988. A study of the sensitivity of rat brain alpha$_2$-adrenoceptors during chronic antidepressants. Naunyn Schmiedebergs Arch. Pharmacol. **320:** 90–92.

30. PILETZ, J. E., D. S. P. SCHUBERT & A. HALARIS. 1986. Evaluation of studies on platelet alpha$_2$-adrenoreceptors in depressive illness. Life Sci. **39:** 1589–1616.

31. DAVIS, M. & D. B. MENKES. 1982. Tricyclic antidepressants vary in decreasing alpha$_2$-adrenoceptor sensitivity with chronic treatment: assessment with clonidine inhibition of acoustic startle. Br. J. Pharmacol. **77:** 217–222.

32. GARCIA-SEVILLA, J. A., A. P. ZIP, P. J. HOLLINGSWORTH, J. F. GREDEN & C. B. SMITH. 1981. Platelet alpha$_2$-adrenergic receptors in major depressive disorder. Arch. Gen. Psychiatry **38:** 1327–1333.

33. GARCIA-SEVILLA, J. A., J. GUIMON, P. GARCIA-VALLEJO & M. J. FUSTER. 1986. Biochemical and functional evidence of supersensitive platelet alpha$_2$-adrenoceptors in major affective disorder. Arch. Gen. Psychiatry **43:** 51–57.

34. GARCIA-SEVILLA, J. A., C. UDINA, E. ALVAREZ & M. CASAS. 1987. Enhanced binding of ^{3}H (−)adrenaline to platelets of depressed patients with melancholia: effect of long-term clomipramine treatment. Acta Psychiatr. Scand. **75:** 150–157.

35. SILVERSTONE, T. 1985. Dopamine in manic depressive illness: a pharmacological synthesis. J. Affective Disorders. **8:** 225.

36. SVENSSON, T. H., B. S. BUNNEY & G. K. AGHAJANIAN. 1975. Inhibition of both noradrenergic and serotonergic neurons in brain by the alpha-adrenergic agonist clonidine. Brain Res. **92:** 291.

37. JOUVENT, R., Y. LECRUBIER, A. J. PUECH, P. SIMON & D. WIDLOCHER. Antimanic effect of clonidine. Am. J. Psychiatry **13:** 1.

38. GOLDING, D. W. & B. A. MAY. 1982. Duality of secretory inclusions in neurons. Acta Zool. Stockh. **63:** 229–238.
39. VIZI, E. S., K. GYIRES, G. T. SOMOGYI & G. UNGVARI. 1983. Evidence that transmitter can be released from regions of the nerve cell other than presynaptic axon terminal axonal release of acetylcholine without modulation. Neuroscience **10:** 967–972.
40. CUELLO, A. C. 1983. Non-classical neuronal communications. Fed. Proc. **42:** 2912–2922.
41. SCHMITT, F. O. 1984. Molecular regulators of brain function. A new view. Neuroscience: 991–1001.
42. AGNATI, L. F., K. FUXE, M. ZOLI, I. ZINI, G. TOFFANO, F. FERRAGUTI. 1986. A correlation analysis of the regional distribution of central enkephalin and beta-endorphin immunoreactive terminals and of opiate receptors in adult and old male rats. Evidence for the existence of two main types of communications in the central nervous system: the volume transmission and the wiring transmission. Acta Physiol. Scand. **128:** 201–207.
43. FUXE, K., L. F. AGNATI, A. CINTRA, K. ANDERSSON, P. ENEROTH, A. HARFSTRAND, M. ZOLI & M. GOLDSTEIN. 1988. Studies on central D1 receptors role in volume transmission, neuroendocrine regulation and release of noradrenaline. Adv. Exp. Med. Biol. **235:** 83–119.

Presynaptic Receptors[a]

Relevance to Psychotropic Drug Action in Man

HERBERT Y. MELTZER

Department of Psychiatry
Case Western Reserve University
School of Medicine
2040 Abingdon Road
Cleveland, Ohio 44106

INTRODUCTION

The ability of presynaptic receptors to modulate the release of neurotransmitters such as serotonin (5-HT), dopamine (DA), norepinephrine (NE), and acetylcholine (ACh) is well established.[1-4] There are two major types of presynaptic receptors: (1) autoreceptors which are involved in negative feedback mechanisms, and (2) heteroreceptors, i.e., presynaptic receptors which respond to neurotransmitters released from adjacent nerve terminals, to cotransmitter neuropeptides, or to other substances that may be hormonal (secreted elsewhere in the body) or the product of the same or nearby neurons, and which may enhance or diminish the calcium-dependent release of the neurotransmitter.[1] This review will consider some of the effects of clinically effective psychotropic drugs on auto- and heteroreceptors and their possible clinical significance. The roles of (1) alpha$_2$-adrenergic, D-2 and 5-HT autoreceptors in the action of antidepressants and lithium; (2) DA and 5-HT autoreceptors and heteroreceptors in the action of clozapine, a novel antipsychotic drug; and (3) DA autoreceptor agonists in the treatment of psychosis will be discussed. Brief mention will also be made of other possible links between psychotropic drug action and presynaptic mechanisms, e.g., the effect of 5-HT$_3$ antagonists on 5-HT heteroreceptors that stimulate DA release and of buspirone-type anxiolytic and antidepressant drugs on 5-HT$_{1A}$ presynaptic receptors.

POSSIBLE EFFECTS OF ANTIDEPRESSANT DRUGS AND LITHIUM ON PRESYNAPTIC RECEPTORS

Alpha$_2$ Presynaptic Receptors

Presynaptic alpha$_2$-adrenergic autoreceptors have been reported to inhibit the release of NE.[1,2] Because of the hypothesis that a decrease in noradrenergic activity may be etiologically important in major depression,[5] it has been suggested that decreasing the inhibitory effect of alpha$_2$-adrenoceptors on the release of NE might have an antidepressant action by enhancing the synaptic availability of NE. This presupposes that alpha$_2$-adrenoceptors regulate NE release in man. There is, however, some evidence challenging the view that presynaptic alpha$_2$-adrenoceptors provide

[a]Supported in part by National Institute of Mental Health grants number MH-41684, MH-41683, MH-41594, grants from the Cleveland Foundation, the Sawyer Foundation, and the Laureate Foundation/NARSAD. Dr. Meltzer is the recipient of a Research Scientist Award grant number MH-47808.

tonic regulation of NE release in man. Thus, Nutt and Molyneux found that the clonidine-induced decrease in plasma 3-methoxy-4-hydroxyphenylglycol (MHPG), the major metabolite of NE in the central nervous system (CNS), was positively correlated with baseline plasma MHPG levels in man.[6] Since greater alpha$_2$-adrenoceptor sensitivity would be expected to lead to lower basal MHPG levels and a greater response to clonidine, i.e., a negative correlation between these two measures, they could not confirm the existence of an alpha$_2$-adrenoceptor-mediated presynaptic mechanism, at least in the group of noradrenergic neurons that contributes most heavily to plasma MHPG. While it seems unlikely that the alpha$_2$-adrenergic autoreceptor mechanism is not relevant to tonic NE release in man, it is prudent not to be overly confident that all such mechanisms present in rodents have the same importance or even presence in man. Furthermore, the evidence for the importance of negative feedback autoreceptors for the regulation of neurotransmitter release in general has been called into question.[7]

The mechanism by which antidepressant drugs have been thought to induce alpha$_2$-adrenoceptor subsensitivity is via an increase in the concentration of NE either in the synapse (uptake blockers) or intracellularly [monoamine oxidase (MAO) inhibitors]. The classical tricyclic antidepressants (e.g., imipramine, desipramine, etc.) increase synaptic concentrations of NE by blocking reuptake. MAO inhibitors do so by blocking the metabolism of NE after synthesis or reuptake. However, these are acute effects and there is evidence that suggests that MAO inhibition does not enhance the release of NE (see Reference 4 for review). Chronic treatment with both tricyclic antidepressants and MAO inhibitors appears to decrease the turnover of NE in the CNS as evidenced by decreased firing of NE neurons in the locus ceruleus[8] as well as by biochemical studies.[9,10] It is possible that antidepressants might increase the concentration of NE at presynaptic receptors irrespective of depolarization-induced release, perhaps due to altered internal compartmentalization of the neurotransmitter.[4] Antidepressants such as specific 5-HT uptake blockers which do not affect the availability of NE also can decrease the sensitivity of the alpha$_2$-adrenergic site, presumably by an indirect mechanism (see Reference 11 for review).

There is extensive pharmacologic evidence which has been reviewed elsewhere[4,12] that some antidepressants down regulate alpha$_2$-adrenoceptors, but this is not necessarily the alpha$_2$-adrenoceptor autoreceptor. Chronic treatment with desipramine has been found to increase NE release in response to electrical stimulation of isolated rat atrial strips and to decrease the ability of clonidine to inhibit this release.[13] Chronic imipramine treatment decreased the ability of clonidine to inhibit locus ceruleus firing[14] as did chronic desipramine and zimelidine but chlorimipramine, mianserin, and iprindole did not.[15] Other indicators of alpha$_2$-adrenoreceptor subsensitivity, e.g., antidepressant-induced diminished decreases in brain MHPG-SO$_4$ following clonidine administration, could be due to an effect of the antidepressant at the somatodendritic presynaptic receptor.

There is also negative evidence for alpha$_2$-adrenoceptor desensitization by antidepressant drugs. Chronic treatment with either clorgyline[16] or desipramine[17] had no effect on clonidine-induced inhibition of the release of NE from cerebral cortex slices or synaptosomes, suggesting no desensitization had occurred. Schoffelmeer *et al.* reported that chronic treatment with desipramine but not maprotiline or chlorimipramine enhanced the electrically evoked release of [^{3}H]NE from neocortical slices whereas chronic zimelidine and iprindol treatment slightly decreased it.[18] None of the treatments affected the inhibitory effect of NE on release of [^{3}H]NE. Ligand binding studies of the effect of antidepressants on alpha$_2$ sites have produced mixed results, e.g., increases,[19] decreases,[20,21] and no effect,[22,23] and in any event do not distinguish between pre- and postsynaptic alpha$_2$-adrenoceptor sites. In fact, if presynaptic alpha$_2$-

adrenoceptor sites were to serve their homeostatic function, they would be expected to up regulate in response to increased release of NE, in order to be able to decrease further NE release,[19,24] and there is some evidence that this may, in fact, occur.

There are recent reports that demonstrate rapid down regulation of cortical beta-adrenoreceptors by the combination of alpha$_2$-antagonists and antidepressants.[25] The alpha$_2$-adrenergic antagonists are believed to promote the further release of NE by blocking terminal autoreceptors, thereby enhancing NE-induced down regulation of the beta-receptor. However, clinical studies show that yohimbine, a potent alpha$_2$-antagonist, did not enhance the antidepressant effect of tricyclic antidepressants in depressed nonresponders.[26]

There is no evidence for up regulation of presynaptic alpha$_2$-adenoceptors in major depression as indicated by plasma NE levels[27,28] or the 24 hour urine levels of metabolites of NE.[29] If anything, the results in major depression suggest increased NE release from the adrenal gland, and peripheral or central nervous systems, more consistent with down regulation of alpha$_2$-adrenoceptors. Since chronic treatment with a variety of antidepressants has been reported to either decrease[30–32] or have no effect[33] on cerebrospinal fluid (CSF) MHPG concentrations, it seems unlikely that increased NE due to alpha$_2$ subsensitivity is a necessary aspect of antidepressant action. Some have suggested a dysregulation of pre- and postsynaptic mechanisms is present in depression and that the mechanism of action of antidepressants is to bring these systems into better coordination.[34]

Chronic desipramine or clorgyline treatment was found not to alter the sensitivity of alpha$_2$-adrenergic heteroreceptors which regulate 5-HT release from rat cortical synaptosomes.[35] This suggests no alteration by antidepressant treatment of the sensitivity of alpha$_2$-adrenoceptors on 5-HT nerve terminals.

In conclusion, alpha$_2$-adrenergic receptor subsensitivity, either at the autoreceptor or the 5-HT-release-modulating heteroreceptor, remains an unproven and perhaps unlikely mechanism of action of antidepressant drug treatments.

Serotonin Autoreceptors

There is extensive evidence that the release of 5-HT from serotonergic neurons is under negative feedback control by inhibitory terminal autoreceptors.[36,37] In an extensive series of studies, DeMontigny, Blier, and colleagues have sought to explain the action of various antidepressants by a shared ability to enhance serotonergic neurotransmission, albeit by at least three different mechanisms (see references 38 and 39 for review). They suggested MAOI act by increasing the amount of 5-HT released per impulse while tricyclic antidepressants and mianserin may act by sensitizing postsynaptic 5-HT$_2$ receptors.[38,39]

DeMontigny and Blier have suggested that chronic treatment with specific 5-HT uptake blockers, e.g., zimelidine, indalpine, citalopram and fluoxetine, enhances 5-HT neurotransmission by desensitizing both the somatodendritic 5-HT autoreceptor and the terminal autoreceptors.[40–44] The evidence for the former effect is inhibition of firing, followed by recovery of the firing rate, of dorsal raphé neurons over a 14-day treatment period. The evidence for the latter is that the effect of electrical stimulation of the ascending 5-HT pathway to the hippocampal pyramidal neurons is increased after 14 days treatment with 5-HT uptake blockers. This enhancement does not appear to be directly due to uptake blockade or to the development of increased postsynaptic receptor sensitivity. Because the effect of an autoreceptor antagonist, methiothepin, was diminished after chronic citalopram or fluoxetine, these results suggested the development of terminal autoreceptor subsensitivity.

The mechanism by which 5-HT uptake blockers may affect autoreceptor sensitivity has been studied. It has been reported that the inhibitory effect of LSD on ^{3}H-5-HT release from hypothalamic slices was antagonized by the antidepressants imipramine, amitriptyline, paroxetine, and citalopram, but not by an MAO inhibitor, pargyline. *Para*-Chlorophenylalanine (PCPA) treatment, to inhibit 5-HT synthesis, had no influence on this effect.[45] Other antidepressants such as maprotiline, buproprion, and mianserin which do not affect 5-HT uptake did not antagonize the inhibitory effect of LSD on 5-HT release in this preparation.[46] These results suggest that the effects on the autoreceptor of the 5-HT uptake blockers are due to an interaction between the uptake site and the autoreceptor rather than via an effect on synaptic 5-HT levels.[46] This hypothesis has been questioned by Raiteri *et al.* who proposed that the 5-HT uptake blockers may act by increasing the concentration of 5-HT in the biophase.[47] Their view is supported by Feuerstein *et al.*, who found an inverse relation between the concentration of 5-HT in the biophase and the effect of exogenous 5-HT agonists to inhibit ^{3}H-5-HT release from hippocampal slices.[48] It should be noted that Galzin *et al.* found that imipramine and amitriptyline, both nonspecific 5-HT uptake blockers, were able to antagonize the effect of LSD on ^{3}H-5-HT release[45] whereas the electrophysiological studies of de Montigny and colleagues did not find an effect of these types of antidepressants, only the 5-HT uptake blockers, on somatodendritic or terminal autoreceptors.[38,39,49]

There is evidence that 5-HT autoreceptors can modulate 5-HT release in human cortex.[50] The human 5-HT autoreceptor has been labeled the 5-HT$_{1D}$ receptor to distinguish it from the rat cortical autoreceptor, which has been labeled as the 5-HT$_{1B}$ subtype.[51] It is uncertain which drug effects at the 5-HT$_{1B}$ autoreceptor apply to the 5-HT$_{1D}$ autoreceptor. Moreover, there is no evidence that 5-HT uptake blockers, either specific or nonspecific, have a differential antidepressant action over drugs such as maprotiline or buproprion which do not affect 5-HT uptake.[52,53] This mechanism could also not account for why specific 5-HT uptake blockers such as fluoxetine are more effective in obsessive-compulsive disorder (OCD) than tricyclic antidepressant drugs such as imipramine.[54]

Specific 5-HT uptake blockers decrease CSF 5-hydroxyindoleacetic acid (5-HIAA) levels[55,56] as do typical antidepressants (see Reference 57 for references). Subsensitive 5-HT autoreceptors might be expected to enhance 5-HT release and increase 5-HIAA levels in CSF. This does not necessarily rule out 5-HT$_2$ receptor subsensitivity because there could be other mechanisms to restrict 5-HT release but increase overall synaptic transmission.[56,57] It is also possible that increased 5-HT release is not reflected in increased CSF 5-HIAA levels because only small areas of brain are involved. However, the decrease in CSF 5-HIAA does not support a singular importance of 5-HT autoreceptor subsensitivity in the action of antidepressants.

Effect of Lithium Treatment on 5-HT Autoreceptors

Despite extensive effort, there is no consensus as to the mechanism of action of lithium carbonate in affective disorder treatment or prophylaxis. However, there is considerable evidence that a variety of effects of lithium treatment on serotonergic neurotransmission may be important.[58] Short-term lithium treatment has been reported to enhance serotonergic neurotransmission by enhancing 5-HT release.[59] This conclusion was based on the ability of lithium treatment for 48 hours to enhance the response of CA$_3$ hippocampal pyramidal neurons to electrical stimulation of the ventromedial 5-HT pathway without increased postsynaptic receptor responsivity. However, no direct electrophysiological evidence for enhanced 5-HT release was obtained and mechanisms other than those involving autoreceptors could have been a factor. There is, however, some biochemical evidence in favor of this hypothesis. It has recently been reported that three days treatment with lithium carbonate attenuated the ability of 5-HT to decrease the release of ^{3}H-5-HT from hippocampal but not frontal cortical slices.[60] It was suggested this increase in 5-HT release may account for the ability of lithium to selectively decrease the density of 5-HT$_1$ receptors in rat hippocampus without affecting their density in cortex. Wang and Friedman[61] demonstrated that chronic lithium treatment decreased the ability of LSD to inhibit the release of preloaded [^{3}H]5-HT from superfused cortical, hippocampal, and hypothalamic brain slices. Lithium treatment also markedly diminished the potentiation by methiothepin, a presynaptic 5-HT antagonist, on [^{3}H]5-HT release. These results are also consistent with an effect on presynaptic receptors. There is, thus intriguing evidence that 5-HT autoreceptor antagonism, subsensitivity, or effects mediated through changes in the biophase of 5-HT$_2$ concentration could play an important role in the antidepressant, antimanic, or prophylactic effects of lithium treatment. However, the effects of lithium on 5-HT release develop rapidly while the clinical effects take weeks to months. The ability of lithium to rapidly potentiate the action of antidepressants in depressed nonresponders might be related to these acute effects on the release of 5-HT.

DA Autoreceptor Drugs

There have been numerous suggestions that antidepressant drugs may act, in part, via an ability of even single doses plus a period of time[62] or chronic treatment[63-65] to induce DA autoreceptor subsensitivity and increase the release of DA, presumably from mesolimbic or mesocortical DA neurons. There is pharmacologic and biochemical evidence for a deficiency in dopaminergic activity in depression so that increasing the release of DA by the development of autoreceptor subsensitivity could be a relevant basis for an antidepressant response.[66] Other biochemical,[67,68] electrophysiological,[69] and behavioral[70] studies have failed to confirm antidepressant-induced subsensitivity of DA autoreceptors. Smialowski and Maj reported the development of a supersensitive postsynaptic response to apomorphine in the hippocampus following chronic imipramine treatment.[71] There is no evidence from studies of DA metabolites in CSF that would indicate that chronic antidepressant treatment increases DA turnover in man,[33,72] as would be expected if DA receptor subsensitivity had developed, at least in the nigrostriatal system.

Autoreceptor Subsensitivity and Antidepressant Action

As reviewed above, some or all of the various types of antidepressant drugs have been suggested to act via induction of noradrenergic, serotonergic, and dopaminergic terminal or somatodendritic autoreceptor subsensitivity. The type of evidence available to support these hypotheses is almost entirely preclinical but of a diverse nature, e.g., release of amines from slices or synaptosomes *ex vivo* after chronic treatment, and electrophysiological, biochemical, or behavioral studies in rodents. Clinical evidence is generally lacking or does not provide support for the development of subsensitive autoreceptors, but it is, of course, uncertain that studies of CSF or plasma concentrations of metabolites could provide definitive tests of these hypotheses. Future studies are therefore needed to test these concepts in a more definitive way. However, when predictions from animal studies fail to be confirmed in man, e.g., the inability of yohimbine to facilitate the action of antidepressant treatment,[26] this must be considered as a serious mark against the hypothesis provided that the clinical studies did provide appropriate tests of the hypothesis and were done in a fashion to provide a valid test of the hypothesis with sufficient power.

It seems reasonable that all antidepressants need not act via the same mechanism. The work of de Montigny and colleagues explicitly relates an effect on 5-HT autoreceptors only to specific 5-HT uptake blockers.[38–41] It is possible that variable effect on combinations of autoreceptors by specific antidepressants, e.g., down regulation of NE, DA, and 5-HT autoreceptors, may contribute to their action. This could be supplemented by effects on postsynaptic beta-adrenergic, 5-HT_{1A}, or 5-HT_2 receptors as well as other mechanisms.

Dopamine Autoreceptors and Psychosis

The hypothesis that modulating dopaminergic neurotransmission is critical to the action of antipsychotic drugs in schizophrenia and other psychoses such as manic-depressive psychosis, psychotic depression, and some drug-induced psychoses remains unchallenged. The most commonly held view is that to achieve an antipsychotic effect it is necessary to reduce dopaminergic neurotransmission in the mesolimbic system and that neuroleptic drugs do so by blocking postsynaptic D-2 receptors, eventually inducing depolarization inactivation of the mesolimbic DA neurons.[73,74] The major modification to the DA hypothesis that has emerged in recent years is the concept that the so-called negative symptoms, e.g., flat affect, withdrawal, anhedonia, and lack of motivation, may be due to decreased dopaminergic activity in the prefrontal or other cortical areas.[75,76] If decreased dopaminergic activity in some regions, is, in fact, part of the pathophysiology of schizophrenia, it is important to develop drug strategies that do not uniformly decrease dopaminergic activity throughout the neuraxis. The importance of maintaining some dopaminergic activity in the striatum to prevent severe parkinsonian symptoms and possibly to lessen the chances of developing tardive dyskinesia must also be considered. The modulatory role of postsynaptic D-1 receptors on dopaminergic neurotransmission and the action of antipsychotic drugs has also been emphasized in many recent stuides (see Reference 77 for review).

Stimulation of DA autoreceptors to decrease the release of DA in some, if not all, DA neurons as a potential means of achieving an antipsychotic effect has been suggested by many authors. Numerous clinical studies reviewed elsewhere[78,79] have suggested that low doses of DA agonists can produce a variety of effects in man that are consistent with autoreceptor-mediated decreases in dopaminergic neurotransmission: (1) induction of relaxation, sedation, and sleep in a variety of excited psychiatric and neurological conditions, as well as in normal volunteers; (2) an antimanic effect; (3) an antipsychotic effect; (4) a reduction in tardive dyskinesia movements; (5) a reduction in Huntington's chorea; (6) a reduction in the symptoms of Tourette's syndrome; (7) the production of Parkinsonian symptoms in schizophrenics when combined with low doses of a DA agonist, although neither have a discernable effect alone; (8) a reduction of symptoms in spasmodic torticollis; and (9) a reduction in craving for alcohol. These studies have been reviewed previously.[78] The evidence that available DA agonists, e.g., apomorphine or bromocriptine, are effective as antipsychotic agents, at any dosage, is unconvincing.[78] This topic will also be reviewed elsewhere in this volume and will not be further discussed here except for the issue of the presence of DA autoreceptors in the mesocortical neurons which bears upon the issue of drugs selectively affecting mesolimbic and nigrostriatal dopaminergic neurotransmission.

Dopamine autoreceptors have at least three distinct regulatory functions: inhibition of synthesis, release, and neuronal firing. Thus, DA nerve terminals in the prefrontal and cingulate cortex have been reported to possess autoreceptors that regulate DA release[80-82] unlike mesolimbic and mesocortical dopaminergic neurons which have few or no autoreceptors that modulate DA synthesis.[83] There is, however, recent biochemical evidence that a subgroup of the mesolimbic DA neurons also do not possess synthesis modulating autoreceptors.[84] This has been related to the absence of a marked effect of low-dose DA agonists on psychotic symptoms. There is also biochemical evidence for DA autoreceptors that regulate DA synthesis in the mesoprefrontal DA neurons.[85] There is general agreement that mesocortical DA neurons possess impulse-modulating somatodendritic autoreceptors.[86,87] The differences in autoreceptors between the mesoprefrontal and mesocingulate and other DA neurons (nigrostriatal, mesolimbic, and mesopyriform) may underlie some of the differences in basal DA activity between these neurons and responses to DA agonists and antagonists (see Reference 86 for review).

At the current time, there is an intensive effort to develop specific DA autoreceptor agonists, drugs that lack significant agonist or antagonist effects at postsynaptic DA receptors. These drugs should be distinguished from the autoreceptor agonists which are D-2 antagonists at the postsynaptic receptor and other partial antagonists that are not even relatively selective for the DA autoreceptor. Further study is needed to clarify the existence of, and relative importance of, these various types of autoreceptor mechanisms in various dopaminergic neurons. There are so many possible selective effects on the gamut of DA autoreceptors: (1) selective effects on different types of autoreceptors; (2) selective effects in various regions; (3) development of subsensitivity of only some types of DA autoreceptors following the administration of the new generation of more specific agonists; (4) differences in the ability of postsynaptic feedback mechanisms to overcome the presynaptic effects of these agonists; (5) differences in postsynaptic DA receptor sensitivity among patients; (6) differences in the extent, if any, of degeneration of DA neurons in various stages of different neuropsychiatric diseases. Because of these and other potentially important variables, it is probably impossible to predict what would be the long-term outcome of treatment with more specific DA autoreceptor agonists. Nevertheless, it is desirable that *specific* agents be clinically tested, preferably without concomitant neuroleptic treatment, in a wide range of disorders in which a role of DA is suspected and that effects on positive and negative symptoms as well as movement disturbances be carefully monitored by experienced clinicians.

5-HT$_3$ RECEPTOR ANTAGONISTS

Another strategy of utilizing presynaptic receptors as the target for psychotropic drugs that has been advanced is based upon the reported ability of 5-HT to increase DA release in various brain regions. This appears to be mediated by the 5-HT$_3$ receptor subtype, but other 5-HT receptor subtypes may also be involved.[88] The relatively selective 5-HT$_3$ agonist 2-methyl-5-HT increases DA release from striatal slices *in vitro*. This is blocked by the 5-HT$_3$ antagonist ICS 205-930.[89] Similarly, the 5-HT$_3$ antagonist odansetron (GR 38032F) inhibits the locomotor hyperactivity caused by the injection of amphetamine or DA into the nucleus accumbens.[88] Odansetron also blocks the hyperactivity in rats following an injection into the ventral tegmental area of the neurokinin receptor agonist [p Glu^5, MePhe^8, Sar^9] SP-5-11 (DiMe-C7) which increases DA metabolism in the mesolimbic and mesocortical DA areas by activating A10 neurons.[90] On the other hand, odansetron does not antagonize nonmesolimbic dopaminergic responses such as amphetamine-induced stereotypy in rats.[91] These results suggest that 5-HT$_3$ antagonists might have a role in any psychiatric condition in which excessive dopaminergic activity may play a role, e.g., psychotic symptoms of schizophrenia, mania, etc. A variety of antidepressant drugs also have a high potency as 5-HT$_3$ antagonists and it has been suggested that an effect at the 5-HT$_3$ receptor may mediate some of their therapeutic or side effects, or both.[92]

The 5-HT$_3$ antagonists BRL 43694 and GR 38032F have also been reported to be as active as diazepam in the social interaction model of anxiety in rats[93] as well as place-avoidance conditioning,[94] but other studies fail to confirm this[95] or to demonstrate significant anxiolytic effects in other animal models.[96] Furthermore, there is still some question as to whether 5-HT$_3$ receptors are present in human brain.[97,98] 5-HT$_3$ receptors are definitely present in the periphery in man and appear to be involved in cardiovascular, gastrointestinal, and pain regulatory mechanisms.[99] Clinical trials with these agents are essential to begin to determine if this strategy is successful.

EFFECT OF CLOZAPINE ON PRESYNAPTIC RECEPTORS

Clozapine has recently been shown to be more effective in treating positive and negative symptoms than typical neuroleptic drugs.[100,101] It also differs in producing fewer extrapyramidal symptoms (EPS), has not been reported to produce tardive dyskinesia (TD), and does not elevate serum prolactin (PRL) levels in man.[102] We have suggested that these advantages may stem from a common mechanism since it seems unlikely that heterogeneous biological mechanisms would underlie these causes.[103] Possible presynaptic receptor mechanisms that may contribute to the unique profile of clozapine will be considered here.

The unique effect of clozapine that is best understood at the current time is its effect on PRL secretion in the rodent. Clozapine is a relatively potent D-2 receptor blocker at the rat pituitary[104] and would be expected to stimulate PRL secretion *in vivo*. Clozapine does, in fact, cause a marked but unusually short-lived increase in rat serum PRL levels.[105] Clozapine at micromolar concentrations can also block the ability of DA to inhibit PRL release from rat hemipituitary glands *in vitro* (Meltzer *et al.*, unpublished data). We have shown that clozapine can increase the turnover of DA in the tuberoinfundibular (TIDA) neurons and hence increase the release of DA.[106] This effect is blocked by 5-HT$_2$ antagonists or interference with 5-HT synthesis by pretreatment with PCPA (Gudelsky *et al.*, in preparation). The increase in DA release from the median eminence is believed to overcome the blockade of pituitary DA receptors by clozapine, leading to an inhibition of PRL secretion.

We have suggested that the ability of clozapine to maintain or enhance DA release may be one of the main reasons for its unique clinical profile.[107,108] This may occur via effects on presynaptic receptors. However, there is conflicting evidence concerning the effect of acute or chronic clozapine on DA release in the striatum and nucleus accumbens. There are also data on the effect of clozapine on 5-HT and acetylcholine (ACh) release which will be reviewed here.

Rebec *et al.* compared the ability of clozapine and haloperidol to antagonize the depression of the firing rate produced by *d*-amphetamine and apomorphine in the neostriatum and nucleus accumbens of anesthetized rats.[109] Clozapine was more effective than haloperidol in blocking the effect of amphetamine but both were comparable in blocking apomorphine-induced depression in both regions. They attributed the greater antiamphetamine effect of clozapine to a possible ability of clozapine to inhibit DA release. No evidence to support this hypothesis was offered. Other early studies suggested that clozapine was not an effective presynaptic receptor antagonist.[110,111] Acute clozapine was reported to increase DA release in the nucleus accumbens but not the caudate nucleus, using an *in vivo* electrochemical method,[112] thus suggesting a selective effect on mesolimbic neurons. Bunney *et al.*, using an extracellular recording technique, also found that clozapine increased the firing rate of DA-containing neurons only in the ventral tegmental area.[113]

Chronic clozapine treatment was reported to inhibit the firing of A10 but not A9 DA neurons by depolarization inactivation, as determined by extracellular recording.[114] These authors believed that the inhibitory effect of chronic but not acute clozapine on cell firing, and presumably DA release, in the A10 neurons explained the delay onset of its antipsychotic action. They suggested that the development of depolarization inactivation in the A9 neurons may be related to EPS and subsequent TD.

Lane and Blaha also reported that acute clozapine treatment significantly increased the release of DA only in the nucleus accumbens using *in vivo* voltammetry.[115] They reported that chronic clozapine decreased basal DA release in the nucleus accumbens but not the striatum. This was accompanied by a significant decrease (60%) in the ability of clozapine to increase DA release in the accumbens after an acute challenge dose. These effects were reversed by apomorphine and were attributed to depolarization inactivation.[116] The regional selectivity of clozapine was attributed to its anticholinergic properties on the basis of pharmacologic studies with haloperidol and the anticholinergic, trihexyphenidyl.[115]

Hand *et al.* also found that clozapine increased the firing rate of A10 but not A9 neurons at lower doses and that even a single higher dose of clozapine could produce depolarization inactivation of A10 neurons since their firing could be restored by apomorphine.[117] They suggested this was due to blockade of somatodendritic autoreceptors. This was supported by the finding that clozapine could reverse the apomorphine-induced inhibition of firing of A9 and A10 neurons, with a more potent effect in the A10 region. Iontophoretic clozapine in the A9 and A10 regions also blocked the inhibitory effect of iontophoretic DA on cell firing. Pharmacologic and lesion studies suggested this selectivity was not due to anticholinergic effects or the influence of forebrain structures, respectively.

Hetey and Drescher reported that clozapine could block the ability of DA to inhibit the K^+-stimulated release of ^3H-DA and ^3H-5-HT from accumbens synaptosomes.[118] The EC_{50} values for these effects were 0.5 and 10.0 μM, respectively. They also found that clozapine could block the ability of 5-HT to inhibit the K^+-stimulated release of ^3H-5-HT and ^3H-DA from the same preparation at EC_{50} values of 1.0 and 7.0 μM, respectively.[119] These findings suggested that clozapine was a potent inhibitor of autoreceptors and heteroreceptors that modulate DA and 5-HT release in the mesolimbic system. There is some evidence that clozapine is a $5\text{-}HT_3$ antagonist.[120] Thus, it could increase DA release by blocking the inhibitory effect of $5\text{-}HT_3$ receptor stimulation on DA release.[88,89]

Compton and Johnson recently reported that chronic clozapine treatment significantly enhanced the ability of amphetamine to stimulate ^3H-DA release from striatal slices, *ex vivo*.[121] Clozapine itself also stimulated ^3H-DA release but only at high concentrations (10 μM). Acute clozapine treatment (20 mg/kg) significantly increased (29%) the electrical field–stimulated DA release in striatal slices *ex vivo* but not in accumbens slices. Clozapine had no effect on the electrically stimulated DA release from striatal slices when added *in vitro*. However, chronic clozapine (21 days) did not affect the release of DA from striatal slices after electrical stimulation, indicating the development of tolerance. These results are analogous to the tolerance to the ability of clozapine to increase DA release that we have observed with *in vivo* dialysis (see below) Clozapine also had complex effects on ACh release.[121] It blocked the ability of the DA agonist TL-99 to inhibit electrical field–stimulated ACh in both regions. No *ex vivo* effect on electrically evoked ACh release or the effect of TL-99 was noted with acute or chronic clozapine treatment. However, clozapine was able to partially reverse the inhibitory effect of TL-99 on ACh release in the accumbens but not the striatum. Compton and Johnson suggested the greater effect of clozapine on the ACh release-modulating presynaptic receptor in the nucleus accumbens might explain why it produces depolarization inactivation of A10 and not A9.[121] The ability of chronic clozapine to enhance amphetamine-stimulated DA release in the striatum was taken as an indication that clozapine may enhance the levels of DA in the synapse in this region via an effect on carrier-mediated DA release.

Using *in vivo* dialysis in freely moving rats, we have found that clozapine produces an acute increase in the release of DA and DOPAC in the striatum and nucleus accumbens[122] (Ichikawa and Meltzer, in preparation). The effect in the nucleus accumbens is approximately twice as large, expressed as percentage of baseline DA efflux. We have also found that chronic clozapine (20 mg/kg in drinking water for 21 days) did not significantly decrease basal DA release in either region.[122] It did attenuate the ability of clozapine to increase DA release in both regions, but especially the nucleus accumbens. Imperato and Angelucci also found that low doses of clozapine acutely increased striatal DA release in the striatum and frontal cortex *in vivo*.[123] Clozapine at higher doses suppressed the ability of the D-2 agonist LY 171555 to decrease DA release. These results are consistent with the ability of clozapine to produce DA autoreceptor blockade. O'Connor *et al.* using *in vivo* dialysis reported that clozapine increased DA release in the nucleus accumbens but not the striatum.[124]

These more recent studies suggest that clozapine is an effective presynaptic receptor antagonist, having the capacity to modulate both impulse-regulating and release-inhibiting presynaptic autoreceptors and heteroreceptors that modulate DA, ACh, and 5-HT release. Since clozapine is a potent alpha$_2$-adrenoceptor antagonist,[125] it most likely enhances NE release as well.

Because of the recent interest in the hypothesis that negative symptoms in schizophrenia are due, in part, to diminished dopaminergic activity in the prefrontal cortex,[75,76] the ability of clozapine to increase DA release in the prefrontal cortex may be of particular interest. Clozapine has been thought to be a potent D-1 antagonist relative to other antipsychotic drugs and to D-2 potency, in both preclinical[126] and positron emission studies[127] in man. However, the importance of D-1 receptor blockade for the action of clozapine and other atypical antipsychotic drugs has been challenged.[128] Increased release of clozapine might be expected to overcome its blockade at either or both D-1 and D-2 receptors. The ability of clozapine to enhance 5-HT release needs to be further studied, especially with *in vivo* dialysis or with appropriate *in vitro* preparations. Clozapine may also increase DA, 5-HT, and NE release by blockade of postsynaptic receptors (D2, 5-HT$_2$, alpha$_1$- and alpha$_2$-adrenoceptor) which would lead to increased presynaptic release by negative feedback mechanisms. Clozapine is a potent 5-HT$_2$ antagonist,[128] and it down regulates 5-HT$_2$ receptors.[129] It is not clear what other 5-HT receptors subtypes it may effectively antagonize, if any. If it does increase 5-HT release *in vivo,* it could increase stimulation of any 5-HT receptor subtypes it does not antagonize, e.g., the 5-HT$_{1A}$ or 5-HT$_3$ receptors. Clozapine is also a moderately potent alpha$_1$ and alpha$_2$ blocker.[125]

It is not possible to predict the net effect of the ability of clozapine to enhance release of neurotransmitters and to block their postsynaptic receptors. The balance may vary over time and between subjects and be relevant to its ability to help some but not all nonresponders and to have a very variable time course in the onset of action.[130] In any event, the importance of the ability of an antipsychotic drug to maintain normal levels of a neurotransmitter by presynaptic mechanisms remains a fruitful area of research to explain the action of clozapine and for the further development of other antipsychotic treatments. If these hypotheses are correct, then the specific design of drugs to enhance DA and 5-HT release while achieving D-2 and 5-HT$_2$ postsynaptic blockade may become a goal of drug development.

5-HT$_{1A}$ AGONIST ANXIOLYTICS AND ANTIDEPRESSANTS

Buspirone and geperone appear to be clinically effective anxiolytic and antidepressant agents.[131-133] These drugs bind with high affinity to the 5-HT$_{1A}$ receptor subtype.[134,135] Acutely, these agents inhibit nerve impulse flow within serotonergic neurons of the dorsal raphé in rats,[136,137] but the firing rate returns to normal after 14 days treatment.[138] The initial inhibition appears to be due to an action at inhibitory somatodendritic autoreceptors, followed by desensitization of these autoreceptors. These agents are partial 5-HT$_{1A}$ agonists[135,139] which decrease 5-HT turnover acutely.[132,136] The actual antidepressant and anxiolytic effects of these agents may be due to stimulation of postsynaptic 5-HT$_{1A}$ receptors after desensitization of inhibitory autoreceptors.[140,141]

CONCLUSIONS

There has been intensive investigation of the possibility that a variety of psychotropic drugs exert their clinical effect or could be targeted to exert their effect, via an action on specific presynaptic receptors that modulate the release of neurotransmitters whose synaptic concentration it would appear desirable to increase or decrease. This brief review could consider only some of this evidence. Insofar as there is clearly a need to consider the availability of neurotransmitters and modulators as well as the full range of postsynaptic mechansisms to understand psychotropic drug action, it is clearly essential to carry out studies of possible effects of acute and chronic treatment with psychotropic drugs on a variety of auto- and heteroreceptors using both *in vitro* and *in vivo* systems, with a range of outcome measures. While the results of such studies are of keen interest in what they would reveal about potential mechanisms, it would be inappropriate to assume they are relevant to man simply because of any apparent consistency with *a priori* hypotheses about possible pathophysiologic deficits in neurotransmission in specific disease states. It will still be necessary to carry out a series of studies to test specific hypotheses based on the suggested effect on autoreceptor mechanisms in animal models both *in vitro* and *in vivo* and to carry out appropriate clinical studies in man whenever possible to put such hypotheses to rigorous testing. These clinical studies should include an integrated examination of effects on multiple pre- and postsynaptic receptors wherever possible, especially if relationship to clinical response is part of the research design. Failure to find evidence suggestive of significant increases or decreases in the release of a neurotransmitter or its metabolite, by measuring its levels in plasma or CSF before and after drug treatment, may not be inconsistent with an effect on autoreceptors or heteroreceptors. Simultaneous changes in postsynaptic mechanisms could have led to markedly different levels of neurotransmitter release in the absence of a drug effect on presynaptic mechanisms, thus masking the presence and significance of an effect on presynaptic mechanism.

Of the many possible effects of psychotropic drugs on presynaptic mechanisms that were reviewed in this article, it is fair to say that there is no convincing evidence that any currently used antidepressant or antipsychotic agent acts mainly via an effect on presynaptic receptors. It would seem likely that 5-HT$_{1A}$ agonists achieve their anxiolytic and possibly their antidepressant action via an effect on postsynaptic 5-HT$_{1A}$

receptors, but only after the development of subsensitivity to their inhibitory effect on somatodendritic 5-HT_{1A} autoreceptors. There have been numerous suggestions from preclinical studies that one or more types of antidepressants act via desensitizing $alpha_2$-adrenergic, serotonergic, or dopaminergic autoreceptors but there is no clinical evidence to support any of these suggestions. The strategy of treating schizophrenia with DA autoreceptor agonists remains unproven but is hopefully heading towards definitive testing. Finally, the importance of the ability of clozapine to enhance 5-HT or DA release via an action at presynaptic receptors remains to be determined by further study but appears to be important.

ACKNOWLEDGMENT

The expert typing of Ms. Diane Mack is gratefully acknowledged.

REFERENCES

1. LANGER, S. Z. 1977. Presynaptic receptors and their role in the regulation of transmitter release. Br. J. Pharmacol. **60:** 481–497.
2. STARKE, K. 1977. Regulation of noradrenaline release by presynaptic receptor systems. Rev. Psychol. Biochem. Pharmacol. **77:** 1–124.
3. LANGER, S. Z. 1987. Presynaptic regulation of monoaminergic neurons. *In* Psychopharmacology: The Third Generation of Progress. H. Y. Meltzer, Ed.: 151–157. Raven Press. New York, N.Y.
4. FINBERG, J. P. M. 1987. Antidepressant drugs and down-regulation of presynaptic receptors. Biochem. Pharmacol. **36:** 3557–3562.
5. SIEVER, L. 1987. Role of noradrenergic mechanisms in the etiology of the affective disorder. *In* Psychopharmacology: The Third Generation of Progress. H. Y. Meltzer, Ed.: 493–504. Raven Press. New York, N.Y.
6. NUTT, D. J. & S. G. MOLYNEUX. 1986. The effect of clonidine on plasma MHPG: evidence against tonic $alpha_2$-adrenoceptor control of noradrenergic function. Psychopharmacology **90:** 509–512.
7. KALSNER, S. 1985. Is there feedback regulation of neurotransmitter release by autoreceptors? Biochem. Pharmacol. **34:** 4085–4097.
8. HUANG, Y. H., J. W. MAAS & G. H. HU. 1980. The time course of noradrenergic pre- and post-synaptic activity during chronic desipramine treatment. Eur. J. Pharmacol. **68:** 41–47.
9. RACAGNI, G., I. MOCCHETTI, G. CALDERINI, A. BATTISTELLA & N. BRUNELLO. 1983. Temporal sequence of changes in central noradrenergic system of rat after prolonged antidepressant treatment: receptor desensitization and neurotransmitter interactions. Neuropharmacology **22:** 415–424.
10. LERNER, P., L. F. MAJOR, D. L. MURPHY, S. LIPPER, C. R. LAKE & W. LOVENBERG. 1979. Dopamine-β-hydroxylase and norepinephrine in human cerebrospinal fluid: effects of monoamine oxidase inhibitors. Neuropharmacology **18:** 423–426.
11. PILC, A. & S. J. ENNA. 1986. Antidepressant administration has a differential effect on rat brain $alpha_2$-adrenoceptor sensitivity to agonists and antagonists. Eur. J. Pharmacol. **132:** 277–282.
12. COHEN, R. M., I. C. CAMPBELL, M. R. COHEN, T. TORDA, D. PICKAR, L. J. SIEVER & D. L. MURPHY. 1980. Presynaptic noradrenergic regulation during depression and antidepressant drug treatment. Psychiatry Res. **3:** 93–105.
13. CREWS, F. T. & C. B. SMITH. 1978. Presynaptic alpha-receptor subsensitivity after long-term antidepressant treatment. Science **202:** 322–324.
14. SVENSSON, T. H. & T. USDIN. 1978. Feedback inhibition of brain noradrenaline neurons by tricyclic antidepressants: alpha-receptor mediation. Science **202:** 1089–1091.

15. SCUVEE-MOREAU, J. J. & T. H. SVENSSON. 1982. Sensitivity *in vivo* of central alpha$_2$- and opiate receptors after chronic treatment with various antidepressants. J. Neural Transm. **54:** 51–63.

16. SCHOFFELMEER, A. N. M. & A. H. MULDER. 1982. [^{3}H] Noradrenaline and [^{3}H] 5-HT release from rat brain slices and its presynaptic alpha-adrenergic modulation after long-term DMI treatment. Naunyn Schmiedebergs Arch. Pharmacol. **318:** 173–180.

17. CAMPBELL, I. C. & R. M. MCKERNAN. 1982. Central and peripheral changes in alpha-adrenoceptors in response to chronic antidepressant drug administration. Adv. Biosci. **40:** 153–160.

18. SCHOFFELMEER, A. N. M., M. D. HOORNEMAN, P. SMINIA & A. H. MULDER. 1984. Presynaptic alpha$_2$- and post-synaptic beta-adrenoceptor sensitivity in slices of rat neocortex after chronic treatment with various antidepressant drugs. Neuropharmacology **23:** 115–119.

19. ASAKURA, M., T. TSUKAMOTO & K. HASEGAWA. 1982. Modulation of rat brain alpha$_2$ and beta-adrenergic receptor sensitivity following long term treatment with antidepressants. Brain Res. **235:** 192–197.

20. SMITH, C. B., J. A. GARCIA-SEVILLA & P. J. HOLLINGSWORTH. 1981. Alpha-adrenoceptors in rat brain are decreased after long term tricyclic antidepressant treatment. Brain Res. **210:** 413–418.

21. CAMPBELL, I. C. & R. M. MCKERNAN. 1986. Clorgyline and desipramine alter the sensitivity of [^{3}H] noradrenaline release to calcium but not to clonidine. Brain Res. **372:** 253–259.

22. SUGRUE, M. F. 1982. A study of the sensitivity of rat brain alpha$_2$-adrenoceptors during chronic antidepressant treatments. Naunyn Schmiedebergs Arch. Pharmacol. **320:** 90–96.

23. PEROUTKA, S. J. & S. H. SNYDER. 1980. Regulation of serotonin$_2$ (5-HT$_2$) receptors labeled with [^{3}H] spiroperidol by chronic treatment with the antidepressant amitriptyline. J. Pharmacol. Exp. Ther. **215:** 582–587.

24. JOHNSON, R. W., T. REISINE, S. SPOTNITZ, N. WIECH, R. URSILLO & H. I. YAMAMURA. 1980. Effect of desipramine and yohimbine on alpha- and beta-adrenoceptor sensitivity. Eur. J. Pharmacol **67:** 123–127.

25. SCOTT, J. A. & F. T. CREWS. 1983. Rapid decrease in rat brain beta adrenergic receptor binding during combined antidepressant alpha-2 antagonist treatment. J. Pharmacol. Exp. Ther. **224:** 640–646.

26. SCHMAUSS, M., G. LAAKMANN & D. DIETERLE. 1988. Effects of alpha$_2$-receptor blockade in addition to tricyclic antidepressants in therapy resistant depression. J. Clin. Psychopharmacol. **8:** 108–111.

27. SIEVER, L. J. & T. W. UHDE. 1984. New studies and perspectives on the noradrenergic receptor system in depression: effects of the alpha$_2$-adrenergic agonist clonidine. Biol. Psychiatry **19:** 131–156.

28. ROY, A., S. GUTHRIE, D. PICKA & M. LINNOILA. 1987. Plasma norepinephrine responses to cold challenge in depressed patients and normal controls. Psychiatry Res. **21:** 161–168.

29. MAAS, J. W., S. H. KOSLOW, J. DAVIS, M. KATZ, A. FRAZER, C. L. BOWDEN, N. BERMAN, R. GIBBONS, P. STOKES & H. LANDIS. 1987. Catecholamine metabolism and disposition in healthy and depressed subjects. Arch. Gen. Psychiatry **44:** 337–344.

30. POTTER, W. Z., M. SCHEINEN, R. N. GOLDEN, M. V. REIDORFER, R. Y. COWDRY, H. M. CALIL, R. J. ROSS & M. LINNOILA. 1985. Selective antidepressants and cerebrospinal fluid. Arch. Gen. Psychiatry **42:** 1171–1177.

31. BERTILSSON, L., J. R. TUCK & B. SUVERS. 1980. Biochemical effects of zimelidine in man. Eur. J. Clin. Pharmacol. **18:** 483–487.

32. BOWDEN, C. L., S. H. KOSLOW, I. HANIN, J. W. MAAS, J. M. DAVIS & E. ROBINS. 1985. Effects of amitriptyline and imiparmine on brain amine neurotransmitter metabolites in cerebrospinal fluid. Clin. Pharmacol. Ther. **37:** 316–324.

33. HSIAO, J. K., H. AGREN, J. J. BARTKO, M. V. RUDORFER, M. LINNOILA & W. Z. POTTER. 1987. Monoamine neurotransmitter interactions and the predictor of antidepressant response. Arch. Gen. Psychiatry **44:** 1078–1083.

34. SIEVER, L. J. & K. L. DAVIS. 1985. Overview: toward a dysregulation hypothesis of depression. Am. J. Psychiatry **142:** 1017–1031.

35. ELLISON, D. W. & I. C. CAMPBELL. 1986. Studies on the role of alpha$_2$-adrenoceptors in the control of synaptosomal [^{3}H]5-hydroxytryptamine release: effects of antidepressant drugs. J. Neurochem. **46:** 218–223.

36. CERRITO, F. & M. RAITERI. 1979. Serotonin release is modulated by presynaptic autoreceptors. Eur. J. Pharmacol. **57:** 427–430.

37. GOTHERT, M. & G. WEINHEIMER. 1979. Extracellular 5-hydroxytryptamine inhibits 5-hydroxytryptamine release from rat brain cortex slices. Naunyn Schmiedebergs Arch. Pharmacol. **310:** 93–96.

38. DEMONTIGNY, C., P. BLIER & Y. CHAPUT. 1984. Electrophysiologically-identified serotonin receptors in the rat CNS. Effect of antidepressant treatment. Neuropharmacology **23:** 1511–1520.

39. BLIER, P., C. DEMONTIGNY & Y. CHAPUT. 1988. Electrophysiological assessment of the effects of antidepressant treatments on the efficacy of 5-HT neurotransmission. Clin. Neuropharmacol. **11**(Suppl. 2): S1–S10.

40. DEMONTIGNY, C., P. BLIER, G. CAILLÉ & E. KOVASSI. 1981. Pre- and post-synaptic effects of zimelidine and norzimelidine on the serotonergic system: single cell studies in the rat. Acta Psychiatr. Scand. **63:** 79–90.

41. BLIER, P. & C. DEMONTIGNY. 1983. Electrophysiological investigations on the effect of repeated zimelidine administration on serotonergic neurotransmission in the rat. J. Neurosci. **3:** 1270–1278.

42. BLIER, P., C. DEMONTIGNY & D. TARDIF. 1984. Effects of the two antidepressant drugs mianserin and indalpine on the serotonergic system: single-cell study in the rat. Psychopharmacology **84:** 242–249.

43. CHAPUT, Y., P. BLIER & C. DEMONTIGNY. 1986. In vivo electrophysiological evidence for the regulatory role of autoreceptors on serotonergic terminals. J. Neurosci. **6:** 2796–2801.

44. BLIER, P., Y. CHAPUT & C. DEMONTIGNY. 1988. Long-term 5-HT reuptake blockade but not MAO inhibition, decreases the function of terminal 5-HT autoreceptors: an electrophysiological study in the rat brain. Naunyn Schmiedebergs Arch. Pharmacol. **337:** 246–254.

45. GALZIN, A. M., C. MORET, B. VERZIER & S. Z. LANGER. 1985. Interaction between tricyclic and nontricyclic 5-hydroxytryptamine uptake inhibitors and the presynaptic 5-hydroxytryptamine inhibitory autoreceptors in the rat hypothalamus. J. Pharmacol. Exp. Ther. **235:** 200–211.

46. MORET, C. & M. BRILEY. 1988. Sensitivity of the response of 5-HT autoreceptors to drugs modifying synaptic availability of 5-HT. Neuropharmacology **27:** 43–49.

47. RAITERI, M., G. BONANNO, M. MARCHI. & G. MAURA. 1984. Is there a functional linkage between neurotransmitter uptake mechanisms and presynaptic receptors? J. Pharmacol. Exp. Ther. **231:** 671–677.

48. FEUERSTEIN, T. J., A. LUPP & G. HERTTING. 1987. The serotonin (5-HT) autoreceptor in the hippocampus of the rabbit: role of 5-HT biophase concentration. Neuropharmacology **26:** 1071–1080.

49. BLIER, P. & C. DEMONTIGNY. 1980. Effect of chronic tricyclic antidepressant treatment on the serotonergic autoreceptor. A microiontophoric study in the rat. Naunyn Schmiedebergs Arch. Pharmacol. **314:** 123–128.

50. SCHICKTER, E., F. BRANDT, K. CLASSEN. & M. GOTHERT. 1985. Serotonin release in human cerebral cortex and its modulations via serotonin receptors. Brain Res. **331:** 337–341.

51. WAEBER, C., P. SCHOEFFTER, J. M. PALACIOS & D. HOYE. 1988. Molecular pharmacology of 5-HT$_{1D}$ recognition sites: radioligand binding studies in human, pig, and calf brain membranes. Naunyn Schmiedebergs Arch. Pharmacol. **337:** 595–601.

52. ABERG-WISTEDT, A. 1982. A double-blind study of zimelidine, a serotonin uptake inhibitor, and desipramine, a nonradrenaline uptake inhibitor, in endogenous depression. Acta Psychiatr. Scand. **66:** 50–65.

53. LADER, M. 1988. Fluoxetine efficacy vs comparative drugs: an overview. Br. J. Psychiatry **153**(Suppl. 3): 51–58.

54. LEVINE, R., J. S. HOFFMAN, E. D. KNEPPLE & M. KENIN. 1989. Long-term fluoxetine treatment of a large number of obsessive-compulsive patients. J. Clin. Psychopharmacol. **9**: 281–283.

55. MELTZER, H. Y., M. LOWY, A. ROBERTSON, P. GOODNICK & R. PERLINE. 1984. Effect of 5-hydroxytryptophan on serum cortisol levels in major affective disorders. III. Effect of antidepressants and lithium. Arch. Gen. Psychiatry **41**: 391–402.

56. ABERG-WISTEDT, A., M. ALVARIZA, L. BERTILSSON, R. MALMGREN & H. WACHMEISTER. 1985. Alaproclate, a novel antidepressant? A biochemical and clinical comparison and zimelidine. Acta Psychiatry. Scand. **71**: 256–268.

57. BJERKENSTEDT, L., G. EDMAN, L. FLYCKT, L. HAGENFELDT, G. SEDVALL & F. A. WIESEL. 1985. Clinical and biochemical effects of citalopram, a selective 5-HT reuptake inhibitor—a dose-response study in depressed patients. Psychopharmacology **87**: 253–259.

58. BOWDEN, C. L., S. H. KOSLOW, I. HANIN, J. W. MAAS, J. M. DAVIS & E. ROBINS. 1985. Effect of amitriptyline and imipramine on brain amine neurotransmitter metabolites in cerebrospinal fluid. Clin. Pharmacol. Ther. **37**: 316–324.

59. BLIER, P. & C. DEMONTIGNY. 1985. Short-term lithium administration enhances serotonergic neurotransmission: electrophysiological evidence in the rat CNS. Eur. J. Pharmacol. **113**: 69–77.

60. HOTTA, I. & S. YAMAWAKI. 1988. Possible involvement of presynaptic 5-HT autoreceptors in effect of lithium on 5-HT release in hippocampus of rat. Neuropharmacology **27**: 987–992.

61. WANG, H-Y. & E. FRIEDMAN. 1988. Chronic lithium: desensitization of autoreceptors mediating serotonin release. Psychopharmacology **94**: 312–314.

62. CHIODO, L. A. & S. M. ANTELMAN. 1980. Repeated tricyclics induce a progressive dopamine autoreceptor subsensitivity independent of daily drug treatment. Nature **287**: 451–454.

63. SERRA, G., A. ARGIOLAS, V. KLIMEK, F. FADDA & G. L. GESSA. 1979. Chronic treatment with antidepressants prevents the inhibitory effect of small doses of apomorphine on dopamine synthesis and motor activity. Life Sci. **25**: 415–424.

64. KOIDE, T. & H. MATSUSHITA. 1981. An enhanced sensitivity of muscarinic cholinergic receptors associated with dopaminergic receptor subsensitivity after chronic antidepressant treatment. Life Sci. **28**: 1139–1145.

65. SCAVONE, C., M. L. AIZENSTEIN, R. DELUCIA & P. C. DASILVA. 1986. Chronic imipramine administration reduces apomorphine inhibitory effects. Eur. J. Pharmacol. **132**: 263–267.

66. JIMERSON, D. C. 1987. Role of dopamine in affective disorders. *In* Psychopharmacology: the Third Generation of Progress. H. Y. Meltzer, Ed.: 505–512. Raven Press. New York, NY.

67. HOLCOMB, H. H., M. J. BANNON & R. H. ROTH. 1982. Striatal dopamine receptors uninfluenced by chronic administration of antidepressants. Eur. J. Pharmacol. **82**: 173–178.

68. MUSCAT, R., A. TOWELL & P. WILLNER. 1988. Changes in dopamine autoreceptor sensitivity in an animal model of depression. Psychopharmacology **94**: 545–550.

69. MACNEIL, D. A. & M. GOWER. 1982. Do antidepressants induce dopamine autoreceptor subsensitivity? Nature **298**: 302–304.

70. SPYRAKI, C. & H. FIBIGER. 1985. Behavioral evidence for supersensitivity of post-synaptic dopamine receptors in the mesolimbic system after chronic administration of dopamine. Eur. J. Pharmacol. **74**: 195–206.

71. SMIALOWSKI, A. & J. MAJ. 1985. Repeated treatment with imipramine potentiates the locomotor effect of apomorphine administered into the hippocampus in rats. Psychopharmacology **86**: 468–471.

72. RISBY, E. D., J. K. HSIAO, T. SUNDERLAND, H. AGREN, M. V. RUDORFER & W. V. POTTER. 1987. The effects of antidepressants on the cerebrospinal fluid homovanillac acid/5-hydroxyindoleacetic acid ratio. Clin. Pharmacol. Ther. **42**: 547–554.

73. MELTZER, H. Y. & S. M. STAHL. 1976. The dopamine hypothesis of schizophrenia. Schizophr. Bull. **2:** 19–76.

74. CHOIDO, L. A. & B. S. BUNNEY. 1983. Typical and atypical neuroleptics: differential effects of chronic administration of the activity of A9 and A10 mid brain dopaminergic neurons. J. Neurosci. **3:** 1607–1619.

75. MACKAY, A. V. P. 1980. Positive and negative symptoms and the role of dopamine. Br. J. Psychiatry **137:** 379–383.

76. MELTZER, H. Y. 1985. Dopamine and negative symptoms in schizophrenia: critique of the type I–type II hypothesis. *In* Controversies in Schizophrenia: Changes and Constancies. M. Alpert, Ed.: 110–136. Guilford Press. New York, N.Y.

77. CLARK, D. & F. J. WHITE. 1987. D1 dopamine receptor—the search for a function: a critical evaluation of the D1/D2 dopamine receptor classification and its functional implications. Synapase **1:** 347–388.

78. MELTZER, H. Y. 1980. Relevance of dopamine autoreceptors for psychiatry: preclinical and clinical studies. Schizophr. Bull. **6:** 456–475.

79. CARLSSON, A. 1988. Dopamine autoreceptors and schizophrenia. *In* Receptors and Ligands in Psychiatry. A. K. Sen & T. Lee, Eds: 1–10. Cambridge University Press. Cambridge, England.

80. GALLOWAY, M. P., M. E. WOLF & R. H. ROTH. 1986. Regulation of dopamine synthesis in medial prefrontal cortex is mediated by release modulating autoreceptors: studies in vivo. J. Pharmacol. Exp. Ther. **236:** 689–698.

81. PLANTUE, J. F., H. W. STEINBUSCH, J. SCHIPPER, F. A. DIJCKS, P. F. VERHEIJDEN & J. C. STOOF. 1981. D-2 dopamine receptors regulate the release of [^{3}H] dopamine in rat cortical regions showing dopamine immunoreactive fibers. Neuroscience **20:** 157–168.

82. TALMACIU, R. K., I. S. HOFFMANN & L. Y. CUBEDDU. 1986. Dopamine autoreceptors modulate dopamine release from the prefrontal cortex. J. Neurochem. **47:** 865–870.

83. WOLF, M. E., M. P. GALLOWAY & R. H. ROTH. 1986. Regulation of dopamine synthesis in the medial prefrontal cortex: studies in brain slices. J. Pharmacol. Exp. Ther. **236:** 699–707.

84. KILTS, C. D., C. M. ANDERSON, T. D. ELY & J. K. NISHITA. 1987. Absences of synthesis-modulating nerve terminal autoreceptors on mesoamygdaloid and other mesolimbic dopamine neuronal populations. J. Neurosci. **7**(12): 3961–3975.

85. FADDA, F., G. L. GESSA, M. MARCOW, E. MOSCA & Z. ROSSETTI. 1984. Evidence for dopamine autoreceptors in mesocortical dopamine neurons. Brain Res. **293:** 67–72.

86. WOLF, M. E., A. Y. DEUTCH & R. H. ROTH. 1987. Pharmacology of central dopaminergic neurons. *In* Handbook of Schizophrenia. Neurochemistry and Neuropharmacology of Schizophrenia. F. A. Henn & L. E. DeLisi, Eds. **2:** 101–147. Elsevier Science Publishers, B. V. Amsterdam, the Netherlands.

87. GARIANO, R. F., J. M. TEPPER, S. F. SAWYER, S. J. YOUNG & P. M. GROVES. 1989. Mesocortical dopaminergic neurons. I. Electrophysiological properties and evidence for soma-dendritic autoreceptors. Brain Res. Bull. **22:** 511–516.

88. COSTALL, B., A. M. DOMENEY, R. J. NAYLOR & M. B. TYERS. 1987. Effects of the 5-HT$_3$ receptor antagonist GR 38032F on raised dopaminergic activity in the mesolimbic system of the rat and marmoset brain. Br. J. Pharmacol. **92:** 881–894.

89. BLANDINA, P., J. GOLDFARB & J. P. GREEN. 1988. Activation of a 5-HT$_3$ receptor releases dopamine from rat striatal slices. Eur. J. Pharmacol. **155:** 349–350.

90. HAGAN, R. M., A. BUTLER, J. M. HILL, C. C. JORDAN, S. J. IRELAND & M. B. TYERS. 1987. Effect of the 5-HT$_3$ receptor antagonist, GR 38032F on responses to injection of a neurokinin agonist into the ventral tegmental area of rat brain. Eur. J. Pharmacol. **138:** 303–305.

91. COSTALL, B., A. M. DOMENEY, M. E. KELLY, R. J. NAYLOR & M. B. TYERS. 1987. Antipsychotic potential of GR 38032F, a selective antagonist of 5-HT$_3$ receptors in the central nervous system. Br. J. Pharmacol. Proc. Suppl. **90:** 89.

92. SCHMIDT, A. W. & S. J. PEROUTKA. 1989. Antidepressant interactions with 5-hydroxytryptamine$_3$ receptor binding sites. Eur. J. Pharmacol. **163:** 397–398.

93. TYERS, M. B., B. COSTALL, A. DOMENEY, B. J. JONES, M. E. KELLEY, R. J. NAYLOR &

N. P. OAKLEY. 1987. The anxiolytic activity of 5-HT$_3$ antagonists in laboratory animals. Neurosci. Lett. **29**(Suppl.): S68.

94. PAPP, M. 1988. Similar effects of diazepam and the 5-HT$_3$ receptor antagonist ICS 205-930 on place avoidance conditioning. Eur. J. Pharmacol. **151**: 321–324.

95. JOHNSTON, A. & S. E. FILE. 1988. Effects of 5-HT$_2$ antagonists in two animal tests of anxiety. Neurosci. Lett. **32**(Suppl.): S44.

96. PIPER, D., N. UPTON, D. THOMAS & J. NICHOLASS. 1988. The effects of the 5-HT$_3$ receptor antagonists BRL 43694 and GR 38032F in animal behavioral models of anxiety. Br. J. Pharmacol. **94**: 314.

97. PEROUTKA, S. J. 1988. Species variations in 5-HT$_3$ recognition sites labelled by [^{3}H] quipazine in the central nervous system. Naunyn Schmiedebergs Arch. Pharmacol. **338**: 472–475.

98. BARNES, J. M., N. M. BARNES, B. COSTALL, J. W. IRONSIDE & R. J. NAYLOR. 1989. Identification and characterization of 5-hydroxytryptamine$_3$ recognition sites in human brain tissue. J. Neurochem. **53**: 1787–1793.

99. ENGEL, G., B. P. RICHARDSON, P. DONATSCH & P. STADLER. 1986. A new class of drugs: 5-HT$_3$ receptor antagonists. Triangle **25**: 123–130.

100. CLAGHORN, J., G. HONIGFELD, F. S. ABUZZAHAB, R. WANG, R. STEINBOOK, V. TUASON & G. KLERMAN. 1987. The risks and benefits of clozapine versus chlorpormazine. J. Clin. Psychopharmacol. **7**: 377–384.

101. KANE, J., G. HONIGFELD, J. SINGER & H. MELTZER. 1988. Clozapine for the treatment-resistant schizophrenic: a double-blind comparison versus chlorpromazine/benztropine. Arch. Gen. Psychiatry **45**: 789–796.

102. MELTZER, H. Y. 1989. Clinical studies on the mechanism of action of clozapine: the dopamine serotonin hypothesis of schizophrenia. Psychopharmacology **99**(Suppl.): S18–S27.

103. MELTZER, H. Y. 1989. Clozapine: clinical advantages and biological mechanisms. *In* Schizophrenia: a Scientific Focus. C. Schulz & C. Tamminga, Eds.: 302–309. Oxford Press. New York, N.Y.

104. MELTZER, H. Y., R. SO, R. J. MILLER & V. S. FANG. 1979. Comparison of the effects of substituted benzamides and standard neuroleptics on the binding of ^{3}H-spiroperidol in the rat pituitary and striatum with *in vivo* effects on rat prolactin secretion. Life Sci. **25**: 573–584.

105. MELTZER, H. Y., S. DANIELS & V. S. FANG. 1975. Clozapine increases rat serum prolactin levels. Life Sci. **17**: 339–342.

106. GUDELSKY, G. A., J. F. NASH, S. A. BERRY & H. Y. MELTZER. 1989. Basic biology of clozapine: the dopamine serotonin hypothesis of schizophrenia. Psychopharmacology **99**:(Suppl.): S13–S17.

107. MELTZER, H. Y., B. BASTANI, L. F. RAMIREZ & S. MATSUBARA. 1989. Clozapine: new research on efficacy and mechanism of action. Eur. Arch. Psychiatry Neurol. Sci. **238**: 332–339.

108. MELTZER, H. Y. 1990. Clozapine: mechanism of action in relation to its clinical advantages. *In* Recent Advances in Schizophrenia. A. Kales, G. N. Stefanos & J. A. Talbott, Eds.: 237–246. Springer Verlag. Heidelberg, New York & Tokyo.

109. REBEC, G. V., T. R. BASHORE, K. S. ZIMMERMAN & K. D. ALLOWAY. 1979. "Classical" and "atypical" antipsychotic drugs: differential antagonism of amphetamine- and apomorphine-induced alterations of spontaneous neuronal activity in the neostriatum and nucleus accumbens. Pharmacol. Biochem. Behav. **11**: 529–538.

110. ANDEN, N. E. & M. GRABOWSKA-ANDEN. 1980. Drug effects on pre- and post-synaptic dopamine receptors. Adv. Biochem. Psychopharmacol. **24**: 57–64.

111. WALTERS, J. R. & R. H. ROTH. 1976. Dopaminergic neurons: an *in vivo* system of measuring drug interactions with presynaptic receptors. Naunyn Schmeidebergs Arch. Pharmacol. **296**: 5–14.

112. HUFF, R. & R. N. ADAMS. 1980. Dopamine release in nucleus accumbens and striatum by clozapine: simultaneous monitoring by *in vivo* electrochemistry. Neuropharmacology **19**: 587–590.

113. BUNNEY, B. S., J. R. WALTERS, R. H. ROTH & G. K. AGHAJANIAN. 1973. Dopaminergic

neurons: effects of antipsychotic drugs and amphetamine on single cell activity. J. Pharmacol. Exp. Ther. **185:** 560–571.

114. CHIODO, L. A. & B. S. BUNNEY. 1983. Typical and atypical neuroleptics differential effects of chronic administration on the activity of A9 and A10 midbrain dopaminergic neurons. J. Neurosci. **3:** 1607–1619.

115. LANE, R. F. & C. D. BLAHA. 1986. Electrochemistry *in vivo:* application to CNS pharmacology. Ann. N.Y. Acad. Sci. **473:** 50–69.

116. BLAHA, C. D. & R. F. LANE. 1987. Chronic treatment with classical and atypical antipsychotic drugs differentially decrease dopamine release in striatum and nucleus accumbens *in vivo.* Neurosci Lett. **78:** 188–204.

117. HAND, T. H., S-T. HU & R. Y. WANG. 1987. Differential effects of acute clozapine and haloperidol on the activity of ventral tegmental (A10) and nigrostriatal (A9) dopamine neurons. Brain Res. **415:** 257–269.

118. HETEY, L., K. DRESCHER & W. OELSSNER. 1982. Different influence of antipsychotics and serotonin antagonists on presynaptic receptors modulating the synaptosomal release of dopamine and serotonin. Wiss. Z. Humbolt Univ. Berl. **131:** 487–489.

119. DRESCHER, K. & L. HETEY. 1988. Influence of antipsychotics and serotonin antagonists on presynaptic receptors modulating the release of serotonin in synaptosomes of the nucleus accumbens of rats. Neuropharmacology **27:** 31–36.

120. ASHBY, C. R., E. EDWARDS, K. L. HARKINS & R. Y. WANG. 1989. Differential effects of typical and atypical antipsychotic drugs on the suppressant action of 2-methylserotonin on medial prefrontal cortical cells: a microiontophoretic study. Eur. J. Pharmacol. **166:** 583–584.

121. COMPTON, D. R. & K. M. JOHNSON. 1989. Effects of acute and chronic clozapine and haloperidol on *in vivo* release of acetylcholine and dopamine from striatum and nucleus accumbens. J. Pharmacol. Exp. Ther. **248:** 521–530.

122. ICHIKAWA, J. & H. Y. MELTZER. The effect of chronic clozapine and haloperidol on basal dopamine release and metabolism in rat striatum and nucleus accumbens studied by *in vivo* microdialysis. Eur. J. Pharmacol. **176:** 371–374.

123. IMPERATO, A. & L. ANGELUCCI. 1988. Effects of the atypical neuroleptics clozapine and fluperlapine on the *in vivo* dopamine release in the dorsal striatum and in the prefrontal cortex. Psychopharmacol. **96:**(Suppl 1. Abstracts of the XVI CINP Congress, Munich): 79.

124. O'CONNOR, W. T., K. L. DREW & V. UNGERSTEDT. 1989. Differences in dopamine release and metabolism in rat striatal subregions following acute clozapine using *in vivo* microdialysis. Neurosci. Lett. **99:** 211–216.

125. RICHELSON, E. 1984. Neuroleptic affinities for human brain receptors and their use in predicting adverse effects. J. Clin Psychiatry **45:** 331–336.

126. ANDERSEN, P. H. & C. BRAESTRUP. 1986. Evidence for different states of the dopamine D-1 receptor: clozapine and fluperlapine may preferentially label an adenylate cyclase–coupled state of the D-1 receptor. J. Neurochem. **47:** 1830–1831.

127. FARDE, L. F.-A. WIESEL, A-L. NORDSTROM & G. SEDVALL. 1989. D-1 and D2 dopamine receptor occupancy during treatment with conventional and atypical neuroleptics. Psychopharmacology **99**(Suppl.): S28–S31.

128. MELTZER, H. Y., S. MATSUBARA & J-C. LEE. 1989. Classification of typical and atypical antipsychotic drugs on the basis of dopamine D-1, D-2 and serotonin$_2$ pKi values. J. Pharmacol. Exp. Ther. **251:** 238–246.

129. MATSUBARA, S. & MELTZER, H. Y. 1989. Effect of typical and atypical antipsychotic drugs on 5-HT$_2$ receptor density in rat cerebral cortex. Life Sci. **45:** 1397–1406.

130. MELTZER, H. Y., B. BASTANI, K. Y. KWON, L. F. RAMIREZ, S. BURNETT & J. SHARPE. 1989. A prospective study of clozapine in treatment-resistant patients. I. Preliminary report. Psychopharmacology **99**(Suppl.): S68–S72.

131. FEIGHNER, J. P., C. H. MERIDETH & G. A. HENDRICKSON. 1982. A double-blind comparison of buspirone and diazepam in outpatients with generalized anxiety disorder. J. Clin. Psychiatry **43:** 103–107.

132. JACOBSON, A. F., P. A. DOMINGUEZ, B. J. GOLDSTEIN & R. M. STEINBOOK. 1985.

Comparison of buspirone and diazepam in generalized anxiety disorder. Pharmacotherapy **5:** 290–296.

133. SCHWEIZER, E. E., J. AMSTERDAM, K. RICKELS, M. KAPLAN & M. DROBA. 1985. Open trials of buspirone in the treatment of major depressive disorder. Psycho. Pharmacol. Bull. **22:** 183–185.

134. PEROUTKA, S. J. 1985. Selective labeling of 5-HT$_{1A}$ and 5-HT$_{1B}$ binding sites in bovine brain. Brain Res. **344:** 167–171.

135. YOCCA, F. D., D. A. HYSLOP, D. P. TAYLOR & S. MAAYANI. 1986. Buspirone and gepirone: partial agonists at the 5-HT$_{1A}$ receptor linked to adenylate cyclase in rat and guinea pig hippocampal preparations. Fed. Proc. **45:** 436.

136. MCMILLEN, B. A., S. M. SCOTT, H. L. WILLIAMS & M. K. SANGHERA. 1987. Effects of gepirone, an aryl-piperazine anxiolytic drug, on aggressive behavior and brain monoaminergic neurotransmission. Naunyn Schmiedebergs Arch. Pharmacol. **335:** 454–464.

137. SPROUSE, J. S. & G. K. AGHAJANIAN. 1987. Electrophysiological responses of serotonergic dorsal raphé neurons to 5-HT$_{1A}$ and 5-HT$_{1B}$ agonists. Synapse **1:** 3–9.

138. BLIER, P. & C. DeMONTIGNY. 1987. Modification of 5-HT neuron properties by sustained administration of the 5-HT$_{1A}$ agonist gepirone: electrophysiological studies in the rat brain. Synapse **1:** 470–480.

139. LUCKI, I. 1986. The nonbenzodiazepine anxiolytics buspirone and ipsaperone antagonize serotonin-mediated behavioral responses. Psychopharmacology **89:** S55.

140. HAMON, M., C.-M. FATTACCINI, J. ADRIEN, M.-C. GALLISSOT, P. MARTIN & H. GOZLAN. 1988. Alterations of central serotonin and dopamine turnover in rats treated with ipsapirone and other 5-hydroxytryptamine$_{1A}$ agonists with potential anxiolytic properties. J. Pharmacol. Exp. Ther. **246:** 745–752.

141. KENNETT, G. A., C. T. DOURISH & G. GURZON. 1987. Antidepressant-like action of 5-HT$_{1A}$ agonists and conventional antidepressants in an animal model of depression. Eur. J. Pharmacol. **134:** 265–274.

Presynaptic Peptide Receptors and Hypertension [a]

THOMAS C. WESTFALL, SONG-PING HAN, XIAOLI
CHEN, KATHERINE DEL VALLE, MELISSA CURFMAN,
ANITA CIARLEGLIO, AND LINDA NAES

Department of Pharmacology
St. Louis University School of Medicine
1402 South Grand Boulevard
St. Louis, Missouri 63104

Vascular smooth muscle is subject to control from a variety of neuronal, hormonal, and autacoidal influences. Among the various chemical mediators that exert important actions on the cardiovascular system are a variety of vasoactive peptides stored and released from perivascular nerves or from components of blood vessels such as smooth muscle and endothelium. Two peptides of great interest include neuropeptide Y (NPY) which has been shown to be colocalized and coreleased with norepinephrine from noradrenergic nerves[1,2] and calcitonin gene-related peptide (CGRP) located in a variety of sensory and perivascular nerves.[3,4] The purpose of this paper is to summarize recent studies on the pre- and postjunctional actions of NPY and CGRP at the vascular neuroeffector junction of normotensive and the spontaneously hypertensive rat (SHR). The perfused mesenteric arterial bed of the rat has been utilized as a model of the vascular neuroeffector junction.

METHODS

Experiments were carried out in various strains of rats including spontaneously hypertensive (SHR), Wistar Kyoto (WKY) (Taconic Farms, Germantown, N.Y.) and Sprague Dawley (SD) (Sasco, St. Louis, Mo.). Animals were maintained under standard laboratory conditions with food and water *ad libitum*. Blood pressure was monitored in conscious rats by tail cuff plethysmography one or two days prior to experimentation.

The perfused mesenteric arterial bed was utilized as described by us previously.[5] Briefly, the mesenteric vascular bed was dissected from the rat, perfused (5 ml/ minute) and superfused (0.5 ml/minute) with modified Krebs bicarbonate buffer bubbled with 95% O_2 and 5% CO_2 at 37°C. A pair of platinum electrodes were placed around the mesenteric artery for periarterial nerve stimulation (various frequencies, 2 mseconds duration supramaximal voltage). The preparation was allowed to equilibrate for 45–60 minutes prior to experimentation. Perfusion pressure was continuously monitored by pressure transducers coupled to chart recorders. In some experiments perfusate effluents were collected in 1-minute fractions into tubes containing 0.1 N perchloric acid, concentrated by alumina chromatography, separated by high performance liquid chromatography (HPLC) and quantified by electrochemical (EC) detection.

[a]These studies have been supported by U.S. Public Health Service grants HL 26319, HL 35202, and NIDA 02668.

To study nerve stimulation–induced vasodilation, the vascular bed was perfused with a buffer containing guanethidine (5×10^{-6} M) and methoxamine (5×10^{-5} M) to block norepinephrine (NE) release and to precontract the vascular bed, respectively.[6] Drugs were added to the perfusion buffer by infusion into the perfusion line or dissolved in the perfusion buffer.

The effect of NPY and NPY fragments on the potassium (K^+) induced release of NE from slices of the posterior hypothalamus (PH) was investigated as described by us previously.[7] K^+ (56 mM) was used during two stimulation periods (S_1 and S_2) to evoke the release of NE measured by HPLC-EC detection or by liquid scintillation spectrometry from slices of the PH obtained from SD, SHR, or WKY. Drugs were added prior to S_2. Other experiments examined the effect of drugs on the release of NPY-like immunoreactivity (NPY-LI) from brain slices and measures by a specific radioimmunoassay.[8]

RESULTS

Neuropeptide Y

NPY was observed to exert both pre- and postjunctional effects on noradrenergic transmission in the perfused mesenteric arterial bed. The peptide exerted a concentration-dependent decrease in the periarterial nerve stimulation induced release of endogenous norepinephrine, which was not altered by addition to the perfusion buffer of the α_1-adrenoceptor antagonist prazosin or the α_2-adrenoceptor antagonist yohimbine.[5] NPY produced very little direct increase in perfusion pressure but significantly enhanced the increase in perfusion pressure produced by a variety of agonists including angiotensin II, phenylephrine, norepinephrine, arginine vasopressin, or periarterial nerve stimulation[9] (FIGURE 1). These pre- and postjunctional effects were shared by intestinal polypeptide (PYY) but not pancreatic polypeptide (PPY) (FIGURE 2). It is unclear whether or not the postjunctional effect of NPY requires the presence of endothelial cells. The potentiation of the contractile response to vasoactive agents by NPY was examined in preparations in which the endothelial cells were removed. Removal of the endothelial cells was accomplished by switching the perfusion buffer from one containing normal Krebs-bicarbonate buffer to one containing distilled water for 2–3 minutes and then returning to normal Krebs buffer. Successful removal of the endothelial cells was verified by the demonstration that acetylcholine no longer could relax the precontracted mesenteric arterial bed. It was observed that removal of the endothelial cells in this fashion did not alter the NPY-induced potentiation of the contractile effect of either NE or vasopressin (FIGURE 3).

The pre- and postjunctional effects of a series of C terminal NPY fragments, synthesized and generously provided by Dr. Jean Rivier of the Salk Institute, were also examined. FIGURE 4 demonstrates the effect of equimolar (10^{-7} M) concentration of a variety of these C terminal fragments in inhibiting the periarterial nerve stimulation induced release of NE (prejunctional response) and the simultaneous effect on perfusion pressure (postjunctional response). The NPY fragments were observed to produce a concentration-dependent decrease in the periarterial nerve stimulation induced release of norepinephrine and a parallel decrease in the increase in perfusion pressure. This is in contrast to NPY and PYY which decreased norepinephrine release but potentiated the nerve stimulation induced increase in perfusion pressure.

The pre- and postjunctional effects of NPY on the perfused mesenteric artery obtained from 8–10 week SHR and normotensive controls (WKY) were also examined. It was observed that the prejunctional inhibitory effect of NPY was attenuated in

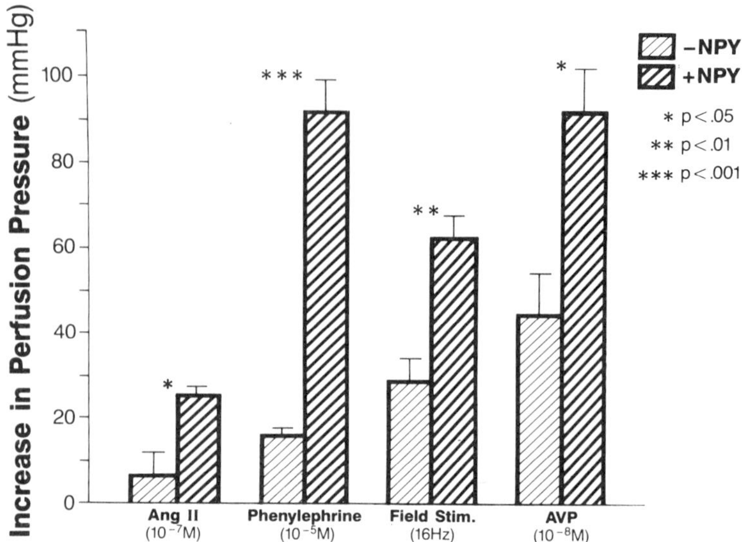

FIGURE 1. The effect of NPY (10^{-7} M) on the increase in perfusion pressure in the perfused mesenteric arterial bed produced by angiotensin II (Ang II) (10^{-7} M), phenylephrine (10^{-5} M), periarterial nerve stimulation (16 Hz, 30 seconds, supramaximal voltage), and vasopressin (AVP, 10^{-8} M). Data are plotted as the increase in perfusion pressure in mmHg produced by each mode of stimulation in the absence or presence of 10^{-7} M NPY. NPY produced a significant potentiation of the response to each agonist.

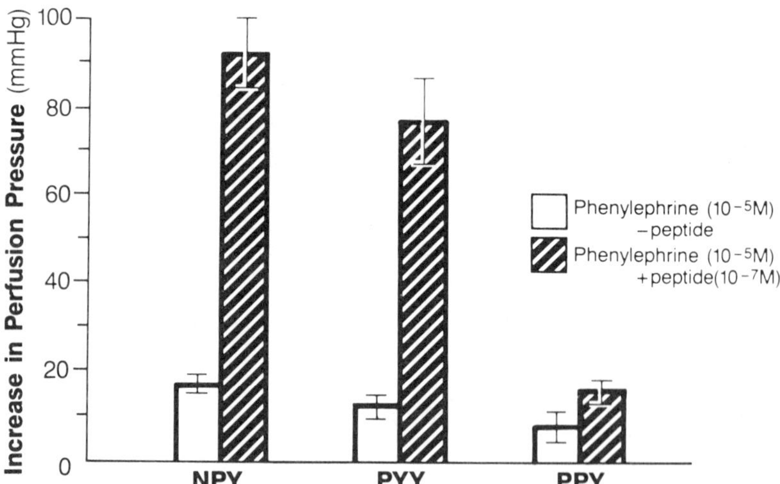

FIGURE 2. The effect of 10^{-7} M NPY, PYY, or PPY (added for 4 minutes) on the increase in perfusion pressure produced by phenylephrine (10^{-5} M). Data are plotted as the increase in perfusion pressure in mmHg. Each bar is the mean ± standard error of the mean (SEM) of five to seven preparations. NPY and PYY produced a significant increase in the response, PPY did not.

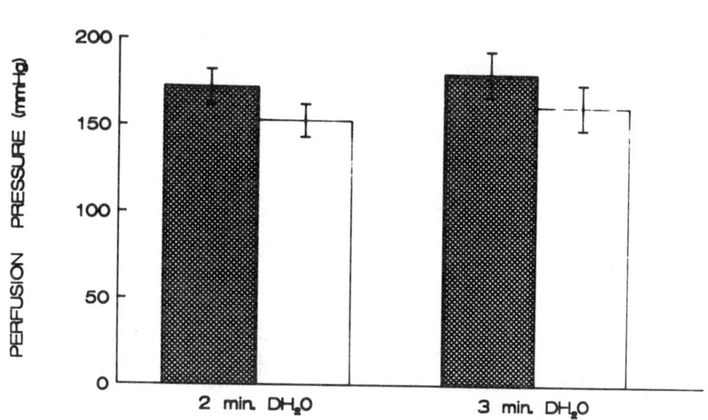

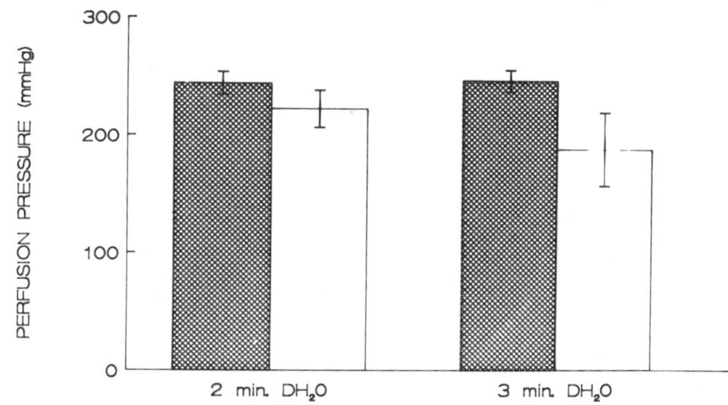

FIGURE 3. NPY induced potentiation of the increase in perfusion pressure to vasopressin (Arg-VPL, 10^{-8} M) or norepinephrine (10^{-6} M) in the perfused mesenteric arterial bed before and after removal of endothelial cells with distilled water.

SHR compared to WKY (FIGURE 5). In contrast the potentiation of the nerve stimulation on norepinephrine-induced increase in perfusion pressure was significantly enhanced (FIGURE 6).

NPY was shown to reduce the K^+-evoked release of norepinephrine from slices obtained from the hypothalamus of Sprague Dawley rats (FIGURE 7). A similar inhibition was observed in slices obtained from 8–10 week SHR or WKY (FIGURE 7). Utilizing a specific radioimmunoassay for NPY, it was observed that K^+ produced a

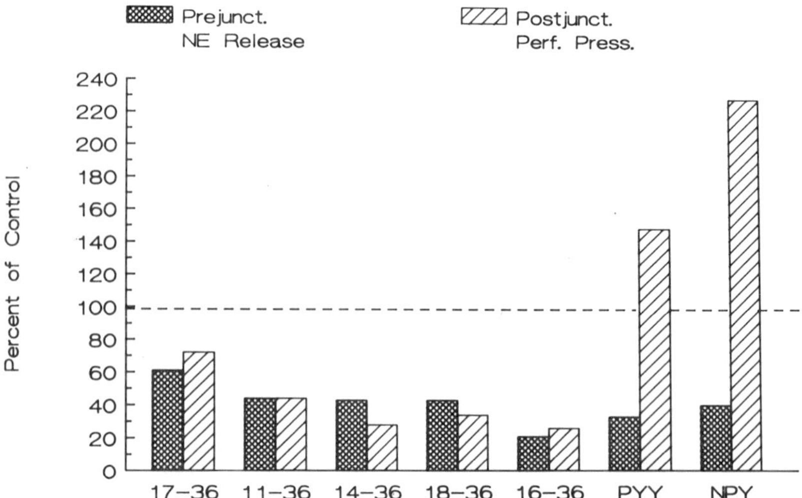

FIGURE 4. The effect of 10^{-7} M concentrations of NPY fragments 11-36, 14-36, 16-36, 17-36, 18-36, or the entire peptide NPY or PYY on the periarterial nerve stimulation induced inhibition of NE release (prejunct.) or increase in perfusion pressure (postjunct.). Data are plotted as percent of control. The fragments produced a parallel decrease in release and perfusion pressure while NPY and PYY produced an inhibition of release but a potentiation of the increase in perfusion pressure.

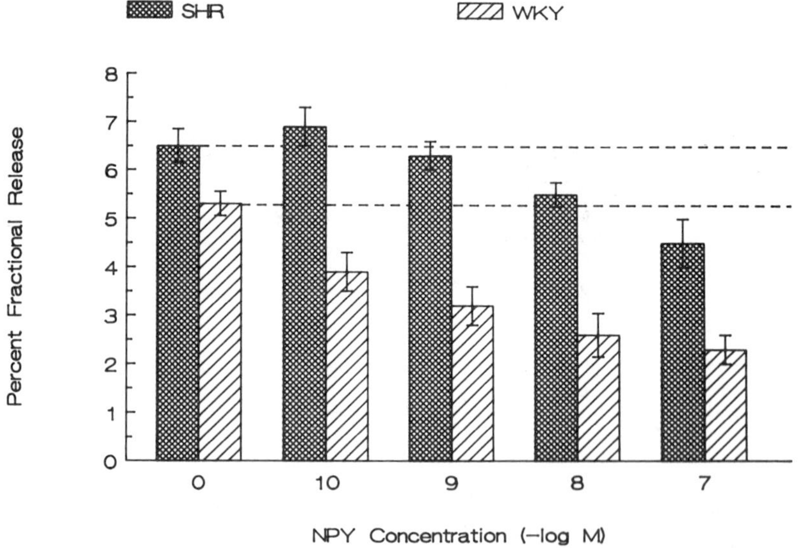

FIGURE 5. The effect of NPY on the periarterial nerve stimulation induced release of endogenous NE from the perfused mesenteric arterial bed of 8–10 week old WKY or SHR. Data are plotted as percent fractional release vs. NPY concentration. Each bar is the mean ± SEM of 5–7 preparations.

calcium-dependent simultaneous release of both norepinephrine and NPY-LI (FIGURE 8). The K^+-induced release of NPY-LI was reduced by administration of the α_2-adrenoceptor agonist clonidine (FIGURE 9) or the β-adrenoceptor antagonist propranolol. (FIGURE 10). The α_2-adrenoceptor antagonist yohimbine significantly enhanced the K^+-induced release of NPY-LI (data not shown).

Calcitonin Gene Related Peptide

As mentioned above periarterial nerve stimulation produces a frequency-dependent increase in perfusion pressure (FIGURE 11A) indicative of vasoconstriction as well as a parallel increase in NE release. In the presence of guanethidine (5×10^{-6} M)

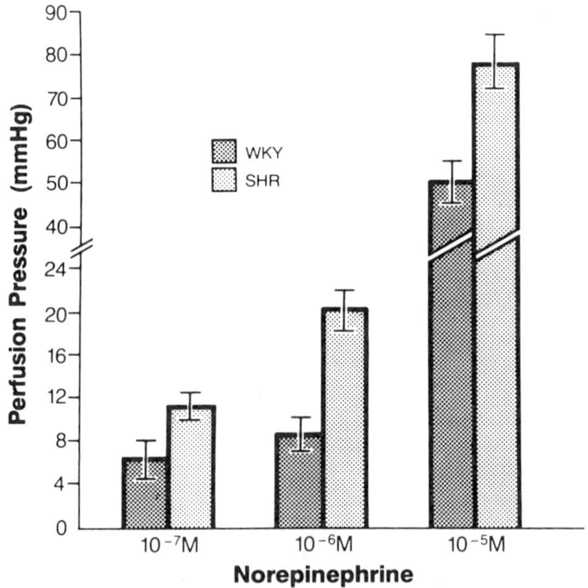

FIGURE 6. The effect of NPY (10^{-7} M) on the increase in perfusion pressure produced by NE on the perfused mesenteric artery of 8–10 week old WKY and SHR. Data are plotted as increase in perfusion pressure vs. NE concentration (M). Each bar is the mean ± SEM of 5–7 experiments. The NPY-induced potentiation of the NE effect was significantly greater in the SHR.

to block NE release and methoxamine (5×10^{-5} M) to precontract the mesenteric bed, periarterial nerve stimulation produced a frequency-dependent vasodilation (FIGURE 11B). The nerve stimulation induced vasodilation was unaffected by pretreatment with propranolol or atropine which abolished the effect of isoproterenol or acetylcholine, respectively, but was abolished by tetrodotoxin (FIGURE 11B). The nerve stimulation induced vasodilation was attenuated by prior treatment with capsaicin or by buffer containing antisera against rat CGRP (1:16,000 dilution) (FIGURE 12A and B). Removal of the endothelial cells in a manner described above in NPY experiments did not alter the nerve stimulation induced vasodilation while abolishing acetylcholine-induced vasodilation (FIGURE 12C). CGRP also attenuated the periarterial nerve

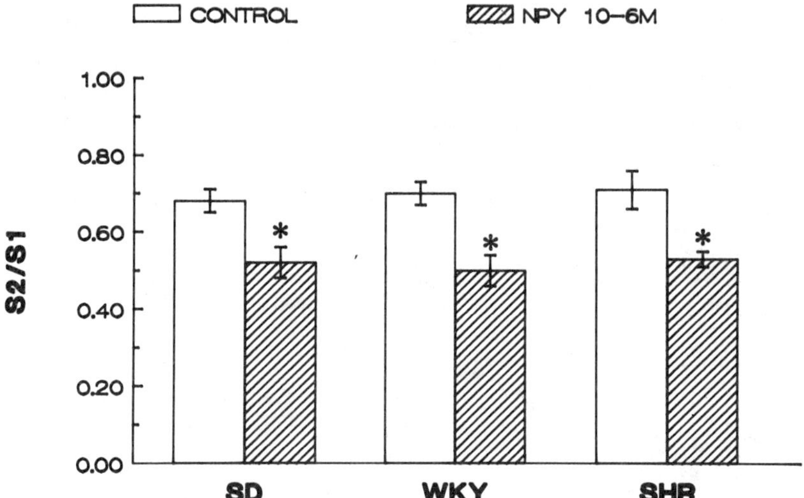

FIGURE 7. Effect of NPY 10^{-6} M on the potassium-evoked release of ^{3}H-NE from slices of the anterior hypothalamus of SD, SHR, and WKY rats. NPY was administered after the S_1 period. Results represent the mean ± SEM of 6 to 10 animals in each group. For the SD rats the S_2/S_1 control = 0.68 ± 0.03 and for the NPY group = 0.52 ± 0.04. WKY control S_2/S_1 = 0.70 ± 0.03 and NPY treated = 0.50 ± 0.04. SHR control S_2/S_1 = 0.71 ± 0.05 and NPY treated = 0.53 ± 0.02. *$p < 0.05$.

stimulation induced vasoconstriction. Basal perfusion pressure was not affected by CGRP but nerve stimulation induced vasoconstriction was attenuated with CGRP for 5 minutes in a concentration-dependent manner (FIGURE 13) with a minimal effective concentration of 10^{-14} M. In contrast, CGRP failed to alter either basal or periarterial nerve stimulation induced release of NE (FIGURE 14).

In vascular beds obtained from SHR, it was observed that the inhibitory effect of exogenous CGRP on the nerve stimulation induced increase in perfusion pressure was reduced (FIGURE 15). In preparations pretreated with guanethidine and precontracted with methoxamine, vasodilation was also produced by nerve stimulation in the vascular bed from SHR. The amplitude of the peak vasodilator response was slightly attenuated (FIGURE 16A) and the duration of the vasodilator response (duration between the time when perfusion pressure started dropping and when it came back to one-half of its peak value) was much shorter in the mesentery from SHR compared to SD rats (FIGURE 16B).

DISCUSSION

These studies show that NPY produces marked effects at the vascular neuroeffector junction. Using the perfused mesenteric arterial bed as a model of the noradrenergic neuroeffector junction, NPY exerted both prejunctional and postjunctional effects. Prejunctionally, NPY decreased the periarterial nerve stimulation induced release of NE while postjunctionally it potentiated the nerve stimulation induced increase in perfusion pressure. Moreover, NPY potentiated the increase in perfusion pressure produced by a variety of vasoactive agents including angiotensin, vasopressin,

as well as α-adrenoceptor agonists like phenylephrine or norepinephrine. These results suggest that NPY is acting at the level of transduction mechanisms common to this diverse group of vasoactive substances. Our results suggest that the postjunctional effect of NPY is independent of endothelial cells.

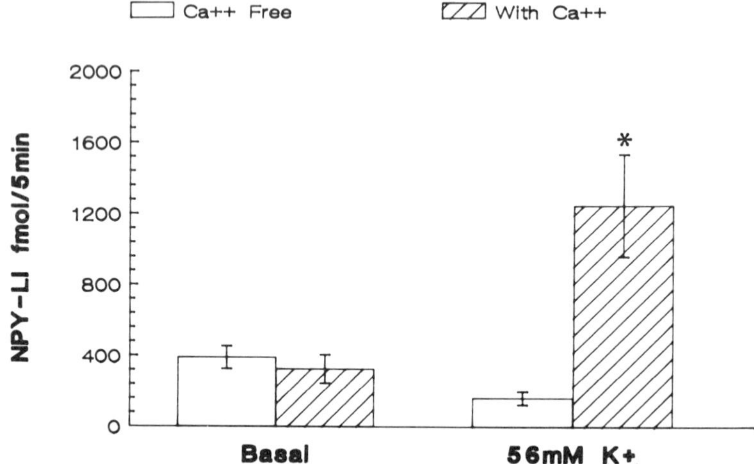

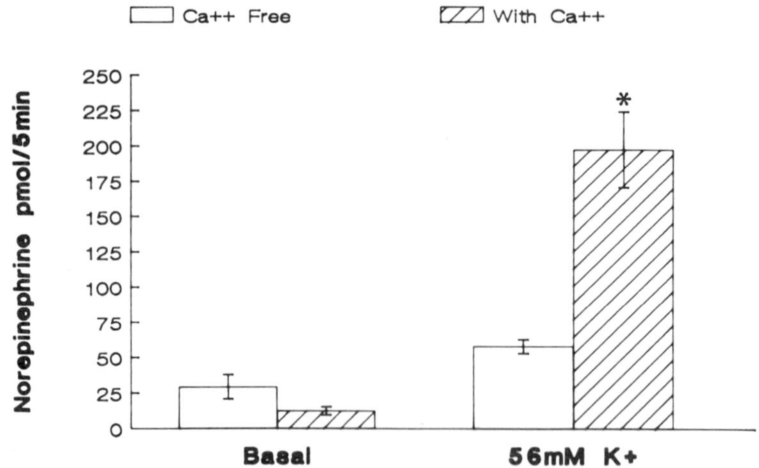

FIGURE 8. Calcium dependence of the potassium-evoked release of NE and NPY from slices of the rat hypothalamus. Slices were exposed to CA^{++}-free buffer during the first basal and stimulation period after which calcium was resupplied during a second basal and stimulation period. The stimulated release of NE and NPY was significantly attenuated in the absence of calcium. Release of NE occurred concurrently with NPY. *$p < 0.05$.

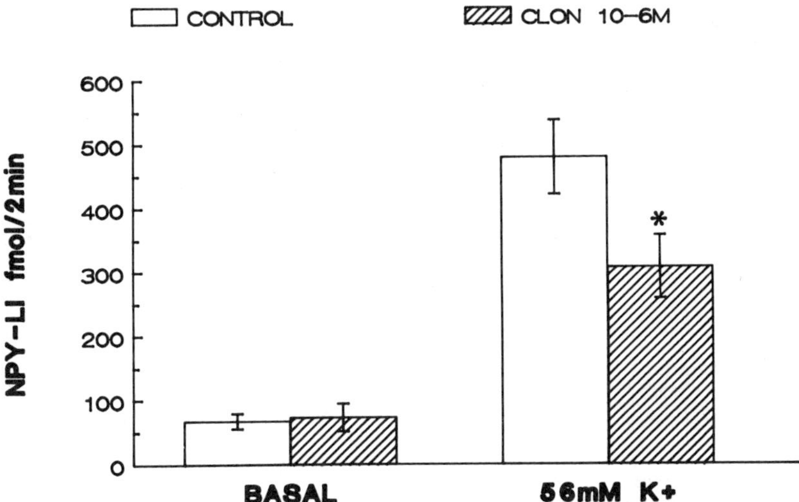

FIGURE 9. Effect of clonidine 10^{-6} M on the potassium-evoked release of NPY from slices of the rat hypothalamus. Clonidine was administered as described in the methods during both the basal and stimulation periods. Results represent the mean $\pm$ SEM of 15 animals in the control group and 8 animals in the clonidine group. Basal release of NPY in the control group = 67.3 fmol $\pm$ 11.9 and for the clonidine group = 72.8 fmol $\pm$ 21.8. Control stimulated release = 480.1 fmol $\pm$ 58.1 and in the presence of clonidine = 307.7 fmol $\pm$ 49.6. *$p < 0.05$.

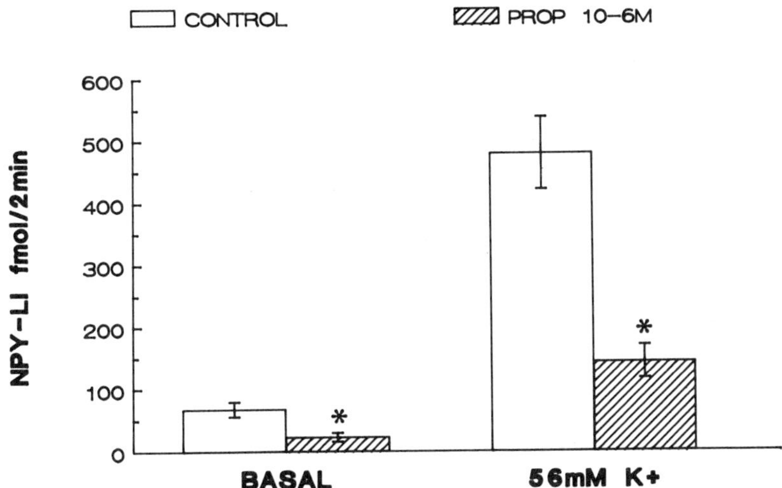

FIGURE 10. Effect of propranolol 10^{-6} M on the potassium-evoked release of NPY from slices of the rat hypothalamus. Propranolol was present during the basal and stimulation periods as described in Methods. Results represent the mean $\pm$ SEM for 15 control animals and 6 propranolol-treated animals. Basal release of NPY for the control group was 67.3 fmol $\pm$ 11.9 and for the propranolol-treated group 22.4 fmol $\pm$ 6.19. Control stimulated release of NPY was 480.1 fmol $\pm$ 58.1 while stimulated release in the propranolol group was 143.1 fmol $\pm$ 26.90. *$p < 0.05$.

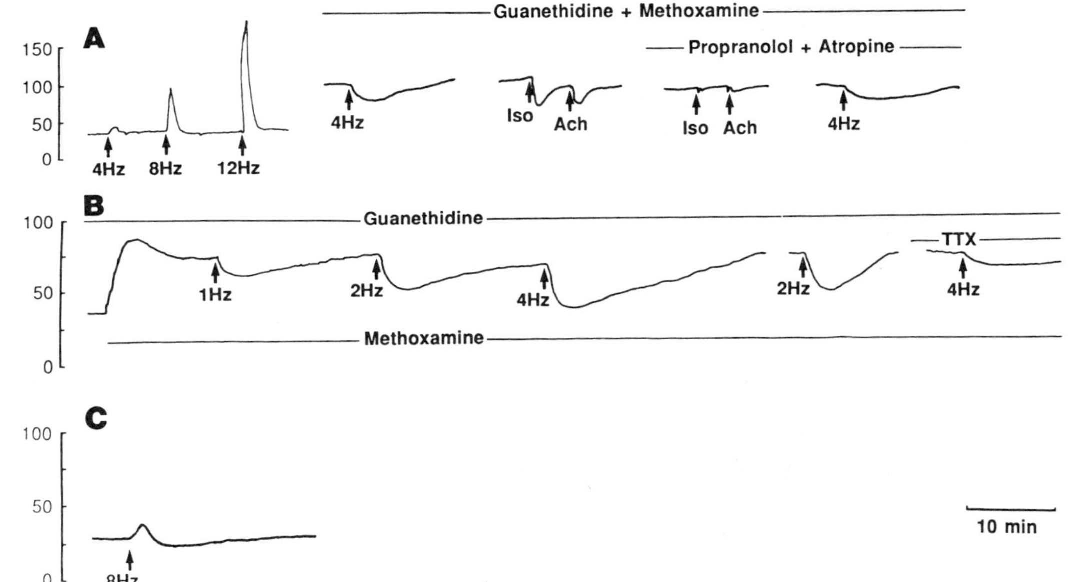

FIGURE 11. (**A**) A representative tracing of periarterial nerve stimulation–induced vasoconstriction in rat mesentery and vasodilation after pretreatment with guanethidine (5×10^{-6} M) and precontracted with methoxamine (5×10^{-5} M) (see Methods). This vasodilation was not blocked by propranolol (10^{-7} M) or by atropine (10^{-6} M) in doses sufficient to block the vasodilation induced by isoproterenol (Iso, 5×10^{-6} M, 20 seconds) or acetylcholine ACh, 10^{-6} M, 20 seconds), respectively. (**B**) The vasodilation was frequency dependent, repeatable, and tetrodotoxin (TTX, 10^{-7} M) sensitive. (**C**) The initial vasoconstriction and the prolonged vasodilation produced by periarterial nerve stimulation in a naive mesenteric vascular bed. Perf. Press.: perfusion pressure.

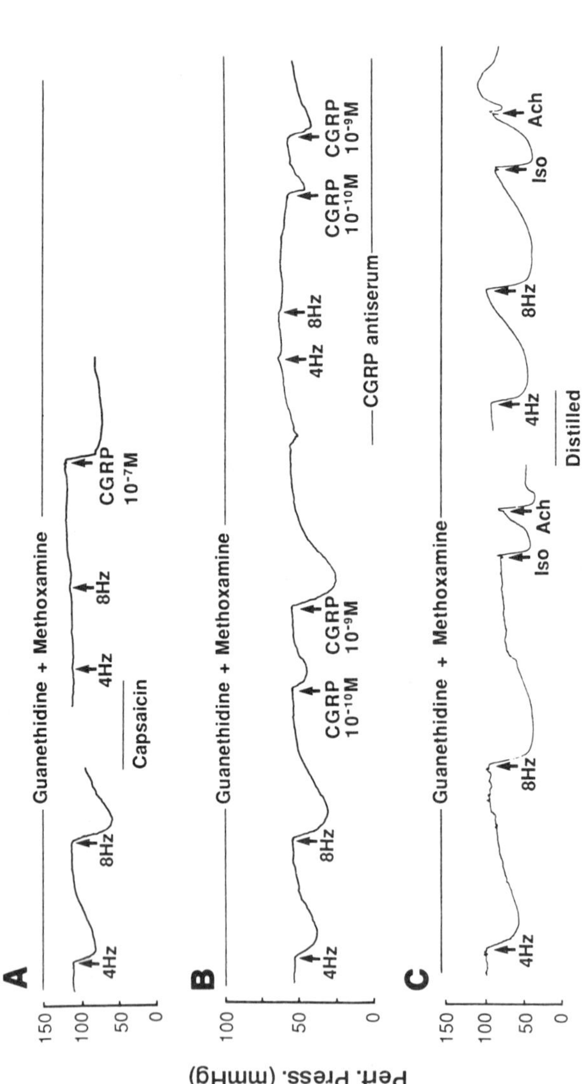

FIGURE 12. (A) Representative tracings showing that periarterial nerve stimulation–induced vasodilation was abolished by capsaicin (10^{-7} M) which did not affect the vasodilator action of exogenous CGRP (10^{-7} M). (B) Antiserum against CGRP prevented the vasodilation produced by nerve stimulation and partially prevented the vasodilation produced by exogenous CGRP. (C) Removal of endothelium from the vascular bed by perfusion with distilled water for 3 minutes changed the effect of acetylcholine (ACh) from vasodilation to vasoconstriction but did not affect vasodilation produced by nerve stimulation or by isoproterenol (Iso). Perf. Press.: perfusion pressure.

There is growing evidence for the existence of multiple subtypes of NPY receptors.[11,12] Studies by Wahstedt and colleagues suggest that the postjunctional response requires the entire 1-36 peptide and have called this receptor subtype Y_1.[10–12] In contrast the prejunctional response could be produced by shorter C terminal fragments (e.g., PPY 13-36 or NPY 13-36) and has been designated the Y_2 receptor.[10–12] Ligand binding studies support the concept of at least two types of NPY receptors that can also be distinguished on the basis of the entire NPY sequence or shorter NPY fragments.[13,14] Our results utilizing NPY, PYY, and a series of NPY C terminal fragments are consistent with the Y_1-Y_2 concept. The strength of our study is that pre- and postjunctional responses were examined simultaneously in a model noradrenergic vascular neuroeffector junction. Our results are consistent with the

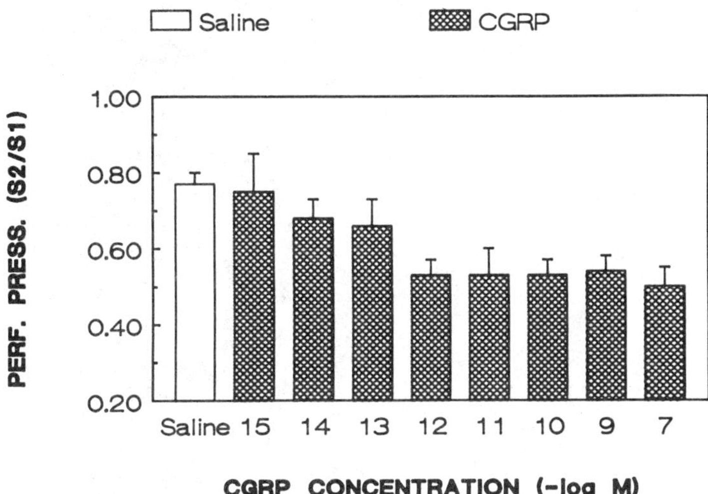

FIGURE 13. Effect of increasing concentrations of exogenous CGRP on periarterial nerve stimulation–induced vasoconstriction. The isolated mesenteric vascular bed was stimulated (60V, 16 Hz, 1 msecond duration for 30 seconds) twice, before (S_1) and during infusion with CGRP for five minutes (S_2). Change in perfusion pressure (perf. press.) is presented as S_2/S_1 ratio. The nerve stimulation–induced vasoconstriction was attenuated by CGRP in a concentration-dependent manner although the basal perfusion pressure was not affected ($n = 6$–8 in each group).

postjunctional response (Y_1) requiring both the N and C terminal while the prejunctional response requires only the C terminal end of the peptide. Although it may be premature or ever simplistic to think that there is a pre–postjunctional distribution of Y_1 and Y_2 receptors, our results do suggest that the predominant postjunctional receptor is of the Y_1 subtype while the predominant form of the prejunctional receptors is of the Y_2 subtype. Because of the known colocalization of NPY and NE in noradrenergic neurons innervating the mesenteric arterial bed, our results reinforce the idea that NPY acts as an important cotransmitter/modulator at this vascular neuroeffector junction.

Our results also implicate NPY in the pathophysiology of hypertension development and/or maintenance in the SHR. At the vascular neuroeffector junction of the

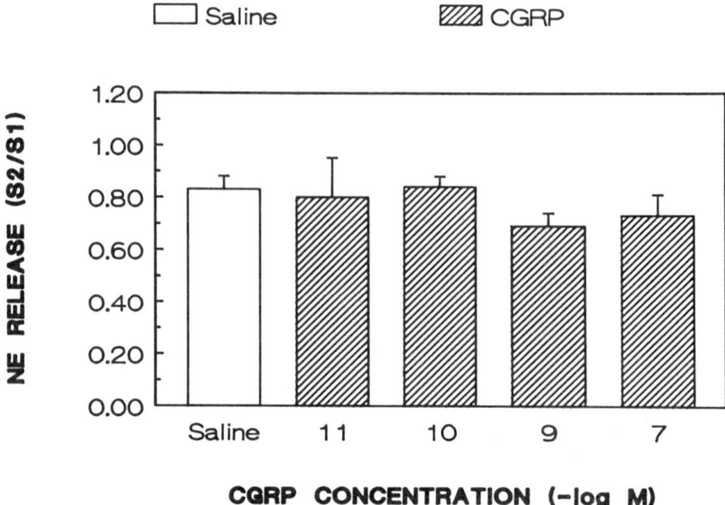

FIGURE 14. Effect of CGRP on NE release evoked by periarterial nerve stimulation (60V, 16 Hz, 1 msecond duration for 30 seconds) in isolated mesenteric arterial bed. Release of NE, expressed as S_2/S_1 ratio, was not affected by CGRP in concentrations effective in inhibiting the vasoconstriction induced by nerve stimulation or exogenous NE ($n = 6$).

mesenteric arterial bed, the postjunctional potentiation of contraction to vasoactive agents and nerve stimulation by NPY was enhanced. In contrast, the prejunctional inhibitory effect of NPY on NE release was attenuated. The net effect of this combination of responses would be a greater increase in blood pressure. The mechanism of these changes is not understood. However, since there is evidence for subtypes of NPY receptors, it is possible that there might be differential up and down regulation of Y_1 and Y_2 receptors. It is also possible that there may be differential effects on the intracellular transduction mechanisms. To date NPY has been shown to inhibit the stimulation of cAMP induced by various agonists[15,16] to increase phosphoinositide turnover[17,18] to produce an increase in intracellular calcium[18,19] or to inhibit intracellular Ca^{2+} transient and depolarization induced transmembrane Ca^{2+} currents.[20–23] Alterations in one or more of these transduction mechanisms may occur in blood vessels or neuronal tissue of the SHR.

The present studies have demonstrated that periarterial nerve stimulation elicits a frequency-dependent nonadrenergic-noncholinergic (NANC) vasodilation. This type of vasodilation has been observed in numerous vascular preparations, however the mediator is not clear.[24–26] Our results suggest that CGRP is the NANC mediator in mesenteric arterial bed. This is consistent with its distribution in most blood vessels. The nerve stimulation induced vasodilation was not affected by pretreatment of the vasculature with propranolol or atropine in concentrations sufficient to antagonize the vasodilation produced by isoproterenol or acetylcholine. The vasodilation was sensitive to TTX and capsaicin indicating that it was not myogenic and likely released from the primary sensory nerve terminals. The nerve stimulation induced vasodilation was mimicked by exogenously applied CGRP and the response of both was attenuated by CGRP antisera. Exogenous CGRP also decreased the increase in perfusion pressure produced by periarterial nerve stimulation or NE administration. Our results suggest that these effects of CGRP were postjunctional and that CGRP had no prejunctional action. Moreover, our studies indicate that the postjunctional effects of CGRP are independent of endothelial cells.

The effect of exogenous CGRP to inhibit the electrical stimulation–induced vasoconstriction in the isolated perfused mesentery was significantly attenuated in the spontaneously hypertensive rat. It is possible that there is a deficiency of the modulatory function of CGRP system at the receptor level in the genetically hypertensive animals. Furthermore, the amplitude of the vasodilation mediated by endogenous CGRP via periarterial nerve stimulation tended to be smaller and the duration of the vasodilation was shortened markedly in spontaneously hypertensive animals. The differences in amplitude and duration of the vasodilation may reflect a difference either in the amount of CGRP released or in the rate of CGRP degradation between the two strains. Consistent with this hypothesis, the plasma CGRP concentration has been found to be lower in SHR than in SD rats.[27] The differences in the vasodilator activity between normotensive and hypertensive animals may contribute to the development or maintenance of hypertension.

In conclusion, our studies have demonstrated that NPY produces prejunctional inhibitory and postjunctional excitatory effects at the noradrenergic neuroeffector junction of the mesenteric arterial bed. The postjunctional effect is nonspecific in that NPY potentiated the increase in perfusion pressure to a number of diverse substances suggesting an effect on a common intracellular transduction mechanism. The postjunctional effect is independent of endothelial cells and appears due to activation of Y_2 receptors while the prejunctional effect is due to activating the Y_1 receptor subtype. The prejunctional effect of NPY was attenuated while the postjunctional effect was enhanced in beds obtained from the SHR. Our results therefore implicate an alteration

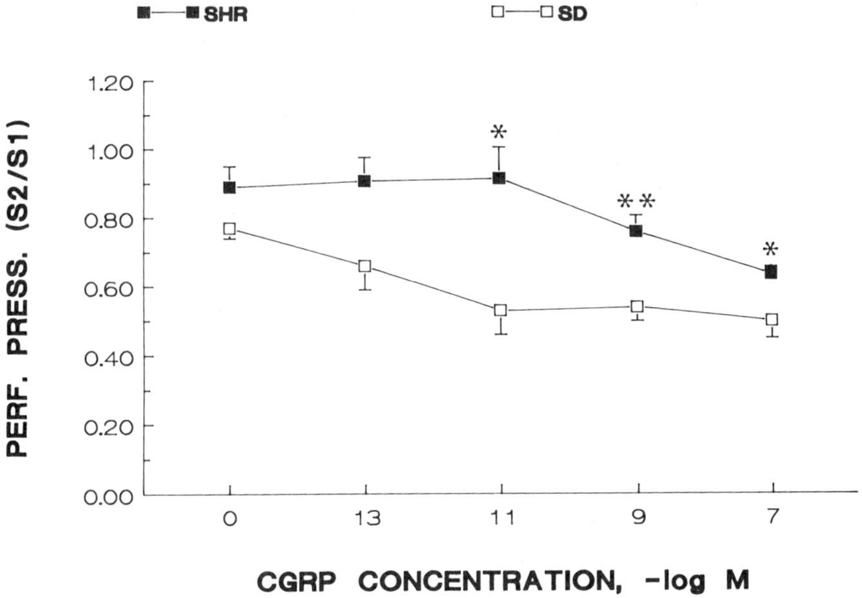

FIGURE 15. The effect of exogenous CGRP on the nerve stimulation-induced vasoconstriction (60V, 16 Hz, 1 msecond duration for 30 seconds) in the vascular beds from spontaneously hypertensive rat (SHR) and Sprague-Dawley (SD) rat. The inhibitory effect of CGRP was attenuated in the vascular bed from SHR as compared to SD ($n = 7$–8).

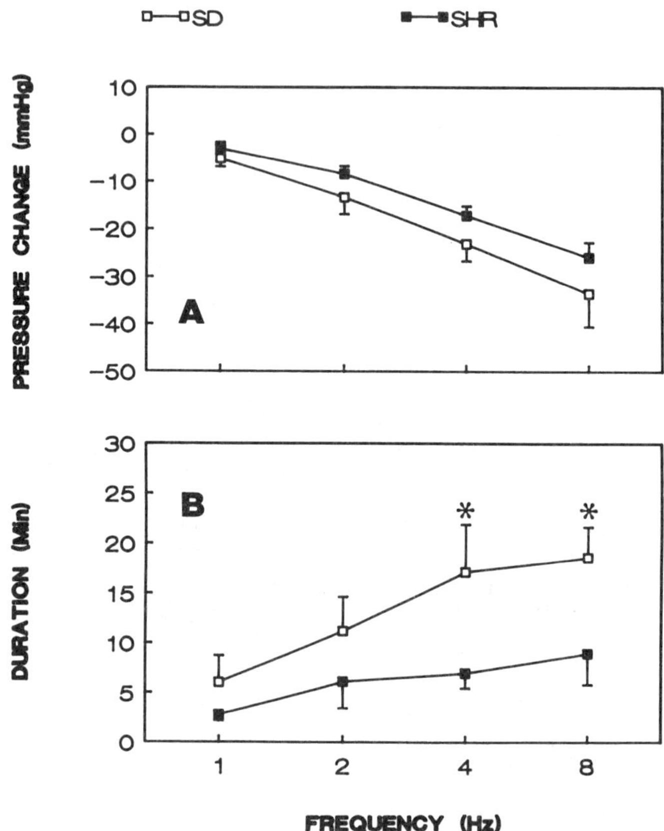

FIGURE 16. The effect of CGRP on amplitude of the peak vasodilator response (A) and duration of the vasodilator response (B) produced by periarterial nerve stimulation in the mesenteric vascular bed from spontaneously hypertensive rat (SHR) and Sprague-Dawley (SD) rat. The amplitude was slightly attenuated while the duration of the response was markedly attenuated in the mesenteric vascular bed from SHR compared to that from SD rat, respectively. The experiment was carried out in a manner similar to that described in the legend to FIGURE 4 ($n = $ 5–6 in each group).

in the NPY system in the development or maintenance of hypertension in the SHR. We have also demonstrated that CGRP can be released from the primary sensory nerve terminals as an endogenous vasodilator following periarterial nerve stimulation. CGRP modulates vasomotor tone in the mesenteric artery through postjunctional mechanisms. A deficiency of the CGRP system in the SHR may play a role in the pathogenesis of hypertension in this species.

REFERENCES

1. EDBLAD, E., L. EDVINSSON, C. WAHLSTEDT, R. UDDMAN, R. HADAMON & F. SUNDLER. 1984. Neuropeptide Y co-exists and cooperates with noradrenaline in perivascular nerve fibers. Regul. Pept. **8:** 225–235.

2. LUNDBERG, J. M. & T. HOKFELT. 1983. Coexistence of peptides and classical neurotransmitters. Trends Neurosci. **6**(8): 325–328.

3. MULDBERRY, P. K., M. A. GHATEI, J. RODRIGO, J. M. ALLEN, M.G. ROSENFELD, J. M. POLAK & S. R. BLOOM. 1985. Calcitonin gene related peptide in cardiovascular tissues of the rat. Neuroscience **14**: 947–954.

4. WHARTON, J., S. GULBENKIAN, P. K. MULDERRY, M. A. GHATER, G. P. McGREGOR, S. R. BLOOM & J. M. POLAK. 1986. Calcitonin gene related peptide immunoreactive nerves in the cardiovascular system of the guinea pig and rat. J. Auton. Nerv. Syst. **16**: 289–309.

5. WESTFALL, T. C., S. CARPENTIER, X. L. CHEN, M. C. BEINFELD, L. NAES & M. J. MELDRUM. 1987. Prejunctional and postjunctional effects of neuropeptide Y at the noradrenergic neuroeffector junction of the perfused mesenteric arterial bed of the rat. J. Cardiovasc. Pharmacol. **10**: 716–722.

6. HAN, S-P., L. NAES & T. C. WESTFALL. 1990. Inhibition of periarterial nerve stimulation–induced vasodilation of the mesenteric arterial bed by CGRP (18-37) and CGRP desensitization. Biochem. Biophys. Res. Commun. **168**: 786–791.

7. CIARLEGLIO, A. E., M. C. BEINFELD & T. C. WESTFALL. 1987. Neuropeptide Y modulation of noradrenergic transmission in the posterior hypothalamus of the rat. Pharmacologist **29**: 128.

8. CIARLEGLIO, A. E., M. C. BEINFELD & T. C. WESTFALL. 1988. Release of neuropeptide Y immunoreactivity from the rat hypothalamus. Soc. Neurosci. Abstr. **14**: 843.

9. WESTFALL, T. C., J. MARTIN, X. CHEN, A. CIARLEGLIO, S. CARPENTIER, K. HENDERSON, M. KNUEPFER, M. C. BEINFELD & L. NAES. 1988. Cardiovascular effects and modulation of noradrenergic neurotransmission following central and peripheral administration of neuropeptide Y. Synapse **2**: 299–307.

10. WAHLESTEDT, C., N. YANAIHARA & R. HAKANSON. 1986. Regul. Pept. **13**: 307–318.

11. WAHLESTEDT, C. & R. HAKANSON. 1986. Med. Biol. **64**: 85–88.

12. WAHLESTEDT, C., R. KOMAN & R. HAKANSON. 1987. Soc. Neurosci. Abstr. **13**: 1271.

13. SHEIKH, S. P., R. HAKANSON & T. W. SCHWARTZ. 1989. FEBS Lett. **245**: 209–214.

14. SCHWARTZ, T. C., J. FAHLENDORFF & J. LANGELAND. 1989. Y_1 and Y_2 receptor for NPY: the evolution of PP fold peptides and their receptors. Nobel Symposium XIV. Neuropeptide Y. Raven Press. New York, N.Y.

15. LUNDBERG, J. M., A. EMSON, O. LARSSON, A. RUDEHILL, A. SARIA & B. FREDHOLM. 1988. Neuropeptide Y receptor in pig spleen: binding characteristics, reduction of cyclic AMP formation and calcium antagonist inhibition of vasoconstriction. Eur. J. Pharmacol. **145**: 21–29.

16. HARFSTRAND, A., B. FREDHOLM & K. FUXE. 1987. Inhibitory effects of cyclic AMP accumulation in slices of the nucleus tractus solitarius. Neurosci. Lett. **76**: 185–190.

17. HINSON, J., C. RANH & J. COUPET. 1988. Neuropeptide Y stimulates inositol phospholipid hydrolysis in rat brain minipressure. Brain Res. **446**: 379–382.

18. PERNEY, T. M. & R. J. MILLER. 1989. Two different G-proteins mediate neuropeptide Y and bradykinin stimulated phospholipid breakdown in cultured rat sensory neurons. J. Biol. Chem. **264**: 7317–7327.

19. MOTINLSKY, H. J. & M. C. MICHEL. 1988. Neuropeptide Y mobilized Ca^{2+} and inhibits adenylate cyclase in human erythroleukemia cells. Am. J. Physiol. **255**: E880–E885.

20. SCHAFIELD, G. G. & R. S. IKEDA. 1988. Neuropeptide Y blocks a calcium current in C-cells of bullfrog sympathetic ganglia. Eur. J. Pharmacol. **151**: 131–134.

21. EWALD, D. A., P. C. STERNWEISS & R. J. MILLER. 1988. Guanine nucleotide binding protein Go-induced coupling of neuropeptide Y receptors to Ca^{2+} channels in sensory neurons. Proc. Nat. Acad. Sci. USA **85**: 3633–3637.

22. WALKER, M. W., D. A. EWALD, T. M. PERNEY & R. J. MILLER. 1988. Neuropeptide Y modulates neurotransmitter release and Ca^{2+} currents in rat sensory neurons. J. Neurosci. **8**: 2438–2446.

23. COLMERS, W. F., K. LUKOWIAK & O. J. PITTMAN. 1988. Neuropeptide Y action in the rat hippocampal slice: site and mechanism of presynaptic inhibition. J. Neurosci. **8**: 3827–3837.

24. KOWASAKI, H., K. TAKASAKI, A. SAITO & K. GOTO. 1988. Calcitonin gene related peptide

       acts as a novel vasodilator neurotransmitter in mesenteric resistance vessels of the rat. Nature **338:** 164–167.

25. DUCKLES, S. P. 1986. Effect of capsaicin on vascular smooth muscle. Naunyn Schmiedebergs Arch. Pharmacodyn. **333:** 59–64.

26. HOLZER, P. 1988. Local effector functions of capsaicin-sensitive sensory nerve endings: involvement of tackykinins, calcitonin-gene related peptides and other neuropeptides. Neuroscience **24:** 739–768.

27. XU, D., X. WANG, J-P. WANG, Q. X. YUAN, R. R. FISCUS, J. K. CHANG & J. THANG. 1989. Calcitonin gene related peptide in normotensive and hypertensive rats. Peptides **10:** 309–312.

Presynaptic Receptors in Hypertension

KAZIMIERZ R. BORKOWSKI

Clinical Pharmacology Group
John P. Robarts Research Institute
Post Office Box 5015
100 Perth Drive
London, Ontario, Canada N6A 5K8

INTRODUCTION

Early observations that alpha-adrenoceptor antagonists increased electrical stimulation–evoked noradrenaline (NA) overflow, in perfused cat spleen and intestine, were considered mainly as postsynaptic events and related to transmitter release and the inhibition of neuronal and extraneuronal uptake.[1-3] Later, the effects of angiotensin, phenoxybenzamine, and propranolol on NA release during sympathetic nerve stimulation[4] and a clonidine-induced inhibition of adrenergic neurotransmission[5] were described. Sympathomimetic amines were found to inhibit nerve-stimulation-evoked NA release in cat spleen.[6] A feedback control of NA secretion in human vasoconstrictor nerves[7] and a frequency dependence of negative feedback control of sympathetic transmitter release in guinea pig vas deferens[8] were reported. These, and other observations on the effects of adrenergic agents on neuronal NA release,[9-11] led to the concept of presynaptic receptors and the hypothesis that presynaptic receptor mechanisms play a role in modulating NA release from sympathetic varicosities.[12]

PRESYNAPTIC ALPHA-ADRENOCEPTORS

On the basis of anatomical location and that of differences in order of potency of alpha-adrenoceptor agonists and antagonists, alpha-adrenoceptors were subdivided into postsynaptic alpha$_1$- and presynaptic alpha$_2$-receptors; the latter mediating negative feedback inhibition of stimulation-evoked NA release.[9,13,14] This subdivision into alpha$_1$- and alpha$_2$-adrenoceptor subtypes, to signify post- and presynaptic alpha-adrenoceptors respectively, no longer appears tenable. Indeed, continued development and use of selective agonists and antagonists has led to the identification of postsynaptic alpha$_2$-adrenoceptors[15-19] and, more recently, presynaptic alpha$_1$-adrenoceptors.[20-22] Postsynaptic alpha$_2$-adrenoceptors inhibit adenylyl cyclase[23] and cause constriction in vascular smooth muscle.[24] Presynaptic alpha$_1$- and alpha$_2$-adrenoceptors both appear to inhibit NA release.[20-21]

PRESYNAPTIC BETA-ADRENOCEPTORS

As the concept of presynaptic alpha-adrenoceptor-mediated negative feedback control of neuronal NA release was being formulated, evidence was accumulating to suggest the existence of presynaptic beta-adrenoceptors mediating a positive feedback facilitation of neuronal NA release.[25-29] These adrenoceptors appear to be of the beta$_2$-subtype,[28,30-34] at which NA is only a weak agonist.[35,36] Thus, presynaptic beta$_2$-adrenoceptors may not represent a true feedback loop, in which the release of endogenous NA would be expected to be capable of affecting its own subsequent release from sympathetic nerve terminals.

Adrenaline appears to be the naturally occurring agonist at these adrenoceptors,[27,31–33,37–41] and elevated plasma adrenaline levels have been reported in patients with essential hypertension.[42–44] Indeed, adrenaline has been shown to have presynaptic effects in man,[45–47] enhancing pressor responses to cold exposure and isometric exercise,[45] causing a sustained elevation in plasma NA and insulin during and following its infusion,[46] and facilitating reflex forearm vasoconstriction to lower body negative pressure.[47]

Most earlier investigators concluded that presynaptic facilitatory beta-adrenoceptors were of the beta$_2$-subtype. Nevertheless, data indicating the possibility that these receptors were of the beta$_1$-subtype have also been reported[39] and the hypothesis that presynaptic facilitatory beta$_1$- and beta$_2$-adrenoceptors coexist in several tissues[48] has been supported recently.[49] Thus, the antihypertensive effects of nonselective and beta$_1$-selective beta-antagonists may, at least in part, be due to blockade of presynaptic facilitatory beta-adrenoceptors.

PRESYNAPTIC ADRENOCEPTORS AND HYPERTENSION

The regulation of sympathetic neurotransmitter release by presynaptic adrenoceptors, under physiological conditions, remains controversial.[50,51] However, if presynaptic adrenoceptors modulating NA release are accepted, it can be argued that a dysfunction in either receptor mechanism may lead to inappropriate sympathetic neurotransmission. Thus, decreased presynaptic alpha$_2$- or increased beta$_2$-adrenoceptor responsiveness would augment the per-pulse release of NA. This facilitation of sympathetic activity will increase peripheral vascular resistance and may play a role in hypertension development and maintenance.

PRESYNAPTIC BETA-ADRENOCEPTORS AND HYPERTENSION

The implantation of slow-release depot preparations containing adrenaline (in doses that were not themselves acutely pressor and, hence, apparently devoid of postsynaptic effect) led to the development of a sustained elevation in blood pressure in normotensive rats.[39] This prohypertensive effect of adrenaline could be attenuated by concomitant treatment with metoprolol,[39] suggesting a possible involvement of beta$_1$-adrenoceptors in the adrenaline-induced blood pressure elevation. However, metoprolol's selectivity for beta$_1$-adrenoceptors is not marked[52] and such an interpretation is complicated by its ability to exert a direct hypotensive effect.

Indeed, subsequent experiments have shown that surgical adrenal enucleation in young (4-week-old) spontaneously hypertensive (SHR) rats, to deplete circulating levels of plasma adrenaline (the purported natural agonist at presynaptic beta$_2$-adrenoceptors), attenuates the development of high blood pressure in this model of hypertension.[53,54] In the adrenal enucleated SHRs, hypertension development could be restored by adrenaline supplementation using slow-release depot implants. Moreover, it was possible to mimic this prohypertensive effect with implants containing the selective beta$_2$-adrenoceptor agonists salbutamol or procaterol, and the adrenaline-induced elevation in blood pressure could be attenuated by concomitant treatment with ICI 118551,[54] a selective beta$_2$-adrenoceptor antagonist[55] which does not exhibit a direct hypotensive effect.

These results indicate that the activation of presynaptic beta$_2$-adrenoceptors, by adrenaline, may play a role in the development of spontaneous hypertension. However, a role in hypertension maintenance is unlikely since reducing presynaptic beta-adrenoceptor stimulation, by adrenal enucleation to deplete circulating adrenaline, did not significantly affect resting blood pressure or heart rate in adult SHRs with established hypertension[56] (Borkowski, unpublished observations). This may indicate that presynaptic facilitatory beta-adrenoceptors are desensitized in established hypertension. Conversely, the multiple mechanisms contributing and serving to maintain hypertension may be capable of masking any reduction in arterial pressure expected as a consequence of eliminating adrenaline's facilitation of sympathetic activity. It is interesting to note that adrenal enucleation induced small but significant reductions in blood pressure and heart rate in adult normotensive rats,[57] suggesting that presynaptic facilitatory beta-adrenoceptors may have a physiological role in normotension.

The ability of adrenaline depletion to attenuate hypertension development in young SHRs,[53,54] but not established hypertension in adults,[56] suggests there may be a period of critical sensitivity to adrenaline's prohypertensive effects during the time course of development of spontaneous hypertension, presumably due to altered presynaptic beta-adrenoceptor acitvity. Attempts to demonstrate differences in presynaptic beta-adrenoceptor properties between hypertensive and normotensive rats and between young and adult hypertensive animals have, however, been largely inconclusive.

In isolated perfused mesenteric beds, isoprenaline's enhancement of periarterial nerve stimulation induced increases in perfusion pressure and ^{3}H-NA overflow was greater in SHRs with established hypertension than in age-matched (14- to 16-week-old) Wistar-Kyoto (WKY) normotensive controls.[58] Salbutamol also caused greater enhancement of pressor responses in SHR compared to WKY mesenteries, and the perfusion pressure effects were attenuated by propranolol, sotalol, and butoxamine, but not by practolol.[58] The results suggest that presynaptic beta$_2$-adrenoceptor-mediated facilitation of sympathetic neurotransmitter release is enhanced in the vasculature of adult SHRs and may play a role in the maintenance of hypertension.

However, in isolated perfused kidneys, no difference was observed between age-matched (18-week-old) SHR and WKY rats in isoprenaline's facilitation of periarterial nerve stimulation induced effects.[59] In addition, in isolated superfused spleen strips, there was no difference between age-matched (15-week-old) SHR and WKY rats in isoprenaline's facilitation of field stimulation induced ^{3}H-NA release.[60] These results appear to indicate that presynaptic beta$_2$-adrenoceptor responsiveness is similar in adult hypertensive and normotensive rats.

Isoprenaline-induced facilitation of field stimulation induced ^{3}H-NA release was, however, greater in isolated superfused spleen strips from young (5-week-old) SHRs compared to age-matched WKY rats,[60] supporting the hypothesis that presynaptic beta-adrenoceptors exhibit increased responsiveness in early hypertension and may be involved in its development rather than maintenance. However, in the isolated portal vein, there was no difference in isoprenaline's facilitation of field stimulation induced 3H-NA release between SHRs and age-matched (6- to 8-week-old) WKY rats.[61] While it could be argued that presynaptic beta-adrenoceptor sensitivity is enhanced only in developing spontaneous hypertension, and even then only in SHRs less than 6 weeks old, these results reinforce the impression of variability between tissues.

In patients with essential hypertension and age-matched normotensive controls, both endogenous (glucagon-stimulated) and exogenous (infused) adrenaline, at physiological plasma concentrations, acted as a sustained stimulator of presynaptic beta-adrenoceptors to enhance NA release from peripheral sympathetic nerve terminals.[62,63] The results support the hypothesis that adrenaline is taken up into sympathetic neurotransmitter stores, activates presynaptic beta-adrenoceptors when subsequently released as a cotransmitter, and enhances neuronal NA release and sympathetic cardiovascular responses. However, an apparent difference could not be demonstrated between normotensive and hypertensive subjects in adrenaline's ability to effect cardiovascular and plasma NA responses.[62,63] This agrees with the majority of animal experiments, which appear to indicate that presynaptic facilitatory beta-adrenoceptor activity in established hypertension is not significantly different from that in normotension.

A possible facilitation of neuronal NA release before the development of essential hypertension has, however, been indicated by the presence of increased sympathetic activity in offspring of hypertensive parents.[64] Compared to age-matched normotensive controls with normotensive parents, young (17- to 26-year-old) offspring of hypertensives demonstrated increases in plasma NA, during submaximal exercise, 75–79% higher than those in the controls.[64] Although it is not known whether the increased sympathetic activity is due to increased presynaptic beta-adrenoceptor-mediated facilitation, or decreased presynaptic alpha-adrenoceptor-mediated inhibition, the results indicate that it is not secondary to hypertension and may, therefore, be important in hypertension development in man.

PRESYNAPTIC ALPHA-ADRENOCEPTORS AND HYPERTENSION

Whether or not changes in presynaptic beta-adrenoceptors occur and play a role in the pathophysiology of hypertension, their facilitatory effect on sympathetic neurotransmitter release is small compared to the inhibitory effect of presynaptic alpha-adrenoceptors. Beta-adrenoceptor agonists increase NA overflow by 30–60%.[65] Blocking presynaptic inhibitory alpha-adrenoceptors with phentolamine, however, increased NA overflow by 300%.[8] Thus, decreased presynaptic alpha-adrenoceptor activity might be expected to play a more important role, in elevating sympathetic activity and hypertension development and maintenance, than does increased presynaptic beta-adrenoceptor activity.

In isolated perfused kidneys from young (7-week-old) SHRs, renal nerve stimulation induced ^{3}H-NA overflow and vasoconstrictor responses were enhanced compared to those in age-matched normotensive rats.[66,67] Blocking NA uptake with cocaine induced similar increases in response in both hypertensive and normotensive kidneys,[67,68] indicating that the enhanced NA overflow in SHRs reflects increased neuronal release, not differences in uptake. The relative ineffectiveness of NA as an agonist at presynaptic facilitatory beta$_2$-adrenoceptors[36] suggests that an increased responsiveness of this facilitatory mechanism is not responsible for the enhanced NA release. Thus, the increased NA release in early hypertension[61] may be due to decreased responsiveness of presynaptic inhibitory alpha-adrenoceptors and may play a role in hypertension development. However, a role in maintaining established hypertension is unlikely since ^{3}H-NA release was lower in kidneys from adult SHRs compared to age-matched (6-month-old) WKY rats,[69] suggesting that presynaptic inhibitory alpha-adrenoceptors may be more, rather than less, sensitive in established hypertension. Indeed, presynaptic alpha-adrenoceptors have been shown to be supersensitive in younger (14-week-old), but still fully hypertensive, SHRs compared to age-matched WKY rats.[59] Despite the presynaptic alpha-adrenoceptor supersensitivity, an increased release of NA in kidneys from the SHR rats was found.[59]

These indications that presynaptic inhibitory alpha-adrenoceptors may be desensitized in prehypertensive SHRs, but not in adults with established hypertension, are not supported by other observations. In isolated arteries and veins from 28-week-old SHRs, presynaptic alpha$_2$-adrenoceptor antagonism with yohimbine elicited a smaller enhancement of field stimulation induced release of both endogenous and ^{3}H-NA compared to age-matched WKY rats.[61,70] The effects of yohimbine were similar in tissues from 6- and 10-week-old SHR and WKY rats,[61,70] suggesting that presynaptic alpha-adrenoceptors may be desensitized in established hypertension and may play a role in its maintenance but not development. Ekas *et al.* used periarterial nerve stimulation in perfused kidneys and tramazoline to stimulate alpha$_2$-adrenoceptors.[59] Westfall and colleagues field stimulated segments of blood vessels and blocked alpha$_2$-adrenoceptors with yohimbine.[61,70] Despite these differences, it is difficult to reconcile their findings and the role, if any, of altered presynaptic alpha-adrenoceptor responsiveness in hypertension development and maintenance is unclear. Moreover, in chronic neurogenic hypertension, induced by sinoaortic baroreceptor denervation, phentolamine caused similar increases in nerve stimulation induced ^{3}H-NA overflow in isolated perfused mesenteric arteries from hypertensive and sham-operated rats.[71] It appears, therefore, that presynaptic alpha-adrenoceptors are unaffected in chronic neurogenic hypertension.

Much of the evidence for an increased sympathetic nerve activity in human essential hypertension is indirect, being based on measurements of plasma catecholamines, and cannot be used readily to distinguish between effects on transmitter release and reuptake, let alone effects of presynaptic alpha- or beta-adrenoceptors. Nevertheless, differences in plasma catecholamines have been identified between hypertensive and normotensive subjects,[42,43] although they have been usually small and inconsistent.[72] Despite mechanisms responsible for the increase being unclear, plasma NA was elevated in 14 out of 15 studies of young hypertensives,[72] a result in agreement with observations of an increased sympathetic neuronal activity in young SHR rats.[73] In a recent comparison of young men with borderline hypertension and age-matched normotensive controls, direct recording from multifiber postganglionic sympathetic nerves to muscle demonstrated increased activity in the hypertensive subjects.[74] This does not, of course, indicate altered presynaptic regulation of neurotransmitter release, but supports suggestions that an elevated sympathetic activity is involved in hypertension development.

CONCLUSIONS

Presynaptic alpha$_2$-adrenoceptors mediate a negative feedback inhibition of neuronal NA release; beta$_2$-adrenoceptors mediate a positive feedback facilitation. NA stimulates alpha$_2$-adrenoceptors and these may represent a true feedback mechanism. However, adrenaline appears to be the natural agonist at beta$_2$-adrenoceptors. Facilitation of neuronal NA release by presynaptic beta$_2$-adrenoceptor stimulation may not, therefore, represent a true feedback loop. Decreased alpha-adrenoceptor-mediated inhibition of NA release, or increased beta-adrenoceptor-mediated facilitation, leading to enhanced per-pulse release of NA and increased vasoconstrictor tone, might be expected to play a role in the development and maintenance of hypertension.

Attempts to identify changes in presynaptic alpha-adrenoceptor responsiveness, in developing or established hypertension, have yielded equivocal results. However, there is evidence that presynaptic facilitatory beta-adrenoceptor responsiveness is increased in young prehypertensive SHRs but not in adults with established hypertension. Thus, presynaptic beta-adrenoceptors may play a role in hypertension development. Their involvement in the development of human essential hypertension is an intriguing, but unproven, possibility.

SUMMARY

The physiological status of presynatic receptors, regulating sympathetic neurotransmitter release, remains subject to debate. Nevertheless, pharmacological techniques have shown presynaptic alpha-adrenoceptors, mediating a negative feedback inhibition of neuronal noradrenaline (NA) release, and presynaptic beta-adrenoceptors mediating a positive feedback facilitation. Decreased presynaptic alpha- or increased beta-adrenoceptor responsiveness might be expected to result in enhanced per-pulse release of NA and may contribute to hypertension development and maintenance. A potential role in hypertension development, but not its maintenance, has been established for presynaptic beta-adrenoceptors. Attempts to identify altered presynaptic adrenoceptor responsiveness in hypertension have, however, been inconclusive.

REFERENCES

1. BROWN, G. L. & J. S. GILLESPIE. 1956. Output of sympathin from the spleen. Nature **178:** 980.
2. BROWN, G. L. & J. S. GILLESPIE. 1957. The output of sympathetic transmitter from the spleen of the cat. J. Physiol. **138:** 81–102.
3. BROWN, G. L., B. N. DAVIES & J. S. GILLESPIE. 1958. The release of chemical transmitter from the sympathetic nerves of the intestine of the cat. J. Physiol. **143:** 41–54.
4. STARKE, K. & H. J. SCHUMANN. 1972. Interactions of angiotensin, phenoxybenzamine and propranolol on noradrenaline release during sympathetic nerve stimulation. Eur. J. Pharmacol. **18:** 27–30.
5. STARKE, K. & K. P. ALTMANN. 1973. Inhibition of adrenergic neurotransmission by clonidine: an action on prejunctional alpha-receptors. Neuropharmacology **12:** 339–347.
6. KIRPEKAR, S. M., R. F. FURCHGOTT, A. R. WAKADE & J. C. PRAT. 1973. Inhibition by sympathomimetic amines of the release of norepinephrine evoked by nerve stimulation in the cat spleen. J. Pharmacol. Exp. Ther. **187:** 529–538.
7. STJARNE, L. & K. GRIPE. 1973. Prostaglandin-dependent and independent feedback control of noradrenaline secretion in vasoconstrictor nerves of normotensive human subjects. Naunyn Schmiedebergs Arch. Pharmacol. **280:** 441–446.
8. STJARNE, L. 1973. Frequency dependence of dual negative feedback control of secretion of sympathetic neurotransmitter in guinea-pig vas deferens. Br. J. Pharmacol. **49:** 358–360.
9. DUBOCOVICH, M. L. & S. Z. LANGER. 1974. Negative feedback regulation of noradrenaline release by nerve stimulation in the perfused cat's spleen: differences in potency of phenoxybenzamine in blocking pre- and postsynaptic adrenergic receptors. J. Physiol. **237:** 505–519.
10. FARNEBO, L. O. & B. HAMBERGER. 1974. Influence of alpha- and beta-adrenoreceptors on the release of noradrenaline from field stimulated atria and cerebral cortex slices. J. Pharm. Pharmacol. **26:** 644–646.
11. STARKE, K. & H. MONTEL. 1974. Influence of drugs with affinity for alpha-adrenoceptors on noradrenaline release by potassium, tyramine and dimethylphenylpiperazinium. Eur. J. Pharmacol. **27:** 273–280.
12. LANGER, S. Z. 1974. Presynaptic regulation of catecholamine release. Biochem. Pharmacol. **23:** 1793–1800.
13. STARKE, K., T. ENDO & H. D. TAUBE. 1975. Relative pre- and postsynaptic potencies of alpha-adrenoceptor agonists in the rabbit pulmonary artery. Naunyn Schmiedebergs Arch. Pharmacol. **291:** 55–78.
14. STARKE, K., T. ENDO & H. D. TAUBE. 1975. Pre- and postsynaptic components in effects of drugs with alpha-adrenoceptor affinity. Nature **254:** 440–441.
15. BERTHELSON. S. & W. A. PETTINGER. 1977. A functional basis for classification of alpha-adrenergic receptors. Life Sci. **21:** 596–606.
16. DREW, G. M. & S. B. WHITING. 1979. Evidence for two distinct types of postsynaptic alpha-adrenoceptors in vascular smooth in vivo. Br. J. Pharmacol. **78:** 489–505.

17. GRANT, J. A. & M. G. SCRUTTON. 1979. Novel alpha$_2$-adrenoceptors primarily responsible for inducing human platelet aggregation. Nature 277: 659–661.

18. HOFFMAN, B. B., A. DELEAN, C. L. WOOD, D. D. SCHOKEN & R. J. LEFKOWITZ. 1979. Alpha-adrenergic receptor subtypes; quantitative assessment by ligand binding. Life Sci. 24: 1739–1746.

19. WICKBERG, J. E. S. 1979. The pharmacological classification of adrenergic alpha$_1$-receptors and their mechanism of action. Acta Physiol. Scand. 468(Suppl.): 1–99.

20. STORY, D. F., C. A. STANDFORD-STARR & M. J. RAND. 1985. Evidence for the involvement of alpha$_1$-adrenoceptors in negative feedback regulation of noradrenergic transmitter release in rat atria. Clin. Sci. 68(Suppl. 10): 111–115.

21. DOCHERTY, J. R. 1986. Prejunctional inhibitory actions of alpha$_1$-adrenoreceptor agonists in the rat. Br. J. Pharmacol. 89: 842P.

22. WARNOCK, P. & J. R. DOCHERTY. 1986. Further investigation of the sites of alpha$_1$ and alpha$_2$ adrenoceptors in the anaesthetized spontaneously hypertensive rat. J. Cardiovasc. Pharmacol. 8: 67–70.

23. NAHORSKI, S. R. & D. B. BARNETT. 1982. Biochemical assessment of adrenoceptor function and regulation: new directions and clinical relevance. Clin. Sci. 63: 97–105.

24. DOCHERTY, J. R. & J. C. MCGRATH. 1980. A comparison of pre- and postjunctional potencies of several alpha-adrenoceptor agonists in the cardiovascular system and anococcygeus muscle of the rat. Naunyn Schmiedebergs Arch. Pharmacol. 312: 107–116.

25. ADLER-GRASCHINSKY, E. & S. Z. LANGER. 1975. Possible role of a beta-adrenoceptor in the regulation of noradrenaline release by nerve stimulation through a positive feedback mechanism. Br. J. Pharmacol. 53: 43–50.

26. LANGER, S. Z., M. A. ENERO, E. ADLER-GRASCHINSKY, M. L. DUBOCOVICH & S. M. CELUCH. 1975. Presynaptic regulatory mechanisms for noradrenaline release by nerve stimulation. *In* Central Action of Drugs in Blood Pressure Regulation. D. S. Davies & J. L. Reid, Eds.: 133–150. Pitman Press. Tunbridge Wells, England.

27. STJARNE, L. & J. BRUNDIN. 1975. Dual adrenoceptor-mediated control of noradrenaline secretion from human vasoconstrictor nerves: facilitation by beta-receptors and inhibition by alpha-receptors. Acta Physiol. Scand. 94: 139–141.

28. STJARNE, L. & J. BRUNDIN. 1976. Beta$_2$-adrenoceptors facilitating noradrenaline secretion from human vasoconstrictor nerves. Acta Physiol. Scand. 97: 88–93.

29. STJARNE, L. & J. BRUNDIN. 1976. Additive stimulating effects of inhibitor of prostaglandin synthesis and of beta-adrenoceptor agonist on sympathetic neuroeffector function in human omental blood vessels. Acta Physiol. Scand. 97: 267–269.

30. WESTFALL, T. C., M. J. PEACH & V. TITTERMARY. 1979. Enhancement of the electrically induced release of norepinephrine from the rat portal vein: mediation by beta$_2$-adrenoceptors. Eur. J. Pharmacol. 58: 67–74.

31. BORKOWSKI, K. R. & P. QUINN. 1984. Facilitation by beta-adrenoreceptors of stimulation-induced vasoconstriction in pithed spontaneously hypertensive rats. J. Auton. Pharmacol. 4: 127–132.

32. BORKOWSKI, K. R. & P. QUINN. 1984. Beta$_2$-adrenoreceptors mediate adrenaline's facilitation of neurogenic vasoconstriction. Eur. J. Pharmacol. 103: 339–342.

33. QUINN, P., K. R. BORKOWSKI & M. G. COLLIS. 1985. Epinephrine enhances neurogenic vasoconstriction in the perfused rat kidney. Hypertension 7: 47–52.

34. BROKOWSKI, K. R., C. Y. KWAN & E. E. DANIEL. 1989. Epinephrine facilitates neurogenic responses in isolated segments of dog mesenteric arteries. J. Cardiovasc. Pharmacol. 13: 760–766.

35. LANDS, A. M., A. ARNOLD, J. P. MCAULIFF, F. P. LUDUENA & T. G. BROWN. 1967. Differentiation of receptor systems activated by sympathomimetic amines. Nature 214: 597–598.

36. DAHLOF, C., B. LJUNG & B. ABLAD. 1978. Increased noradrenaline release in rat portal vein during sympathetic nerve stimulation due to activation of presynaptic beta-adrenoceptors by noradrenaline and adrenaline. Eur. J. Pharmacol. 50: 75–78.

37. MAJEWSKI, H., M. W. MCCULLOCH, M. J. RAND & D. F. STORY. 1980. Adrenaline activation of prejunctional beta-adrenoceptors in guinea-pig atria. Br. J. Pharmacol. 71: 435–444.

38. MAJEWSKI, H., M. J. RAND & L-H. TUNG. 1981. Activation of prejunctional beta-adrenoceptors in rat atria by adrenaline applied exogenously or released as a cotransmitter. Br. J. Pharmacol. **73:** 669–679.

39. MAJEWSKI, H., L-H. TUNG & M. J. RAND. 1981. Adrenaline-induced hypertension in rats. J. Cardiovasc. Pharmacol. **3:** 179–185.

40. MAJEWSKI, H., L-H. TUNG & M. J. RAND. 1982. Adrenaline activation of prejunctional beta-adrenoceptors and hypertension. J. Cardiovasc. Pharmacol. **4:** 99–106.

41. MISU, Y., M. KUWAHARA, H. AMANO & T. KUBO. 1989. Evidence for tonic activation of prejunctional beta-adrenoceptors in guinea-pig pulmonary arteries by adrenaline derived from the adrenal medulla. Br. J. Pharmacol. **98:** 45–50.

42. DECHAMPLAIN, J., L. FARLEY, D. COUSINEAU & M. R. VAN AMERINGEN. 1974. Circulating catecholamine levels in human and experimental hypertension. Circ. Res. **38:** 109–114.

43. FRANCO-MORSELLI, R., J. L. ELGHOZI, E. JOLY, S. DIGUILIO & P. MEYER. 1977. Increased plasma adrenaline concentrations in benign essential hypertension. Br. Med. J. **2:** 1251–1254.

44. BERTEL, O., F. R. BUHLER, W. KIOWSKI & B. E. LUTOLD. 1980. Decreased beta-adrenoreceptor responsiveness as related to age, blood pressure, and plasma catecholamines in patients with essential hypertension. Hypertension **2:** 130–138.

45. VICENT, J. H., F. BOOMSMA, A. J. MAN IN'T VELD & M. A. D. H. SCHALEKAMP.1983. Beta-receptor stimulation by adrenaline elevates plasma noradrenaline and enhances the pressor responses to cold exposure and isometric exercise. J. Hypertens. **1:** 74–76.

46. BROWN, M. J., A. D. STRUTHERS, J. M. BURRIN, L. DISILVIO & D. C. BROWN. 1985. The physiological and pharmacological role of presynaptic alpha- and beta-adrenoceptors in man. Br. J. Clin. Pharmacol. **20:** 649–658.

47. FLORAS, J. S., P. E. AYLWARD, R. G. VICTOR, A. L. MARK & F. M. ABBOUD. 1988. Epinephrine facilitates neurogenic vasoconstriction in humans. J. Clin. Invest. **81:** 1265–1274.

48. MISU, Y. & T. KUBO. 1986. Presynaptic beta-adrenoceptors. Med. Res. Rev. **6:** 197–225.

49. HEIMBURGER, M., M. J. MONTERO, V. FOUGERES, F. BESLOT, M. DAVY, M. MIDOL-MONNET & Y. COHEN. 1989. Presynaptic beta-adrenoceptors in rat atria: evidence for the presence of stereoselective beta$_1$-adrenoceptors. Br. J. Pharmacol. **98:** 211–217.

50. BLAKELEY, A. G. H., T. C. CUNNANE & S. A. PETERSEN. 1982. Local regulation of transmitter release from rodent sympathetic nerve terminals? J. Physiol. **325:** 93–109.

51. KALSNER, S. & M. QUILLAN. 1984. A hypothesis to explain the presynaptic effects of adrenoceptor antagonists. Br. J. Pharmacol. **82:** 515–522.

52. HARMS, H. H. 1976. Cardioselectivity of beta-adrenoreceptor blocking agents: in vitro studies with human and guinea-pig preparations. *In* Beta-Adrenoreceptor Blocking Agents. P. R. Saxena & R. P. Forsyth, Eds.: 311–315. Elsevier/North Holland Biomedical Press. Amsterdam, The Netherlands.

53. BORKOWSKI, K. R. & P. QUINN. 1983. The effect of bilateral adrenal demedullation on vascular reactivity and blood pressure in spontaneously hypertensive rats. Br. J. Pharmacol. **80:** 429–437.

54. BORKOWSKI, K. R. & P. QUINN. 1985. Adrenaline and the development of spontaneous hypertension in rats. J. Auton. Pharmacol. **5:** 89–100.

55. BILSKI, A. J., S. E. HALLIDAY, D. J. FITZGERALD & J. L. WALE. 1983. The pharmaology of a beta$_2$-selective adrenoceptor antagonist (ICI 118,551). J. Cardiovasc. Pharmacol. **5:** 430–437.

56. FINCH, L. & G. D. H. LEACH. 1970. Does the adrenal medulla contribute to the maintenance of experimental hypertension? Eur. J. Pharmacol. **11:** 388–391.

57. BORKOWSKI, K. R. & E. KELLY. 1986. The effect of adrenal demedullation on cardiovascular responses to environmental stimulation in conscious rats. Br. J. Pharmacol. **88:** 943–945.

58. KAWASAKI, H., W. H. CLINE & C. SU. 1982. Enhanced presynaptic beta adrenoceptor-mediated modulation of vascular adrenergic neurotransmission in spontaneously hypertensive rats. J. Pharmacol. Exp. Ther. **223:** 721–728.

59. EKAS, R. D., M. L. STEENBERG, M. S. WOODS & M. F. LOKHANDWALA. 1983. Presynaptic

alpha- and beta-adrenoceptor stimulation and norepinephrine release in the spontaneously hypertensive rat. Hypertension **5:** 198–204.

60. KUBO, T., M. KUWAHARA & Y. MISU. 1984. Effect of isoproterenol on vascular adrenergic neurotransmission in prehypertensive and hypertensive spontaneously hypertensive rats. Jpn J. Pharmacol. **36:** 419–421.

61. WESTFALL, T. C. & M. J. MELDRUM. 1985. Alterations in the release of norepinephrine at the vascular neuroeffector junction in hypertension. Annu. Rev. Pharmacol. Toxicol. **25:** 621–641.

62. NEZU, M., Y. MIURA, M. ADACHI, M. ADACHI, S. KIMURA, S. TORIYABE, Y. ISHIZUKA, H. OHASHI, T. SUGAWARA, M. TAKAHASHI, T. NOSHIRO & K. YOSHINAGA. 1983. The role of epinephrine in essential hypetension. Jpn Circ. J. **47:** 1242–1246.

63. NEZU, M., Y. MIURA, M. ADACHI, M. ADACHI, S. KIMURA, S. TORIYABE, Y. ISHIZUKA, H. OHASHI, T. SUGAWARA, M. TAKAHASHI, T. NOSHIRO & K. YOSHINAGA. 1985. The effects of epinephrine on norepinephrine release in essential hypertension. Hypertension **7:** 187–195.

64. ROKKEDAL NIELSEN, J., L. F. GRAM & P. K. PEDERSEN. 1988. Plasma noradrenaline levels in young subjects at increased risk of developing essential hypertension. Response to a multistage exercise test. Pharmacol. Toxicol. **63**(Suppl. 1): 32–34.

65. DAHLOF, C. 1981. Studies on beta-adrenoceptor mediated facilitation of sympathetic neurotransmission. Acta Physiol. Scand. **500**(Suppl.): 1–147.

66. COLLIS, M. G., C. DE MEY & P. M. VANHOUTTE. 1979. Enhanced release of noradrenaline in the kidney of the young spontaneously hypertensive rat. Clin. Sci. **57**(Suppl. 5): 233–234.

67. COLLIS, M. G., C. DE MEY & P. M. VANHOUTTE. 1980. Renal vascular reactivity in the young spontaneously hypertensive rat. Hypertension **2:** 45–52.

68. COLLIS, M. G. & P. M. VANHOUTTE. 1977. Vascular reactivity of isolated perfused kidneys from male and female spontaneously hypertensive rats. Circ. Res. **41:** 759–767.

69. VANHOUTTE, P. M., D. BROWNING, E. COEN, T. J. VERBEUREN, L. ZONNEKEYN & M. G. COLLIS. 1982. Decreased release of norepinephrine in the isolated kidney of the adult spontaneously hypertensive rat. Hypertension **4:** 251–256.

70. GALLOWAY, M. P. & T. C. WESTFALL. 1982. The release of endogenous norepinephrine from the coccygeal artery of spontaneously hypertensive and Wistar-Kyoto rats. Circs. Res. **51:** 225–232.

71. GRANATA, A. R., M. A. ENERO, E. M. KRIEGER & S. Z. LANGER. 1983. Norepinephrine release and vascular response elicited by nerve stimulation in rats with chronic neurogenic hypertension. J. Pharmacol. Exp. Ther. **227:** 187–193.

72. GOLDSTEIN, D. S. 1983. Plasma catecholamines and essential hypertension: an analytical review. Hypertension **5:** 86–99.

73. GROBECKER, H., M. F. ROIZEN, V. WEISE, J. M. SAAVEDRA & I. J. KOPIN. 1975. Sympathoadrenal medullary activity in young, spontaneously hypertensive rats. Nature **258:** 267.

74. FLORAS, J. S. & B. L. MORRIS. 1989. Blood pressure and sympathetic nerve activity fall after exercise in borderline hypertensive but not normotensive men: direct evidence for increased sympathetic outflow in borderline hypertension. Clin. Invest. Med. **12**(Suppl.): R315.

The Physiological Operation of Presynaptic Inhibitory Autoreceptors[a]

THOMAS C. WESTFALL

*Department of Pharmacology
St. Louis University School of Medicine
1402 South Grand Boulevard
St. Louis, Missouri 63104*

INTRODUCTION AND DEFINITIONS

There is now overwhelming evidence for the existence of receptors located on the soma, dendrites, or axons of neurons which respond to transmitters or modulators released from adjacent neurons or cells or transmitters released from the same neuron. Somadendritic receptors are those receptors located on or near the cell body and dendrites and when activated primarily modify the function of the somadendritic region (e.g., protein synthesis; generation of action potentials). Presynaptic receptors, by convention, have been used to describe receptors presumed to be located on, in, or near the axon terminals or varicosities and when activated modify the function of the terminal region (transmitter synthesis and release). Two types of presynaptic receptors have been defined. Heteroreceptors are presynaptic receptors that respond to transmitters or modulators released from adjacent neurons or cells. Presynaptic autoreceptors on the other hand are receptors located on or close to those axon terminals of a neuron through which the neuron's transmitter can and, under appropriate conditions, may modify transmitter biosynthesis or depolarization-evoked transmitter release.

The basic evidence for presynaptic autoreceptors has been extensively summarized and discussed in numerous reviews.[1–15] Briefly, agonists inhibit or enhance the release of the neurotransmitter, while antagonists counteract the effect of the agonists, and when given alone produce opposite effects of the agonists. The location of these receptors is thought to be presynaptic because the effects are obtained in the absence of cell bodies, can be seen in synaptosomal preparations, and can be seen in neurons grown in cell culture and occasionally by ligand binding studies where binding sites disappear upon denervation.[1–15] Evidence exists for the presence of presynaptic autoreceptors on many types of neurons including noradrenergic, adrenergic, dopaminergic, serotonergic, cholinergic, histaminergic, GABAergic, glutamergic, glycinergic, and enkephalinergic. It is possible that presynaptic autoreceptors may be a basic feature of all neurons and only time will reveal if this is in fact the case. Although there is general agreement that presynaptic receptors play a pharmacological role in influencing transmitter synthesis and/or release, there is a lack of agreement as to whether or not presynaptic autoreceptors play a physiological role in regulating the synthesis and/or release of neurotransmitters.[16] This has led to open discussion and debate in the literature of the arguments for and against the hypothesis (for example see References 17–22). The purpose of this essay is to summarize data consistent with the physiological relevance of autoreceptor inhibitory modulation of neurotransmitter release. The

[a]Research carried out in the author's laboratory has been supported by U.S. Public Health Service grants HL 26319, HL 35202, and NIDA 02668.

reader is referred to numerous recent reviews as well as articles in this volume that also discuss evidence for and against a physiological role of presynaptic autoreceptors.[1-15]

Evidence for the physiological relevance of presynaptic inhibitory autoreceptors is supported by three general lines of evidence as discussed in great detail in two recent reviews:[12,14] *firstly, in vitro* studies demonstrating that the effect of agonists and antagonists varies with the biophase concentration of the neurotransmitter; *secondly,* studies in which the postsynaptic response is influenced by presynaptic modulation of transmitter release; and *thirdly,* studies demonstrating the existence of presynaptic autoreceptors using *in vivo* techniques. Most of the evidence presented in this paper will utilize presynaptic α_2 inhibitory autoreceptors on noradrenergic neurons as the model since this system has been the most thoroughly studied. However, similar arguments and data exist for numerous other neurotransmitters.

IN VITRO STUDIES DEMONSTRATING THAT THE EFFECT OF AGONISTS AND ANTAGONISTS VARIES WITH THE BIOPHASE CONCENTRATION OF TRANSMITTER

Studies with Antagonists In Vitro

Both proponents and opponents of the presynaptic inhibitory autoreceptor hypothesis suggest that an important argument for an interaction of both agonists and antagonists with endogenous transmitter is a dependence of their effect on the transmitter concentration in the biophase.[12,14,16] Accordingly the agonist effect should

TABLE 1. Studies with Antagonists *In Vitro*

---

1. α_2 antagonists fail to increase the release of norepinephrine (NE) elicited by a single pulse
2. The release-enhancing effect of α_2 antagonists increases with increasing frequency of stimulation
3. The release-enhancing effect of α_2 antagonists increases in the presence of uptake blockers
4. The release-enhancing effect of α_2 antagonists was reduced or abolished when NE was depleted

---

decrease and the antagonist effect increase with an increase in the biophase concentration of the endogenous neurotransmitter. TABLE 1 summarizes four lines of evidence utilizing α_2-adrenoceptor antagonists *in vitro* that are consistent with this general premise (TABLE 1).

Investigators have utilized numerous ways of altering the biophase concentration of the neurotransmitter including changes in the electrical parameters (nerve frequency),[23,24] train length or pulse series,[24,25] current strength,[26,27] addition of compounds that retard removal of the transmitter,[28,29] alterations in the perfusion rate[30] or depletion of the transmitter.[31-36]

Evidence for the existence of presynaptic autoreceptors has been obtained by all of these means of varying the biophase concentration of norepinephrine. As mentioned above there should be no or little enhancement of transmitter release by antagonists when the biophase concentration of transmitter is zero or very low. Such a condition may exist when an experimental preparation is stimulated with a single pulse or very few pulses. There are numerous published studies that report that antagonists do not enhance transmitter release from peripheral[24,25,29,37,38] or central tissue[39–43] when stimulated with a single pulse or very low frequencies of nerve stimulation.

In those studies in which α-adrenoceptor antagonists such as phenoxybenzamine were observed to increase the stimulation induced overflow of transmitter,[47,48] it is likely that the concentration of antagonists was too high and produced nonspecific increases in the concentration of transmitter in the biophase. In other studies it is also possible that the biophase concentration of transmitters was also higher than thought since the transmitter value is in fact unknown and can be increased by multiple factors. As stated above, the presynaptic inhibitory autoreceptor hypothesis suggests that the release-enhancing effect of α_2-adrenoceptor antagonists should be greater with increasing frequencies of nerve stimulation. Data supporting this point are equivocal. There are many studies demonstrating that such antagonists do result in greater release of transmitter with increasing frequencies or stimulus trains of different lengths.[24–26,32,37,39,40,49–52] In TABLE 2, which has been adapted from experiments carried out by Kahan,[52] it can be seen that the phenoxybenzamine-induced increase in ^{3}H-norepinephrine or endogenous norepinephrine overflow from the skeletal muscle vasculature *in situ* increased with increasing frequencies of stimulation.

In contrast, there are many other studies reporting that there is an inverse or constant release of transmitter with increasing frequency of nerve stimulation.[12,14,47,53,54] It may be argued that this is inconsistent with the presynaptic inhibitory receptor hypothesis. However, there are several explanations for the lack of α_2-adrenoceptor antagonists producing an increase in the release of norepinephrine with increasing frequency. First, it is possible that the endogenous neurotransmitter, norepinephrine, may surmount the effect of the antagonist, especially in the case of competitive antagonists. Second, the noradrenergic neuroeffector junction is far more complex than previously appreciated. There is now convincing evidence for multiple neurotransmitters. FIGURE 1 depicts a summary of the situation at the vascular noradrenergic neuroeffector junction. At this junction there is evidence for the existence of multiple neurotransmitters or neuromediators including ATP, neuropeptide Y, as well as norepinephrine. All three mediators have been shown to be released following appropriate stimulation parameters and all three have both prejunctional inhibitory effects on transmitter release as well as postjunctional effects.[55–58] All three mediators exert the biological effects by acting on their own discrete receptors: NPY on Y_1 (postsynaptic) or Y_2 (presynaptic) receptors, ATP on P_{2x} (postsynaptic) and P_3 (presynaptic) receptors and norepinephrine, of course, on α_1- or α_2-adrenoceptors. An explanation for the lack of α_2 antagonists enhancing norepinephrine release at increasing frequencies of stimulation is that at higher frequencies there is release of NPY and ATP which, by activating Y_2 and P_3 receptors, inhibit norepinephrine release and therefore nullify the enhancing effect of α_2-adrenoceptor antagonists.

A final explanation has been advanced by Starke,[14] namely, that increasing frequencies of nerve stimulation results in an increase in the intra axoplasmic Ca^{2+}

TABLE 2. Phenoxybenzamine-Induced Increase in Norepinephrine Overflow (Skeletal Muscle Vasculature *In Situ*)[a]

	Endogenous NE		^{3}H-NE	
Frequency (Hz)	Control	PBZ	Control	PBZ
1	1.5 ± 4	10.4 ± 2.4	3.1 ± 0.5	21.8 ± 0.4
2	2.4 ± 0.5	27.2 ± 0.8	4.5 ± 1.3	35.9 ± 1.3
4	3.1 ± 0.5	32.7 ± 8	5.9 ± 1.4	39.6 ± 1.2

[a]Adopted from Table I in Reference 52. PBZ is phenoxybenzamine.

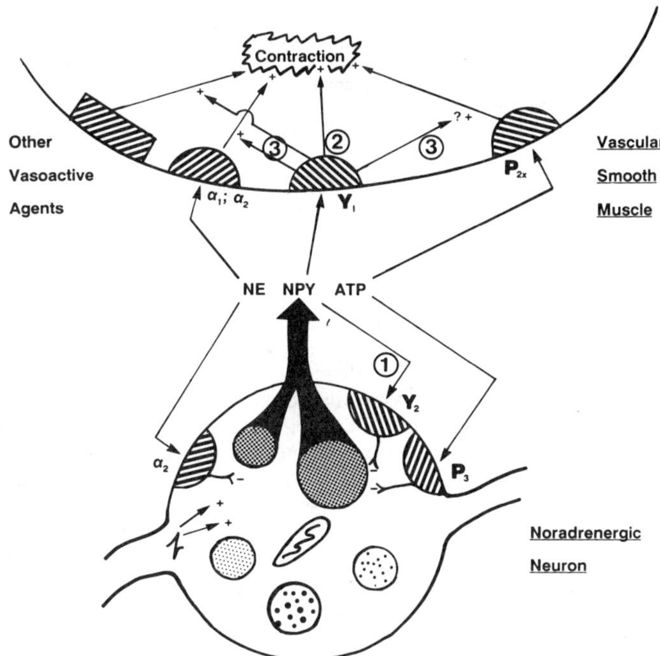

FIGURE 1. A summary of a noradrenergic varicosity and effector cell.

concentrations. As proposed by Starke,[14] a change in Ca^{2+} rather than a change in transmitter concentration may modify the sensitivity of autoreceptor agonists and antagonists. According to this argument, the effect of autoreceptor agonists on transmitter release should decline because axoplasmic Ca^{2+} levels become too high and lead to release that is no longer modulated via presynaptic autoreceptors. If this situation does exist, then the effect of autoreceptor antagonists on neurotransmitter release would decrease with increasing frequency of nerve stimulation.

Another way of increasing the biophase concentration of norepinephrine is to interfere with neuronal uptake of the amine (uptake 1). Numerous investigators have demonstrated that the release-enhancing effect of α_2-adrenoceptor antagonists increases in the presence of norepinephrine uptake blockers;[28–30,40,51,59,60] an effect consistent with the biophase argument. Finally, several investigators have utilized measurements of dopamine β hydroxylase as an index of transmitter release after depletion of norepinephrine stores.[31–33] It is reasoned that following depletion of norepinephrine stores, there would be little amine released upon nerve stimulation while one would still be able to release dopamine β hydroxylase, and it would now become an index of neurotransmission. Under such conditions the release-enhancing effect of α_2-antagonists on dopamine β hydroxylase was reduced or abolished; presumably because there was little norepinephrine available to activate presynaptic receptors. The release-enhancing effect of α-adrenoceptor antagonists was also reduced when the vesicular marker ³H-amezinime[36] or the cotransmitter NPY[35] was measured as a marker of transmitter release. The biophase concentration has also been changed by altering the rate of perfusion of the perfused rat heart.[30] When the rate was reduced from 7 to 1.8 ml/minute and presumably increasing the biophase concentration of NE, the α_2-

TABLE 3. Studies with Agonists *In Vitro*

1. α_2 agonists are most effective when there is little NE in the vicinity of the autoreceptor
2. α_2 agonist inhibition declines with increasing frequency of nerve stimulation
3. α_2 agonist inhibition declines in the presence of NE uptake blockers
4. α_2 agonist inhibition is enhanced when the endogenous transmitter is depleted

adrenoceptor antagonist yohimbine more than doubled the evoked overflow of ^{3}H-norepinephrine.

Studies with Agonists In Vitro

TABLE 3 summarizes four lines of evidence utilizing α-agonists *in vitro*. In contrast to the situation with antagonists and as would be expected, α-adrenoceptor agonists are most effective when there is little norepinephrine in the vicinity of the autoreceptor. Moreover the inhibitory effect of α-adrenoceptor agonists declines with increasing frequency of nerve stimulation, an effect observed in numerous peripheral[61,62] or central preparations.[63] A representative experiment is depicted in FIGURE 2 which shows the results of experiments carried out utilizing the guinea pig vas deferens.[1] Concentrations of three agonists including clonidine, acting on α_2-adrenoceptor, methacholine, acting on muscarinic receptors, and prostaglandin E_2, acting on prostaglandic receptors, were chosen that produced a 50% reduction in the field stimulation induced release of norepinephrine from the vas deferens when stimulated at a frequency of 2 Hz. The inhibitory effect was reduced as the frequency of nerve stimulation was increased.

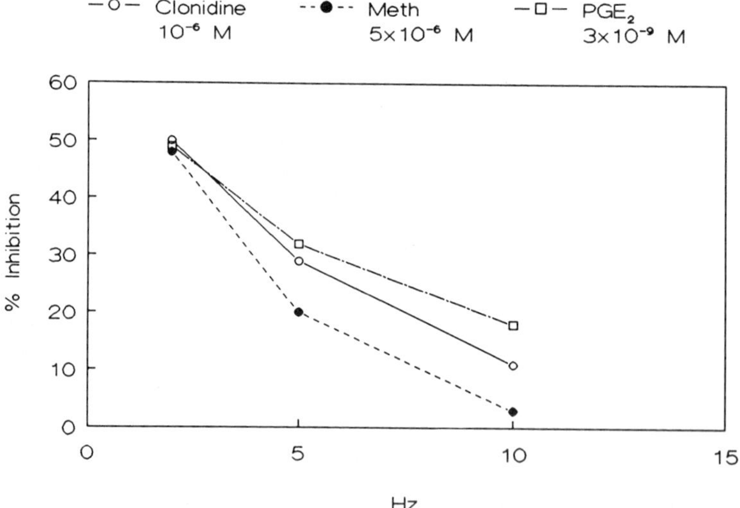

FIGURE 2. Effect of clonidine, methacholine, and prostaglandin E_2 (PGE_2) on inhibition of electrically induced release of ^{3}H-norepinephrine from superfused guinea pig vas deferens at different frequencies of stimulation (Hz). Data are plotted as percent inhibition produced by 10^{-6} M clonidine, 5×10^{-6} M methacholine or 3×10^{-9} M PGE_2 at 2, 5, or 10 Hz.

Additional experiments utilizing adrenoceptor agonists reveal that if the biophase concentration of norepinephrine is increased by the addition of neuronal uptake blockers there is a decline in the release inhibitory effect of α-adrenoceptor agonists.[27,30] Finally, the inhibitory effect of α-adrenoceptor agonist on transmitter release is enhanced when norepinephrine is depleted,[31–33] an effect opposite what was seen with α-adrenoceptor antagonists.

STUDIES IN WHICH THE POSTSYNAPTIC RESPONSE IS INFLUENCED BY PRESYNAPTIC MODULATION OF TRANSMITTER RELEASE

Evidence for the physiological role of presynaptic inhibitory autoreceptors requires the demonstration that α-adrenoceptor agonists decrease and α-adrenoceptor antagonists increase the response of the effector cell to nerve stimulation. There are, in fact,

TABLE 4. α_2-Adrenoceptor Antagonist Induced Enhancement of the Effector Cell Response to Noradrenergic Nerve Stimulation

Species	Tissue or Preparation	Reference
Dog	Blood vessels	64
Dog	Mesenteric artery	65
Dog	Heart	66–73
Cat	Heart	74, 75
Rabbit	Pulmonary artery	67, 76, 77, 78
Rabbit	Aorta	79
Rabbit	Ileocolic artery	80
Rabbit	Hindlimb	81
Rabbit	Mesenteric artery	82
Rabbit	Heart	83
Rat	Vasculature	84–88
Rat	Heart	60
Pithed rat	Heart	89–93
Rat	Hindlimb	94
Rat	Vas deferens	95–99
Guinea pig	Saphenous artery	100
Guinea pig	Heart	25, 29, 101–105

numerous examples of such responses. When using antagonists, of course, it is necessary that the effector cell response due to increased release of transmitter is not neutralized by the blocking effect of the antagonists at the postsynaptic receptor. TABLE 4 summarizes the results of representative studies in which it has been demonstrated that α-adrenoceptor antagonists enhance the response of the effector cell to noradrenergic nerve stimulation. If one examines the data from the studies listed in TABLE 4, it can be concluded that an enhancement of the effector cell response following appropriate antagonists has been demonstrated in a wide variety of various tissues and organs and in numerous species including, the dog, cat, rabbit, rat, and guinea pig.

In contrast to the enhancement of the effector cell response to nerve stimulation produced by α-adrenoceptor antagonists, a decrease in the effector cell response has been observed for numerous α-adrenoceptor agonists. A representative example is depicted in FIGURE 3 which demonstrates that clonidine produces a concentration-dependent decrease in the field stimulation induced twitch response of the guinea pig vas deferens. The neurogenic nature of the twitch response is verified by the fact that tetrodotoxin completely prevents the response to nerve stimulation.

In Vivo *Studies Consistent with Presynaptic Autoreceptors*

A final criterion for the physiological role of presynaptic inhibitory autoreceptors is that they be demonstrated *in vivo*. In addition to the large number of studies demonstrating that α-adrenoceptor antagonists enhance and α-adrenoceptor agonists decrease the nerve stimulation induced response of the effector cell, there are also a large number of studies demonstrating that these drugs produce an increase or

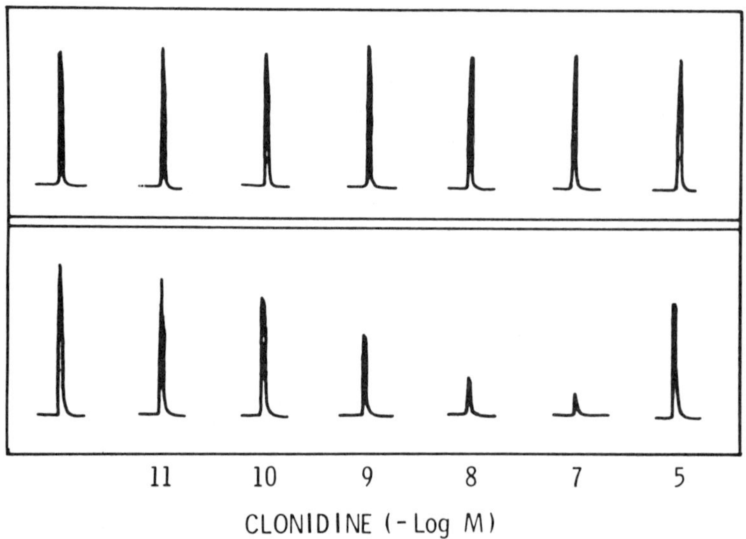

FIGURE 3. Effect of cumulative concentrations of clonidine on the field stimulation (5 Hz, 3 seconds) induced twitch contraction of the guniea pig vas deferens. Field stimulation was applied at 2.5 minute intervals. Clonidine was added immediately after the end of electrical stimulation. (From Sax and Westfall, with permission.)[121]

decrease in the *in vivo* or *in situ* overflow of norepinephrine into the perfusate from perfused organs, into the plasma, or into centrally placed microdialysis cannulae following activation of the respective nerves. TABLE 5 summarizes representative studies carried out in different species or preparations. Appropriate increases or decreases in transmitter release or overflow have been demonstrated in many different species and utilizing a variety of procedures including conscious, freely moving animals. These studies are consistent with the interpretation that the agonists are decreasing and the antagonists increasing the release of transmitter in response to nerve stimulation via their activating or blocking presynaptic autoreceptors and add to the tremendous amount of data consistent with the physiological operation of presynaptic autoreceptors.

TABLE 5. Representative *In Vivo* Studies Consistent with Physiological Role of Presynaptic α_2-Autoreceptors

Species and Preparation	Reference
Coronary overflow of NE to sympathetic nerve stimulation of the dog heart	66, 68, 71–73, 106, 107
Dog hindlimb and gracilis muscle	52, 108, 109
Dog liver	110
Pig spleen	111, 112
Pithed rabbit	113, 114
Conscious rabbit	113, 115, 116
Pithed rat	84, 85
Human spillover	117–120

COMMENTS AND SUMMARY

When evaluating all of the available information it would seem that there is an overwhelming amount of evidence that supports not only a pharmacological but physiological role for presynaptic inhibitory autoreceptors in modulating noradrenergic neurotransmission. There are data, nevertheless, from individual experiments that would not seem to support a physiological role for presynaptic autoreceptors. It is likely that some investigators, including opponents of the theory, have placed unrealistic demands on various experimental protocols, and have viewed the system as if a single presynaptic receptor is the only system modulating release. This is clearly not the case and it needs to be appreciated that neurotransmission and its modulation are a very complex process. There are numerous explanations as to why individual experiments may provide results not consistent with the hypothesis. First, the tools and techniques used in individual experiments are often imperfect. TABLE 6 provides a list of some of these that have contributed to differing results among investigators.

Moreover, as just mentioned, it is clear that neurotransmission is a complex process that involves multiple neurotransmitters each with the ability to act on discrete pre- and postsynaptic receptors, as well as other processes such as facilitation which also influence neurotransmission. The response of the effector cell is also influenced by multiple factors in addition to a specific neurotransmitter. For instance, FIGURE 4 depicts how complicated the situation is at the vascular noradrenergic neuroeffector junction. The effector cell can be influenced by noradrenergic and nonadrenergic, noncholinergic neurons. The noradrenergic neuron can release as many as three

TABLE 6. Factors Complicating Interpretation of Results Concerning Presynaptic Autoreceptors

Experimental conditions (Species, age, temperature, stimulation parameters, perfusion rate)
Drugs—nonspecific effects
Presence of heteroreceptors
Cotransmitters or multiple transmitters
Other factors influencing transmitter release
In vitro studies—may be artificial
In vivo studies—multiple complications

neurotransmitter neuromodulators (norepinephrine; NPY; ATP) each with their own pre- (α_2, Y_2, P_3) and postsynaptic (α_1, α_2, Y_1, P_{2x}) receptors. Moreover, there are heteroreceptors on noradrenergic neurons that may be activated from mediators released from adjacent neurons, as well as nonneuronal cells such as endothelial or smooth muscle cells. The effector cell itself may be influenced by mediators released from noradrenergic neurons as well as from endothelial cells, or nonnoradrenergic neurons. This very great complexity must be taken into consideration when interpreting experiments designed to study presynaptic autoreceptors.

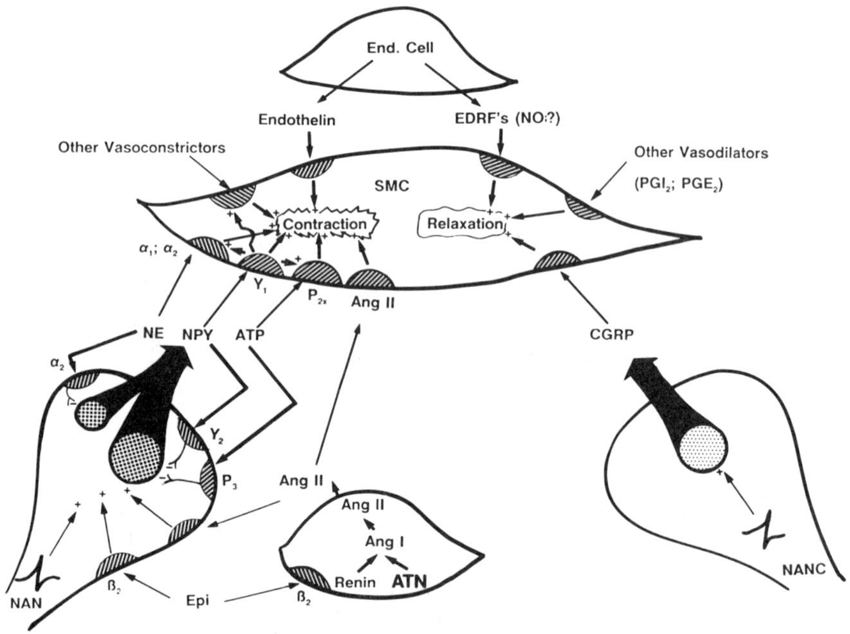

FIGURE 4. A summary of the vascular noradrenergic neuroeffector junction. Abbreviations: NAN, noradrenergic nerve; NANC, nonadrenergic, noncholinergic nerve; CGRP, calcitonin gene related peptide; Ang, angiotensin; ATN, angiotensinogen; NPY, neuropeptide Y; NE, norepinephrine; SMC, smooth muscle cell; End. cell, endothelial cell; EDRF, endothelium derived relaxant factor; NO, nitric oxide; PGI$_2$, prostacyclin; PGE$_2$, prostaglandin E$_2$.

It would appear prudent to examine the evidence for a physiological role of presynaptic receptors in its totality and not to dismiss the concept because results in one or another experiment seem not to support or fit.

It must be admitted that the precise physiological role of presynaptic autoinhibitory receptors is still not clear and much more information is required. It is possible that autoregulation may not exist at all junctions or synapses. It is also not clear if autoregulation takes place at each varicosity, upstream of terminal varicosities, or by lateral regulation (one varicosity to another). Possible physiological functions may include pulse to pulse regulation of transmitter release during nerve stimulation; upstream or lateral regulation of release from various varicosities; modulation of release during periods of rest or during periods in which the frequency of nerve stimulation is low; modulation of release during periods of very intense stimulation; and finally as a physiological antagonist to facilitate transmitter release that is known to take place in some neurons during nerve stimulation.

A challenge for the future will be to unravel which of these are most important at discrete synapses or junctions. Despite a lack of understanding of the precise role it would appear that there is an overwhelming amount of evidence that supports a physiological role of presynaptic autoinhibitory receptors in the process of noradrenergic neurotransmission.

REFERENCES

1. WESTFALL, T. C. 1977. Local regulation of adrenergic neurotransmission. Physiol. Rev. **57:** 659–728.
2. STARKE, K. 1977. Regulation of noradrenaline release by presynaptic receptor systems. Rev. Physiol. Biochem. Pharmacol. **77:** 1–124.
3. VIZI, S. 1979. Presynaptic modulation of neurochemical transmission. Prog. Neurobiol. **12:** 181–292.
4. WESTFALL, T. C. 1980. Neuroeffector mechanisms. Annu. Rev. Physiol. **42:** 383–397.
5. RAND, M. J., M. W. McCULLOCH & D. F. STORY. 1980. Catecholamine receptors on nerve terminals. *In* Handbook of Experimental Pharmacology. Adrenergic Activators and Inhibitors. L. Szekeres, Ed. **54(I):** 223–266. Springer-Verlag. Berlin, FRG.
6. GILLESPIE, J. S. 1980. Presynaptic receptors in the autonomic nervous system. *In* Handbook of Experimental Pharmacology. Adrenergic Activators and Inhibitors. L. Szekeres, Ed. **54:** 353–425. Springer-Verlag. Berlin, FRG.
7. LANGER, S. Z. 1981. Presynaptic regulation of the release of catecholamines. Pharmacol. Rev. **32:** 337–362.
8. STARKE, K. 1981. Presynaptic receptors. Annu. Rev. Pharmacol. Toxicol. **21:** 7–30.
9. CHESSELET, M. F. 1984. Presynaptic regulation of neurotransmitter release in the brain: facts and hypothesis. Neuroscience **12:** 347–375.
10. WESTFALL, T. C., S-Q. ZHANG, S. CARPENTIER, L. NAES & M. J. MELDRUM. 1986. Local modulation of noradrenergic neurotransmission in blood vessels of normotensive and hypertensive animals. *In* Central and Peripheral Mechanisms of Cardiovascular Regulation. A. Margo, A. Osswald, D. Reis & P. Vanhoutte, Eds.: 111–133. Plenum Publishing Corporation. New York, N.Y.
11. TIMMERMANS, P. B. M. W. M. & M. J. M. C. THOOLEN. 1987. Autoreceptors in the central nervous system. Med. Res. Rev. **7:** 307–332.
12. STARKE, K. 1987. Presynaptic α-autoreceptors. Rev. Physiol. Biochem. Pharmacol. **107:** 74–146.
13. LANGER, S. Z. & J. LEHMANN. 1988. Presynaptic receptors on catecholamine neurons. *In* Handbook of Experimental Pharmacology. Catecholamines I. U. Trendelenbury & N. Werner, Eds. **85:(I)** 419–507. Springer-Verlag. Berlin, FRG.
14. STARKE, K., M. GÖTHERT & H. KILBUNGER. 1989. Modulation of neurotransmitter release by presynaptic receptors. Physiol. Rev. **69:** 865–989.
15. WESTFALL, T. C. & J. R. MARTIN. 1990. Presynaptic autoreceptors in the peripheral and central nervous system. *In* Presynaptic Regulation of Neurotransmitter Release. J. J. Feigenbaum & M. M. Hanani, Eds. Freund Publishing Co. Ltd. London, England.
16. KALSNER, S. 1985. Is there feedback regulation of neurotransmitter release by autoreceptors? Biochem. Pharmacol. **34:** 4085–4097.
17. ANGUS, J. A. & P. I. KORNER. 1981. Presence and physiological role of presynaptic inhibitory α-adrenoceptors in guinea pig atria. Nature London **294:** 671–672.
18. LANGER, S. Z. 1981. Presence and physiological role of presynaptic inhibitory α_2 adrenoceptors in guinea pig atria. Nature London **294:** 671–672.
19. KALSNER, S. 1982. The presynaptic receptor controversy. Trends Pharmacol. Sci. **3:** 11–16.

20. RAND, M. J., M. W. McCULLOCH & D. F. STOREY. 1982. Comments on the presynaptic receptor controversy. Trends Pharmacol. Sci. **3:** 16–18.

21. KALSNER, S. 1984. Limitations of presynaptic theory: no support for feedback control of autonomic effectors. Fed. Proc. **43:** 1358–1364.

22. WESTFALL, T. C. 1984. Evidence that noradrenergic transmitter release is regulated by presynaptic receptors. Fed. Proc. **43:** 1352–1357.

23. STARKE, K., H. MONTEL, W. GAYK & R. MERKER. 1974. Comparison of the effects of clonidine on pre and postsynaptic adrenoceptors in the rabbit pulmonary artery. Naunyn Schmeidebergs Arch. Pharmacol. **285:** 133–150.

24. AUCH-SCHWELK, W., K. STARKE & A. STEPPELER. 1983. Experimental conditions required for the enhancement by α-adrenoceptor antagonists of noradrenaline release in the rabbit ear artery. Br. J. Pharmacol. **78:** 543–551.

25. STORY, D. F., M. W. McCULLOCH, M. J. RAND & C. A. STANDFORD-STAN. 1981. Conditions required for the inhibitory feedback loop in noradrenergic transmission. Nature **293:** 62–65.

26. BAUMANN, P. A. & W. P. KOELLA. 1980. Feedback control of noradrenaline release as a function of noradrenaline concentration in the synaptic cleft in cortical slices of the rat. Brain Res. **189:** 437–448.

27. REICHENBACHER, D., W. REIMANN & K. STARKE. 1982. α-Adrenoceptor mediated inhibition of norepinephrine release in rabbit brain cortex slices. Naunyn Schmidebergs Arch. Pharmacol. **319:** 71–77.

28. SCHLICKER, E., M. GOTHERT, F. KOSTERMANN & R. CLAUSING. 1983. Effects of α-adrenoceptor antagonists on the release of serotonin and noradrenaline from rat brain cortex slices. Naunyn Schmiedeberg Arch. Pharmacol. **323:** 106–113.

29. ANGUS, J. A., A. BOBIK, G. P. JACKMAN, I. J. KOPIN & P. I. KORNER. 1984. Role of auto-inhibitory feedback in cardiac sympathetic transmission assessed by simultaneous measurements of changes in ^{3}H-efflux and atrial rate in guinea pig atrium. Br. J. Pharmacol. **81:** 201–214.

30. FUDER, H., F. BATH, H. NIEBELT & E. MUNSCHOTT. 1984. Autoinhibition of noradrenaline release from the rat heart as a function of the biophase concentration. Effects of exogenous α-adrenoceptor agonists, cocaine and perfusion rate. Naunyn Schmidebergs Arch. Pharmacol. **325:** 25–33.

31. ENERO, M. A. & S. Z. LANGER. 1973. Influence of reserpine-induced depletion of noradrenaline on the negative feedback mechanism for transmitter release during nerve stimulation. Br. J. Pharmacol. **49:** 214–225.

32. CUBEDDIN, L. & N. WEINER. 1975. Nerve stimulation mediated overflow of norepinephrine and dopamine β hydroxylase. III. Effects of norepinephrine depletion in the alpha presynaptic regulation of release. J. Pharmacol. Exp. Ther. **192:** 1–14.

33. DIXON, W. R., W. F. MOSIMANN & N. WERNER. 1979. The role of presynaptic feedback mechanisms in regulation of norepinephrine release by nerve stimulation. J. Pharmacol. Exp. Ther. **209:** 196–204.

34. LUNDBERG, J. M., A. RUNDEHILL, A. SOLLEVI & B. HAMBUGER. 1989. Evidence for cotransmitter role of neuropeptide Y in the pig spleen. Br. J. Pharmacol. **96:** 675–687.

35. LUNDBERG, J. M., J. PERNOW, G. FRIED & A. ANGGARD. 1987. Neuropeptide Y and noradrenaline mechanisms in relation to reserpine induced impairment of sympathetic neurotransmission in the cat spleen. Acta Physiol. Scand. **131:** 1–10.

36. STARKE, K. 1983. Amezinum: a novel pattern of effects on the sympathetic nervous system. Trends Pharmacol. Sci. **4:** 269–272.

37. MARSHALL, I. 1983. Stimulation-evoked release of ^{3}H-noradrenaline by 1, 10 or 100 pulses and its modification through presynaptic α_2-adrenoceptors. Br. J. Pharmacol. **78:** 221–231.

38. RAND, M. J., D. F. STORY, G. S. ALLEN, A. B. GLOVER & M. W. McCULLOCH. 1973. Pulse to pulse modulation of noradrenaline release through a prejunctional α-receptor autoinhibitory mechanism. Front. Catecholamine Res.: 579–581.

39. LIMBERGER, N. & K. STARKE. 1984. Further study of prerequisites for the enhancement by α_2-adrenoceptor antagonists of the release of noradrenaline. Naunyn Schmiedebergs Arch. Pharmacol. **325:** 240–246.

40. HEEPE, P. & K. STARKE. 1985. α-Adrenoceptor antagonists and the release of noradrenaline in rabbit cerebral cortex slices: support for the α-autoreceptor hypothesis. Br. J. Pharmacol. **84:** 147–155.

41. MAYER, A., N. LIMBERGER & K. STARKE. 1988. Transmitter release patterns of noradrenergic, dopaminergic and cholinergic axons in rabbit brain slices during short pulse trains and the operation of presynaptic autoreceptors. Naunyn Schmiedebergs Arch. Pharmacol. **338:** 632–643.

42. LIMBERGER, N., E. A. SINGER & K. STARKE. 1988. Only activated but not non-activated presynaptic α_2-autoreceptors interfere with neighboring presynaptic receptor mechanisms. Naunyn Schmiedebergs Arch. Pharmacol. **338:** 62–67.

43. VALENTA, B., H. DROBNY & E. A. SINGER. 1988. Presynaptic autoinhibition of central noradrenaline release *in vitro:* operational characteristics and effects of drugs acting at alpha$_2$ adrenoceptors in the presence of uptake inhibition. J. Pharmacol. Exp. Ther. **245:** 944–949.

44. KALSNER, S. 1979. Single pulse stimulation of guinea pig vas deferens and the presynaptic receptor hypothesis. Br. J. Pharmacol. **66:** 343–349.

45. CHAIN, C. C. & S. KALSNER. 1979. An examination of the negative feedback function of presynaptic adrenoceptors in vascular tissue. Br. J. Pharmacol. **67:** 401–407.

46. KALSNER, S. 1981. The role of calcium in the effects of noradrenaline and phenoxybenzamine on adrenergic transmitter release from atria: no support of negative feedback of release. Br. J. Pharmacol. **73:** 363–371.

47. KALSNER, S. 1985. Is there a feedback regulation of neurotransmitter release by autoreceptors? Biochem. Pharmacol. **34:** 4085–4097.

48. WAKADE, A. R. & T. D. WAKADE. 1981. Release of noradrenaline by one pulse: modulation of such release by alpha-adrenoceptor antagonists and uptake blockers. Naunyn Schmidebergs Arch. Pharmacol. **317:** 302–309.

49. MCCULLOCH, M. W., M. PAPANICOLAOU & M. J. RAND. 1985. Evidence for autoinhibition of stimulation-induced noradrenaline release from vas deferentia of the guinea pig and rat. Br. J. Pharmacol. **86:** 455–464.

50. HIEBLE, J. P., R. M. DEMARINIS, P. J. FOWLER & W. O. MATTHEWS. 1986. Selective alpha$_2$-adrenoceptor blockade by SKF 86466: *in vitro* characterization of receptor selectivity. J. Pharmacol. Exp. Ther. **236:** 90–96.

51. NEDERGAARD, D. A. 1986. Presynaptic α-adrenoceptor control of transmitter release from vascular sympathetic neurons *in vitro. In* New Aspects of the Role of Adrenoceptors in the Cardiovascular System. H. Grobecker, A. Philippu & K. Starke, Eds.: 24–32. Springer-Verlag. Berlin, FRG.

52. KAHAN, T. 1987. Prejunctional adrenergic receptors and sympathetic neurotransmission: studies in canine skeletal muscle vasculature *in situ.* Acta Physiol. Scand. **129**(Suppl): 560.

53. LANGER, S. Z., M. L. DUBOCOVICH & S. M. CELUCH. 1975. Prejunctional regulatory mechanisms for noradrenaline release elicited by nerve stimulation. *In* Regulation of Catecholamine Turnover. Chemical Tools in Catecholamine Research. O. Almgen, A. Carlsson & J. Engel, Eds. **11:** 183–191. Elsevier-North Holland. Amsterdam, the Netherlands.

54. ALBERTS, P., T. BARTFAI & L. STARNE. 1981. Sites and ionic basic of α autoinhibition and facilitation of ^{3}H-noradrenaline secretion in guinea pig vas deferens. J. Physiol. London **312:** 297–334.

55. SNEDDON, P. & D. P. WESTFALL. 1984. Pharmacological evidence that adenosine triphosphate and noradrenaline are co-transmitters in the guinea-pig vas deferens. J. Physiol. London **347:** 561–580.

57. WESTFALL, T. C., S. CARPENTIER, X. CHEN, M. C. BEINFELD, L. NAES & M. J. MELDRUM. 1987. Prejunctional and postjunctional effects of neuropeptide Y at the noradrenergic neuroeffector junction of the perfused mesenteric arterial bed of the rat. J. Cardiovasc. Pharmacol. **10:** 716–722.

58. WESTFALL, T. C., J. R. MARTIN, X. CHEN, A. CIARLEGLIO, S. CARPENTIER, K. HENDERSON, M. M. KNUEPFER, M. C. BEINFELD & L. NAES. 1988. Cardiovascular effects and

modulation of noradrenergic neurotransmission following central and peripheral administration of neuropeptide Y. Synapse **2:** 299–307.

59. BAKER, D. J., G. M. DREW & A. HILDITCH. 1984. Presynaptic α-adrenoceptors: Do exogenous and neuronally released noradrenaline act at different sites? Br. J. Pharmacol. **81:** 457–464.

60. ENERO, M. A. 1984. Influence of neuronal uptake on the presynaptic α-adrenergic modulation of noradrenaline release. Naunyn Schmiedebergs Arch. Pharmacol. **328:** 38–40.

61. STARKE, K. 1972. Alpha sympathomimetic inhibition of adrenergic and cholinergic transmission in the rabbit heart. Naunyn Schmidebergs Arch. Pharmacol. **274:** 18–45.

62. MEDGETT, I. C., M. W. McCULLOCH & M. J. RAND. 1978. Partial agonist atria of clonidine on prejunctional and postjunctional α-adrenoceptor. Naunyn Schmiedebergs Arch. Pharmacol. **304:** 215–221.

63. CICHINI, G., H. LASSMANN, P. PLACHETA & E. A. SINGER. 1986. Effects of clonidine on the stimulation-evoked release of ^{3}H-noradrenaline from superfused rat brain slices as a function of the biophase concentration. Naunyn Schmiedebergs Arch. Pharmacol. **333:** 36–42.

64. TODA, N., T. OKAMURA, M. NAKAJIMA & M. MIYAZAKI. 1984. Modification by yohimbine and prazosin of the mechanical response of isolated dog mesenteric, renal and coronary arteries to transmural stimulation and norepinephrine. Eur. J. Pharmacol. **98:** 69–78.

65. MURAMATSU, I., T. OHMURA & M. OSHITA. 1989. Comparison between sympathetic adrenergic and aminergic transmission in the dog mesenteric artery. J. Physiol. **411:** 227–243.

66. KANDA, A., H. KOYANAGAWA, R. YORIKANE, T. KIMURA & S. SATOH. 1987. Enhancement by yohimbine of nicotine and DMPP induced release of norepinephrine from cardiac sympathetic nerves of the dog. J. Pharmacol. Exp. Ther. **243:** 1095–1100.

67. CONSTANTINE, J. W., R. A. WEEKS & W. K. McSHANE. 1978. Prazosin and presynaptic α-receptors in the cardioaccelerator nerve of the dog. Eur. J. Pharmacol. **50:** 51–60.

68. CAVERO, I., T. DENNIS, F. LEFEVRE-BORG, P. PERROT, A. G. ROACH & B. SCATTON. 1979. Effects of clonidine, prazosin and phentolamine on heart rate and coronary sinus catecholamine concentration during cardioaccelerator nerve stimulation in spinal dogs. Br. J. Pharmacol. **67:** 283–292.

69. SHEPPERSON, N. B., N. DUVAL, R. MASSINGHAM & S. Z. LANGER. 1981. Pre- and postsynaptic alpha adrenoceptor selectivity studies with yohimbine and its two diastereoisomers rauwolscine and corynanthine in the anesthetized dog. J. Pharmacol. Exp. Ther. **219:** 540–546.

70. DABIRE, H., P. MOUILLE, M. ANDREJAK, B. FOURNIER & H. SCHMITT. 1981. Pre- and postsynaptic α-adrenoceptor blockade by (imidazolinyl-2)-2-benzodioxane 1-4 (170 150): antagonistic action on the central effects of clonidine. Arch. Int. Pharmacodyn. Ther. **254:** 252–270.

71. HEYNDRICKX, G. R., J. P. VILAINE, E. J. MOERMAN & I. LEUSEN. 1984. Role of prejunctional α_2-adrenoceptors. J. Cardiovasc. Pharmacol. **7:** 167–173.

72. SAEED, M., J. HOLTZ, D. ELSNER & E. BASSENGE. 1985. Sympathetic control of myocardial oxygen balance in dogs mediated by activation of coronary vascular α_2-adrenoceptors. J. Cardiovasc. Pharmacol. **7:** 167–173.

73. YORIKANE, R., A. KANDA, T. KIMURA & S. SATOH. 1986. Effects of epinephrine, isoproterenol and IPS-3390 on sympathetic transmission to the dog heart: evidence against the facilitatory role of presynaptic beta adrenoceptors. J. Pharmacol. Exp. Ther. **238:** 334–340.

74. ILHAN, M., J. P. LONG & J. G. CANNON. 1976. The ability of pimozide to prevent inhibition by dopamine analogs of cardioaccelator nerves in cat hearts. Arch. Int Pharmacodyn. Ther. **222:** 70–80.

75. ABDEL RAHMAN, A. R. A. & F. M. SHARABI. 1981. Presynaptic alpha receptors in relation to the cardiovascular effect of yohimbine in the anesthetized cat. Arch. Int. Pharmacodyn. Ther. **252:** 229–240.

76. STARKE, K., T. ENDO & H. D. TAUBE. 1975. Relative pre- and postsynaptic potencies of

  α-adrenoceptor agonists in the rabbit pulmonary artery. Naunyn Schmiedebergs Arch. Pharmacol. **291:** 55–78.

77. STARKE, K., E. BOROWSKI & T. ENDO. 1975. Preferential blockade of presynaptic α-adrenoceptors by yohimbine. Eur. J. Pharmacol. **34:** 385–388.

78. LATTIMER, N., R. P. MCADAMS, K. F. RHODES, S. SHARMA, S. J. TURNER AND J. N. WATERFALL. 1984. Alpha$_2$-adrenoceptor antagonism and other pharmacological antagonist properties of some substituted benzyquinolizines and yohimbine *in vitro*. Naunyn Schmiedebergs Arch. Pharmacol. **327:** 312–318.

79. DOCHERTY, J. R. & K. STARKE. 1982. An examination of the pre- and postsynaptic α-adrenoceptors involved in neuroeffector transmission in rabbit aorta and portal vein. Br. J. Pharmacol. **76:** 327–335.

80. VON KÜGELGEN, I. & K. STARKE. 1985. Noradrenaline and adenosine triphosphate as co-transmitters of neurogenic vasoconstriction in rabbit mesenteric artery. J. Physiol. London **367:** 435–455.

81. MADJAR, H., J. R. DOCHERTY & K. STARKE. 1980. An examination of pre- and postsynaptic α-adrenoceptors in the autoperfused rabbit hindlimb. J. Cardiovasc. Pharmacol. **2:** 619–627.

82. RAMME, D., J. T. REGENOLD, K. STARKE, R. BUSSE & P. ILLES. 1987. Identification of the neuroeffector transmitter in jejunal branches of the rabbit mesenteric artery. Naunyn Schmiedebergs Arch. Pharmacol. **336:** 267–273.

83. CARR, S. R. & J. R. FOZARD. 1981. Lack of modulation by presynaptic α_2-adrenoceptors of adrenergic transmitter release evoked by activation of 5-hydroxytryptamine and nicotine receptors. Eur. J. Pharmacol. **72:** 27–34.

84. YAMAGUCHI, I. & I. J. KOPIN. 1980. Differential inhibition of alpha-1 and alpha-2 adrenoceptor-mediated pressor responses in pithed rats. J. Pharmacol. Exp. Ther. **214:** 275–281.

85. ZUKOWSKA-GROJEC, Z., M. A. BAYORH & I. J. KOPIN. 1983. Effect of desipramine on the effects of α-adrenoceptor inhibitors on pressor responses and release of norepinephrine into plasma of pithed rats. J. Cardiovasc. Pharmacol. **5:** 297–301.

86. EIKENBURG, D. C. 1984. Functional characterization of the pre- and postjunctional α-adrenoceptors in the *in situ* perfused rat mesenteric vascular bed. Eur. J. Pharmacol. **105:** 161–165.

87. NILLSON, H., B. LJUNG. N. SJÖBLOM & B. G. WALLIN. 1985. The influence of the sympathetic impulse pattern on contractile responses of rat mesenteric arteries and veins. Acta Physiol. Scand. **123:** 303–309.

88. NILSSON, H., N. SJÖBLOM & B. FOLKOW. 1985. Interaction between prejunctional α_2-receptors and neuronal transmitter reuptake in small mesenteric arteries from the rat. Acta Physiol. Scand. **125:** 245–252.

89. ALGATE, D. R. & J. F. WATERFALL. 1978. Action of indoramin on pre- and postsynaptic α-adrenoceptors in pithed rats. J. Pharm. Pharmacol. **30:**651–652.

90. CAVERO, I., R. GOMENI, F. LEFEVRE-BORG & A. G. ROACH. 1980. Comparison of mianserin with desipramine, maprotiline and phentolamine on cardiac presynaptic and vascular postsynaptic α-adrenoceptors and noradrenaline reuptake in pithed normotensive rats. Br. J. Pharmacol. **68:** 321–332.

91. DREW, G. M. 1980. Presynaptic modulation of heart rate responses to cardiac nerve stimulation in pithed rats. J. Cardiovasc. Pharmacol. **2:** 843–856.

92. DEMICHEL, P., P. GOMOND & J. ROQUEBERT. 1982. α-Adrenoceptor blocking properties of raubasine in pithed rats. Br. J. Pharmacol. **77:** 449–454.

93. DEJONGE, A., P. N. SANTING, P. B. M. W. M. TIMMERMANS & P. A. VAN ZWIETEN. 1983. Effect of age on the prejunctional α-adrenoceptor-mediated feedback in the heart of spontaneously hypertensive rate and normotensive Wistar Kyoto rats. Naunyn Schmiedebergs Arch. Pharmacol. **323:** 33–36.

94. MEDGETT, I. C. & R. R. RUFFALO, JR. 1987. Characterization of α-adrenoceptors mediating sympathetic vasoconstriction in rat autoperfused hindlimb. Br. J. Pharmacol. **144:** 393–397.

95. VIZI, E. S., G. T. SOMOGYI, P. HADHAZY & J. KNOLL. 1973. Effect of duration and

frequency of stimulation on the presynaptic inhibition by α-adrenoceptor stimulation of the adrenergic transmission. Naunyn Schmiedebergs Arch. Pharmacol. **280:** 79–91.

96. BROWN, C. M., J. C. McGRATH & R. J. SUMMERS. 1979. The effects of α-adrenoceptor agonists and antagonists on responses of transmurally stimulated prostatic and epididymal portions of the isolated vas deferens of the rat. Br. J. Pharmacol. **66:** 553–564.
97. FRENCH, A. M. & N. C. SCOTT. 1983. Feedback inhibition of responses of rat vas deferens to twin pulse stimulation. Eur. J. Pharmacol. **86:** 379–383.
98. PENNEFATHER, J. N. 1983. A study of stimulation-evoked activation of α_2-adrenoceptors in the rat isolated vas deferens. Clin. Exp. Pharmacol. Physiol. **10:** 381–393.
99. STJÄRNE, L. & P. ÅSTRAND. 1985. Relative pre- and postjunctional roles of noradrenaline and adenosine 5′-triphosphate as neurotransmitters of the sympathetic nerves of guinea-pig and mouse vas deferens. Neuroscience **14:** 929–946.
100. FUJIOKA, M. & P. W. CHEUNG. 1987. Autoregulation of neuromuscular transmission in the guinea pig saphenous artery. Eur. J. Pharmacol. **139:** 147–153.
101. DYKE, A. & J. A. ANGUS. 1988. Comparative assay of neuronal uptake and autoinhibitory feedback in guinea pig and rat atria. J. Auton. Pharmacol. **8:** 219–228.
102. LANGER, S. Z., E. ADLER-GRASCHINSKY & O. GIORGI. 1977. Physiological significance of α-adrenoceptor-mediated negative feedback mechanisms regulating noradrenaline release during nerve stimulation. Nature **265:** 648–650.
103. McCULLOCH, M. W., M. J. RAND, D. F. STORY & I. SUTTON. 1981. Prejunctional α-adrenoreceptors subserve a physiological role in cardiac noradrenergic transmission. J. Auton. Pharmacol. **1:** 407–412.
104. BEALLEAU, B., B. G. BENFEY & C. MELCHIORRE. 1982. Presynaptic effect of clonidine antagonized by the tetramine disulphide, benextramine. Br. J. Pharmacol. **75:** 617–621.
105. LEDDA, F. & L. MANTELLI. 1984. Differences between the prejunctional effects of phenylephrine and clonidine in guinea-pig isolated atria. Br. J. Pharmacol. **81:** 491–497.
106. YAMAGUCHI, N., J. deCHAMPLAIN & R. A. NADEAU. 1977. Regulation of norepinephrine release from cardiac sympathetic fibers in the dog by presynaptic α- and β-receptors. Cir. Res. **41:** 108–117.
107. COUSINEAU, D., C. A. GORESKY, G. G. BACH & C. P. ROSE. 1984. Effect of β-adrenergic blockade on *in vivo* norepinephrine release in canine heart. Am. J. Physiol. **246:** H283–H292.
108. ELSNER, D., M. SAEED, O. SOMMER, J. HOLTZ & E. BASENGE. 1984. Sympathetic vasoconstriction sensitive to α_2-adrenergic receptor blockade. Hypertension **6:** 915–925.
109. KAHAN, T., P. HJEMDAHL & C. DAHLÖF. 1984. Relationship between the overflow of endogenous and radiolabelled noradrenaline from canine blood perfused gracilis muscle. Acta Physiol. Scand. **122:** 571–582.
110. YAMAGUCHI, N. 1982. Evidence supporting the existence of presynaptic α-adrenoceptors on the regulation of endogenous noradrenaline release upon hepatic sympathetic nerve stimulation in the dog liver *in vivo*. Naunyn Schmiedebergs Arch. Pharmacol. **321:** 177–184.
111. LUNDBERG, J. M., A. RUDEHILL, A. SOLLEVI, E. THEODORSSON-NORHEIM & B. HAMBERGER. 1986. Frequency- and reserpine-dependent chemical coding of sympathetic transmission: differential release of noradrenaline and neuropeptide Y from pig spleen. Neurosci. Lett. **63:** 96–100.
112. LUNDBERG, J. M., A. RUDEHILL, A. SOLLEVI & B. HAMBERGER. 1989. Evidence for co-transmitter role of neuropeptide Y in the pig spleen. Br. J. Pharmacol. **96:** 657–687.
113. MAJEWSKI, H., L. HEDLER & K. STARKE. 1983. Modulation of noradrenaline release in the conscious rabbit through α-adrenoceptors. Eur. J. Pharmacol. **93:** 255–264.
114. MAJEWSKI, H., L. HEDLER, C. SCHURR & K. STARKE. 1985. Dual effect of adrenaline on noradrenaline release in the pithed rabbit. J. Cardiovasc. Pharmacol. **7:** 251–257.
115. BROWN, M. J. & D. HARLAND. 1984. Evidence for a peripheral component in the sympatholytic effect of clonidine in rats. Br. J. Pharmacol. **83:** 657–665.
116. SZABO, B. & L. HEDLER. 1989. Effect of yohimbine rauwolscine and clonidine on postganglionic sympathetic nerve activity, plasma noradrenaline, blood pressure and heart rate in anesthetized rabbits. Naunyn-Schmiedebergs Arch. Pharmacol. **339:** R89.
117. MURPHY, M. B., M. J. BROWN & C. T. DOLLERY. 1984. Evidence for a peripheral

       component in the sympatholytic actions of clonidine and guanfacine in man. Eur. J. Clin. Pharmacol. **27:** 23–27.

118. BROWN, M. J., A. D. STRUTHERS, J. M. BURRIN, L. DiSILVIO & D. C. BROWN. 1985. The physiological and pharmacological role of presynaptic α- and β-adrenoceptors in man. Br. J. Clin. Pharmacol. **20:** 649–658.

119. ELLIOTT, H. L., C. R. JONES, J. VINCENT, C. B. LAWRIE & J. L. REID. 1984. The alpha adrenoceptor antagonist properties of idazoxan in normal subjects. Clin. Pharmacol. Ther. **36:** 190–196.

120. PERNOW, J., J. M. LUNDBERG & L. KAIJSER. 1988. α-Adrenoceptor influence on plasma levels of neuropeptide Y-like immunoreactivity and catecholamines during rest and sympathoadrenal activation in humans. J. Cardiovasc. Pharmacol. **12:** 593–599.

121. SAX, R. D. & T. C. WESTFALL. 1981. Alterations in prejunctional adrenoceptor function in guinea pig and rat vas deferens after chronic ganglionic blockade. J. Pharmacol. Exp. Ther. **219:** 21–26.

The Problem with Autoreceptors

STANLEY KALSNER

Department of Physiology
City University of New York Medical School
138th Street and Convent Avenue
New York, New York 10031

THE QUESTION OF AUTOREGULATION

Whereas neuronal homoreceptors are defined simply by their responsivity to exogenous transmitter the definition of autoreceptors, involving as it does endogenous feedback regulation of transmitter release, is more complex. An earlier report by me in this volume discusses the expectations for agonist performance in the presence of feedback, and the present study deals with the actions of antagonists and the pivotal role these compounds play in certifying the operation of feedback. An analysis of their presynaptic actions leads to the inescapable conclusion that they act in ways unrelated to feedback.

METHODS

Tissue preparations were done as described previously, and were dependent on species and organ. For example, renal arteries were removed from beef cattle at the slaughterhouse and taken promptly to the laboratory in cold previously oxygenated Krebs-Henseleit solution [composition (millimolar concentrations): NaCl, 115.3; KCl, 4.6; $CaCl_2$, 2.3; $MgSO_4$, 1.1; $NaHCO_3$, 22.1; KH_2PO_4, 1.1; glucose, 7.8; and disodium EDTA, 0.03]. After trimming of fat and connective tissue the vessels were cut into spiral strips of about 4 mm × 30 mm. Rats, guinea pigs, and rabbits of either sex were killed by cervical dislocation or decapitation and appropriate tissues were immediately removed and placed in oxygenated (95% O_2 and 5% CO_2) Krebs-Henseleit (Krebs) solution. The left atria of rabbits was cut into four longitudinal strips and that of the rat into two strips. The two vasa deferentia were desheathed and left intact; the ureters were cut into distal and proximal halves and the two distal portions considered as one matched pair and the two proximal another; capsular strips of the rat spleen were used, with two adjoining strips (1.7 cm long) taken from each spleen. In experiments with phenoxybenzamine and tetraethylammonium (TEA) one single strip (1.0 cm long) was taken from each spleen. The two central ear arteries of the rabbit were opened longitudinally and each cut into two equal segments. All tissues were trimmed of adherent fat and connective tissue.

All tissues were maintained for approximately 30 minutes in oxygenated Krebs' solution before incubation with tritiated norepinephrine. The desired preparations were then incubated for 60 minutes in 4.0 ml of oxygenated Krebs' solution containing l-[7,8-^{3}H]norepinephrine (10 μCi/ml; 7.0–10 × 10^{-7} M) at 37°C. The tissues were then dipped in fresh Krebs' solution, suspended under 1 or 2 g of tension in a pump-driven superfusion apparatus, and superfused continuously at 4 ml/minute with warmed (37°C) and oxygenated Krebs' solution containing cocaine (1 or 3 × 10^{-5} M) or desmethylimipramine (1.1×10^{-6} M), to inhibit neuronal reuptake, and normetanephrine (1×10^{-5} M), to inhibit extraneuronal uptake of released amine.

Stimulation Parameters

The tissues were mounted between platinum wire electrodes fixed vertically (or for heart tissue, horizontally) on opposite sides and stimulated transmurally. Pulses, usually of 0.5-msec and duration, and usually at supramaximal voltage (50–70 V) were delivered at the desired frequency using Grass stimulators. A 90-minute equilibration period under superfusion conditions was allowed for all tissues before the onset of stimulation protocols. One typical protocol would be as follows. After the equilibration period, which included one primer stimulation (10 pulses at 1 Hz given at 60 minutes), each tissue received, at 9-minute intervals, a set of 100 pulses of 0.5-msecond duration, twice at each of the test frequencies, e.g., 2 and 5 Hz. In a typical protocol, after completion of the initial set of stimulations (S_1) the experimental tissues, but not the controls, were exposed to the desired agonist or antagonists, dissolved in the superfusate for the indicated time, and the identical stimulation cycle was repeated (S_2) in both the untreated and treated tissues. Other stimulation conditions are as described under Results.

Drugs and Radiochemicals

The drugs used and their sources were as follows: cocaine hydrochloride (May & Baker Ltd., Manchester, England), normetanephrine hydrochloride (Calbiochem Biochemicals, San Diego, Calif.), desmethylimipramine (Merrell Dow Pharmaceuticals, Cincinnati, Ohio), phenoxybenzamine hydrochloride (Smith Kline & French, Philadelphia, Pa.), yohimbine hydrochloride (Sigma Chemical Co., St. Louis, Mo.), l-norepinephrine bitartrate (Calbiochem), carbamyl choline chloride (carbachol) (Aldrich), veratradine (Sigma). The radioisotope l-[7,8-^{3}H]noradrenaline hydrochloride (specific activity about 15 Ci/mmol) was usually obtained from Amersham International (Bucks, England). It was diluted to a stock concentration of 100 μCi/ml (7.0–10.0×10^{-6} M) in ascorbic acid (50 μg/ml) and stored at 4°C in 10-ml aliquots, under nitrogen gas. To obtain a final concentration of 10 μCi/ml (7.0–10.0×10^{-7} M) in the incubation medium, 0.4 ml of this stock solution was added to 3.6 ml of Krebs' solution.

Efflux of Transmitter

The basal and stimulation-induced efflux of tritium from the tissues was determined by assaying 1.0-ml aliquots of the superfusate, collected in vials by a fraction collector which rotates at 3-minute intervals. The aliquots are subsequently transferred to vials containing 10 ml of Aqueous Counting Scintillant (Amersham/Searle Corp., Des Plaines, Ill.) and counted to a 1% error in a Wallace Rackbeta LKB scintillation system with automatic external standardization to determine efficiency or in a Beckman L5-5801. Basal efflux is expressed as disintegrations per minute and is referred to as the total radioactivity detected in the 3-minute sample collected immediately before each stimulation. Stimulation-evoked efflux is expressed as disintegrations per minute and was determined by subtracting the basal efflux from the total disintegrations per minute detected in the 3-minute sample collected during, and immediately after, stimulation. Transmural stimulation, when given, was always administered at the onset of a collection period and collection of superfusate was continued until tritium levels returned to basal levels.[1]

In preparations in which variable pulse durations were used after a 90-minute equilibration period, including one initial stimulation (10 pulses, 2 Hz) at 30 minutes following suspension, each pair of tissues was stimulated transmurally with a train of 100 monophasic pulses at 2 Hz and supramaximal voltage. The pulse durations utilized were 50, 100, 200, 500, 1000, 2000, and 5000 μseconds with 9 minutes between each stimulation. One of each pair of tissues was exposed to yohimbine (3×10^{-6} M), phenoxybenzamine (3×10^{-6} M), or tetraethylammonium (TEA) (3×10^{-3} M) for 30 minutes followed without washout by the identical stimulation cycle done simultaneously in both the control and treated preparations. No tissue was used for more than one stimulation cycle.

Data Evaluation

Data are routinely expressed either as absolute disintegrations per minute collected from individual tissues during a stimulation period (minus basal efflux) or as the ratio of stimulation-induced efflux during a second period of stimulation (usually in the presence of the desired test drug) to that obtained in the first period of stimulation (usually in the absence of drug).[1] This latter procedure eliminates any differences between preparations or animals in the absolute amounts of tritium released because the effect of a drug is determined by a same-tissue (intratissue) comparison of stimulation-induced efflux, first in the absence and then in the presence of drug, and a ratio is expressed. A matching group of control tissues taken from the same animals and treated identically except for the drug under study serves to correct and account for (normalize) any changes in efflux due to time alone. This is achieved by comparison of S_2/S_1 or other desired ratio in treated strips with that of the comparable ratio in untreated preparations. Additionally, it should be noted that no material decline in stimulation-induced efflux was noted in control preparations, with any of the tissues used here, when second vs. first stimulation runs were compared.

The above-described procedure, now widely used, is a precise and desirable way of recording efflux events and avoids a correction related to total tissue content of tritium, a factor of which the full implications are not yet clear.[2] A preliminary study by my laboratory, however, confirmed that data expressed by the fractional release method[3] and the ratio method[1] give essentially similar results.

A procedure even more sensitive than the intergroup comparison of ratios were also used, as deemed appropriate. The ratio of the absolute output of tritium in the second vs. the first (or third or fourth vs. the first) period of stimulation of any given treated tissue was calculated as a percentage of the similarly obtained ratio for the matching control (untreated) tissue, taken from the same animal. Thus, e.g., a ratio of 1.2 for a control tissue and 0.9 for its drug-treated mate yields a value of 75.0% of the control output of tritium, for that particular experiment. Mean percentage values for each treatment condition then allow a precise assessment of the comparative effects of a given drug on the stimulation-induced efflux of tritium by nulling out the sometimes encountered spontaneous variations in the absolute output of radioactivity between first and second stimulation periods, which are reflected in ratio determinations.

Data on efflux are presented as mean $\pm$ standard error (SE). Student's paired t test was used for all intrastrip comparisons with the unpaired test used for comparisons between groups; p values of <0.05 were considered significant.[4]

RESULTS

Effects of Antagonists on Transmitter Release

When feedback is operative an increase in the intensity of stimulation should increase the effect of the antagonist to enhance transmitter release. This is because enhanced potentiation is presumed to represent blockade by the antagonist of the concentration of transmitter at the neuronal receptor sites and these would be increased during conditions of heightened stimulation. As shown in FIGURE 1b, however, the enhancement of stimulation-induced ^{3}H-norepinephrine release by yohimbine in cortical slices from rat brain did not increase with increasing stimulation frequency; instead it tended to decline. In additional experiments increasing simultaneously the frequency of stimulation and the pulse train number, from 50 to 400 (FIGURE 1a), did not increase the effectiveness of yohimbine to enhance release.

These observations were not restricted to central systems but were routinely noted in a variety of peripheral preparations.[5] For example, the stimulation-induced release of ^{3}H-norepinephrine in cattle circular iris preparations was enhanced to a similar extent by yohimbine during stimulation with pulse trains (FIGURE 1c) of varying number from 10 to 100 despite the different biophase transmitter levels achieved, and this latter point was confirmed by monitoring the size of the circular muscle contractions. In detailed studies, published elsewhere by my laboratory, nine preparations collected from four species were examined under a range of stimulation conditions, utilizing a variety of pulse train lengths and frequencies.[5] The extent to which norepinephrine release was enhanced by yohimbine did not correlate at all with the anticipated biophase transmitter levels. Similar discordancies were obtained with other test conditions,[6] and using several different antagonists, including phenoxybenzamine, a highly reactive blocker which forms a covalent chemical link with the alpha-receptor site and consequently is not in competition with the agonist.

Observations of a reduced antagonist effect under conditions of increased stimulation, opposite to theoretical expectations for interruption of a negative feedback system, have also been reported by other investigators (see TABLE 1), and appear to be unquestionable. For example, Rand *et al.* found that enhancement of ^{3}H-norepinephrine release by phenoxybenzamine, from blood vessel preparations stimulated distally via propagated nerve impulses, was most pronounced at the lowest frequencies and least at the higher frequencies,[7] not at all in keeping with a mechanism of enhancement linked to interruption of a feedback loop. Similarly, De Potter *et al.* found that the release of endogenous norepinephrine and the enzyme marker dopamine B hydroxylase, from the sympathetic nerves of cat spleen, was most increased by phenoxybenzamine at the lowest frequencies and least at the highest test frequency.[8] Similarly, Richards noted that the release of labeled 5-hydroxytryptamine from rat hypothalamic slices was increased most by the 5-HT antagonists methiothepin and quipazine at the low frequency of 1 Hz and increased least at 5 Hz.[9] In experiments with rat cerebral cortex slices, it was noted that atropine enhanced the stimulation-induced release both of endogenous and of tritiated acetylcholine, and did so most at the lowest frequency (0.25 Hz) and least at the highest test frequency (10 Hz).[10] These kinds of observations make clear that increases in stimulation-induced transmitter release by antagonists, however they occur, are not linked to a process that is intimately associated with the perineuronal level of transmitter.

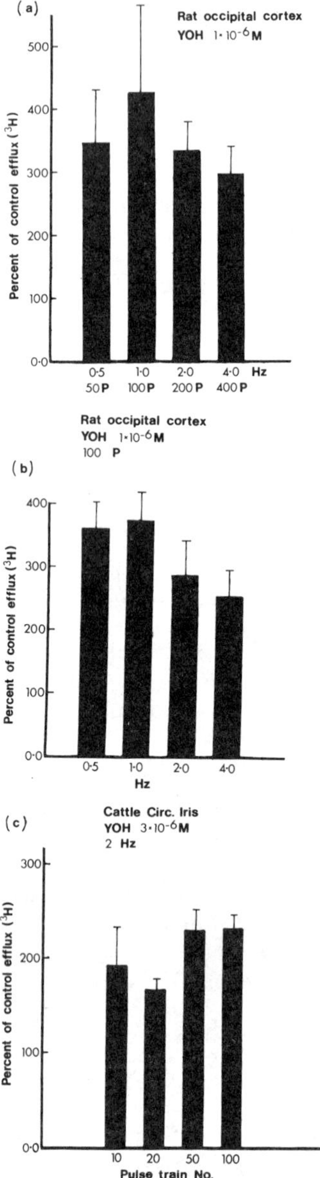

FIGURE 1. Effect of yohimbine on stimulation-induced efflux of ^{3}H-norepinephrine in rat occipital cortex and cattle iris. Tissues were exposed to yohimbine (1×10^{-6} M) after the initial period of stimulation (S_1) followed 20 minutes later, without washout of the yohimbine, by a repetition of the stimulations (S_2). (**a**) Stimulation of cortical slices for 100 seconds at 0.5, 1.0, 2.0, and 4.0 Hz; (**b**) Stimulation of cortical preparations with 100 pulses at 0.5, 1.0, 2.0, and 4.0 Hz; (**c**) Stimulation of cattle iris preparations at 2 Hz with 10, 20, 50, and 100 pulses. Control tissues were not exposed to yohimbine between S_1 and S_2. See Methods for details.

Discordancies between Per Pulse Release and Antagonist Enhancement

The discrepancy between biophase transmitter level and magnitude of antagonist effect is not limited to conditions of changing frequency and pulse number. As shown in FIGURE 2, when extracellular calcium is reduced to 20% of the control value, and the per pulse release of transmitter curtailed in this way, the effectiveness of yohimbine to increase stimulation-induced release in guinea pig ureter is not at all diminished. Similarly, the enhancing effect of yohimbine was not diminished by a reduction in the applied stimulation voltage, which reduces the number of varicosities contributing to

TABLE 1. Antagonists—Inverse Effect or No Change with Frequency[a]

Dzielak *et al.* (1983)[23]	Bovine splenic vein; efflux—NA; (1–10 Hz); phentol.
Abrahamsen & Nedergaard (1981)[24]	Rabbit pulmonary artery; efflux—NA/CPZ; (1–30 Hz).
Arbilla *et al.* (1985)[25]	Rabbit caudate nucleus response; dopamine/sulpiride; (1–3 Hz).
Carr & Fozard (1981)[26]	Rabbit pulmonary artery; NA/yohimbine; (0.3–10 Hz).
Bourdois *et al.* (1974)[10]	Rat cortex; Ach/atropine; (0.25–64 Hz).
Stjarne (1980)[27]	Human omental artery; NA/phentol; (1 and 10 Hz).
Stjarne (1973)[28]	Guinea pig vas; NA/phentol.; (1–30 Hz).
Richards (1985)[9]	Rat hypothalamus slices; 5-HT; methiothepin, quipazine; (1–5 Hz).
Story *et al.* (1981)[29]	NA; 4 pulses phentolamine (0.5–10 Hz).
Kalsner (1979–85)[1,4–6,15–16,21]	Many tissues from 5 species; NA-efflux; phentolamine, POB, yohimbine.
Daly *et al.* (1989)[30]	Rat heart; NA response/yohimbine. ER$_{20–80}$.
Wakade (1980)[31]	Rat vas; NA/POB; (freq. range).
Guimaraes *et al.* (1978)[32]	Dog saphenous vein; efflux—NA/phentol; (1–5 Hz).
Dart *et al.* (1984)[33]	Rat heart; NA/yohimbine; (1–4 Hz).
Rand *et al.* (1988)[7]	Rabbit ear artery; efflux—NA/POB; (10–50 Hz).
Kahan *et al.* (1984)[34]	Dog muscle bed; efflux—NA/POB; (1–4 Hz).
Saelens & Williams (1983)[35]	Dog saph. vein; NA/phentolamine; (1–5 Hz).
Medgett & Rand (1981)[36]	Rat tail artery; NA/phentol.

See also Bell & Vogt (1971);[37] Nedergaard & Abrahamsen (1988);[38] Vizi *et al.* (1973);[39] McCulloch *et al.* (1975);[40] Kirpekar & Cervoni (1963);[41] Hughes (1972);[2] De Potter *et al.* (1976).[8]

[a]NA is noradrenaline. CPZ is chlorpromazine. 5-HT is serotonin. POB is phenoxybenzamine.

the response, and correspondingly the total amount of transmitter released from the stimulated issue (FIGURE 3a). Also, shortening the duration of each stimulation pulse, from 500 to 50 μseconds such as to diminish the number of recruited terminal fibers, and thereby the amount of transmitter released from the tissues, did not reduce but potentiated the capacity of yohimbine to enhance release (FIGURE 3b).

The Amount of Enhancement by Antagonists

A study was done in guinea pig atria to examine a number of presynaptic alpha-receptor blocking agents so as to assess their relative abilities to enhance transmitter release. It was reasoned that the maximal amount of enhancement should be similar for each established antagonist if enhancement was due to a common

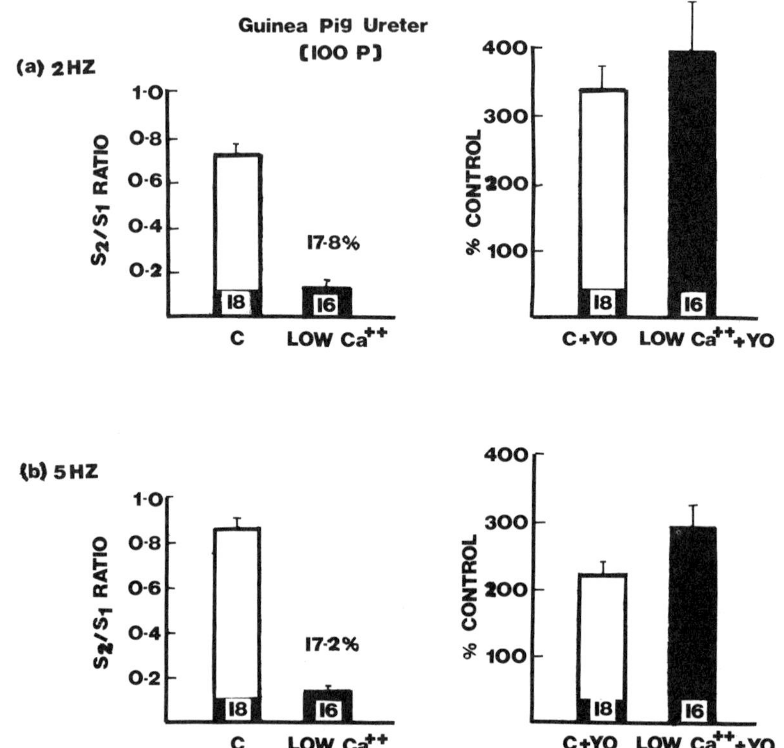

FIGURE 2. The effect of low extracellular calcium and yohimbine on the efflux of ^{3}H-norepinephrine in guinea pig ureter. Preparations were stimulated twice with 100 pulses at 2 Hz (a) and 5 Hz (b) in standard calcium (2.3 mM) Krebs and then exposed to low calcium Krebs (0.23 mM) followed 30 minutes later by restimulation with 100 pulses. Control ureter segments (c), taken from the same animals, were maintained in standard calcium Krebs in the interval between S_1 and S_2. Yohimbine (3×10^{-6} M), when needed, was added 20 minutes prior to the second stimulation period. See Methods for details.

capacity to occupy a finite number of presynaptic alpha-receptor sites. Tissues were stimulated with 100 pulses at two different test frequencies, and the effects on transmitter release were examined using at least a 100-fold concentration range for each antagonist. The effects of the antagonists varied widely (FIGURE 4), with some concentrations causing inhibition of release. As shown in FIGURE 5, the maximum enhancement of stimulation-induced transmitter release by each of the test agonists differed greatly. Whereas phentolamine increased efflux more than 2-fold, priscoline (tolazoline) had only a very modest effect on the efflux of ^{3}H-norepinephrine.

In light of such findings how are we to decide which of the antagonists serves as the appropriate index of feedback inhibition? Is it phentolamine which increased transmitter output the most? Or is it estimated best through priscoline, which had the least effect? Or should we compromise and use yohimbine, which had an intermediate effect, and is regarded as a highly specific alpha₂-receptor antagonist? If we examine the effects of presynaptic receptor antagonists on other tissues, the magnitude of the problem becomes more sharply defined.

In experiments with guinea pig ureter, using a range of antagonists and concentrations, norepinephrine release was increased most by phenoxybenzamine and by priscoline (about 6-fold) (FIGURE 6). The latter compound, however, as was noted above, increased efflux the least in guinea pig atria (FIGURE 5). Some other alpha-receptor antagonists had only relatively modest effects to enhance release in the ureters (e.g., chlorpromazine). Overall the antagonist effects ranged from as much as 6-fold (600% of controls) to as little as an 80% increase in transmitter overflow. Yohimbine produced an intermediate level of enhancement, raising again the plausibility of its unique suitability as the determinant of feedback regulation through presynaptic alpha sites. In still another studied tissue, however, namely, cattle radial artery, yohimbine did not enhance but, in fact, inhibited transmitter efflux over a 100-fold range of test concentrations, questioning the validity of such a deduction (FIGURE 7). Yohimbine-induced inhibition of transmitter release has been observed in other cattle vascular tissue, as well, and even in other species including the cerebral vessels of the cat.

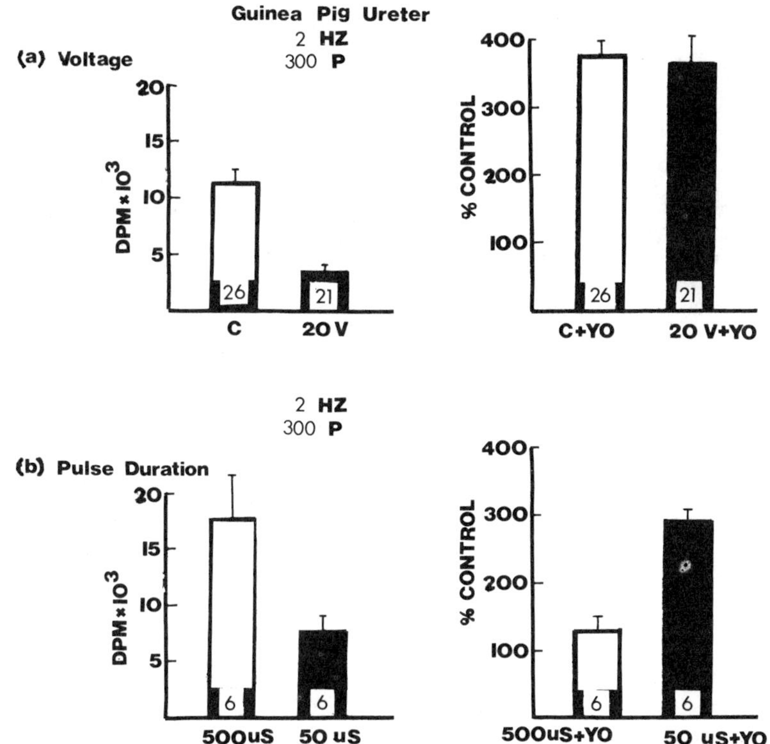

FIGURE 3. The effect of yohimbine on the efflux of ^{3}H-norepinephrine in guinea pig ureter under conditions of altered stimulation parameters. Preparations were stimulated with 300 pulses at 2 Hz under conditions of standard voltage (60 V) and standard pulse durations (0.5 msecond) and also stimulated in the presence of yohimbine (3×10^{-6} M) with the voltage reduced to 20 V (panel a, right) or the pulse duration reduced to 0.05 msecond only (panel b, left). Matched control tissues (c) were exposed to yohimbine and maintained under standard conditions of voltage and pulse duration for both S_1 and S_2. Panel a and panel b, left, show the per pulse efflux of ^{3}H-norepinephrine under conditions of altered voltage and pulse duration respectively.

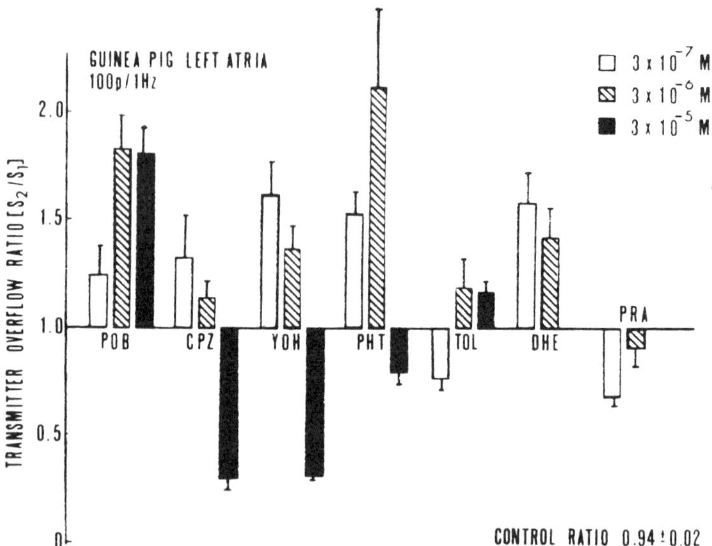

FIGURE 4. Ratios of transmitter efflux in the second to the first periods of field stimulation in the presence of various adrenergic antagonists in guinea pig atria stimulated with 100 pulses at 1 Hz. The antagonists were given, one per experiment, in the period between the first and second stimulation sets. The antagonists were phenoxybenzamine (POB), chlorpromazine (CPZ), yohimine (YOH), phentolamine (PHT), tolazoline (priscoline)(TOL), dihydroergotamine (DHE), and prazosin (PRA). The concentration of each antagonist used was 3×10^{-7} M (open bar), 3×10^{-6} M (hatched bar), and 3×10^{-5} M (filled bar). Each bar represents between 4 and 12 values.

Obviously even yohimbine is not a suitable compound to assess the presence of feedback, quantitatively or even qualitatively.

One Pulse Stimulation and Antagonist Enhancement

About 12 years ago I reported that both the efflux of ^{3}H-norepinephrine and the motor response to stimulation with a single brief pulse were increased significantly in guinea pig vas deferens. This was observed regardless of the presence or absence of blockers of neuronal and extraneuronal uptake.[1] It is not possible for such data to be explained through interruption of a feedback loop since the transmitter released by a single brief pulse cannot retroactively inhibit its own release. Some workers subsequently reported the lack of enhancement of release by a single pulse in their own test systems. Others, however (TABLE 2), have sustained my initial observations and described enhancement of transmitter release, or of the effector response, following a single stimulation pulse.

The electrophysiological studies by Holman and Surprenant involved several antagonists,[11] whereas those by Blakeley and associates showed that piperoxan and yohimbine enhanced the amplitude of postsynaptic discrete events initiated by a single pulse to the vas deferens preparation.[12,13] The notion that transmitter is released, in the absence of any stimulation, sufficient to activate presynaptic autoinhibitory receptors and inhibit substantially stimulation-induced release can be dismissed. Firstly it is well established that the trickle of transmitter released under such conditions is almost all in the form of deaminated metabolites.[14] Secondly, should the minimal amount of transmitter released during quiescent conditions be sufficient to activate the feedback loop, such activation must be to an extent comparable to that achieved during conditions of intense stimulation. This is so since the magnitudes of enhancement by the antagonist are not widely discrepant in the different circumstances. If enhancement with one pulse is due only to feedback blockade, then the system is operative under absurd test conditions, and unable to discriminate between incidental stray transmitter, insufficient to initiate even a minimal effector motor response, and moderate or heavy neuronal traffic. More likely the antagonists act directly on the nerve terminal to alter neurosecretion.

Mechanism of Antagonist Action

Yohimbine is an effective antagonist of presynaptic $alpha_2$-receptors but it has numerous other actions as well (TABLE 3). These are too often dismissed, without regard, when considering the interactions of yohimbine with neurons that might result in increased transmitter release. A body of evidence indicates that yohimbine increases stimulation-induced transmitter release not as a consequence of alpha-receptor blockade but due to some other action of the alkaloid at the nerve terminals.

This point of view is validated by my finding that a concentration of yohimbine producing near maximal to maximal enhancement of transmitter release in autonomic

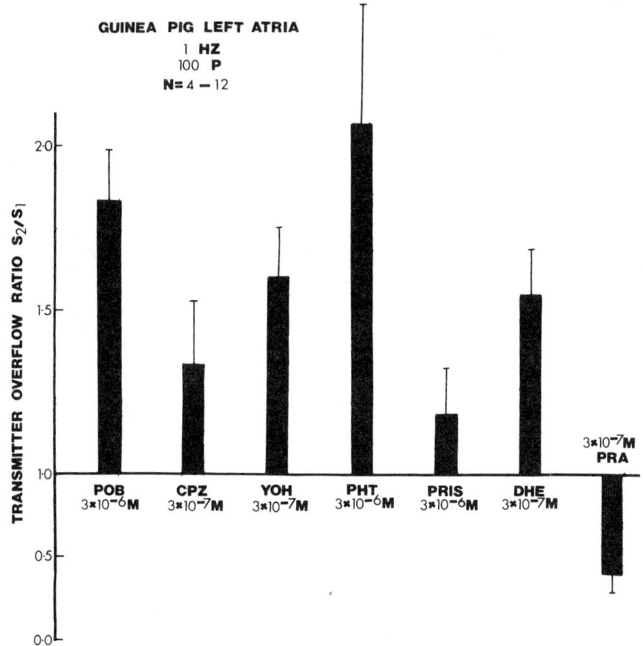

FIGURE 5. Peak enhancing effects of several adrenergic antagonists on the efflux of ³H-norepinephrine in response to 100 pulses at 1 Hz. See legend to FIGURE 4 for details.

effectors does not significantly antagonize the inhibition of transmitter release by exogenous transmitter[15] (TABLE 4). A similar dissociation between antagonist enhancement of transmitter release and the blockade of inhibition by exogenous agonists was obtained in experiments with the imidazoline derivative oxymetazoline[15] (TABLE 5). As can be seen, inhibition of ^{3}H-norepinephrine efflux by oxymetazoline was not significantly attenuated by a concentration of yohimbine that markedly enhanced release of ^{3}H-norepinephrine during stimulation. I also reported previously that inhibition of

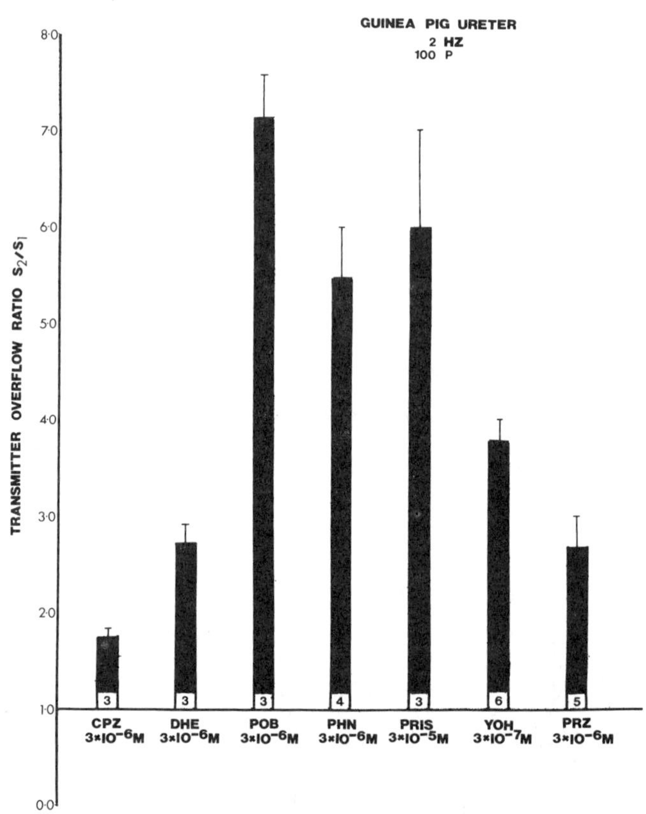

FIGURE 6. Peak enhancing effects of several adrenergic antagonists on the efflux of ^{3}H-norepinephrine in guinea pig ureter stimulated with 100 pulses at 2 Hz. A 100-fold concentration range was tested for each antagonist and that concentration producing the most enhancement is shown in the Figure. See legend to FIGURE 4 for other details.

norepinephrine release by clonidine in guinea pig atria was not antagonized by a concentration of yohimbine that produced near maximal enhancement of ^{3}H-norepinephrine release.[16] Similar observations were made by Baker *et al.* in their studies in guinea pig atria.[17]

The above-described studies demonstrate that the enhancement of overflow by antagonists, presumed to represent blockade of endogenously released transmitter, occurs at antagonist concentrations that are often magnitudes lower than the concen-

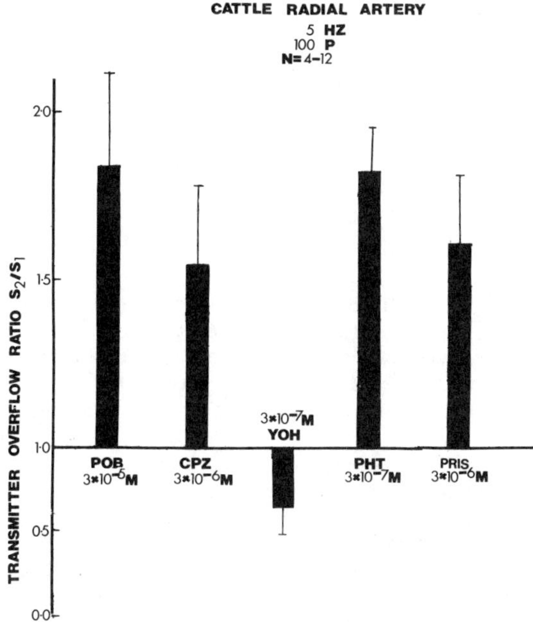

FIGURE 7. Peak enhancing effects of several adrenergic antagonists on the efflux of ^{3}H-norepinephrine in cattle radial artery stimulated with 100 pulses at 5 Hz. Yohimbine did not enhance efflux in this preparation at any of the test concentrations (3×10^{-7} M, 3×10^{-6} M, 3×10^{-5} M). See legend to FIGURE 4 for other details.

trations needed to block out the inhibitory effects on transmitter release of even moderate concentrations of exogenous agonists. This is so although peak transmitter concentrations following release from point sources, presumably blocked by the antagonist, are manyfold greater than the concentrations of exogenous transmitter used in such experiments. It appears that increased transmitter release and block of the presynaptic effects of exogenous agonists are not expressions of a unitary event, namely, passive occupation of alpha$_2$-receptor sites, but likely result from separate presynaptic actions of the antagonist.

An orientation towards accepting the idea that antagonists act directly on nerve terminals to enhance transmitter release is provided by discussion of the schematic

TABLE 2. Enhancement by Single Pulse

Blakeley (1981, 1984)[12,13]—mouse vas def—yohimbine, piperoxan; excitatory junction potentials (EJP)

Holman and Surprenant (1979)[11]—rabbit ear artery, saphenous artery, rat tail artery—phenoxybenzamine (POB), phentolamine; EJP

Kalsner (1979)[1]—guinea pig vas—POB; efflux and response.

Kuriyama & Makita (1982)[42]—guinea pig mesenteric artery—phentolamine; EJP

Suzuki (1984)[43]—dog mesenteric vein—prazosin; EJP

Wakade & Wakade (1981)[44]—guinea pig heart—yohimbine, phentol, POB; efflux (TEA).

TABLE 3. Known Actions of Yohimbine[a]

1. Alpha$_2$ antagonist	8. Alpha$_2$ agonist
2. Local anesthetic	9. Ca^{++} channel antagonist
3. Alpha$_1$ blocker	10. Monoamine oxidase inhibitor
4. Na$^+$ channel gating modifier	11. 5-HT receptor antagonist
5. 5-HT uptake blocker	12. Cholinergic (muscarinic) receptor antagonist
6. Dopamine receptor blocker	13. Alpha$_1$ agonist
7. 5-HT agonist	14. Cholinesterase inhibitor

[a]5-HT is serotonin (5-hydroxytryptamine).

shown in FIGURE 8. The schematic shows the intraneuronal transport of extracellular ^{3}H-norepinephrine via the catecholamine transport system. The membrane pump is inhibited by substances such as cocaine or desmethylimipramine, thereby increasing the efflux of transmitter. Are these substances appropriately described as antagonists of uptake? If so is the tritiated norepinephrine acting as an agonist during transport? This is clearly not the case.

Certain alpha-receptor blocking agents also block intraneuronal transport. Are substances such as phentolamine and phenoxybenzamine to be described as antagonists in achieving such blockade, since they are already known as alpha antagonists, or is this an inappropriate use of the term? The problem of "names" is brought to a head when we consider blockade of ^{3}H-norepinephrine transport by other agonists, which themselves compete for the transport sites. Should epinephrine be described as an antagonist of ^{3}H-norepinephrine transport? Finally, should the transmitter substance itself, unlabeled norepinephrine, in competition with the uptake of ^{3}H-norepinephrine, as schematized in FIGURE 8, be classified, under the described circumstances, as an uptake antagonist?

The limits of such constraining terminology are clear and apply also when other presynaptic systems are considered. It is suggested here that yohimbine and other

TABLE 4. Actions and Interactions of Yohimbine (3×10^{-7} M) and Noradrenaline (1.8×10^{-6} M) on Stimulation-Induced ^{3}H-Transmitter Efflux in Guinea Pig Ureter

		Transmitter Efflux ($\times 10^3$ d/minute)			Percent Inhibition of
Experimental Group (n)	Hz	First Run (S$_1$)	Second Run (S$_2$)	Efflux Ratio S$_2$/S$_1$	Efflux by Noradrenaline
a. Control (6)	1	3.69 ± 0.71	4.12 ± 0.78	1.12 ± 0.08	
Noradrenaline (6)	1	4.54 ± 0.72	0.48 ± 0.18	0.12 ± 0.04[a]	(a) 88.6 ± 5.4
Control (6)	5	5.36 ± 1.02	5.21 ± 1.08	0.96 ± 0.06	
Noradrenaline (6)	5	5.47 ± 0.90	2.11 ± 0.70	0.42 ± 0.04[a]	(b) 56.0 ± 3.4
b. Yohimbine (8)	1	4.48 ± 0.23	23.38 ± 1.56	5.26 ± 0.34	
Yohimbine + noradrenaline (8)	1	5.19 ± 0.39	4.09 ± 0.91	0.89 ± 0.24[a]	(c) 84.3 ± 3.5
Yohimbine (8)	5	5.86 ± 0.26	25.52 ± 1.76	4.36 ± 0.24	
Yohimbine + noradrenaline (8)	5	6.78 ± 0.71	10.84 ± 0.95	1.81 ± 0.30[a]	(d) 59.5 ± 5.4

[a]$p < 0.001$ compared with the ratio for the corresponding group without noradrenaline. Ureters were exposed to noradrenaline (1.8×10^{-6} M) or to yohimbine (3×10^{-7} M) or to both drugs together after the initial period of stimulation at 1 Hz and 5 Hz with 100 pulses. This was followed, without washout, by a repetition of the stimulations (S$_2$). (a) vs. (c) not significant; (b) vs. (d) NS. Note enhanced efflux ratios in the presence of yohimbine.

antagonists act on membrane systems to increase efflux of transmitter in a way not appropriately described as "antagonistic," and the justification for such an assertion is provided below.

A Direct Action of Presynaptic Antagonists to Increase Transmitter Release

Experimental evidence was obtained that yohimbine acts directly to increase transmitter release and this is unrelated to passive occupancy of presynaptic receptor sites mediating feedback. The stimulation-induced efflux of ^{3}H-norepinephrine is not materially altered by a 60-minute exposure of rabbit aortic strips to acetylcholine $(1.4 \times 10^{-5}$ M) (FIGURE 9), when tested at two different test freqencies, namely, 2 and 5 Hz. However, the enhanced efflux in the presence of yohimbine is virtually entirely blocked by acetylcholine (FIGURES 9a and b).[18] This antagonism is completely reversed

TABLE 5. Actions and Interactions of Yohimbine $(3 \times 10^{-7}$ M) and Oxymetazoline $(1.0 \times 10^{-7}$ M) on Stimulation-Induced ^{3}H-Transmitter Efflux in Guinea Pig Ureter

| | | Transmitter Efflux ($\times 10^3$ d/minute) | | | Percent Inhibition of |
| | | First Run (S_1) | Second Run (S_2) | Efflux Ratio S_2/S_1 | Efflux by |
Experimental Group (n)	Hz				Noradrenaline
a. Control (2)	1	10.06 ± 2.92	9.95 ± 3.72	0.96 ± 0.09	
Oxymetazoline (3)	1	12.33 ± 2.96	3.65 ± 1.47	0.28 ± 0.06[a]	(a) 70.1 ± 4.2
Control (2)	5	12.16 ± 2.86	12.05 ± 2.54	1.00 ± 0.03	
Oxymetazoline (3)	5	15.60 ± 2.20	13.42 ± 2.73	0.85 ± 0.05[a]	(b) 23.3 ± 2.3
b. Yohimbine (4)	1	7.13 ± 1.25	22.19 ± 3.26	3.16 ± 0.21	
Yohimbine + oxymetazoline (4)	1	8.07 ± 1.46	9.20 ± 2.41	1.11 ± 0.12[a]	(c) 64.7 ± 3.1
Yohimbine (4)	5	11.18 ± 1.94	27.99 ± 3.18	2.58 ± 0.17	
Yohimbine + oxymetazoline (4)	5	11.62 ± 2.96	20.43 ± 5.34	1.76 ± 0.05[a]	(d) 30.9 ± 3.7

[a]$p < 0.05$ compared with the ratio for the corresponding group without oxymetazoline. Ureters were exposed to oxymetazoline $(1.0 \times 10^{-7}$ M) or to yohimbine $(3 \times 10^{-7}$ M) or to both drugs together after the initial period of stimulation (S_1) at 1 Hz and 5 Hz with 100 pulses. This was followed, without washout, by a repetition of the stimulations (S_2). (a) vs. (c) NS; (b) vs. (d) NS. Note enhanced efflux ratios in the presence of yohimbine.

by a moderate concentration of atropine, confirming an action of acetylcholine at discrete muscarinic sites. Atropine itself does not alter efflux[18] (FIGURE 9).

A similar antagonism of yohimbine-induced enhancement of ^{3}H-norepinephrine efflux by acetylcholine was noted in guinea pig ureter and also in atria.[19] As shown in FIGURE 9c, acetylcholine slightly reduced the stimulation-induced efflux of norepinephrine in ureter preparations and completely antagonized the enhancing effect of yohimbine. In contrast, the inhibitory effect of exogenous norepinephrine on stimulation-induced transmitter release was much less compromised by acetylcholine. All the effects of acetylcholine were entirely antagonized by atropine,[18] including the block of yohimbine-induced enhancement. How does a compound (acetylcholine) antagonize the effects of a supposed antagonist (yohimbine) without itself producing effects similar to the antagonist (yohimbine)? It appears, more likely, that yohimbine acts directly (at a neuronal site) to increase norepinephrine release and that acetylcholine acts at a separate muscarinic site to conformationally distort, or otherwise modify, the site for yohimbine. In this model, then, atropine blocks the muscarinic sites through which acetylcholine acts, and thus restores the enhancing effect of yohimbine.

Presynaptic Regulation in the Cold

The several types of experiments described above indicate that presynaptic receptor antagonists increase transmitter overflow by an action distinct from blockade of presynaptic receptor sites. To assess if metabolic components might possibly be

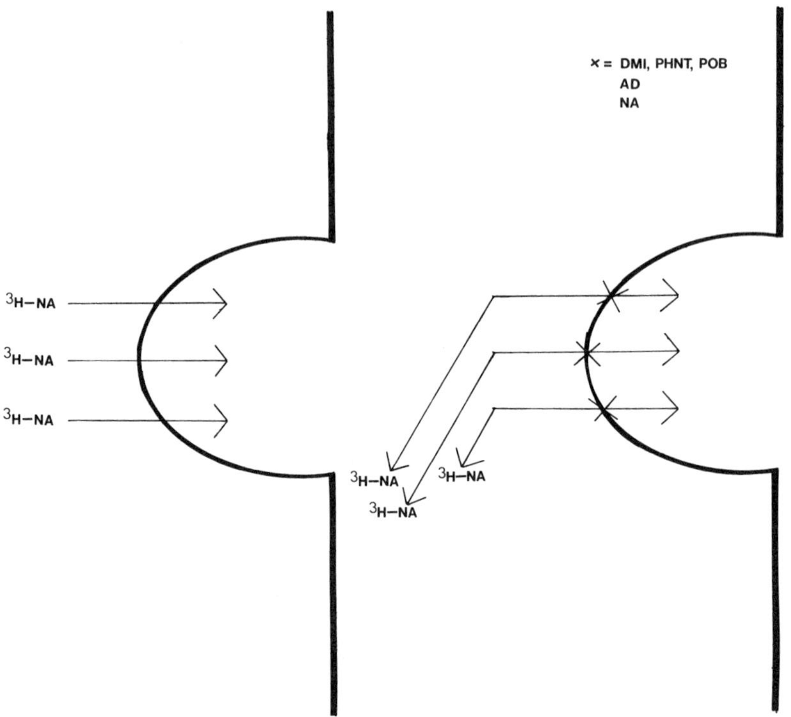

FIGURE 8. Schematic of neuronal transport sites for norepinephrine. Desmethylimipramine (DMI), phentolamine (PHNT), phenoxybenzamine (POB), epinephrine (AD), and norepinephrine (NA) are all indicated as blockers of uptake ($\times$) of ^{3}H-norepinephrine (^{3}H-NA).

involved in enhancement or inhibition of transmitter release and be selectively affected by the cold, experiments were performed at 11°C, using refrigerated circulating water baths and chilled Krebs superfusate solution.

The per pulse release of norepinephrine was much reduced in the cold, but the values were well maintained during four successive stimulation periods over the course of several hours (FIGURE 10a). The effectiveness of norepinephrine to inhibit stimulation-induced ^{3}H-norepinephrine efflux was not at all diminished in the cold, despite the reduced per pulse release of transmitter; in fact, agonist effectiveness was slightly

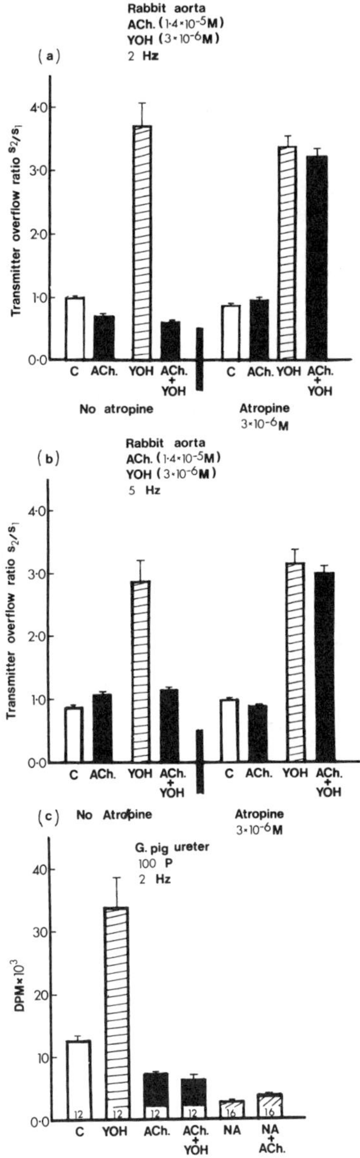

FIGURE 9. Actions and interactions of acetylcholine (Ach), yohimbine (Yoh), and atropine on the stimulation-induced efflux of ^{3}H-norepinephrine in rabbit aorta and guinea pig ureter. Effects of Ach (1.4×10^{-5} M) and yohimbine (3×10^{-6} M) on efflux in the absence and presence of atropine (3×10^{-6} M), at 2 Hz (panel a) and 5 Hz (panel b). **Panel c:** the effect of yohimbine, acetylcholine, and norepinephrine on the efflux of ^{3}H-norepinephrine with 100 pulses at 2 Hz in guinea pig ureter. See Kalsner and Quillan[18] for details.

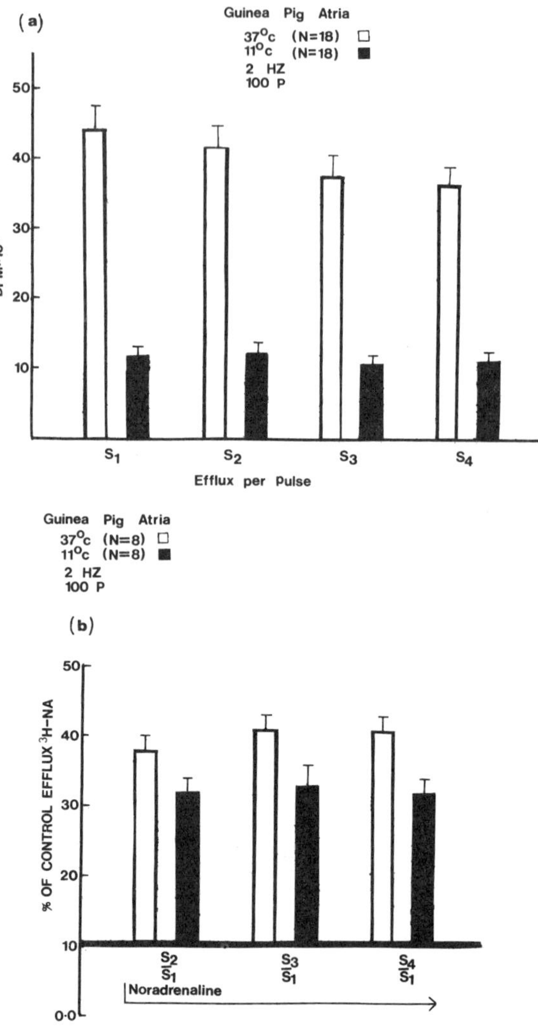

FIGURE 10. Effect of cold on the efflux of ³H-norepinephrine in guinea pig atria stimulated with 100 pulses at 2 Hz. Open bars (37°C) and solid bars (11°C). Tissues were exposed to the cold for 90 minutes prior to the onset of the first stimulation period and then maintained at 11°C for the duration of the experiment, consisting of four stimulation sets, at least 30 minutes apart. Matching control atrial halves were maintained at 37°C during S_1–S_4. **Panel a:** Per pulse efflux with 100 pulses (200 μseconds) during S_1 to S_4 at 37°C and at 11°C. **Panel b:** Inhibition of stimulation-induced efflux at 11°C and at 37°C by norepinephrine (1.8×10^{-6} M), administered after S_1 and maintained during S_2, S_3, and S_4.

enhanced (FIGURE 10b). The inhibitory effect of the agonist was well sustained during the course of the experiment both at 37°C and over four stimulation runs in the cold. The full retention in the cold of the inhibitory effect of the transmitter indicates that second messenger systems are not likely to be intimately involved in the presynaptic effects of norepinephrine.

In contrast, the enhancement of stimulation-induced transmitter release by yohimbine was somewhat reduced during the first stimulation test at 11°C and then subsequently disappeared (FIGURE 11). For example, during the fourth stimulation set, norepinephrine continued to exert its usual inhibition of transmitter release, but yohimbine had no or only minimal effect to enhance release. This is in contrast to the maintenance of the yohimbine effect at 37°C (FIGURE 11).

Raising the per pulse efflux of transmitter at 11°C to the levels seen at 37°C, by increasing the extracellular calcium concentration threefold, did not alter the characteristics of the rapid disappearance of the yohimbine effect in the cold (FIGURE 12). It appears that an essential component of yohimbine action fatigues rapidly in the cold although the mechanism through which exogenous agonists inhibit release remains intact.

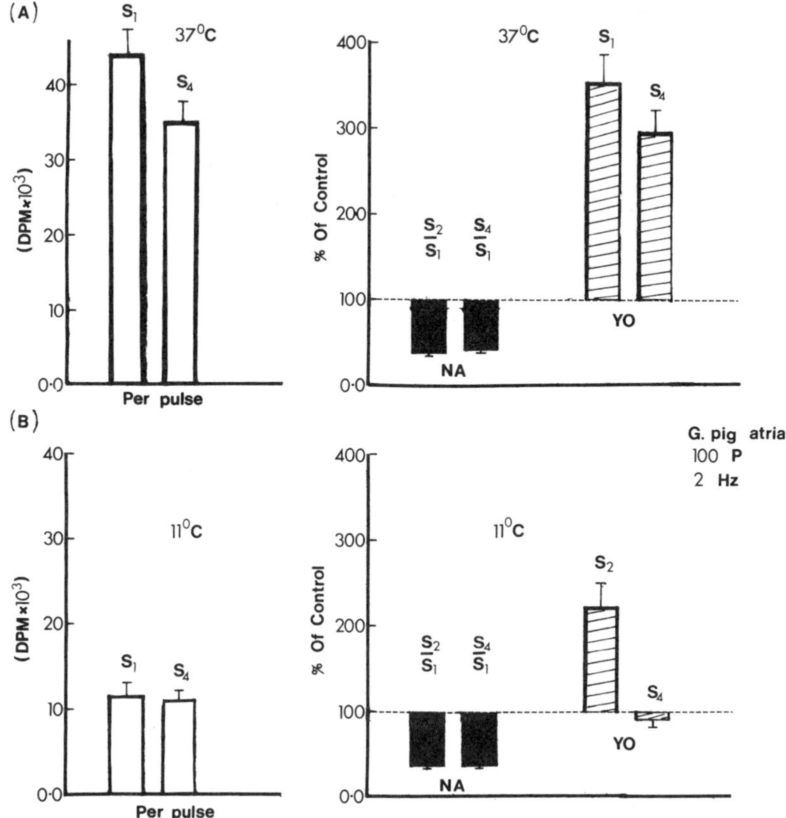

FIGURE 11. The effect of yohimbine and norepinephrine on the stimulation-induced efflux of ^{3}H-norepinephrine in guinea pig atria at 11°C and at 37°C. Preparations were stimulated with 100 pulses at 2 Hz. **Left panels:** Per pulse efflux of ^{3}H-norepinephrine at 37°C (A) and 11°C (B) during the first (S_1) and the fourth (S_4) stimulation period. **Right panels:** Inhibitory effect of norepinephrine (NA) (1.8×10^{-6} M) (solid bars) and enhancing effect of yohimbine (yo) (3×10^{-6} M) (hatched bars) during the second (S_2) and the fourth (S_4) stimulation period at 37°C (A) and at 11°C (B).

Mechanism of Antagonist Action

The action of yohimbine to increase the release of norepinephrine has negligible lag time, not in keeping with an action linked to accumulated biophase levels of transmitter from prior neuronal activity. The data with a single pulse, about which there is disagreement, have already been discussed. Numerous laboratories, however, agree that enhancement is readily obtained for the second of two pulses, or after several stimulation pulses.[20] Further, the extent of the potentiation after only a few pulses is comparable to what is seen under moderate or even extreme stimulation conditions.[4,6]

For example, in my own work, it was routinely noted that efflux of ^{3}H-norepinephrine following stimulation with 5 pulses in guinea pig vas deferens was

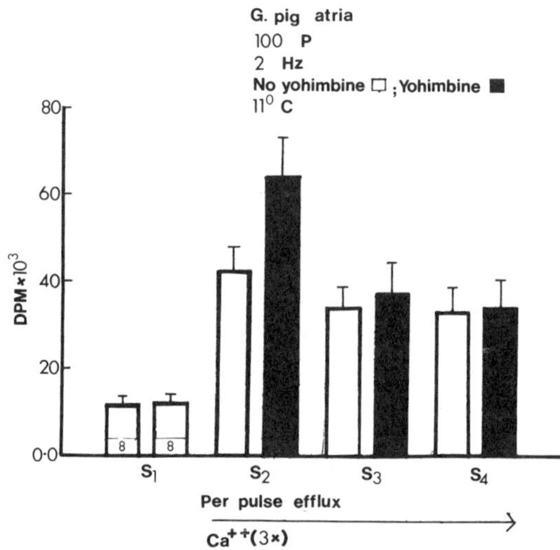

FIGURE 12. Effect of yohimbine on stimulation-induced transmitter efflux at 11°C in standard and in elevated calcium Krebs. Tissues were maintained at 11°C and superfused with chilled Krebs prior to the start of S_1. Guinea pig left half atrial preparations were exposed to elevated calcium (6.9 mM) after completion of S_1. One member of each pair of half atria (solid bars) was additionally exposed to yohimbine (3×10^{-6} M) 20 minutes prior to S_2 and S_2 to S_4 conducted in the presence of yohimbine.

enhanced about threefold by yohimbine, comparable to that seen with much more extensive stimulation. This pronounced enhancement was seen even when the frequency of stimulation was 20 and 50 Hz. These latter two frequencies restrict the total stimulation period to a very brief 200 mseconds and 80 mseconds, respectively. The presynaptic action of yohimbine is obviously linked to changes exerted by the alkaloid on some membrane component of the nerve terminals affected by the electrical impulse, and is not associated with the time-dependent development of accumulated free and active transmitter.

The locus of antagonist action was defined more clearly in a series of experiments in which the duration of the stimulation pulse was varied from 50 μseconds to 2000 μseconds[21] (FIGURE 13). The per pulse release of transmitter increased in guinea pig

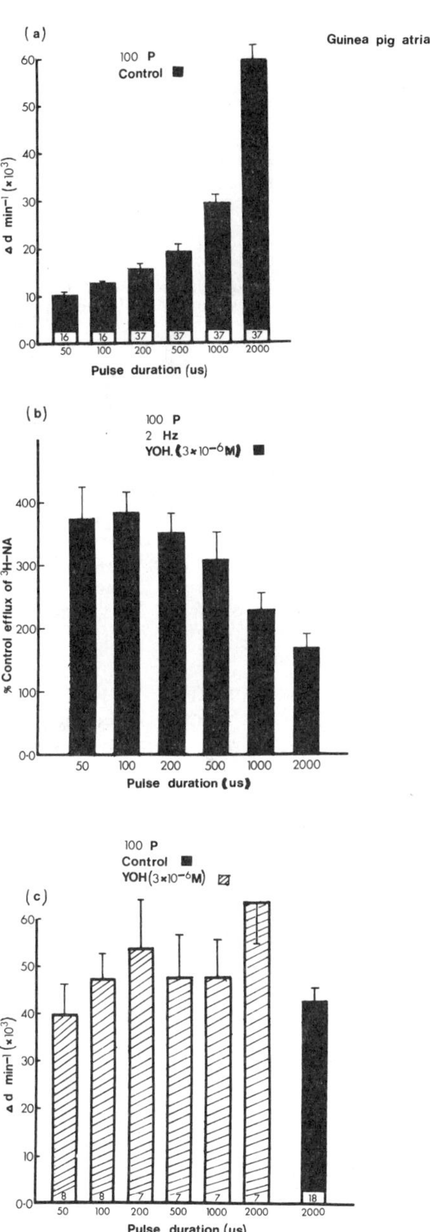

FIGURE 13. The efflux of ^{3}H-norepinephrine from guinea pig atria during stimulation with 100 pulses at varying pulse durations in the presence and absence of yohimbine. (a) Control per pulse efflux. Number of values in each group is shown inside each column. (b) Efflux in the presence of yohimbine (3×10^{-6} M) during stimulation with 100 pulses at varying pulse duration. Percentage values of efflux, relative to controls, were determined as described in Methods. (c) Per pulse efflux of ^{3}H-norepinephrine in the presence of yohimbine (solid bars) at the indicated pulse durations. Solid bar indicates control per pulse value with 2000 μsec pulse duration. See Kalsner[21] for details.

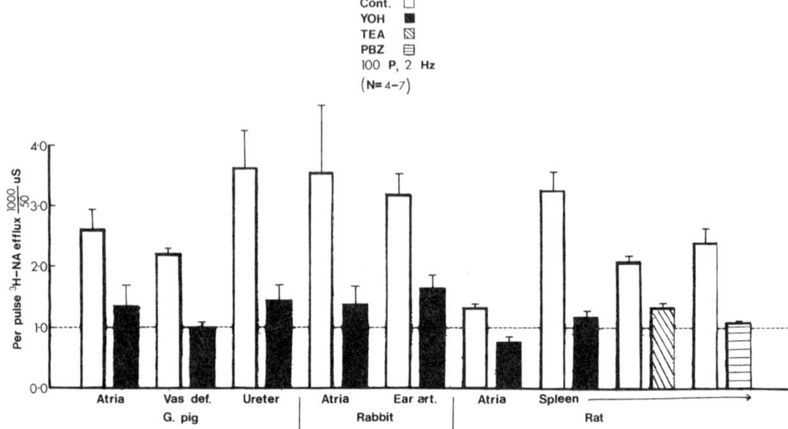

FIGURE 14. The effects of pulse duration changes, yohimbine, phenoxybenzamine and tetraethylammonium on the per pulse efflux of ³H-norepinephrine in several tissues. Ratios shown (efflux at 100 μseconds vs. efflux at 50 μseconds) are for untreated controls (open bars), yohimbine-treated (solid bars), tetraethyl-ammonium treated (3 mM) (slanted hatched bars), and phenoxybenzamine treated (horizontal hatched bars) tissues. Drugs, when used, were added about 30 minutes before stimulation to one of each set of tissues. See Kalsner and Quillan[19] for details.

atria in proportion to the increased duration of the stimulation pulse, when 100 stimulation pulses was delivered. Since the individual action potential is an all or none event, the increase in per pulse release is attributable to the increased recruitment of terminal filaments, through the increasingly prolonged depolarizations and the associated opening (activation) of sodium channels.

Although the amount of transmitter liberated into the extracellular space increased with increasing pulse duration in control tissues, the amount of enhancement of efflux by yohimbine decreased, rather than increased, under the same conditions (FIGURE 13b). This again is contrary to expectations for a mechanism linked to extracellular transmitter levels. Examination revealed that the declining effectiveness of yohimbine derived from the fact that the per pulse release in the presence of the antagonist did not change materially when the pulse duration was changed. That is, the per pulse release in the presence of yohimbine was essentially the same at 50 μseconds and at 1000 and 2000 μseconds.

Since per pulse output increased in control tissues as pulse duration increased, but not in yohimbine-treated tissues, the two groups converged at approximately 2000 to 5000 μsecond pulse durations. Yohimbine has a minimal effect to enhance transmitter output at long pulse durations.[21] These observations were confirmed and extended to a variety of tissues taken from four species.[19] The findings are summarized in FIGURE 14. The per pulse efflux of transmitter in the presence of yohimbine approached a constant value at all test pulse durations, whereas in control tissues the per pulse efflux increased progressively.

It appears that prolongation of the pulse duration in untreated tissues, which involves primarily the recruitment of additional terminal filaments, and the effect of yohimbine are somehow similar. Increasing the pulse duration allows sufficient current to reach the tissue sites such that the threshold for the initiation of the all or none action potential is achieved in an increasing number of fibers. Since depolarization to the critical threshold for the action potential is linked to the opening of sodium channels, the possibility was examined that yohimbine acts to lower the voltage threshold for the opening of the sodium channels and the initiation of the action potential and neurosecretion. Yohimbine by altering the voltage dependency of sodium channel gating could cause pulses of only 50 μseconds duration to activate (with action potentials) the same population of terminal filaments as does 1000 μsecond pulses under control conditions. It would explain the per pulse efflux values recorded in the presence and absence of yohimbine and also the declining differences between control and antagonist-treated preparations as the pulse duration lengthens. Phenoxybenzamine, another alpha antagonist, appeared to act like yohimbine (FIGURE 15).[19]

Veratridine and Yohimbine and the Sodium Pump

High concentrations of veratridine are known to attach to sites on the sodium channel and cause massive discharge of transmitter through changes in the voltage dependency of sodium channel activation.[22] The possibility that a low concentration of veratridine might simulate the enhancement of release seen with alpha-receptor antagonists was explored in preparations of guinea pig ureter stimulated with 100 pulses at 2 Hz and at 5 Hz. Veratridine (4.5×10^{-7}) was added to the superfusate for 45 minutes following a control set of stimulations at 2 Hz and 5 Hz (S_1), followed by repetition of the stimulation cycle (S_2). Matching tissues from the same ureters were either not exposed to veratridine, or were exposed to yohimbine or to both yohimbine plus veratridine in the interval between stimulation sets (FIGURE 15).

Veratridine, in low concentration, increased stimulation-induced efflux of ^{3}H-norepinephrine, and did so in a surprisingly specific way that did not increase the basal efflux of tritium. The effect was similar to that produced by yohimbine, and the two

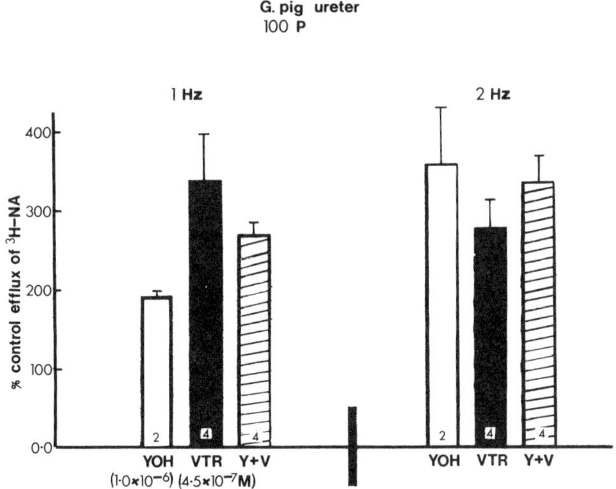

FIGURE 15. The actions and interactions of veratradine and yohimbine on stimulation-induced efflux of ^{3}H-norepinephrine in guinea pig ureter. Preparations were stimulated first in the absence and then in the presence of veratradine (VTR) (4.5×10^{-7} M) or yohimbine (YOH) (1.0×10^{-6} M), or the combination (Y + V).

compounds together did not have an additive effect (FIGURE 15). It seems likely that yohimbine and veratridine (in low concentration) are both acting on a sodium channel site to alter the voltage dependency for activation and thereby promote transmitter release. The lack of additivity in the effects is not due to a ceiling or limit on the magnitude of efflux since it was determined that tetraethylammonium (3 mM), which acts on K^+ channels, produces a substantially greater increase in transmitter release than does yohimbine together with the low concentration of veratridine under similar test conditions.

SUMMARY

There are numerous problems with the concept that antagonists enhance transmitter release by blockade of feedback. It was shown that antagonist enhancement of transmitter release does not correlate satisfactorily with the intensity of stimulation or with other indices of biophase transmitter concentration. Wide variations were shown to exist between antagonists in the amount of enhancement of release they induce. Also, antagonists enhance transmitter release or the effector response with a single stimulation pulse, a condition under which no feedback is possible. A study of agonist/antagonist relationships indicates different sites of action, and it was determined that the antagonist effect has negligible or minimal latency and that enhancement by antagonists is maximal under minimal condition of stimulation.

Antagonists were shown to enhance release by a direct action, not by passive occupancy of agonist sites. Experiments were described in which acetylcholine and cold selectively antagonized antagonist but not agonist effects. Further, experiments with pulse duration shifts and with veratridine pointed to a direct action of antagonists on Na^+ (also Ca^{++}?) channel gating mechanisms, which results in a shift in the voltage dependence of activation.

If antagonists, in some particular instances, enhance release by blockade of sites involved in negative feedback this is likely lost or mired in their more prominent direct actions on neurosecretion—and these must be sorted out. The acceptance of the fact that antagonists act directly to alter transmitter release (and not only as passive occupiers of presynaptic receptors), as the present study shows, both in the central nervous system and in the periphery, opens a new area for future investigation, and may be exploitable for therapeutic purposes and to gain an enriched understanding of the mechanisms of neurosecretion.

REFERENCES

1. KALSNER, S. 1979. Single pulse stimulation of guinea-pig vas deferens and the presynaptic receptor hypothesis. Br. J. Pharmacol. **66:** 343–349.
2. HUGHES, J. 1972. Evaluation of mechanisms controlling the release and inactivation of the adrenergic transmitter in the rabbit portal vein and vas deferens. Br. J. Pharmacol. **44:** 472–491.
3. STARKE, S., T. ENDO & H. D. TAUBE. 1975. Relative pre- and postsynaptic potencies of a α-adrenoceptor agonists in the rabbit pulmonary artery. Naunyn Schmiedebergs Arch. Pharmacol. **291:** 55–78.
4. KALSNER, S. 1980. Limitations of presynaptic adrenoceptor theory: the characteristics of the effects of noradrenaline and phenoxybenazmine on stimulation-induced efflux of [^{3}H]noradrenaline in vas deferens. J. Pharmacol. Exp. Ther. **212:** 232–239.
5. KALSNER, S. 1985. Is there feedback regulation for neurotransmitter release by autoreceptors? Biochem. Pharmacol. **34:** 4085–5097.

6. KALSNER, S. 1982. Feedback regulation of neurotransmitter release through adrenergic presynaptic receptors: time for a reassessment. *In* Trends in Autonomic Pharmacology. S. Kalsner, Ed. **2:** 385–425. Urban & Schwarzenberg, Inc. Baltimore, Md.

7. RAND, M. J., D. F. STORY & I. SUTTON. 1988. Comparison of the effects of phenoxybenzamine and uptake blockade on noradrenaline efflux from rabbit ear arteries evoked by field stimulation and propagated nerve impulses. J. Auton. Pharmacol. **8:** 287–295.

8. DE POTTER, W. P., I. W. CHUBB, A. PUT, *et al.* 1971. Facilitation of the release of noradrenaline and dopamine-hydroxylase at low stimulation frequencies by blocking agents. Arch. Int. Phamacodyn Ther. **193:** 191–197.

9. RICHARDS, M. H. 1985. Efflux of ^{3}H-5-hydroxytryptamine from rat hypothalamic slices by continuous electrical stimulation: frequency-dependent responses to serotonergic antagonists and 5-hydroxytryptamine. Naunyn Schmiedebergs Arch. Pharmacol. **329:** 359–366.

10. BOURDOIS, P. S., J. F. MITCHELL, G. T. SOMOGYI & J. C. SZERB. 1974. The output per stimulus of acetylcholine from cerebral cortical slices in the presence or absence of cholinesterase inhibition. Br. J. Pharmacol. **52:** 509–517.

11. HOLMAN, M. E. & A. M. SURPRENANT. 1979. Some properties of the excitatory junction potentials recorded from saphenous arteries of rabbits. J. Physiol. London **287:** 337–351.

12. BLAKELEY, A. G. H., T. C. CUNNANE & S. A. PETERSEN. 1981. The packeted release of sympathin from single release sites in the rodent vas deferns. Is release subject to feedback control? *In* Chemical Transmission 75 years. L. Stjarne, P. Hedqvist, H. Lagercrantz & Å. Wennmalm, Eds. (Chapter 19): 201–210. Academic Press. New York, N.Y.

13. BLAKELEY, A. G. H., A. MATHIE & S. A. PETERSEN. 1984. Facilitation at single release sites of a sympathetic neuroeffector junction in the mouse. J. Physiol. London **349:** 57–71.

14. CUBEDDU, L., E. M. BARNES, S. Z. LANGER & N. WEINER. 1974. Release of norepinephrine and dopamine-β-hydroxylase by nerve stimulation. I. Role of neuronal and extraneuronal uptake and alpha presynaptic receptors. J. Pharmacol. Exp. Ther. **190:** 431–450.

15. KALSNER, S. 1982. Evidence against the unitary hypothesis of agonist and antagonist action at presynaptic adrenoceptors. Br. J. Pharmacol. **77:** 375–380.

16. KALSNER, S. 1985. Clonidine and presynaptic adrencoceptor theory. Br. J. Pharmacol. **85:** 143–147.

17. BAKER, D. J., G. M. DREW & A. HILDITCH. 1984. Presynaptic α-adrenoceptors: Do exogenous and neuronally released noradrenaline act at different sites? Br. J. Pharmacol. **81:** 457–464.

18. KALSNER, S. & M. QUILLAN. 1988. Presynaptic interactions between acetylcholine and adrenergic antagonists on norepinephrine release. J. Pharmacol. Exp. Ther. **244:** 879–891.

19. KALSNER, S. & M. QUILLAN. 1984. A hypothesis to explain the presynaptic effects of adrenoceptor antagonists. Br. J. Pharmacol. **82:** 515–522.

20. MARKIEWICZ, M., I. MARSHALL & P. A. NASMYTH. 1980. Lack of feedback via presynaptic α-adrenoceptors by noradrenaline released by a single pulse. Br. J. Pharmacol. **69:** 343.

21. KALSNER, S. 1983. Yohimbine and prolongation of stimulation pulse duration alter similarly 3H-transmitter efflux in heart: an alternative to the negative feedback hypothesis. Br. J. Pharmacol. **79:** 985–992.

22. HILLE, B. 1984. Ionic Channels of Excitable Membranes. Sinauer Associates Inc. Sunderland, Mass.

23. DZIELAK, D. J., Å. THURESON-KLEIN & R. L. KLEIN. 1983. Local modulation of neurotransmitter release in bovine splenic vein. Blood Vessels **20:** 122–134.

24. ABRAHAMSEN, J. & O. A. NEDERGAARD. 1981. Br. J. Pharmacol. 943P.

25. ARBILLA, S., J. Z. NOWAK & S. Z. LANGER. 1985. Rapid desensitization of presynaptic dopamine autoreceptors during exposure to exogenous dopamine. Brain Res. **337:** 11–17.

26. CARR, S. R. & J. R. FOZARD. 1981. Lack of modulation by presynaptic α_2-adrenoceptors of adrenergic transmitter release evoked by activation of 5-hydroxytryptamine and nicotine receptors. Eur. J. Pharmacol. **72:** 27–34.

27. STJARNE, L. 1980. Presynaptic regulation of the innervation of blood vessels. *In* Vascular Neuroeffector Mechanisms. J. A. Bevan, *et al.,* Eds.: 129–137. Raven Press. New York, N.Y.

28. STJARNE, L. 1973. Frequency dependence of dual negative feedback control of secretion of sympathetic neurotransmitter in guinea-pig vas deferens. Br. J. Pharmacol. **49:** 358–360.
29. STORY, D. F., M. W. MCCULLOCH, M. J. RAND & C. A. STANDFORD-STARR. 1981. Conditions required for the inhibitory feedback loop in noradrenergic transmission. Nature **293:** 62–65.
30. DALY, R. N., P. B. GOLDBERG & J. ROBERTS. 1989. The effect of age on presynaptic alpha 2 adrenoceptor autoregulation of norepinephrine release. J. Gerontol. **44:** 859–866.
31. WAKADE, A. R. 1980. A maximum contraction and substantial quantities of tritium can be obtained from tetraethylammonium-treated [^{3}H]-noradrenaline preloaded, rat vas deferens in response to a single electrical shock. Br. J. Pharmacol. **68:** 425–436.
32. GUIMARÃES, S., F. BRANDÃO & M. Q. PAIVA. 1978. A study of adrenoceptor feedback mechanisms by using adrenaline as a false transmitter. Naunyn Schmiedebergs Arch. Pharmacol. **305:** 185–188.
33. DART, A. M., R. DIETZ, K. HIERONYMUS, W. KÜBLER, E. MAYER, A. SCHÖMIG & R. STRASSER. 1984. Effects of α- and β-adrenoceptor blockade on the neurally evoked overflow of endogenous noradrenaline from the rat isolated heart. Br. J. Pharmacol. **81:** 475–478.
34. KAHAN, T., P. HJEMDAHL & C. DAHLOF. 1984. Relationship between the overflow of endogenous and radiolabelled noradrenaline from canine blood perfused gracilus muscle. Acta Physiol. Scand. **122:** 571–582.
35. SAELENS, D. A. & P. B. WILLIAMS. 1983. Evidence for prejuctional α- and β-adrenoceptors in the canine saphenous vein: influence of frequency of stimulation and external calcium contraction. J. Cardiovasc. Pharmacol. **5:** 598–603.
36. MEDGETT, I. C. & M. J. RAND. 1981. Dual effects of clonidine on rat prejuctional α-adrenoceptors. Clin. Exp. Pharmacol. Physiol. **8:** 503–507.
37. BELL, C. & M. VOGT. 1971. Release of endogenous noradrenaline from an isolated muscular artery. J. Physiol. London **215:** 509–520.
38. NEDERGAARD, O. A. & J. ABRAHAMSEN. 1988. Effect of chlorpromazine on sympathetic neurofeffector transmission in the rabbit isolated pulmonary artery and aorta. Br. J. Pharmacol. **93:** 23–24.
39. VIZI, E. S., G. T. SOMOGYI, P. HADHAZY & J. KNOLL. 1973. Effect of duration and frequency of stimulation on the presynaptic inhibition by α-adrenoceptor stimulation of the adrenergic transmission. Naunyn Schmiedebergs Arch. Pharmacol. **280:** 79–91.
40. MCCULLOCH, M. W., J. A. BEVAN & C. SU 1975. Effects of phenoxybenzamine and norepinephrine on transmitter release in the pulmonary artery of the rabbit. Blood Vessels **12:** 122–133.
41. KIRPEKAR, S. M. & P. CERVONI. 1963. Effect of cocaine, phenoxybenzamine and phentolamine on the catecholamine output from spleen and adrenal medulla. J. Pharmacol. Exp. Ther. **142:** 59–70.
42. KURIYAMA, H. & Y. MAKITA. 1982. Modulation of neuromuscular transmission by endogenous and exogenous prostaglandins in the guinea-pig mesenteric artery. J. Physiol. London **327:** 431–448.
43. SUZUKI, H. 1984. Adrenergic transmission in the dog mesenteric vein and its modulation by α-adrenoceptor antagonists. Br. J. Pharmacol. **81:** 479–489.
44. WAKADE, A. R. & T. D. WAKADE. 1981. Release of noradernaline by one pulse: modulation of such release by alpha-adrenoceptor antagonists and uptake blockers. Naunyn Schmiedebergs Arch. Pharmacol. **317:** 302–309.

Conditions for the Operation of Presynaptic Receptors[a]

D. F. STORY, M. J. RAND, C. A. STANDFORD-STARR,
AND M. A. WIDODO

Department of Pharmacology
University of Melbourne
Victoria 3052, Parkville
Australia

Autoregulation of neurotransmitter release refers to control of the transmission process by the action of substances released from the nerve terminals at receptors located on the terminals or on the preterminal axons. When the concept of autoregulation was first expounded, the mediator was taken to be the primary transmitter; however, other substances released from the nerve terminals may also act prejunctionally and thereby influence subsequent release. In this expanded concept of autoinhibition the role of cotransmitters must be considered.[1,2] The term autoreceptor was introduced by Carlsson to define the prejunctional receptors subserving autoregulation.[3]

It is not our intention here to review the evidence for the existence of autoregulation at various synapses and neuroeffector junctions. Interested readers should refer to the following reviews: References 4–9. We will deal mainly with autoinhibitory regulation of norepinephrine release at sympathetic neuroeffector sites. In addition, experimental findings from two studies in our laboratory which bear on the conditions under which prejunctional receptor control of norepinephrine release operates will be presented. The first study is concerned with the activation of autoinhibitory regulation of cardiac noradrenergic transmission from the onset of trains of neuronal stimulation, and the second, to a set of experimental conditions under which we observed prejunctional receptor modulation of nonexocytotic release of norepinephrine. At the outset, some general issues will be raised relating to the conditions of operation of the inhibitory feedback loop in noradrenergic transmission.

CONDITIONS OF OPERATION OF INHIBITORY AUTOREGULATION OF NORADRENERGIC TRANSMISSION

Inhibitory autoregulation of transmitter norepinephrine release through prejunctional α-adrenoceptors seems to be a universal characteristic of noradrenergic neuroeffector and synaptic transmission. However, it is important to recognize that under certain conditions the feedback system may not be operational. The usual criterion for postulating that the release of a neurotransmitter is subject to autoinhibitory feedback regulation through a prejunctional receptor mechanism is that its release is enhanced by specific antagonists of the prejunctional receptors involved. Conversely, failure to enhance transmitter release by blockade of prejunctional receptors for the putative transmitter may be taken as evidence that autoinhibition is not operational, at least

[a]The research from the authors' laboratory reported in this publication was supported by a National Health and Medical Research Council of Australia Program Grant. Dr. Widodo was sponsored by The Australian International Development Assistance Bureau.

under the particular experimental conditions. We have recently reviewed the conditions of operation of autoregulation of noradrenergic transmission.[10] The principal conditions under which autoinhibition of norepinephrine release is not experessed are stimulation with single pulses; stimulation with short trains of pulses; stimulation at very low frequencies; stimulation at high frequencies; and stimulation when the transmitter pool is depleted.[10]

Two of the conditions under which feedback regulation of norepinephrine release from sympathetic nerves does not operate were noted by Story et al.[11] The conclusions arose from a study that was undertaken to seek an explanation for the findings of Angus and Korner that, when the intramural sympathetic nerves on guinea pig atrial preparations were stimulated by brief trains of field pulses, the chronotropic response of the atria, which was taken as an index of transmitter release, was not enhanced by α-adrenoceptor antagonists.[12] We demonstrated that there was a time delay before the autoinhibitory feedback effect through prejunctional α-adrenoceptors is mainfested, and that the duration of the trains of stimulation employed in the experiments of Angus and Korner were within this latent period. Thus, in guinea-pig atria in which the noradrenergic transmitter stores had been labeled with ^{3}H-norepinephrine, phentolamine failed to enhance the efflux of the radioactive label evoked by field stimulation with trains of 4 pulses delivered at a frequency of 2 Hz (corresponding to a 1.5 second train duration). However, when stimulation was with a longer train of pulses, (16 pulses at 4 Hz, corresponding to a 7.5 second train duration), phentolamine increased the stimulation-evoked efflux more than 2.5-fold.[11] It was also noted that phentolamine did not enhance the efflux of radioactivity evoked from the atria by four pulses of field stimulation if there was a long interval between pulses (24 seconds). It was concluded from these observations that there is a latent period, from the onset of neuronal activity, before autoinhibition of noradrenergic transmission is expressed, and that once activated, the inhibition will decay, such that there is a limited period during which transmitter release evoked by a subsequent nerve impulse may be influenced. We pointed out that these are general characteristics of many receptor-mediated events and that the magnitude of the latency and the period of persistence of autoinhibition of noradrenergic transmission in guinea pig atria suggest the involvement of an intracellular mediator.[11] Thus, the latent period can be attributed to the time required for the formation of a second messenger and the expression of its effects resulting in modulation of the transmitter release mechanisms, and the persistence reflects the time for the decay of the effects.

A delay in the onset of autoinhibition of transmitter release from the noradrenergic nerves in the rabbit ear artery has also been reported.[13] Thus, from the onset of field stimulation of the periarterial sympathetic nerves of isolated artery preparations with trains of pulses at a frequency of 2 Hz, autoinhibition did not operate until about 4.5 seconds after delivery of the first pulse. The similar magnitudes of the latencies of autoinhibition of transmitter release and of the development of the vasoconstrictor response to stimulation were pointed out, and it was suggested that the delayed onset of autoinhibition may be a physiologically relevant factor in the rate of development of the postjunctional response.[13]

A delay in the onset of autoinhibition was also observed in experiments in which the cardiac sympathetic spinal outflow in pithed rats was stimulated with 8 pulses at frequencies ranging from 2 to 32 Hz. In this case the selective α_2-adrenoceptor antagonist yohimbine was found to increase the chronotropic response when stimulation was at 2 or 4 Hz but not that with 8, 16, or 32 Hz, indicating a delay in onset of autoinhibition of about 1 second.[14]

It is likely that the latency of autoregulation of norepinephrine release is influenced by the rate of activation of the prejunctional α_2-adrenoceptors subserving the inhibition of release and, hence, on the frequency of neuronal activity. In line with this, the tachycardia response to stimulation of the spinal sympathetic outflow between C7 and T1 with 6 pulses at 0.5 Hz was not enhanced by phentolamine, whereas that elicited by the same period of stimulation at 2 Hz (10 seconds) was enhanced.[15] Presumably, with the higher frequency, the concentration of norepinephrine in the vicinity of the prejunctional receptors was maintained at a level sufficient to produce autoinhibition of transmitter release.

The persistence of feedback inhibition of transmitter release has been extensively explored in rat isolated vasa deferentia preparations. It has been reported that a train of pulses delivered at a frequency of 10 Hz within 10 seconds of subsequent stimulation with a single pulse inhibited the contractile response of the rat vas deferens to the single pulse, whereas if the interval was extended to 1 minute, there was no inhibition of the response to the single pulse.[16] Moreover, the inhibitory effect of the conditioning train of stimulation was prevented by phentolamine. However, these findings, although relevant to the time course of feedback action of norepinephrine released from nerves, must be interpreted in the light of evidence that the motor response of the rat vas deferens to sympathetic nerve stimulation is predominantly due to the release of a nonnoradrenergic transmitter.[17,18] Thus, the experiments of Vizi et al.[16] presumably demonstrate inhibition of release of one transmitter by a cotransmitter.[10]

More recently, the persistence of the inhibitory effect of norepinephrine released from sympathetic nerves in the rat vas deferens has been explored systematically by French and Scott.[19–22] These studies involved either the application of conditioning trains of stimulation before the application of a single test pulse, or the application of a single pulse before a train of test pulses, or the application of two single pulses with various intervals between them. Essentially, the findings of these studies confirm the concepts of latency and decay of autoinhibition.

A ROLE FOR AUTOINHIBITION IN NORADRENERGIC TRANSMISSION?

The concept of autoinhibition of noradrenergic transmission provides that norepinephrine released from the nerve terminals activates receptors associated with the terminal axons to inhibit subsequent transmitter release. Given that there is a latency in onset of autoinhibition, it follows that after a delay, the feedback mechanism should activate progressively with the arrival of successive impulses until an equilibrium condition is attained. However, electrophysiological studies in which the excitatory junction potential (ejp) of smooth muscle cells in the mouse vas deferens was taken as an index of transmitter release indicate that the release increases with the first few nerve impulses in a train, up to a plateau level.[23–25] The rate and extent of the growth of the ejp from the onset of stimulation increases with the frequency of stimulation. The phenomenon is referred to as frequency-dependent facilitation of transmission.[23,25] Thus, early in a train of stimulation, transmitter release from noradrenergic nerves is subject to two opposing influences: autoinhibition and frequency-dependent facilitation.

We have explored the interaction between autoinhibition and frequency-dependent facilitation by investigating the changes in transmitter release from the sympathetic nerves in rat isolated atria. The release of radioactivity after labeling transmitter stores with ^{3}H-norepinephrine was taken as the primary index of transmitter release; in addition, the stimulation-induced efflux of radioactivity was correlated with the peak chronotropic responses of the atria.

Rat isolated atria were set up under a diastolic tension of 1 g wt, in an organ bath containing 2 ml of physiological salt solution (PSS) of the following composition (mM): NaCl, 118; KCl, 4.7; NaHCO$_3$, 25; MgSO$_3$, 0.45; KH$_2$PO$_4$, 1.03; CaCl$_2$, 2.5; D-(+)-glucose, 11.1; sodium edetate, 0.07; ascorbic acid, 0.142. Atropine (1 μM) was present in the PSS to block the pre- and postjunctional actions of acetylcholine at muscarinic cholinoceptors. The solution in the organ bath and in the reservoirs supplying it was maintained at 37°C and continuously gassed with 5% CO$_2$ in O$_2$. Two platinum wire electrodes were located, one on each side of the atria, for applying field stimulation. After a 15 minute period of equilibration, the noradrenergic transmitter in the atria was labeled by incubating the preparations with (−)-[7,8-^{3}H]-norepinephrine (10 μCi/ml, specific activity 8–14.5 Ci/mmol) for 30 minutes, followed by a 60 minute washout with repeated exchanges of the PSS in the organ bath with norepinephrine-free solution. After 30 minutes of washing, a 30 second train of 1 msecond field pulses at 1 Hz was applied to the atria to assist in clearing unspecifically bound radiolabeled compounds.

Each atria preparation was subjected to 5 periods of field stimulation at 30 minute intervals with trains of 1, 2, 4, 8, and 16 pulses delivered in random sequence, using a latin square design with 5 atria in each block of experiments. The stimulating pulses were 1 msecond square waves of supramaximal voltage (15 V/cm). The stimulation-induced (S-I) release of norepinephrine was deduced from the efflux of radioactivity from the atria which was measured in 30 second collections of the PSS in contact with the atria. The S-I efflux was taken as the sum of amounts of radioactivity exceeding the prestimulation level in 5 successive 30 second collections taken from the beginning of stimulation. When the duration of stimulation exceeded 30 seconds, the collection period during which stimulation was applied was extended and an appropriate correction for the basal efflux was applied. Contractions of the atria were monitored using a strain gauge transducer and a pen recorder.

S-I Efflux per Train

As shown in FIGURE 1a, with stimulation at 0.25 and 2 Hz, the total S-I efflux of radioactivity from the labeled atria increased linearly with the number of pulses in the train of stimulation. The slope of the relationship between S-I efflux per train and the number of pulses was significantly greater ($p < 0.05$, unpaired t-test) with stimulation at 2 Hz than with 0.25 Hz.

Mean S-I Efflux per Pulse

Since with both frequencies of stimulation, a linear relationship of the form $y = ax + b$ exists between the evoked efflux (y) and the number of pulses in the train of stimulation (x), the relationship between average evoked efflux per pulse (y/x) with each train of stimulation and the number of pulses must be a hyperbola of the form $y/x = a/x + b$. Hence, the data plotted in FIGURE 1a may be transformed into plots of average efflux per pulse *versus* the number of pulses in the trains of stimulation. The transformed data for control experiments with stimulation at 0.25 and 2 Hz are shown in FIGURE 1b.

The mean efflux per pulse rose rapidly with the number of pulses in the train and reached a plateau level when stimulation was with trains of 4–8 pulses: the plateau level of efflux was greater with the higher frequency of stimulation. The findings are thus consistent with frequency-dependent facilitation of noradrenergic transmission. However, if the ratio of the efflux evoked by a single pulse to the mean efflux per pulse in trains of 16 pulses is taken as a measure of facilitation of transmitter release, there was no significant difference between the ratio determined for the two frequencies of stimulation. Thus the "facilitation ratio" was 1.8 [standard error of the mean (SEM) = 0.2] for both 0.25 and 2 Hz stimulation.

Effect of Phentolamine

The effect of phentolamine (0.3 μM) on the mean efflux of radioactivity per pulse with trains of 1 to 16 pulses of stimulation at 0.25 and 2 Hz is shown in FIGURE 2.

Phentolamine (3 μM) markedly increased the plateau levels of the relationship between the S-I efflux of radioactivity per pulse and the number of pulses per train for stimulation at 0.25 and 2 Hz. Thus, there was enhanced facilitation of transmitter

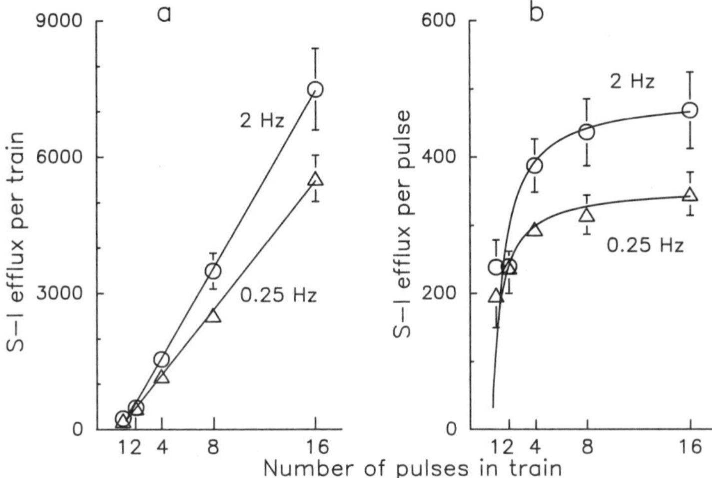

FIGURE 1. Stimulation-induced (S-I) effluxes of radioactivity as a function of the number of pulses in trains of stimulation with frequencies of 0.25 Hz ($n = 5$) and 2 Hz ($n = 5$). The symbols are means and standard errors; in some cases, the standard error is smaller than the size of the symbols. (a) S-I effluxes per train of stimulation. The lines are calculated regression lines. (b) Effluxes per pulse. Transformation of data on S-I effluxes per train of stimulation (see text). The hyperbola are derived from the regression lines.

efflux in the presence of phentolamine and this was much greater with the higher frequency of stimulation (FIGURE 2). The findings suggest that frequency-dependent facilitation of norephinephrine release is normally restrained by autoinhibition, subserved by prejunctional α-adrenoceptors.

With both frequencies of stimulation, there were significant linear relationships between the positive chronotropic response to stimulation (peak increase in rate of atrial beating) and the logarithm of the S-I efflux of radioactivity. In control experiments, the correlation coefficients (r) were 0.88 and 0.86 ($p < 0.001$ in each case) for stimulation with 0.25 and 2 Hz respectively. The linear correlations between the atrial response to stimulation and log efflux were maintained when phentolamine (3 μM) was present, the slope being slightly increased ($p < 0.05$) with stimulation at 0.25 Hz, but not with stimulation at 2 Hz.

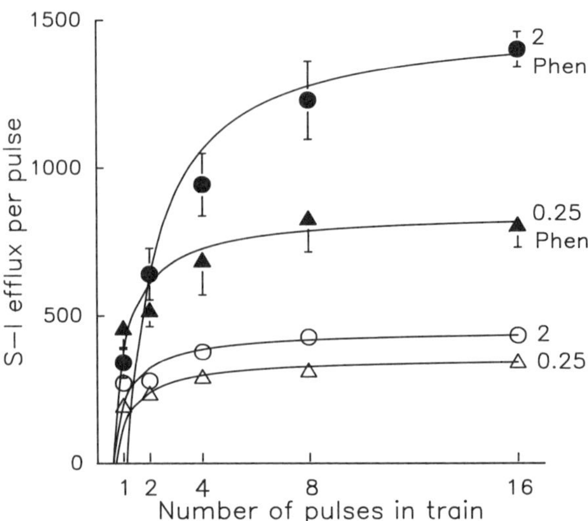

FIGURE 2. Effect of phentolamine (3 μM) on the relationship between stimulation-induced (S-I) efflux per pulse (see text) and the number of pulses of stimulation with trains of pulses at 0.25 Hz and 2 Hz. The data from experiments in the absence of phentolamine are from FIGURE 1. The symbols are means and standard errors; in some cases, the standard error is smaller than the size of the symbols. In each case $n = 5$.

Effect of Blockade of Neuronal Uptake

Blockade of neuronal uptake with desipramine (0.1 μM) altered the relationship between mean S-I efflux per pulse and the number of pulses in a manner strikingly different from that of phentolamine. With stimulation at 2 Hz, desipramine enhanced the mean efflux per pulse for each train of stimulation with 1 to 16 pulses. Thus, as shown in FIGURE 3, instead of a rising hyperbola, the relationship between efflux per pulse and the number of pulses was almost a constant. Thus, there was no evidence of facilitation of transmitter release with trains of increasing number of pulses, up to 16. The upward displacement of the relationship is presumably due to blockade of neuronal reuptake of released norepinephrine. The lack of facilitation is probably due to enhanced autoinhibition of release due to higher concentrations of norepinephrine in the biophase of the prejunctional autoreceptors. This suggestion is supported by the effects of a combination of phentolamine (3 μM) and desipramine (0.1 μM) on the relationship between S-I efflux per pulse and the number of pulses with stimulation at 2 Hz. In this case, the additional presence of phentolamine restored the relationship to a rising hyperbola and increased the plateau level over that obtained with phentolamine alone (FIGURE 3). The marked facilitation observed in the presence of phentolamine and desipramine, in contrast to the apparent absence of facilitation when desipramine was present alone, indicates the very considerable role of autoinhibition in restraining facilitation.

The evidence for the marked influence of autoinhibition of norepinephrine release in rat atria with short trains of pulses of stimulation is difficult to reconcile with the conclusion of Dyke and Angus[26] that it is of little consequence. These authors used the chronotropic responses of the atrial preparations as the index of transmitter release, whereas we monitored the S-I efflux. However, in our study there was excellent correlation between effluxes and responses.

The major conclusion we draw from our studies is that autoinhibition of norepinephrine release, at least in the rat atria, provides restraint on frequency-dependent facilitation. The view that these two mechanisms interact as determinants of norepinephrine release has previously been proposed.[25]

MODULATION OF NONEXOCYTOTIC RELEASE OF ³H-NOREPINEPHRINE BY PREJUNCTIONAL RECEPTOR MECHANISMS

Norepinephrine in noradrenergic nerve terminals is predominantly located within the "small" and "large" dense-cored amine storage vesicles with only a small proportion present free in the neuronal cytoplasm (see review by Stjärne).[27] It is generally considered that nerve impulse–evoked release of neurotransmitters, including norepinephrine, is through exocytosis of the contents of synaptic vesicles. The process is triggered by influx of Ca^{2+} through specific voltage-dependent channels in the neuronal membrane, the calcium channels being activated by depolarization of the terminals caused by invading action potentials (see review by Zimmerman).[29] Although norepinephrine may be released from nerve terminals by mechanisms not involving depolarization, and hence Ca^{2+} entry, for example, "displacement" release by indirectly acting sympathomimetic amines, the ability of prejunctional receptor mechanisms to modulate release is restricted to that involving depolarization.[30] Although the mechanisms by which prejunctional receptors modulate transmitter release are as yet unknown,

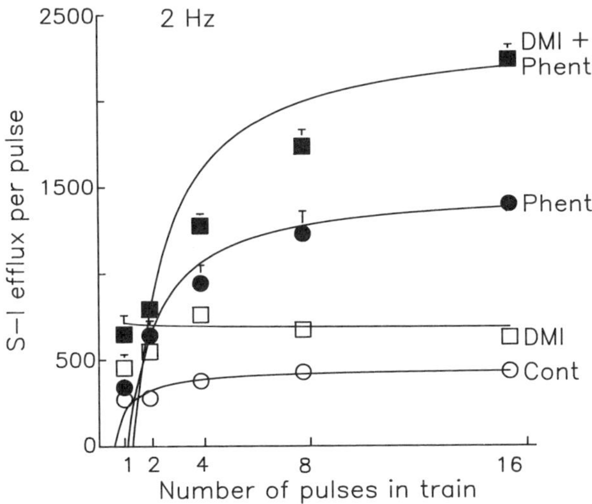

FIGURE 3. Effects of desipramine (0.1 μM) and phentolamine (3 μM) on the relationship between stimulation-induced (S-I) efflux per pulse (see text) and the number of pulses of stimulation with trains of pulses at 2 Hz. The data from experiments in the absence of phentolamine are from FIGURE 1 and those from experiments in the presence of phentolamine are from FIGURE 2. The symbols are means and standard errors; in some cases, the standard error is smaller than the size of the symbol. In each case $n = 5$.

modulation of Ca^{2+} entry into nerve terminals has often been proposed as a primary or sequential consequence of activation of the control systems.[1,27,30] One possibility that has been frequently porposed is that changes in the amount of Ca^{2+} entering the nerve terminals are secondary to altered K^+ conductance, rather than being due to a primary effect on Ca^{2+} conductance.[1] There are considerable technical difficulties in studying the intracellular mechanisms of presynaptic receptor control of transmitter release. Not the least of these results from the complexity and uncertainty of the mechanism coupling Ca^{2+} entry to exocytosis, and of the mechanism of exocytosis. We have recently developed a model system that isolates the membrane ionic conductance changes evoked by action potential invasion of varicosities from the process of exocytosis. Preliminary findings suggest that the model may be useful in the study of presynaptic receptor control of transmitter release.

Kostyuk et al.[31] demonstrated that the voltage-dependent calcium channels of certain molluscan neurones acquire the ability to admit monovalent cations when exposed to a calcium-chelating agent such as ethylenediamine tetraacetic acid (EDTA) in a Ca^{2+}-free medium. We have now demonstrated that, under similar conditions, the calcium channels of noradrenergic nerve terminals in rat atria admit Na^+ when they are depolarized.[32] If norepinephrine is present in the cytoplasm of the terminals, the increase in $[Na^+]_i$ leads to activation of the neuronal amine pump, resulting in the cotransport out of the terminals of norepinephrine and Na^+. Under these conditions, the amount of norepinephrine released is presumably dependent on the conductance (of Na^+) through the modified calcium channels, and on the activity of the amine carrier.

Atria were taken from rats that had been pretreated with reserpine [5 mg/kg, subcutaneously (sc), 20 hours and 2.5 mg/kg, intraperitoneally (ip), 3 hours previously] and pargyline (100 mg/kg, ip, 3 hours previously). These treatments were to inhibit the storage of norepinephrine in neuronal vesicles and to inhibit monoamine oxidase, respectively. The rats were killed and their atria dissected and set up in 4 ml of PSS from which Ca^{2+} had been omitted. The PSS contained the catechol-O-methyltransferase inhibitor U-0521 (5 μM) which remained present throughout, and pargyline (0.5 μM) was also present for the first 30 minutes only. The PSS was gassed with 5% CO_2 in O_2 and maintained at 37°C throughout. A low concentration of EDTA (0.07 mM) was present initially in the PSS and this was increased to 0.3 mM during the course of the experiment (see below). After 30 minutes equilibration, ^{3}H-norepinephrine was incorporated into the cytoplasm of noradrenergic terminals by incubating the atria for 30 minutes with the labeled amine (40 nCi/ml, 3.2 nM). After the incubation, the atrial preparations were washed for 100 minutes with norepinephrine- and Ca^{2+}-free PSS. The above procedures are essentially the same as those described by Bönisch in studies in which he investigated carrier-facilitated exchange of indirectly acting sympathomimetic amines with ^{3}H-norepinephrine present in the cytoplasm of noradrenergic nerves.[33]

After the period of washing, when the efflux of radioactivity from the atria had declined to a relatively constant level, we have established that the radioactivity retained in the atria is present almost entirely as ^{3}H-norepinephrine, and is dependent upon the presence of intact noradrenergic nerve terminals (Widodo and Story, unpublished).

The intramural noradrenergic nerves of the atria were subjected to two periods of field stimulation with a 30 minute interval between them, and the resting and S-I effluxes of radioactivity from the atria were determined in 1 minute collections of the bathing solution.

Figure 4 shows the effect of the two periods of field stimulation (2 Hz, 10 second trains of pulses) on the efflux of radioactivity from atrial preparations from reserpine-

and pargyline-treated rats. Note that the concentration of EDTA in the bathing solution was increased from 0.07 mM to 0.03 mM, 20 minutes before the second period of stimulation.

In the presence of 0.07 mM EDTA, the first period of stimulation consistently evoked a small efflux of radioactivity; the increase in the 1 minute period in which the stimulation was delivered over the prestimulation resting level amounted to about 20%. In the presence of the higher concentration of EDTA, the S-I efflux was markedly enhanced (FIGURE 4). The S-I efflux in the presence of 0.3 mM EDTA amounted to about 11% of the tissue content of radioactivity.

When, in addition to increasing the concentration of EDTA, the neuronal uptake inhibitor desipramine (1 μM) was introduced into the PSS bathing the atria before the second period of stimulation, the S-I efflux of radioactivity was almost abolished

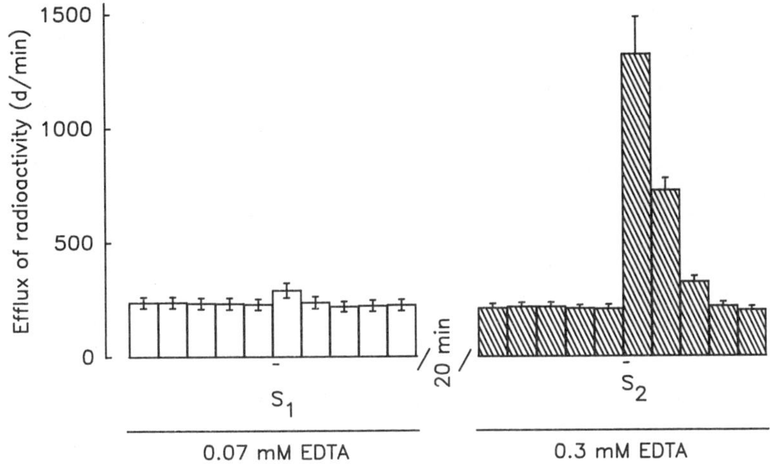

FIGURE 4. Efflux of radioactivity in 1 minute periods from rat atria in which ^{3}H-norepinephrine had been incorporated into the cytoplasm of the atrial noradrenergic nerves (see text). The atrial noradrenergic nerves were subjected to two 10 second periods of field stimulation (S$_1$ and S$_2$) at a frequency of 2 Hz, with a 30 minute interval. Note that Ca^{2+} was not present in the solution bathing the atria and that EDTA was present in concentrations of 0.07 mM and 0.3 mM for the first and second periods, respectively.

(Table 1). Indicating that the release was due to outward transport of ^{3}H-norepinephrine from noradrenergic nerves by the amine carrier mechanism. The S-I efflux in the second period was also almost abolished by blockade of voltage-dependent Na$^+$ channels with 1 μM tetrodotoxin (TABLE 1). Thus, the release of cytoplasmic ^{3}H-norepinephrine in the presence of EDTA would appear to depend on the generation and propagation of action potentials in the atrial sympathetic nerves.

The involvement of modified voltage-dependent Ca^{2+} channels in the S-I efflux is also suggested by the data summarized in TABLE 1. Thus, the S-I efflux with the second period of stimulation was markedly reduced when CdCl$_2$ (0.1 mM) was introduced into the PSS bathing the atria. Moreover, the potent inhibitor of voltage-dependent N-type calcium channels omega-conotoxin (30 nM)[34,35] also strongly inhibited S-I efflux. Thus, despite the complete absence of Ca^{2+} in the atrial bathing solution, it appears that the S-I efflux of cytoplasmic ^{3}H-norepinephrine involves a transfer of ions into the nerve terminals through voltage-dependent Ca^{2+} channels. That the release is nonexocytotic is indicated by the finding that introduction of Ca^{2+} (2.5 mM) before the second period of stimulation did not enhance efflux but almost abolished it (TABLE 1). The involvement of Ca^{2+} channels in the S-I release is also suggested by the finding that tetraethylammonium (TEA, 3 mM) enhanced release (TABLE 1). It is well known that TEA enhances transmitter release from noradrenergic nerves. The effect is generally attributed to blockade of K^+ channels at nerve terminals, thereby delaying repolarization after invasion of the terminals by an action potential, resulting in enhanced Ca^{2+} entry and, consequently, increased transmitter release. Presumably, under the present experimental conditions, TEA enhanced the conductance of Na^+ through the EDTA-modified Ca^{2+} channels.

Taken together, the findings summarized in TABLE 1 strongly suggest that the S-I efflux of cytoplasmic norepinephrine results from the invasion of varicosities of the terminal sympathetic nerves of the atria by action potentials; the resulting depolarization of the varicosities activates Ca^{2+} channels which have been modified in such a way that they conduct monovalent cations (predominantly Na^+) into the terminals, as proposed by Kostyuk.[31] The increased levels of Na^+ on the inner side of the neuronal membrane provide for the outward transport of cytoplasmic ^{3}H-norepinephrine by the amine transport carrier.

The mechanism of norepinephrine release proposed above is dependent on the ion channels that underly the process of transmitter release and on the amine transport system. It does not involve exocytosis or the mechanisms normally coupling Ca^{2+} entry with exocytosis. Hence, the model may provide a means of investigating the consequences of activation and blockade of prejunctional receptor systems to the entry of ions into the nerve terminals through voltage-dependent calcium channels. Accordingly, experiments were undertaken to investigate if S-I release of cytoplasmic norepinephrine in this model was subject to modulation by activation of inhibitory α_2-adrenoceptors, muscarinic cholinoceptors, or facilitatory angiotension II receptors.

As shown in FIGURE 5, introduction of the selective α_2-adrenoceptor agonist UK 14304 (0.3 μM), 10 minutes before the second period of stimulation significantly reduced the S-I efflux of radioactivity. The inhibitory effect of UK 14304 was prevented by the selective α_2-adrenoceptor antagonist idazoxan (1 μM) (FIGURE 5), indicating that the effect of the agonist was a consequence of α_2-adrenoceptor activation. Indazoxan alone did not alter S-I release, indicating that there was no appreciable activation of the inhibitory α_2-adrenoceptors by endogenously released norepinephrine.

TABLE 1. Effects of Various Drugs on the Stimulation-Induced (S-I) Release of Radioactivity from the Cytoplasm of Sympathetic Nerves in Rat Atria

Drug Present for Second Period of Stimulation	S-I Efflux with Second Stimulation (percent of tissue radioactivity)		
	Mean	SEM	n
None	10.6	1.0	9
Desipramine (1 μM)	1.1[a]	0.3	5
Tetrodotoxin (1 μM)	0.8[a]	0.4	4
$CdCl_2$ (0.1 mM)	1.9[a]	0.9	4
Omega-conotoxin (0.1 nM)	0.2[a]	0.02	4
Ca^{2+} (2.5) mM	0.2[a]	0.07	4
Tetraethylammonium (3 mM)	14.0[a]	0.3	3

[a]Significantly different from control ($p < 0.05$, t-test).

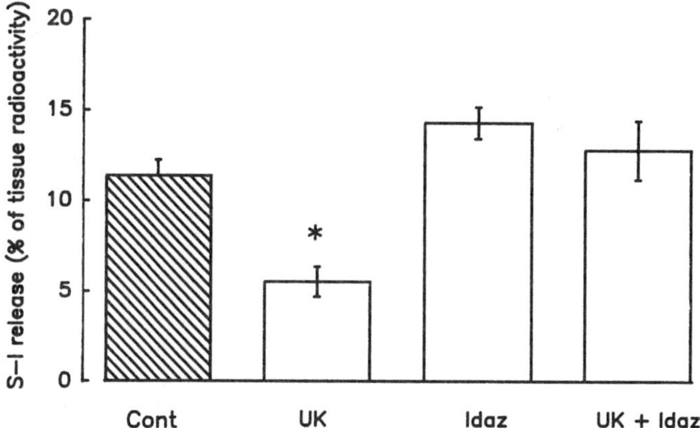

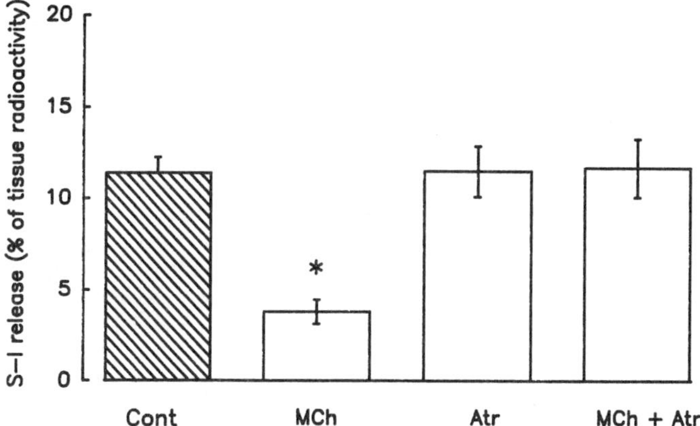

FIGURE 5. Top panel: Effect of UK 14304 (UK, 0.3 μM), idazoxan (Idaz, 1 μM), and UK 14304 in the presence of idazoxan on the stimulation-induced (S-I) efflux of radioactivity from rat atria in which ^{3}H-norepinephrine had been incorporated into the cytoplasm of the atrial noradrenergic nerves (see text). The atrial noradrenergic nerves were subjected to two 10 second periods of field stimulation (S_1 and S_2) at a frequency of 2 Hz, with a 30 minute interval. Note that Ca^{2+} was not present in the solution bathing the atria and that EDTA was present in concentration of 0.07 mM and 0.3 mM for the first and second periods, respectively. The vertical bars show the mean content of radioactivity ($\pm$ SEM) in 1 minute collections of the atrial bathing solution. There were 3 to 9 atria in each group. The asterisk indicates a significant difference ($p < 0.05$, t-test) from control (Cont). **Lower panel:** Effects of methacholine (30 μM), atropine (0.3 μM), and methacholine in the presence of idazoxan on the S-I efflux of radioactivity from atrial preparations radiolabeled with ^{3}H-norepinephrine as above. The experimental conditions are as specified above.

The muscarinic cholinoceptor agonist methacholine (30 μM) also inhibited S-I release. The effect of methacholine was abolished by atropine (0.3 μM), but atropine alone did not alter S-I efflux (FIGURE 5).

In contrast to the findings with the α_2-adrenoceptor and muscarinic agonists, angiotensin II (0.1 μM) did not significantly alter S-I efflux of cytoplasmic ^{3}H-norepinephrine, the S-I efflux for the second period of stimulation in the presence of angiotensin II being 12.3% (SEM = 1.6, $n = 5$) of the tissue content of radioactivity, compared with the control value of 10.6% (SEM = 1.0, $n = 9$). It has been shown in our laboratory that angiotension II enhances the S-I release of transmitter ^{3}H-norepinephrine in "normal" atrial preparations, that is, in preparations in which transmitter release was by exocytosis.[36]

The EDTA-modified voltage-dependent calcium channels of noradrenergic nerves of rat atria appear to retain the pharmacological characteristics of unmodified calcium channels, as has been shown for similarly modified channels of molluscan neurones.[31] This would appear to extend to the ability of TEA to prolong the open state of the channels by prolonging terminal depolarization.

The inhibitory effects of UK 14304 and of methacholine on S-I efflux are clearly exerted as a result of the activation of α-adrenoceptors and muscarinic cholinoceptors, respectively. In this model, these agonists presumably directly or indirectly inhibit the influx of extracellular Na^+ into the neurone terminals through modified calcium channels after they have been activated by depolarization. The findings do not rule out the possibility of the primary effect of prejunctional receptor agonists being to enhance a repolarizing K^+ efflux. Nor do they allow us to differentiate between a primary action at the nerve terminal varicosities and an effect on action potential propagation. However, the inhibitory effects of UK 14304 and methacholine on the release of cytoplasmic ^{3}H-norepinephrine are presumably ultimately due to the modulation of conductance of an ion channel in the terminal or preterminal membrane of the atrial noradrenergic nerves. The technique has the potential to provide insight into receptor-activated mechanisms modulating transmitter release. The failure of angiotensin II to enhance S-I efflux is puzzling in view of the well-established facilitatory effect of angiotensin II on depolarization-evoked exocytotic release of norepinephrine. The lack of effect of angiotensin II in the present experiments, in contrast to the abilities of UK 14304 and methacoline to modulate the release of cytoplasmic ^{3}H-norepinephrine, may indicate that there are fundamental differences in the mechanisms in which different prejunctional receptor agonists modulate transmitter release of noradrenergic terminals.

Finally, the inability of both idazoxan and atropine to enhance the S-I release of cytoplasmic ^{3}H-norepinephrine would indicate that under these experimental conditions there was no activation of either the prejunctional α-adrenoceptors or muscarinic cholinoceptors by endogenous agonists.

REFERENCES

1. STARKE, K., M. GÖTHERT & H. KILBINGER. 1989. **69:** 864–989.
2. RAND, M. J., M. W. McCULLOCH, H. WONG-DUSTING. D. F. STORY, R. E. LOICONO & J. ZIOGAS. 1987. J. Cardiovasc. Pharmacol. **10**(Suppl. 12): S33–S44.
3. CARLSSON, A. 1975. Dopaminergic autoreceptors. *In* Chemical Tools in Catecholamine Research. O. Almgren, A. Carlsson & J. Engel, Eds. **11:** 219–225. Elsevier-North Holland. Amsterdam, the Netherlands.
4. LANGER, S. Z. 1977. Br. J. Pharmacol. **60:** 481–497.
5. STARKE, K. 1977. Rev. Physiol. Biochem. Pharmacol. **77:** 1–124.
6. STARKE, K. 1987. Rev. Physiol. Biochem. Pharmacol. **107:** 73–146.
7. WESTFALL, T. C. 1977. Physiol. Rev. **57:** 659–728.
8. GILLESPIE, J. S. 1980. Presynaptic Receptors in the Autonomic Nervous System. *In* Adrenergic Activators and Inhibitors. L. Szekeres, Ed. (Part I): 353–425. Springer-Verlag. Berlin, FRG.
9. RAND, M J., M. W. McCULLOCH & D. F. STORY. 1980. Catecholamine receptors on nerve

terminals. *In* Adrenergic Activators and Inhibitors. L. Szekeres, Ed. (Part I): 223–266. Springer-Verlag. Berlin, FRG.

10. RAND, M. J. & D. F. STORY. Conditions required for the inhibitory feedback loop in noradrenergic transmission. *In* Presynaptic Regulation of Neurotransmitter release: a Handbook. J. Feigenbaum, Ed. Freund Publishing House. Tel Aviv, Israel. (In press.)

11. STORY, D. F., M. W. McCULLOCH, M. J. RAND & C. A. STANDFORD-STARR. 1981. Nature **293:** 62–65.

12. ANGUS, J. A. & P. I. KORNER. 1980. Nature **286:** 288–291.

13. AUCH-SCHWELK, W., K. STARKE & A. STEPPELER. 1983. Br. J. Pharmacol. **78:** 543–551.

14. DE LUCA, A. & M. J. RAND. 1987. Clin. Exp. Pharmacol. Physiol. **15:** 33–41.

15. RAND, M. J., M. W. McCULLOCH, C. STANDFORD-STARR, D. F. STORY & C. YANG. The time course of the development and persistence of the autoinhibitory effect in noradrenergic transmission. *In* Neurotransmitter Receptors. H. Yoshida, Y. Hagihara & S. Ebashi, Eds.: 121–130. Pergamon Press. Oxford, England.

16. VIZI, E. S., G. T. SOMOGYI, P. HADHAZY & J. KNOLL. 1973. Naunyn Schmiedebergs Arch. Pharmacol. **280:** 79–91.

17. PENNEFATHER, J. N., L. VARDOLOV & P. HEATH. 1974. Clin. Exp. Pharmacol. Physiol. **1:** 451–462.

18. ANTON, P. G., M. E. DUNCAN & J. D McGRATH. 1977. J. Physiol. London. **273:** 23–43.

19. FRENCH, A. M. & N. C. SCOTT. 1981. Br. J. Pharmacol. **74:** 182P.

20. FRENCH, A. M. & N. C. SCOTT. 1981. Br. J. Pharmacol. **74:** 183P.

21. FRENCH, A. M. & N. C. SCOTT. 1983. Eur. J. Pharmacol. **86:** 379–383.

22. FRENCH, A. M. & N. C. SCOTT. 1981. Br. J. Pharmacol. **74:** 616.

23. BENNETT, M. R. 1973. J. Physiol. London. **229:** 515–531.

24. BENNETT, M. R. & J. MIDDLETON. 1975. Br. J. Pharmacol. **55:** 87–95.

25. STJÄRNE, L. 1975. Basic mechanisms and local feedback control of secretion of adrenergic and cholinergic neurotransmitters. *In* Handbook of Psychopharmacology. L. L. Iversen, S. D. Iversen & S. H. Snyder, Eds.: 179–223. Plenum Press. New York, N.Y.

26. DYKE, A. & J. A. ANGUS. 1988. J. Auton. Pharmacol. **8:** 219–228.

27. ALBERTS, P., T. BARTFAI & L. STJÄRNE. 1981. J. Physiol. London. **312:** 297–334.

28. STJÄRNE, L. 1989. Rev. Physiol. Biochem. Pharmacol. **112:** 1–137.

29. ZIMMERMANN, H. 1979. Neuroscience **4:** 1773–1804.

30. LANGER, S. Z. 1981. Pharmacol. Rev. **32:** 337–362.

31. KOSTYUK, P. G., S. L. MIRONOV & YA. M. SHUBA. 1983. J. Membr. Biol. **76:** 83–93.

32. WIDODO, M. A., D. F. STORY & M. J. RAND. 1988. Proc Aust. Physiol. Pharmacol. Soc. **19:** 207P.

33. BÖNISCH, H. 1984. Naunyn Schmiedebergs Arch. Pharmacol. **323:** 233–244.

34. AUGUSTINE, G. J., M. P. CHARLTON & S. J. SMITH. 1987. Annu. Rev. Neurosci. **10:** 633–693.

35. MILLER, R. J. 1987. Science **235:** 46–52.

36. STORY, D. F. & J. ZIOGAS. 1987. Trends in Pharmacol. Sci. **8:** 269–271.

Is the Release of Dopamine from Medial Prefrontal Cortex Modulated by Presynaptic Receptors?

Comparison with Nigrostriatal and Mesolimbic Terminals

LUIGI X. CUBEDDU,[a] IRENE S. HOFFMANN,
AND REIZA K. TALMACIU

Division of Pharmacology
Pharmacy Faculty
Central University of Venezuela
Apartado Nueva Granada
Caracas, Venezuela

INTRODUCTION

This study summarizes our work on the control of dopamine (DA) release via presynaptic DA receptors (autoreceptors) in central DA neurons, with emphasis on the dopaminergic terminals of the medial prefrontal cortex (PFC). Data on the stimulus-release characteristics of the different mesotelencephalic DA terminal fields and the competence of presynaptic release modulatory mechanisms at striatal, limbic, and cortical DA terminals are presented.

DA neuronal systems in the brain are diverse in distribution and function.[1] The largest group of DA neurons have cell bodies in the ventral tegmental region and axons that project to various forebrain regions. Nigrostriatal DA neurons originate in the pars compacta of the substantia nigra and terminate in the striatum (putamen and caudate nucleus). Mesolimbic-mesocortical DA neurons originate from cell bodies in the substantia nigra and the ventromedial tegmentum; they project to limbic forebrain regions (e.g., nucleus accumbens, olfactory tubercle) and to the cerebral cortex (e.g., cingulate, frontal, and entorhinal cortices). Another group of DA neurons is located in the mediobasal hypothalamus. Autoreceptor mechanisms in hypothalamic DA neurons will not be discussed.

The olfactory tubercle (OT) occupies the part of the basal hemisphere wall just caudal to the point of attachment of the olfactory tract. Heimer and coworkers proposed that the OT and the nucleus accumbens constitute the main components of the ventral striatum,[2] whereas the caudate-putamen forms the dorsal striatum. Functionally, the OT is an integral part of the mesolimbic system; the caudate-putamen, on the other hand, is considered part of the extrapyramidal system. The OT and the nucleus caudate have similar DA concentrations.[3,4] The PFC is defined as the cortical area receiving a projection from the mediodorsal nucleus of the thalamus. This region is innervated by DA neurons originating in the ventral mesencephalic tegmentum.[3,5] Most of the DA present in the PFC is located in nerve terminals.[6] Three DA terminal

[a] Address for correspondence: Calle Los Samanes, Res. Florida Palace, #8, La Florida, Caracas, Venezuela.

systems can be demonstrated in the PFC cortex: the anteromedial (medial PFC) which originates from the medial part of the ventral mesencephalic tegmentum; the suprarhinal, which originates from the dorsal-lateral part of the ventral mesencephalic tegmentum; and the supragenual system (cingulate), originating from the lateral part of the ventral mesencephalic tegmentum and the mediolateral extent of the substantia nigra.[7] In general, cortical DA exerts inhibitory effects. Administration of DA into the PFC inhibits the firing rate of the PFC cells.[9]

Electrophysiological and biochemical studies have indicated that the mesocortical DA projections are functionally and pharmacologically different from the better known nigrostriatal and mesolimbic DA projections.[11] Comparatively with the mesolimbic and nigrostriatal DA neurons, the mesocortical DA neurons have faster rates of synthesis, turnover, and firing,[12,13] have different firing patterns characterized by more frequent and intense burst activity,[13] do not exhibit tolerance to haloperidol-induced increases in DOPAC levels, do not develop depolarization inactivation after chronic treatment with neuroleptics,[14,15] are less susceptible to inhibition by DA agonists (firing rates and synthesis rates),[12,13] and can sustain high release at rapid stimulation rates.[16]

It thus appears that the mesocortical DA neurons are functionally different from the striatal and limbic DA neurons. Their faster firing, synthesis, turnover, and release rates suggest that the mesocortical neurons are under less tonic inhibition than the nigrostriatal and mesolimbic DA neurons. It is also possible that their higher activity may result from poorer autoreceptor-control mechanisms in mesocortical DA neurons. Results from investigations conducted in our laboratories in the last years, on the role of presynaptic receptors modulating DA release from PFC DA terminals, will be presented.

METHODS

DA release was studied from superfused rabbit brain slices of the PFC, OT, and nucleus caudate (striatum). The slices were prelabeled with ^{3}H-DA (0.2 μM) for 30 minutes and, after several washes, placed in glass superfusion chambers containing platinum electrodes for electrical stimulation. The slices were superfused with Krebs Ringer bicarbonate at 37°C, pH 7.4, bubbled with 95% oxygen and CO_2. In general, two electrical stimulations were applied to each slice, per experiment. Often, the first stimulation (S_1) was used as control, and drugs were applied 20–30 minutes prior to the second stimulation (S_2). In some experiments, drugs were applied from 20 minutes prior to S_1 and continued until the end of the experiment. In control slices both stimulations were applied in the absence of drugs. Five minute samples of the perfusate were collected throughout the experiment. At the end of the experiment, the slices were placed in scintillation vials, and the radioactivity present in the slices quantified. Results were expressed as the percentage of tissue radioactivity released by each period of electrical stimulation, and the ratio S_2/S_1 was calculated. Results were also expressed as percentage of inhibition and percentage of increase above control values.

RESULTS AND DISCUSSION

Characteristics of DA Release from Prefrontal Cortex: Comparative Studies with Striatal and Olfactory Tubercle DA Terminals

We have compared the release of DA from the nucleus caudate, OT, and PFC of the rabbit (TABLES 1 and 2). In the three structures, DA release was elicited by

TABLE 1. Electrically Evoked Release of DA from the Striatum, OT, and PFC[a]

	1 Hz, 120 Pulses	10 Hz, 120 Pulses	10 Hz, 1200 Pulses
Striatum	1.7 ± 0.1	1.3 ± 0.1	4 ± 0.6
OT	1.8 ± 0.2	1.5 ± 0.2	—
PFC	2.3 ± 0.1^b	4.7 ± 0.2^c	14 ± 0.9^c

[a]DA release was evoked by electrical stimulation. Slices from the three regions were prelabeled with ^{3}H-DA. Significantly different from striatum and OT at $^bp < 0.05$ and $^cp < 0.01$.

electrical stimulation and was abolished by reduction of extracellular calcium. The release of DA per pulse and the frequency-release relationships in the OT were qualitatively and quantitatively similar to those observed in the nucleus caudate. Little change in the release of DA per pulse from the OT and nucleus caudate occurs real when the frequencies of stimulation were increased from 0.3 to 10 Hz (TABLE 1).

DA release per pulse from PFC was higher than from striatal and OT slices (TABLE 1). At low frequencies (0.3 Hz, 120 pulses), DA release from the PFC was 60% higher than from the striatum, whereas at higher frequencies (10 Hz and 120 pulses or 1200 pulses), release from the PFC was 550% higher than release from the striatum. These differences in the release per pulse could not be fully reverted by blockade of the autoreceptors with haloperidol, suggesting that the lesser release from striatal slices is not due to feedback inhibition of DA release.

In order to investigate the reason for the differences in release, the following comparative experiments were performed in superfused slices of the striatum and the PFC of the rabbit:

1. Calcium dependence of electrically evoked release.
2. Potassium-evoked DA release.
3. Tetraethylammonium (TEA) induced facilitation of electrically evoked release.
4. Phorbol ester–induced facilitation of electrically evoked release.

TABLE 2. Comparative Effects of Low Calcium, High Potassium, TEA, and Phorbol Ester on DA Release from the Striatum and the PFC[a]

Electrically evoked DA release (as % of tissue 3H released)	Calcium (1.3 mM)	Calcium (0.33 mM)	TEA (3 mM)
Striatum	3.6 ± 3	4.2 ± 5	13.3 ± 1.0
PFC	$13.8 \pm .6^b$	0.1 ± 0^b	13.0 ± 0.8
Electrically-evoked DA release (as % increase above controls)	PDBu (0.1 μM)	PDBu (0.3 μM)	PDBu (1 μM)
Striatum	52.3 ± 9	98 ± 12	25 ± 12
PFC	166.5 ± 18^b	189 ± 8^b	245 ± 19^b
Potassium-evoked DA release (as % tissue 3H released)	K^+ (30 mM)	K^+ (40 mM)	K^+ (60 mM)
Striatum	0.4 ± 0.2	5.8 ± 1.2	21.6 ± 4
PFC	4.2 ± 0.3^b	12.5 ± 1.1^b	28.1 ± 2

[a]The effects of phorbol ester (PDBu) and TEA were studied with stimulations at 1 Hz, 120 pulses. The effects of low calcium were studied at 10 Hz, 1200 pulses. Potassium was applied as a 1 minute pulse. Significantly different from striatum at $p < 0.05$ and $p < 0.01$.

Extracellular Calcium and DA Release

Gradual reduction of extracellular calcium concentrations from 1.3 mM to 0.13 mM, reduced the electrically evoked release of ^{3}H (10 Hz, 1200 pulses) from striatum and PFC. However, perfusion for 20 minutes with 0.33 mM extracellular calcium abolished the electrically evoked release from the striatum, whereas, nearly 4% of tissue radioactivity was released from PFC (TABLE 2).

Extracellular Potassium and DA Release

Perfusion with high potassium enhanced DA release from striatum and PFC. Maximal release was obtained with 60 mM potassium (TABLE 2). There were no differences in the maximal amount of DA release (percent of tissue ^{3}H released) with 60 mM potassium in both brain regions. However, at lower concentrations of potassium, release from PFC was greater than release from the striatum. The ratio DA release PFC/striatum was 10.3 ± 0.5 at 30 mM potassium and 2.2 ± 0.1 for 40 mM potassium.

Facilitation of Electrically Evoked DA Release by TEA

Addition of TEA increased in a concentration-dependent manner (from 0.1 to 3 mM) the electrically evoked release of DA from the PFC and the striatum. There were no significant differences in the magnitude of increase produced by TEA in both brain regions (TABLE 2). Maximal increases in DA release were obtained with 3 mM TEA, and ranged from 400 to 600% above the release observed in control stimulations (no TEA).

Facilitation of DA Release by Phorbol Ester

Perfusion for 20 minutes with 0.1 μM 4-beta-phorbol 12,13 dibutyrate, enhanced the electrically evoked release of DA from striatum and PFC (TABLE 2). However, the release from prefrontal cortex was facilitated to a greater extent. A higher concentration of phorbol ester (1 μM) further enhanced DA release from the PFC, whereas it produced lesser facilitation of DA release from the striatum.

In summary, these observations indicate that the lower release observed at high frequencies from the striatum than from the PFC cannot be attributed to different E_{max} for release in both structures. Striatal terminals, similarly to the cortical terminals, were able to release nearly 13% and 25% of their contents when electrically stimulated in the presence of 3 mM TEA, and when stimulated with 60 mM potassium, respectively. It was also observed that at high stimulation rates, DA release from the PFC is more resistant to reductions in extracellular calcium. In addition, DA release from PFC is very sensitive to non-action-potential-dependent depolarization (potassium 20–40 mM). Finally, the greater efficacy of phorbol esters in the PFC may be related to a more efficient excitation-release coupling in these dopaminergic terminals.

Modulation of DA Release by Presynaptic D2 DA Receptors

All the requisites to establish that the release of a neurotransmitter is modulated by D2 DA autoreceptors have been fulfilled for the release of DA from the striatum of several animal species.[17,18] It appears that the agonists, the antagonists, and the endogenously released DA act at a similar presynaptic DA receptor to modulate DA release. DA, apomorphine, bromocriptine, LY-171555, pergolide, and other DA agonists have been shown to inhibit DA release, and this inhibition is prevented by haloperidol, sulpiride, metoclopramide, thioridazine, and others. Under appropriate experimental conditions, these DA receptor antagonists have been shown to facilitate DA release. The behavior of the DA agonists and antagonists on DA release has been consistent with the expected interactions between the endogenous transmitter and the exogenously applied compounds.

Effects of DA Receptor Agonists and Antagonists on the Release of DA from the Olfactory Tubercle and the PFC

Similarly to the striatum, DA receptor agonists (apomorphine and LY 171555) induced a concentration-dependent inhibiiton of DA release from the OT and PFC (TABLE 3). The inhibitory effect of both agonists was antagonized by sulpiride or haloperidol.

TABLE 3. Potency and Maximal Inhibitory Effects of D2 DA Agonists on DA Release[a]

	APOMORPHINE (1 Hz, 120 pulses)		LY-171555 10 Hz, 1200 pulses		APOMORPHINE (10 Hz, 1200 pulses)	
	IC_{50}	E_{max}	IC_{50}	E_{max}	IC_{50}	E_{max}
Striatum	17 nM	98%	38 nM	98%	55 nM	48%
OT	14 nM	96%	28 nM	92%	—	—
PFC	25 nM	60%[b]	29 nM	61%[b]	48 nM	24%[b]

[a]DA release was elicited by electrical stimulation. The agonists were added 20 minutes prior to S_2. Significantly different from striatum and OT at $p < 0.05$ and [b]$p < 0.01$.

In the OT, the concentration-effect curves for apomorphine and LY-171555 in the OT overlapped with those obtained in the nucleus caudate. Similar IC_{50} and maximal degree of inhibition were obtained for both agonists in the striatum and the OT (TABLE 3). In the absence of neuronal uptake inhibitors, sulpiride enhanced the stimulation-evoked release of DA; this effect was similar to that observed in the nucleus caudate. Metoclopramide, a selective D2 antagonist, antagonized the inhibitory effect of apomorphine on the release of DA from the OT. The pA_2 obtained for metoclopramide on release modulatory autoreceptors of the OT [19] was similar to the pA_2 obtained for DA autoreceptors in the nucleus caudate.[20]

These observations suggest that a similar extent of modulation of DA release occurs in DA terminals of the striatum and the OT. The equal potency and efficacy of apomorphine and LY 171555 on both structures suggests that the number and affinity of the autoreceptors may be similar despite the different origin of the DA cell bodies.

In the PFC, DA receptor agonists (apomorphine, bromocriptine, and LY-171555) also reduced in a concentration-dependent manner the release of DA evoked by electrical stimulation (TABLES 3 and 5). The EC_{50} were similar to those observed in the striatum and OT; however, lower E_{max} was obtained in the PFC (TABLE 3). The effect of the agonists was antagonized by haloperidol and sulpiride.

In summary, under appropriate experimental conditions, it is possible to demonstrate that DA receptor agonists inhibit the release of DA from medial PFC; however, under similar conditions of stimulation, the maximal degree of inhibition produced by the agonists was greater in the striatum and the OT than in the PFC. The effects of the antagonists are less impressive in the PFC, and in this structure, the experimental conditions ought to be set appropriately in order to observe increases in DA release of a magnitude comparable to those seen in the striatum.[16,21]

Frequency and Duration of Stimulation as Determinants of Drug Effects on Presynaptic DA Receptors

In general in the three brain regions, D2 DA antagonists produced greater facilitation of the electrically evoked release of DA from rabbit striatal slices at higher than at lower stimulation frequencies. Little facilitation of release is observed at frequencies lower than 1 Hz (30–360 pulses). Optimal facilitation was observed between 3 and 10 Hz. In the absence of any drug that alters DA disposition, maximal facilitation of DA release by DA antagonists is achieved when the DA terminals are depolarized with short trains of rapidly delivered pulses (3–10 Hz).

The effects of DA agonists on DA release are also highly dependent on the parameters of stimulation (TABLE 3). In the striatum, OT, and PFC the inhibitory effects of DA agonists on DA release were greatest at lower stimulation rates (TABLE 3). Greatest inhibitory effects of apomorphine and LY-171555 on DA release, were obtained at 0.3 and 1 Hz and the inhibition was reduced at 3 and even more at 10 Hz. In addition, the inhibitory effects of apomorphine and bromocriptine were attenuated as the number of pulses was increased. When both parameters were altered simultaneously (large number of pulses at high frequencies) the effects of the agonists were markedly reduced. As we previously reported,[22] a 50% reduction in efficacy and an eightfold reduction in the potency of bromocriptine for inhibition of DA release from the striatum were observed when the conditions of stimulation were changed from 120 pulses at 0.3 Hz to 360 pulses at 3 Hz.

In the PFC, the maximal degree of inhibition produced by apomorphine was reduced from 66% at 1 Hz and 120 pulses to 24% when DA release was evoked by stimulations at 10 Hz and 1200 pulses. Haloperidol enhanced DA release by 20% at 1 Hz (120 pulses), 49% at 10 Hz (120 pulses), and by 13% at 10 Hz (1200 pulses) (52). Sulpiride enhanced DA release by 43% at 1 Hz, 120 pulses and by 68% at 10 Hz, 120 pulses.

Interactions between Neuronal Uptake Inhibitors and Drugs Acting at Presynaptic DA Receptors

Further evidenc in favor of the existence of modulation of DA release by presynaptic receptors derives from experiments where the synaptic concentrations of the transmitter are increased by inhibition of the neuronal uptake pump. If endogenous DA modulates its own release, inhibition of neuronal uptake should enhance the facilitation of DA release produced by DA antagonists and reduce the inhibitory action of the agonists.[23,24] Both postulates have been demonstrated in the striatum and the OT.[19] However, different findings were obtained in the PFC.

At 1 Hz, 120 pulses, and in the absence of other drug treatments, nomifensine (3 μM) enhanced DA overflow by 60% in the striatum and by 95% in the PFC ($p < 0.05$) (TABLE 4). Similar greater effects on overflow were observed with other structurally different uptake inhibitors, bupropion and cocaine, in PFC than in striatal slices. In nomifensine-treated slices, addition of sulpiride (1 and 10 μM) enhanced DA release by 40–50% in the PFC and by 200 to 255% in the striatum, respectively ($p < 0.01$) (TABLE 4). Interestingly, in the absence of nomifensine, sulpiride (10 μM) enhanced the electrically evoked release of DA (at 10 Hz, 120 pulses) by 80% in the striatum and by 70% in the PFC ($p > 0.1$) (TABLE 4).

In summary, the lack of synergistic interaction between uptake inhibitors and D2 DA receptor antagonists in the PFC is evidence against modulation of DA release from PFC DA terminals by endogenous DA. This is also supported by the observation that neuronal uptake inhibitors induce a much greater facilitation of DA overflow in the PFC than in the striatum or the OT.

Pharmacological Characteristics of Presynaptic DA Receptors

There is a large amount of evidence to suggest that the presynaptic receptors that modulate DA release from the striatum are of the D2 subtype. In the OT and PFC, the autoreceptors that modulate the release of DA are also akin to the DA D2 receptor subtype (TABLES 3–5). This conclusion is based on the following observations:

1. LY 171555 a selective D2 agonist inhibited DA release, and SKF 38393, a selective D1 agonist, had no effect per se on DA release from OT and PFC DA terminals. Sulpiride, a selective D2 antagonist, antagonized the inhibition of DA release produced by LY 171555 and apomorphine. In addition, sulpiride enhanced the DA release and this effect was enhanced in the presence of nomifensine. On the other hand, SCH 23390, a selective D1 antagonist, did not modify DA release and did not antagonize the inhibitory effect of LY 171555 and apomorphine both in the PFC and the OT.[16,19,24]
2. Metoclopramide, a selective D2 antagonist, blocked apomorphine-induced inhibition of DA release. The pA_2 for metoclopramide was similar to that reported for DA D2 receptors.[19,20] Similar studies have not been conducted in the PFC.

Interaction between Phorbol Esters and D2 DA Autoreceptors

Pretreatment of slices for 20 minutes with 0.1 μM 4-beta-phorbol 12,13 dibutyrate (PDBu), but not with its 4-alpha isomer (does not activate protein kinase C), antagonized the inhibitory effect of apomorphine and LY 171555 on the electrically

TABLE 4. Interaction between Neuronal Uptake Inhibitors and D2 DA Receptor Antagonists[a]

	S_2/S_1 Control	S_2/S_1 Nomifensine	Percent Increase
Effects of nomifensine 3 μM on DA release[d]			
Striatum	1.0 ± 0.07	1.6 ± 0.2	58 ± 7
PFC	1.0 ± 0.03	2.1 ± 0.1^b	94 ± 19^b

	S_2/S_1 Control	S_2/S_1 Sulpiride	Percent Increase
Effects of 10 μM sulpiride in the presence of 3 μM nomifensine[e]			
Striatum	0.83 ± 0.15	2.5 ± 0.3	254 ± 47^c
PFC	0.97 ± 0.03	1.4 ± 0.1^b	54 ± 10

[a]Significantly different from striatum at $^bp < 0.05$ and $^cp < 0.01$.
[d]Nomifensine was added 20 minutes prior to S_2.
[e]Nomifensine was present from 20 minutes prior to S_1 and sulpiride was added 20 minutes prior to S_2.

TABLE 5. Inhibition of DA Release from PFC and Striatum by Bromocriptine: Antagonism by 4-beta-Phorbol 12,13 Dibutyrate[a]

PDBu	BROMO	S_2/S_1	Percent Inhibition
BA release from striatum			
0	0.03	0.43 ± 0.06	63 ± 5
0	0.3	0.10 ± 0.03	89 ± 4
0.1	0.03	0.83 ± 0.08	17 ± 6
0.1	0.3	0.47 ± 0.03	52 ± 3
DA release from prefrontal cortex			
0	0.03	0.60 ± 0.04	39 ± 3
0	0.3	0.42 ± 0.04	59 ± 4
0.1	0.03	0.82 ± 0.05	6 ± 1
0.1	0.3	0.71 ± 0.03	26 ± 6

[a]PDBu was perfused from 20 minutes prior to S_1 until the end of the experiment. Bromocriptine was perfused from 20 minutes prior to S_2 until the end of the experiment. All values in the presence of PDBu were significantly different ($p < 0.01$) from their respective controls. PDBu: 4-beta phorbol 12,13 dibutyrate. BROMO: bromocriptine.

evoked release of DA from striatum and PFC (TABLE 5). In the striatum, where a kinetic evaluation of this interaction was performed, we observed that both the E_{max} and the IC_{50} of apomorphine were affected by the 4-beta-phorbol ester. In the absence of 4-beta phorbol ester, apomorphine E_{max} was 99% and its IC_{50} was 16 nM. In the presence of 0.1 μM 4-beta-phorbol ester, apomorphine E_{max} was 69% ($p < 0.01$) and its IC_{50} was 99 nM ($p < 0.01$).

We cannot be certain that the effects of the 4-beta-phorbol ester are mediated by protein kinase C. However, since its 4-alpha isomer, which does not activate the enzyme, did not enhance DA release and did not antagonize the effects of the D2 DA agonists on DA release, our findings suggest that the effects of 4-beta-phorbol 12,13 dibutyrate may be due to activation of protein kinase C. Our observations support the view that protein kinase C may be involved in the regulation of the amount of transmitter release and on the density or coupling of release modulatory D2 DA receptors.[25]

SUMMARY

Results obtained from our *in vitro* studies employing superfused slices obtained from three functionally different brain regions rich in DA axon terminals were discussed. Striking qualitative and quantitative similarities were found for the modulation of DA release from the nucleus caudate and the OT of the rabbit. However, the PFC DA terminals showed important differences from the nigrostriatal and mesolimbic DA terminals. Although release modulatory D2 DA autoreceptors could also be demonstrated in superfused slices of the PFC, our results suggest that the cortical nerve terminals may have a lower number of functional autoreceptors or a reduced efficiency of coupling between receptors and inhibition of release. Either possibility could explain (a) the poor inhibitory efficacy of the agonists, (b) the small facilitatory effect of the antagonists, (c) the disproportionate increase in transmitter overflow produced by neuronal uptake inhibitors, and (d) the lack of synergism between uptake inhibitors and DA antagonists.

When the efficacy of the autoreceptor mechanisms was evaluated at stimulation frequencies comparable to the *in vivo* firing rates reported for each of the three neuronal groups, it was found that DA release from the striatum and the OT was tightly modulated by presynaptic D2 DA receptors; whereas release from PFC was not. We propose that the autoreceptor-mediated control of DA release from PFC may not function *in vivo,* even though modulation of release by presynaptic D2 DA receptors from PFC terminals could be demonstrated under specific experimental conditions *in vitro.* However, it is envisaged that if *in vivo* firing rate of the PFC DA neurons is reduced, the inhibitory actions of DA agonists on DA release may be regained.

From these and other studies it is apparent that drug effects on autoreceptors are highly dependent on the rate and duration of stimulation applied to a specific neuronal group. We propose that the basal status of activity of a specific neuronal target could determine the type and magnitude of the effect produced by a therapeutic agent acting at release modulatory receptors. The neuronal activity (firing rate and pattern) may be affected by physiological status, disease, and by current or previous drug treatments.

The mechanisms by which PFC DA terminals release a larger proportion of their storage pool compared to other mesotelencephalic DA terminals is unknown and may represent a compensatory mechanism to the continuous rapid firing rates at which these neurons are exposed *in vivo.* The lesser sensitivity to reductions in extracellular calcium, their greater sensitivity to extracellular potassium, and the greater facilitation of release produced by active phorbol esters (protein kinase C activators) may represent adaptive changes in neuronal function required to sustain a high output of transmitter under *in vivo* conditions. For example, PFC DA terminals may have adapted to develop a greater number of protein kinase C molecules (or greater enzyme activity) or a more efficient coupling between protein kinase C and transmitter secretion. Such a mechanism may account for its release characteristics and for the greater facilitation of release produced by active phorbol ester in PFC DA terminals. Adaptive changes in the activity of protein kinase C may determine the level of transmitter release, as well as the receptor densities in the neuronal membrane. Consequently, one could postulate that the reduced efficacy of autoreceptor mechanisms observed in the presynaptic DA terminals of the PFC could be mediated through adaptive changes in protein kinase C activity.

REFERENCES

1. BJORKLUND, A. & O. LINDVALL. 1978. The mesoencephalic dopamine neuron system: a review of its anatomy. *In* Limbic Mechanisms. K. E. Livingston, & O. Hornykiewicz, Eds.: 307–331. Plenum Press. New York, N.Y.
2. HEIMER, L., G. ALHEID & L. ZABORSZKY. 1985. Basal ganglia. *In* The Rat Nervous System. G. Paxinos, Ed. 1: 37–86. Academic Press. Orlando, Fla.
3. PALKOVITS, M., L. ZABORSZKY, M. J. BROWNSTEIN, M. I. K. FEKETE, J. P. HERMAN & B. KANYCSKA. 1979. Distribution of norepinephrine and dopamine in cerebral cortical areas of the rat. Brain Res. Bull. **4:** 593–601.
4. MOORE, R. Y. & F. E. BLOOM. 1978. Central catecholamine neuron system: anatomy and physiology of the dopaminergic system. Annu. Rev. Neurosci. **1:** 129–169.
5. BERGER, B., A. M. THIERRY, J. P. TASSIN & M. A. MOYNE. 1976. Dopaminergic innervation of the rat prefrontal cortex: a fluorescence histochemical study. Brain Res. **106:** 133–145.
6. THIERRY, A. M., J. P. TASSIN, G. BLANC & G. GLOWINSKI. 1974. Presence of dopaminergic terminals and absence of dopaminergic cell bodies in the cerebral cortex of the cat. Brain Res. **79:** 77–88.
7. LINDVALL, O., A. BJORKLUND & I. DIVAC. 1978. Organization of catecholamine neurons projecting to the frontal cortex of the rat. Brain Res. **142:** 1–24.

9. BUNNEY, B. S. & G. K. AGHAJANIAN. 1976. Dopamine and norepinephrine innervated cells in the rat prefrontal cortex: pharmacological differentiation using microiontophoretic techniques. Life Sci. **19:** 1783–1792.

11. BANNON, M. J. & R. H. ROTH. 1983. Pharmacology of mesocortical dopamine neurons. Pharmacol. Rev. **35:** 53–68.

12. BANNON, M. J., E. B. BUNNEY & R. H. ROTH. 1981. Mesocortical dopamine neurons: rapid transmitter trunover compared to other brain catecholamine systems. Brain Res. **218:** 376–382.

13. CHIODO, L. A., M. J. BANNON, A. A. GRACE, R. H. ROTH & B. S. BUNNEY. 1984. Evidence of the absence of impulse-regulating somatodendritic and synthesis-modulating nerve terminal autoreceptors on subpopulations of mesocortical dopamine neurons. Neuroscience **12:** 1–16.

14. SCATTON, B. 1977. Differential regional development of tolerance to increase in dopamine turnover upon repeated neuroleptic administration. Eur. J. Pharmacol. **46:** 363–369.

15. CHIODO, L. A. & B. S. BUNNEY. 1983. Typical and atypical neuroleptics: differential effects of chronic administration on the activity of A9 and A10 midbrain dopaminergic neurons. J. Neurosci. **3:** 1607–1619.

16. HOFFMANN, I. S., R. K. TALMACIU, C. P. FERRO & L. X. CUBEDDU. 1988. Sustained high release at rapid stimulation rates and reduced functional autoreceptors characterize prefrontal cortex dopamine terminals. J. Pharmacol. Exp. Ther. **245:** 761–772.

17. CUBEDDU, L. X. & I. S. HOFFMANN. 1982. Operational characteristics of the inhibitory feedback mechanism for regulation of dopamine release via presynaptic receptors. J. Pharmacol. Exp. Ther. **223:** 497–501.

18. CUBEDDU, L. X. Modulation of dopamine release from corpus striatum, limbic regions and prefrontal cortex by presynaptic receptors. *In* Presynaptic Regulation of Neurotransmitter Release: a Handbook. J. J. Feigenbaum & M. Hannani, Eds. Freund Publishing House Ltd. London, England. (In press.)

19. SUAREZ-ROCA, H., T. LOVENBERG & L. X. CUBEDDU. 1987. Comparative dopamine-cholinergic mechanisms in the olfactory tubercle and the striatum: effects of metoclopramide. J. Pharmacol. Exp. Ther. **243:** 840–851.

20. HELMREICH, I., W. REIMAN, G. HERTTING & K. STARKE. 1982. Are presynaptic dopamine autoreceptors and postsynaptic dopamine receptors in the rabbit caudate nucleus pharmacologically different? Neuroscience **7:** 1559–1566.

21. TALMACIU, R. K., I. S. HOFFMANN & L. X. CUBEDDU. 1986. DA autoreceptors modulate DA release from the prefrontal cortex. J. Neurochem. **47:** 865–870.

22. HOFFMANN, I. S. & L. X. CUBEDDU. 1984. Differential effects of bromocriptine on dopamine and acetylchonline release modulatory receptors. J. Neurochem. **42:** 278–282.

23. CUBEDDU, L. X., I. S. HOFFMANN & M. K. JAMES. 1983. Frequency-dependent effects of neuronal uptake inhibitors on the autoreceptor-medicated modulation of dopamine and acetylcholine release from the rabbit striatum. J. Pharmacol. Exp. Ther. **226:** 89–94.

24. HOFFMANN, I. S., R. K. TALMACIU & L. X. CUBEDDU. 1986. Interactions between endogenous dopamine and dopamine agonists at release modulatory receptors: multiple effects of neuronal uptake inhibitors on transmitter release. J. Pharmacol. Exp. Ther. **238:** 437–446.

25. CUBEDDU, L. X., T. W. LOVENBERG, I. S. HOFFMANN & R. K. TALMACIU. 1989. Phorbol esters and D2-dopamine receptors. J. Pharmacol. Exp. Ther. **251:** 687–693.

Axonal Transport of Receptors

A Major Criterion for Presynaptic Localization

PIERRE M. LADURON AND MARIE-NOËLLE CASTEL

Department of Biology
Rhône-Poulenc Santé
Centre de Recherche de Vitry-Alportville
13, quai Jules Guesde
94403 Vitry sur Seine, France

INTRODUCTION

Interneuronal communication through chemical transmission involves the release of transmitters from nerve terminals and their action on postsynaptic receptors. Owing to the distance of nerve terminals from the neuronal cell body, traffic systems or transport mechanisms are required in order to allow neuronal constituents or informative molecules to reach all the intracellular compartments, including nerve terminals. It is interesting to note that the sequence of intracellular events (FIGURE 1) that had been firmly established for numerous neurotransmitters, and on which the concept of neurotransmission is founded, also seems to be valid for presynaptic receptors.

The five different stages, or dynamic events, in neurotransmission are very similar for both transmitters and receptors:

1. Biosynthesis (exclusively in the cell body for receptors) and storage in vesicles.
2. Anterograde axonal transport (but also retrograde for receptors).
3. Release of transmitters at nerve terminals by exocytosis, a process that externalizes the presynaptic receptors.
4. Binding of transmitters to postsynaptic receptors, whereas the presynaptic receptors receive signal molecules from the postsynaptic side. Such binding with an agonist can lead to the internalization of presynaptic sites leading to a retrograde axonal transport of receptors.
5. Degradation by enzymes, in the case of receptors after fusion with lysosomes presumably in the cell body.

In addition to this, another analogy concerns the recycling process which has been firmly established for certain classical neurotransmitters in nerve terminals, and has also been suggested for receptors in the cell body.[1] All these intracellular processes have to be considered in order to understand the regulation of receptors at the cellular level.

Like neurotransmitters, very few receptors are operational at a given moment, which means that they have to be in an externalized state in the synapse to allow the interaction with the neurotransmitter (FIGURE 1). Since a very low number of sites (less than 1–10%) exist in an externalized form, hypersensivity and desensitization processes must be analyzed mainly in terms of functional receptors.

The aim of the present paper is to show that axonal transport can be used to demonstrate the presynaptic nature of receptors.

AXONAL TRANSPORT OF PRESYNAPTIC HETERORECEPTORS IN PERIPHERAL NERVES

Axonal transport was first studied in peripheral nerves because the axonal processes of these nerves extend a large distance from the cell body (about 1 meter for the sciatic nerve in man), and also because the axon is easy to access in ligature experiments. Hence, the experimental model consists of analyzing the accumulation of

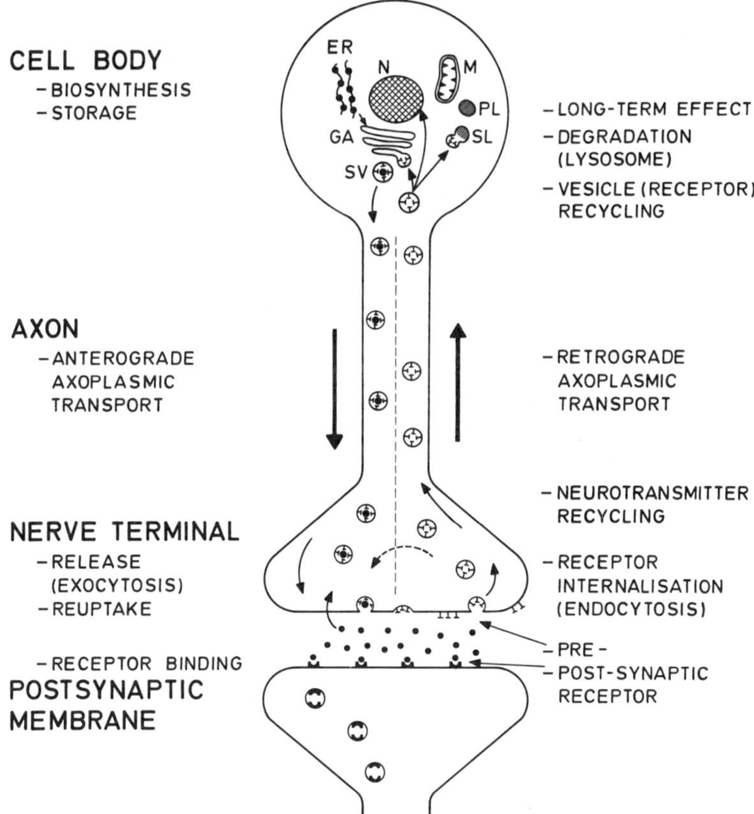

FIGURE 1. Schema illustrating the axonal transport of receptors in the dynamic sequence of events that occurs through the neurotransmission process. N, nucleus; M, mitochondrion; GA, Golgi apparatus; SV, synaptic vesicle; PL, primary lysosome; and SL, secondary lysosome.

neuronal constituents in nerve segments, proximal or distal to a ligature, at different time intervals. Using this approach, noradrenaline, acetylcholinesterase, dopamine-β-hydroxylase, labeled proteins, and neuropeptides were all found to move along the axon at a fast rate.[2–7]

The first evidence for the existence of axonal transport of receptors was found in dog splenic nerves,[8] where the presence of muscarinic receptors (identified by the *in vitro* binding assay using ^{3}H-dexetimide) was demonstrated in the axon and even in the cell body (celiac ganglion) of splenic nerves. These muscarinic sites were found to accumulate on both sides of a ligature up to 48h after constriction, indicating the existence of bidirectional axoplasmic transport of muscarinic receptors. This finding was confirmed not only in different species but also for other receptors: opiate, β-adrenergic, nicotinic, cholecystokinin, neurotensin, and ion channels (TABLE 1). Although only anterograde transport of opiate receptors was initially reported,[9] we have also demonstrated a retrograde process when opiate sites were labeled *in vivo* with ^{3}H-lofentanil.[10–12] It is interesting that the accumulation of opiate receptors on both sides of a ligature in the vagus nerve was almost completely abolished when rats were chronically treated with capsaicin, thus indicating that opiate sites are associated with sensory neurones, in particular those containing substance P.[11] The fast rate of transport for muscarinic receptors (1.2 mm/hour in splenic nerves and 2.7 mm/hour in sciatic nerves) implies that this transport may occur within organelles such as synaptic vesicles. Fractionation using differential and isopycnic centrifugation[13] and structure-latency[14] studies has provided more evidence that muscarinic receptors in noradrenergic nerves are associated with vesicles, and even possibly with vesicles containing noradrenaline and dopamine-β-hydroxylase, thus suggesting the coexistence of neurotransmitter and heteroreceptor in the same vesicle.[1]

In brief, the axonal transport of presynaptic heteroreceptors is bidirectional, fast, and microtubule dependent: receptors are associated with vesicles that are possibly recycled in the cell body, and are externalized and internalized at the nerve terminals. This latter process was clearly demonstrated for muscarinic receptors in cell culture:[15] following only minutes of exposure to carbachol, this neurotransmitter acted as signal for inducing the internalization of the receptors; when carbachol was removed, the receptors rapidly reappeared at the cell surface, therefore providing further evidence for an internalization/externalization process of receptors.

In vivo experiments in rats with ^{3}H-lofentanil also substantiate a process of internalization of the labeled opiate receptors at the nerve terminals, previous to retrograde transport in the vagus nerve.[10–12]

TABLE 1. Axonal Transport of Receptors

Receptor	Nerve or Pathway	Neurone	Species	Reference
Muscarinic	Splenic	Noradrenergic	Dog	8
	Hypogastric	Noradrenergic	Cat	18
	Sciatic	Noradrenergic	Rat	1, 14, 19, 20
	Vagus		Rat	21
Opiate	Vagus	Sensory	Rat	9, 10, 11, 12
	Nigrostriatal	Dopaminergic	Rat	22
β-Adrenergic	Sciatic	—	Rat	23
Nicotinic	Dorsal root	Sensory	Rat	24
	Sciatic	—	Rat	25
	Optic	—	Rat	26
CCK	Vagus	—	Rat	27
Na$^+$ Channel	Sciatic	—	Rat	28
Digitalis	Sciatic	—	Rat	29
Angiotensin	Vagus	—	Rat	30
Neurotensin	Nigrostrital	Dopaminergic	Rat	31, 32
	Vagus	Sensory, motor	Rat	33
NMDA	Vagus	—	Rat	34

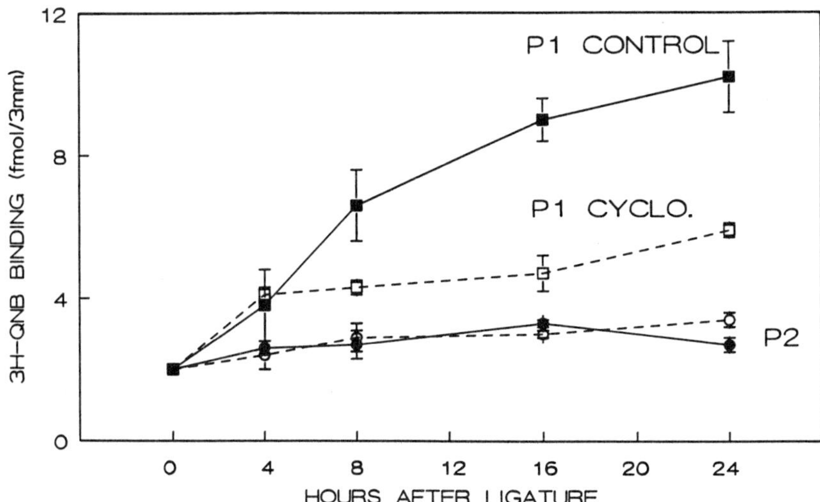

FIGURE 2. Effect of cycloheximide on the time course of accumulation of muscarinic receptors in proximal segments of ligated sciatic nerves. Cycloheximide (1.5 mg/kg) was administered 2 hours before ligation and the nerve segments (P_1, proximal segment adjacent to the ligature; P_2, more proximal segment) were removed at different times after ligation. [³H]QNB binding was assayed in controls (solid line, dark symbols) and cycloheximide-treated (cyclo) rats (dashed line, open symbols). Each point represents the mean value obtained from four different pools of five segments assayed in duplicate [± standard error of the mean (SEM)] (for detail cf. Reference 16).

Recent experiments in the rat showed that the anterograde accumulation of muscarinic receptors in ligated sciatic nerves was reduced after treatment with protein synthesis inhibitors.[16] FIGURE 2 shows that the reduction in the accumulation began to be detectable 10 hours after treatment with cycloheximide (note that cycloheximide was given 2 hours before ligation). Such a delay was compatible with the time necessary for receptor sites to move from the cell body to the ligature. Therefore the inhibition of protein synthesis only affects the biosynthesis of muscarinic receptors in the cell body, but not in the axon where receptor sites move in vesicles. The fact that cycloheximide may have also partially affected the machinery involved in the transport mechanisms must not be overlooked.

TABLE 1 shows that numerous receptors other than muscarinic and opiate were found to move in peripheral nerves through fast axonal transport mechanisms. However, although this has been clearly demonstrated for presynaptic heteroreceptors, there is no evidence of such a process for autoreceptors.[17]

AXONAL TRANSPORT OF RECEPTORS IN THE BRAIN

Since axonal transport of receptors is now firmly established in peripheral nerves, what is the evidence for such a process in the central nervous system (CNS)? Clearly, because ligature or constriction experiments are not feasible, another experimental approach must be developed. Lesion experiments may be useful, if all the criteria defined in periphery are fulfilled in the CNS. This was not exactly the case when an accumulation of β-adrenergic and α_2-adrenergic binding sites were reported in rat brain after 6-hydrodopamine lesion;[35,36] the time course of their accumulation (up to 4 days) was not compatible with a fast axonal transport. Moreover, the neurons were not clearly identified and the transport was not shown to be bidirectional. Other attempts were reported with ^{3}H ketanserin,[37] which is known to label the serotonin ($5HT_2$) or S_2 receptors.[38] However, in the striatum this ligand was found to bind preferentially to the amine transporter of synaptic vesicles;[39,40] therefore these experiments have to be reinterpreted in terms of retrograde axonal transport of the amine transporter associated with vesicles containing dopamine rather than axonal transport of $5HT_2$ receptors.

Since nigrostriatal dopaminergic neurones are one of the well-documented pathways in the brain, they could provide an excellent model to study the axonal transport of receptors in the CNS. Indeed, we recently demonstrated that ^{125}I-neurotensin injected into the rat striatum was transported retrogradely to the substantia nigra.[30,31] TABLE 2 shows that the labeling started to accumulate in the ipsilateral, but not the contralateral, substantia nigra, about 2 hours after injection of ^{125}I-neurotensin into the striatum, and that the maximum was reached 3 or 4 hours after injection. This is entirely compatible with a fast axonal transport mechanism, and excludes a diffusion

TABLE 2. Labeling in Substantia Nigra after Injection of I^{125} Neurotensin in Rat Caudate Putamen[a]

	Time after Injection			
	1 Hour	2 Hours	3 Hours	4 Hours
I^{125} neurotensin	0	205 ± 28	456 ± 31	431 ± 31
+ unlabeled neurotensin	0	0	0	0
+ unlabeled fragment 8-13	—	—	—	—
+ unlabeled fragment 1-8	—	—	—	427 ± 95
+ pretreatment				
6-hydroxydopamine	—	—	—	0
colchicine	—	—	—	0

[a]Mean disintegrations per minute ± standard error of the mean.

phenomenon which was observed in other brain regions immediately after the injection of labeled peptide. TABLE 2 also shows that the accumulation of radioactivity in the substantia nigra was prevented when a 1000-fold excess of unlabeled neurotensin, or its active 8-13 fragment, was injected together with labeled neurotensin, but not when the inactive 1-8 fragment was used. Colchicine and 6-hydroxydopamine pretreatment provided more direct evidence that the dynamic process was microtubule dependent, and occurred in dopaminergic neurones. It is important to note that in order to measure a detectable signal in the experiment it was necessary to inject kelatorphan (a peptidase inhibitor) which prevents the degradation of labeled neurotensin. Interestingly, the neurotensin receptors associated with dopaminergic terminals in the caudate nucleus (for references cf Reference 41) may be regarded as true presynaptic heteroreceptors, because there is as yet no proof for colocalization of neurotensin and dopamine in the nigrostriatal pathway.

CAN AUTORECEPTORS UNDERGO AXONAL TRANSPORT?

The question Can autoreceptors undergo axonal transport? was first asked in 1985,[17] and we wonder whether the answer is still the same today.

As previously stated, axonal transport has up until now been reported for presynaptic heteroreceptors. TABLE 1 shows a list of these heteroreceptors. To this, one may also add NGF which has been reported to move retrogradely in different neurones, prior to the identification of NGF receptors (cf. References 42 and 43). As for the nicotinic receptor, it is not clear whether bungarotoxin binding sites are only associated with sensory or with motoneurones. Moreover one cannot be sure that the binding site in this case represents true receptors rather than acceptor sites.

Among autoreceptors, the most interesting candidate still remains the presynaptic α_2-adrenergic receptor. Unfortunately, numerous groups have tried to identify α_2 binding sites in noradrenergic nerves and also to demonstrate their axonal transport, but without success.[17,18,44] One cannot argue that this is due to a problem concerning the binding assay, because different ligands effective for α_2-adrenergic sites have been tested under various conditions. These negative results are compatible with the fact that α_2-adrenergic receptors increase, rather than decrease, in number after noradrenergic denervation as is the case for most postsynaptic sites. Under the same conditions of denervation, presynaptic muscarinic receptors in sympathetic nerves (heteroreceptors) were reduced. As stated previously,[17] three major criteria must be fulfilled for assessing the presynaptic nature of receptors, thus including autorecptors if they exist. Firstly, the presence of receptor sites should be demonstrated in the cell body, the axon, and the nerve terminals of the neuron that contains the transmitter for this receptor. For instance, α_2-adrenergic autoreceptors should be detected in noradrenergic neurons. Secondly, axonal transport of receptors should be proved; for the autoreceptors, such a process should occur in the neurons that contain the endogenous ligand. Therefore, for those, one might expect a parallel accumulation above the ligature of the receptor and its own neurotransmitter. Lastly, a physiological response at the nerve terminal should be demonstrated; for instance, presynaptic muscarinic receptors in noradrenergic terminals are involved in the regulation of noradrenaline release. To this, a fourth criterion may be proposed: the demonstration in the same neuron of the mRNA of receptors and the endogenous ligand. For this, a clear-cut localization must be shown because lesion experiments can also affect postsynaptic sites.

As a rule, indirect evidence is generally provided to suggest the presynaptic nature of receptors. For the autoreceptors, direct evidence is still lacking; as far as we know, there are no available data to indicate that autoreceptors are transported axonally. The answer to the question raised in 1985 remains the same: autoreceptors do not fulfill the conditions of presynaptic localization. The most severe criterion for proving the existence of autoreceptors is the axonal transport; this has never been taken into consideration. Therefore, we suggest a return to the first proposal that a transynaptic process should be responsible for the regulation of presynaptic events through postsynaptic receptor sites.

In fact, this represents the first idea that was developed to try to explain the intriguing results obtained in the field of α_2-adrenergic receptors with phenoxybenzamine and phentolamine. Reconsideration of such an explanation has the advantage of opening up a new line of research for isolation of the factor (peptide or not) that should be released from the postsynaptic site after stimulation of α_2-receptors and which could act on the presynaptic side. This implies the existence of a bidirectional mechanism of communication between the pre- and postsynaptic sides. NGF is already an example of such a substance released by target cells.

REFERENCES

1. LADURON, P. M. 1984. Axonal transport of receptors: coexistence with neurotransmitter and recycling. Biochem. Pharmacol. **33:** 897, 903.

2. DAHLSTRÖM, A. 1965. Observations on the accumulation of noradrenaline in the proximal and distal part of peripheral adrenergic nerves after compression. J. Anat. **99:** 677–689.

3. LUBINSKA, L. & S. NIEMERKO. 1971. Velocity and intensity of bidirectional migration of acetylcholinesterase in transected nerves. Brain Res. **27:** 329–342.

4. LADURON, P. & F. BELPAIRE. 1968. Transport of noradrenaline and dopamine-β-hydroxylase in sympathetic nerves. Life Sci. **7:** 1–7.

5. OCHS, S. 1972. Fast transport of materials in mammalian nerve fibers. Science **176:** 252–260.

6. LASEK, R. J. 1968. Axoplasmic transport of labeled proteins in rat ventral motoneurons. Exp. Neurol. **21:** 41–51.

7. LUNDBERG, J. M., T. HÖKFELT, G. NILSSON, L. TERENIUS, J. REHFELD, R. ELDE & R. SAID. 1978. Peptide neurons in the vagus, splanchnic and sciatic nerves. Acta Physiol. Scand. **104:** 499–501.

8. LADURON, P. 1980. Axoplasmic transport of muscarinic receptors. Nature **286:** 287–288.

9. YOUNG, W. S., J. K. WAMSLEY, M. A. ZARBIN & M. J. KUHAR. 1980. Opioid receptors undergo axonal flow. Science **210:** 76–78.

10. LADURON, P. M. & P. F. M. JANSSEN. 1982. Axoplasmic transport and possible recycling of opiate receptors labeled with ^{3}H-lofentanil. Life Sci. **31:** 457–462.

11. LADURON, P. M. 1984. Axonal transport of opiate receptors in capsaicin-sensitive neurones. Brain Res. **294:** 157–160.

12. LADURON, P. M. & P. F. M. JANSSEN. 1985. Retrograde axonal transport of receptor-bound opiate in the vagus and delayed accumulation in the nodose ganglion. Brain Res. **333:** 389–392.

13. LADURON, P. M. 1984. Axonal transport of muscarinic receptors in vesicles containing noradrenaline and dopamine-β-hydroxylase. FEBS Lett. **165:** 128–132.

14. LADURON, P. M. & T. SCHOLTUS. 1986. Structure-linked latency of muscarinic receptors in axonal transport. FEBS Lett. **206:** 83–86.

15. MALOTEAUX, J. M., A. GOSSUIN, P. J. PAUWELS & P. M. LADURON. 1983. Short-term disappearance of muscarinic cell surface receptors in carbachol-induced desensitization. FEBS Lett. **156:** 103–107.

16. LEVY, C., D. SCHERMAN & P. M. LADURON. 1990. Axonal transport of synaptic vesicles and muscarinic receptors: role of synthesis and recycling. J. Neurochem. **54:** 880–885.

17. LADURON, P. M. 1985. Axonal transport of presynaptic receptors. *In* Trends in Autonomic Pharmacology. S. Kalsner, Ed. **3:** 113–127. Raven Press. New York, N.Y.

18. ALONSO, F. G., V. CENA, A. G. GARCIA, S. M. KIRPEKAR & P. SANCHEZ-GARCIA. 1982. Presence and axonal transport of cholinoceptor but not adrenoceptor sites on a cat noradrenergic neurone. J. Physiol. London **333:** 595–618.

19. GULYA, K. & P. KASA. 1984. Transport of muscarinic cholinergic receptors in the sciatic nerve of rat. Neurochem. Int. **6:** 123–126.

20. WAMSLEY, J. K., M. A. ZARBIN & M. J. KUHAR. 1981. Muscarinic cholinergic receptors flow in sciatic nerves. Brain Res. **217:** 155–161.

21. ZARBIN, M. A., J. K. WAMSLEY & M. J. KUHAR. 1982. Axonal transport of muscarinic cholinergic receptors in rat vagus nerve: high and low affinity agonist receptors move in opposite directions and differ in nucleotide sensitivity. J. Neurosci. **2:** 934–941.

22. VAN DER KOOY, W. P. & J. I. NAGY. 1986. Dopamine and opiate receptors: localization in the striatum and evidence for their axoplasmic transport in the nigrostriatal and striatonigral pathways. Neuroscience **19:** 139–146.

23. ZARBIN, M. A., J. M. PALACIOS, J. K. WAMSLEY & M. J. KUHAR. 1983. Axonal transport of β-adrenergic receptors. Antero- and retrogradely transported receptors differ in agonist affinity and nucleotide sensitivity. Mol. Pharmacol. **24:** 341–348.

24. NINOVIC, M. & S. P. HUNT. 1983. α-Bungarotoxin binding sites on sensory neurones and their axonal transport in sensory afferents. Brain Res. **272:** 57–69.

25. MILLINGTON, W. R., E. AIZENMAN, G. G. BIERKAMPER, M. A. ZARBIN & M. J. KUHAR. 1985. Axonal transport of α-bungarotoxin binding sites in rat sciatic nerve. Brain Res. **340:** 269–276.

26. HENLEY, J. M., J. M. LINDSTROM & R. E. OSWALD. 1986. Acetylcholine receptor synthesis in retina and transport to optic tectum in goldfish. Science **232:** 1627–1629.

27. ZARBIN, M. A., J. K. WAMSLEY, R. B. INNIS & M. J. KUHAR. 1981. Cholecystokinin receptors: presence and axonal flow in the rat vagus nerve. Life Sci. **29:** 697–705.

28. LOMBET, A., P. LADURON, C. MOURRÉ, Y. JACOMET & M. LAZDUNSKI. 1985. Axonal transport of the voltage-dependent Na+ channel protein identified by its tetrodotoxin binding site in rat sciatic nerves. Brain Res. **345:** 153–158.

29. LOMBET, A., P. LADURON, C. MOURRE, Y. JACOMET & M. LAZDUNSKI. 1986. Axonal transport of Na$^+$, K$^+$, ATPase identified as a ouabain binding site in rat sciatic nerve. Neurosci. Lett. **64:** 177–183.

30. DIZ, D. I. & C. M. FERRARIS. 1988. Bidirectional transport of angiotensin II binding sites in the vagus nerve. Hypertension **11**(Suppl. 2): I-139–143.

31. CASTEL, M. N., C. MALGOURIS, J. C. BLANCHARD & P. M. LADURON. 1989. Retrograde axonal transport of neurotensin in the rat brain. Eur. J. Pharmacol. **166:** 353–354.

32. CASTEL, M. N., C. MALGOURIS, J. C. BLANCHARD & P. M. LADURON. Retrograde axonal transport of neurotensin in the dopaminergic nigrostriatal pathway in the rat. Neuroscience. (In press.)

33. KESSLER, J. P. & A. BEAUDET. 1989. Association of neurotensin binding sites with sensory and visceromotor components of the vagus nerve. J. Neurosci. **9**(2): 446–472.

34. CINCOTTA, M., P. M. BEART, R. J. SUMMERS & D. LODGE. 1989. Bidirectional transport of NMDA receptor and ionophore in the vagus nerve. Eur. J. Pharmacol. **160:** 167–171.

35. LEVIN, B. E. 1984. Retrograde axonal transport of β-adrenoreceptors in rat brain: effect of reserpine. Brain Res. **300:** 103–112.

36. LEVIN, B. E. 1984. Axonal transport and presynaptic location of α_2 adrenoreceptors in locus coeruleus neurons. Brain Res. **321:** 180–182.

37. SNOWHILL, E. W. & J. K. WAMSLEY. 1983. Serotonin type-2 receptors undergo axonal transport in the medical forebrain bundle. Eur. J. Pharmacol. **95:** 325–327.

38. LEYSEN, J. E., C. J. E. NIEMEGEERS, J. M. VAN NUETEN & P. M. LADURON. 1982. (^{3}H) ketanserin, a selective ^{3}H ligand for serotonin$_2$ receptor binding sites. Mol. Pharmacol. **21:** 301–314.

39. LEYSEN, J. E., A. EENS, W. GOMMEREN, P. VAN GOMPEL, J. WYNANTS & P. A. JANSSEN. 1987. Non serotonergic (^{3}H) ketanserin binding sites in striatal membranes are associated with a dopac release system on dopaminergic nerve endings. Eur. J. Pharmacol. **134:** 373–375.

40. DARCHEN, F., D. SCHERMAN, P. M. LADURON & J. P. HENRY. 1988. Ketanserin binds to the monoramine transporter of chromaffin granules and of synaptic vesicles. Mol. Pharmacol. **33:** 672–677.

41. SCHOTTE, A., W. ROSTENE & P. M. LADURON. 1988. Different subcellular localization of neurotensin-receptor and neurotensin-acceptor sites in the rat brain dopaminergic system. J. Neurochem. **50:** 1026–1031.

42. THOENEN, H. & Y. BARDE. 1980. Physiology of nerve growth factor. Physiol. Rev. **60:** 1284–1335.

43. LADURON, P. M. 1987. Axonal transport of neuro-receptors: possible involvement in long-term memory. Neuroscience, **22:** 767–779.

44. PIMOULE, C., B. SCATTON & S. Z. LANGER. 1983. (^{3}H) RX.78 10 94: a new antagonist ligand labels α_2 adrenoreceptors in the rat brain cortex. Eur. J. Pharmacol. **95:** 79–85.

In Vivo Electrophysiology of Central Nervous System Terminal Autoreceptors[a]

JAMES M. TEPPER[b] AND PHILIP M. GROVES[c]

[b]Center for Molecular and Behavioral Neuroscience
Rutgers, The State University of New Jersey
195 University Avenue
Newark, New Jersey 07102

[c]Department of Psychiatry
University of California, San Diego
School of Medicine
La Jolla, California 92093

INTRODUCTION

Evidence both for and against the existence of presynaptic autoreceptors and the related question of, if they exist, do they play a significant role in the "normal" or physiological functioning of synaptic transmission, has been presented throughout this volume (for a recent excellent and comprehensive review, see Reference 1). Some of the controversy regarding the existence and functioning of autoreceptors derives from the techniques and conditions under which their existence and functioning have been measured. In this chapter we will review data derived from an electrophysiological technique for examining the neurophysiological consequences of autoreceptor stimulation and blockade at the terminals of central nervous system neurons *in vivo* that support the existence of such autoreceptors, and give some clues to the conditions under which they operate *in situ*. The data to be discussed concern changes in the electrical excitability of single central nervous system (CNS) axon terminals as a function of the local and systemic application of drugs, and the rate of impulses reaching the terminal fields. Thus far, the technique has been used to study apparent autoreceptor-mediated changes in terminal excitability in rat mesencephalic dopaminergic neurons projecting to neostriatum,[2–7] nucleus accumbens[8] prefrontal, cingulate, and entorhinal cortices,[9] in rat noradrenergic locus ceruleus neurons projecting to frontal cortex,[10–12] in rat serotonergic dorsal raphé neurons projecting to neostriatum,[13] in presumed glutamatergic rat corticostriatal neurons,[14] and in monkey nigral dopaminergic neurons projecting to the putamen.[15] In addition, this technique has been used to study the electrophysiological consequences of activation or blockade of presynaptic heteroreceptors in several different systems, e.g., opioid receptors on cortical noradrenergic terminal of locus ceruleus neurons,[16] dopamine receptors on hippocampal and cortical axons projecting to ventral striatum,[14,17] and dopamine receptors on the terminals of presumed GABAergic (γ-aminobutyric acid) striatonigral neurons.[18]

Despite the fact that neurotransmitter release itself is not directly measured in these studies, the nature of the changes in terminal excitability and the conditions under which these changes are observed suggest rather strongly that many different classes of CNS neurons possess at their terminal fields neurotransmitter receptors

[a]Some of the research and the writing of this manuscript were supported by the National Institute of Mental Health and the National Institute on Drug Abuse, grants MH-45286, DA-02854, and DA-00079.

sensitive to the neurotransmitter that these neurons release, and that these receptors can and do operate under more or less normal physiological conditions *in vivo* to modify the electrophysiological properties of the nerve terminals in a manner consistent with the observed modulation of impulse-dependent neurotransmitter release *in vitro* and *in vivo*.

TERMINAL EXCITABILITY TESTING PARADIGM

The details of the methods for measuring autoreceptor-mediated changes in terminal excitability resulting from local drug application and changes in the rate of impulses reaching the terminal fields have been published previously.[4,5,10] The paradigm is illustrated schematically for the nigrostriatal dopaminergic system in FIGURE 1A. Briefly, terminal excitability is indexed by the current intensity necessary to elicit antidromic responses from single units recorded extracellularly. A number of trials (50–100) at a variety of different stimulating currents are presented, ranging from the minimum current needed to evoke antidromic responses to each stimulus delivery (excluding collisions) to the maximum current that fails to evoke any antidromic responses. The entire procedure is repeated several times, in a counterbalanced fashion, until successive determinations at the same currents yield the same proportion of antidromic responding ±10%. These points are taken as baseline, and an excitability curve relating percent antidromic response to stimulus current can be plotted as shown in FIGURE 1B. Drugs or vehicle is then administered, either systemically via an intravenous catheter or locally, directly into the terminal fields or preterminal axons via infusion cannulae coupled to the stimulating electrode. Local drug administration consists of relatively low concentrations (0.1–10 μM) in small volumes (325 nL). Post drug measures of terminal excitability are performed, and the pre- and postdrug excitability curves compared. Shifts to the right of the excitability curve signify decreases in terminal excitability whereas shifts to the left indicate increased terminal excitability. In some cases, instead of administering drugs, preterminal regions of the axon are stimulated at different rates to alter the synaptic concentrations of released neurotransmitter at the terminal regions and the excitability reassessed.

AUTORECEPTOR-MEDIATED CHANGES IN TERMINAL EXCITABILITY

When appropriate autoreceptor agonists are administered systemically or applied directly to the terminal regions of dopaminergic, noradrenergic, or serotonergic neurons, the terminal excitability of each is reduced, as illustrated in FIGURE 2.[2–4,8,10,13] The decreased excitability is dose dependent and obtains only in the terminal fields; if drugs are locally infused into and excitability tested from preterminal regions of monoaminergic axons (e.g., the medial forebrain bundle or the dorsal noradrenergic pathway), no changes in excitability are observed. In addition to increasing the threshold for antidromic activation, autoreceptor agonists produce slight increases in the antidromic latency, and in the variability of the antidromic latency.[4,10] These decreases in terminal excitability can be reversed by subsequent infusions of appropriate autoreceptor antagonists, or blocked by pretreatment with these drugs.[4,8,10,19]

In addition, appropriate autoreceptor antagonists can not only reverse the effects of prior administration of agonists, but, when administered alone, produce increased terminal excitability.[4,8,10] This is an important observation indicating that *in vivo,* there is a sufficient concentration of endogenous agonist in the terminal regions to activate terminal autoreceptors, thus suggesting a physiological role for these autoreceptors.

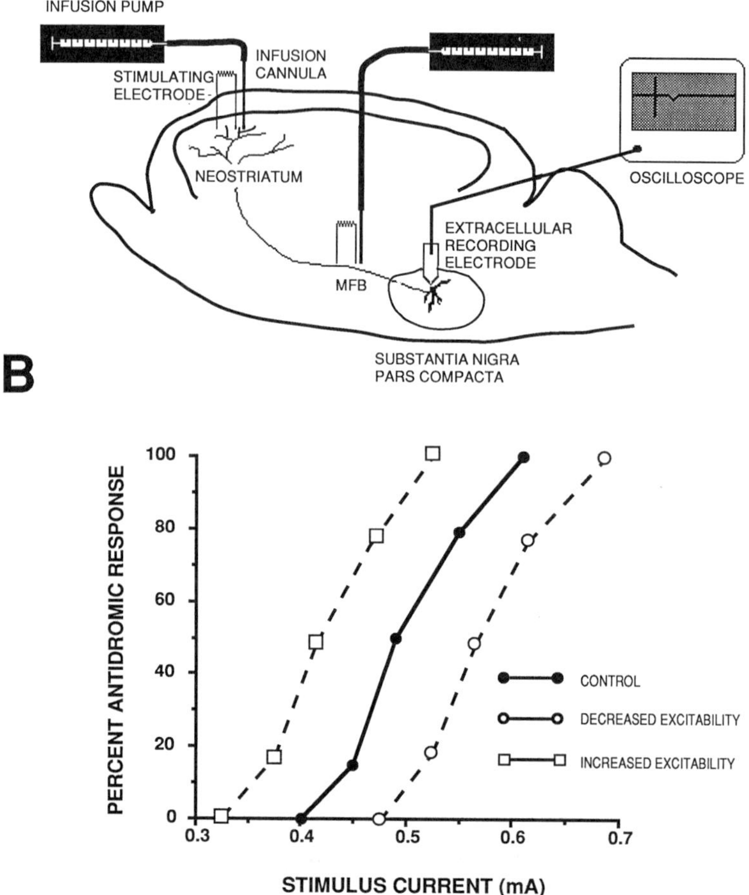

FIGURE 1. (**A**) Schematic illustration of method for measuring autoreceptor-mediated changes in dopaminergic terminal excitability *in vivo*. Excitability is indexed by the stimulating currents required to elicit varying proportions of antidromic responding from terminal fields in neostriatum or preterminal axons in the medial forebrain bundle (MFB). Excitability is compared before and after local infusions of drugs or spontaneous or stimulation-induced alterations in firing rate. (**B**) Idealized sample excitability curves resulting from application of the method in A. Shifts to the right of the excitability curve indicate decreased excitability, whereas shifts to the left indicate increased excitability.

As a general rule, the nature of the terminal autoreceptor mediating the decreases in terminal excitability seems to be of the same general class as that of the somadendritic autoreceptor which inhibits the firing rate of monoamine neurons and the axonal autoreceptor that inhibits transmitter release. Thus, for example, noradrenergic terminal excitability is decreased by clonidine, and increased by phentolamine or

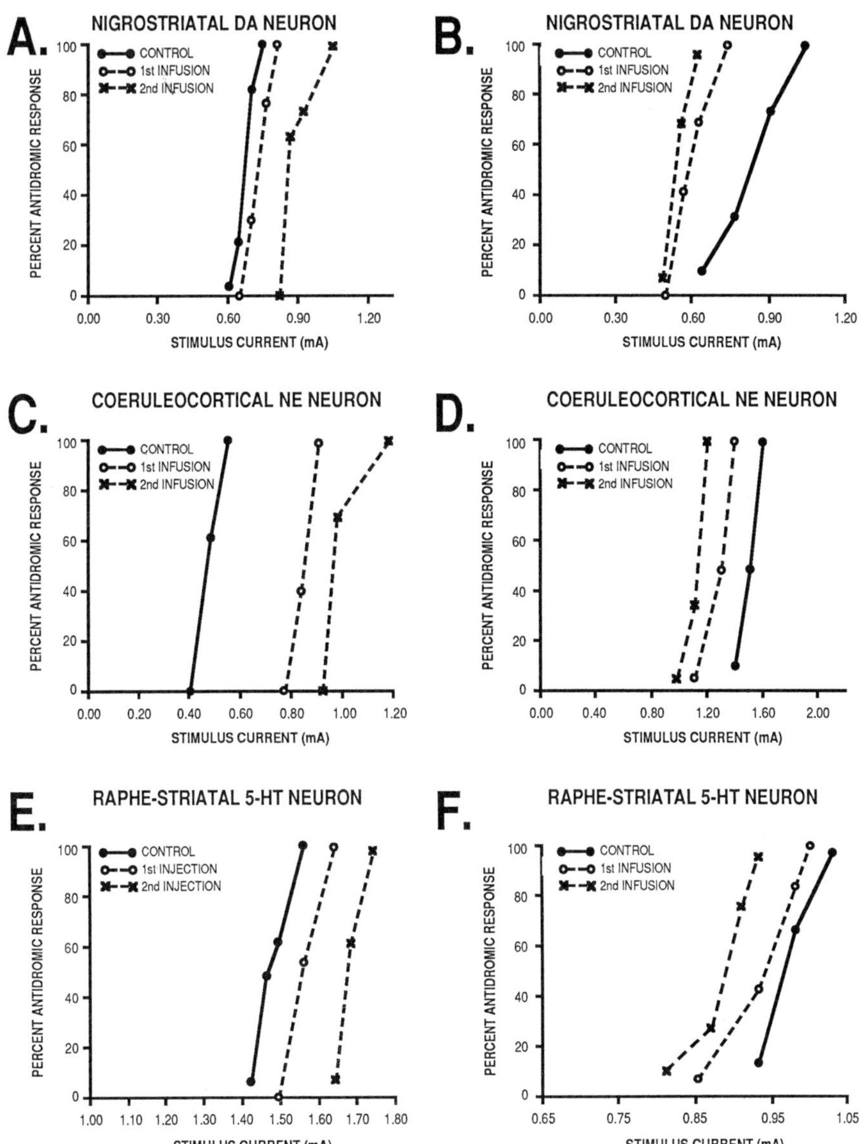

FIGURE 2. Sample excitability curves illustrating autoreceptor-mediated changes in terminal excitability *in vivo* in dopaminergic nigrostriatal (A, B), noradrenergic ceruleocortical (C, D) and serotonergic raphé-striatal neurons (E, F). In all cases, local or systemic application of the appropriate autoreceptor agonist produces a dose-dependent decrease in terminal excitability, whereas the appropriate autoreceptor antagonist produces increased terminal excitability. (A) 10 μM apomorphine. (B) 10 μM sulpiride. (C) 10 μM clonidine. (D) 10 μM phentolamine. (E) 20 μg/kg 5-methoxy dimethyl tryptamine. (F) 10 μM methiothepin. A–D, and F are local infusions of drugs into the stimulating site in the terminal fields. E is intravenous injection.

yohimbine, suggesting that the receptor may be pharmacologically defined as an α_2-receptor,[10] consistent with the receptor subtype that mediates the inhibition of firing of locus ceruleus neurons *in vivo* and *in vitro*[20–22] and the autoinhibition of norepinephrine release from cortical noradrenergic terminals *in vitro*.[1,23,24] Similarly, based on the results obtained with haloperidol and sulpiride,[2,4,8] the dopaminergic terminal autoreceptor appears to be of the D_2 variety, as does the somadendritic autoreceptor[25–27] as well as the receptor responsible for autoinhibition of dopamine release both *in vitro* and *in vivo*.[1,28,29] It should be noted however, that recent terminal excitability experiments indicate the possible existence of D_1 receptors on the terminals of nigrostriatal neurons that, like the D_2 receptors, serve to decrease terminal excitability.[30] It remains to be seen whether these receptors operate as functional autoreceptors, i.e., are sensitive to endogenously released dopamine or modify terminal excitability depending on changes in impulse flow, and whether they function to modulate dopamine release. Finally, serotonergic terminal excitability is decreased by the serotonin ($5HT_1$) agonist 5-methoxy dimethyl tryptamine.[13] These effects can be reversed by the relatively nonselective serotonergic antagonist methiothepin,[19] consistent with the identification of the serotonergic somadendritic autoreceptor modulating serotonergic neuron firing rate[31] as well as the axonal autoreceptor mediating autoinhibition of serotonin release[32] as a $5HT_1$ receptor.

It should be noted that not all terminal autoreceptors mediate feedback inhibition of terminal excitability and neurotransmitter release. Although this appears to be the case for all monoaminergic terminal autoreceptors,[33] recent data on drug and impulse dependent changes in the excitability of presumably glutamatergic corticostriatal terminals suggest that while these axons do possess terminal autoreceptors, these autoreceptors appear to mediate increases in the excitability of corticostriatal terminals,[14] as well as increases in evoked glutamate release from hippocampal slices *in vitro*.[34]

BIOPHYSICAL INTERPRETATION OF CHANGES IN TERMINAL EXCITABILITY AND THEIR RELATION TO TRANSMITTER RELEASE

The most direct way to uncover the membrane events resulting in presynaptic receptor–mediated changes in terminal excitability would be to record intracellularly from the presynaptic terminals while applying receptor agonists and antagonists. While this has been achieved in certain invertebrate preparations (e.g., Reference 35), the small size of central monoamine presynaptic axons has, so far, precluded this approach. Classically, synaptic potentials are associated with an alteration in conductance to an ion or ions that leads to a change in the membrane potential. At the nerve terminal, if the membrane were hyperpolarized, more current would be required to depolarize it to threshold for initiation of an antidromic spike. It is also true that an increase in membrane conductance alone could raise the threshold.[36] Either or both could result in the decrease in terminal excitability observed following terminal monoamine autoreceptor stimulation. However, there are several indirect lines of evidence suggesting that what is monitored by terminal excitability measurements in monoamine neurons is in fact the membrane potential or level of polarization of the terminal fields.[37] For example, when dopaminergic and noradrenergic terminal fields are depolarized by local infusion of potassium chloride (which probably also produces a slight increase in membrane conductance), their terminal excitability is increased,[4,10] suggesting that decreases in terminal excitability are associated with a terminal hyperpolarization and increases in terminal excitability are associated with a terminal depolarization. This is consistent with the results from *in vitro* intracellular recordings from central monoaminergic neurons which have revealed that although the autoreceptor recognition site is distinct for dopaminergic, noradrenergic, and serotonergic neurons, all three autoreceptors act to hyperpolarize the neuron by increasing its conductance to potassium.[21,26,38] Further evidence that terminal excitability measurements index membrane potential rather than membrane conductance comes from studies of presynaptic inhibition in the mammalian spinal cord where the presynaptic inhibition is known to be associated with a depolarization of the primary afferent terminals mediated by an increase in chloride conductance.[39] Despite the increased conductance, terminal excitability measurements similar to those described in this chapter revealed an increase in terminal excitability rather than a decrease.[40]

Thus, it is likely that decreases in monoamine terminal excitability seen following autoreceptor stimulation reflect a hyperpolarization of the terminal membranes. Since these same agonists produce a reduction in evoked release of monoamines, it follows that autoinhibition of monoamine release occurs when these nerve terminals are hyperpolarized. It is important to note that this does not necessarily suggest that it is the hyperpolarization *per se* that produces the inhibition of release. Indeed, such a hypothesis would be inconsistent with the mechanism of presynaptic inhibition in the spinal cord which, as described above, is associated with a depolarization of the primary afferent terminals. It is more likely that it is the conductance increase underlying the change in potential that serves to shunt presynaptic action currents out through the terminal membrane that is responsible for the inhibition, as has been demonstrated in the crayfish claw opener system.[35]

It must be noted that the above arguments regarding the interpretation of autoreceptor-mediated changes in terminal excitability and their relation to modulation of transmitter release hold only for receptors and their associated ionophores that function by altering conductance to ions other than Ca^{2+}, as seems to be the case with monoamine autoreceptors.[21,26,38] Since it is well-established that the only absolute requisite for evoking neurotransmitter release from nerve terminals is an increase in the availability of intracellular calcium (e.g., Reference 41), presynaptic receptors that act to directly modify Ca^{2+} permeability may be expected to lead to altered transmitter release regardless of other concurrent biophysical changes at the terminal. Such may be the case with the N-methyl-D-aspartate (NMDA) autoreceptor postulated to exist on the terminals of glutamatergic corticostriatal neurons,[14] as has been shown for NMDA receptors on other central neurons.[42]

NEURONAL LOCALIZATION OF THE TERMINAL AUTORECEPTOR

Attempts to localize the neuronal site of the receptors mediating changes in dopaminergic terminal excitability were made in dopaminergic nigrostriatal neurons by using kainic acid to destroy postsynaptic neostriatal neurons in the region from which drug infusions and terminal excitability testing were performed.[4] Three to six days following kainate-induced lesions, the effects of local infusions of apomorphine and haloperidol were tested, and found to cause changes in nigrostriatal terminal excitability indistinguishable from those in intact animals. Histological examination of the neostriata from these animals indicated that the lesions extended for at least 0.5–1.0 mm beyond the site of drug infusion and terminal excitability testing, verifying that the drug infusions and excitability changes occurred at sites devoid of intrinsic neurons. A similar lack of effect of ibotenic acid lesions of the nucleus accumbens on agonist-induced changes in mesoaccumbens terminal excitability[8] and of kainate lesions of striatum on changes in corticostriatal terminal excitability[14] has also been reported. Thus, the receptors mediating drug-induced changes in dopaminergic and corticostriatal terminal excitability would seem to be located directly on their respective axons and not on postsynaptic striatal neurons, and drug-induced changes in terminal excitability do not require the participation of postsynaptic neurons. Further, since local infusions or even intravenous administration of agonists and antagonists failed to alter excitability from preterminal regions of monoamine axons in the medial forebrain bundle, the axonal autoreceptors appear to be constrained to terminal regions of the axonal field.[2–4,10,13]

EVIDENCE FROM TERMINAL EXCITABILITY TESTING SUPPORTING A PHYSIOLOGICAL ROLE FOR TERMINAL AUTORECEPTORS

The data discussed thus far provide electrophysiological evidence that *in vivo*, the terminal excitability of central monoamine neurons can be altered by the local application of drugs that directly stimulate or block receptors located on the axon terminal regions that are sensitive to the transmitter that the neuron itself releases. However, with the exception of the observation that terminal excitability can be increased by the local infusion of antagonists by themselves, the data cannot be used to argue convincingly that these presynaptic receptors function as autoreceptors.

One piece of evidence to support the operation of these presynaptic receptors as true autoreceptors derives from the use of amphetamine as a probe. Amphetamine increases extracellular levels of catecholamines by simultaneously promoting catecholamine release and blocking catecholamine uptake.[43] When amphetamine is locally infused into neostriatal, ventral striatal, or neocortical terminal fields, the excitability of dopaminergic or noradrenergic terminals is markedly reduced, just as with infusions of exogenously applied direct-acting agonists,[4,8,11] as illustrated in FIGURE 3. In both catecholaminergic systems, the effects of amphetamine were completely eliminated by prior treatment with the catecholamine synthesis inhibitor α-methyl-p-tyrosine, and could be reversed by haloperidol or sulpiride in the dopaminergic system or phentolamine or yohimbine in the noradrenergic system.[4,8,11,12] Thus, the axonal presynaptic receptors of both systems can be considered autoreceptors since they are responsive to their endogenously released neurotransmitter.

In a recent review, Kalsner suggested several operational criteria that should be fulfilled before accepting the hypothesis that autoreceptors modulate impulse- or stimulus-evoked transmitter release.[44] Two of these criteria had to do with the relation between the magnitude of the effects of agonists and antagonists on evoked release of transmitters and the level of ongoing autoinhibition. If the synaptic concentration of endogenous transmitter is high, for example, due to a high rate of neuronal activity and consequent transmitter release, then antagonists would be expected to exert relatively large effects due to the occupation of many autoreceptor sites by the endogenous neurotransmitter. By similar logic, the effects of exogenously applied agonists would be expected to be relatively small at high rates of firing, since a fixed concentration (dose) of the applied agonist corresponds to a progressively smaller proportion of the total agonist available for binding to the autoreceptor. The converse should be true at low rates of impulse activity and release, when agonists would be expected to exert relatively large effects and antagonists relatively small effects. Precisely this relation was observed with dopaminergic nigrostriatal terminal excitability with a variety of autoreceptor agonists and antagonists infused into the terminal fields,[4] as illustrated in FIGURE 4. When the magnitude of the drug-induced changes in terminal excitability was correlated with the baseline firing rate of the neuron, a significant negative correlation was observed with the agonists amphetamine and apomorphine, and a significant positive correlation with the antagonists haloperidol and sulpiride. Similar

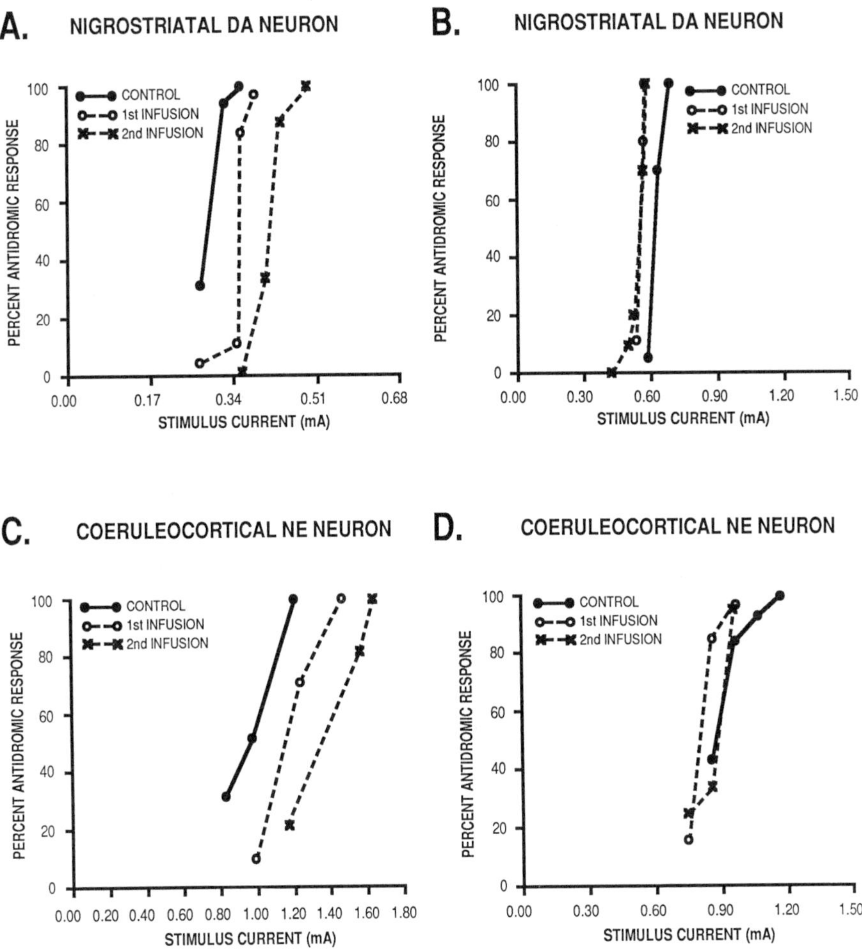

FIGURE 3. Sample excitability curves illustrating effects of local infusion of amphetamine on terminal excitability of nigrostriatal dopaminergic (A, 1 μM) and ceruleocortical noradrenergic (C, 10 μM) neurons. In both systems, increasing endogenous levels of catecholamines by amphetamine infusion leads to decreased terminal excitability. The effects of amphetamine can be prevented in both systems by pretreatment with the catecholamine synthesis inhibitor α-methyl-p-tyrosine. (**B**) 10 μM amphetamine after α-methyl-p-tyrosine, nigrostriatal neuron. (**D**) 10 μM amphetamine after α-methyl-p-tyrosine, ceruleocortical neuron.

effects were observed with 5-methoxy-dimethyl tryptamine induced decreases in serotonergic terminal excitability.[13] A similar relation, i.e., that between the rate of stimulation and the strength of the inhibitory and facilitatory effects of autoreceptor agonists and antagonists *in vitro* has also been described for dopamine, norepinephrine, and serotonin release.[45–50] It is interesting to note that a similar negative correlation obtains between the baseline firing rate of monoaminergic neurons and the extent to which exogenously applied autoreceptor agonists depress the firing rate of the neuron by acting on soma-dendritic autoreceptors.[51–54]

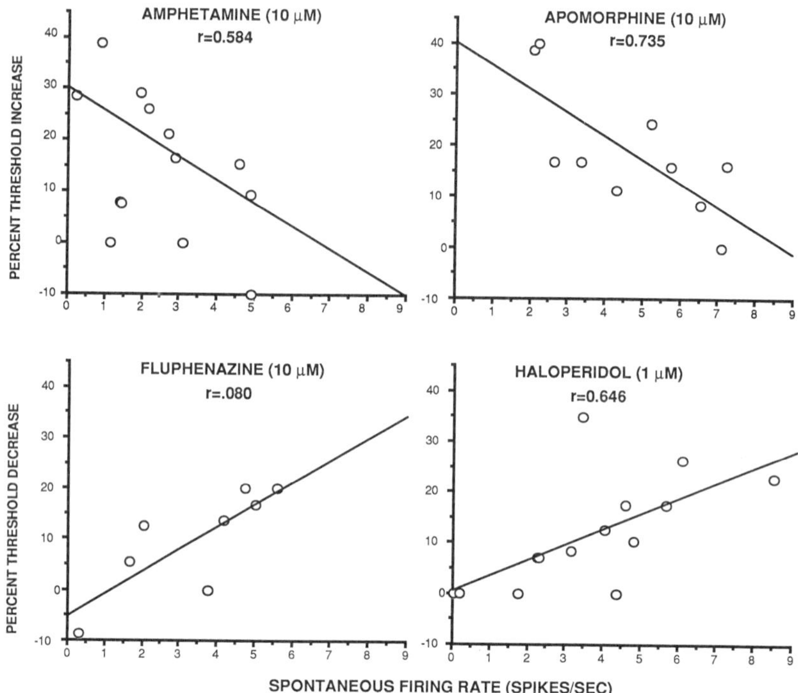

FIGURE 4. Correlations between the magnitude of the change in dopaminergic terminal excitability resulting from local neostriatal infusions of amphetamine, apomorphine, fluphenazine, and haloperidol and the predrug spontaneous firing rate. Autoreceptor agonists are most effective at decreasing terminal excitability in slowly firing neurons, whereas autoreceptor antagonists are most effective at increasing terminal excitability in rapidly firing neurons. $p <$ 0.05 for all regressions. (Data replotted and modified from Tepper *et al.*)[4]

Another criterion suggested by Kalsner was that changes in the rate of neuronal activity, in the absence of exogenous drug application, should in some way reflect changes in ongoing autoinhibition.[44] Terminal excitability measurements *in vivo* in the nigrostriatal and mesolimbic dopaminergic system,[5,9] ceruleocortical noradrenergic projections,[10] and presumed glutamatergic corticostriatal system[14] have all afforded direct evidence for this phenomenon. This is illustrated for a noradrenergic ceruleocortical neuron and a dopaminergic mesoprefrontal cortical neuron in FIGURE 5. Although the firing rate of individual central monoamine neurons *in vivo* is usually relatively constant, occasionally neurons are recorded whose firing rate varies significantly over the course of a few minutes. FIGURE 5A illustrates the results of repeatedly measuring the spontaneous firing rate and the antidromic threshold from the frontal cortex of a noradrenergic locus ceruleus neuron whose firing rate is varying over the course of the measurements.[10] The antidromic threshold current is seen to exhibit a tight direct relationship to the firing rate. FIGURE 5B illustrates the same relation in a slightly different way for a dopaminergic ventral tegmental area neuron projecting to the prefrontal cortex.[37] In this case, the antidromic stimulating current was adjusted to a subthreshold value, and the percent antidromic responding and spontaneous firing rate of the neuron were measured every 30 seconds for several minutes. The terminal excitability exhibits a significant inverse correlation to the firing rate ($r = -0.84$, df $= 22$, $p < 0.001$) which is abolished by intravenous administration of haloperidol (0.05 mg/kg).

In addition to responding to these relatively prolonged changes in firing rate (over the course of several seconds), dopaminergic terminal excitability is sensitive to considerably briefer events, such as bursts consisting of 2 or 3 spikes lasting on the order of 150–200 mseconds, or even the occurrence of single spikes just outside the antidromic collision interval. In these cases, terminal excitability is transiently depressed by increased spiking that occurs within 5–10 mseconds of the delivery of the antidromic stimulus.[3,5]

Impulse-dependent changes in terminal excitability of noradrenergic and dopaminergic systems have also been demonstrated by altering the frequency of impulses along the axon by the application of conditioning stimuli to preterminal regions of the axon. The range of effective frequencies employed (brief trains of pulses at 1–10 Hz) are well within the range of the rates of spontaneous activity for these neurons. Three important characteristics of stimulus-dependent decreases in terminal excitability are illustrated in FIGURES 6–8. FIGURE 6 illustrates decreased terminal excitability of cortical locus ceruleus terminals following suprathreshold stimulation of the preterminal axons in the dorsal noradrenergic pathway (DP).[10] It was possible to determine the threshold for the conditioning stimulation by simply observing the antidromic response to the conditioning pulses. When conditioning was applied at suprathreshold currents, terminal excitability was reduced. If however, the dorsal pathway stimulus current was slightly reduced to a value that failed to elicit antidromic responding, there was no alteration in terminal excitability. This illustrates the point that the reduced terminal excitability results from some consequence of impulse flow along the axon whose excitability is being tested. That this phenomenon occurs only at the terminal regions of the axon is

A. COERULEOCORTICAL NE NEURON **B.** MESOPREFRONTAL DA NEURON

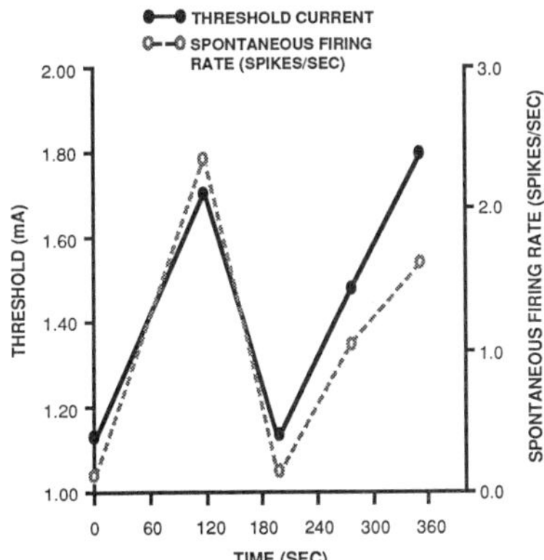

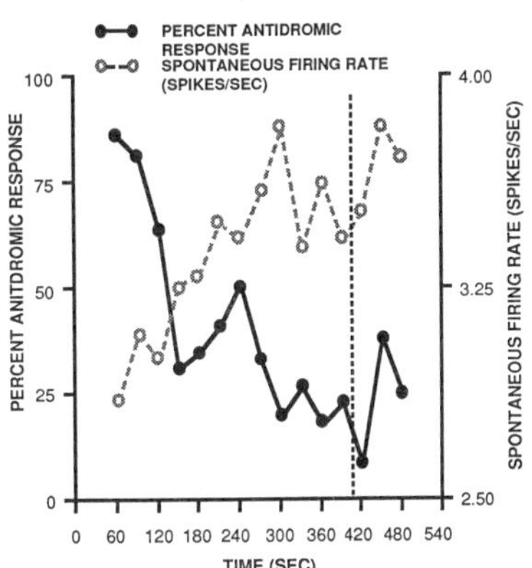

FIGURE 5. Changes in terminal excitability associated with spontaneous changes in neuronal firing rate. (**A**) Threshold current for a noradrenergic ceruleocortical neuron varies directly with spontaneous changes in firing rate (Redrawn from Nakamura *et al.,* with permission)[10] (**B**) Antidromic excitability of a dopaminergic mesocortical neuron exhibits significant negative correlation with spontaneous firing rate ($r = -0.84$, $p < 0.05$). This relation is abolished by intravenous administration of haloperidol (50 µg/kg) at the double dashed line. (Redrawn from Tepper et al., with permission.)[37]

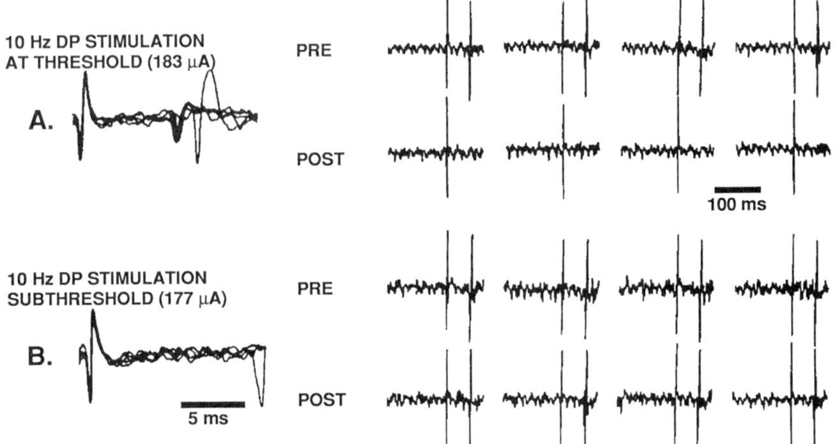

FIGURE 6. (A) High frequency conditioning stimulation applied at threshold (183 µA) to preterminal axon (DP) of a noradrenergic ceruleocortical neuron elicits antidromic responses and produces decreased terminal excitability that outlasts the conditioning stimulation (compare PRE versus POST traces at same stimulating current). (B) Similar conditioning stimulation applied to DP just below threshold (177 µA) does not evoke antidromic responses and does not affect terminal excitability. (Modified from Nakamura et al., with permission.)[10]

illustrated for a nigrostriatal dopaminergic neuron in FIGURE 7. The roles of the conditioning and testing electrodes can be reversed within a single experiment so that in one case the preterminal axon electrode (MFB electrode) can be used to condition the axon for excitability testing from the terminal regions. The terminal field electrode can then be used to send an identical train of stimuli down through the medial forebrain bundle, and the medial forebrain bundle electrode can be used to test excitability from this preterminal site.[5] The results from this experiment show that although increased impulse flow decreases terminal excitability, there is little or no effect along preterminal regions of the axon, which are also insensitive to the effects of locally or systemically administered autoreceptor agonists and antagonists. Finally, FIGURE 8 illustrates that the effects of conditioning stimulation of the medial forebrain bundle on dopaminergic terminal excitability can be blocked by local infusion of haloperidol into the neostriatal stimulating site.[5] Thus, the reduction in terminal excitability is not simply due to the increased impulse flow, but to stimulation of terminal autoreceptors by increased levels of transmitter evoked by the conditioning stimuli.

In the monoaminergic systems, as discussed above, autoreceptor stimulation results in decreased terminal excitability and decreased transmitter release. However, as mentioned previously, corticostriatal neurons also appear to possess autoreceptors, but these autoreceptors act in an opposite fashion to those on monoamine terminals. Thus the terminal excitability of (presumed) glutamatergic corticostriatal neurons is directly correlated with the firing rate of these cortical neurons, a relationship that, like that described above for mesocortical dopaminergic neurons, can be abolished by the local application of an appropriate autoreceptor antagonist, in this case, an NMDA antagonist.[14]

PRESYNAPTIC HETERORECEPTOR-MEDITATED CHANGES IN TERMINAL EXCITABILITY

In addition to the presynaptic autoreceptor systems discussed above, terminal excitability testing has also been employed to investigate the electrophysiological consequences of stimulation and blockade of presynaptic heteroreceptors. Local infusions of opioid agonists have been shown to decrease the terminal excitability of cortical terminals of noradrenergic locus ceruleus neurons in a naloxone-reversible manner.[16] The decreased excitability is consistent with *in vitro* intracellular recording studies of locus ceruleus neurons showing that opioids hyperpolarize locus ceruleus neurons by activating a potassium conductance identical to that activated by α_2-autoreceptor agonists.[55,56] The fact that naloxone alone produced increases in noradrenergic terminal excitability suggests that the presynaptic opioid receptors may have physiological relevance since they appear to be at least partially occupied by an endogenous agonist *in vivo*. It is even possible that these presynaptic opioid

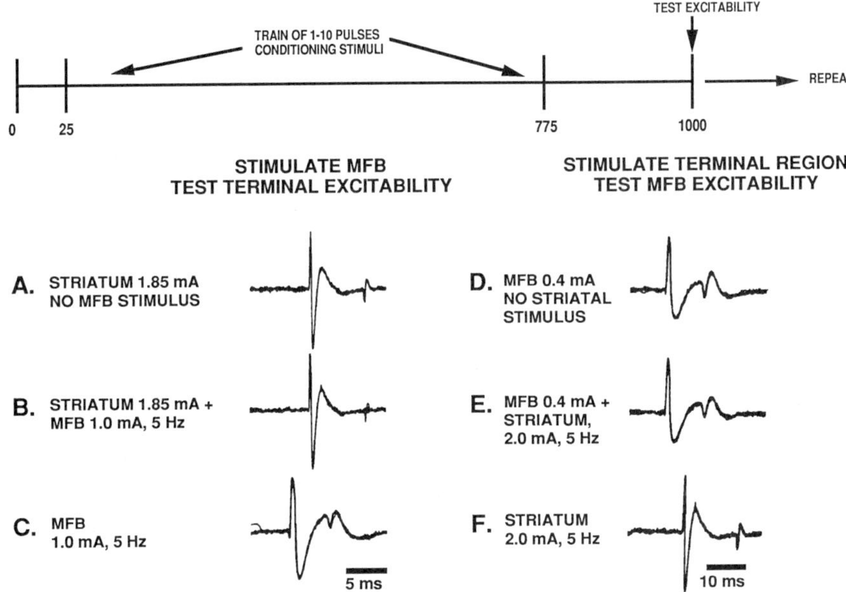

FIGURE 7. Increasing the rate of impulse flow in a dopaminergic nigrostriatal neuron decreases excitability, but only at the terminal. A train of conditioning stimuli is delivered in a 750 msecond window that terminates 225 mseconds prior to excitability testing. Entire sequence is repeated once per second. (**A**) Threshold for antidromic responding from striatum is 1.85 mA. (**B**) When preterminal axon in medial forebrain bundle (MFB) is stimulated at 5Hz, 1.85 mA stimulus to terminal field evokes antidromic spike on only one of 5 trials. (**C**) Antidromic response elicited by the MFB conditioning pulses at 5 Hz. (**D**) Threshold for antidromic responding from MFB is 0.4 mA. (**E**) When axon is antidromically conditioned by suprathreshold stimulation of terminal at 5 Hz, there is no change in antidromic responses elicited from MFB. (**F**) Antidromic responses elicited by terminal conditioning pulses at 5 Hz proving that the conditioning pulses pass through the region of the MFB from which excitability was tested, and demonstrating that increases in impulse flow only produce decreased excitability at the terminal region. Each trace comprised of superimposition of 5 consecutive sweeps. (Modified from Tepper et al., with permission).[5]

"hetereoreceptors" are in fact autoreceptors (see Reference 57), since at least in some species, there is evidence for colocalization of norepinephrine and enkephalin in locus ceruleus neurons.[58] In any event, the decreased noradrenergic terminal excitability resulting from opioid receptor stimulation *in vivo* is consistent with the decreases in evoked norepinephrine release seen with presynaptic opioid receptor stimulation *in vitro*.[1,23,24]

In contrast to its apparent hyperpolarizing action at striatal dopaminergic terminals, at least one study has reported increased terminal excitability of hippocampal afferents to ventral striatum elicited by stimulation of the ventral tegmental area or by

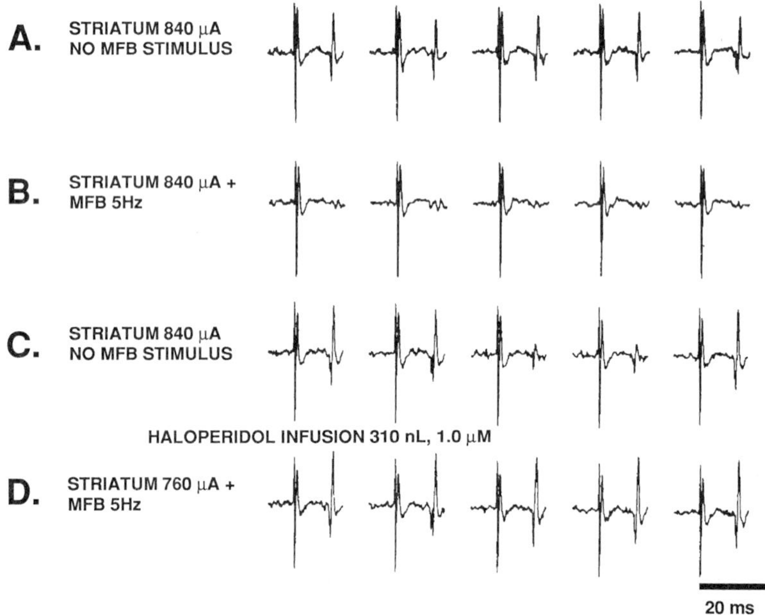

FIGURE 8. Decreased terminal excitability resulting from stimulation-induced increases in impulse flow in dopaminergic nigrostriatal neuron are blocked by local infusion of haloperidol into the terminal fields (**A**) Predrug, preconditioning threshold is 0.84 mA. (**B**) 5 Hz conditioning stimulation of MFB decreases threshold and eliminates antidromic responding. (**C**) Recovery baseline terminal excitability after cessation of MFB conditioning pulses. (**D**) After local neostriatal infusion of 310 nL haloperidol (1 μM), antidromic threshold decreases slightly and MFB conditioning no longer affects terminal excitability. (Modified from Tepper et al.)[33]

iontophoretic application of dopamine to these terminals. These changes were mimicked by iontophoresis of a D_2 agonist (quinpirole), but not by a D_1 agonist (SKF38393), suggesting that they were mediated by dopamine D_2 heteroreceptors.[17] The interpretation of these findings is complicated by a seeming lack of correspondence between the actions of dopamine at hippocampal terminals *in vivo* and dopamine's action on hippocampal cell bodies *in vitro*, where it causes a membrane hyperpolarization and decreases excitability.[59,60] In addition, some aspects of the dopamine response in the hippocampal slice preparation may be mediated by β-adrenergic receptors;[61] it is at present unclear if a similar situation may also obtain at hippocampal terminals.

Two different studies have been conducted on dopamine receptor–mediated effects on corticostriatal terminal excitability.[14,62] In one, conditioning stimulation of the substantia nigra led to prolonged increases in corticostriatal terminal excitability.[62] The increased terminal excitability could be blocked by systemic administration of sulpiride, suggesting D_2 receptor mediation. In a second study, corticostriatal terminal excitability was decreased by local infusion of amphetamine or apomorphine, or electrical stimulation of substantia nigra,[14] consistent with the ability of D_2 receptor stimulation to decrease evoked release of glutamate from corticostriatal terminals.[63] These effects could be blocked by administration of haloperidol or sulpiride, and obtained even in animals with kainate lesions of the striatal terminal fields. Perhaps even more interesting, local infusions of haloperidol or sulpiride alone produced increased terminal excitability, suggesting that *in vivo,* the presynaptic dopamine heteroreceptors on corticostriatal afferents are stimulated by endogenously released dopamine. It is difficult to reconcile the contradictory findings of these two studies. Despite the success of sulpiride at antagonizing the opposite changes in terminal excitability in both experiments, it is conceivable that transmitters other than dopamine may have been involved. For example, since cholecystokinin has been shown to be colocalized in some ascending dopaminergic pathways,[64] it is possible that the increased terminal excitability seen by Mogenson and Yang may be due to release of this tachykinin, similar to effects observed on dopaminergic mesoaccumbens terminals.[65] Cholecystokinin has also been shown to exert an excitatory effect on dopamine-sensitive cortical neurons, which, incidentally, appear to possess dopamine receptors that are not easily classified as D_1 or D_2, perhaps further contributing to the disparate results.[66] Clearly, further experiments are necessary to clarify these issues.

Finally, evidence from terminal excitability testing also reveals the presence of dopaminergic receptors on the terminals of presumed GABAergic striatonigral neurons,[18] consistent with previous observations suggesting a role for presynaptic dopamine receptors in the modulation of nigral GABA release.[67] Interestingly, in contrast to all of the autoreceptors and most of the presynaptic heteroreceptor systems described above, careful examination of the presynaptic dopamine heteroreceptor on striatonigral terminals in vivo suggests that these receptors may not have a physiological function since although they can be activated by local infusions of a D_1 agonist, they appear insensitive to systemic administration of amphetamine, even at high doses. Furthermore, neither haloperidol nor the D_1 antagonist SCH23390, administered alone either systemically or locally, produced a significant change in striatonigral terminal excitability. These last two observations suggest that the levels of endogenously released dopamine, even when augmented by amphetamine, are not sufficient to stimulate these receptors.[18] Alternatively, perhaps the appropriate physiological conditions necessary for endogenous activation of these receptors have not yet been determined.

CONCLUSIONS

The data reviewed in this chapter consist largely of electrophysiological measurements of autoreceptor-mediated changes in terminal excitability in vivo induced by direct application of drugs or by changes in the rate of impulses reaching the axon terminal fields of several neurochemically and topographically distinct neuron types. These data are consistent with the presence of autoreceptors on the nerve terminals of a variety of different CNS neurons. At least within the class of monoaminergic neurons, presynaptic autoreceptor stimulation appears to be associated with a hyperpolarization of the terminal regions. As mentioned at the outset, the actual release of neurotransmitter is not measured in these experiments, but these electrophysiological techniques do offer the advantages that they are performed *in vivo* in intact brains, under more or less "normal" physiological conditions, and can be correlated with the neuronal activity of the systems under study with temporal resolution in the millisecond range. These characteristics are particularly important when trying to address questions like "Are presynaptic autoreceptors physiologically functional *in vivo?*" and "Under what conditions of neuronal activity might they operate?". The results obtained following local infusions of autoreceptor antagonists alone into various axon terminal regions suggest quite strongly that the autoreceptors are physiologically functional *in vivo,* as do the observed correlations between the magnitude of the changes in terminal excitability observed after application of both agonists and antagonists. The nature of the correlations, coupled with the observations that physiologically relevant increases in impulse flow decrease monoaminergic terminal excitability through an autoreceptor-mediated mechanism, suggests that the presynaptic autoreceptors are most likely to be activated in vivo during prolonged periods of high activity. However, these systems also have the capacity to respond to very transient changes in neuronal activity, such as during brief bursts that may last less than 100 mseconds. These data paint a picture of an autoregulatory system that may be constantly changing its tone, or level of control perhaps on a spike by spike basis, to fine tune neurotransmitter output at individual sites within an axonal arbor in response to local conditions, at least partially independently of events occurring at the cell body.

ACKNOWLEDGMENTS

Thanks to F. Trent for assistance in preparing the figures and to Dr. D. H. Schwartz for helpful comments on the manuscript.

REFERENCES

1. STARKE, K., M. GOTHERT & H. KILBINGER. 1989. Modulation of neurotransmitter release by presynaptic autoreceptors. Physiol. Rev. **69:** 864–989.
2. GROVES, P. M., G. A. FENSTER, J. M. TEPPER, S. NAKAMURA & S. J. YOUNG. 1981. Changes in dopaminergic terminal excitability induced by amphetamine and haloperidol. Brain Res. **221:** 425–431.
3. TAKEUCHI, H., S. J. YOUNG & P. M. GROVES. 1982. Dopaminergic terminal excitability following arrival of the nerve impulse: the influence of amphetamine and haloperidol. Brain Res. **245:** 47–56.
4. TEPPER, J. M., S. NAKAMURA, S. J. YOUNG & P. M. GROVES. 1984. Autoreceptor-mediated changes in dopaminergic terminal excitability: effects of striatal drug infusions. Brain Res. **309:** 317–333.
5. TEPPER, J. M., S. J. YOUNG & P. M. GROVES. 1984. Autoreceptor-mediated changes in dopaminergic terminal excitability: effects of increases in impulse flow. Brain Res. **309:** 309–316.
6. TEPPER, J. M., S. F. SAWYER, S. J. YOUNG & P. M. GROVES. 1986. Autoreceptor-mediated changes in dopaminergic terminal excitability: effects of potassium channel blockers. Brain Res. **367:** 230–237.
7. CHAVEZ-NORIEGA, L., P. PATINO & M. GARCIA-MUNOZ. 1986. Excitability changes induced in the striatal dopamine-containing terminals following frontal cortex stimulation. Brain Res. **379:** 300–306.
8. MEREU, G., T. C. WESTFALL & R. Y. WANG. 1985. Modulation of terminal excitability of mesolimbic dopaminergic neurons by D-amphetamine and haloperidol. Brain Res. **359:** 88–96.

9. GARIANO, R. F., S. F. SAWYER, J. M. TEPPER, S. J. YOUNG & P. M. GROVES. 1989. Mesocortical dopaminergic neurons. 2. Consequences of terminal autoreceptor activation. Brain Res. Bull. **22:** 517–523.

10. NAKAMURA, S., J. M. TEPPER, S. J. YOUNG & P. M. GROVES. 1981. Neurophysiological consequences of presynaptic receptor activation: changes in noradrenergic terminal excitability. Brain Res. **226:** 155–170.

11. NAKAMURA, S., J. M. TEPPER, S. J. YOUNG & P. M. GROVES. 1982. Changes in noradrenergic terminal excitability induced by amphetamine and their relation to impulse traffic. Neuroscience **7:** 2217–2224.

12. RYAN, L. J., J. M. TEPPER, S. J. YOUNG & P. M. GROVES. 1985. Amphetamine's effect on terminal excitability of noradrenergic locus coeruleus neurons are impulse dependent at low but not high doses. Brain Res. **341:** 155–163.

13. SAWYER, S. F., J. M. TEPPER, S. J. YOUNG & P. M. GROVES. 1985. Antidromic activation of dorsal raphé neurons from neostriatum: physiological characterization and effects of autoreceptor activation. Brain Res. **332:** 15–28.

14. GARCIA-MUNOZ, M., S. J. YOUNG, & P. M. GROVES. 1989. Modification of excitability of corticostriatal afferents: evidence for auto and heteroreceptor regulation. Soc. Neurosci. Abstr. **15:** 584.

15. ROMO, R. & W. SCHULTZ. 1985. Prolonged changes in dopaminergic terminal excitability and short changes in dopaminergic neuron discharge rate after short peripheral stimulation in monkey. Neurosci. Lett. **62:** 335–340.

16. NAKAMURA, S., J. M. TEPPER, S. J. YOUNG, N. LING & P. M. GROVES. 1982. Noradrenergic terminal excitability: effects of opioids. Neurosci. Lett. **30:** 57–62.

17. YANG, C. R. & G. J. MOGENSON. 1986. Dopamine enhances terminal excitability of hippocampal-accumbens in presynaptic inhibition. J. Neurosci. **6:** 2470–2478.

18. RYAN, L. J., M. DIANA, S. J. YOUNG & P. M. GROVES. 1989. Dopamine D1 heteroreceptors on striatonigral axons are not stimulated by endogenous dopamine either tonically or after amphetamine: evidence from terminal excitability. Exp. Brain Res. **77:** 161–165.

19. TEPPER, J. M., S. F. SAWYER, S. NAKAMURA & P. M. GROVES. Dopaminergic and serotonergic terminal excitability: effects of autoreceptor stimulation and blockade. *In* The Presynaptic Regulation of Neurotransmitter Release. J. J. Feigenbaum & M. Hanai, Eds. Freund Publishing Co. Tel Aviv, Israel. (In press.)

20. CEDARBAUM, J. M. & G. K. AGHAJANIAN. 1977. Catecholamine receptors on locus coeruleus neurons: pharmacological characterization. Eur. J. Pharmacol. **44:** 375–385.

21. AGHAJANIAN, G. K. & C. P. VANDERMAELEN. 1982. Alpha2-adrenoceptor-mediated hyperpolarization of locus coeruleus neurons: intracellular studies in vivo. Science **215:** 1394–1396.

22. WILLIAMS, J. T., G. HENDERSON & R. A. NORTH. 1985. Characterization of alpha2-adrenoceptors which increase potassium conductance in rat locus coeruleus neurones. Neuroscience **14:** 95–101.

23. TAUBE, H. D., K. STARKE & E. BOROWSKI. 1977. Presynaptic receptor systems on the noradrenergic neurones of rat brain. Naunyn Schmiedebergs Arch. Pharmacol. **299:** 123–141.

24. LANGER, S. Z. 1977. Presynaptic receptors and their role in the regulation of transmitter release. Br. J. Pharmacol. **60:** 481–497.

25. AGHAJANIAN, G. K. & B. S. BUNNEY. 1977. Dopamine 'autoreceptors': pharmacological characterization by microiontophoretic single cell recording studies. Naunyn Schmiedebergs Arch. Pharmacol. **297:** 1–7.

26. LACEY, M. G., N. B. MERCURI & R. A. NORTH. 1987. Dopamine acts on D2 receptors to increase potassium conductance in neurons of the rat substantia nigra zona compacta. J. Physiol. London **392:** 497–516.

27. MORELLI, M., T. MENNINI & G. DI CHIARA. 1988. Nigral dopamine autoreceptors are exclusively of the D2 type: quantitative autoradiography of [^{125}I]iodosulpiride and [^{125}I]SCH 23982 in adjacent brain sections. Neuroscience **27:** 865–870.

28. HELMREICH, I., W. REIMANN, G. HERTTING & K. STARKE. 1982. Are presynaptic dopamine autoreceptors and postsynaptic dopamine receptors in the rabbit caudate nucleus pharamcologically different? Neuroscience **7:** 1559–1566.

29. BOYAR, W. C. & C. A. ALTAR. 1987. Modulation of in vivo dopamine release by D2 but not D1 receptor agonists and antagonists. J. Neurochem. **48:** 824–831.
30. DIANA, M., S. J. YOUNG & P. M. GROVES. 1989. Modulation of dopaminergic terminal excitability by D1 selective agents. Neuropharmacology **28:** 99–101.
31. VANDERMAELEN, C. P. & G. K. AGHAJANIAN. 1983. Electrophysiological and pharmacological characterization of serotonergic dorsal raphé neurons recorded extracellularly and intracellularly in rat brain slices. Brain Res. **289:** 109–119.
32. MARTIN, L. L. & E. SANDERS-BUSCH. 1982. Comparison of the pharmacological characteristics of 5HT1 and 5HT2 binding sites with those of serotonin autoreceptors which modulate serotonin release. Naunyn Schmiedebergs Arch. Pharmacol. **321:** 165–170.
33. TEPPER, J. M., P. M. GROVES & S. J. YOUNG. 1985. The neuropharmacology of the autoinhibition of monoamine release. Trends Pharmacol. Sci. **6:** 251–256.
34. CONNICK, J. H. & T. W. STONE. 1988. Excitatory amino acid antagonists and endogenous aspartate and glutamate release from rat hippocampal slices. Br. J. Pharmacol. **93:** 863–867.
35. BAXTER, D. A. & G. D. BITTNER. 1981. Intracellular recordings from crustacean motor axons during presynaptic inhibition. Brain Res. **223:** 422–428.
36. HUBBARD, J. I., R. LLINAS & D. M. J. QUASTAL. 1969. Electrophysiological Basis of Synaptic Transmission: 15–84, 112–173.50. Williams & Wilkins Co. Baltimore, Md.
37. TEPPER, J. M., R. F. GARIANO & P. M. GROVES. 1987. The neurophysiology of dopamine nerve terminal autoreceptors. Neurophysiology of Dopaminergic Systems: Current Status and Clinical Perspectives. *In* L. A. Chiodo & A. S. Freeman, Eds.: 93–127. Lakeshore Press. Grosse Pt., Mich.
38. AGHAJANIAN, G. K. & J. M. LAKOSKI. 1984. Hyperpolarization of serotonergic neurons by serotonin and LSD: studies in brain slices showing increased K + conductance. Brain Res. **305:** 181–185.
39. ECCLES, J. C., R. F. SCHMIDT & W. D. WILLIS. 1963. The mode of operation of the synaptic mechanism producing presynaptic inhibition. J. Neurophysiol. **26:** 523–536.
40. WALL, P. D. 1958. Excitability changes in afferent fibre terminations and their relation to slow potentials. J. Physiol. London **142:** 1–21.
41. ZUCKER, R. S. & L. LANDO. 1986. Mechanism of transmitter release: voltage hypothesis and calcium hypothesis. Science **231:** 574–579.
42. DINGLEDINE, R. 1983. N-Methyl aspartate activates voltage-dependent calcium conductance in rat hippocampal pyramidal cells. J. Physiol. London **343:** 385–405.
43. KUCZENSKI, R. 1983. Biochemical actions of amphetamine and other stimulants. *In* Stimulants: Neurochemical, Behavioral and Clinical Perspectives. I. Creese, Ed.: 31–61. Raven Press. New York, N.Y.
44. KALSNER, S. 1985. Is there feedback regulation of neurotransmitter release by autoreceptors? Biochem. Pharmacol. **34:** 4085–4097.
45. CUBEDDU, L. X. & I. S. HOFFMANN. 1982. Operational characteristics of the inhibitory feedback mechanism for regulation of dopamine release via presynaptic receptors. J. Pharmacol. Exp. Ther. **223:** 497–501.
46. HOFFMANN, I. S. & L. X. CUBBEDDU. 1982. Rate and duration of stimulation determine presynaptic effects of haloperidol on dopaminergic neurons. J. neurochem. **39:** 585–588.
47. LEHMANN, J. & S. Z. LANGER. 1982. The pharmacological distinction between central pre- and postsynaptic dopamine receptors: implications for the pathology and therapy of schizophrenia. Adv. Dopamine Res. Adv. Biosci. **37:** 25–39.
48. DUBOCOVICH, M. L. & J. G. HENSLER. 1986. Modulation of [^{3}H]-dopamine release by different frequencies of stimulation from rabbit retina. Br. J. Pharmacol. **88:** 51–61.
49. MAYER, A., N. LIMBERGER & K. STARKE. 1988. Transmitter release patters of noradrenergic, dopaminergic and cholinergic axons in rabbit brain slices during short pulse trains and the operation of presynaptic autoreceptors. Naunyn Schmiedebergs Arch. Pharmacol. **338:** 632–643.
50. BLIER, P., R. RAMDINE, A.-M. GALZIN & S. Z. LANGER. 1989. Frequency-dependence of serotonin autoreceptor but not α_2-adrenoceptor inhibition of [^{3}H]-serotonin release in rat hypothalamic slices. Naunyn Schmiedebergs Arch. Pharmacol. **339: 60–64.**
51. STAUNTON, D. A., S. J. YOUNG & P. M. GROVES. 1980. The effect of long-term amphet-

amine administration on the activity of dopaminergic neurons of the substantia nigra. Brain Res. **188:** 107–117.

52. TEPPER, J. M., S. NAKAMURA, S. J. YOUNG & P. M. GROVES. 1982. Subsensitivity of catecholaminergic neurons to direct acting agonists after single or repeated electroconvulsive shock. Biol. Psychiatry –17: 1059–1070.

53. JACOBS, B. L., J. HEYM & K. RASMUSSEN. 1983. Raphé neurons: firing rate correlates with size of drug response. Eur. J. Pharmacol. **90:** 275–278.

54. WHITE, F. J. & R. Y. WANG. 1984. A10 dopamine neurons: role of autoreceptors in determining firing rate and sensitivity to dopamine agonists. Life Sci. **34:** 1161–1170.

55. PEPPER, C. M. & G. HENDERSON. 1980. Opiates and opioid peptides hyperpolarize locus coeruleus neurons in vitro. Science **209:** 394–396.

56. NORTH, R. A., & J. T. WILLIAMS. 1985. On the potassium conductance increased by opioids in rat locus coeruleus neurones. J. Physiol. London **364:** 265–280.

57. NAKAMURA, S., J. M. TEPPER & P. M. GROVES. Noradrenergic terminal excitability: effects of presynaptic receptor stimulation and blockade. *In* The Presynaptic Regulation of Neurotransmitter Release. J. J. Feigenbaum & M. Hanani, Eds. Freund. London & Tel Aviv. (In press.)

58. CHARNAY, Y., L. LÉGER, F. DRAY, A. BEROD, M. JOUVET, J. F. PUJOL & P. M. DUBOIS. 1982. Evidence for the presence of enkephalin in catecholaminergic neurones of cat locus coeruleus. Neurosci. Lett. **30:** 147–151.

59. BERNARDO, L. S. & D. A. PRINCE. 1982. Dopamine action on hippocampal pyramidal cells. J. Neurosci. **2:** 415–423.

60. STANZIONE, P., P. CALABRESI, N. MERCURI & G. BERNARDI. 1984. Dopamine modulates CA1 hippocampal neurons by elevating the threshold for spike generation: an in vitro study. Neuroscience **13:** 1105–1116.

61. MALENKA, R. C. & R. A. NICOLL. 1986. Dopamine decreases the calcium-activated hyperpolarization in hippocampal CA1 pyramidal cells. Brain Res. **379:** 210–215.

62. MOGENSON, G. J. & C. R. YANG. 1987. Dopamine modulation of limbic and cortical input to striatal neurons. *In* Neurophysiology of Dopaminergic Systems: Current Status and Clinical Perspectives. L. A. Chiodo & A. S. Freeman, Eds.: 237–251. Lakeshore Press. Grosse Pt., Mich.

63. MAURA, G., A. GIARDI & M. RAITERI. 1988. Release-regulating D-2 dopamine receptors are located on striatal glutamatergic nerve terminals. J. Pharmacol. Exp. Ther. **247:** 680–684.

64. HOKFELT, T., J. F. REHFELD, L. SKIRBOLL, B. IREMARK, M. GOLDSTIEN & K. MARKEY. 1980. Evidence for coexistence of dopamine and CCK in mesolimbic neurons. Nature London **285:** 476–478.

65. GARCIA-MUNOZ, M. & P. PATINO. 1987. Excitability changes of the dopamine containing terminals in the nucleus accumbens induced by cholecystokinin. Neurosci. Lett. *Suppl.* **29:** 24.

66. BUNNEY, B. S. & S. R. SEESACK. 1987. Electrophysiological identification and pharmacological characterization of dopamine sensitive neurons in the rat prefrontal cortex. *In* Neurophysiology of Dopaminergic Systems: Current Status and Clinical Perspectives. L. A. Chiodo & A. S. Freeman, Eds.: 129–140. Lakeshore Press. Grosse Pt., Mich.

67. REUBI, J.-C., L. L. IVERSEN & T. M. JESSELL. 1977. Dopamine selectively increases ^{3}H-GABA release from slices of rat substantia nigra in vitro. Nature London **268:** 652–654.

Early Evidence of Presynaptic Receptors

GEORGE B. KOELLE

Department of Pharmacology
School of Medicine
University of Pennsylvania
36th Street and Hamilton Walk
Philadelphia, Pennsylvania 19104-6084

Probably the first concrete evidence of presynaptic receptors was published in a classical paper by Masland and Wigton in 1940.[1] The authors reasoned that the fasciculation, rather than fibrillation, that follows the intraarterial injection of acetylcholine (ACh) or an anticholinesterase (anti-ChE) agent (physostigmine, neostigmine) into a skeletal muscle must reflect the synchronous firing of entire motor units rather than activation of individual muscle fibers. In support of this concept they quoted a much earlier study by Langley and Kato in 1915, in which it had been shown that injection of physostigmine does not produce fasciculation in chronically denervated muscle.[2] Richard Masland (a University of Pennsylvania neurologist, who subsequently became director of the then National Institute of Neurological Diseases) and Wigton monitored both muscle action potentials and antidromic firing along the corresponding motor nerve; following the injection of ACh or an anti-ChE agent, they recorded synchronous bursts of activity at both. Furthermore, both were blocked by curare. Therefore, they concluded that cholinergic nicotinic receptors are present in both presynaptic motor nerve terminals and the postsynaptic portion of the motor end plate.

A number of confirming reports followed. Abdon and Bjarke concluded that tubocurarine exerts its blocking effect predominantly at the presynaptic motor terminal, probably by interfering with the release of ACh.[3] Walter Riker and associates examined a number of analogues of neostigmine, with and without anti-ChE activity, and arrived at similar conclusions.[4,5]

The first suggestion of presynaptic receptors in autonomic ganglia was published in 1950 by Laporte and Lorenté de No, who showed that transmission in a turtle ganglion was blocked by tubocurarine at a predominantly presynaptic site.[6] Unlike what occurs at motor end plates of skeletal muscle, ACh and anti-ChE agents fail to induce retrograde preganglionic firing when injected into normal mammalian sympathetic ganglia.[7] This is probably due to the difference in the geometry of the two presynaptic sites: whereas motor nerves terminate in sizable structures at the motor end plate, the overwhelmingly axodendritic synapses of preganglionic terminals are extremely small twigs.[8] However, following infection with pseudorabies virus, such an action is apparently unmasked in ganglia. In the late 1950s John Dempsher and associates showed that when pseudorabies virus is introduced into the eye of an anesthetized cat, it passes via the postganglionic sympathetic fibers to the superior cervical ganglion (SCG).[9–11] After a few days' latency, spontaneous, continuous bursts of firing develop synchronously in the postganglionic (anterograde) and preganglionic (retrograde) trunks. Simultaneously, show waves of localized depolarization, probably representing generator potentials, were detected at both poles of the ganglion. All this activity was suppressed by tubocurarine, suggesting that it was due to the action of ACh at both pre- and postsynaptic receptors.

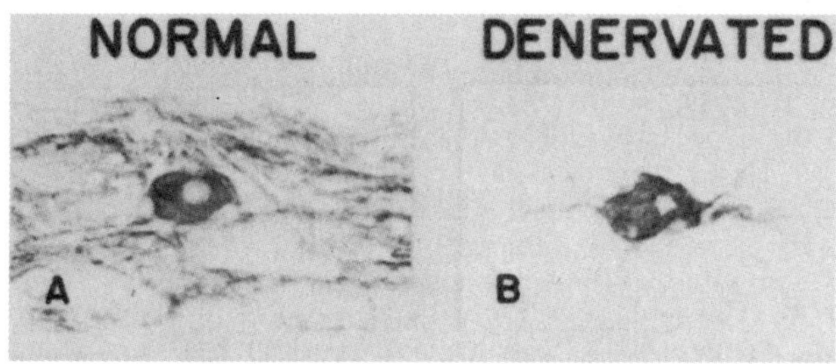

FIGURE 1. Light micrographs of normal (A) and preganglionically denervated (B) superior cervical ganglia of the cat stained by CuThCh method for AChE. Incubated at 22°C, 60 minutes, ×270. (Reference 14, 1959. Reproduced with permission.)

At this time Burn and Rand published their hypothesis of a cholinergic link in adrenergic transmission.[12] According to this concept adrenergic nerves release first ACh, which then acts at presynaptic receptors on the same terminals to release norepinephrine. The hypothesis never gained wide acceptance, although it seemed to explain a number of observations that could not be accounted for otherwise. In discussing the hypothesis with British colleagues I found that they had two major

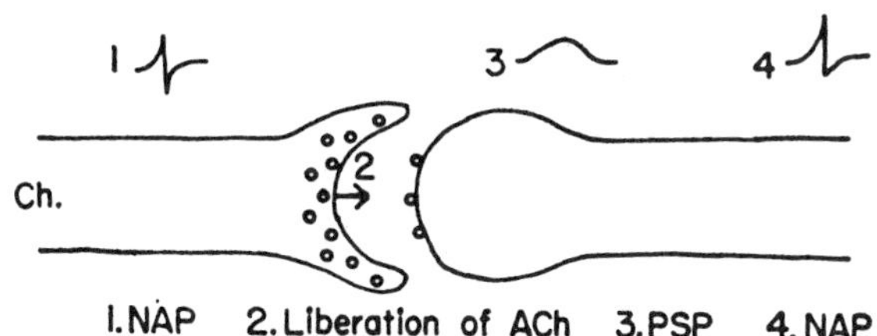

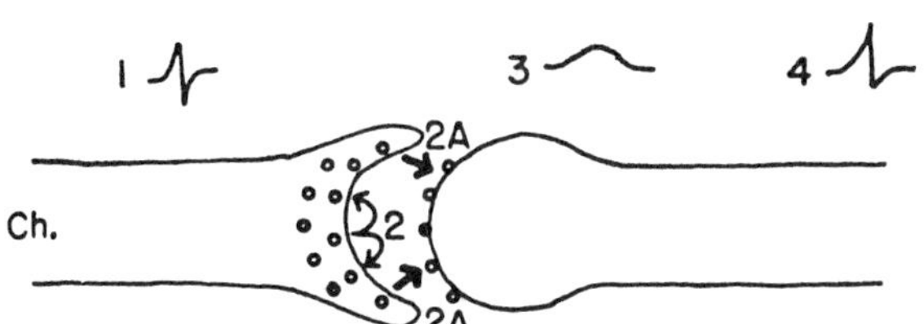

FIGURE 2. Functions of ACh in cholinergic transmission. **A** (upper): Standard concept. (1) Nerve action potential (NAP) causes axonal terminals to liberate (2) ACh, which diffuses across the synaptic cleft and combines with postsynaptic receptors, resulting in (3) localized depolarization, the postsynaptic potential (PSP), which initiates electrogenically (4) NAP in second axon. **B** (lower): Proposed presynaptic function. Acetylcholine acts first at terminal from which liberated to activate release of (2A) additional quanta of ACh, which produce the PSP. (Reference 17, 1962. Reproduced with permission.)

objections: (1) you don't need it, and (2) we're tired of hearing Burn talk about it. Neither point seemed especially apposite. Only Sir Henery Dale took a definite opposite stand. When he was 89 I visited him at the Wellcome Foundation in London. We discussed the hypothesis at length, then he leaned back and said, "I think that young fellow Burn [then in his 70s] has a pretty good idea!"

My colleagues and I became involved with presynaptic receptors through histochemical findings. By means of the copper thiocholine method,[13] acetylcholinesterase (AChE) in the normal cat SCG was found by light microscopy (LM) to be present in high concentration in the perikarya of occasional ganglion cells (those giving rise to cholinergic postganglionic fibers) and in the neuropil (the intertwined presynaptic terminals and dendrites, which cannot be distinguished by LM), and in very low concentration in the perikarya of the overwhelmingly predominant adrenergic gan-

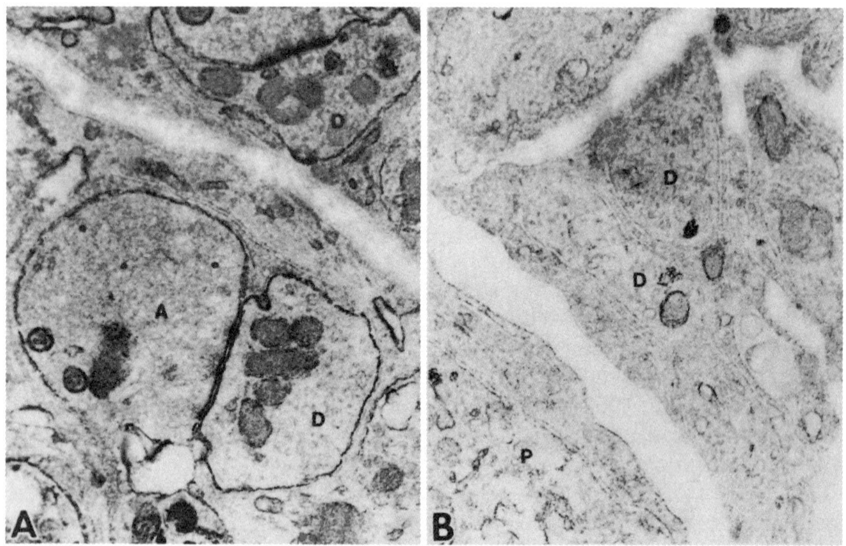

FIGURE 3. Electron micrographs of normal (A) and preganglionically denervated (B) superior cervical ganglia of the cat stained by Au(TA)$_2$ method for AChE. Six hours of incubation, 5°C. A, axon; D, dendrite; P, perikaryon. (References 19, 1978 and 20, 1981. Reproduced with permission.)

glion cells (FIGURE 1A). Following preganglionic denervation, the enzyme disappeared from the neuropil but remained unchanged in the ganglion cells (FIGURE 1B). Therefore, it was concluded that nearly all of the ganglionic AChE was normally located presynaptically, and that it loss following denervation resulted from degeneration of the sectioned fibers.[14] The earlier work from other laboratories, quoted above, provided a clue to its probable function. In contrast to the accepted sequence of cholinergic transmission (FIGURE 2A), it was hypothesized that the ACh released initially by the preganglionic fibers acted at receptors on the same terminals to amplify the release (i.e., a positive feedback mechanism) and that the secondarily released ACh effected transmission (FIGURE 2B). The apparently presynaptically located AChE would then be most strategically situated to terminate this process.[15] Bob Volle and I[16] then conducted a pharmacological study which appeared to support the proposal. We found, among other things, that preganglionic denervation of the SCG resulted in an increase in the threshold dose of intraarterially injected ACh required to activate the ganglion cells; thus it seemed that the presynaptic terminals were more sensitive to ACh than the postsynaptic membranes.

In 1962 an extensive review was published, containing three pages of references, of the evidence for presynaptic receptors at that time.[17] The presentations in this volume have by and large fully confirmed and greatly amplified the concept.

Ironically, we realized subsequently that we had entered the field of presynaptic receptors for the wrong reason. As noted above, the axonal terminals and ganglion cell dendrites of the neuropil cannot be distinguished by LM. With the development of the bis-thioacetoxy aurate method, it became possible to localize AChE accurately by electron microscopy (EM).[18] It was then found that in the normal cat SCG, AChE is present at both the presynaptic and postsynaptic membranes (FIGURE 3A).[10] (We still endorse the function of the presynaptic enzyme postulated above; the added identification of the postsynaptic enzyme is consistent with its termination of the transmitter action of ACh at that site.) Following preganglionic denervation, it was found by EM, as earlier by LM, that AChE disappears from both pre- and postsynaptic sites (FIGURE 3B).[20] This observation raised the question of why the integrity of the presynaptic terminals is essential for the maintenance of postsynaptic AChE. It was postulated that the preganglionic fibers release a neurotrophic factor that regulates the synthesis of postsynaptic AChE.[21] Subsequent studies have indicated that glycylglutamine or a similar small peptide functions in such a capacity.[22,23]

REFERENCES

1. MASLAND, R. L. & R. S. WIGTON. 1940. J. Neurophysiol. **3:** 269–275.
2. LANGLEY, J. N. & T. KATO. 1915. J. Physiol. **49:** 410–431.
3. ABDON, N.-O. & T. BJARKE. 1945. Acta Pharmacol. Toxicol. Kbh. **1:** 1–17.
4. RIKER, W. F., JR., J. ROBERTS. F. G. STANDAERT & H. FUJIMORI. 1957. J. Pharmacol. Exp. Ther. **121:** 286–312.
5. RIKER, W. F., JR., G. WERNER, J. ROBERTS & A. S. KUPERMAN. 1959. J. Pharmacol. Exp. Ther. **125:** 150–158.
6. LAPORTE, Y. & R. LORENTE DE NÒ. 1950. J. Cell. Comp. Physiol. (Suppl. 2) **35**(Suppl. 2): 61–106.
7. DOUGLAS, W. W., D. W. LYWOOD & R. W. STRAUB. 1960. J. Physiol. **153:** 250–264.
8. ELFVIN, L.-G. 1963. J. Ultrastruct. Res. **8:** 403–440, 441–476.
9. DEMPSHER, J., M. G. LARRABEE, F. B. BANG & D. BODIAN. 1955. Am. J. Physiol. **182:** 203–216.
10. DEMPSHER, J. & W. K. RIKER. 1957. J. Physiol. **139:** 145–156.
11. DEMPSHER, J. & J. ZABARA. 1960. J. Physiol. **151:** 217–224.
12. BURN, J. H. & M. J. RAND. 1960. Br. J. Pharmacol. **15:** 56–66.
13. KOELLE, G. B. & J. S. FRIEDENWALD. 1949. Proc. Soc. Exp. Biol. NY **70:** 617–622.
14. KOELLE, W. A. & G. B. KOELLE. 1959. J. Pharmacol. Exp. Ther. **126:** 1–8.
15. KOELLE, G. B. 1961. Nature London **190:** 208–211.
16. VOLLE, R. L. & G. B. KOELLE. 1961. J. Pharmacol. Exp. Ther. **133:** 223–240.
17. KOELLE, G. B. 1962. J. Pharm. Pharmacol. **14:** 65–90.
18. KOELLE, G. B., R. DAVIS, E. G. SMYRL & A. V. FINE. 1974. J. Histochem. Cytochem. **22:** 252–259.
19. DAVIS, R. & G. B. KOELLE. 1978. J. Cell. Biol. **78:** 785–809.
20. DAVIS, R. & G. B. KOELLE. 1981. J. Cell. Biol. **88:** 581–590.
21. KOELLE, G. B. & G. A. RUCH. 1983. Proc. Nat. Acad. Sci. USA **80:** 3106–3110.
22. KOELLE, G. B. 1988. Trends Pharmacol. Sci. **9:** 318–321.
23. KOELLE, G. B., J. J. O'NEILL, N. S. THAMPI, M. S. HAN & R. CACCESE. 1989. Proc. Nat. Acad. Sci. USA **86:** 10153–10155.

In Vivo Evidence for the Existence of Autoreceptors on Dopaminergic, Serotonergic, and Cholinergic Neurons in the Brain

B. H. C. WESTERINK, P. DE BOER, W. TIMMERMAN,
AND J. B. DE VRIES

*Department of Medicinal Chemistry
Center for Pharmacy
University of Groningen
A. Deusinglaan 2
9713 AW Groningen, the Netherlands*

INTRODUCTION

Most of the studies on autoreceptors in the central nervous system (CNS) were thus far carried out with indirect methods. *In vitro* experiments using brain slices or synaptosomes and analysis of postmortem tissue levels have much contributed to our knowledge of the role of autoreceptors in the modulation of neurotransmitter release.[1] The use of the push-pull technique able to record the release of a radiolabeled neurotransmitter was a more direct approach to analyze the properties of autoreceptors in the brain.[2] A relatively simple method to record the endogenous release of neurotransmitters in conscious rats is the brain microdialysis technique. Microdialysis is becoming increasingly established as a useful method for repeated sampling of the extracellular fluid of brain regions *in vivo*.[3–5] Here we report on the use of the microdialysis technique for the investigation of autoreceptors in the striatum of the rat.

We have studied the effect of the specific D-2 antagonist sulpiride and the specific D-2 agonist N-0437 on the *in vivo* release of dopamine (DA). Infusion experiments were carried out with both enantiomers of the compounds. In an attempt to localize the release-modulating autoreceptors, sulpiride or N-0437 was administered systemically or was infused into the striatum. Finally rats received a combined treatment of intraperitoneally (ip) administered sulpiride and infused sulpiride via the dialysis membrane. To eliminate the possible contribution of postsynaptic connections, sulpiride or N-0437 was administered ip to rats that had received a unilateral injection of kainic acid into the striatum.

Autoreceptors are also involved in the modulation of the synthesis rate of neurotransmitters such as dopamine (DA) and serotonin (5-HT).[6–8] To analyze the autoreceptor-synthesis interaction we have applied a microdialysis method able to monitor the synthesis rate of DA and 5-HT. In this method a decarboxylase inhibitor is infused via the perfusion fluid and the activity of tyrosine hydroxylase or tryptophan hydroxylase is estimated by analysis of dialysate levels of DOPA and 5-HTP respectively.[9] D-2 specific drugs [(−)-N-0437 and (−)-sulpiride] were infused as well as administered systemically and their effects on tyrosine hydroxylase were determined *in vivo*. The presence of autoreceptors on serotonergic neurons involved in the synthesis of 5-HT was investigated by systemic as well as intrastriatal application of the 5-HT-IA agonist 8-hydroxy-2-(di-*n*-propylamino)-tetralin (8-OH-DPAT).

The presence of autoreceptors on cholinergic neurons was evaluated by infusing a muscarinic agonist (oxotremorine) and a muscarinic antagonist (atropine) during microdialysis of acetylcholine (ACh). The possible interference of coinfusion of an esterase inhibitor (neostigmine) was investigated. The muscarinic autoreceptor involved in the release of ACh was identified as an M_3-type.[10]

It is concluded that microdialysis is a valuable tool to study the properties of autoreceptors *in vivo*. The presented data support the concept that dopaminergic, serotonergic, and cholinergic neurons possess autoreceptors that are involved in a negative feedback control of the release and/or synthesis rate of the respective transmitter.

MATERIALS AND METHODS

Animals and Drug Treatment

Male albino rats of a Wistar-derived strain (175–200 g weight, C.D.L., Groningen) were used. The following drugs were used: ±-sulpiride (Dogmatil), (−)-sulpiride and (+)-sulpiride (Ravizza, Milan), (−)-N-0437, (+)-N-0437, 8-OH-DPAT (synthesized in our laboratory by known methods), atropine, oxotremorine sesquioxalaat, 3-hydroxybenzylhydrazine hydrochloride (NSD 1015, Janssen, Beerse).

Lesions

The kainic acid lesions were made unilaterally in the left striatum of the brain of the rat under chloral hydrate anesthesia (400 mg/kg).[11] The rat was positioned in the stereotaxic frame and an injection of 2 μg kainic acid (dissolved in 0.9% NaCl) was given at coordinates A 8.8, L −3.0, and V 0.0.[12] The lesions were controlled by measuring the level of ACh in the striatum by means of microdialysis.[13] The rats were used about 20 days after lesioning.

Surgery and Brain Dialysis

Most of the experiments were performed with a transstriatal cannula, essentially as described earlier.[4] In brief, one hole was drilled on each side of the temporal bone at the level of the head of the caudate nucleus (coordinates A 7.4, V 5.5 from temporal bone).[12] A dialysis tube bearing a tungsten wire was fastened in a transverse position to a stereotaxic holder mounted on the right bar of a stereotaxic apparatus. The dialysis tube (inside diameter, 0.22 mm; outside diameter, 0.27 mm) was prepared from saponified cellulose ester (artificial kidney for human use; Cordis Dow Medical International, Oosterwolde, the Netherlands); the molecular weight cutoff of the membrane was 10,000 Daltons. In the case of experiments depicted in FIGURE 1, a U-shaped cannula[14] was used. The probes were implanted during chloral hydrate anesthesia (400 mg/kg).

The perfusion experiments were carried out 24–168 hours after implantation of the cannula using an on-line high-performance liquid chromatography (HPLC) system. The experiments were done with caged but otherwise unrestrained animals. The rat was directly connected to the HPLC equipment. Two polyethylene tubes (inner diameter, 0.22 mm) were connected to the outlets of the dialysis tube. One tube (length 45 cm) was connected to the perfusion pump, and the second tube (length 45 cm) to the injection valve of the HPLC equipment. In case of recording of DOPA, the polyethylene tube connected with the HPLC was replaced weekly. The striatum was perfused with a Ringer solution at 5.5–6 μl/minute (Perfusor VI, B. Braun). The composition of the Ringer solution was NaCl, 147 mmol/l; KCl, 4 mmol/l; $CaCl_2$, 3.4 mmol/l (FIGURES 1–3 and 8) or 1.2 mmol/l (FIGURES 4–7); $MgCl_2$, 1.0 mmol/l. With the help of an electronic timer the injection valve (Valco) was held in the load position for 15 minutes, in which the sample loop (50 μl) was filled with dialysate. The valve was then switched automatically to the injection position for 20 seconds. This procedure was repeated every 15 minutes, which was the time needed to record a complete chromatogram.

Chemical Assays

DOPA

As DOPA was not detectable in the dialysates, a decarboxylase inhibitor (NSD 1015) was added to the perfusate in a concentration of 10 μmol/l. Two hours after infusion of NSD 1015, a stable DOPA output was achieved.[9] DOPA was quantitated by HPLC with electrochemical detection. In brief, a Perkin Elmer (series 10) pump was used in conjunction with a rotating disc electrochemical detector. The detector potential was set at 600 mV vs. an Ag/AgCl reference electrode. A reverse-phase column (250 × 4.7 mm) filled in our laboratory with Nucleosil 5C18 (Mächerey-Nagel, Düren, FRG) was used. The mobile phase consisted of a mixture of 0.1 mol/l trichloroacetic acid adjusted to pH 3.2 with sodium acetate and 0.01 mmol/l Na_2 EDTA.

5-HTP

As 5-HTP could not be reliably detected in the dialysate, a decarboxylase inhibitor (NSD 1015) was added to the perfusate in a concentration of 10 μmol/l. Two to three hours after start of the infusion of NSD 1015 a stable 5-HTP output was reached. 5-HTP was quantitated by HPLC with electrochemical detection. A Perkin-Elmer (series 10) pump was used in conjunction with a rotating disc electrochemical detector. The detector potential was set at 600 mV vs. an Ag/AgCl reference electrode. An Altech-RSL cartridge (150 × 4.6 mm) reverse phase column was used. The mobile phase consisted of 0.1 mol/l sodium acetate adjusted to pH 4.1 with acetic acid, 0.3 mmol/l Na_2EDTA, and 2% methanol.

Dopamine

DA was quantified by HPLC with electrochemical detection. A Perkin-Elmer (series 10) pump was used in conjunction with a rotating disc electrochemical detector. The detector potential was set at 650 mV vs. an Ag/AgCl reference electrode. A reverse-phase column (150 × 4.7 mm) filled with Nucleosil 5 C18 (Machery-Nagel, Düren, FRG) was used. The mobile phase consisted of a mixture of 0.1 mol/l sodium acetate adjusted to a pH of 4.1 with concentrated acetic acid, 2 mmol/l 1-heptanesulfonic acid, 0.01 mmol/l Na_2EDTA, and 60–80 ml of methanol/l. The flow was set at 1.0 ml/minute. The detection limit of the assay was about 10 fmol/injection.

Acetylcholine

ACh was separated on a cation exchange column.[13] This column was prepared by loading a homemade reverse phase column (100 × 2.1 mm) containing Chromspher C18 material (Chrompack, Middleburg, the Netherlands) with a lauryl sulfate solution. An enzymatic postcolumn reactor (10 × 2.1 mm) containing acetylcholinesterase (EC 3.1.1.7, type VI-S, Sigma) and choline oxidase (EC 1.1.3.17, Sigma), covalently attached to Lichrosorb NH_2 (Merck) and activated with glutaraldehyde, was used to convert ACh to hydrogen peroxide. The hydrogen peroxide was electrochemically detected by a rotating platinum electrode at +500 mV. The mobile phase was delivered by a Perkin Elmer series 10 pump (Perkin Elmer, Norwalk, Conn.) at 0.6 ml/minute and contained a 0.1–0.2 mol/l phosphate buffer of pH 8.0, 0.5 mmol/l EDTA and 1 mmol/l tetrametylammoniumchloride. The detection limit of the ACh assay was 50 fmol/injection.

RESULTS

Basal Output of ACh and DA

Baseline dialysis output of ACh and DA showed variations from rat to rat. Therefore the average amount of neurotransmitter output of the last three preinjection samples was taken as 100% and all subsequent postinjection samples were expressed relative to the basal values. The effect of each drug treatment was expressed graphically by taking the mean of the relative value of the corresponding time points

Effect of Infusion of (−)-Sulpiride or (+)-Sulpiride on the Dialysate Content of DA

The two enantiomers of the D-2 antagonist sulpiride were dissolved in the perfusion fluid. Both enantiomers increased the output of DA. The maximal increase in the dialysate content of DA for various concentrations is depicted in FIGURE 1. From the two dose-effect curves it can be calculated that the $ED_{50\%}$ of (−)-sulpiride is about 8 × 10^{-8} mol/l, whereas the $ED_{50\%}$ of (+)-sulpiride is about 2 × 10^{-6} mol/l.

Effect of Infusion of (−)-N-0437 or (+)-N-0437 on the Dialysate Content of DA

The two enantiomers of the D-2 agonist N-0437 were dissolved in the perfusion fluid. The maximal change in the dialysate content of DA for various concentrations is depicted in FIGURE 2. (−)-N-0437 decreased the dialysate content of DA. The $ED_{50\%}$ of this effect was close to 10^{-8} mol/l. (+)-N-0437 caused an increase in the output of DA. The latter effect was observed at higher concentrations: $>10^{-6}$ mol/l.

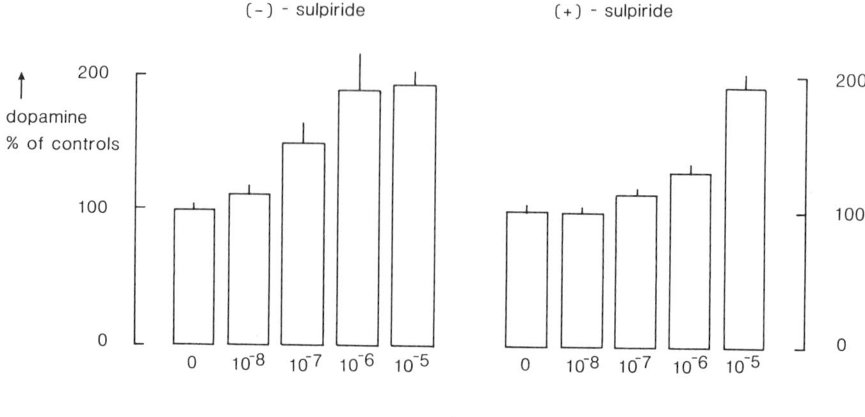

FIGURE 1. Effect of different concentrations of ($-$)-sulpiride and ($+$)-sulpiride, infused via the dialysis membrane, on the striatal dialysate content of DA. Basal values for DA were [±standard error of the mean (SEM)] 6.2 ± 0.4 fmol/minute ($n = 25$). A U-shaped cannula was used.

Effect of Different Routes of Administration of ($-$)-Sulpiride or ($-$)-N-0437 on the Dialysate Content of DA in Control and Kainic Acid Pretreated Animals

Ip administration of a high dose of ($\pm$)-sulpiride (40 mg/kg) caused an increase in the dialysate content of DA that was very similar to the maximal effects noticed in infusion experiments (180–190% of controls) (FIGURE 3). A similar elevation of the output of DA was seen after combined treatment of ip administered ($\pm$)-sulpiride (40 mg/kg) and infused sulpiride (10 μmol/L). Pretreatment of kainic acid did not modify the effect of ip administered ($\pm$)-sulpiride (40 mg/kg) on the dialysate content of DA (FIGURE 3).

Ip administration of high doses of ($-$)-N-0437 (1 μmol/kg) caused a decrease in the dialysate content of DA that was very similar to the maximal effects noticed in infusion experiments (40–50% of controls) (FIGURE 4). Pretreatment of kainic acid did not modify the effect of ip administered N-0437 (1 μmol/kg) on the dialysate content of DA (FIGURE 4).

Effect of Infusion of ($-$)-Sulpiride and ($-$)-N-0437 on the Formation of DOPA in Striatal Dialysates

The tyrosine hydroxylase activity in the striatum was monitored by continuous infusion of 10 μmol/l of the decarboxylase inhibitor NSD 1015, whereas the extracellular levels of DOPA were recorded by microdialysis. After 2 hours infusion of NSD 1015 a stable output of DOPA was obtained, and at that time point the infusion of ($-$)-sulpiride or ($-$)-N-0437 was started. ($-$)-Sulpiride (10 μmol/l) caused an increase in the formation of DOPA (FIGURE 5), whereas ($-$)-N-0437 (1 μmol/l) decreased the output of DOPA (FIGURE 6).

Effect of Infusion of 8-OH-DPAT or Intraperitoneal Administration of 8-OH-DPAT on the Formation of 5-HTP in Striatal Dialysates

The tryptophan hydroxylase activity in the striatum was monitored by continuous infusion of 10 1 μmol/l of the decarboxylase inhibitor NSD 1015, whereas the extracellular levels of 5-HTP were recorded by microdialysis. After 2 hours infusion of NSD 1015 a stable output of 5-HTP was established, and at this time point 8-OH-DPAT was infused in a concentration of 1 μmol/l, or administered ip in a dose of 1.5 μmol/kg. Infusion of 8-OH-DPAT was without effect on the output of 5HTP, but ip administered 8-OH-DPAT clearly decreased the formation of 5-HTP (FIGURE 7).

Effect of Infusion of Atropine or Oxotremorine on the Dialysate Content of ACh

The muscarinic antagonist atropine was infused into the striatum dissolved in the perfusion fluid in a concentration of 1 μmol/l. During infusion of atropine the dialysate levels of ACh were not affected (FIGURE 8). The muscarinic agonist oxotremorine, infused in a concentration of 100 μmol/l, decreased the ACh output to about 50% of controls. When the experiments were carried out in the presence of neostigmine (0.1 μmol/l), different results were obtained. Infused atropine (1 μmol/l) increased the

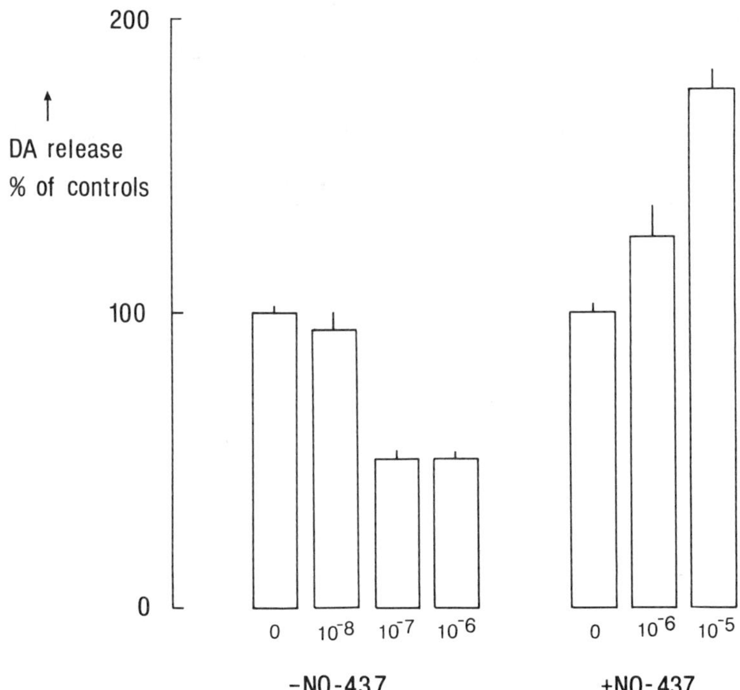

FIGURE 2. Effect of different concentrations of ($-$)-N-0437 and ($+$)-N-0437 infused via the dialysis membrane on the striatal dialysate content of DA. Basal values for DA were ($\pm$SEM): 27.5 $\pm$ 1.6 fmol/minute ($n = 25$).

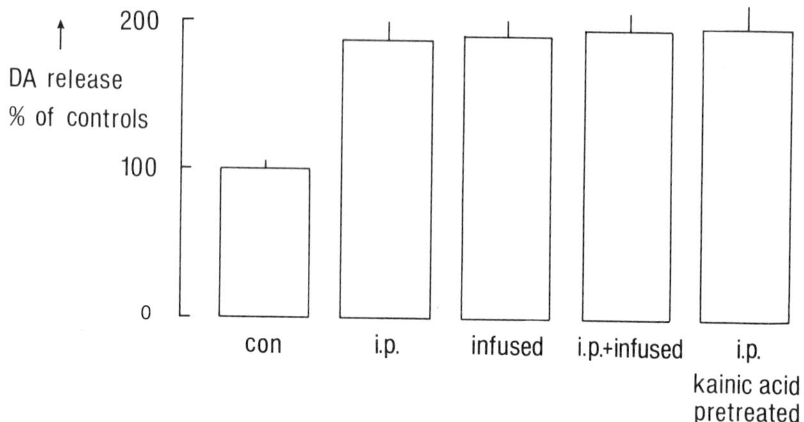

FIGURE 3. Effect of ip administered ($\pm$)-sulpiride (40 mg/kg), infused sulpiride (10 μmol/kg), ip administered + infused sulpiride, and sulpiride administered (40 mg/kg ip) to rats that had received a unilateral kainic acid lesion on the DA content of striatal dialysates. Basal values for DA ($\pm$SEM) were 5.2 $\pm$ 0.7 ($n = 5$) fmol/minute, for unlesioned rats, and 2.7 $\pm$ 0.4 fmol/minute ($n = 4$), for lesioned rats. A U-shaped cannula was used.

output of ACh to about 250% of controls, whereas oxotremorine (100 μmol/l) was without effect.

In an attempt to classify the muscarinic receptor, 4 anticholinergics (atropine, pirenzipine, 4-DAMP, and AF-DX 116) were infused into the striatum at different concentrations.[10] ACh was recorded in the presence of neostigmine (0.1 μmol/l). All anticholinergics increased the output of ACh. $ED_{50\%}$ values are given in TABLE 1.

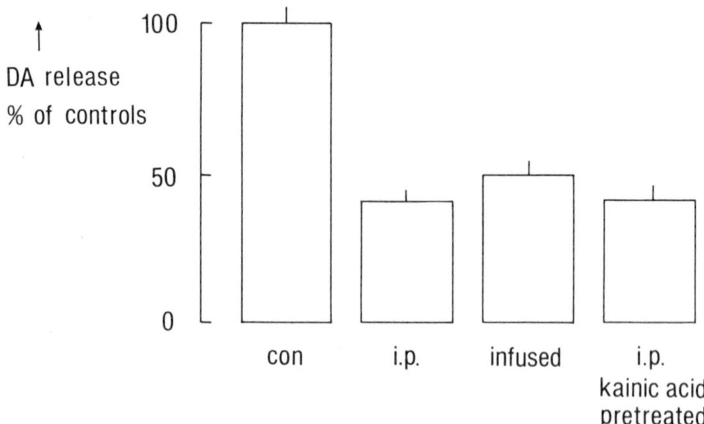

FIGURE 4. Effect of ip administered ($-$)-N-0437 (1 μmol/kg), infused ($-$)-N-0437 (1 μmol/l), and ip administered ($-$)-N-0437 (1 μmol/kg) to rats that had received a unilateral kainic acid lesion on the DA content of striatal dialysates. Basal values for DA were 21.1 $\pm$ 2.9 ($n = 3$) fmol/minute, for unlesioned rats and 12.2 $\pm$ 1.1 ($n = 4$) fmol/minute for lesioned rats.

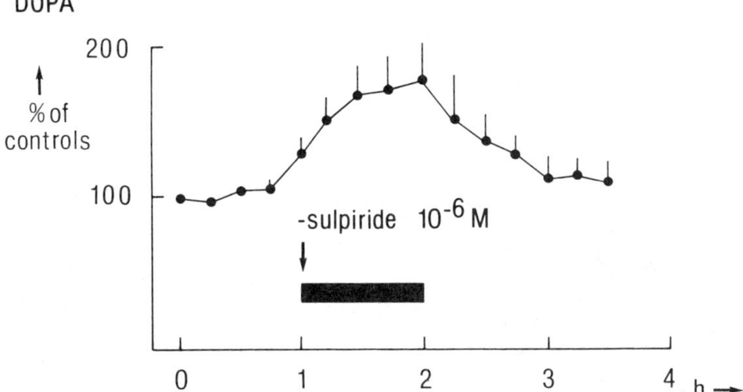

FIGURE 5. Effect of ($-$)-sulpiride infusion (1 μmol/l) on the DOPA accumulation in striatal dialysates. Control values of DOPA were 0.24 $\pm$ 0.02 pmol/minute ($n = 6$).

DISCUSSION

Intrastriatal infusion as well as systemic administration of the selective D-2 antagonist sulpiride caused a similar increase in the release of DA of about 180–190% of controls.[15] The ($+$)-isomer was about 25 times less active in this respect. A comparable conclusion was drawn when the selective D-2 agonist ($-$)-N-0437 was infused intrastriatally or administered systemically: both routes of administration caused a maximal decrease in the release of DA to about 40–50% of controls.[16,17] The ($+$)-isomer of N-0437 increased the release of DA at relatively high concentrations ($>10^{-6}$ mol/l). The latter effect suggest that ($+$)-N-0437 has antagonistic properties towards autoreceptors.[17]

The fact that both routes of administration of the D-2 agonist and the D-2 antagonist induced nonadditive changes in the release of DA, and the additional finding that a similar change of the release of DA was seen in kainic acid–treated rats

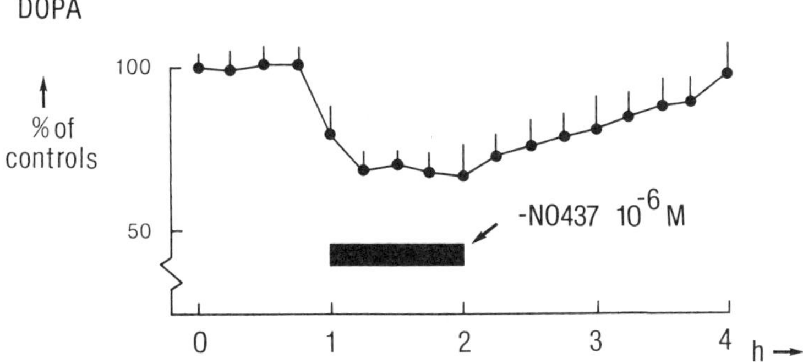

FIGURE 6. Effect of ($-$)-N-0437 infusion (1 μmol/l) on the DOPA accumulation in striatal dialysates. Control values of DOPA were ($\pm$ SEM) 0.23 $\pm$ 0.02 pmol/minute ($n = 8$).

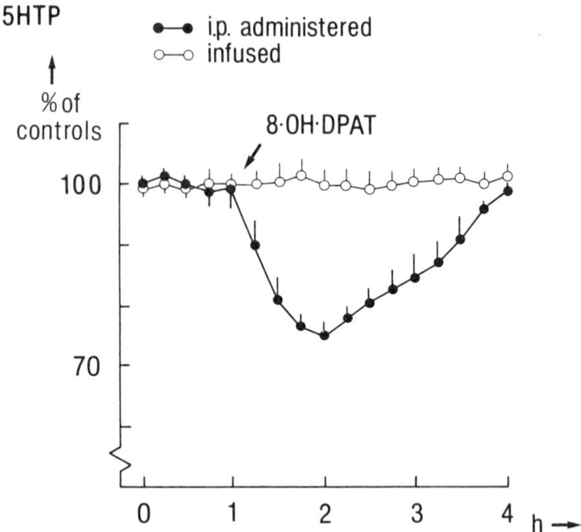

FIGURE 7. Effect of 8-OH-DPAT ip administered (1.5 μmol/kg) or infused via the dialysis membrane (1 μmol/l) on the 5-HTP accumulation in striatal dialysates. Control values of 5-HTP were ($\pm$ SEM) 26.5 $\pm$ 1.2 fmol/minute ($n = 8$).

(FIGURE 3), strongly suggest that the mechanism responsible for the D-2 receptor–mediated changes in the release of DA acts at the level of autoreceptors localized at nerve terminals. These data support earlier conclusions that were drawn based on postmortem tissue analysis.[11,18]

In order to study the D-2 autoreceptors that regulate the synthesis of DA, we used a

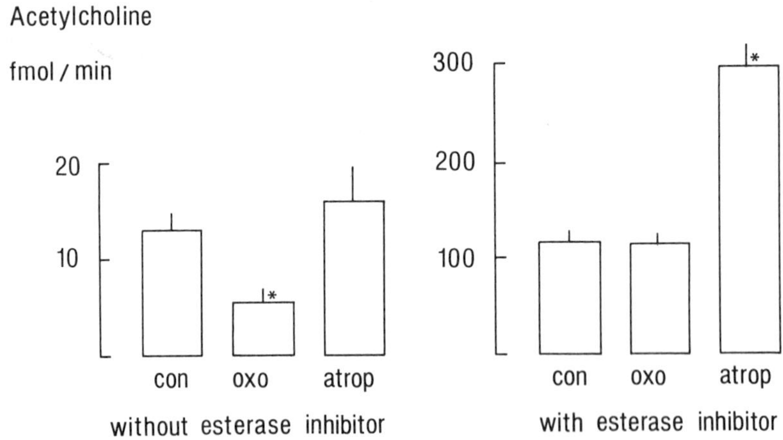

FIGURE 8. Effect of oxotremorine (100 μmol/l) and atropine (1 μmol/1), infused via the dialysis membrane, with and without esterase inhibitor (neostigmine 0.1 μmol/l), on the striatal dialysate content of ACh. Data shown are the average $\pm$ SEM of 4 experiments. *$p < 0.01$ vs. controls (Student's t-test).

microdialysis method in which DOPA formation was monitored during continuous infusion of a decarboxylase inhibitor.[9] Infusion of $(-)$-N-0437 or $(-)$-sulpiride indicated that the D-2 receptors controlling the synthesis of DA are localized on nerve terminals. Such findings are in line with earlier studies using pharmacological models of autoreceptors.[7,16]

The conclusion that the autoreceptors involved in the release and synthesis of DA are localized on nerve terminals means that various other mechanisms that have been suggested to explain DA-receptor-mediated changes in the release of DA are less likely. Firstly, it implies that striatal neurons that are localized postsynaptically to the dopaminergic terminals (the so-called feedback loop theory[19,20]) are not involved in the neuroleptic-induced increase in DA release. Secondly, the autoreceptors present on dopaminergic cell bodies[2,21] apparently do not contribute to the neuroleptic-induced increase in the release of DA. Thirdly the increase in nerve impulse activity of nigral dopaminergic neurons, as has been described after neuroleptics,[19] does not contribute to the rise in the release of DA. The latter conclusion questions the relation between nerve impulse activity and transmitter release, although it is emphasized that impulse activity of dopaminergic neurons is a prerequisite for the effects of neuroleptics on the release of DA.[22]

In order to evaluate the properties of synthesis-modulating autoreceptors on serotonergic neurons we have estimated the synthesis rate of 5-HT by monitoring

TABLE 1. Potency of Four Anticholinergic Agents to Increase the ACh Content of Striatal Dialysates[a]

$EC_{50\%}$	Values
Atropine	31 nmol/l
4-DAMP	54 nmol/l
Pirenzepine	280 nmol/l
AF-DX 116	6300 nmol/l

[a]ACh values were recorded in the presence of neostigmine (0.1 μmol/l).[10]

5-HTP formation in the dialysates during infusion of a decarboxylase inhibitor. Administration of the selective 5-HT-IA agonist 8-hydroxy-dipropyl-aminotetraline (8-OH-DPAT) resulted in a decrease in the synthesis rate of 5-HT. When 8-OH-DPAT was infused via the dialysis membrane, the agonist was unable to modify the release of 5-HT. This suggests that 8-OH-DPAT preferentially acts on somatodendritic autoreceptors, which have indeed been reported to belong to the 5HT-IA class. Similar conclusions were drawn in a recent study based on postmortem analysis of accumulated DOPA.[8]

Finally we have investigated the existence of autoreceptors on striatal cholinergic neurons. Infusion of the muscarinic agonist oxotremorine decreased the output of ACh, whereas infusion of atropine was without effect on the release of ACh. These results are somewhat in conflict with earlier microdialysis studies, which demonstrated that peripheral administration of anticholinergic drugs caused a clear increase of the release of ACh.[23-25] However, in the latter studies, ACh was protected from degradation by coinfusion of an esterase inhibitor. When atropine is infused in the presence of the esterase inhibitor neostigmine (FIGURE 6), it is evident that under such conditions atropine is very effective in stimulating the release of ACh. The finding that oxotremorine is unable to modify the output of ACh in the presence of neostigmine is explained by the occupation of muscarinic autoreceptors by endogenous ACh. Taken together these data suggest that muscarinic autoreceptors in the brain are not fully occupied under physiological conditions. Similar results have been obtained in *in vitro* studies.[1] When the effect of a series of anticholinergic drugs on the release of ACh was determined in the presence of neostigmine, evidence was provided (TABLE 1) that the autoreceptor involved in this mechanism is of the M-3 type.[10]

In conclusion, microdialysis of neurotransmitters is a valuable tool for the study of autoreceptors *in vivo*. The presented studies provided evidence for the existence of autoreceptors controlling the synthesis and/or release of DA, 5-HT, as well as ACh in the striatum. The properties of the three studied autoreceptors differed to some extent. The autoreceptor on dopaminergic neurons regulates both the release and synthesis of DA. The receptors are localized on nerve endings and are of the D-2 type. The receptor is functional during normal conditions. The 5-HT-IA autoreceptor controlling the synthesis rate of 5-HT is localized on cell bodies and/or dendrites. The autoreceptor controlling the release of ACh is of the M_3 type and is not fully occupied during normal conditions.

SUMMARY

Intrastriatal infusions as well as systemic administration of the selective D-2 antagonist ($-$)-sulpiride caused similar increases in the dialysate levels of dopamine (DA) to about 180% of controls. A similar conclusion was drawn when the selective D-2 agonist ($-$)-N-0437 was infused intrastriatally or administered systemically: both routes of administration caused a decrease in the release of DA to about 40–50% of controls. In order to evaluate the properties of synthesis-modulating autoreceptors on dopaminergic and serotonergic neurons we have estimated the synthesis rate of serotonin (5-HT) or DA by monitoring the 5-HTP or DOPA formation in the dialysates during infusion of a decarboxylase inhibitor. Infusion of ($-$)-N-0437 decreased the DOPA formation, whereas infusion of ($-$)-sulpiride increased the dialysate levels of DOPA; these results indicate that the D-2 receptors controlling the synthesis of DA are localized on nerve terminals. Administration of the selective 5-HT-IA agonist 8-hydroxy-dipropyl-aminotetraline (8-OH-DPAT) resulted in a decrease in the synthesis rate of 5-HT. When 8-OH-DPAT was infused via the dialysis membrane, the agonist was unable to modify the release of 5-HT. The effects of infusion of the muscarinic agonist oxotremorine and the muscarinic antagonist atropine were dependent on the presence of the esterase inhibitor neostigmine in the perfusion fluid. In the absence of neostigmine, oxotremorine caused a pronounced decrease in the output of acetylcholine (ACh), whereas atropine was without effect. In the presence of neostigmine oxotremorine was without effect but infusion of atropine or other anticholinergics caused a pronounced increase in the dialysate levels of ACh. It is concluded that the autoreceptor controlling the release of ACh is of the M_3-type and that the receptor is not fully occupied during normal conditions. In conclusion, microdialysis of neurotransmitters is a valuable tool for the study of autoreceptors *in vivo*. The presented studies provided evidence for the existence of autoreceptors controlling the synthesis and/or release of DA, 5-HT, as well as ACh in the striatum.

REFERENCES

1. STARKE, K., M. GÖTHERT & H. KILBINGER. 1989. Modulation of neurotransmitter release by presynaptic receptors. Physiol. Rev. **69:** 865–989.

2. CHERAMY, A., V. LEVIEL & J. GLOWINSKI. 1981. Dendritic release of dopamine in the substantia nigra. Nature **289:** 537–542.

3. UNGERSTEDT, U. 1984. Measurement of neurotransmitter release by intracranial dialysis. *In* Measurement of Neurotransmitter Release In Vivo. C. A. Marsden, Ed.: 81–105. Wiley, Chichester, England.

4. IMPERATO, A. & G. DI CHIARA. 1984. Trans-striatal dialysis coupled to reverse phase high performance liquid chromatography with electrochemical detecion: a new method for the study of the in vivo release of endogenous dopamine and metabolites. J. Neurosci. **4:** 966–977.

5. WESTERINK, B. H. C., G. DAMSMA, H. ROLLEMA, J. B. DE VRIES & A. S. HORN. 1987. Scope and limitations of in vivo brain dialysis: a comparison of its application to various neurotransmitter systems. Life. Sci. **41:** 1763–1776.

6. KEHR, W., A. CARLSSON, M. LINDQVIST, T. MAGNUSSON & C. V. ATACK. 1972. Evidence for a receptor mediated feedback control of striatal tyrosine hydroxylase activity. J. Pharm. Pharmacol. **24:** 744–747.

7. WALTERS, J. & R. H. ROTH. 1976. Dopaminergic neurons: an in vivo system for measuring drug-interactions with presynaptic receptors. Naunyn Schmiedebergs Arch. Pharmacol. **296:** 5–14.

8. HJORTH, S. & T. MAGNUSSON. 1988. The 5-HT-IA receptor agonist, 8-OH-DPAT, preferentially activates cell body 5-HT autoreceptors in rat brain in vivo. Naunyn Schmiedebergs Arch. Pharmacol. **338:** 463–471.

9. WESTERINK, B. H. C., J. B. DE VRIES & R. DURAN. 1990. Use of microdialysis for monitoring tyrosine hydroxylase activity in the brain of conscious rat. J. Neurochem. **54:** 381–387.

10. DE BOER, P., B. H. C. WESTERINK, H. ROLLEMA, J. ZAAGSMA & A. S. HORN. 1990. An M-3 like muscarinic receptor regulates the in vivo release of acetylcholine in the striatum. Eur. J. Pharmacol. **179:** 167–172.

11. DI CHIARA, G., M. L. PORCEDDU, P. F. SPANO & G. L. GESSA. 1977. Haloperidol increases and apomorphine decreases striatal dopamine metabolism after destruction of striatal dopamine-sensitive adenylate-cyclase by kainic acid. Brain Res. **130:** 374–382.

12. KÖNIG, J. F. R. & R. A. KLIPPEL. 1963. The Rat Brain: a Stereotaxic Atlas of the Forebrain and Lower Parts of the Brainstem: 162. Williams and Wilkins. Baltimore, MD.

13. DAMSMA, G., B. H. C. WESTERINK, J. B. DE VRIES, C. J. VAN DEN BERG & A. S. HORN. 1987. Measurement of acetylcholine release in freely moving rats by means of automated intracerebral dialysis. J. Neurochem. **48:** 1523–1528.

14. KORF, J. & K. VENEMA. 1985. Amino acids in rat striatal dialysates: methodological aspects and changes after electroshock. J. Neurochem. **45:** 1341–1347.

15. WESTERINK, B. H. C. & J. B. DE VRIES. 1989. On the mechanism of neuroleptic induced increase in striatal dopamine release: brain dialysis provides direct evidence for mediation by autoreceptors localized on nerve terminals. Neurosci. Lett. **99:** 197–202.

16. TIMMERMAN, W., B. H. C. WESTERINK, J. B. DE VRIES, P. G. TEPPER & A. S. HORN. 1989. Microdialysis and striatal dopamine release: stereoselective actions of the enantiomers of N-0437. Eur. J. Pharmacol. **162:** 143–150.

17. TIMMERMAN, W., M. L. DUBOCOVICH, B. H. C. WESTERINK, J. B. DE VRIES, P. G. TEPPER & A. S. HORN. 1989. The enantiomers of the dopamine agonist N-0437: in vivo and in vitro effects on the release of striatal dopamine. Eur. J. Pharmacol. **166:** 1–11.

18. MURRIN, L. C. & R. H. ROTH. 1987. Nigrostriatal dopamine neurons: modulation of impulse-induced activation of tyrosine hydroxylase by dopamine autoreceptors. Neuropharmacology **26:** 591–595.

19. BUNNEY, B. A., J. R. WALTERS, R. H. ROTH & G. K. AGHAJANIAN. 1973. Dopaminergic neurons: effect of antipsychotic drugs and amphetamine on single cell activity. J. Pharmacol. Exp. Ther. **195:** 560–571.

20. CARLSSON, A. & M. LINDQVIST. 1963. Effect of chlorpromazine or haloperidol on formation of 3-methoxytyramine and normetanephrine in mouse brain. Acta Pharmacol. **20:** 140–144.

21. GROVES, P. M., C. J. WILSON, S. J. YOUNG & G. V. REBEC. 1975. Self-inhibition of dopaminergic neurons. Science **190:** 522–529.

22. IMPERATO, A., & G. DI CHIARA. 1985. Dopamine release and metabolism in awake rats after systemic neuroleptics as studied by transstriatal dialysis. J. Neurosci. **5:** 297–306.

23. CONSOLO, S., C. FU WU, F. FIORENTINI, H. LADINSKY & A. VEZZANI. 1987. Determination of endogenous acetylcholine release in freely moving rats by transstriatal dialysis coupled to a radioenzymatic assay: effect of drugs. J. Neurochem. **48:** 1459–1465.

24. DAMSMA, G., B. H. C. WESTERINK, P. DE BOER, J. B. DE VRIES & A. S. HORN. 1988. Basal acetylcholine release in freely moving rats detected by on-line trans-striatal dialysis: pharmacological aspects. Life Sci. **43:** 1161–1168.

25. TOIDE, K. 1989. Effects of scopolamine on extracellular acetylcholine and choline levels and on spontaneous motor activity in freely moving rats measured by brain dialysis. Pharmacol. Biochem. Behav. **33:** 109–113.

Modulation of Serotonin Release[a]

Interactions between the Serotonin Transporter and Autoreceptors

WILLIAM A. WOLF[b] AND DONALD M. KUHN

Laboratory of Neurochemistry
Lafayette Clinic
and
Cellular and Clinical Neurobiology Program
Department of Psychiatry
Wayne State University School of Medicine
Detroit, Michigan 48207

The release of neurotransmitter from the presynaptic neuron could be considered among the most important processes in brain function since it is the primary means through which nerves communicate with one another. The synaptic availability of a neurotransmitter is therefore an important determinant of the operation of a particular neurochemical system. For some time now, the focus of clinical psychiatry, at least from a neurobiological viewpoint, has been on discovering means through which the synaptic availability of neurotransmitters can be altered. For example, many of the more effective antidepressants are inhibitors of biogenic amine uptake.

It has proved difficult to stimulate the release of neurotransmitters *in vivo,* especially that of serotonin (5-HT). It now seems clear that increases in the synthesis of 5-HT through the administration of its substrate tryptophan do not result in increased *release* of 5-HT.[1-3] Agents such as fenfluramine and *para*-chloroamphetamine (PCA) are indeed potent releasers of 5-HT, and they are of some effectiveness in the treatment of eating disorders, yet the neurotoxic and other nonspecific effects of these compounds[4] place limits on their usefulness. Inhibition of the metabolism of neurotransmitters, for example with inhibitors of monoamine oxidases, is also clinically effective in certain affective disorders. While inhibition of the metabolism of neurotransmitters will certainly increase their concentration in brain, it is not certain that this results in an increase in the synaptic availability of the transmitter substance.

The intraneuronal levels of 5-HT, like those of almost any transmitter, represent neurochemical potential which is realized only upon release of the transmitter into the synapse. Developments in two related areas of neurochemical research have led to new approaches to altering the synaptic availability of 5-HT so that its function can be more accurately and specifically manipulated. First, it is now recognized that a large variety of receptors for 5-HT exist in brain.[5] Second, certain receptors for 5-HT seem to be important in regulating its release. According to the feedback regulation hypothesis of neurotransmitter release,[6-8] liberated transmitter can act upon receptors located on the presynaptic side of the synapse to modulate subsequent release: increases in the synaptic concentration of a transmitter such as 5-HT "feed back" on the release mechanism through an autoreceptor to decrease release. The existence of

[a]Supported by the Department of Mental Health, State of Michigan and Grant MH44873 from the National Institute of Mental Health.
[b]Supported in part by a research fellowship from the Office of the Dean, Wayne State University School of Medicine.

release-regulating receptors for 5-HT was established long ago,[9] and recent research has identified the receptor subtype as either 5-HT_{1B}^{10} or 5-HT_{1D}^{11}. Thus, a variety of compounds are being developed that are targeted at specific 5-HT receptors and many of these can interact with 5-HT autoreceptors which, in turn, could alter the synaptic availability or function of 5-HT.

THEORETICAL AND TECHNICAL ASPECTS OF 5-HT AUTORECEPTORS

A number of variables interact to determine the function of the autoreceptors that regulate 5-HT release. For example, the extent to which release is regulated by autoreceptors and the pharmacological subtypes of those receptors that modulate release can be influenced by the methods used to stimulate release (e.g., electrical stimulation or KCl-induced depolarization) from brain tissue *in vitro*. Considering electrical stimulation, the interval between pulses appears to be an important determinant of how much release takes place and how much endogenous autoinhibition occurs.[12,13] Of equal importance is the brain area used for studies of autoreceptor function and some controversy exists over whether slices or synaptosomes are better models for studying the regulation of release.[14] The most commonly employed technique for measuring transmitter release involves superfusing the brain tissue and measuring the 5-HT that is released into the superfusate. With this technique, uptake of released neurotransmitter is minimized by the superfusion which sweeps the transmitter away.

Limits on the sensitivity for detecting released 5-HT in superfusion methods have been overcome through preloading tissue with isotopically labeled 5-HT, and release is assessed by measuring tritium overflow. It should be remembered that while the preloading technique labels vesicle stores of 5-HT, little consideration is given to the pool of newly synthesized 5-HT (i.e., endogenous), and studies with dopamine release[15] suggest that differences exist in the regulation of endogenous versus preloaded pools of neurotransmitter. As long as tritium overflow accurately reflects 5-HT release, conclusions on the modulation of release by autoreceptors can be made. Some studies that have determined how much of the tritium released is actually 5-HT have reported values as low as 50%.[16] Care must be taken to ensure that the experimental conditions are measuring release and not metabolism of the preloaded $^3\text{H-5-HT}$ as well.

In superfusion experiments, inhibitors of 5-HT uptake are commonly employed to prevent the rapid reuptake of released 5-HT. In addition, 5-HT uptake inhibitors are required when studying the effect of exogenous 5-HT on its own release in order to prevent the rapid uptake of the exogenously applied 5-HT. However, if the uptake mechanism (i.e., the transporter) is involved with the release process or if it is linked to the receptors regulating release, inhibition of the transporter could confound results on release. The requirement for an uptake inhibitor in studies of 5-HT release (even those involving superfusion) focuses attention on the important role of uptake in determining the synaptic availability and activity of 5-HT.

It has been proposed that the uptake mechanism and the receptors regulating 5-HT release are related. Agonists at 5-HT autoreceptors inhibit release, and 5-HT uptake inhibitors are known to shift the inhibitory effect of the agonists to the right.[17] As mentioned above, 5-HT cannot be used as an agonist at the presynaptic receptor since the amine is rapidly taken up into tissue. An additional difficulty exists in tissue preloaded with $^3\text{H-5-HT}$ in that the exogenous 5-HT will displace the isotope label upon uptake, giving the appearance of a large efflux of transmitter. Thus, 5-HT inhibits its own release only when the uptake mechanism is blocked. On the other hand, autoreceptor-mediated inhibition of 5-HT release by various 5-HT agonists is antagonized by uptake inhibitors. The fact that an uptake inhibitor is necessary to observe the inhibition of 5-HT release by 5-HT itself, while at the same time the uptake inhibitors *reduce* the ability of other agonists at the 5-HT autoreceptor to inhibit release, seems paradoxical. Perhaps 5-HT autoreceptor agonists such as LSD or 5-methoxy-tryptamine (5-MeOT) act at different receptors than endogenous 5-HT (plus an uptake inhibitor) to inhibit release but this has not received much attention experimentally. On the other hand, a functional or molecular link may exist between the 5-HT transporter and the 5-HT autoreceptor and this could account for some of the controversial or paradoxical results observed thus far regarding the 5-HT autoreceptor.

MECHANISM OF 5-HT RELEASE

Perhaps the first step in understanding the autoreceptor regulation of 5-HT release involves understanding the mechanisms by which 5-HT release occurs. The release of 5-HT follows most criteria that would lead to the conclusion that exocytosis is involved. Cytoplasmic stores of 5-HT can also be released by a carrier-mediated process which is inhibited by 5-HT uptake inhibitors (see Reference 1 for discussion). Whether 5-HT is released from vesicles, from the cytoplasm, or a combination of both may determine how the release process is regulated. Thus, modulation of the synaptic availability of 5-HT could differ for each mechanism of release. The release process itself, the transporter, or the autoreceptor may also interact with ion channels (directly or indirectly) and this could influence the intracellular levels of calcium. Since the release of 5-HT is sensitive to the frequency of stimulation,[12,13,18,19] it is also possible that release could be associated with different levels of intracellular calcium and the free calcium levels could also be regulated by the autoreceptor. By modulating the levels of calcium, activation of an autoreceptor could inhibit release of 5-HT but lead to an uncoupling of the stimulus secretion process as described for gamma-aminobutyric acid (GABA) or adrenergic systems.[8,20] In essence, if the cytoplasmic levels of free calcium were to increase in the presence of an autoreceptor agonist, a pool of transmitter could be released that otherwise would be under tighter autoreceptor control. Thus, autoreceptor control of release could involve the interaction of numerous factors that influence free calcium levels. If intraneuronal pools of 5-HT differed in their sensitivity to calcium, the effects of autoreceptor agonists on release might appear to function inappropriately or they may not exert their normal control over a certain pool of transmitter.

INTERACTION OF 5-HT₁ AGONISTS WITH THE 5-HT TRANSPORTER

In an attempt to determine how the transporter and the autoreceptor may interact to influence release of 5-HT, the ability of several 5-HT$_{1B}$ agonists to modulate 5-HT uptake was determined in synaptosomes. It was observed initially that both *m*-trifluoromethylphenyl-piperazine (TFMPP) and RU 24969 were potent inhibitors of 5-HT uptake into synaptosomes. TABLE 1 presents results from the uptake inhibition studies with these compounds and also includes data from an expanded analysis of the effects of other 5-HT$_{1B}$ agonists on uptake. It can be seen in TABLE 1 that the agonists TFMPP, RU 24969, mCPP, and MK-212 concentration dependently inhibited the uptake of exogenous 5-HT by purified synaptosomes.

In order to further characterize the interaction of these agents with the 5-HT

TABLE 1. Effects of 5-HT$_{1B}$ Agonists on 5-HT Uptake[a]

Agonist	Concentration (μM)	Uptake (% Control)
TFMPP	0.05	93
	0.10	87
	0.30	64
	0.60	47
	1.00	36
RU 24969	0.01	84
	0.02	71
	0.05	53
	0.10	34
	0.30	14
mCPP	0.05	70
	0.10	53
	0.30	28
	0.60	10
MK-212	0.10	91
	0.30	76
	0.60	60
	1.00	47
	3.00	22

[a]Synaptosomes from rat forebrain were incubated with the indicated concentration of the 5-HT agonists in the presence of 50 nM exogenous 5-HT. Uptake of 5-HT was assessed as previously described.[24]

transporter, their ability to displace ^{3}H-paroxetine from neuronal membranes was tested. Paroxetine is a potent 5-HT uptake inhibitor, and the binding of ^{3}H-paroxetine is used as a marker for the neuronal 5-HT transporter.[21] TABLE 2 presents the results of the effects of various 5-HT$_{1B}$ agonists on the binding of ^{3}H-paroxetine. As with 5-HT uptake, it was observed that the 5-HT$_{1B}$ agonists (with 5-HT uptake inhibiting properties) also displaced ^{3}H-paroxetine from its binding site. The order of potency of uptake inhibition and displacement of ^{3}H-paroxetine binding were closely correlated, confirming that the binding of ^{3}H-paroxetine is an effective marker of the 5-HT transporter. This relationship between uptake inhibition and displacement of ^{3}H-paroxetine from membranes also further establishes that many of the 5-HT$_{1B}$ agonists that are thought to bind to the autoreceptor and inhibit 5-HT release also interact closely with the 5-HT transporter.

The relationship between the 5-HT$_{1B}$ site and the 5-HT transporter was approached from another avenue. The autoreceptor in rat brain tissue has been classified pharmacologically as the 5-HT$_{1B}$ subtype.[10] Thus, the published binding constants for these agents at either the 5-HT$_{1A}$ or the 5-HT$_{1B}$[22] binding sites were compared to their IC$_{50}$ values for inhibition of 5-HT uptake. The results of this comparison are presented in FIGURE 1. It can be seen that a close correlation exists between potency of uptake inhibition and binding to the 5-HT$_{1B}$ site ($r = 0.89$). On the other hand, a correlation between binding to the 5-HT$_{1A}$ site and uptake inhibitory properties was not evident ($r = 0.01$; data not shown). These results establish that a relationship indeed exists between the 5-HT$_{1B}$ site and the 5-HT transporter. Based on the range of 5-HT agonists included in this analysis, one could also predict that the higher the affinity of a compound for the 5-HT$_{1B}$ binding site, the greater is the likelihood that the compound will inhibit 5-HT uptake on a functional level. This relationship is unidirectional since uptake inhibitors such as fluoxetine and paroxetine do not bind to the 5-HT$_{1B}$ site.[22]

INFLUENCE OF SELECTED 5-HT$_{1B}$ AGONISTS ON 5-HT RELEASE FROM BRAIN SLICES

Having established that at least TFMPP and RU 24969 interact directly with the 5-HT transporter, the previously discussed interaction between the autoreceptor and the uptake site for 5-HT can be viewed in a different light. While certain interpretational problems may exist in this regard,[17,23] these results provide a molecular basis for the interaction. Furthermore, the properties of some of the 5-HT$_{1B}$ agonists prompted a reexamination of their ability to modulate 5-HT release. In these studies, slices from rat hippocampus were superfused and the release of endogenous 5-HT was measured by high performance liquid chromatography (HPLC) with fluorescence detection.[24] The 5-HT uptake mechanism was not inhibited in these superfusion experiments. In an effort to confirm that the release model was functioning as previously described, it was first determined that the KCl-induced release of 5-HT from the slices was calcium dependent. Second, the fractional release of 5-HT produced by KCl was in the range of 0.8–1.5% of the tissue content of 5-HT. Third, 5-MeOT at a concentration of 500 nM was also capable of reducing the evoked release of 5-HT. When either TFMPP (1.0 μM) or RU 24969 (100 nM) was added to the superfusion medium prior to release, the effects on release were somewhat surprising. Basal release of 5-HT was increased twofold by TFMPP but was not altered by RU 24969. Upon stimulation of release by KCl (20 mM), each agonist augmented the release of 5-HT. TFMPP increased fractional release from 1.14% to 3.40% and RU 24969 produced a similar effect, increasing the fractional release from 1.36% to 2.26%. These results would not have been predicted based solely on the 5-HT$_{1B}$ properties of TFMPP or RU 24969, yet their

TABLE 2. Effects of 5-HT$_{1B}$ Agonists on ^{3}H-Paroxetine Binding[a]

Agonist	Concentration (μM)	^{3}H-Paroxetine Bound (% Control)
TFMPP	0.30	79
	0.60	66
	1.00	53
	2.00	40
	5.00	23
RU 24969	0.02	71
	0.05	53
	0.10	37
	0.30	18
5-HT	0.30	68
	0.60	55
	1.00	41
	2.00	25
Fluoxetine	0.005	62
	0.01	40
	0.03	20
	0.06	12

[a]Membranes were prepared from rat brain synaptosomes and were incubated with the indicated concentrations of the 5-HT agonists in the presence of 0.05 nM ^{3}H-paroxetine. After a 60 minute incubation at 25°C, membranes were filtered and washed, and the amount of radioactivity trapped on the filters was quantified by liquid scintillation spectrometry. Binding refers to specific binding as defined as the difference between the amount of ^{3}H-paroxetine bound in the absence of 100 μM 5-HT and the amount of ^{3}H-paroxetine bound in its presence.

effects on the release of *endogenous* 5-HT, with the uptake system functioning normally, were more consistent with that of a 5-HT uptake inhibitor/releasing agent. Similar results were observed when the release of ^{3}H-5-HT from preloaded slices was measured. It is difficult to ascribe the effects of TFMPP or RU 24969 to the method of measuring release, to the brain area selected for study, or to the method by which release was induced since all other aspects of the superfusion system functioned as previously described for 5-HT, from very different methods.

Certain 5-HT autoreceptor agonists such as TFMPP or RU 24969 clearly interact with the 5-HT transporter while others such as 5-MeOT do not. Perhaps the additional property of interaction with the 5-HT transporter by certain 5-HT_{1B} agonists is

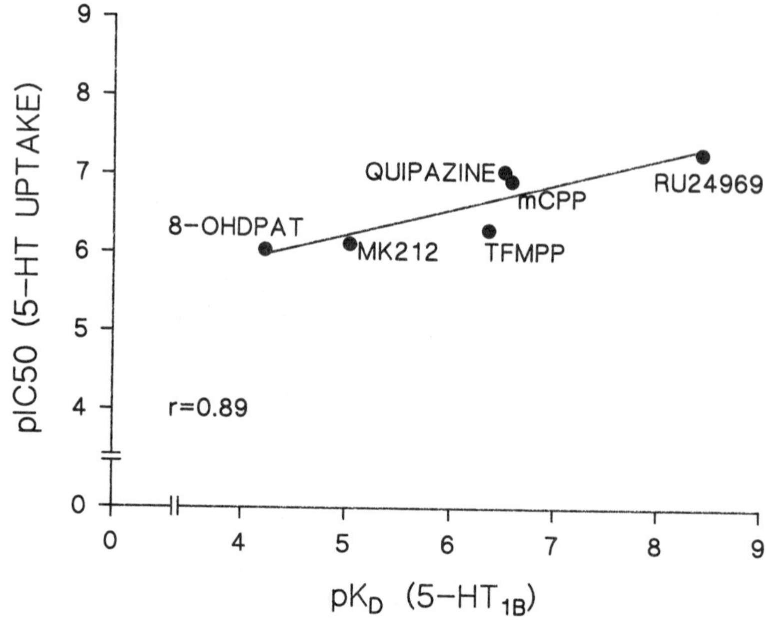

FIGURE 1. Relationship between the affinity of 5-HT agonists for the 5-HT_{1B} binding site and 5-HT uptake inhibiting properties. The published affinities of the various agonists for the 5-HT_{1B} binding site (see Reference 22 for values) were compared to their IC_{50} for inhibiting the uptake of exogenous 5-HT into rat brain synaptosomes.

responsible for the somewhat anomalous augmentation of 5-HT release. Additional study will be required in order to substantiate this possibility. A closer inspection of previous experiments with TFMPP or RU 24969 indicates that these compounds have only minor effects on 5-HT release. Their classification as 5-HT_{1B} agonists is based largely on binding experiments.

NATURE OF THE 5-HT_{1B} AUTORECEPTOR

The results presented above along with others further establish that a number of significant differences exist among various agonists at the 5-HT_{1B} receptor. For example, 5-HT does not appear to inhibit its own release in the absence of an uptake inhibitor, suggesting that reuptake predominates over the autoregulation of 5-HT release by newly released transmitter. Agonists that do not interact with the 5-HT transporter (e.g., 5-MeOT) do inhibit release. Finally, those 5-HT_{1B} agonists that interact with the 5-HT transporter (e.g., TFMPP or RU 24969) do not appear to inhibit release. How can these widely differing results on release by compounds that interact with the 5-HT_{1B} receptor site explained? First, 5-HT_{1B} receptor sites are not necessarily restricted to the presynaptic portion of the neuron. Combined interaction with pre- and postsynaptic receptors could determine whether "autoreceptor" agonists will actually inhibit release. Second, the 5-HT transporter and the autoreceptor may well be linked through functional or molecular events, in which case certain agonists will produce anomalous results with regard to autoreceptor function.

Our results with measures of the release of endogenous 5-HT indicate that the release of this transmitter can be detected without interfering with the uptake mechanism. However, the question of endogenous autoregulation of neurotransmitter release for both noradrenaline[8] and 5-HT[18] has been approached in a slightly different manner recently. Considering 5-HT, Blier *et al.* demonstrated that the release of 5-HT (per pulse, when electrical stimulation is used to induce release) was inversely proportional to the time interval between pulses.[18] In essence, at lower stimulation frequencies (longer interpulse intervals), more transmitter is released. Blier *et al.* concluded that at higher stimulation frequencies, the biophase concentration of 5-HT is increased to a sufficient extent to inhibit release.[18] Thus, while the fractional release of 5-HT is lower at 3 Hz than at 1 Hz, the increased levels of 5-HT in the biophase that are responsible for the autoinhibition of release were nevertheless not substantiated. Although the same number of pulses are delivered in these experiments, the length of time that the tissue is stimulated is necessarily different for 1 or 3 Hz. Thus, it seems possible that other regulatory processes could come to influence 5-HT release given different times of stimulation. The relationship between frequency of stimulation and 5-HT release appears to be different *in vivo*. Sharp *et al.* found that higher frequency stimulation of the dorsal raphe nucleus resulted in the release of greater amounts of 5-HT into hippocampal dialyzates.[2]

Another aspect of the experiment of Blier *et al.* points out the differences between the regulation of release by the endogenous ligand 5-HT and that of other 5-HT agonists.[18] These investigators observed that the inhibitory efficacy of 5-MeOT was much greater at lower stimulation frequencies. For example, at 1 Hz, the IC_{50} for inhibition of 5-HT release by 5-MeOT was 34 nM compared to 560 nM at 3 Hz. The differences in the synaptic concentration of 5-HT induced by 1 Hz or 3 Hz stimulation are thought to lead to the large differences in IC_{50} values for 5-MeOT. What is not as clear is how the stimulation at 1 Hz versus 3 Hz results in less than a 2-fold difference in the amount of 5-HT that is released yet the same stimulation parameters (and presumably the same synaptic concentrations of 5-HT) result in more than a 16-fold difference in the efficacy of the 5-HT autoreceptor agonist 5-MeOT to inhibit release.

IS THERE SUFFICIENT EVIDENCE FOR THE OPERATION OF CENTRAL 5-HT AUTORECEPTORS?

This is not an easy question to answer in reality. It is clear that a wide variety of compounds can modulate the release of 5-HT *in vitro*. Furthermore, a great deal of evidence leads to the conclusion that the pharmacological profile of this receptor which regulates 5-HT release is of the 5-HT_{1B} subtype.[10] Although many 5-HT_{1B} agonists also interact with the 5-HT_{1A} receptor, selective 5-HT_{1A} agonists such as 8-OH-DPAT do not inhibit 5-HT release.[25] However, several other observations regarding the autoreceptor and its agonists should also be discussed in order to place it in some perspective. As discussed above, a selective autoreceptor agonist could possibly be useful in the treatment of psychiatric disorders if it were important to modulate the release and synaptic availability of 5-HT. Compounds like LSD or 5-MeOT are potent inhibitors of 5-HT release *in vitro* yet these agents will not find clinical usefulness as autoreceptor agonists for obvious reasons. The newer 5-HT_1 agonists such as TFMPP or RU 24969 appear to have many actions in addition to their abilities to interact with the 5-HT autoreceptor and at least RU 24969 interacts with the 5-HT_2 receptor.[26] This is not intended to imply that they will not have clinical uses—only to focus on the complexity of their mechanisms of action. It must also be considered that the 5-HT_{1B} binding site does not apparently exist in human brain[27] and thought should be given to the possibility that the autoreceptor in human brain, assuming that one exists, is pharmacologically distinct from that seen in rodent brain.

Finally, the apparent differences in the modulation of 5-HT release by 5-HT itself or by other 5-HT_{1B} agonists, such as 5-MeOT, lead to the question of whether the biophase concentration of 5-HT regulates its own release. It appears that the uptake mechanism for 5-HT plays a very important role in determining the synaptic activity of 5-HT after stimulation of release. In fact, apart from the recent results of Blier *et al.*,[18] the ability of 5-HT to inhibit its own release is restricted to conditions where 5-HT uptake is inhibited. In view of the ability of 5-HT_{1B} agonists to interact with the 5-HT transporter and inhibit 5-HT uptake, the modulation of release by 5-HT in the presence of a 5-HT uptake inhibitor could reflect more than an interaction of 5-HT with the autoreceptor.

The regulation of 5-HT release by an autoreceptor can be substantiated from a pharmacological viewpoint, but the operation of such a mechanism from a physiological basis is not so clear. In the final analysis, it may be of little importance if an autoreceptor regulates 5-HT release as long as the 5-HT transporter functions normally. It is also possible that a dysfunction at the level of the 5-HT autoreceptor could occur and this might lead to altered function in 5-HT neurotransmission. In this case, it would be important to have a pharmacological agent that could interact with the receptor(s) that regulates 5-HT release and compounds such as TFMPP or RU 24969 might serve in this function.

REFERENCES

1. KUHN, D. M., W. A. WOLF & M. B. H. YOUDIM. 1986. Neurochem. Int. **8:** 141–154.
2. SHARP, T., S. R. BRAMWELL, D. CLARK & D. G. GRAHAME-SMITH. 1989. J. Neurochem. **53:** 234–240.
3. KALEN, P., R. E. STRECKER, E. ROSENGREN & A. BJORKLAND. 1988. J. Neurochem. **51:** 1422–1435.
4. JACOBY, J. H. & L. D. LYTLE, Eds. 1978. Serotonin Neurotoxins. New York Academy of Sciences. New York, N.Y. (Ann. N.Y. Acad. Sci. **305.**)
5. PEROUTKA, S. J. 1988. Annu. Rev. Neurosci. **11:** 45–60.
6. KALSNER, S. 1985. Biochem. Pharmacol. **34:** 4085–4097.
7. CHESSELET, M.-F. 1984. Neuroscience **12:** 347–375.
8. STARKE, K. 1987. Rev. Physiol. Biochem. Pharmacol. **107:** 73–146.
9. CERRITO, F. & M. RAITERI. 1979. Eur. J. Pharmacol. **57:** 427–430.
10. MORET, C. 1985. *In* Neuropharmacology of Serotonin. A. R. Green, Ed.: 21–49. Oxford University Press. Oxford, England.
11. SCHLICKER, E., M. FINK, M. GOTHERT, D. HOYER, G. MOLDERINGS, I. ROSCHKE & P. SCHOFFTER. Naunyn Schmiedebergs Arch. Pharmacol. **340:** 45–51.
12. GOTHERT, M. 1980. Naunyn Schmiedebergs Arch. Pharmacol. **314:** 223–230.

13. CHAPUT, Y. & C. DE MONTIGNY. 1988. J. Pharmacol. Exp. Ther. **246:** 359–370.
14. RAITERI, M., M. MARCHI & G. MAURA. 1984. *In* Handbook of Neurochemistry. A. Lajtha, Ed. 2nd edit. **6.** Plenum Press. New York, N.Y.
15. HERNDON, H. & S. R. NAHORSKI. 1987. Naunyn Schmiedebergs Arch. Pharmacol. **335:** 238–242.
16. BAUMANN, P. A. & P. C. WALDMEIER. 1981. Naunyn Schmiedebergs Arch. Pharmacol. **317:** 36–43.
17. PASSARELLI, F., A. M. GALZIN & S. Z. LANGER. 1987. J. Pharmacol. Exp. Ther. **242:** 1056–1063.
18. BLIER, P., R. RAMDINE, A.-M. GALZIN & S. Z. LANGER. 1989. Naunyn Schmiedebergs Arch. Pharmacol. **339:** 60–64.
19. RICHARDS, M. H. 1985. Naunyn Schmiedebergs Arch. Pharmacol. **329:** 359–366.
20. WALDMEIER, P. C., P. WICKI, J.-J. FELDTRAUER & P. A. BAUMAN. 1989. Naunyn Schmiedebergs Arch. Pharmacol. **340:** 372–378.
21. HABERT, E., D. GRAHAM, L. TAHRAOUI, Y. CLAUSTRE & S. Z. LANGER. 1985. Eur. J. Pharmacol. **118:** 107–114.
22. HOYER, D. 1988. J. Recept. Res. **8:** 59–81.
23. BONANNO, G. & M. RAITERI. 1987. Naunyn Schmiedebergs Arch. Pharmacol. **335:** 219–225.
24. WOLF, W. A. & D. M. KUHN. 1986. J. Neurochem. **46:** 61–67.
25. MIDDLEMISS, D. N. 1984. Naunyn Schmiedebergs Arch. Pharmacol. **327:** 18–22.
26. VAN DE KAR, L. D., M. CARNES, R. J. MASLOWSKI, A. M. BONADONNA, P. A. RITTENHOUSE, K. KUNIMOTTO, R. A. PIECHOWSKI & C. L. BETHEA. 1989. J. Pharmacol. Exp. Ther. **251:** 428–434.
27. HOYER, D., A. PAZOS, A. PROBST & J. M. PALACIOS. 1986. Brain Res. **376:** 85–96.

Operation of Presynaptic α_2-Adrenoceptors *In Vivo* in the Rat Brain

BERNARD SCATTON

Synthélabo Research (LERS)
Biochemical Pharmacology Group
31 avenue Paul Vaillant-Couturier
92220 Bagneux, France

INTRODUCTION

In the autonomic nervous system, a host of evidence has suggested the existence of presynaptic α_2-adrenoceptors that control the release of noradrenaline (NA) via a negative feedback mechanism.[1-4] The location of α_2-adrenoceptors on nerve endings in the periphery has been clearly demonstrated by (1) the diminution of the binding of α_2-adrenoceptor ligands after chemical sympathectomy,[5] and (2) the fact that the regulation of NA release by α_2-adrenoceptors is independent of the α or β nature of the postsynaptic adrenoceptors that mediate the response of the effector organ and is not affected by the destruction of the postsynaptic effector cells.[1,2]

A similar modulation of NA release by presynaptic α_2-adrenoceptors is believed to operate in the central nervous system (CNS). Much of the evidence for this view is derived from *in vitro* studies on tissue slices or synaptosomes which demonstrate that α_2-adrenoceptor agonists inhibit whereas α_2-adrenoceptor antagonists stimulate the electrically or chemically evoked release of ^{3}H-NA.[1-3,6-9] Moreover, anterograde transport of α_2-adrenoceptors from NA perikarya to terminals has been demonstrated,[10] suggesting that α_2-adrenoceptors are delivered to the presynaptic region. However, the existence of presynaptic α_2-adrenoceptors in the CNS is controversial since neurotoxic destruction of ascending noradrenergic pathways fails to reduce ^{3}H-idazoxan or ^{3}H-clonidine binding in the corresponding projection fields,[11,12] even after ibotenate-induced lesion of postsynaptic elements.[13]

In spite of extensive *in vitro* studies devoted to the role of presynaptic α_2-adrenoceptors in the control of NA release, little is known of the operation of presynaptic α_2-adrenoceptors *in vivo*. Several investigations have shown that after systemic injection, α_2 antagonists increase, whereas α_2 agonists decrease, NA turnover and release in various NA-rich brain areas.[14-18] However, the relative contribution of presynaptic, postsynaptic, or somatodendritic α_2-adrenoceptors or α_2-heteroreceptors to this effect is not known.

We have recently attempted to investigate whether presynaptic α_2-adrenoceptors play a significant role in the control of NA release *in vivo* by studying the influence of α_2-adrenoceptor antagonists and agonists on NA release (as assessed by dialysis) and metabolism in NA-rich rat brain areas, using experimental manipulations that allow an exclusive interaction of these drugs with presynaptic α_2-adrenoceptors.[19,20] We presently provide evidence that in spite of their low density, presynaptic α_2-adrenoceptors operate *in vivo* and play a cardinal role in the regulation of NA turnover in the CNS.

CONTROL OF CORTICAL NA RELEASE *IN VIVO* BY PRESYNAPTIC α_2-ADRENOCEPTORS IN THE RAT

The control of NA release by α_2-adrenoceptors in the CNS is rather complex due to the different synaptic localization of these receptors (cf. FIGURE 1). Thus, apart from α_2-adrenoceptors theoretically located on noradrenergic nerve terminals, there exist postsynaptic α_2-adrenoceptors that may regulate NA release by local (to terminal) or

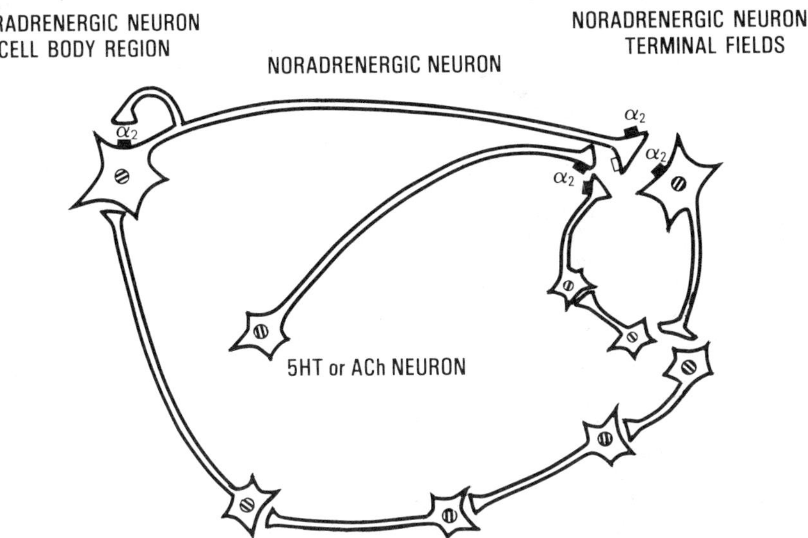

FIGURE 1. Schematic representation of differently located α_2-adrenoceptors in the central nervous system and possible mechanism of action. Activation of postsynaptic α_2-adrenoceptors by released NA could trigger a negative feedback mediated either via long multisynaptic pathways projecting to noradrenergic cell bodies or via local circuits to noradrenergic terminals. A further alternative could implicate the release of a humoral agent (e.g., prostaglandin E), which influences the noradrenergic terminal and inhibits NA release. α-2-Adrenoceptors may exist on terminals of neurons (e.g., serotonergic or cholinergic) afferent to noradrenergic nerve endings (heteroreceptors). Activation of these receptors would lead to the liberation of an inhibitory neurotransmitter thus inhibiting the release of NA from noradrenergic terminals. Inhibitory somatodendritic α_2-adrenoceptors appear to be innervated by recurrent axon collaterals. Activation of such α_2-adrenoceptors by released NA triggers a compensatory feedback inhibition of noradrenergic neuron cell firing. Finally, the activation by released NA of presynaptic α_2-adrenoceptors located on noradrenergic nerve endings inhibits the excitation-secretion coupling leading to a reduction of NA release. 5-HT, serotonin; ACh, acetylcholine.

distant (to cell bodies) polysynaptic feedback circuits or by controlling the release of a postsynaptically produced modulatory factor (e.g., prostaglandins).[6] Somatodendritic α_2-adrenoceptors may also regulate NA release via alterations of noradrenergic neuronal activity.[21] Finally α_2-heteroreceptors located on the nerve terminals of other neuronal systems (e.g., cholinergic or serotonergic)[22-25] may modulate NA release by controlling the release of these transmitters. To investigate the relative contribution of each of these α_2-adrenoceptor populations, we studied the effects of idazoxan on cortical NA release, using transcortical dialysis,[18] in awake unrestrained rats before and after surgical or experimental manipulations that eliminated specific feedback loops. The dialysis fiber was used to simultaneously administer drugs locally and measure extracellular NA concentrations.

The local infusion of idazoxan excluded an involvement of somatodendritic α_2-adrenoceptors. The influence of postsynaptic α_2-adrenoceptors was examined by comparing the effects of systemic or local perfusion of idazoxan before and after lesion of cortical cell bodies with ibotenic acid, administered via the dialysis fiber. Finally, we investigated the possible contribution of α_2-heteroreceptors by studying the effects of idazoxan on cortical NA release in rats deprived of their serotonergic or cholinergic innervation.

Destruction of Postsynaptic α_2-Adrenoceptors

Previous studies from our laboratory have shown that the technique of transcortical dialysis coupled to a sensitive radioenzymatic assay is a suitable and sensitive means to investigate the effects of centrally acting drugs on the release of endogenous NA in the rat cerebral cortex.[18] The basal levels of NA measured in the extracellular space of the cerebral cortex ($\simeq 7$ pg/20 minute fraction) are fairly constant for a period of at least 4 days in the awake, freely moving rat. The neurogenic origin of the amine measured in the perfusate is supported by the observation that tetrodotoxin (TTX) injected into the medial forebrain bundle or calcium-free perfusion medium almost totally abolished NA output, whereas electrical stimulation of the ascending noradrenergic pathways dramatically increased NA output in the dialysate.[18]

As shown in FIGURE 2A, systemic injection of idazoxan [20 mg/kg intraperitoneally (ip)] to naive rats produced a marked increase in extracellular cortical NA levels (maximum increase +230% over basal levels at 1 hour postinjection). The effect of idazoxan was presumably due to its antagonistic action at α_2-adrenoceptors as it was fully antagonized by systemic administration of clonidine (0.3 mg/kg ip).[20] Idazoxan (20 mg/kg ip) produced a similar augmentation of NA output on day 4 after implantation of the dialysis fiber (FIGURE 2A) showing that the responsiveness to the drug does not change over this time period.

The possible contribution of postsynaptic α_2-receptors to the idazoxan-induced NA release was investigated. After the acute administration of idazoxan (above), rats received an infusion of ibotenic acid (10 μg/μl for 20 minutes) via the cortically implanted dialysis fiber; perfusion with Ringer's medium was continued until 3 days postlesion at which time further experiments were performed. The neurotoxin by itself failed to affect basal NA efflux. In ibotenate-lesioned rats, systemic administration of idazoxan (20 mg/kg ip) caused a maximal increase in NA output which was similar in magnitude to that seen in the sham-lesioned rats; however, this increase in NA release was prolonged in the lesioned animals, as compared to naive or sham-lesioned rats (FIGURE 2A).

Similar results were obtained when desipramine (20 mg/kg ip) was injected prior to idazoxan (FIGURE 3A). Indeed, after desipramine pretreatment, idazoxan (20 mg/kg ip) produced a 34-fold increase in extracellular cortical NA concentrations in naive rats and a 32-fold increase in this parameter in ibotenate-lesioned rats (at its peak of effect). Desipramine by itself caused only a 2-to-3-fold increase in cortical NA output.

Histological examination of the extent of the ibotenate lesioned area revealed that the neurotoxin produced a zone of total neuronal depopulation in the cerebral cortex with an average radius of 2.5 mm from the dialysis fiber.[20]

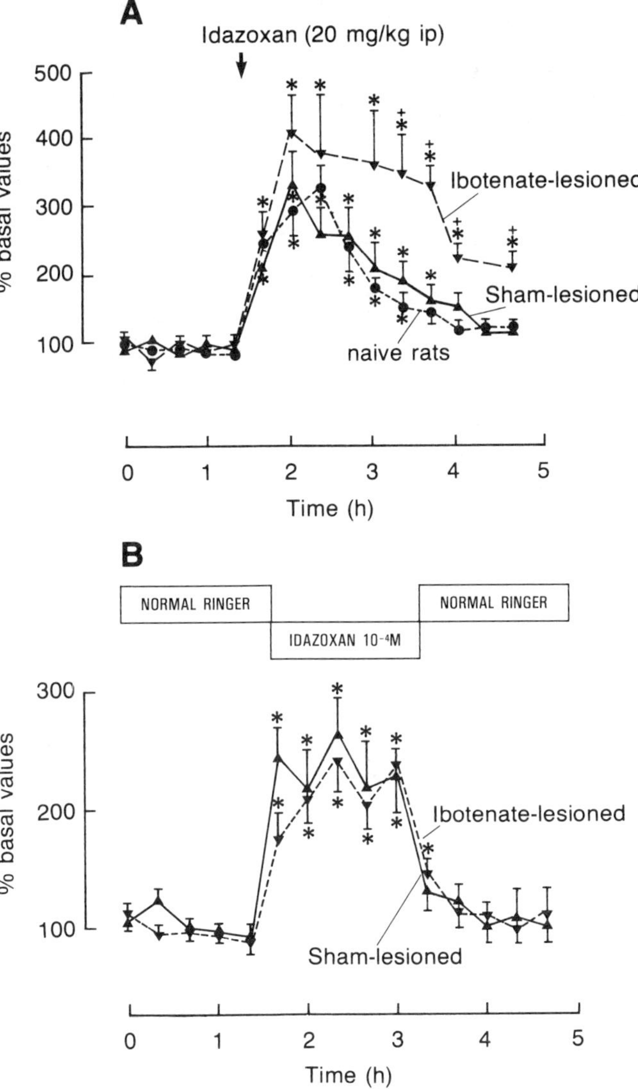

FIGURE 2. Effect of systemic injection (A) or intracortical perfusion (B) of idazoxan on extracellular NA concentrations, in the cerebral cortex or normal and ibotenate-lesioned awake unrestrained rats. Rats were implanted transcortically with a dialysis tube (Diaflo, 200 μm internal diameter) under halothane/N_2:O_2 anesthesia according to L'Heureux *et al.*[18] The inlet tube was connected via a swivel connector to a Hamilton syringe driven by a microinjection pump and the outlet tube was positioned in a collecting tube on ice. The dialysis fiber was perfused with Ringer's medium at a rate of 2 μl/minute. NA was measured by a radioenzymatic technique[36] in 20-minute samples. Naive rats were perfused with normal Ringer's medium on day 1 after implantation of the dialysis fiber. Sham-lesioned rats were perfused on day 4 after fiber implantation. In the ibotenate-lesioned group, after the experiment on day 1, rats were perfused with ibotenic acid (10 μg/μl, 2 μl/minute for 20 minutes) and a systemic injection (20 mg/kg ip) or local perfusion (10^{-4} M) of idazoxan was carried out on day 4 (e.g., 3 days after ibotenate lesion). Results are means with standard errors of the mean (SEM) of data obtained from 6 to 9 rats per experimental group and are expressed as the percentage of respective basal values. *$p <$ 0.01 vs. basal values, ⁺$p <$ 0.05 vs. sham-lesioned rats.

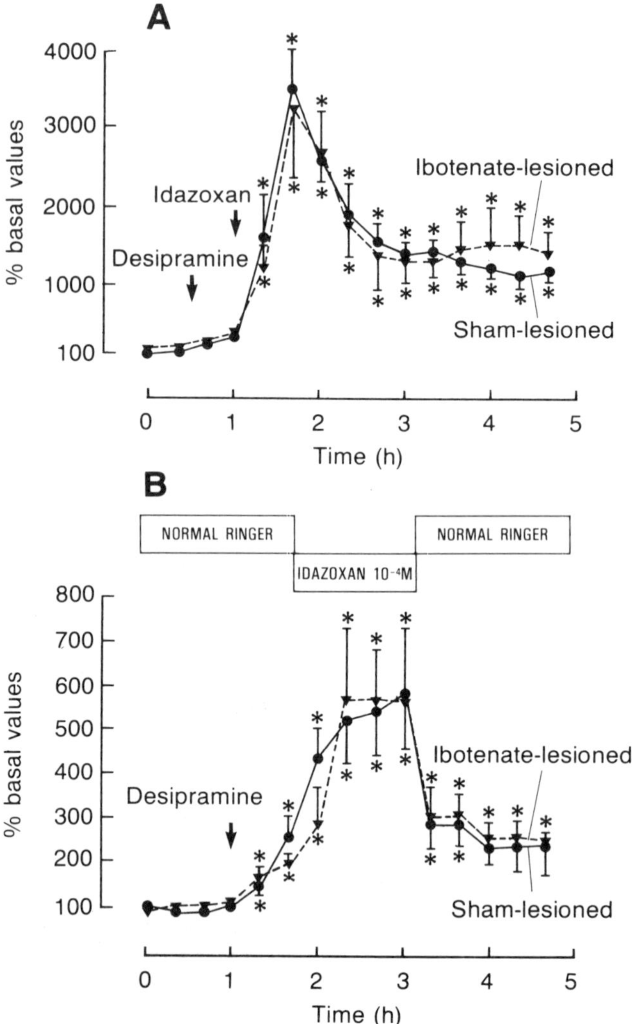

FIGURE 3. Effect of systemic (20 mg/kg ip) (A) or intracortical (10^{-4} M) (B) injection of idazoxan and systemic injection of desipramine (20 mg/kg ip) on extracellular NA concentrations in the cerebral cortex or normal and ibotenate-lesioned unrestrained rats. For experimental conditions see FIGURE 2 legend. Desipramine was injected 40 minutes before idazoxan. Results are means with SEM of data obtained from 6 to 8 rats and are expressed as the percentage of respective basal values. *$p < 0.01$ vs. basal values.

These results indicate that the effect of idazoxan on cortical NA release does not depend upon the integrity of postsynaptic neurons within a close proximity to the dialysis fiber.

To investigate whether presynaptic α_2-auto- or heteroreceptors are involved in the regulation of cortical NA release *in vivo*, idazoxan was directly infused into the cerebral cortex (so as to eliminate the contribution of somatodendritic α_2-autoreceptors) after destruction of postsynaptic elements by ibotenic acid.

Local infusion of idazoxan (10^{-4} M) via the dialysis fiber in both sham- and ibotenate-lesioned rats produced a sharp increase in NA efflux reaching a maximal of 150% above basal levels (FIGURE 2B). NA output returned rapidly to basal values when normal Ringer's solution was reintroduced. Likewise, in sham- or ibotenate-lesioned rats pretreated with desipramine (20 mg/kg ip), intracortical infusion of idazoxan (10^{-4} M) provoked a fivefold increase in NA efflux during the period of application (FIGURE 3B).

The extent of diffusion and effective concentration of idazoxan after its local administration via the dialysis fiber was evaluated by perfusing sham- and ibotenate-lesioned rats with ^{3}H-idazoxan (10^{-4} M) under standard conditions and by processing the brain for autoradiography. The zone of ^{3}H-idazoxan diffusion was contained within the lesioned area in the cerebral cortex and was comparable in sham- and ibotenate-lesioned rats.[20] Measurement of optical densities at different radii from the dialysis fiber center indicated that the extracellular concentrations of idazoxan ranged from 4 to 48 μM. However, due to diffusion barriers and drug compartmentation, the tissue concentration of idazoxan is likely to be lower. Nonetheless, these values are close to the range of idazoxan concentrations needed to elicit a clear-cut increase in the electrically evoked release of ^{3}H-NA from brain tissue slices.[25]

These results thus demonstrate that idazoxan has an action within the cerebral cortex that does not necessarily depend upon the integrity of the postsynaptic elements and of the operation of somatodendritic α_2-adrenoceptors.

Destruction of α_2-Heteroreceptors Located on Cortical Serotonergic and Cholinergic Afferents

α_2-Adrenoceptors localized on the terminals of neurotransmitter systems afferent to noradrenergic nerve endings (heteroreceptors) may also take part in the regulation of cortical NA release. Indeed, the existence of α_2-heteroreceptors on serotonergic and cholinergic nerve endings in the cerebral cortex exerting an inhibitory control over serotonin and acetylcholine release has been proposed[22–25]; it is thus possible that such α_2-heteroreceptors indirectly modulate the release of cortical NA. To investigate this possibility, idazoxan was infused (via the dialysis fiber) into the cerebral cortex of rats deprived of their cortical serotonergic or cholinergic innervation.

As shown in FIGURE 4, neither 5,7-dihydroxytryptamine (5,7-DHT) lesions of serotonergic neurons nor electrolytic lesions of the nucleus basalis magnocellularis had any significant effect on the stimulation of cortical NA output produced by intracortical infusion of idazoxan (10^{-4} M). The effectiveness of the 5,7-DHT lesion was attested by a 77% depletion of cortical serotonin levels.[20] Histological examination revealed that the electrolytic lesion of the nucleus basalis magnocellularis caused a total destruction of the substantia innominata, ventral pallidum and globus pallidus.[20] Moreover, nucleus basalis lesion reduced cortical choline acetyltransferase activity by 47%.[20]

Altogether, these results indicate that α_2-heteroreceptors located on serotonergic and cholinergic fibers probably do not contribute to the effect of idazoxan on cortical NA release.

Discussion

The majority of α_2-adrenoceptors in the CNS appear to be located post-synaptically.[11–13,26] However, the fact that ibotenate-induced lesion of postsynaptic elements within the cerebral cortex did not reduce (but rather enhanced) the effects of either systemically or locally administered idazoxan on cortical NA release strongly

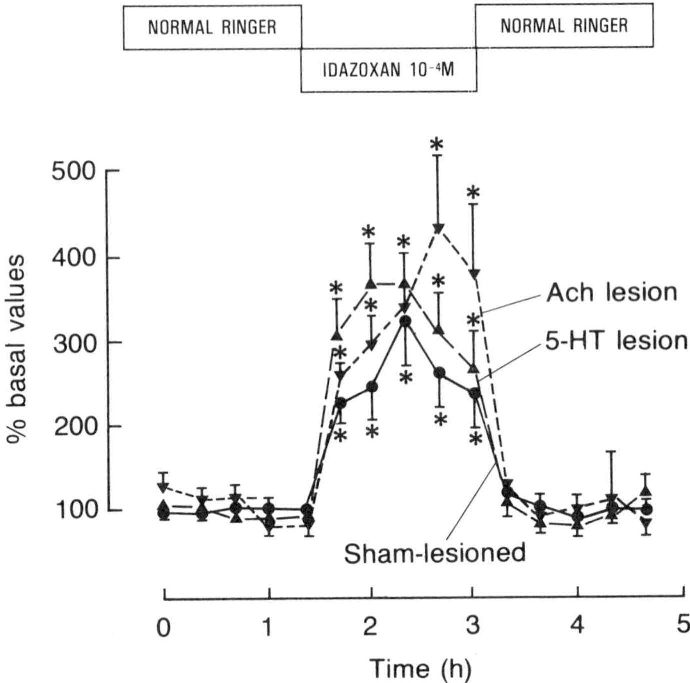

FIGURE 4. Effect of local perfusion of idazoxan (10^{-4} M) on cortical NA release after lesion of serotonergic or cholinergic afferents to the cerebral cortex. For the lesion of serotonergic neurons, rats pretreated with desipramine (25 mg/kg ip) and nomifensine (10 mg/kg ip) received an intracerebroventricular (icv) injection of 200 μg of 5,7-dihydroxytryptamine or vehicle 15 days before the perfusion experiment. Cholinergic neurons projecting to the cortex were destroyed by bilateral electrolytic lesion of the nucleus basalis magnocellularis 7 days before the perfusion experiment.[20] Results are means with SEM of data obtained from 5–15 rats and are expressed as the percentage of respective basal values. *$p < 0.01$ vs. respective basal values. There was no statistically significant difference between sham, 5-HT-, and ACh-lesioned rats. Abbreviations: 5-HT, serotonin; ACh, acetylcholine.

argues against the involvement of postsynaptic α_2-adrenoceptors in the effects of idazoxan on NA release. Thus, neuronal circuits from the cerebral cortex, either to noradrenergic nerve terminals or to cell body regions, appear to contribute very little to the α_2-receptor-mediated control of NA release. It also seems unlikely that α_2-adrenoceptors regulate NA release through the action of a humoral agent released by postsynaptic neurons which in turn influences the noradrenergic nerve terminal.

Intracortical infusion of idazoxan via the dialysis fiber in ibotenate-lesioned rats caused an increase in cortical NA output almost similar in magnitude to that seen after systemic injection of the α_2-antagonist. This demonstrates that even when somatodendritic α_2-adrenoceptors are bypassed and postsynaptic elements are destroyed, the effects of idazoxan on cortical NA release are maintained. In this situation, both presynaptic α_2-autoreceptors and presynaptic α_2-heteroreceptors located on afferents to noradrenergic nerve terminals may be targets for idazoxan. However, as lesions of serotonergic and cholinergic afferents to the cerebral cortex failed to modify the effects of intracortically administered idazoxan on cortical NA release, the contribution of α_2-heteroreceptors on these fibers appears to be minimal.

Thus, despite the fact that presynaptic α_2-adrenoceptors constitute a relatively small portion of central α_2-adrenoceptors (to date undetectable in binding studies), it seems by a process of elimination that they are very much involved in the regulation of cortical NA release *in vivo*. This does not rule out an influence of postsynaptic or somatodendritic α_2-adrenoceptors on cortical NA release (in fact, postsynaptic or somatodendritic α_2-adrenoceptors are also probably involved in the control of NA release as idazoxan increased cortical NA release to a larger extent after systemic than after intracortical injection, expecially when animals were pretreated with desipramine) but does demonstrate the functional importance of presynaptic α_2-receptors in regulating cortical NA release *in vivo*.

The effects of both systemically and locally administered idazoxan on NA release were potentiated markedly by systemic pretreatment with the NA uptake inhibitor desipramine. NA uptake blockade per se would be expected to increase the extracellular concentrations of NA, but this phenomenon, which results in only a doubling of the amount of NA recovered in the dialysis tube, cannot solely account for the dramatic potentiation of the effect of idazoxan on cortical NA release. A further explanation for the effects of desipramine might be that when NA uptake is inhibited, a higher fraction of the transmitter released becomes available for the activation of presynaptic α_2-adrenoceptors. Consequently, the effects of α_2-blockade are better expressed against a background of tonic α_2-adrenoceptor stimulation provided by high levels of NA in the synapse. This view is consistent with *in vitro* studies that showed that the facilitatory effect of α_2-adrenoceptor antagonists on ^{3}H-NA release is directly related to the frequency of electrical stimulation and is enhanced in the presence of NA uptake blockers.[2,27,28]

INVOLVEMENT OF PRESYNAPTIC α_2-ADRENOCEPTORS IN THE REGULATION OF NA METABOLISM IN THE RAT BRAIN

Previous studies have shown that α_2-adrenoceptor agonists and antagonists diminish and increase, respectively, the levels of free 3,4-dihydroxyphenylethyleneglycol (DOPEG), a major NA metabolite, in brain areas receiving noradrenergic afferents.[16,17,19] This biochemical index is directly related to impulse flow in noradrenergic neurons[16] and can therefore be used as a reflection of rapid variations in noradrenergic neuronal activity.

To clarify further the role of presynaptic α_2-adrenoceptors in controlling brain noradrenergic nerve transmission, we studied the effects of clonidine or idazoxan on cerebral free DOPEG levels in the anesthetized rat under conditions that eliminated somatodendritic and postsynaptic α_2-adrenoceptor-mediated influences (FIGURE 5).[19] Cortical noradrenergic terminals were isolated from their cell bodies by injection of the sodium channel blocker TTX into the ascending noradrenergic bundle and were subsequently driven artificially by electrical stimulation of the nerve fibers proximally to the site of TTX injection (FIGURE 5). Under these conditions, an effect of systemically administered compounds is unrelated to (1) interaction with somatodendritic receptors; (2) control via polysynaptic feedback pathways projecting to noradrenergic cell bodies. Postsynaptic α_2-adrenoceptor-mediated indirect control of noradrenergic terminals by local neuronal circuits, or by local release of some humoral agent, was then obviated by ibotenic acid lesion of the cell bodies located within the noradrenergic terminal fields under study (FIGURE 5). We reasoned that any effects of α_2-adrenergic agents on free DOPEG levels under such conditions probably reflect an interaction with presynaptic α_2-adrenoceptors.

Initial studies validated the experimental procedure. As shown in TABLE 1, local application of TTX (50 ng) in the ascending noradrenergic bundle markedly decreased free DOPEG levels in the corresponding projection fields at 1 hour postinjection. That

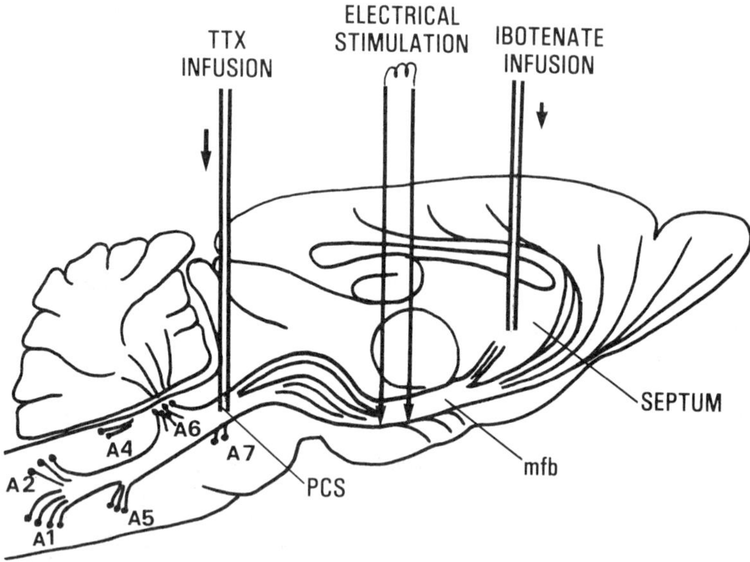

FIGURE 5. Schematic drawing illustrating the experimental model used. Abbreviations: mfb, medial forebrain bundle; PCS, pedunculus cerebellaris superior.

impulse flow in noradrenergic neurons was effectively interrupted in these animals is demonstrated by the failure of electrical stimulation of the locus ceruleus or systemic injection of idazoxan to increase cortical free DOPEG levels.[19] Electrical stimulation of the medial forebrain bundle (400 μA, 2 msecond, 10 minutes) proximal to the TTX injection site, at a frequency of 8 Hz, totally reversed the effect of TTX on free DOPEG levels in the cerebral cortex, hypothalamus, and medial septum, bringing the NE metabolite back to control levels (TABLE 1). These results suggest that electrical stimulation of noradrenergic fibers releases NA from the corresponding nerve terminals.

In naive rats, systemic injection of idazoxan (20 mg/kg ip) increased free DOPEG levels by 49, 66, and 48% in the hypothalamus, cerebral cortex, and medial septal area, respectively (TABLE 1). After infusion of TTX into, and electrical stimulation of, the

TABLE 1. Effect of Systemic Injection of Clonidine or Idazoxan on Free DOPEG Levels in Rat Brain Areas after Combined Local Infusion of TTX into, and Electrical Stimulation of, the Ascending Noradrenergic Bundle[a]

Local Injection	Experimental Procedure	Drug	Dose (mg/kg ip)	Free DOPEG (ng/g)		
				Medial Septum	Hypothalamus	Cerebral Cortex
Vehicle	Sham	Saline	—	21.2 ± 1.1 (100)	26.8 ± 2.6 (100)	9.7 ± 1.1 (100)
Vehicle	Sham	Idazoxan	20	31.4 ± 1.0^c (148)	39.9 ± 2.0^c (149)	16.0 ± 1.0^c (166)
Vehicle	Sham	Clonidine	0.3	15.5 ± 0.6^b (73)	14.5 ± 1.0^c (54)	6.8 ± 0.2^b (70)
TTX	Sham	Saline	—	10.5 ± 0.6^c (50)	15.5 ± 1.9^c (42)	4.7 ± 0.8^b (52)
TTX	Electrical stimulation	Saline	—	22.0 ± 1.6 (103)	25.7 ± 2.2 (96)	9.8 ± 1.7 (101)
TTX	Electrical stimulation	Idazoxan	20	34.1 ± 2.0^{c,e} (155)	38.0 ± 2.0^c (148)	18.3 ± 1.8^{c,d} (186)
TTX	Electrical stimulation	Clonidine	0.3	17.6 ± 1.3^{b,d} (80)	15.4 ± 2.0^{b,d} (60)	7.3 ± 0.4^{c,d} (74)

[a]Rats anesthetized with chloral hydrate (400 mg/kg ip) received a unilateral infusion of TTX (50 ng) laterally to the pedunculus cerebellaris superior. Electrical stimulation of the medial forebrain bundle (400 μA, 2 mseconds, 8 Hz, 10 minutes), performed on the side ipsilateral to TTX injection, was initiated 1 hour after TTX injection. Idazoxan and/or clonidine was injected ip 55 minutes after TTX infusion. Rats were decapitated immediately after the completion of the 10 minute stimulation period (i.e., 15 minutes following idazoxan or clonidine injection). Free DOPEG levels were measured according to the radioenzymatic method of Dennis and Scatton.[37] Results are means with standard errors of the mean of data obtained on 6 to 10 rats per group. Numbers in parentheses, percentage of respective controls.

[b]$p < 0.05$; [c]$p < 0.01$ vs. respective controls.
[d]$p < 0.05$; [e]$p < 0.01$ vs. TTX + electrical stimulation.

noradrenergic bundle, systemic injection of the same dose of the α_2-antagonist provoked comparable increases in the NA metabolite levels in these structures (TABLE 1). Moreover, clonidine (0.3 mg/kg ip) decreased free DOPEG levels in these brain areas to a similar extent in normal rats and in rats subjected to TTX infusion and electrical stimulation (TABLE 1).

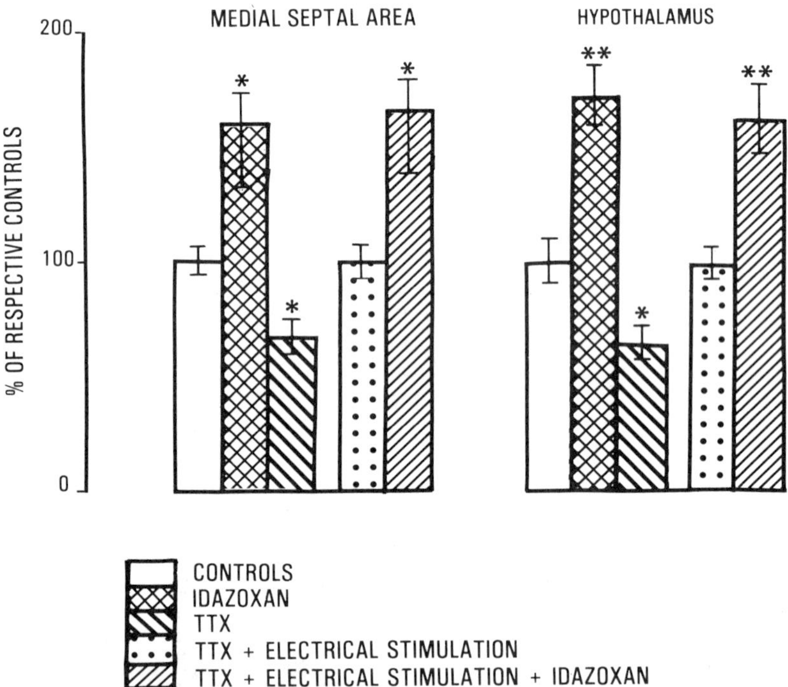

FIGURE 6. Effect of idazoxan on free DOPEG levels in medial septal area and hypothalamus after ibotenate-induced lesion of these areas and local injection of TTX into, and electrical stimulation of, the ascending noradrenergic bundles in the rat. One month after unilateral infusion of ibotenate (10 μg/0.5 μl) into the medial septal area or the hypothalamus, rats were anesthetized with chloral hydrate (400 mg/kg ip) and received a unilateral injection of TTX (50 ng) laterally to the pedunculus cerebellaris superior (on the side ipsilateral to the ibotenate lesion) 15 minutes thereafter. Idazoxan (20 mg/kg ip) was injected 55 minutes after TXX infusion. Five minutes after idazoxan injection, electrical stimulation of the medial forebrain bundle (400 μA, 2 mseconds, 8 Hz, 10 minutes) was performed on the side ipsilateral to TTX infusion. Rats were sacrificed immediately after the end of the stimulation period and free DOPEG levels were measured radioenzymatically according to Dennis and Scatton.[37] Results are means with SEM of data obtained on 6 to 9 animals per group. They are expressed as a percentage of respective control values. *$p < 0.05$; **$p < 0.01$ vs. respective controls.

To eliminate the possible contribution of postsynaptic α_2-adrenoceptors to the effects of α_2-adrenoceptor agents in this situation, the same experiments were repeated after ibotenate-induced lesion of postsynaptic elements in the medial septum and hypothalamus. As shown in FIGURE 6, one month after ibotenate (10 μg) induced lesion of the hypothalamus or medial septum, local injection of TTX (50 ng) into the ascending noradrenergic bundle caused a marked diminution of free DOPEG levels in these areas. Subsequent electrical stimulation (400 μA, 2 mseconds, 8 Hz) of the noradrenergic pathways for 10 minutes restored normal levels of free DOPEG. These biochemical alterations were similar to those observed previously in unlesioned rats (TABLE 1). In ibotenate-lesioned rats that received a local injection of TTX into, and an electrical stimulation of, the ascending noradrenergic pathways, systemic administration of idazoxan (20 mg/kg ip) still increased the levels of free DOPEG in both medial septum and hypothalamus; the magnitude of this effect being comparable to that seen in normal rats (FIGURE 6).

Histological examination revealed a large neuronal loss with glial proliferation throughout the medial septal area and the hypothalamus after ibotenate infusion.[19] The effectiveness of the ibotenate lesion was attested by a 70% decrease in choline acetyltransferase activity in the hippocampus (which receives its cholinergic afferents mainly from the medial septum)[29] one month after the infusion of ibotenic acid into the septal area. The failure of ibotenate lesion to totally reduce the enzyme activity in the hippocampus can be ascribed to the fact that those cholinergic afferents to this structure that are situated more laterally and caudally in the horizontal limb of the diagonal band and nucleus basalis were spared by the lesion.

The present study shows that, when noradrenergic neuron activity is interrupted with TTX, and the activity of the terminals controlled artificially by electrical stimulation of the ascending noradrenergic bundle, idazoxan and clonidine maintain their effects on free DOPEG levels in noradrenergic projection fields. Thus, somatodendritic α_2-adrenoceptors are not essential for this action. Furthermore, postsynaptic actions of the α_2-adrenergic agents probably do not contribute as idazoxan still increases free DOPEG levels in the terminal fields of such artificially driven neurons after destruction of postsynaptic neurons with ibotenic acid. Thus, (1) elimination of postsynaptic α_2-adrenoceptors involved either in the control of the release of a humoral substance or in the initiation of polysynaptic feedback regulation of noradrenergic neurons (ibotenate-induced destruction of postsynaptic elements) and (2) isolation of the terminals from the function of somatodendritic α_2-adrenoceptors and of polysynaptic feedback from noradrenergic nerve terminal to cell body regions (TTX + electrical stimulation + ibotenate lesion) essentially have no effect on the responses produced by idazoxan or clonidine. This suggests that postsynaptic or somatodendritic α_2-adrenoceptors, although constituting a major proportion of the central α_2-receptor population, play a minimal role in the actions of these agents in the control of central noradrenergic neuronal activity. One has to conclude therefore that, as is the case for NA release, presynaptic α_2-adrenoceptors play a predominant role in the regulation of NA metabolism in the CNS *in vivo*. Although these results cannot completely exclude the possibility that α_2-adrenergic agents modify NA metabolism via an action on α_2-heteroreceptors, the overall contribution of this population to the regulation of NA release is likely to be comparatively minor as the affinity of clonidine for the α_2-heteroreceptor seems to be lower than that for the presynaptic α_2-autoreceptor in the rat brain *in vitro*.[30]

CONCLUDING REMARKS

The two different experimental approaches used in the present study clearly demonstrate that although postsynaptic or somatodendritic α_2-adrenoceptors can play a part in regulating noradrenergic activity under control conditions, under conditions in which postsynaptic α_2-adrenoceptors have been destroyed, or the means for somatodendritic α_2-autoceptor-mediated and for polysynaptic feedback control of noradrenergic neuron are removed, drugs affecting α_2-adrenoceptors continue to modify NA release and metabolism, presumably via an interaction with α_2-adrenoceptors located on noradrenergic terminals. This suggests that, as in the periphery, CNS presynaptic α_2-adrenoceptors play an important role in regulating noradrenergic transmission *in vivo*.

Binding studies after noradrenergic denervation[11–13,26] have failed to demonstrate the existence of presynaptic α_2-autoceptors probably because they represent a small fraction of the total α_2-adrenoceptor population. It might therefore appear paradoxical that in spite of their low density presynaptic α_2-autoceptors play such a cardinal role in the regulation of NA release and metabolism *in vivo*. However, the density of a given receptor population is not related necessarily to its functional importance. Although present in a low density, presynaptic α_2-autoceptors may be in a key position to influence noradrenergic activity. A similar situation occurs for the presynaptic dopamine autoreceptors as binding studies have failed to reveal their existence[31,32] whereas functional studies[33–35] fully support their involvement in the control of cerebral dopamine metabolism.

REFERENCES

1. LANGER, S. Z. 1977. Presynaptic receptors and their role in the regulation of transmitter release. Br. J. Pharmacol. **60:** 481–497.
2. LANGER, S. Z. 1981. Presynaptic regulation of the release of catecholamines. Pharmacol. Rev. **32:** 337–362.
3. STARKE, K. 1977. Regulation of noradrenaline release by presynaptic receptor systems. Rev. Physiol. Biochem. Pharmacol. **77:** 1–124.
4. WESTFALL, T. C. 1977. Local regulation of adrenergic neurotransmission. Physiol. Rev. **57:** 659–728.
5. STORY, D. F., M. S. BRILEY & S. Z. LANGER. 1979. The effects of chemical sympathectomy with 6-hydroxydopamine on α-adrenoceptor and muscarinic cholinergic binding in rat heart ventricle. Eur. J. Pharmacol. **57:** 423–426.
6. TAUBE, H. D., K. STARKE & E. BOROWSKI. 1977. Presynaptic receptor systems on the noradrenergic neurones of rat brain. Naunyn Schmiedebergs Arch. Pharmacol. **299:** 123–141.
7. DE LANGEN, C. D. J., F. HOGENBOOM & A. H. MULDER. 1979. Presynaptic noradrenergic α-receptors and modulation of ^{3}H-noradrenaline release from rat brain synaptosomes. Eur. J. Pharmacol. **60:** 79–89.
8. DUBOCOVICH, M. L. 1984. Presynaptic alpha-adrenoceptors in the central nervous system. Ann. N.Y. Acad. Sci. **430:** 7–25.
9. MAURA, G., A. PITTALUGA, A. RICCHETTI & M. RAITERI. 1984. Noradrenaline uptake inhibitors do not reduce the presynaptic action of clonidine on ^{3}H-noradrenaline release in superfused synaptosomes. Naunyn Schmiedebergs Arch. Pharmacol. **327:** 86–89.
10. LEVIN, B. E., 1984. Axonal transport and presynaptic location of α_2-adrenoceptors in locus coeruleus neurons. Brain Res. **321:** 180–182.
11. U'PRICHARD, D. C., T. D. REISINE, S. T. MASON, H. C. FIBIGER & H. I. YAMAMURA. 1980. Modulation of rat brain α- and β-adrenergic receptor populations by lesion of the dorsal noradrenergic bundle. Brain Res. 187: 143–154.
12. PIMOULE, C., B. SCATTON & S. Z. LANGER. 1983. ^{3}H-RX 781094: a new antagonist ligand labels α_2-adrenoceptors in the rat brain cortex. Eur. J. Pharmacol. **95:** 79–85.
13. CURET, O. 1985. Contribution à l'étude du rôle des récepteurs α_2-adrénergiques présynaptiques dans la régulation du métabolisme de la noradrénaline dans le système nerveux central de rat. Ph.D. Thesis. R. Descartes University. Paris, France.
14. ANDEN, N.E., M. GRABOWSKA & U. STROMBOM. 1976. Different alpha-adrenoceptors in the central nervous system mediating biochemical and functional effects of clonidine and receptor blocking agents. Naunyn Schmiedebergs Arch. Pharmacol. **292:** 43–53.
15. DIETL, H., J. N. SINHA & A. PHILIPPU. 1981. Presynaptic regulation of the release of catecholamines in the cat hypothalamus. Brain Res. **208:** 213–218.
16. SCATTON, B. 1982. Brain 3,4-dihydroxyphenylethyleneglycol levels are dependent on central noradrenergic neuron activity. Life Sci. **31:** 495–504.

17. SCATTON, B., J. DEDEK & B. ZIVKOVIC. 1983. Lack of involvement of α_2-adrenoceptors in the regulation of striatal dopaminergic transmission. Eur. J. Pharmacol. **86:** 427–433.

18. L'HEUREUX, R., T. DENNIS, O. CURET & B. SCATTON. 1986. Measurement of endogenous noradrenaline release in the rat cerebral cortex in vivo by transcortical dialysis: effects of drugs affecting noradrenergic transmission. J. Neurochem. **46:** 1794–1801.

19. CURET, O., T. DENNIS & B. SCATTON. 1987. Evidence for the involvement of presynaptic alpha-2 adrenoceptors in the regulation of norepinephrine metabolism in the rat brain. J. Pharmacol. Exp. Ther. **240:** 327–336.

20. DENNIS, T., R. L'HEUREUX, C. CARTER & B. SCATTON. 1987. Presynaptic alpha-2 adrenoceptors play a major role in the effects of idazoxan on cortical noradrenaline release (as measured by in vivo dialysis) in the rat. J. Pharmacol. Exp. Ther. **241:** 642–649.

21. CEDARBAUM, J. M. & G. K. AGHAJANIAN. 1977. Catecholamine receptors on locus coeruleus neurons: pharamacological characterization. Eur. J. Pharmacol. **44:** 375–385.

22. BEANI, L., C. BIANCHI, A. GIACOMELLI & F. TAMBERI. 1978. Noradrenaline inhibition of acetylcholine release from guinea-pig brain. Eur. J. Pharmacol. **48:** 179–183.

23. VIZI, E. S. 1980. Modulation of cortical release of acetylcholine by noradrenaline released from nerves arising from the rat locus coeruleus. Neuroscience **5:** 2139–2144.

24. GOTHERT, M., H. HUTH & E. SCHLICKER. 1981. Characterization of the receptor subtype involved in alpha-adrenoceptor-mediated modulation of serotonin release from rat brain cortex slices. Naunyn Schmiedebergs Arch. Pharmacol. **317:** 199–203.

25. GALZIN, A. M., C. MORET & S. Z. LANGER. 1984. Evidence that exogenous but not endogenous norepinephrine activates the presynaptic alpha-2 adrenoceptors on serotonergic nerve endings in the rat hypothalamus. J. Pharmacol. Exp. Ther. **228:** 725–732.

26. GROSS, G., M. GOTHERT, U. GLAPA, G. ENGEL & H. T. SCHUMANN. 1985. Lesioning of serotoninergic and noradrenergic nerve fibres of the rat brain does not decrease binding of ^{3}H-clonidine and ^{3}H-rauwolscine to cortical membranes. Naunyn Schmiedebergs Arch. Pharmacol. **328:** 229–235.

27. REICHENBACHER, D., W. REIMAN & K. STARKE. 1982. α-Adrenoceptor-mediated inhibition of noradrenaline release in rabbit brain cortex slices. Receptor properties and role of biophase concentration of noradrenaline. Naunyn Schmiedebergs Arch. Pharmacol. **319:** 71–77.

28. CHESSELET, M. F. 1984. Presynaptic regulation of neurotransmitter release in the brain: facts and hypothesis. Neuroscience **12:** 347–375.

29. FIBIGER, H. C. 1982. The organization and some projections of cholinergic neurons of the mammalian forebrain. Brain Res. Rev. **4:** 327–388.

30. MAURA, G., A. GEMIGNANI & M. RAITERI. 1985. α_2-Adrenoceptors in rat hypothalamus and cerebral cortex: functional evidence for pharmacologically distinct subpopulations. Eur. J. Pharmacol. **116:** 335–339.

31. LEFF, S. E. & I. CREESE. 1983. Dopaminergic D-3 binding sites are not presynaptic autoreceptors. Nature **306:** 586–589.

32. HALL, M., P. JENNER, E. KELLY & C. D. MARSDEN. 1983. Differential anatomical location of [^{3}H]-N-n-propylnorapomorphine and [^{3}H]-spiperone binding sites in the striatum and substantia nigra of the rat. Br. J. Pharmacol. **79:** 599–610.

33. CARLSSON, A. 1975. Receptor-mediated control of dopamine metabolism. *In* Pre- and Postsynaptic Receptors. E. Usdin & W. E. Bunney, Jr., Eds.: 49–55. Marcel Dekker. New York, N.Y.

34. NOWYCKY, M. & R. H. ROTH. 1978. Dopaminergic neurons: role of presynaptic receptors in the regulation of transmitter biosynthesis. Prog. Neuro-Psychopharmacol. **2:** 139–158.

35. LEHMANN, J., M. BRILEY & S. Z. LANGER. 1983. Characterization of dopamine autoreceptor and ^{3}H-spiperone binding sites in vitro with classical and novel dopamine receptor agonists. Eur. J. Pharmacol. **88:** 11–26.

36. DA PRADA, M. & G. ZURCHER. 1976. Simultaneous radioenzymatic determination of plasma and tissue adrenaline, noradrenaline and dopamine within the femtomole range. Life Sci. **19:** 1161–1174.

37. DENNIS, T. & B. SCATTON. 1982. A radioenzymatic technique for the measurement of free and conjugated 3, 4-dihydroxyphenylethyleneglycol in brain tissue and biological fluids. J. Neurosci. Methods **6:** 369–382.

Modulation of Noradrenaline Release by Activation of Presynaptic β-Adrenoceptors in the Cardiovascular System[a]

OVE A. NEDERGAARD AND JAN ABRAHAMSEN

Department of Pharmacology
School of Medicine
Odense University
J. B. Winsløws Vej 19
DK-5000 Odense C, Denmark

INTRODUCTION

Depolarization-evoked transmitter release from central and peripheral sympathetic neurons is subject to modulation mediated by presynaptic receptors which probably are located on the varicosity membrane of the nerve terminals (FIGURE 1). Many endogenous substances, including the transmitter itself, and drugs can modulate the exocytotic release by either decreasing or enhancing the amount of released noradrenaline. Several receptors are involved in the inhibition of transmitter release, such as, e.g., α_2-adrenoceptors. Presynaptic receptors that are linked to a facilitation of depolarization-evoked noradrenaline release include β_2-adrenoceptors, nicotine receptors, and angiotensin II receptors. Adler-Graschinsky and Langer first suggested that presynaptic β-adrenoceptors may play an important physiological role.[1] This facilitatory system has been reviewed by several authors: Langer,[2] Starke,[3] Westfall,[4] Gillespie,[5] Vizi,[6] Rand *et al.*,[7] Westfall,[8] Patel *et al.*,[9] Vanhoutte *et al.*,[10] Langer,[11] Dahlöf,[12] Majewski,[13] Göthert,[14] Misu and Kubo,[15] and Borkowski.[16] Kalsner has questioned the concept of the positive feedback system itself,[17] when applied to neurotransmitter secretion, since it is an inappropriate application of cybernetic considerations regarding the operation of closed loops. The present communication will primarily concern itself with the *in vitro* pharmacology of presynaptic β-adrenoceptors in the peripheral cardiovascular system.

BASIC EVIDENCE FOR THE EXISTENCE OF PRESYNAPTIC β-ADRENOCEPTORS

Effect of β-Adrenoceptor Agonists

One main prerequisite for the existence of presynaptic β-adrenoceptors is that β-adrenoceptor agonists must enhance transmitter release from sympathetic nerve terminals evoked by depolarization. The effects of meveral β-adrenoceptor agonists

[a]Work in the authors' laboratory was supported by the Danish Medical Research Council, P. Carl Petersen's Fund, The Research Foundation of the Danish Medical Association, and The Fund for the Advancement of Medical Science.

(adrenaline, dobutamine, dopamine, isoprenaline, prenalterol, procaterol, salbutamol, tazolol, terbutaline, and zinterol) on the electrical stimulation–evoked release of transmitter from noradrenergic neurones *in vitro* are summarized in TABLE 1. The β-adrenoceptor agonists alone enhanced the ^{3}H-overflow from many tissues preloaded with ^{3}H-noradrenaline. The maximal enhancement as a rule was rather small (range: 13–133%) compared to that seen in the presence of α_2-adrenoceptor antagonists (up to four- to fivefold). Cat aorta[18] represents an exceptional case since the enhancement was very large (up to 593%). The enhancements were usually caused by low concentrations (10^{-10}–10^{-7} M) of the β-adrenoceptor agonists. The employed frequency of stimulation was commonly 1–10 Hz. The number of pulses in each train of stimuli was most often 60–600.

Noradrenaline, an agonist for both α- and β-adrenoceptors ($\beta_1 > \beta_2$), decreased the

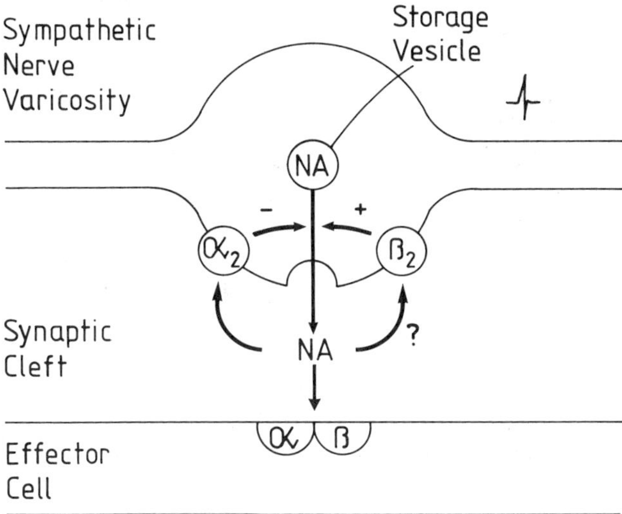

FIGURE 1. Autoregulation of stimulation-evoked noradrenaline release from sympathetic nerve terminals mediated via presynaptic inhibitory α_2-adrenoceptors and presynaptic facilitatory β_2-adrenoceptors. NA, noradrenaline; +, facilitation; −, inhibition.

transmitter release from numerous tissues evoked by electrical stimulation of postganglionic sympathetic nerves.[3,19] Apparently nobody has described enhancement of transmitter release induced by exogenous noradrenaline alone.

Majewski and Rand proposed that in order to observe the effect of activating facilitatory β-adrenoceptors, it might be necessary to block presynaptic inhibitory α-adrenoceptors.[20] The latter are present in all tissues that have been examined so far.[3,19] According to this proposal, blockade of presynaptic α_2-adrenoceptors should unmask the presence of β-adrenoceptors. Thus, only in the presence of the nonselective α_2-adrenoceptor antagonist phentolamine did isoprenaline have a facilitatory effect in the rabbit pulmonary artery[21] and rabbit ear artery.[20] However, using the same experimental conditions, isoprenaline had no effect in either the rabbit pulmonary artery[22] or the ear artery.[23]

TABLE 1. Effect of Some β-Adrenoceptor Agonists on ^{3}H-Noradrenaline Release Evoked by Electrical Stimulation of Sympathetic Nerves *In Vitro*

Species and Tissue	Pretreatment	Agonist	^{3}H-NA Release Maximal Change (%)	Reference
Mouse				
Atrium	None	Isoprenaline	$-13; 7$	57
	Atropine	(+)-Isoprenaline	70^b	21
Rat				
Atrium	Cocaine	(−)-Isoprenaline	81	20
	Cocaine + phentolamine	(−)-Isoprenaline	35^b	
	Cocaine	(−)-Adrenaline	$-55^b; 46^b$	40
	Cocaine + phenoxyben-zamine	(−)-Adrenaline	42^b	
	Atropine	(−)-Isoprenaline	$-39^b; 24$	74
	Atropine + yohimbine	(−)-Isoprenaline	44^b	
Portal vein	None	(−)-Isoprenaline	$—^c$	75
	Cocaine + metanephrine	(−)-Adrenaline	41	25
		(±)-Isoprenaline	41^b	
		Terbutaline	41^b	
		Dobutamine	6	
	Cocaine + metanephrine + phenoxyben-zamine	(−)-Adrenaline	$-23, 22$	
	Desipramine + metanephrine	(±)-Isoprenaline	29	76
Vena cava	Desipramine + corticosterone	(−)-Adrenaline	$-72^b; 13^b$	26
		(−)-Isoprenaline	56^b	
		(±)-Procaterol	51^b	
		(±)-Prenalterol	18^b	
Perfused kidney	Cocaine	Adrenaline	-79^b	27
	Cocaine + phentolamine	Adrenaline	$-21; 15$	
	Cocaine + corticosterone + phentolamine	Adrenaline	$-58; 34$	
	None	Salbutamol	133^b	
		Isoprenaline	94^b	
	None	Salbutamol	28^b	28
Mesenteric artery	None	(−)-Isoprenaline	5	77
	None	Isoprenaline	15^b	78
Renal artery	None	Isoprenaline	4	
Splenic artery	None	Isoprenaline	21^b	
Guinea-pig				
Atrium	Atropine	(±)-Isoprenaline	50^b	1
	Cocaine + normetanephrine	(±)-Isoprenaline	36^b	44

TABLE 1. (Continued)

Species and Tissue	Pretreatment	Agonist	³H-NA Release Maximal Change (%)	Reference
Atrium (cont.)	Cocaine	(−)-Adrenaline	-61^b; 10^b	37
	None	Isoprenaline	56^b	79
Papillary muscle	None	(−)-Isoprenaline	17^b	41
	Phentolamine	(−)-Isoprenaline	48	
		Zinterol	39	
Pulmonary artery	None	(−)-Isoprenaline	39^b	43
	None	(−)-Isoprenaline	56^b	35
		Salbutamol	45	
	None	(−)-Isoprenaline	67	36
		Salbutamol	58	
		Prenalterol	12	
		Tazolol	—	
		(−)-Adrenaline	-22^b	
	Phentolamine	(−)-Adrenaline	49^b	
Rabbit				
Ear artery	Cocaine	(−)-Adrenaline	-49^b	65
		(±)-Isoprenaline	−13; 15	
		Dopamine	-59^b	
	Cocaine + corticosterone	(−)-Adrenaline	-51^b	23
	Cocaine + corticosterone + rauwolscine	(−)-Adrenaline	-51^b	
	Cocaine + corticosterone	(−)-Isoprenaline	−4; 4	
	Cocaine + corticosterone + rauwolscine	(−)-Isoprenaline	−7; 2	
	Cocaine + corticosterone + phentolamine	(−)-Isoprenaline	−11	
	None	(−)-Isoprenaline	9	20
	Phentolamine	(−)-Isoprenaline	28^b	
Aorta	Cocaine + corticosterone	(−)-Adrenaline	-86^b	64
	Cocaine + corticosterone + rauwolscine	(−)-Adrenaline	-63^b; 6	
	Cocaine + corticosterone + phenoxybenzamine	(−)-Adrenaline	−12; 5	
	Cocaine + corticosterone + phentolamine	(−)-Adrenaline	-85^b	

TABLE 1. (Continued)

Species and Tissue	Pretreatment	Agonist	^{3}H-NA Release Maximal Change (%)	Reference
Pulmonary artery	Cocaine + corticosterone	(±)-Isoprenaline	-29^b; 8	80
	Cocaine + corticosterone	(±)-Isoprenaline	-27^b; 9	55
	None	(±)-Isoprenaline	4	21
	ICI 63 197[a]	(±)-Isoprenaline	114^b	
	Phentolamine	(±)-Isoprenaline	28^b	
	None	(±)-Isoprenaline	-13	81
	Phentolamine	(±)-Isoprenaline	26^b	
	Cocaine + corticosterone	(−)-Isoprenaline	-39^b	22
	Cocaine + corticosterone + phentolamine	(−)-Isoprenaline	-9; 3	
	Cocaine + corticosterone + rauwolscine	(−)-Isoprenaline	-21^b; 4	
	ICI 63 197	(−)-Isoprenaline	-25^b; 6	
	Cocaine + corticosterone	(+)-Isoprenaline	23^b	
	Cocaine + corticosterone rauwolscine	(+)-Isoprenaline	10	
Cat				
Aorta	None	Isoprenaline	593	18
Perfused spleen	None	(−)-Isoprenaline	62^b	42
		(+)-Isoprenaline	-33	
	None	(−)-Isoprenaline	51^b	59
	None	(−)-Isoprenaline	63^b	82
Dog				
Saphenous vein	Cocaine + hydrocortisone + U-0521	(−)-Isoprenaline	23^b	60
	None	(±)-Isoprenaline	35^b	61
		Isoprenaline	44^b	62
Man				
Omental blood vessels	Desipramine + normetanephrine	(−)-Isoprenaline	67^b	63
		(−)-Adrenaline	-71; 26^b	
	Desipramine + metanephrine	(−)-Isoprenaline	69^b	29
		Terbutaline	58^b	
		Salbutamol	96^b	
		Tazolol	-17^b	
Digital artery	None	(−)-Adrenaline	35^b	30
		(±)-Dobutamine	88^b	
		(−)-Isoprenaline	71^b	
		(−)-Salbutamol	77^b	

TABLE 1. (Continued)

Species and Tissue	Pretreatment	Agonist	^{3}H-NA Release Maximal Change (%)	Reference
Pulmonary artery	Cocaine + corticosterone	(−)-Isoprenaline	63[b]	31
		(±)-Procaterol	104[b]	
		(±)-Prenalterol	18[b]	
		(−)-Adrenaline	−69[b]	
	Cocaine + corticosterone + rauwolscine	(−)-Adrenaline	71[b]	
Saphenous vein	None	Isoprenaline	129[b]	62
	Desipramine + corticosterone	(−)-Isoprenaline	56[b]	33
		(±)-Procaterol	59[b]	
		(±)-Prenalterol	−8	
	Cocaine + corticosterone	(−)-Adrenaline	−39[b]	
	Cocaine + corticosterone + rauwolscine	(−)-Adrenaline	152[b]	
	Desipramine + corticosterone	(−)-Isoprenaline	38	34
		(±)-Procaterol	95	
Metatarsal vein	None	Salbutamol	60[b]	32

[a]ICI 63 197, 2-amino-6-methyl-5-oxo-4-n-propyl-4,5-dihydro-s-triazolol[1,5-a]pyrimidine; U-0521, 3,4-dihydroxy-2-methyl propiophenone.
[b]Significantly different from control.
[c]—, no quantitative data given.

Effect of β-Adrenoceptor Antagonists

Another prerequisite for the presence of presynaptic β-adrenoceptors is that β-adrenoceptor antagonists must attenuate the facilitatory effect of the β-adrenoceptor agonists. The ability of β-adrenoceptor antagonists to reduce the β-adrenoceptor agonist–induced facilitation of stimulation-evoked ^{3}H-noradrenaline release is summarized in TABLE 2. $β_2$-Adrenoceptor antagonists abolished the enhancement. This was not the case for $β_1$-adrenoceptor antagonists. The block was seen with such low concentrations of $β_2$-adrenoceptor antagonists that it most likely is selective. Nonspecific effects of these antagonists were ruled out.[13]

Subtypes of Presynaptic β-Adrenoceptors

Postsynaptic β-adrenoceptors can be divided into two subtypes ($β_1$- and $β_2$-adrenoceptors).[24] Evidence has accumulated to suggest that presynaptic β-adrenoceptors are of the $β_2$ subtype. This view is supported by the following. Firstly, some relatively selective $β_2$-adrenoceptor agonists (procaterol, salbutamol, and terbutaline) enhanced the stimulation-evoked ^{3}H-noradrenaline release in rat portal vein,[25] rat vena cava,[26] rat perfused kidney,[27,28] human omental vessels,[29] human digital artery,[30] human pulmonary artery,[31] human metatarsal vein,[32] and human saphenous vein[33,34] (TABLE 1). Lack of absolute specificity is indicated by the finding that prenalterol, a

TABLE 2. Effect of β-Adrenoceptor Antagonists on the Enhancing Effect of β-Adrenoceptor Agonists on the Stimulation-Evoked Release of ^{3}H-Noradrenaline *In Vitro*

Species and Tissue	Pretreatment	Agonist	Antagonist[a]	Effect[b]	Reference
Mouse					
Atrium	Atropine	(±)-Isoprenaline	(±)-Propranolol	+	21
Rat					
Atrium	Cocaine	(−)-Isoprenaline	Metoprolol	+	20
	Cocaine + phentolamine	(−)-Isoprenaline	Metoprolol	+	
	Atropine + yohimbine	(−)-Isoprenaline	(−)-Propranolol	+	74
Portal vein	Cocaine + metanephrine	(±)-Isoprenaline	Propranolol	+	25
		Terbutaline	Propranolol	+	
		(±)-Isoprenaline	Butoxamine	+	
		Terbutaline	Butoxamine	+	
		(±)-Isoprenaline	Practolol	0	
		Terbutaline	Practolol	0	
Vena cava	Desipramine + corticosterone	(−)-Isoprenaline	(±)-Propranolol	+	26
		(−)-Isoprenaline	(±)-ICI 118 551	+	
		(−)-Isoprenaline	(±)-Atenolol	+	
Perfused kidney	None	Salbutamol	(−)-Propranolol	+	27
Guinea pig					
Atrium	Cocaine + normetanephrine	(±)-Isoprenaline	(−)-Propranolol	+	44
		(±)-Isoprenaline	(+)-Propranolol	+	
	Cocaine	(−)-Adrenaline	Metoprolol	+	37
	None	Isoprenaline	Propranolol	+	79
Papillary muscle	Phentolamine	(−)-Isoprenaline	ICI 118 551	+	41
		(−)-Isoprenaline	ICI 89 406	0	
Pulmonary artery	None	(−)-Isoprenaline	(−)-Propranolol	+	43
		(−)-Isoprenaline	(+)-Propranolol	0	
	Phentolamine	None	(−)-Propranolol	+	
		None	(+)-Propranolol	0	
	None	(−)-Isoprenaline	(−)-Metoprolol	+	35
		(−)-Isoprenaline	(+)-Metoprolol	0	
		Salbutamol	Practolol	0	
		Salbutamol	Acebutalol	0	
		Salbutamol	Bevantolol	0	
		Salbutamol	Metoprolol	+	
		Salbutamol	Butoxamine	+	
		Salbutamol	IPS 339	+	
		Salbutamol	H 35/25	+	
	None	(−)-Isoprenaline	Atenolol	0	36
		(−)-Isoprenaline	Practolol	0	
		(−)-Isoprenaline	Butoxamine	+	
		(−)-Isoprenaline	H 35/25	+	

TABLE 2. (Continued)

Species and Tissue	Pretreatment	Agonist	Antagonist[a]	Effect[b]	Reference
Rabbit					
Ear artery	Phentolamine	(−)-Isoprenaline	Metoprolol	+	20
Pulmonary artery	ICI 63 197	(±)-Isoprenaline	(±)-Propranolol	+	21
	Phentolamine	(±)-Isoprenaline	(±)-Propranolol	+	
	Cocaine + corticosterone	(+)-Isoprenaline	(±)-Propranolol	+	22
Cat					
Perfused spleen	None	(−)-Isoprenaline	(−)-Propranolol	+	42
	Papaverine	None	(−)-Propranolol	+	
	None	(−)-Isoprenaline	(−)-Propranolol		59
	None	(−)-Isoprenaline	(−)-Propranolol	+	82
Dog					
Saphenous vein	None	(±)-Isoprenaline	(±)-Propranolol	+	61
	None	Isoprenaline	Propranolol	+	62
Human					
Omental vessels	Desipramine + normetanephrine	(−)-Isoprenaline	Propranolol	+	29
		(−)-Isoprenaline	Practolol	0	
		(−)-Isoprenaline	H35/25	+	
Digital arteries	None	(±)-Salbutamol	(±)-Propranolol	+	30
		Adrenaline	(±)-Propranolol	+	
Metatarsal vein	None	Salbutamol	Propranolol	+	32
Pulmonary artery	Cocaine + corticosterone	(±)-Procaterol	(±)-ICI 118 551	+	31
		(±)-Procaterol	(±)-Atenolol	0	
		(−)-Isoprenaline	(±)-Propranolol	+	
		(−)-Isoprenaline	(±)-Atenolol	0	
	None	Isoprenaline	Propranolol	+	62
Saphenous vein	Cocaine + corticosterone	(−)-Isoprenaline	(±)-Propranolol	+	33
	Desipramine + corticosterone	(−)-Isoprenaline	(±)-ICI 118 551	+	
			(±)-Atenolol	0	

[a]ICI 118 551, erythro-*dl*-1-(7-methylindane-4-yloxy)-3-isopropylamino-butan-2-ol hydrochloride; IPS 339, (*t*-butyl-amino-3-ol-2-propyl)oximino-9 fluorene hydrochloride; H 35/25, (±)-erythro-4-methyl-α-(1-isopropylaminoethyl)-benzylalcohol hydrochloride. ICI 89 406, 1-(2-cyanophenoxy)-3β-(3-phenyl-ureido)-ethylamino-2-propanol.

[b]+, block; 0, no effect.

selective β_1-adrenoceptor agonist, caused a slight enhancement of stimulation-evoked ^{3}H-noradrenaline release in rat vena cava[26] and human pulmonary artery.[31] Another β_1-adrenoceptor agonist dobutamine also facilitated the ^{3}H-noradrenaline release in human digital artery.[30] On the other hand, the selective β_1-adrenoceptor agonist tazolol reduced the transmitter release.[29] This was also the case with noradrenaline,[3,19] which has greater affinity for the β_1-adrenoceptors than the β_2 subtype. Secondly, selective β_2-adrenoceptor antagonists (butoxamine, ICI 118 551, IPS 339, H 35/25) reduced the β_2-adrenoceptor agonist–induced enhancement of stimulation-evoked ^{3}H-noradrenaline release in rat atrium,[25] rat vena cava,[26] guinea pig pulmonary artery,[35,36] human omental vessels,[29] and human pulmonary artery.[31] Consistent with the proposal that the presynaptic β-adrenoceptors are of the β_2-subtype, β_1-adrenoceptor antagonists (acebutalol, atenolol, bevantolol, metoprolol, and practolol) did not attenuate the β-adrenoceptor agonist–induced facilitation of transmitter release in rat portal vein,[25] guinea pig pulmonary artery,[35,36] human omental vessels,[29] and human pulmonary artery.[31] Some results, however, suggest that putative presynaptic β_1-adrenoceptors are involved. Thus, the supposedly selective β_1-adrenoceptor antagonist metoprolol blocked the facilitatory effect of β-adrenoceptor agonist-induced enhancement in rat atrium,[20] guinea pig atrium,[37] guinea pig pulmonary artery,[35] and rabbit ear artery.[20] It should be noted, though, that selectivity is a matter of degree and that at least in the guinea pig and in man, metoprolol's selectivity for β_1-adrenoceptors is not marked.[38] In the concentrations that are generally used, metoprolol is, in fact, an effective β_2-adrenoceptor antagonist.[39] This may also explain why metoprolol (and to a lesser degree practolol) also blocked the facilitatory effect of neuronally released adrenaline in guinea pig atrium[37] and rat atrium.[40] The exact characterization of which subtype of presynaptic β-adrenoceptors are involved is difficult to decide with certainty, since classical pharmacological techniques for determining pA_2-values for β-adrenoceptor antagonists have not been carried out except in a few cases.[33,41]

Stereospecificity of Presynaptic β-Adrenoceptors

It would be expected that presynaptic β-adrenoceptors showed the same stereospecificity as that of postsynaptic β-adrenoceptors, i.e., the $(-)$-isomer is more active than the $(+)$-isomer. However, the stereospecificity of presynaptic β-adrenoceptors is somewhat controversial (cf., TABLE 1). Thus, $(-)$-isoprenaline (10^{-10}–10^{-8} M), but not $(+)$-isoprenaline (10^{-8}–10^{-7} M), enhanced the stimulation-evoked ^{3}H-noradrenaline release in the cat perfused spleen.[42] In contrast, $(+)$-isoprenaline (10^{-7}–3×10^{-5} M) caused an enhancement in the rabbit pulmonary artery, while the $(-)$-isomer had no effect.[22] $(-)$-Propranolol was more potent than $(+)$-propranolol as an antagonist against the presynaptic facilitatory effect of β-adrenoceptor agonists in the guinea pig pulmonary artery.[43] On the other hand, both $(-)$- and $(+)$-propranolol blocked the facilitation of ^{3}H-noradrenaline release in guinea pig atria.[44] (TABLE 2). More work is obviously needed to settle the issue of the stereospecificity of presynaptic β-adrenoceptors.

Location of Presynaptic β-Adrenoceptors

Presynaptic β-adrenoceptors are generally assumed to be located on noradrenergic nerve terminals. This view is largely based on the finding that β-adrenoceptor agonists facilitate the stimulation-evoked release of ^{3}H-noradrenaline from axonal sprouts derived from rat cultured superior cervical ganglia which have no postsynaptic structures.[45] In this context, the β-adrenoceptors are most likely localized on the outer surface of the nerve terminals. This is supported by the finding that when neuronal uptake was blocked, β-adrenoceptor agonists, e.g., adrenaline and isoprenaline, enhanced stimulation-evoked ^{3}H-noradrenaline release (TABLE 1). Contrary to earlier conclusions, isoprenaline is taken up by vascular adrenergic nerve terminals to a small degree by a cocaine-sensitive uptake mechanism.[46]

An alternate possibility is that the facilitatory β_2-adrenoceptors are located postsynaptically. Evidence has been obtained to support the hypothesis that β-adrenoceptor activation in the blood vessel wall may lead to local stimulation of angiotensin II synthesis and that this peptide in turn after diffusion across the synaptic cleft activates presynaptic angiotensin II receptors and thereby facilitates the release of noradrenaline, i.e., transsynaptic modulation.[26,34,47,48] This β-adrenoceptor/angiotensin hypothesis appears limited in its application, since several studies have now yielded negative results.[49–52]

Noradrenaline Activation of Presynaptic β-Adrenoceptors

The suggestion that noradrenaline released from sympathetic nerve terminals activates presynaptic β-adrenoceptors and thereby initiates a positive feedback system causing further release of transmitter (FIGURE 1), is controversial and is not supported by convincing evidence.

One of the prerequisites for this hypothesis is that noradrenaline should enhance stimulation-evoked transmitter release. However, this was not the case whenever tissues from normotensive animals were used (see above). Dahlöf et al. reported that high concentrations (10^{-5} and 5×10^{-5} M) of noradrenaline in the presence of phenoxybenzamine caused a small enhancement (maximally 41%) of ^{3}H-noradrenaline release from rat portal vein derived from spontaneously hypertensive rats.[53,54] This facilitation may possibly be a peculiarity of tissue obtained from hypertensive rats. Thus, when portal vein was obtained from normotensive rats, noradrenaline in the presence of phenoxybenzamine decreased transmitter release.[25]

If neurogenic noradrenaline plays a role in the positive feedback loop, then β-adrenoceptor antagonists should decrease the stimulation-evoked ^{3}H-noradrenaline release. The effects of β-adrenoceptor antagonists on stimulation evoked ^{3}H-noradrenaline release from isolated cardiovascular tissues differ (TABLE 3). This was the case even when low concentrations ($<10^{-6}$ M) of the antagonists were used. The discordant results may possibly be explained by differences in the experimental protocol (number of pulses, stimulation frequency, concentration of antagonists, presence of uptake inhibitors). High concentrations of β-adrenoceptor antagonists (propranolol and oxprenolol) surprisingly caused an enhancement of stimulation-evoked ^{3}H-noradrenaline release in rat perfused kidney,[27] guinea pig atrium,[37] and rabbit pulmonary artery.[55] The enhancement may possibly be due to inhibition of the neuronal uptake mechanism (uptake-1) independently of the β-adrenoceptor blocking action of these antagonists.[56] In other experiments, propranolol in high concentrations caused an inhibition of transmitter release in mouse atrium,[57] rat vena cava,[26] guinea pig atrium,[1,58] guinea pig pulmonary artery,[43] rabbit pulmonary artery,[22] cat perfused spleen,[59] and human omental blood vessels.[29] This is probably due to either adrenergic neurone blocking activity or local anesthesia.

Even in tissues where presynaptic β-adrenoceptors have been demonstrated, β-adrenoceptor antagonists per se did not reduce the stimulation-evoked release of ^{3}H-noradrenaline from sympathetic nerves. This was the case in rat perfused kidney,[27] rat atrium,[20,40] dog saphenous vein,[60–62] human digital arteries,[30] human metatarsal vein,[32] human pulmonary artery,[31] and human saphenous vein.[33] This strongly indicates that noradrenaline released from sympathetic nerves does not play a role in the proposed positive feedback loop.

Activation of Presynaptic β-Adrenoceptors by Adrenaline

Stjärne and Brundin were the first to suggest that adrenaline may be the physiological activator of presynaptic β-adrenoceptors[63] (FIGURE 2). Adrenaline is a more potent β-adrenoceptor agonist than noradrenaline.[24] Exogenously applied adrenaline caused a

TABLE 3. Effect of β-Adrenoceptor Antagonists on ^{3}H-Noradrenaline Release Evoked by Electrical Stimulation of Noradrenergic Nerves *In Vitro*

Species and Tissue	Pretreatment	Antagonist	^{3}H-NA Release Maximal Change (%)	Reference
Mouse				
Atrium	None	Propranolol	−27[a]	57
	Atropine	(±)-Propranolol	11	21
Rat				
Atrium	Cocaine	Metoprolol	16	20
	Cocaine + phentolamine	Metoprolol	3	
	Cocaine	Metoprolol	−14	40
	Cocaine + phenoxybenzamine	Metoprolol	−13	
	Atropine + yohimbine	(−)-Propanolol	—[b]	74
Vena cava	Desipramine + corticosterone	(±)-Propranolol	−10[a]	26
Perfused kidney	None	(−)-Propranolol	35[a]	27
Portal vein	Cocaine + metanephrine	Propranolol	−6	25
		Practolol	−11	
		Butoxamine	+6	
Guinea-pig				
Atrium	Atropine	(±)-Propranolol	−43[a]	1
	Cocaine + normetanephrine	(−)-Propranolol	—	44
			—	
		(+)-Propranolol	—	
	Cocaine	Metoprolol	26	37
		Practolol	12	
		Propranolol	−5; 22	
		Oxprenolol	40[a]	
	Cocaine + phentolamine	Oxprenolol	−11; 7	
	None	Metoprolol	17	79
		Propranolol	−16	
	None	(−)-Propranolol	−32[a]	58
	Desipramine	(−)-Propranolol	102[a]	
Pulmonary artery	None	(−)-Propranolol	−7	43
		(+)-Propranolol	3	
	Phentolamine	(−)-Propranolol	−34[a]	
		(+)-Propranolol	−12	
	None	(±)-Carteolol	−15[a]	83
		(±)-Carteolol	−3; 3	
Papillary muscle	None	ICI 118 551	−6	41
		ICI 89 406	−2	
		Pindolol	−7	
		Celiprolol	−8	

TABLE 3. (Continued)

Species and Tissue	Pretreatment	Antagonist	^{3}H-NA Release Maximal Change (%)	Reference
Rabbit				
Aorta	Cocaine + corticosterone	($\pm$)-Propranolol	9	64
	Cocaine + corticosterone	Metroprolol	1	84
	Cocaine + corticosterone	Atenolol	-8	
Pulmonary artery	None	($\pm$)-Propranolol	$-5; 8$	80
	None	($\pm$)-Propranolol	-15	21
	ICI 63 197	($\pm$)-Propranolol	33	
	Cocaine + corticosterone	($\pm$)-Propranolol	-53^a	22
	Cocaine + corticosterone	($\pm$)-Propranolol	$-8; 39^a$	55
Cat				
Perfused spleen	None	($-$)-Propranolol	-17^a	42
	None	($-$)-Propranolol	-17^a	59
	None	($-$)-Propranolol	-8	82
Dog				
Saphenous vein	Cocaine + hydrocortisone + U-0521	($\pm$)-Propranolol	-6	60
	None	($\pm$)-Propranolol	—	61
	None	Propranolol	—	62
Human				
Omental vessels	Desipramine + normetanephrine	Propranolol	—	63
	Desipramine + normetanephrine	Propranolol	-9^a	29
Digital artery	None	($\pm$)-Propranolol	12	30
	Phentolamine	($\pm$)-Propranolol	-7	
Metatarsal vein	None	($\pm$)-Propranolol	-10	32
Pulmonary artery	Cocaine + corticosterone	($\pm$)-Propranolol	9	31
Saphenous vein	None	($\pm$)-Propranolol	—	33
		($\pm$)-ICI 118 551	—	
		($\pm$)-Atenolol	—	

^{a}Significantly different from control.

^{b}—, no quantitative information given.

concentration-dependent dual effect on stimulation-evoked ^{3}H-noradrenaline release in most tissues (TABLE 1). Thus, low concentrations (10^{-10}–10^{-8} M) of adrenaline facilitate the stimulation-evoked transmitter release in rat atrium,[39] rat vena cava,[26] guinea pig atrium,[37] human omental blood vessels,[63] and human digital arteries.[30] Higher concentrations ($>10^{-8}$ M) of adrenaline produced an attenuation of the release in most of these tissues. Adrenaline solely reduced the stimulation-evoked release in rat perfused kidney,[27] guinea pig pulmonary artery,[36] rabbit aorta,[64] rabbit ear artery,[23,65]

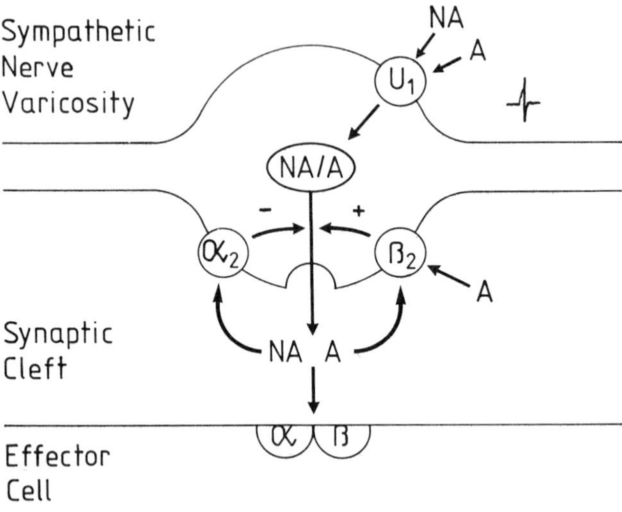

FIGURE 2. Autoregulation of stimulation-evoked noradrenaline release from sympathetic nerves via presynaptic α_2- and β_2-adrenoceptors. Adrenaline is derived originally from the adrenal medulla; this circulating hormone is taken up via the neuronal transport mechanism (uptake-1) and is then released by electrical stimulation from the varicosity as a cotransmitter with noradrenaline. NA, noradrenaline; A, adrenaline; U_1, uptake-1.

and human pulmonary artery.[31] The reduction is most likely due to activation of presynaptic inhibitory α_2-adrenoceptors.[64]

The proposed inhibitory link between presynaptic β-adrenoceptors and presynaptic α-adrenoceptors may explain why in some tissues facilitatory β-adrenoceptors could not be demonstrated.[20] Thus, only in the presence of phentolamine and rauwolscine, a selective α_2-adrenoceptor antagonist, did adrenaline have a facilitatory effect in the guinea pig artery[36] and human pulmonary artery,[31] respectively. In contrast, adrenaline did not enhance stimulation-evoked ³H-noradrenaline release from rabbit aorta either in the presence of rauwolscine, phentolamine, and phenoxybenzamine[64] or rabbit ear artery in the presence of rauwolscine and phentolamine.[23]

The plasma concentration of adrenaline derived from the adenal medulla may be sufficiently high to activate presynaptic β-adrenoceptors. A low concentration (10^{-10} M) of adrenaline was sufficient to activate depolarization-evoked release of dopamine-β-hydroxylase in rat perfused spleen.[66] Stress-induced increase in plasma adrenaline concentration was high enough for activation of presynaptic β-adrenoceptors in rats[67,68] and in man.[69] The direct activation of presynaptic β-adrenoceptors by elevated concentrations of circulatory adrenaline from the adrenal medulla is probably limited by a rather short plasma half-life (about one minute).[40]

The second alternative for the supply of adrenaline that may activate presynaptic β-adrenoceptors is adrenaline that has been taken up from the circulation by noradrenergic neurones and subsequently released (FIGURE 2). Adrenaline can be taken up by cardiovascular sympathetic neurones[70,71] from where this amine can be released as a cotransmitter with noradrenaline.[30,37,40,60,72,73] The neurogenic cotransmitter adrenaline has been shown to activate presynaptic β-adrenoceptors in dog saphenous vein,[60] guinea pig atria,[37] and rat atria.[40] This, however, was not the case with rabbit aorta,[73] human digital arteries,[30] and human saphenous vein.[33] These conclusions were based mainly on the findings that β-adrenoceptor antagonists either reduced or did not alter the stimulation-evoked ^{3}H-overflow from tissues incubated with adrenaline plus ^{3}H-noradrenaline (TABLE 4).

SUMMARY

Peripheral sympathetic nerve terminals in many tissues, but not all, are endowed with β-adrenoceptors. Activation of these result in an enhancement of noradrenaline release evoked by electrical nerve stimulation. These so-called presynaptic β-adrenoceptors are possibly located on the outer surface of the varicosity of the noradrenergic nerves. A postsynaptic location, however, is also a possibility. The presynaptic β-adrenoceptors appear to be of the β_2-adrenoceptor subtype. However, specific classification is lacking. The stereospecificity of the β-adrenoceptors is controversial. These receptors are not activated by noradrenaline released from sympathetic nerves. Adrenaline derived from the adrenal medulla may be the physiological activator. Either circulating adrenaline or adrenaline taken up by sympathetic nerve terminals and then

TABLE 4. Effect of β-Adrenoceptor Antagonists on Stimulation-Evoked ^{3}H-Noradrenaline Release *In Vitro:* Influence of Adrenaline Released as a Cotransmitter

Species and Tissue	Pretreatment	Antagonist	^{3}H-NA Release Maximal Change (%)	Reference
Rat				
Atrium	None	Metoprolol	-28^a	40
Guinea pig				
Atrium	None	Atenolol	-12	39
		Pindolol	17^a	
		Propranolol	-50^a	37
		Metoprolol	-50^a	
		Oxprenolol	-30^a	
	Phentolamine	None	314^a	
		Oxprenolol	-46^a	
Rabbit				
Aorta	Cocaine + corticosterone	(±)-Propranolol	-8	73
	Cocaine +	None	371^a	
	corticosterone +	(±)-Propranolol	-2	
	rauwolscine	(±)-Metoprolol	-9	
Human				
Digital artery	None	(±)-Propranolol	-2	30
		(±)-Metoprolol	-10	
Saphenous vein	Corticosterone	(±)-Propranolol	4	33

^{a}Significantly different from control.

released as a cotransmitter with noradrenaline activates the presynaptic β-adrenoceptors. In the latter case, a "positive" feedback" loop may be formed.

REFERENCES

1. ADLER-GRASCHINSKY, E. & S. Z. LANGER. 1975. Br. J. Pharmacol. **53:** 43–50.
2. LANGER, S. Z. 1977. Br. J. Pharmacol. **60:** 481–497.
3. STARKE, K. 1977. Rev. Physiol. Biochem. Pharmacol. **77:** 1–124.
4. WESTFALL, T. C. 1977. Physiol. Rev. **57:** 659–728.
5. VIZI, E. S. 1979. Prog. Neurobiol. **12:** 181–290.
6. GILLESPIE, J. S. 1980. *In* Adrenergic Activators and Inhibitors: Handbook of Experimental Pharmacology. L. Szekeres, Ed. **54**(part 1): 353–425. Springer Verlag. Berlin & Heidelberg.
7. RAND, M. J., M. W. McCULLOCH & D. F. STORY. 1980. *In* Adrenergic Activators and Inhibitors: Handbook of Experimental Pharmacology. L. Szekeres, Ed. **54**(part 1): 223–266. Springer Verlag. Berlin & Heidelberg.
8. WESTFALL, T. C. 1980. Annu. Rev. Physiol. **42:** 383–397.
9. PATEL, S., U. PATEL, D. VITHALANI & S. C. VERMA. 1981. Gen. Pharmacol. **12:** 405–422.
10. VANHOUTTE, P. M., T. J. VERBEUREN & R. C. WEBB. 1981. Physiol. Rev. **61:** 151–247.
11. LANGER, S. Z. 1981. Pharmacol Rev. **81:** 337–362.
12. DAHLÖF, C. 1981. Acta Physiol. Scand. Suppl. **500:** 1–147.
13. MAJEWSKI, H. 1983. J. Auton. Pharmacol. **3:** 47–60 (Corrigenda: 155).
14. GÖTHERT, M. 1985. Drug Res. **35:** 1909–1916.
15. MISU, Y & T. KUBO. 1986. Med. Res. Rev. **6:** 197–225.
16. BORKOWSKI, K. R. 1988. J. Auton. Pharmacol. **8:** 153–171.
17. KALSNER, S. 1982. Can. J. Physiol. Pharmacol. **60:** 737–743.
18. LANGER, S. Z., M. A. ENERO, E. ADLER-GRASCHINSKY, M. L. DUBOCOVITCH & S. M. CELUCH. 1975. *In* Central Action of Drugs in Blood Pressure Regulation. D. S. Davies & J. L. Reid, Eds.: 133–150. Pitman Medical Publishing. Kent, England.
19. STARKE, K. 1987. Rev. Physiol. Biochem. Pharmacol. **107:** 73–146.
20. MAJEWSKI, H. & M. J. RAND. 1981. Eur. J. Pharmacol. **69:** 493–498.
21. JOHNSTON, H. & H. MAJEWSKI. 1986. Br. J. Pharmacol. **87:** 553–562.
22. NEDERGAARD, O. A. 1987. Naunyn Schmiedebergs Arch. Pharmacol. **336:** 176–182.
23. ABRAHAMSEN, J. & O. A. NEDERGAARD. 1988. Br. J. Pharmacol. **94:** 328P.
24. LANDS, A. M., A. ARNOLD, J. P. McAULIFF, F. P. LUDUENA & T. G. BROWN, JR. 1967. Nature **214:** 597–598.
25. WESTFALL, T. C., M. J. PEACH & V. TITTERMARY. 1979. Eur. J. Pharmacol. **58:** 67–74.
26. GÖTHERT, M. & P. KOLLECKER. 1986. Naunyn Schmiedebergs Arch. Pharmacol. **334:** 156–165.
27. STEENBERG, M. L., R. D. EKAS & M. F. LOKHANDWALA. 1983. Eur. J. Pharmacol. **93:** 137–148.
28. EKAS, R. D., JR., M. L. STEENBERG, M. S. WOODS & M. F. LOKHANDWALA. 1983. Hypertension **5:** 198–204.
29. STJÄRNE, L. & J. BRUNDIN. 1976. Acta Physiol. Scand. **97:** 88–93.
30. STEVENS, M. J., R. E. RITTINGHAUSEN, R. L. MEDCALF & R. F. W. MOULDS. 1982. Eur. J. Pharmacol. **83:** 263–270.
31. GÖTHERT, M. & F. HENTRICH. 1985. Br. J. Pharmacol. **85:** 933–941.
32. MOULDS, R. F. W. & M. J. STEVENS. 1983. Gen. Pharmacol. **14:** 81–83.
33. MOLDERINGS, G. J., J. LIKUNGU, H. R. ZERKOWSKI & M. GÖTHERT. 1988. Naunyn Schmiedebergs Arch. Pharmacol. **337:** 408–414.
34. MOLDERINGS, G. J., J. LIKUNGU, F. HENTRICH & M. GÖTHERT. 1988. Naunyn Schmiedebergs Arch. Pharmacol. **338:** 228–233.
35. MISU, Y., KUWAHARA, M. KAIHO & T. KUBO. 1983. Eur. J. Pharmacol. **91:** 287–290.
36. MISU, Y., M. KAIHO, G. YASUDA, M. KUWAHARA & T. KUBO. 1984. Jpn. J. Pharmacol. **36:** 329–337.

37. MAJEWSKI, H., M. W. MCCULLOCH, M. J. RAND & D. F. STORY. 1980. Br. J. Pharmacol. **71:** 435–444.
38. HARMS, H. H. 1976. *In* Beta-Adrenoceptor Blocking Agents. P. R. Saxena & R. P. Forsyth, Eds.: 311–315. North-Holland Publishing Company. Amsterdam, the Netherlands.
39. RAND, M. J., H. MAJEWSKI, M. W. MCCULLOCH & D. F. STORY. 1979. Adv. Biosci. **18:** 263–269.
40. MAJEWSKI, H., M. J. RAND & L.-H. TUNG. 1981. Br. J. Pharmacol. **73:** 669–679.
41. VALENTA, B., H. PITTNER & E. A. SINGER. 1989. J. Cardiovasc. Pharmacol. **14:** 846–850.
42. CELUCH, S. M., M. L. DUBOCOVITCH & S. Z. LANGER. 1978. Br. J. Pharmacol. **63:** 97–109.
43. MISU, Y., K. KAIHO, K. OGAWA & T. KUBO. 1981. J. Pharmacol. Exp. Ther. **218:** 242–247.
44. KALSNER, S. 1980. Br. J. Pharmacol. **70:** 491–498.
45. WEINSTOCK, M., N. B. THOA & I. J. KOPIN. 1978. Eur. J. Pharmacol. **47:** 297–302.
46. OSSWALD, W. 1986. *In* Central and Peripheral Mechanisms of Cardiovascular Regulation. A. Magro, W. Osswald, D. Reis & P. Vanhoutte, Eds.: 1–31. Plenum Press. New York & London.
47. KAWASAKI, H., W. H. CLINE & C. SU. 1984. J. Pharmacol. Exp. Ther. **231:** 23–32.
48. NAKAMURA, M., E. K. JACKSON & T. INAGAMI. 1986. Am. J. Physiol. **250:** H144–H148.
49. RUMP, L. C. & H. MAJEWSKI. 1987. J. Pharmacol. Exp. Ther. **243:** 1107–1112.
50. LI, C., H. MAJEWSKI & M. J. RAND. 1988. Br. J. Pharmacol. **95:** 385–392.
51. ZIOGAS, J. & D. F. STORY. 1987. *In* Proceedings of the 10th International Congress of Pharmacology P1395. IUPHAR. Sydney, Australia.
52. RAJANAYAGAM, M. A. S., I. F. MUSGRAVE, M. J. RAND & H. MAJEWSKI. 1989. Arch. Int. Pharmacodyn. **299:** 185–199.
53. DAHLÖF, C., B. LJUNG & B. ÅBLAD. 1978. Eur. J. Pharmacol. **50:** 75–78.
54. DAHLÖF, C., B. LJUNG & B. ÅBLAD. 1978. *In* Recent Advances in the Pharmacology of Adrenoceptors. E. Szabadi, C. M. Bradshaw & P. Bevan, Eds.: 355–356. Elsevier/North-Holland Biomedical Press. Amsterdam, the Netherlands.
55. STARKE, K., T. ENDO H. D. TAUBE & E. BOROWSKI. 1975. *In* Chemical Tools in Catecholamine Research. O. Almgren, A. Carlsson & J. Engel, Eds. **2:** 193–200. North-Holland Publishing Company. Amsterdam & Oxford.
56. WERNER, U., J. WAGNER & H. J. SCHÜMANN. 1971. Naunyn Schmiedebergs Arch. Pharmacol. **268:** 102–113.
57. FARNEBO, L. O. & B. HAMBERGER. 1974. J. Pharm. Pharmacol. **26:** 644–646.
58. ADLER-GRASCHINSKY, E. & A. H. CARRARA. 1982. J. Cardiovasc. Pharmacol. **4:** 942–948.
59. LANGER S. Z. & M. L. DUBOCOVITCH. 1978. *In* Recent Advances in the Pharmacology of Adrenoceptors. E. Szabadi, C. M. Bradshaw & P. Bevan, Eds.: 181–189. Elsevier/North-Holland Biomedical Press. Amsterdam, the Netherlands.
60. GUIMARÃES, S., F. BRANDÃO & M. Q. PAIVA. 1978. Naunyn Schmiedebergs Arch. Pharmacol. **305:** 185–188.
61. SAELENS, D. A. & P. B. WILLIAMS. 1983. J. Cardiovasc. Pharmacol. **5:** 598–603.
62. VERBEUREN, T. J., R. R. LORENZ, L. L. AARHUS, J. T. SHEPHERD & P. M. VANHOUTTE. 1983. J. Auton. Nerv. Syst. **8:** 261–271.
63. STJÄRNE, L. & J. BRUNDIN. 1975. Acta Physiol. Scand. **94:** 139–141.
64. ABRAHAMSEN, J. & O. A. NEDERGAARD. 1989. Naunyn Schmiedebergs Arch. Pharmacol. **339:** 281–287.
65. HOPE, W., M. LAW, M. W. MCCULLOCH, M. J. RAND & D. F. STORY. 1976. Clin. Exp. Pharmacol. Physiol. **3:** 15–28.
66. DIXON, W. R., W. F. MOSIMANN & N. WEINER. 1979. J. Pharmacol. Exp. Ther. **209:** 196–204.
67. POPPER, C. W., C. C. CHIUEH & I. J. KOPIN. 1977. J. Pharmacol. Exp. Ther. **202:** 144–148.
68. KVETNANSKY, R., V. K. WEISE, N. B. THOA & I. J. KOPIN. 1979. J. Pharmacol. Exp. Ther. **209:** 287–291.
69. SCHALEKAMP, M. A. D. H., H. H. VINCENT & A. J. MAN IN'T VELD. 1983. Lancet i: 362.
70. ANDÉN, N. E. 1964. Acta Pharmacol. Toxicol. **21:** 59–75.
71. ABRAHAMSEN, J. & O. A. NEDERGAARD. 1985. Blood Vessels **22:** 32–46.

72. ABRAHAMSEN, J. & O. A. NEDERGAARD. 1986. Acta Pharmacol. Toxicol. **59:** 416–424.
73. ABRAHAMSEN, J. & O. A. NEDERGAARD. 1989. J. Auton. Pharmacol. **9:** 337–346.
74. KAZANIETZ, M. G. & M. A. ENERO. 1989. Naunyn Schmiedebergs Arch. Pharmacol. **340:** 274–278.
75. ENERO, M. A. 1979. Adv. Biosci. **18:** 321–325.
76. WESTFALL, T. C., M. J. MELDRUM, L. BADINO & J. T. EARNHARDT. 1984. Hypertension **6:** 267–274.
77. KAWASAKI, H., H. CLINE, JR. & C. SU. 1982. J. Pharmacol. Exp. Ther. **223:** 721–728.
78. KUBO, T., M. KUWAHARA & Y. MISU. 1984. Jpn. J. Pharmacol. **36:** 419–421.
79. RAND, M. J., M. LAW, D. F. STORY & M. W. MCCULLOCH. 1976. Drugs **11**(Suppl. 1): 134–143.
80. ENDO, T., K. STARKE, A. BANGERTER & H. D. TAUBE. 1977. Naunyn Schmiedebergs Arch. Pharmacol. **296:** 229–247.
81. COSTA, M. & H. MAJEWSKI. 1988. Br. J. Pharmacol. **95:** 993–1001.
82. LANGER, S. Z., M. L. DUBOCOVITCH & S. M. CELUCH. 1975. *In* Chemical Tools in Catecholamine Research. O. Almgren, A. Carlsson & J. Engel, Eds. **2:** 183–191. Elsevier/North-Holland Biomedical Press. Amsterdam, the Netherlands.
83. KUWAHARA, M., H. AMAND, T. KUBO & Y. MISU. 1986. Arch. Int. Pharmacodyn. Ther. **284:** 225–230.
84. ABRAHAMSEN, J. & O. A. NEDERGAARD. Unpublished.

Monitoring Acetylcholine Release from the Neuromuscular Junction

M. ALEXANDER, S. A. SLOMOWITZ, AND
R. J. STORELLA

Department of Anesthesiology
Hahnemann University
Philadelphia, Pennsylvania 19102

Train-of-four (TOF) fade and tetanic fade represent decreasing acetylcholine (ACh) release mediated by prejunctional ACh receptors. In contrast to results with nicotinic ACh receptor antagonists less selective for the neuromuscular junction, α-bungarotoxin (αBT) does not produce fade. This suggests that αBT does not alter ACh release and therefore that the receptors affecting ACh release are not identical to muscle ACh receptors.

We examined whether postjunctional monitoring of ACh release is less sensitive to changes in ACh in the presence of αBT than in the presence of d-tubocurarine (dTC) by blocking transmission with either dTC or αBT and then altering junctional ACh levels with different agents. ACh release was increased by posttetanic potentiation. The duration of ACh in the junction was increased by inhibiting acetylcholinesterase with edrophonium. ACh release was decreased by train-of-four stimulation and by adding Mg^{++}.

The mouse phrenic nerve diaphragm preparation was maintained in a modified Krebs solution. The nerve was stimulated supramaximally (0.2 msecond). Isometric

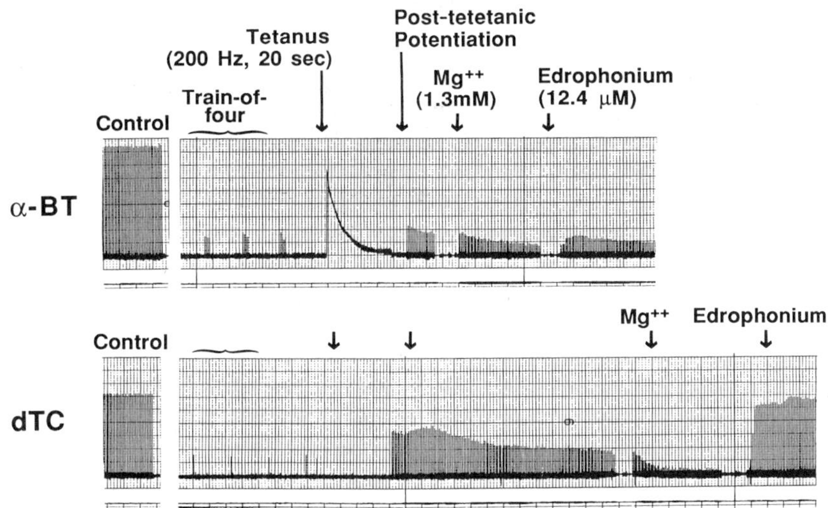

FIGURE 1. Representative experiments with αBT and dTC. Note, the paper speed is not constant, but varied with the agents.

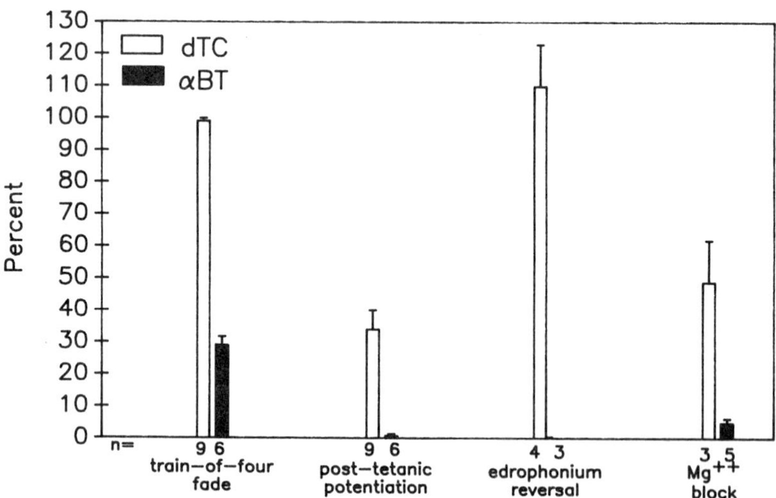

FIGURE 2. The postjunctional response to 4 agents in the presence of dTC or αBT block. The number of preparations tested is listed below each bar (mean + standard error of the mean). TOF fade was calculated as precent decrease in tension of the last pulse compared to the first. Posttetanic potentiation was calculated as peak percent reversal of block. Following edrophonium, percent reversal of block was measured. Percent block due to Mg^+ was measured after 90 seconds. In all cases, the response in the presence of αBT is significantly less than with dTC ($p < 0.05$, t-tests).

tension was recorded. Resting tension adjusted to yield maximum contractile tension at 0.2 Hz.

Preparations were stimulated at 0.2 Hz and 70–80% blocked with either dTC or αBT. All preparations were tested with train-of-four stimulation (4 pulses at 2 Hz) followed by a tetanizing stimulation of 200 Hz for 20 seconds. Immediately after the tetanus, stimulation was returned to 0.2 Hz and posttetanic potentiation was monitored. After block returned to 70–80%, either edrophonium (12.4 μM) or Mg^{++} (0.3 mM $MgCl_2$) was added. FIGURE 1 represents typical experiments except that both Mg^{++} and edrophonium were added.

Following either tetanus or edrophonium, dTC block was reversed. Also train-of-four fade was considerable and $MgCl_2$ (0.3 mM) increased block in the presence of dTC. In contrast, these same agents had minimal effects on αBT block. The response to all agents was significantly less in the αBT group (FIGURE 2). There was no significant differences between the αBT and dTC preparations in the percent block of twitch (0.2 Hz) preceding each test.

We have demonstrated that alterations of ACh release and duration are not as effective in altering αBT block as dTC block. Although the alterations in ACh levels induced by these agents may differ in the presence of dTC and αBT, the variety of mechanisms involved suggests that it is the responsiveness of the end plate to changes in ACh levels that is altered. This may be due to the slower dissociation of αBT from the ACh receptor.

The results suggest that the absence of fade during αBT neuromuscular block cannot be taken as the absence of a toxin-induced decreases in ACh release. It may be that the decrease in ACh release during train-of-four stimulation is not large enough to be reflected by postjunctional (e.g., end plate or force) monitoring in the presence of αBT.

The present investigation supports a simple generalization: postjunctional monitoring of transmitter release should be considered a bioassay and the responsiveness of the system to transmitter should be determined under appropriate conditions.

Evidence that Postsynaptic Effects of *d*-Tubocurarine or α-Toxin Cause Fade at the Neuromuscular Junction[a]

RONALD J. BRADLEY, RAIMUND STERZ, KLAUS PEPER,
WAI-CHUNG CHAU, AND GUOZHU ZHANG

Department of Psychiatry
School of Medicine
University of Alabama at Birmingham
UAB Station
Birmingham, Alabama 35294

Repetitive nerve stimulation does not normally invoke fade or run-down in the amplitude of end plate currents (EPCs) at the mammalian neuromuscular junction. In the past it has been reported that *d*-tubocurarine (d-TC) causes such fade but α-toxins from snake venom reduced the response evenly and without fade.[1] It was assumed that α-toxins do not cause fade because they do not bind to the pre-synaptic terminal[2] and therefore do not interact with presynaptic receptors (AChRs). As d-TC does cause fade it was assumed that this was due to d-TC acting at presynaptic AChRs of the neuronal type which by definition do not bind α-toxins. These AChRs are assumed to increase acetylcholine (ACh) release by a positive feedback mechanism during repetitive stimulation.[1] Others have claimed that fade is not voltage dependent so cannot be due to drug blocking of the open channel of the postsynaptic AChR.[3,4] However, Bradley *et al.* found that low concentrations of α-toxin cause severe fade,[5] which indicates that fade can be caused by high-affinity binding to the postsynaptic AChR. These postsynaptic effects can be probed by using ionophoresis of ACh which bypasses the presynaptic release mechanism.[4] FIGURE 1 shows the fade produced by 1 μM d-TC on the muscle compound action potential and tetanus after repetitive nerve stimulation. The effects of the same concentration of d-TC are shown in FIGURE 2 when repetitive pulses of ACh were released by ionophoresis. When high frequency pulses were delivered through the ionophoresis pipette, the ACh concentration at the AChR is very low because the pulses by definition must be very short. However, if the frequency is low, the pulses can be very long and the ACh concentration is therefore very high and close to saturation. A significant fade response was found only when very high ACh concentrations close to saturation were released at a low frequency but fade was not observed when low ACh concentrations were released at a high frequency (FIGURE 2). During such low ACh concentrations there is a reduced probability that d-TC will bind, if activation by ACh is required for d-TC binding, and there will be a reduced probability that the same AChR will be activated in consecutive pulses. Therefore fade is only observed if the ACh concentration is high which is the case for nerve-induced EPCs[8] or after long ionophoretic pulses (low frequency stimulation). This is a use-dependent effect but it is not voltage dependent and therefore cannot be due to channel blocking.[3,4] The same effect is caused by the purified α-toxin erabutoxin b

[a]Supported by National Institutes of Health grant ES04295 and the Alabama Chapter of the Myasthenia Gravis Foundation.

which under the conditions of the experiments also cannot be due to channel blocking.[6] Paradoxically, we found that high concentrations of α-toxin produced less fade than low concentrations.[5,6] Other investigators who claimed that α-toxins produced no fade were probably correct because they uniformly used high concentrations ($\sim$150 nM). Our conclusion is that low concentrations of α-toxins or d-TC bind to a high affinity site on the postsynaptic AChR and accelerate AChR failure (densensitization?) thereby producing fade of EPCs. Higher concentrations may bind to an additional low affinity

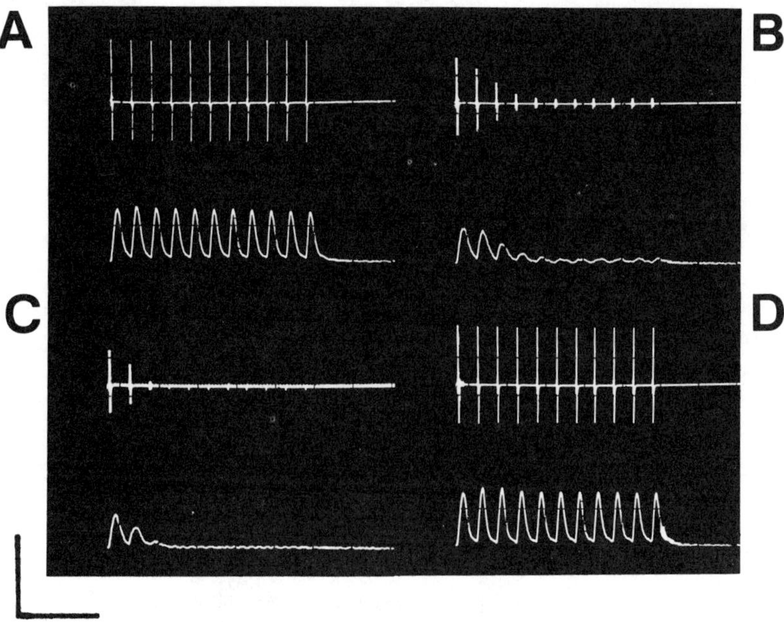

FIGURE 1. Fade of compound action potential (CAP) and contraction in the rat diaphragm caused by 1 μM d-TC at 10 Hz stimulation for 1 second. (**A**) Control recording after equilibrium in the physiological solution for 60 minutes. Upper signal is CAP and lower signal is contraction. (**B**) After 10 minutes incubation with 1 μM *d*-tubocurarine. (**C**) After 30 minutes incubation with 1 μM *d*-tubocurarine. (**D**) After washing out for 1 hour. The vertical bar represents 20 mV for the compound action potential and 50 g for contraction; the horizontal bar represents 400 mseconds. For all experiments the bathing solution consisted of (mM) 135 NaCl, 5 KCl, 2 CaCl$_2$, 1 MgCl$_2$, 1 Na$_2$HPO$_4$, 15 NaHCO$_3$, 11 glucose, pH 7.4 and bubbled with 95% O$_2$ and 5% CO$_2$. The temperature was maintained at 30°C.

site which inactivates the AChR, therefore desensitization or fade is not observed.[6] The theory predicts that these sites have similar on-rate constants but differ widely in their off-rate constants[5,6] and is supported by binding studies with iodoinated purified α-toxins[7] which demonstrate the existence of these two binding sites on the AChR. Our findings do not support the theory that d-TC causes fade by blocking pre-synaptic AChRs and suggest that fade may be due to an allosteric effect at the postsynaptic AChR.

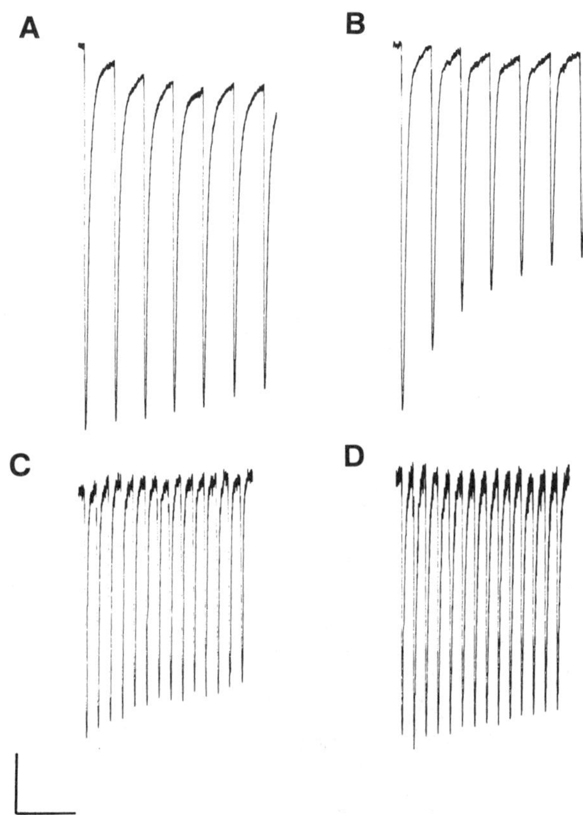

FIGURE 2. Effects of d-TC on end plate currents (EPCs) produced by trains of ionophoretic pulses at single voltage-clamped end plates of rat omohyoideus muscle. The technology was as previously described.[8] The ionophoresis pipette was placed as close as possible to the end plate in order to mimic nerve release. The ionophoresis pipette contained 2 M ACh and had a resistance of 100–200 megohms because such pipettes require only 2–4 nA braking current thus avoiding dilution of ACh at the tip. The pipettes were tested before use and had a constant resistance in the range of 40–500 nA current. (**A**) Control train of ionophoresis induced EPCs at 1 Hz. The current pulse at the ionophoresis pipette was 400 nA for 10 mseconds. (**B**) The same end plate after incubation with 1 μM d-TC for 10 minutes. (**C**) Control train of ionophoresis induced EPCs at 10 Hz. The current pulse at the ionophoresis pipette was 300 nA for 0.22 mseconds. (**D**) The same end plate after incubation with 1 μM d-TC for 10 minutes. The calibration bars for **A** are 15 nA and 2 seconds; **B** is 10 nA and 2 seconds; **C** is 5 nA and 500 mseconds; **D** is 2.5 nA and 500 msecond. The amplitude of the EPC was reduced by d-TC but significant fade was only observed at 1 Hz stimulation. Membrane potential was clamped at -50 mV in all cases.

REFERENCES

1. BOWMAN, W. C., A. G. GIBB, A. L. HARVEY & I. G. MARSHALL. 1986. Prejunctional effects of cholinoceptor agonists and antagonists, and of anticholinesterase drugs, *in* Neuromuscular Blocking Agents, ed. D. A. Kharkevich, Ed.: 141. Springer-Verlag, Berling, FRG.

2. JONES, S. W. & M. M. SALPETER. 1983. Absence of (^{125}I) α-bungarotoxin binding to motor nerve terminals of frog, lizard and mouse muscle. J. Neurosci. **3:** 326.
3. MAGLEBY, K. L., B. S. PALLOTTA & D. A. TERRAR. 1981. The effect of (+)-tubocurarine on neuromuscular transmission during repetitive stimulation in the rat, mouse, and frog. J. Physiol. London **311:** 307.
4. GIBB, A. J. AND I. G. MARSHALL. 1984. Pre- and post-junctional effects of tubocurarine and other nicotinic antagonists during repetitive stimulation in the rat. J. Physiol. (London) **351:** 275.
5. BRADLEY, R. J., M. K. PAGALA & M. T. EDGE. 1987. Multiple effects of α-toxins on the nicotinic acetylcholine receptor. FEBS Lett. **224:** 277.
6. BRADLEY, R. J., M. T. EDGE & W. CHAU. 1990. The α-toxin erabutoxin b causes fade at the rat end-plate. Eur. J. Pharmacol. **176:** 11.
7. MARCHOT, P., P. FRACHON & P. E. BOUGIS. 1988. Selective distinction at equilibrium between the two α-neurotoxin binding sites of Torpedo acetylcholine receptor by microtitration. Eur. J. Biochem. **174:** 537.
8. PEPER, K., R. J. BRADLEY & F. DREYER. 1982. The acetylcholine receptor at the neuromuscular junction. Physiol. Rev. **62:** 1271.

Prejunctional Inhibition by Opioids

α_2-Adrenoceptor Activation and External Calcium

SUE PIPER DUCKLES AND DÉNES BUDAI

Department of Pharmacology
College of Medicine
University of California
Irvine, California 92717

As is true for many presynaptic receptors,[1] inhibition of norepinephrine release by opioid agonists is inversely related to the stimulus intensity. Thus in rabbit arteries, the effects of activation of either presynaptic delta or kappa opioid receptors are critically dependent on the number of pulses in the stimulation train with frequency held constant.[2] For example increasing the train length from 8 to 120 pulses (8 Hz) resulted in a decrease of the inhibitory effect of met-enkephalin (10^{-7} M) from 49 to 12% of control.

It has been suggested that a mutual interaction between presynaptic release inhibiting α_2-adrenoceptors and other types of presynaptic receptors, perhaps due to a common G protein link, may account for this dependence on stimulus intensity. However, neither the activation of the α_2-adrenoceptor-mediated negative feedback by clonidine nor its inhibition by yohimbine nor the increase in biophase norepinephrine produced by addition of cocaine to block neuronal amine uptake altered the neuroinhibitory effect of dynorphin$_{1-13}$ or met-enkephalin at any stimulus train length.[3,4] These results suggest a lack of direct interaction between activation of opioid receptors and α_2-adrenoceptors.

An alternative hypothesis suggests that accumulation of calcium in the nerve terminal with high stimulus intensity serves to counteract the inhibitory effect of presynaptic receptor activation.[4] Indeed conditions known to increase the entry of calcium into the varicosities mimicked the effect of an increase in stimulation train length on the modulation of norepinephrine release by opioids. For example, increasing the extracellular calcium concentration (from 1.6 to 5 or 8 mM) diminished the inhibitory effect of either delta or kappa opioids. Conversely, lowering the calcium concentration (from 1.6 to 1 mM) increased the inhibition by opioids and enhanced the dependence on train length. This suggests that opioid receptor activation involves a primary modulation of Ca^{2+} influx which becomes masked when high levels of axoplasmic calcium are achieved during a continued stimulation train.

The key to this relationship between stimulation intensity, extracellular calcium, and presynaptic receptor activation may be the calcium buffering capacity of the cell, which is thought to reside in calcium sinks, such as mitochondria and endoplasmic reticulum, or in membrane calcium transport.[5] If calcium buffering becomes more important with high intracellular calcium levels (such as achieved with high stimulus intensity), one would predict a nonlinear relationship between intracellular calcium and inward calcium current. At high stimulation intensity when intracellular calcium levels are more efficiently buffered, activation of presynaptic inhibitory receptors may not cause much change in the level of intracellular calcium even though inward calcium current is decreased. This common property of many presynaptically acting neuromodulators implies that the pattern of nerve activation will be critical in determining the degree of effect produced.

REFERENCES

1. ILLES, P. 1986. Neuroscience **17:** 909–928.
2. BUDAI, D. & S. P. DUCKLES. 1988. J. Pharmacol. Exp. Ther. **247:** 839–843.
3. GAN, E. & S. P. DUCKLES. 1988. Eur. J. Pharmacol. **158:** 21–28.
4. BUDAI, D. & S. P. DUCKLES. 1989. J. Pharmacol. Exp. Ther. **251:** 497–501
5. BLAUSTEIN, M. P. 1988. Trends Neurosci. **11:** 438–443.

Presynaptic Regulation of Acetylcholine Release by Endogenous Somatostatin in Ciliary Ganglion Terminals in the Chick Choroid

D. B. GRAY, D. ZELAZNY, N. MANTHAY, AND G. PILAR

Department of Physiology and Neurobiology
The University of Connecticut
Storrs, Connecticut 06269

In this report, we show that a classical transmitter, acetylcholine (ACh), and a neuropeptide, somatostatin, are both released by K^+ depolarization that involves Ca^{++} entry via pharmacologically different Ca^{++} channels, ACh preferentially through the dihydropyridine (DHP)-resistant (putatively N-type) channels, and somatostatin via the DHP-sensitive (L-type) channels. Furthermore, we provide evidence that endogenous somatostatin, acting on presynaptic receptors coupled to a guanosine triphosphate (GTP)-binding protein, inhibits transmitter release. Previously we reported that exogenous somatostatin inhibits K^+-evoked ^{3}H-ACh release in terminals in the vascular smooth muscle of the chick choroid and that this inhibition is blocked by pertussis toxin (PTX), indicating mediation by a GTP-binding protein.[1,2] Experiments with a Ca^{++} ionophore indicated that modulation occurred at the level of Ca^{++} entry.[2] Since somatostatin is colocalized with ACh in the ciliary ganglion terminals of the choroid coat,[2,3] we decided to investigate whether endogenous somatostatin is released and what mechanisms are involved in the modulation of ACh release. Specifically, we were interested in determining if somatostatin is released during high-K^+ incubation, if this release is Ca^{++} dependent, and if the flux responsible for peptide release is through voltage-dependent channels distinct from those that trigger ACh release. This work answers these questions in the affirmative with the use of two pharmacological tools: a somatostatin antagonist cyclo (7-aminoheptanoyl-phe-D-trp-lys-thr(BZL) (CyCam) (Sigma) and dihydropyridine (DHP) compounds, a class of drugs affecting L-type Ca^{++} channels. Earlier, we demonstrated that the K^+-stimulated release of ACh is not antagonized by nifedipine, a DHP antagonist,[2] indicating that Ca^{++} flux through L-type Ca^{++} channels does not trigger this Ca^{++}-dependent event. Previous studies also revealed no enhancement of ACh release in the initial 3 minutes of K^+ stimulation in the presence of CyCam, which might be expected if endogenous somatostatin was cosecreted and subsequently inhibited ACh release. No enhancement was seen for longer periods as well, up to 40 minutes after high-K^+ stimulation. Conversely, ACh release in the first minute after exposure to high K^+ was also examined. FIGURE 1 compares those initial values of labeled ACh release with values averaged over the first 4 minutes of release (the first minute plus the next 3 minute period) for both choroid and iris preparations. There is significant enhancement of ACh release in the first minute in the choroid, but not in subsequent release periods, and not in the iris, which does not contain somatostatin.[2] This indicates that endogenous somatostatin may be released and may modulate ACh release only during the initial 60 seconds of evoked release. Interestingly, nifedipine, a DHP antagonist, also enhances labeled ACh release in the first minute of K^+ stimulation in the choroid but not in the iris. Although

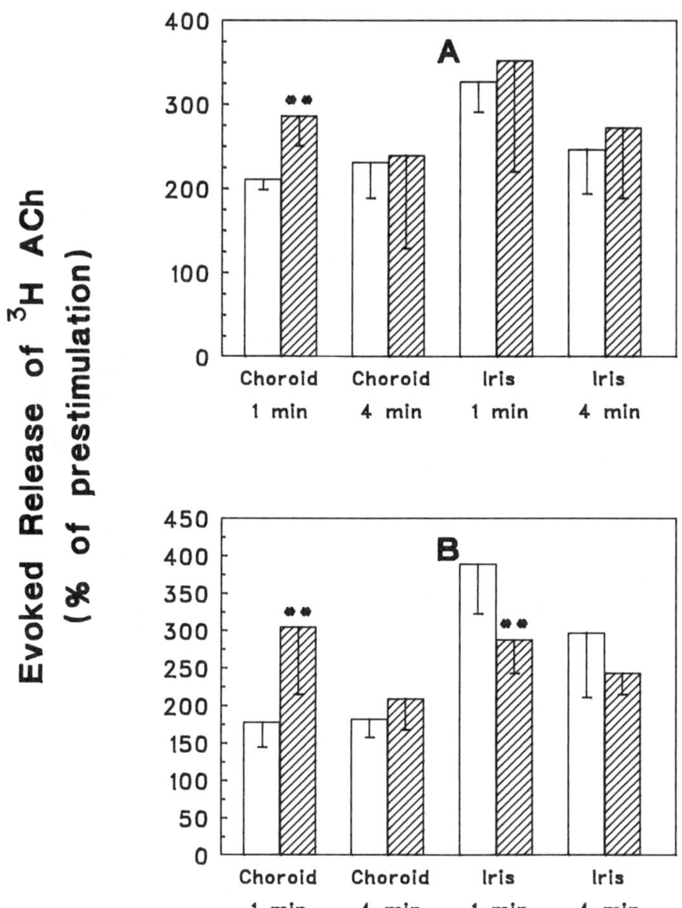

FIGURE 1. Effect of CyCam and nifedipine on the initial one and four minutes of K⁺-evoked ACh release in choroid and iris. Clear bars are controls and hatched bars are in the presence of either 5 μM CyCam (A) or 10 μM nifedipine (B). Hatchling choroids and irises were excised, incubated in ³H-Ch, washed and assayed for labeled ACh release in normal and 55 mM KCl Tyrodes as described previously,[2] except that release samples were taken in the initial minute of exposure to high potassium Tyrodes and the following 3 minute period as well. Values represent the percent increase in labeled ACh release over prestimulation values in the first minute or first 4 minutes (the first and the next 3 minutes) calculated on a per minute value. $n = 3$ and ** $= p < 0.02$.

it appears paradoxical that an antagonist of Ca⁺⁺ entry would enhance transmitter release, it is possible that Ca⁺ flux through DHP-sensitive channels mediates somatostatin release. Thus, nifedipine may act to inhibit somatostatin release, thereby enhancing ACh release. Again, in the iris, no enhancement of transmitter release is seen with nifedipine. This hypothesis was further tested with the DHP agonist Bay K 8644 in FIGURE 2. If L-type channels allow Ca⁺⁺ flux that mediates somatostatin release, then a channel agonist should increase somatostatin release and thus inhibit ACh release. In

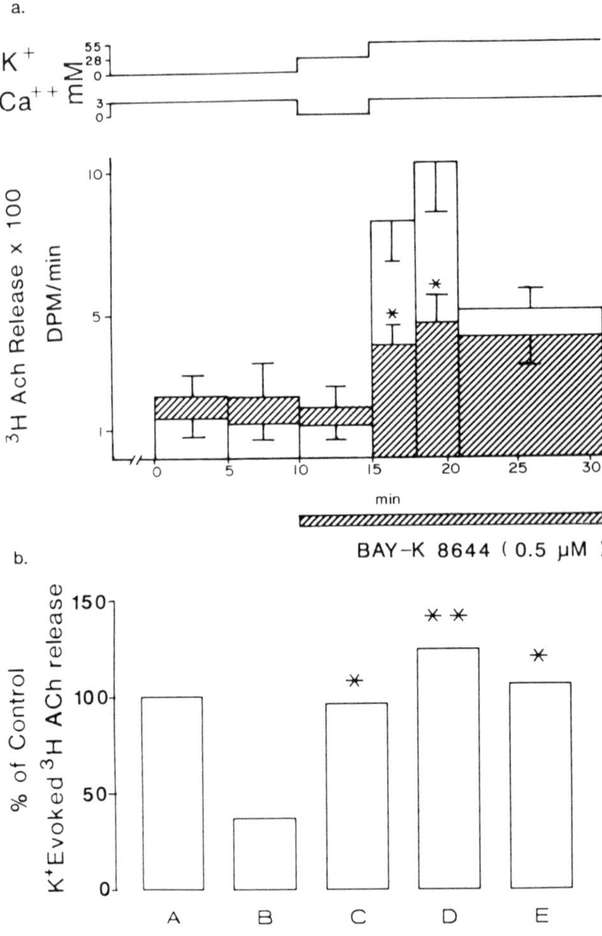

FIGURE 2. (a) Effect of a DHP agonist, Bay K 8644, on release of ^{3}H-ACh from choroid wedges. Clear bars represent controls; hatched bars, tissues exposed to Bay K 8644 (hatched horizontal bar indicates time of application). In these experiments, a 25 mM K$^+$ predepolarization step was included prior to evoked release to maximize Bay K 8644 binding to the DHP receptor. However the Bay K 8644-induced inhibition of ^{3}H-ACh release does not require this step (data not shown). Upper horizontal lines indicate the concentrations of Ca^{++} and K$^+$ in the superfusate Tyrodes. * = significantly different from control evoked release at $p < 0.01$. **(b)** Effect of PTX, CyCam (somatostatin antagonist), and 10 μM nifedipine on the Bay K 8644–induced inhibition of evoked ^{3}H-ACh release as described in (2a) (hatched bars). **(A)** Control K$^+$-evoked ^{3}H-ACh (clear bar). **(B)** 5-minute preincubation with 0.50 μM Bay K 8644. **(C)** Bay K 8644 plus 90 minute preincubation with PTX (200 units/ml). **(D)** Five-minute preincubation with Bay K 8644 plus 10 μM CyCam. We selected CyCam because of its somatostatin antagonist properties, and we chose 10 μM from a dose-response curve. At this concentration, 80% of the somatostatin-induced inhibition of ACh release from choroid terminals was reversed. **(E)** Five-minute preincubation with 10 μM nifedipine. Experimental conditions are similar to those in (a) except that there is no partial depolarization step. Values are expressed as percentage of K$^+$-evoked control for the first 3 minute evoked ^{3}H-ACh release period in each experiment. Note that conditions C, D, and E are significantly different from B, using Student's t test in each individual group of experiments. In all cases, $n \geq 4$. * = $p < 0.05$. ** = $p < 0.02$.

FIGURE 2a, clear bars represent control values and demonstrate a large stimulation in ACh release with high K^+ over 6 minutes, while shaded bars represent values in the presence of 0.5 μM Bay K 8644 and a depressed response to high K^+. FIGURE 2b demonstrates that the Bay K–induced inhibition of evoked ACh release can be reversed by PTX and CyCam, which block somatostatin's effect, and nifedipine, which indicates that Bay K 8644 is acting as a dihydropyridine.

This study provides the first pharmacological evidence for specific roles of different voltage-dependent Ca^{++} channels in controlling the release of a neurotransmitter and its modulator from the same population of terminals. Furthermore, these results suggest distinct secretory pathways in choroid terminals, one for ACh and the other for its modulator, somatostatin, allowing their selective activation and regulation by action potential frequency modalities and second messenger or hormones. It will be important to demonstrate the generality of these findings to more physiological stimulation paradigms and to synapses in the central nervous system.

REFERENCES

1. GRAY, D. B. & G. PILAR. 1988. Differential role of Ca^{++} channels in the release of ^{3}H-ACh and its modulation. Soc. Neurosci. Abstr. **14:** 646.
2. GRAY, D. B., G. R. PILAR & M. J. FORD. 1989. Opiate and peptide inhibition of transmitter release in parasympathetic nerve terminals. J. Neurosci. **9:** 1683–1692.
3. EPSTEIN, M. L., J. P. DAVIS, L. E. GELLMAN, J. R. LAMB, & J. L. DAHL. 1988. Cholinergic neurons of the chicken ciliary ganglion contain somatostatin. Neuroscience **25:** 1053–1060.

Cyclic 3'5'-Guanosine Monophosphate Inhibits Release of Norepinephrine in Vascular Smooth Muscle

STAN S. GREENBERG, FRIEDRICH P. J. DIECKE,
KEITH PEEVY,[a] AND TOSHIOKO P. TANAKA

Department of Physiology
University of Medicine & Dentistry of New Jersey
New Jersey Medical School
185 South Orange Avenue
Newark, New Jersey 07103

[a]Department of Pediatrics
College of Medicine
University of South Alabama
Mobile, Alabama 36688

Cyclic 3'5'-adenosine monophosphate (cAMP) facilitates catecholamine release from sympathetic nerves,[1] whereas the role of cyclic 3'5'-guanosine monophosphate (cGMP) in this process is unclear. Elevated extracellular potassium ion and electrical stimulation increase cGMP in rat superior cervical ganglion.[2] Acetylcholine release from brain slices is inhibited by 8-bromo-cGMP.[3] Inhibitors of type II cyclic GMP phosphodiesterase (PDEI) and stimulators of guanylate cyclase inhibit the contractions of various smooth muscles to sympathetic nerve stimulation.[4] Drewett *et al.* showed that atrial natriuretic peptide inhibited the release of radiolabeled norepinephrine (NE) from sympathetic nerves innervating rabbit vas deferens and rat PC12 cells.[5] Thus, intraneuronal cGMP may inhibit NE release from sympathetic nerves innervating vascular smooth muscle (VSM).

We examined the effects of modulators of cAMP and cGMP on transmural sympathetic neurotransmission (SNS) to isolated canine mesenteric arteries with superfusion and measurement of responses to exogenous NE and efflux of radiolabeled 2-^{14}C-NE during SNS at 0.5 to 32 Hz (calcium-dependent release) or with tyramine (10 μM) (calcium-independent release). Stimulation of adenylate cyclase with forskolin, prostacyclin, and iloprost, a stable prostacyclin analogue, and inhibition of type I, III, and IV cAMP phosphodiesterase with neural specific rolipram and milrinone and isobutylmethylxanthine (IBMX) did not enhance the efflux of NE from sympathetic nerves innervating the VSM (FIGURE 1A). Isoproterenol enhanced NE efflux during SNS. This effect was antagonized by propranolol but was unaffected by concurrent incubation with cAMP PDEI (FIGURE 1B). SIN-1, nitroglycerin, sodium nitroprusside and nitroglycerin (GTN), and inhibitors of type II cGMP phosphodiesterase (M & B-22948 and verofyllin) inhibited the responses to nerve stimulation and NE efflux due to SNS but not tyramine (FIGURE 2A and B).

Modulators of cAMP did not affect SNS. The inability of cAMP PDEI to augment isoproterenol-induced enhancement of NE efflux during SNS suggests that β-adrenoceptor mediated enhancement of NE release efflux may also be independent of cAMP. Thus, cAMP appears not to be a primary modulator of NE release from sympathetic nerves in VSM. The data support the conclusion that cGMP is an inhibitory modulator

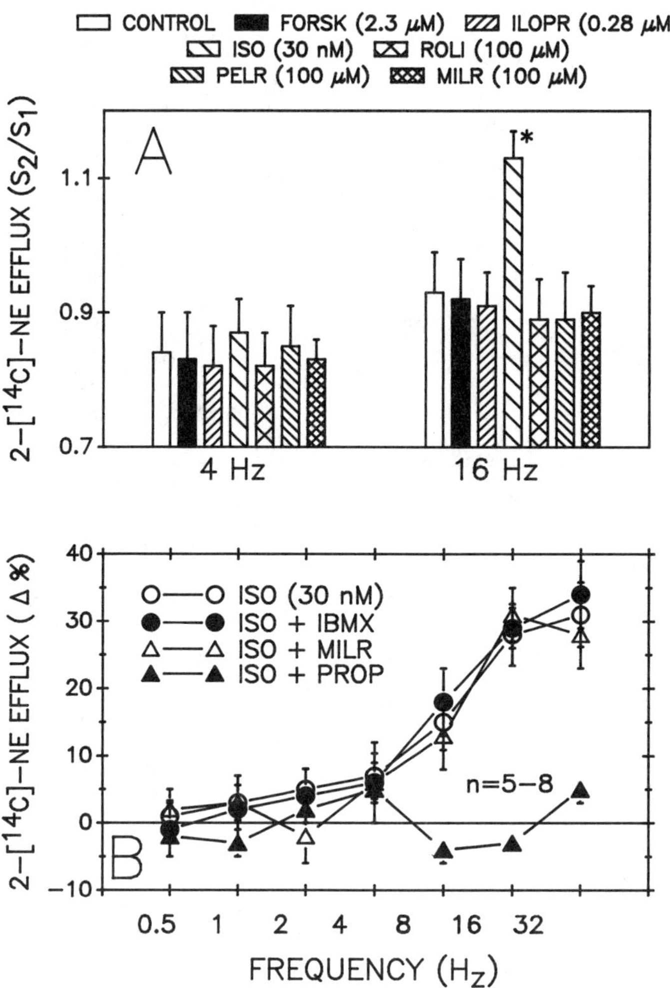

FIGURE 1. (**A**) Preincubation with stimulators of adenylate cyclase and inhibitors of type I, III, and IV cAMP PDE do not affect the efflux of ^{14}C-NE during sympathetic nerve stimulation at 4 Hz or 16 Hz (2 mseconds duration and delay, 9–10 V). (**B**) The efflux of ^{14}C-NE is enhanced at 16 Hz by isoproterenol. Propranolol (1 μM) blocks the response, but IBMX and milrinone do not affect the isoproterenol response.

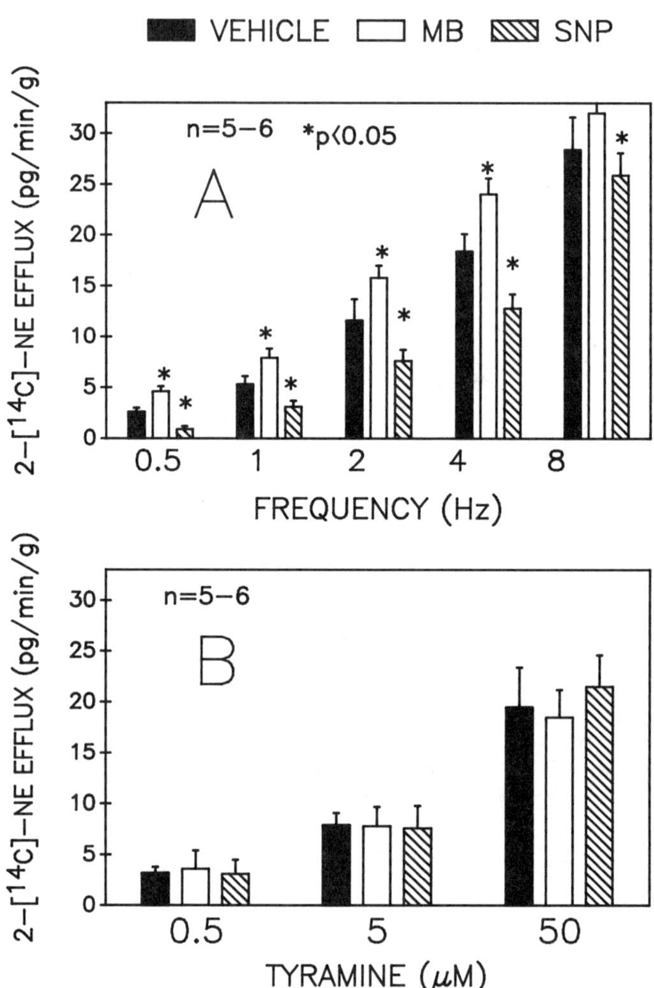

FIGURE 2. Preincubation with stimulators of guanylate cyclase and inhibitors of type II cGMP PDE inhibit the efflux of ^{14}C-NE during SNS (A) but not with tyramine (B).

of calcium- and depolarization-dependent NE release from sympathetic nerves. cGMP does not inhibit calcium-independent NE release from sympathetic nerves. The source of cGMP may be both smooth muscle as well neuronal in origin.

REFERENCES

1. WOOTEN, G. F., N. B. THOA, I. J. KOPIN & J. AXELROD. 1973. Mol. Pharmacol. **9:** 178–193.
2. KALIX, P. 1976. J. Neurochem. **27:** 1563–1564.
3. NORDSTROM, O. & T. BARTFAI. 1981. Brain Res. **213:** 467–471.
4. GOTHERT, M. 1984. Blood Vessels **21:** 117–125.
5. DREWETT, J. G., G. J. TRACHTE & G. R. MARCHAND. 1989. J. Pharmacol. Exp. Ther. **248:** 135–141.

Bradykinin-Evoked Acetylcholine Release via Inositol Trisphosphate–Dependent Calcium Rise in Neuroblastoma x Glioma Hybrid NG108-15 Cells

HARUHIRO HIGASHIDA AND AKIHIKO OGURA[a]

Department of Biophysics
Neuroinformation Research Institute
Kanazawa University School of Medicine
Kanazawa 920, Japan
and
[a]*Department of Neuroscience*
Mitsubishi Kasei Institute of Life Sciences
Machida 194, Japan

The mechanism underlying the bradykinin (BK) induced increase of acetylcholine (ACh) release was studied in NG108-15 mouse neuroblastoma x rat glioma hybrid cells and their synapses formed onto mouse muscle cells. Application of BK produced characteristic membrane potential changes (hyperpolarization followed by depolarization) in the hybrid cells and an increase in the frequency of miniature end plate potentials (MEPPs) in paired myotubes.[1] When the BK-induced depolarization of the

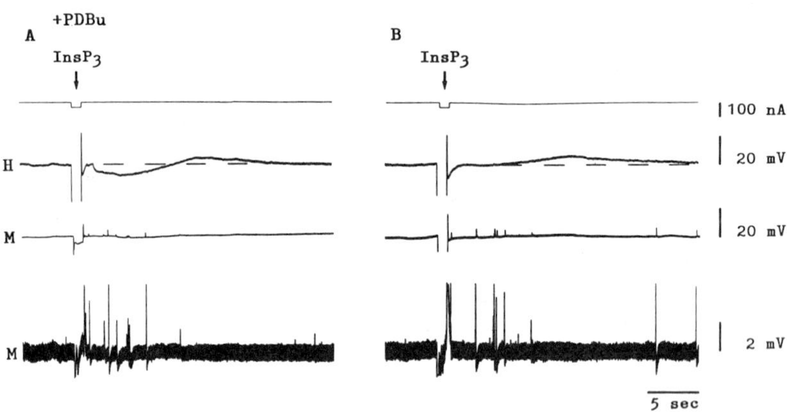

FIGURE 1. MEPPs generated in response to iontophoretic injections of $InsP_3$ in NG108-15 cells. The membrane potentials of a hybrid cell (H) and a paired myotube (M) were simultaneously monitored. The injection of $InsP_3$ (-30 nA, 1 second; uppermost trace) produced a hyperpolarization in **A,** whereas it did not in **B.** Note the independence of the facilitation of MEPPs on the membrane potential of NG108-15 cells. The resting potentials of the hybrid cells and of the muscle cells ranged between -47 and -55 mV. The cell pair in A and B had been pretreated with 1 μM PDBu. Turning off of injection current triggered an action potential in each case, indicating that the injection was intracellular.

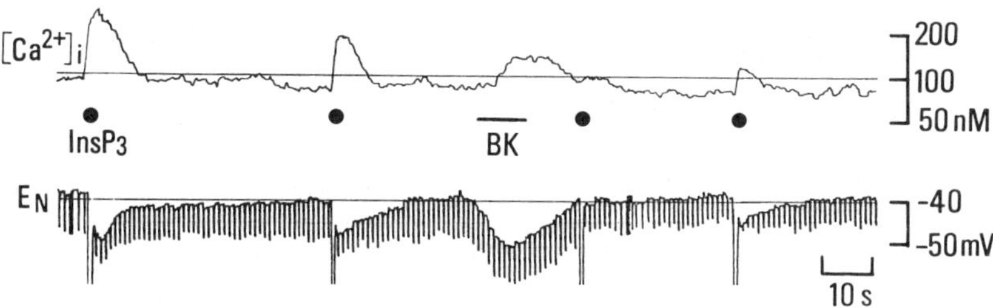

FIGURE 2. Simultaneous recordings of $[Ca^{2+}]_i$ and membrane potential (E_N) from an NG108-15 hybrid cell injected with $InsP_3$. $InsP_3$ injection elicited membrane hyperpolarization (lower trace) and $[Ca^{2+}]_i$ rise (upper trace). Both responses were reversibly suppressed 10 seconds after application of BK. The responses partially recovered after 40 seconds. $InsP_3$ was injected (-100 nA, 0.1 second; indicated by dots) repeatedly. BK was applied by perfusion at the period indicated by a bar (1 μM, 10 seconds). Pulses (-0.2 nA, 0.1 second) were applied at 1 second intervals. Fluoromicroscopic determination of $[Ca^{2+}]_i$ with fura-2 was performed as described (Reference 3). $[Ca^{2+}]_i$ rise in response to injections of $InsP_3$ and BK was monitored 2 μm beside the tip of the $InsP_3$-containing electrode.

hybrid cells was prevented by prior depolarization by phorbol dibutyrate (PDBu), faciliation of MEPPs in the myotubes occurred in the period of hyperpolarization. Ba^{2+} (4 mM) blocked the hyperpolarization in response to BK, but facilitation of MEPPs was still observed. BK is known to accelerate membrane phospholipid metabolism leading to the liberation of inositol 1,4,5-trisphosphate ($InsP_3$) into cytoplasm.[2] Iontophoretic injection of $InsP_3$ into the cytoplasm of the NG108-15 cell briefly evoked MEPPs during the $InsP_3$-induced hyperpolarizing phase in most cases (FIGURE 1). However, $InsP_3$-dependent facilitation of MEPPs was also observed in cells where $InsP_3$ injections produced no detectable hyperpolarization or induced even depolarization. These results indicate that the membrane potential changes are not the primary cause for the BK-induced ACh release.

Real-time quantitative monitoring[3] of the intracellular free Ca^{2+} concentration ($[Ca^{2+}]_i$) with fura-2 in a single NG108-15 cell showed that BK application or intracellular $InsP_3$ injection induced an elevation of $[Ca^{2+}]_i$ which coincided in time with membrane hyperpolarization simultaneously recorded from the same cell (FIGURE 2). The $[Ca^{2+}]_i$ rise produced by $InsP_3$ injection started from the single site of injection while that produced by BK began from a deep compartment of the cytoplasm of the NG108-15 cells. The BK or $InsP_3$ appeared to share the same intracellular Ca^{2+} store site(s) which was also sensitive to caffeine.[4] The BK- and $InsP_3$-evoked facilitation of MEPPs and the $[Ca^{2+}]_i$ rise were relatively independent of extracellular Ca^{2+}. Application of norepinephrine and enkephalin whose receptors are not coupled with phospholipid metabolism in the hybrid cell did not produce hyperpolarization, $[Ca^{2+}]_i$ rise or ACh secretion. These findings suggest that the BK-induced ACh release results from transient $InsP_3$-dependent elevation of $[Ca^{2+}]_i$.

REFERENCES

1. HIGASHIDA, H. 1988. J. Physiol. **397**: 209–222.
2. YANO, K., H. HIGASHIDA, R. INOUE & Y. NOZAWA. 1984. J. Biol. Chem. **259**: 10201–10207.
3. KUDO, Y. & A. OGURA. 1986. Br. J. Pharmacol. **89**: 191–198.
4. OGURA, A., Y. MYOJO & H. HIGASHIDA. 1990. J. Biol. Chem. **265**: 3577–3584.

The Sigma Receptor Ligand (+)3PPP Inhibits Neuronal [³H]Norepinephrine Accumulation in the Rat Tail Artery

T. MASSAMIRI AND S. P. DUCKLES

Department of Pharmacology
College of Medicine
University of California
Irvine, California 92717

Because of its high affinity to the sigma binding site, (+)3PPP (3-[3-hydroxyphenyl]-N-(1-propyl)-piperidine), has been used extensively to investigate the anatomical distribution as well as the biochemical and functional properties of the sigma receptor. This receptor has now been characterized as a unique class of binding sites distinct from dopamine, opioid, and phencyclidine (PCP) receptors although little is known about the mechanism of action or functional role of sigma receptor activation. Nevertheless, it has been suggested that the sigma receptor might play a prominent role in modulating affective states.

Therefore, in order to further explore the mechanism of action of sigma receptors and their ligands, we have developed a functional assay, the perfused rat tail artery, as a model for studying the effects of (+)3PPP and other sigma ligands. Our results have shown that (+)3PPP acting via dopamine D_2 receptors inhibits vascular contraction induced *in vitro* by transmural nerve stimulation. However, when D_2 receptors are blocked with sulpiride, (+)3PPP potentiates the contractile response to norepinephrine to a maximum of 350% with an EC_{50} of 0.96 μM. This potentiation was blocked by 1 μM cocaine, implying an action of (+)3PPP at the monoamine uptake₁ site. We now demonstrate that (+)3PPP blocks neuronal [³H]norepinephrine accumulation in the rat tail artery with a potency and maximal inhibitory effect comparable to cocaine. Segments of the rat tail artery were incubated in [3H]norepinephrine for 30 minutes to determine norepinephrine accumulation. The maximum inhibition as well as the concentration necessary to block 50% of [3H]norepinephrine accumulation (IC_{50}) are similar for (+)3PPP and cocaine (see TABLE 1).

Additional sigma ligands, previously shown to either potentiate or inhibit the contractile response to norepinephrine in the rat tail artery, also exhibited a comparable inhibition of [³H]norepinephrine accumulation. These ligands include (+)SKF 10047, (+)-N-allyl-N-normetazocine; haloperidol; BMY 14802, [α-(4-fluorophenyl)-4-(5-fluoro-2-pyrimidinyl)-1-piperazine butanol]; and rimcazole, *cis*-9-[3-(3,5-dimethyl-1-piperazinyl)propyl]carbazole dihydrochloride. Thus, even ligands that do not potentiate contractile responses of the rat tail artery inhibit neuronal norepineph-

TABLE 1

	IC_{50} (μM)	Maximum Inhibition at 10^{-4} M (%)
(+)3PPP	6.7 ± 1.6	91%
Cocaine	3.6 ± 0.8	90%

rine uptake. We attribute this discrepancy between the observed inhibition of the contractile response to norepinephrine by haloperidol, BMY 14802, and rimcazole and their potency at blocking [^{3}H]norepinephrine accumulation to their greater relative efficacy at a distinct postjunctional site compared to the norepinephrine uptake$_1$ site.

Our observations suggest a mode of action similar to cocaine for $(+)$3PPP and other sigma receptor ligands on the neuronal catecholamine uptake system. This effect should be taken into consideration in studies of sigma receptor effects and could make an important contribution to the psychoactive properties of sigma and PCP receptor ligands.

Two Furan Analogues of Muscarine as Selective Agonists at the Presynaptic Muscarinic Receptors of the Guinea Pig Longitudinal Ileal Muscle[a]

B. V. RAMA SASTRY, L. K. OWENS, AND R. F. OCHILLO

Department of Pharmacology
Vanderbilt University School of Medicine
Nashville, Tennessee 37232-6600

Two mechanisms have been proposed for the regulation of acetylcholine (ACh) release in the nervous sytem: a negative feedback mechanism and an ACh-amplification or a positive feedback mechanism.[1,2] The level of ACh available for chemical transmission

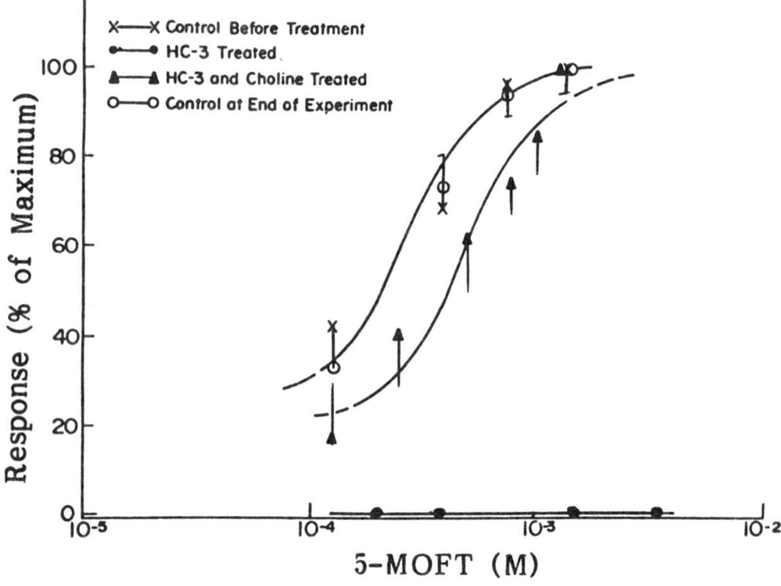

FIGURE 1. Effects of hemicholinium-3 (HC-3) and choline on the concentration-response curve of 5-MOFT on the longitudinal muscle of the guinea pig ileum. Note that the agonist action of this agent in the presence of HC-3 after the muscle has been incubated with HC-3 is not significant within the concentration range tested. Vertical lines denote one standard error (SE) from a minimum of 3 values. Concentrations of various agents: HC-3, 16×10^{-6} M (incubation time, 3 hours); choline 4.8×10^{-7} M (incubation time, 15 minutes). All are cumulative response curves.

[a]This work was supported by grants from the Smokeless Tobacco Research Council Inc. and United States Public Health Service—National Institute of Health ES-03172.

in the synaptic gaps will initiate one or the other feedback mechanism. If the level of ACh is too little to effect synaptic transmission, a group of presynaptic muscarinic receptors (Ms) are activated to initiate the positive feedback mechanism resulting in the release of further quantities of ACh.[3] If the ACh level in the synaptic gap is too high, a group of presynaptic receptors (Mi) is activated resulting in the inhibition of ACh release. A question arises as to whether the presynaptic receptors initiating the two feedback mechanisms belong to two different subtypes (Ms and Mi) of muscarinic receptors. In order to develop selective agonists, we have synthesized 5-methoxyfurfuryl-trimethylammonium (5-MOFT) and 5-hydroxyfurfuryltrimethylammonium (5-HMFT[4] and studied their effects of the guinea pig longitudinal ileal muscle–Auerback plexus preparation.

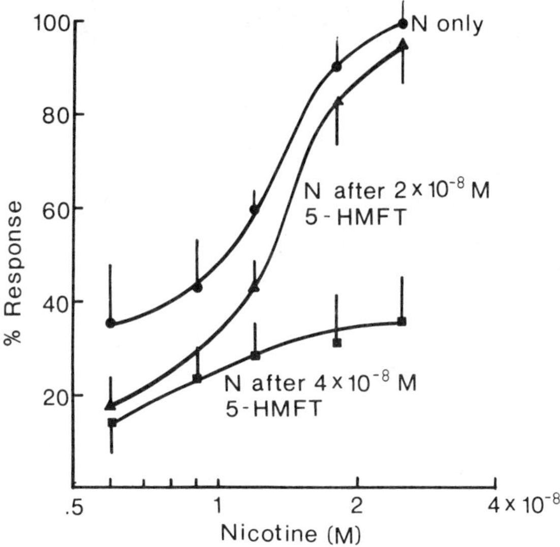

FIGURE 2. Effects of 5-HMFT on the concentration-response curve of nicotine in the guinea pig longitudinal ileal muscle. The bath contained physostigmine sulfate (4×10^{-7} M) but no hexamethonium. Each point is a mean $\pm$ SE from 6 values. In this tissue, nicotine releases ACh from Auerbach plexus and causes contraction of the muscle. 5-HMFT blocked nicotine-induced release of ACh as well as contraction of the muscle due to release of ACh. The nicotine-induced release of ACh was also blocked by morphine (0.75 μM) and procaine (1.1 mM).

METHODS AND RESULTS

The preparation of the guinea pig longitudinal muscle, the composition of Tyrode solution, measurement of contractions of the ileal muscle when challenged with agonists, and the determination of ACh content by gas chromatography have been described in the author's papers.[5,6]

The longitudinal muscle contained about 119–120 nmol ACh/g wet tissue. Incubation of this tissue with hemicholinium-3 (HC-3) (16 μM) depleted 85–90% of ACh which was replenished by reincubation with choline (48 μM). 5-MOFT caused contraction of the muscle which was blocked by atropine (1.0 μM) or morphine (0.76

μM). It was not blocked by hexamethonium (0.4 mM) or procaine (1.1 mM). 5-MOFT did not cause contraction of the muscle that was depleted of ACh (FIGURE 1). These observations indicate that 5-MOFT activates a presynaptic muscarinic receptor, Ms, in the Auerbach plexus and releases ACh to cause contraction of the muscle. At high concentrations (10^{-6} to 10^{-4} M), 5-HMFT induced contraction of the muscle with or without HC-3 treatment. Physostigmine potentiated these responses. It noncompetitively blocked nicotine-induced contractions of the muscle at a relatively low concentration (4×10^{-8} M) (FIGURE 2). These observations indicate that 5-HMFT activates a presynaptic muscarinic receptor, Mi, and inhibits ACh release.

SUMMARY

5-MOFT activates presynaptic Ms receptors and releases ACh while 5-HMFT activates Mi receptors and inhibits ACh release. Thus, two subtypes of muscarinic receptors regulate ACh release in Auerbach plexus.

REFERENCES

1. SASTRY, B. V. R. & O. S. TAYEB. 1982. Adv. Biosci. **38:** 165–172.
2. SASTRY, B. V. R., V. E. JANSON, N. JAISWAL & O. S. TAYEB. 1983. Pharmacology **26:** 61–72.
3. KOELLE, G. B. 1971. Ann. N.Y. Acad. Sci. **183:** 5–20.
4. CHATURVEDI, A. K., P. P. ROWELL & B. V. R. SASTRY. 1981. Pharmacol. Res Commun. **13:** 829–845.
5. CHENG, H. C. & B. V. R. SASTRY. 1972. J. Pharmacol. Exp. Ther. **180:** 326–339.
6. OCHILLO, R., P. P. ROWELL & B. V. R. SASTRY. 1978. Pharmacology **16:** 121–130.

Effects of Nicotinic Agonists and Antagonists on Acetylcholine Release from the Rat Hemidiaphragm

R. J. STORELLA[a] AND G. G. BIERKAMPER[b]

[a]Department of Anesthesiology
Hahnemann University
Philadelphia, Pennsylvania 19102-1192

[b]Department of Pharmacology
University of Nevada School of Medicine
Reno, Nevada

Recent experiments[1] suggest that prejunctional acetylcholine receptors mediate increases in ACh release from the neuromuscular junction (positive feedback receptors). This work used a radiolabel method of monitoring ACh release in preparations treated with a choline uptake inhibitor (hemicholinium). However, initial results using a radioenzymatic assay for ACh in anticholinesterase (neostigmine) treated preparations indicated that nicotine decreases ACh release and that low doses of d-tubocurarine increase ACh release.[2] Using the radioenzymatic assay, we tested the hypothesis that there are negative feedback prejunctional ACh receptors.

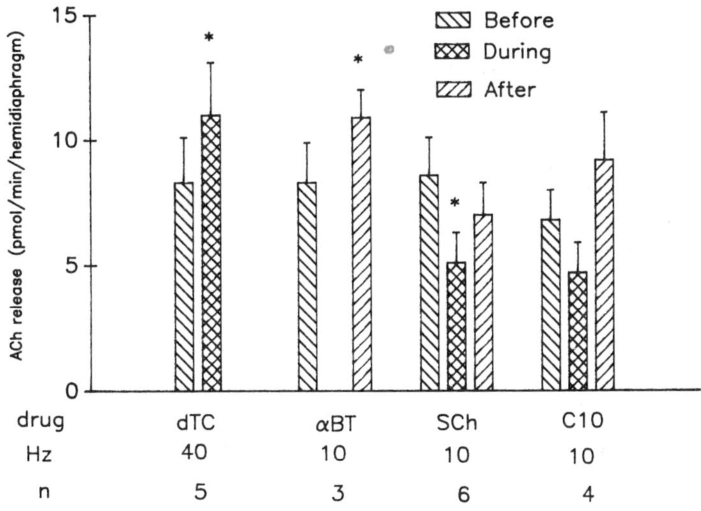

FIGURE 1. Effects of cholinergic drugs on ACh release. ACh release was measured during 15 minute periods before, during, and after drug infusion. Drugs given were d-tubocurarine (dTC, 0.01 μM), α-bungarotoxin (αBT, 50 μg, single bolus), suberyldicholine (SCh, 10 μM), and decamethonium (C10, 10 μM). The frequency of stimulation (Hz), number of preparations (n), and significance* are indicated as appropriate.

Total ACh release from the vascularly perfused rat phrenic nerve hemidiaphragm was measured radioenzymatically as previously described.[2] The preparation was bathed in a standard HEPES buffered physiological saline (30–32°C) to which 10 μM neostigmine and 0.01 mM choline were added. After 60 minutes of indirect simulation, ACh release reaches a plateau.[2] Drugs were infused during this plateau. ACh was measured in perfusates collected for 15 minute periods right *before* drug, *during* the period 30 minutes after the start of the drug, and 15 minutes *after* discontinuing the drug. Figures show means and standard errors. Significance ($p < 0.05$) was tested with one-tailed paired t-tests.

Cholinergic antagonists d-tubocurarine and α-bungarotoxin increased ACh release while cholinergic agonists suberyldicholine and decamethonium decreased ACh release (FIGURE 1). The amount of ACh released per minuted does not increase linearly with stimulation frequency (FIGURE 2). Thus, ACh release per stimulus decreases.

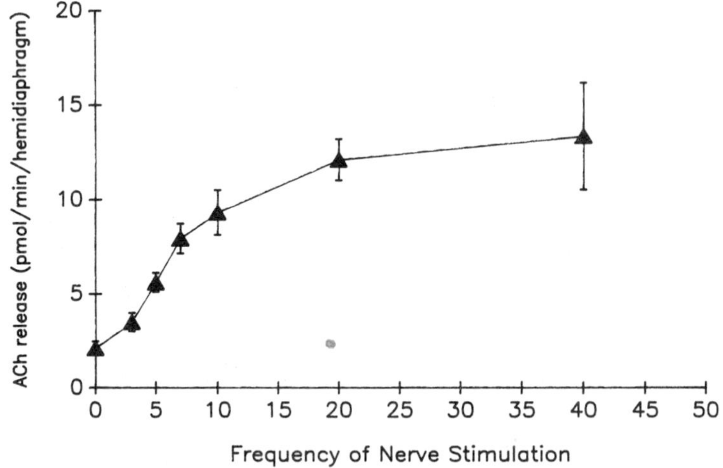

FIGURE 2. ACh release at different stimulation frequencies. ACh was measured for 15 minute periods during the plateau of release at 90, 105, and 120 minutes and averaged. The number of preparations per point at 0, 3, 5, 7, 10, 20, and 40 Hz was 14, 4, 6, 12, 5, and 6 respectively.

The results clearly demonstrate that cholinergic drugs alter total ACh release from the neuromuscular junction. The most straightforward interpretation of the results is that there is a negative feedback ACh receptor. However, the use of the anticholinesterase complicates the interpretation of the results since elevated junctional ACh may produce a predrug baseline in which ACh receptors (both nicotinic and muscarinic) might already be activated or blocked (depolarization/desensitization). Similarly, the work of others suggesting positive feedback receptors[1] is complicated by the potential for selective labeling of ACh and hemicholium actions on ACh mobilization and release. Thus, the elucidation of the physiological and clinical roles of prejunctional receptors on ACh release awaits a better understanding of the complications imposed by the assay techniques as well as direct comparisons between results with ACh release and postjunctional monitoring.

ACKNOWLEDGMENT

This is a posthumous presentation of work from the laboratory of Dr. Bierkamper. Dr. Storella is appreciative of the kind and scholarly mentorship he received from Dr. Bierkamper and is honored to make this presentation.

REFERENCES

1. WESSLER, I. 1989. Trends Pharmacol. Sci. **10:** 100–114.
2. BIERKAMPER, G. G., E. AIZENMAN & W. R. MILLINGTON. 1986. Dynamics of Cholinergic Function. I. Hanin, Ed.: 447–457. Plenum Press. New York, N.Y.

Actions of Angiotensin II on Sympathetic Noradrenergic and Purinergic Transmission in the Guinea Pig Vas Deferens

THOMAS C. CUNNANE, CLAIRE F. WARDELL, AND
JAMES ZIOGAS[a]

University Department of Pharmacology
South Parks Road
Oxford OX1 3QT, United Kingdom

Bell reported that angiotensin II (AII) enhanced the amplitude of excitatory junctional potentials (EJPs) evoked by sympathetic nerve stimulation without affecting spontaneous excitatory junction potentials (SEJPs) in the guinea pig vas deferens,[1] suggesting that AII enhanced noradrenergic transmission. However, it is now believed that EJPs in this preparation are not mediated by noradrenaline, but by the cotransmitter adenosine triphosphate (ATP).[2] AII enhances noradrenaline release in a variety of tissues,[3] which is consistent with the idea that noradrenaline and ATP are coreleased. Recently, Ellis and Burnstock reported that AII may enhance stimulation-evoked ^{3}H-noradrenaline release without affecting ATP release in the guinea pig vas deferens.[4] In the present study the actions of AII on the noradrenergic and purinergic components of the response to sympathetic nerve stimulation in the guinea pig vas deferens have been investigated.

Vasa deferentia removed from male Duncan-Hartley guinea pigs (300–400 g) were bathed in Krebs solution of the following composition (mM): NaCl 118, $NaHCO_3$ 25, NaH_2PO_4 1.13, $CaCl_2$ 1.8, KCl 4.7, $MgCl_2$ 1.3, and glucose 11.1. The solution was bubbled with 95% O_2/5% CO_2 (to pH 7.4) and maintained at 36–37°C. Tissues were mounted vertically in a 15 ml organ bath and perfused with Krebs solution at 5 ml/minute and the proximal end of the vasa was electrically stimulated through Ag/AgCl ring electrodes. Tension was recorded on a Grass 7D pen recorder using Grass FT03 isometric transducers. Stimulation at 10 Hz, 200 pulses, 0.5 msecond, 40 V evoked biphasic contractions. The peaks of the initial rapid and secondary slow contraction were 4.9 ± 0.9 and 3.4 ± 0.5 g ($n = 6$), respectively. These were enhanced 35–65% by AII (30 nM). Contractions obtained in the presence of the α-adrenoceptor antagonist prazosin or the P_2-purinoceptor desensitizing agent α,β-methylene ATP were similarly enhanced by AII.

For biochemical studies of endogenous noradrenaline release, vasa were mounted vertically in a 1 ml organ bath and perfused at 2 ml/minute with Krebs solution which contained 2.5 mM Ca^{2+} and 10 mg/l EDTA. The perfusate was collected at 1 minute intervals in ice-cold vials for subsequent determination of noradrenaline content by high performance liquid chromatography (HPLC). Prior to injection onto the column catecholamines were separated by alumina extraction. Resting levels of noradrenaline release were below detectable levels. In the absence of AII, stimulation (3 or 10 Hz, 200 pulses, 1.0 msecond, 30 V) evoked noradrenaline release was 44.5 ± 5.7 ($n = 6$)

[a]C. J. Martin Fellow of the Australian NH & MRC and author for correspondence.

and 76.2 ± 7.1 (n = 6) pg/mg of tissue. The total tissue noradrenaline content was 10.4 ± 1.1 µg/g of tissue. AII (30 nM) enhanced noradrenaline release by 38.2 ± 11.1 and 73.0 ± 19.8% at 3 and 10 Hz, respectively.

For the electrophysiological studies vasa were pinned to the Sylgard (Dow Corning) covered base of a 3 ml organ bath and perfused with Krebs solution at 2 ml/minute. The hypogastric nerve was electrically stimulated through Ag/AgCl electrodes. Membrane potentials of smooth muscle cells were recorded using conventional glass microelectrodes filled with 5 M potassium acetate, tip resistances 30–70 MΩ. Signals were recorded through a Neurolog NL102 d.c. amplifier and stored on magnetic tape. As shown in FIGURE 1, both in a single cell and a population of cells, AII (30 or 100 nM) enhanced EJP amplitude. As reported by Bell,[1] SEJPs appeared unaffected by angiotensin II (FIGURE 1A) indicating a prejunctional site of action.

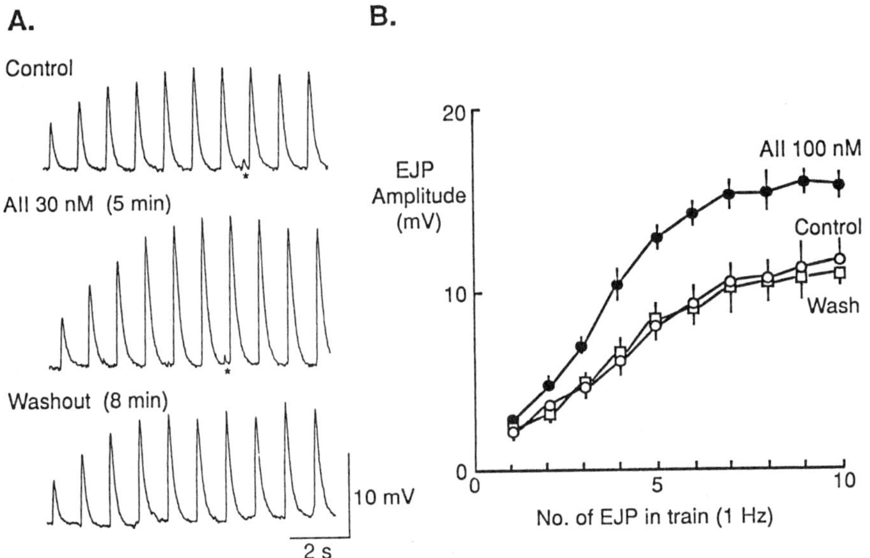

FIGURE 1. Enhancement of EJPs by AII (10 pulses, 1 Hz, 0.1 msecond, 20 V). (**A**) The effect of AII (30 nM) in a single cell. SEJPs are indicated by *. (**B**) The effect of AII (100 nM) in a population of cells (n = 6 cells for each group).

However, in two experiments, AII (100 nM) inhibited EJP amplitude, an effect that was prevented by the cylooxygenase inhibitor indomethacin (3 µM) (FIGURE 2). In contrast, the enhancement by AII (1–100 nM) of contractions evoked by short trains of stimulation (12 p, 10 Hz, 0.1 msecond, 30 V) was unaffected by indomethacin. These contractions were abolished by tetrodotoxin (0.3 µM) and α,β-methylene ATP (1 µM), but only partly (10%) reduced by prazosin (0.3 µM).

Thus, AII enhanced both noradrenergic and purinergic neurotransmission in the guinea pig vas deferens, a finding consistent with corelease of the two transmitters. However, the role of prostaglandins as transjunctional modulators that may be formed in the presence of AII requires further investigation.

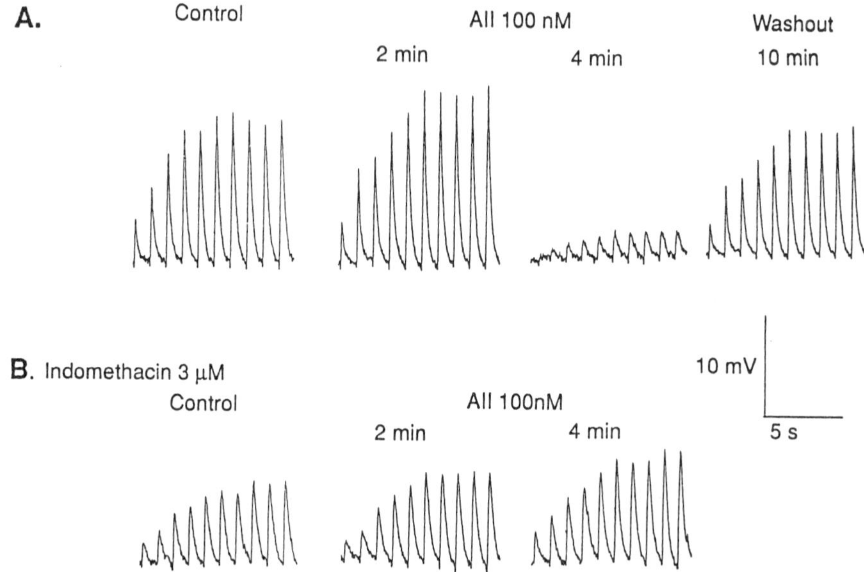

FIGURE 2. (A) Single cell showing inhibition of EJPs (10 pulses, 1 Hz, 0.1 msecond, 15 V) by AII (100 nM). (B) Another cell in the same preparation showing prevention of the inhibitory effect of AII (100 nM) by indomethacin (3 μM).

REFERENCES

1. BELL, C. 1972. Mechanism of enhancement by AII of sympathetic adrenergic transmission in the guinea-pig. Circ. Res. **31:** 348–355.
2. WHITE, T. D. 1988. The role of adenine compounds in autonomic neurotransmission? Pharmacol. Ther. **38:** 129–168.
3. STARKE, K. 1977. Regulation of noradrenaline release by presynaptic receptor systems. Rev. Physiol. Pharmacol. **77:** 1–124.
4. ELLIS, J. L. & G. BURNSTOCK. 1989. Angiotensin neuromodulation of adrenergic and purinergic co-transmission in the guinea-pig vas deferens. Br. J. Pharmacol. **97:** 1157–1164.

Striatal Dopamine Release[a]

In Vivo Evidence for Local Initiation

ELIZABETH D. ABERCROMBIE[b]
AND MICHAEL J. ZIGMOND[b,c]

*[b]Department of Behavioral Neuroscience
[c]Department of Psychiatry
Center for Neuroscience
University of Pittsburgh
Pittsburgh, Pennsylvania 15260*

Transmitter release generally is thought to occur in response to the arrival of an action potential and the consequent initiation of calcium-dependent exocytosis. Over the past two decades it has become clear that release is not governed by action potentials alone but can be influenced by local factors, including the synaptic concentration of the transmitter under investigation. Several recent observations from our research group and others have suggested that, in addition to this impulse-dependent modulation of release, there must exist an additional mechanism for transmitter release: the local *triggering* of release via a heterosynaptic input.

We have focused on the release of endogenous dopamine (DA) from neurons of the nigrostriatal bundle (NSB) of adult, male rats. Where indicated selective lesions of the NSB were made using 6-hydroxydopamine (6-HDA) administered either along the bundle or into the cerebrospinal fluid. Extracellular fluid was sampled in awake rats via chronically implanted microdialysis probes placed in the striatum, and DA analyses were carried out by high performance liquid chromatography (HPLC) with electrochemical detection. Observations are summarized in FIGURE 1.

DA RELEASE FROM RESIDUAL NEURONS IS INCREASED AFTER 6-HDA IN THE ABSENCE OF AN INCREASE IN FIRING RATE

One to two months after 6-HDA administration we observed little or no decline in extracellular DA concentration despite the apparent loss of up to 80% of the dopaminergic NSB neurons.[1] This could not be explained in terms of an increase in the proportion of surviving nigral cells that were active or in the firing rate of those cells,[2] but instead appeared to reflect an increase in the amount of DA released per pulse by residual DA terminals together with a decrease in DA reuptake.[3]

DA RELEASE IS SUSTAINED AFTER DEPOLARIZATION BLOCK OF NSB NEURONS

In rats lesioned with 6-HDA the acute administration of haloperidol [0.5 mg/kg, intraperitoneally (ip)] causes an initial increase in the firing rate of nigral neurons

[a]Research summarized in this review was carried out as part of NINDS Program Project NS19608. Additional support was provided by U.S. Public Health Service Grants MH43947, MH42217, MH00058, MH09658, and EY05283 and the National Alliance for Research on Schizophrenia and Depression.

followed by a cessation of their activity.[4] However, this "depolarization block" of NSB is accompanied by little or no reduction in the concentration of DA in extracellular fluid.[5]

ENVIRONMENTAL STRESS CAUSES A SUSTAINED INCREASE IN DA RELEASE IN STRIATUM DESPITE LITTLE OR NO CHANGE IN NEURONAL ACTIVITY

Although novel stimuli can cause transient increases in the firing rate of DA neurons, little or no sustained increase in activity is seen during stress.[6] Nonetheless, exposure of rats to 30 minutes of tail shock caused a significant increase in extracellular DA in the striatum of intact animals.[7] This response was even larger if animals were previously lesioned with 6-HDA.[8]

STRESS MAY EVEN INCREASE DA RELEASE FROM NIGRAL TRANSPLANTS PLACED IN THE STRIATUM

Neonatal rats were lesioned bilaterally with 6-HDA and given a unilateral mesencephalic transplant. When examined as adults such rats showed extensive

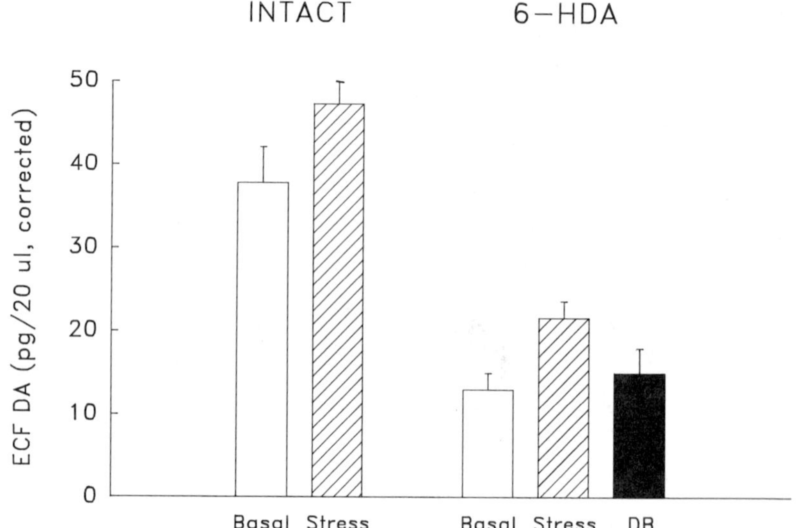

FIGURE 1. Evidence favoring the local initiation of DA release via heterosynaptic input. Extracellular DA level in striatum was assessed under a variety of conditions using *in vivo* microdialysis. 6-HDA-induced lesions reduced the basal level of extracellular DA in striatum by only 67% despite a >85% loss of nigrostriatal DA neurons and no apparent increase in electrophysiological measures of the activity of surviving DA neurons (open bars). Environmental stress produced increases in extracellular DA level in striatum in both intact and 6-HDA-treated rats whereas there is no evidence for increases in nigral DA cell activity under these circumstances (hatched bars). Inactivation of nigral DA neurons by the induction of depolarization block (DB) was not associated with any significant change in extracellular DA level in striatum (solid bar).

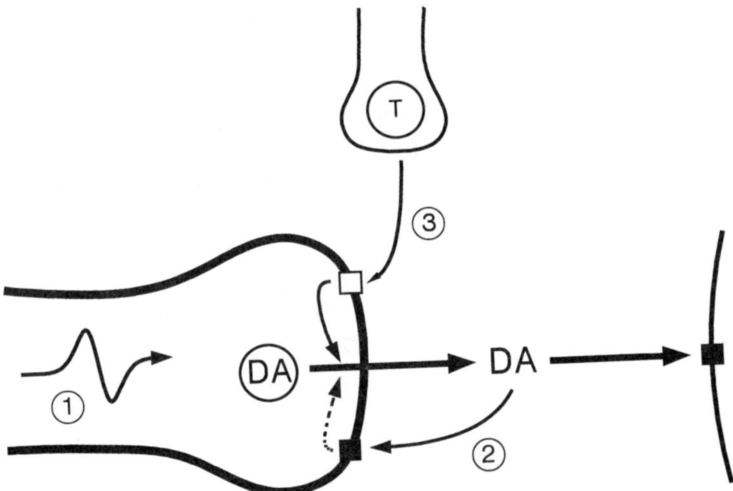

FIGURE 2. A model for multiple influences on transmitter release. (1) Dopamine (DA) is released by classical, Ca^{2+}-dependent exocytosis initiated by the arrival of an action potential. (2) Impulse-dependent DA release can be modulated by local influences such as DA itself acting on presynaptic autoreceptors, as well as other, heterosynaptic influences (not shown). (3) Recent evidence suggests that local factors also can trigger the release of DA in an impulse-independent process. DA released in this manner might involve exocytosis of DA stored in vesicles or the reverse transport of cytoplasmic DA.

sprouting of transplanted DA neurons into the host striatum. Not surprisingly, such animals turned contralaterally when given amphetamine (2 mg/kg, ip). However, the animals also turned contralaterally in response to the stress of tail pinch or cold water.[9]

Collectively these observations suggest that DA release in striatum can be initiated in response to local afferent input independent of the arrival of action potentials (FIGURE 2). DA release has been reported in response to a variety of influences presumably exerted directly on DA terminals. One such influence may be exerted via glutamate released by corticostriatal neurons.[10] It seems probable that the locally initiated release of DA plays an important role in the normal functioning of striatum and may be a key to certain disease states, as well.[11]

REFERENCES

1. BONATZ, A. E., H. J. MORRIS, M. J. ZIGMOND & E. D. ABERCROMBIE. 1989. Soc. Neurosci. Abstr. **15:** 124.
2. HOLLERMAN, J. R., T. W. BERGER & A. A. GRACE. 1986. Soc. Neurosci. Abstr. **12:** 872.
3. STACHOWIAK, M. K., R. W. KELLER, JR., E. M. STRICKER & M. J. ZIGMOND. 1987. J. Neurosci. **7:** 1648–1654.
4. HOLLERMAN, J. R. & A. A. GRACE. 1989. Neurosci. Lett. **96:** 62–88.
5. ABERCROMBIE, E. D., J. R. HOLLERMAN & A. A. GRACE. 1989. Soc. Neurosci. Abstr. **15:** 1002.
6. STRECKER, R. E. & B. L. JACOBS. 1985. Brain Res. **361:** 339–350.
7. ABERCROMBIE, E. D., K. A. KEEFE, D. S. DiFRISCHIA & M. J. ZIGMOND. 1989. J. Neurochem. **52:** 1655–1658.

8. KEEFE, K. A., E. M. STRICKER, M. J. ZIGMOND & E. D. ABERCROMBIE. 1989. Soc. Neurosci. Abstr. **15:** 558.
9. CARDER, R. K., A. M. SNYDER-KELLER & R. D. LUND. 1987. Dev. Brain Res. **33:** 315–318.
10. LONART, G. & M. J. ZIGMOND. Ann. N.Y. Acad. Sci. (This volume.)
11. GRACE, A. A., E. D. ABERCROMBIE & M. J. ZIGMOND. 1989. Soc. Neurosci. Abstr. **15:** 1002.

Relative Importance of Somatodendritic and Terminal Dopaminergic Autoreceptors for the Effects Produced by Systemic (−)3-PPP Administration on Rat Locomotor Activity

SVEN AHLENIUS,[a] VIVEKA HILLEGAART,[a] AND
LENNART SVENSSON[b]

[a]Department of Neuropharmacology
Astra Research Center AB
S-151 85 Södertälje, Sweden

[b]Department of Pharmacology
University of Göteborg
Box 33 031
S-400 33 Göteborg, Sweden

INTRODUCTION

Dopamine autoreceptors exist on cell bodies in the mesencephalon and on terminals in the forebrain.[1,2] In the present experiments we have compared the relative potencies of 3-(3-hydroxyphenyl)-N-n-propylpiperidine (3-PPP) enantiomers on rat locomotor activity after (1) systemic, subcutaneous (sc), administration and (2) intracerebral, nucleus accumbens, administration. (+)3-PPP behaves in most pharmacological models as a full agonist at central dopamine (DA) receptors, whereas (−)3-PPP behaves as a partial agonist.[3] Differences in the potency of (−)3-PPP as compared to (+)3-PPP, at these two administration routes, should be related to different access to somatodendritic and terminal DA autoreceptors.

METHODS

Adult male Sprague-Dawley rats, 280–320 g, were used (ALAB, Laboratorie-tjänst, Sollentuna, Sweden). The animals were kept under standard laboratory conditions on a reversed 12:12 hour light-dark cycle (lights on 18.00). Under deep barbiturate anesthesia, some of the animals were provided with guide cannulas (21G) fixed to the skull by means of acrylic dental cement. These animals were allowed one week of postoperative recovery before the local application of 3-PPP enantiomers into the nucleus accumbens. Injections were made by means of 25G cannulas and the volume was 1 μl, infused at a rate of 0.67 μl minute^{-1}. Locomotor activity was observed in an open field (approx. 0.5 m^2), as described elsewhere.[4]

RESULTS AND DISCUSSION

In comparison with $(-)$3-PPP, $(+)$3-PPP initially produced a stronger inhibition of locomotor activity after systemic administration. This observation, together with the stimulation of activity by high $(+)$3-PPP doses, probably reflects a higher efficacy of $(+)$3-PPP as compared to $(-)$3-PPP at both DA presynaptic autoreceptors and postsynaptic receptors (FIGURE 1). The fact that the relative potency of the 3-PPP enantiomers was reversed after local application into the nucleus accumbens (FIGURE 2) is best explained by less sensitive autoreceptors in the terminal region as compared to the somatodendritic autoreceptors. Furthermore, and probably related to less sensitivity also at the postsynaptic DA receptor, the suppression of activity produced by the local application of $(-)$3-PPP should to a great extent be due to inhibition of postsynaptic DA receptors rather than stimulation of presynaptic autoreceptors in this region.

In conclusion, as indicated by these observations, the suppression of locomotor activity induced by systemically administered $(-)$3-PPP is due to the combined action of *stimulation* of somatodendritic DA autoreceptors and *blockade* of postsynaptic DA receptors, whereas stimulation of terminal DA autoreceptors by $(-)$3-PPP appears to be of less importance under these conditions.

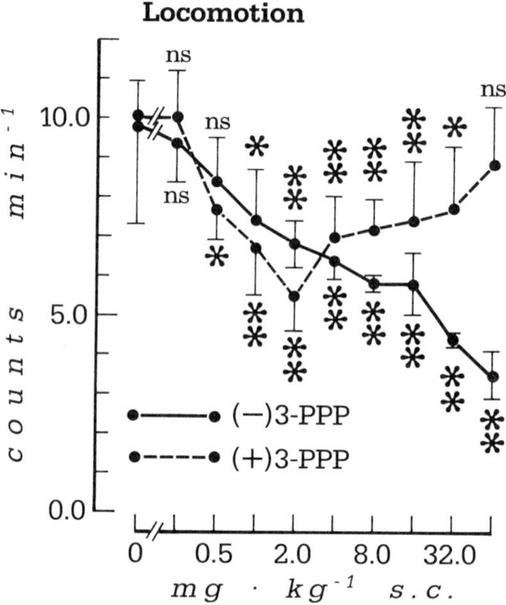

FIGURE 1. Effects of systemic administration of 3-PPP enantiomers on rat locomotor activity. The 3-PPP enantiomers were administered sc 20 minutes before the start of a 15 minute open field session. Shown are the means ± standard deviations (SD) based on 5 observations per group. Statistical evaluation by means of a two-way ANOVA, followed by t-tests for comparisons with saline treated controls as indicated in the figure.[5] *Pretreatment:* $F_{1,80} = 20.53, p < 0.001$. *Dose:* $F_{9,80} = 16.32, p < 0.001$. *Pretreatment × Dose:* $F_{9,80} = 7.96, p < 0.001$. Not significant (ns), $p > 0.05$; *, $p < 0.05$; **, $p < 0.01$.

Locomotion

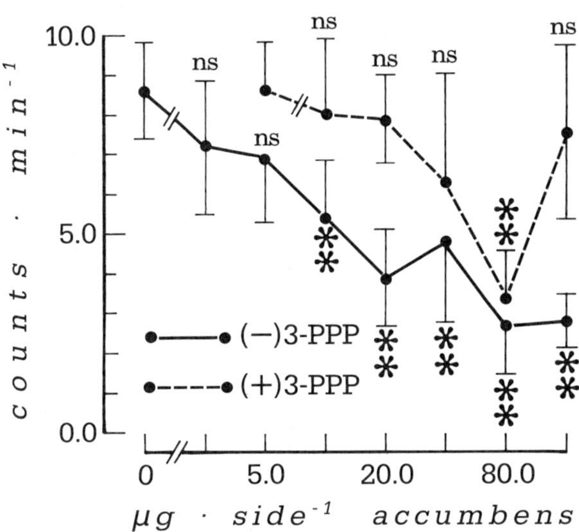

FIGURE 2. Effects of intracerebral (nucleus accumbens) administration of 3-PPP enantiomers on rat locomotor activity. The 3-PPP enantiomers were locally applied into the nucleus accumbens 6 minutes before the start of a 9 minute open field session. Shown are the means ± SD based on 4–8 observations per group. Statistical analysis by means of a one-way ANOVA, followed by the Dunnett's *t*-test for comparisons with saline-treated controls, as indicated in the figure.[5] *(−)3-PPP:* $F_{7,35} = 13.68$, $p < 0.001$. *(+)3-PPP:* $F_{5,22} = 5.29$, $p < 0.01$. ns, $p > 0.05$; **$p < 0.01$.

REFERENCES

1. CARLSSON, A. 1975. *In* Chemical Tools in Catecholamine Research. O. Almgren *et al.*, Eds: 219–225. North Holland. Amsterdam, the Netherlands.
2. ROTH, R. H. 1984. Ann. N.Y. Acad. Sci. **430:** 27–53.
3. CLARK, D., S. HJORTH & A. CARLSSON. 1985. J. Neural Transm. **62:** 1–52.
4. AHLENIUS, S. & V. HILLEGAART. 1986. Pharmacol. Biochem. Behav. **24:** 1409–1415.
5. WINER. 1971. Statistical Principles in Experimental Design. McGraw-Hill. New York, N.Y.

Ultrastructural Immunocytochemical Evidence for Presynaptic Localization of Beta-Adrenergic Receptors in the Striatum and Cerebral Cortex of Rat Brain[a]

CHIYE AOKI AND VIRGINIA M. PICKEL

Division of Neurobiology
Department of Neurology & Neuroscience
Cornell University Medical College
New York, New York 10021

INTRODUCTION

Two important factors determine the sites for catecholaminergic action within the central nervous system (CNS): (1) the pattern of innervation by catecholaminergic axons; and (2) the position of catecholaminergic receptors. Our interests have been to gain precise knowledge of the sites within CNS that are modulated by catecholamines through activation of β-adrenergic receptors (BAR). Both the neocortex and striatum exhibit high densities of BAR, even though the 2 regions receive different degrees of noradrenergic innervation.[1,2] These results suggest that the ultrastructural relation between catecholaminergic axon terminals and BAR may also differ greatly between the two regions. One possibility is that the two regions differ with regard to the relative prevalence of pre- versus postsynaptic BAR.[3] Thus, we used ultrastructural immunocytochemical methods to differentiate pre- from postsynaptic BAR within the neocortex and striatum, then compared the results with previously observed distribution of BAR in the nuclei of the solitary tracts (NTS).[4]

METHODS

The mouse monoclonal antibody against BAR was produced, characterized for its specificity in previous studies,[5] and generously given to us by Dr. C. D. Strader at the Merck, Sharp & Dohme Research Laboratories. Adult male Sprague-Dawley rats were anesthetized with Nembutal, then perfused through the aortic arch with isotonic buffer containing acrolein (3.75%) and paraformaldehyde (2%). Coronal Vibratome sections from brain containing dorsal and ventral striatum (i.e., caudate-putamen nuclei and accumbens nuclei) and cerebral cortex were immunolabeled for BAR using the ABC method.[6] Other Vibratome sections were immunolabeled simultaneously for the catecholamine-synthesizing enzyme tyrosine hydroxylase (TH) by immunoautoradiography and for BAR by the ABC method as previously described.[4]

[a]This work was supported by National Institutes of Health Grants EY08055 and HL18974.

RESULTS

By light microscopy, reddish brown colored reaction product of peroxidase reflecting BAR immunoreactivity (ir) was evident within many of the neuronal *perikarya* in laminae I through VI of the cerebral cortex and in the striatum. Examination of these perikarya by electron microscopy revealed a patchy distribution of BAR-ir (see FIGURE 1) over perikaryal organelles known to be involved in the turnover of mem-

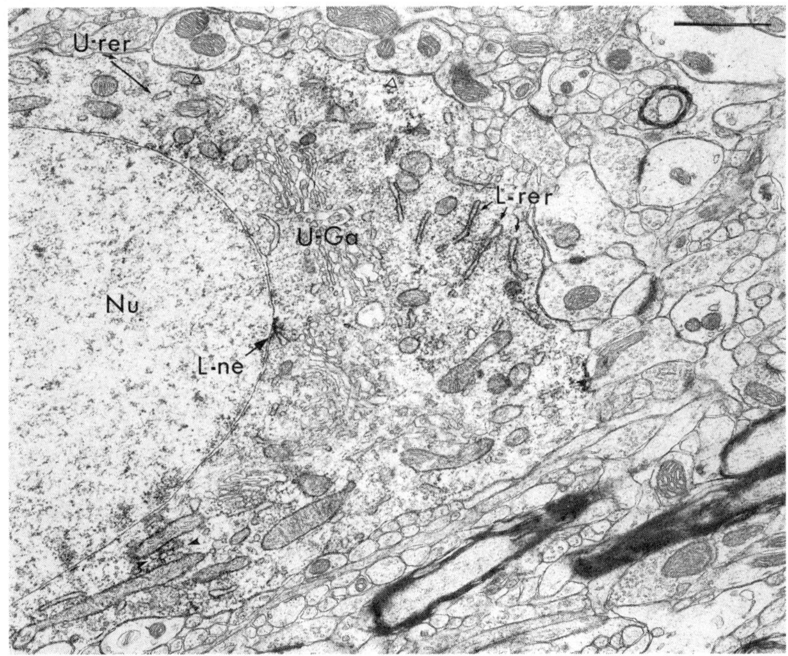

FIGURE 1. Electron micrograph showing perikaryal BAR-ir in layer IV of visual cortex. Electron-dense peroxidase reaction product reflecting BAR is associated with a group of rough endoplasmic reticulum (L-rer), a small portion of smooth endoplasmic reticulum that is continuous with the nuclear envelope (L-ne), and some of the mitochondria (between arrowheads). That these organelles are labeled is evident when compared to other less electron-dense (unlabeled) organelles in the vicinity, such as the Golgi apparatus (U-Ga) and other rough endoplasmic reticulum (U-rer). Open triangles point to unlabeled axosomatic junctions with this neuron. Bar = 1 μm.

brane proteins.[7] These included endoplasmic reticulum, nuclear envelope, mitochondrial outer membrane, and plasma membrane which, when immunolabeled, occurred clustered near other labeled organelles. In contrast, the Golgi apparatus rarely was labeled. BAR-ir in *dendritic shafts* within striatum and cerebral cortex occurred as discrete clumps near the plasma membrane, some of which overlay saccules away from synapses and others that were associated discretely with postsynaptic densities. Within striatum, BAR-ir was also evident along intracellular surfaces of plasma membranes and vesicles within *axon terminals* (see FIGURE 2). BAR-ir axons formed both symmetric and asymmetric types of synapses with unlabeled dendrites. Analyses of dually immunolabeled ultrathin sections revealed that most TH-containing catecholaminergic synapses do not exhibit BAR-ir presynaptically.

DISCUSSION

The localization of BAR-immunoreactivity along cytoplasmic surfaces of neuronal membranes indicates that BAR protein within rat brain neurons contains a portion that is nearly homologous with a peptide sequence within the third intracellular loop of hamster lung BAR. Since a significant portion of the antigenicity resided over

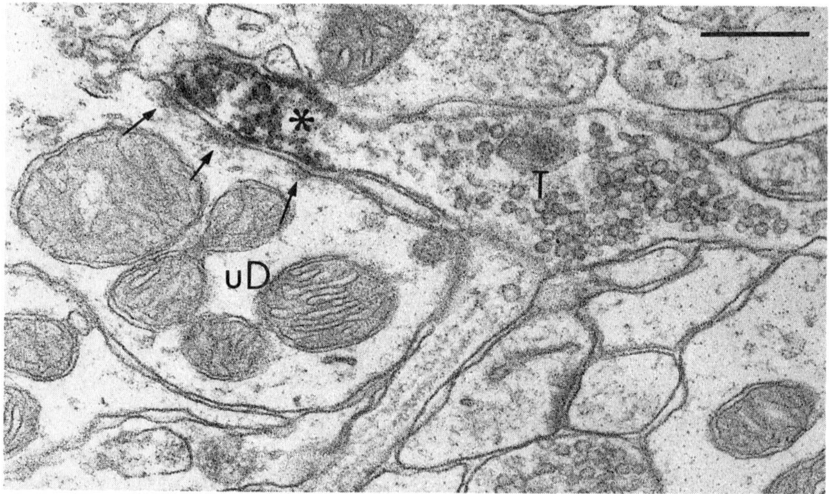

FIGURE 2. Electron micrograph showing presynaptic BAR-ir within the caudate putamen nucleus. The electron-dense peroxidase reaction product (near asterisk) reflecting BAR is evident within a process identifiable as an axon terminal (T) by the cluster of clear vesicles and parallel alignment of the plasma membrane juxtaposed to an unlabeled dendrite (uD). BAR-ir is associated with the presynaptic membrane and nearby vesicles but not with the postsynaptic densities (arrows). Bar = 0.25 μm.

perikaryal organelles involved in the turnover of membrane protein, BAR that are newly synthesized, desensitized and degraded may all be recognized by the antibody. The presence of BAR-ir within nonsynaptic as well as synaptic membranes within dendrites further supports the idea that the antibody immunoreacts with functionally different forms of the receptor protein.

While the perikaryal and dendritic immonolabeling occurred both in the striatum and cerebral cortex, BAR-ir within axons was more frequently encountered within the striatum. The presynaptic BAR in striatum differed from those seen in the NTS, where labeling was evident along both pre- and postsynaptic membranes but not presynaptic vesicles. These ultrastructural observations suggest regional specificity in BAR's presynaptic modulation of neurotransmission. Specifically, since dual labeling for TH and BAR failed to reveal colocalization of the two antigens within axon terminals of the striatum, presynaptic BAR-ir may reflect sites for axoaxonic interactions between catecholamines and noncatecholamines in this area.

REFERENCES

1. RAINBOW, T. C., PARSON B. & B. B. WOLFE. 1984. Proc. Nat. Acad. Sci. USA **81:** 1585.
2. FALLON, J. H. & S. E. LOUGHLIN. 1987. *In* Cerebral Cortex. E. G. Jones & A. Peters, Eds. **6:** 41. Plenum Press.
3. REISINE, T. D., J. I. NAGY, K. BEAUMONT, H. C. FIBIGER & H. I. YAMAMURA. 1979. Brain Res. **177:** 241.
4. AOKI, C., B. A. ZEMCIK, C. D. STRADER & V. M. PICKEL. 1989. Brain Res. **493:** 331.
5. ZEMCIK, B. A. & C. D. STRADER. 1988. BIOCHEM. J. **251:** 333.
6. HSU, S. M., L. RAINE & H. FANGER. 1981. J. Histochem. Cytochem. **29:** 577.
7. VERNER, K. & G. SCHATZ. 1988. Science **241:** 1307.

The Serotonin Autoreceptor in Hypothalamus as Studied by Microdialysis in Awake Rats[a]

S. B. AUERBACH AND J. J. RUTTER

Department of Biological Sciences
Nelson Laboratory
Rutgers University
Piscataway, New Jersey 08855

Numerous *in vitro* studies have suggested the presence of an autoreceptor on serotonergic (5-HT) nerve terminals. A putative 5-HT_1 antagonist methiothepin increases stimulated release of 5-HT *in vitro*.[1] Reciprocally, agonists of the 5-HT_1 subtype such as RU24969 inhibit 5-HT release from superfused synaptosomes and slices prepared from rat brain.[2] Activation of the terminal autoreceptor appears to decrease the availability of calcium for stimulation of exocytosis and also to decrease synthesis of 5-HT.[3,4] This may be a mechanism for negative feedback regulation of 5-HT release. We have used microdialysis and high performance liquid chromatography-electrochemical detection (HPLC-EC) to measure extracellular 5-HT in rat hypothalamus to determine *in vivo* (1) if the putative terminal autoreceptor can be activated by local perfusion of 5-HT_1 agonists, and (2) by local perfusion of an antagonist, whether ambient levels of 5-HT are great enough to activate the autoreceptor.

Male Sprague-Dawley rats were stereotaxically implanted with guide tubes aimed at the ventral medial hypothalamus (VMH). Following at least one week for recovery, on the evening before an experiment, a concentric style microdialysis probe was cemented into the guide tube so that the cellulose tip (4 mm) extended into the VMH. Subjects were attached to a fluid swivel that allowed free movement in a 10 gallon aquarium while being perfused with artificial cerebrospinal fluid at the rate of 1 μl/minute. Early the next morning, 30 minute sample collections were begun about 90 minutes prior to lights out in a 12:12 hour reversed light-dark cycle. Monoamines were analyzed by HPLC with an EG&G PARC dual potentiostat electrochemical detector.[5] The detection limit for 5-HT was approximately 300 fg based on a signal to noise ratio of 3 to 1. The experimental protocol involved administration of drugs in the dialysis solution for two hours. Changes across time were compared to three hours of baseline and two hours of posttreatment sampling. Statistical significance was tested by one-way ANOVA with repeated measures followed by Duncan's multiple range test.

Without uptake inhibition, basal levels of 5-HT analyzed the morning after implantation of the dialysis probe were stable and averaged 1.08 ± 0.09 pg/30 minute sample. This was approximately three times the detection limit of the HPLC-EC assay. After systemic administration of the somatodendritic autoreceptor agonist 8-OH-DPAT [250 μg/kg subcutaneously (SC)] 5-HT was reduced to $36 \pm 8\%$ of baseline [$F(13,39) = 6.90$ $p < 0.0001$ $n = 4$). This suggests that a large percentage of the 5-HT in our microdialysis experiments was neuronal in origin since this dose of 8-OH-DPAT produces a sustained inhibition of serotonergic neuronal discharge.[6]

During perfusion with the putative 5-HT terminal autoreceptor agonist RU24969

[a]This work was supported in part by National Science Foundation Grant BNS-8708014.

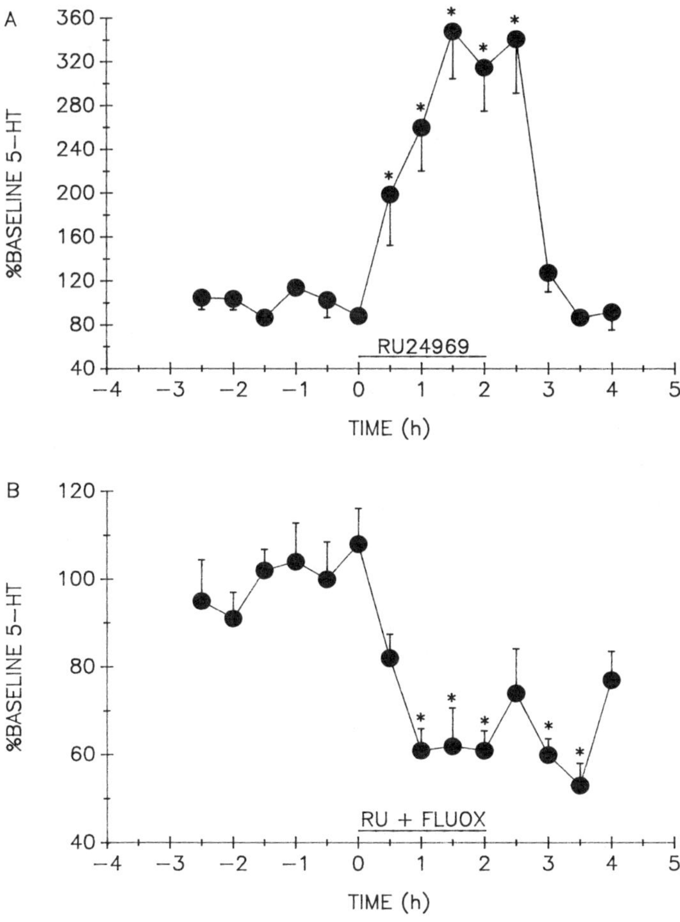

FIGURE 1. (A) Effect of local infusion of RU24969 into the hypothalamus on 5-HT. Beginning at time 0, RU24969 at a concentration of 1 μM was included in the dialysis solution for the two hour period indicated by the horizontal bar. (B) Effect of local infusion of RU24969 into the hypothalamus when uptake was inhibited. Fluoxetine (10 μM) was present in the dialysis solution throughout the experiment to block the 5-HT uptake carrier. Beginning at time 0, RU24969 (1 μM) was included in the dialysis solution for the two hour period indicated by the horizontal bar. Mean values with vertical bars representing standard errors of the mean (SEM) are shown. Asterisks indicate values that are significantly different from baseline according to Duncan's multiple range test ($p < 0.05$).

at a concentration of 1 μM in the dialysis solution, 5-HT levels were increased at 1.5 hours to 348 ± 43% of baseline (FIGURE 1A, $F(13,69) = 14.74$ $p < 0.0001$ $n = 5$). Since uptake inhibitors were used in most previous *in vitro* studies of 5-HT terminal autoreceptor agonists, we reexamined the *in vivo* effects of RU24969 during local perfusion with the selective 5-HT uptake blocker fluoxetine at a concentration of 10 μM in the dialysis solution. With fluoxetine present, baseline 5-HT levels were increased over sixfold to 6.8 ± 0.3 pg/30 minute sample. With the uptake inhibitor

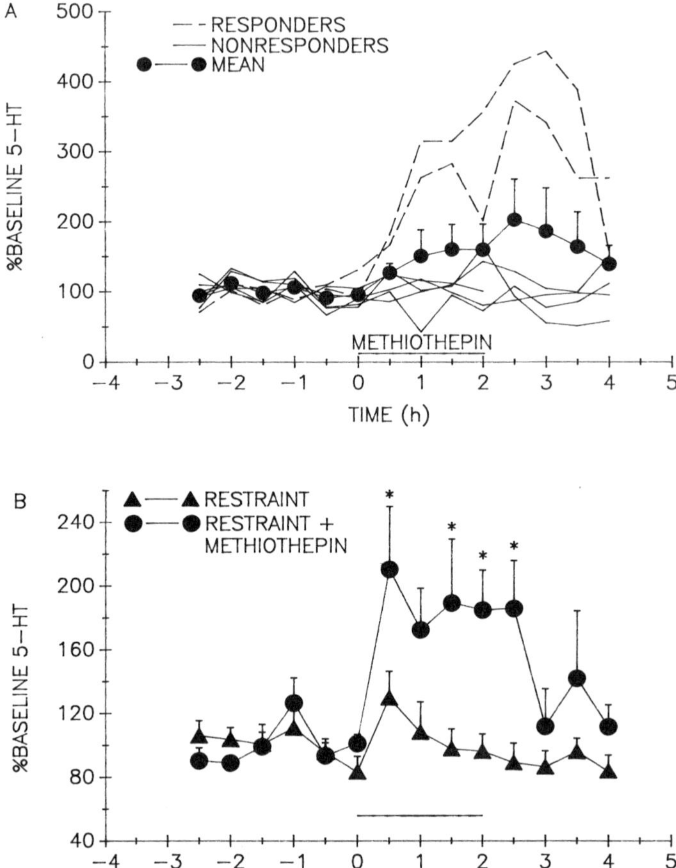

FIGURE 2. (A) Effect of local infusion of methiothepin into the hypothalamus on 5-HT. Beginning at time 0, methiothepin at a concentration of 50 μM was included in the dialysis solution for the two hour period indicated by the horizontal bar. Data from seven individual rats are plotted separately. In addition, the means from these data are also shown with vertical bars representing SEM. (B) Effect of restraint, and of local infusion of methiothepin during restraint. Six rats were manually restrained for the two hour period indicated by the horizontal bar. Five other rats were similarly restrained while methiothepin was infused into the hypothalamus at a concentration of 50 μM. Mean values with vertical bars representing SEM. are shown. Asterisks indicate values that are significantly different from baseline according to Duncan's multiple range test ($p < 0.05$).

present throughout the experiment 5-HT levels were decreased to 61 ± 2% of baseline levels at one hour after the beginning of perfusion with 1 μM RU24969 [FIGURE 1B $F(14,56) = 9.01$ $p < 0.0001$ $n = 5$].

The putative 5-HT$_1$ antagonist methiothepin was infused into the hypothalamus at a concentration of 50 μM in the dialysis solution to determine if the biophase concentration of 5-HT is sufficient to activate the terminal autoreceptor. Serotonin was increased in only two of seven trials and the overall effect of methiothepin was not significant when undisturbed animals were tested (FIGURE 2A). Electrophysiological studies suggest that 5-HT release is increased during wakefulness.[7] In a second series of experiments with methiothepin, rats were subjected to two hours of manual restraint to determine if the terminal autoreceptor is preferentially activated by ambient levels of 5-HT during behavioral arousal. As shown in FIGURE 2B, 5-HT was not significantly elevated during restraint alone ($n = 6$) but was reliably increased when methiothepin (50 μM) was infused into the hypothalamus during the period of stress [$F(13,52) = 3.74$ $p < 0.0003$ $n = 5$).

In conclusion, 5-HT release was inhibited by the terminal autoreceptor agonist RU24969 only when uptake was blocked by fluoxetine. This suggests that there are two distinct nerve terminal sites of action: (1) the 5-HT uptake carrier with some 5-HT$_1$ agonists acting either to produce a reversal of 5-HT transport or to block 5-HT uptake, (2) the autoreceptor, with agonists acting to inhibit 5-HT release. The ability of RU24969 to reduce 5-HT when extracellular levels were elevated sixfold by uptake inhibition indicates that binding of ambient 5-HT to the terminal autoreceptor was not saturated even at levels much higher than the normal physiological range. In support of this conclusion, 5-HT levels were not increased during undisturbed behavior by infusion of the antagonist methiothepin. Methiothepin only produced a reliable increase in 5-HT release during the two hours of restraint stress. These results suggest that the negative feedback effect of 5-HT acting at the terminal autoreceptor is important only during prolonged behavioral arousal.

REFERENCES

1. CERRITO, F. & M. RAITERI. 1979. Eur. J. Pharmacol. **57:** 427–430.
2. MIDDLEMISS, D. N. 1985. J. Pharm. Pharmacol. **37:** 434–437.
3. GOTHERT, M. 1980. Naunyn Schmiedebergs Arch. Pharmacol. **314:** 223–230.
4. CARLI, M., R. INVERNIZZI, L. CERVO & R. SAMANIN. 1988. Psychopharmacology **94:** 359–364.
5. AUERBACH, S. B., M. MINZENBERG & L. O. WILKINSON. 1989. Brain Res. **499:** 281–290.
6. SPROUSE, J. S. & G. K. AGHAJANIAN. 1987. Synapse **1:** 3–9.
7. JACOBS, B. J. 1985. *In* Neuropharmacology of Serotonin. A. R. Green, Ed.: 196–217. Oxford University Press. Oxford, England.

Effects of Endogenously Released Dopamine on [³H]γ-Aminobutyric Acid Efflux from Rat Striatal Slices

SANDOR BERNATH AND MICHAEL J. ZIGMOND

Department of Behavioral Neuroscience
Center for Neuroscience
University of Pittsburgh
570 Crawford Hall
Pittsburgh, Pennsylvania 15260

Dopamine (DA) containing terminals of the nigrostriatal bundle make synaptic contacts with GABAergic cells.[1] However, there are contradictory findings regarding the impact of these afferents. Recently, there have been reports suggesting that DA can influence striatal γ-aminobutyric acid (GABA) release through both DA receptor subtypes, the D1-mediated response being excitatory, and the D2 response being inhibitory.[2] Similar observations in the substantia nigra also have been reported.[3] Our investigations suggest that DA has a third effect on GABA efflux in the striatum, one that is not mediated through either type of receptor. Using slices from rat striatum we have observed that DA accelerates resting GABA efflux as well as electrically evoked GABA overflow through a process that is insensitive to D1 and D2 antagonists but can be blocked by nipecotic acid, an inhibitor of GABA uptake.[4]

In the current study, the impact of electrically and chemically released DA on [³H]GABA efflux from rat striatal slices was measured. Male, Sprague-Dawley rats, weighing 200–300 g, were used throughout these experiments. Animals were killed by decapitation, the brain was quickly removed, and coronal slices (350 μm) of the rostral part of the striatum were made. Slices were preincubated at 25°C in the presence of [³H]GABA (0.1 μM). After 60 minutes of preperfusion, 5-minute fractions were collected and analyzed for tritium by liquid scintillation spectroscopy. Slices were stimulated twice for 3 minutes using biphasic pulses (20 mA, 3-msecond pulse duration with 1-msecond delay, 10 Hz for 9 seconds). These stimulation periods coincided with the beginning of the third (S_1) and twelfth (S_2) 5-minute fractions.

Spontaneous tritium efflux gradually decreased throughout the course of the

TABLE 1. Effect of SCH 23390 and Sulpiride on Spontaneous and Electrically Evoked GABA Efflux[a]

	Concentration (μM)	B_2/B_1	S_2/S_1	(n)
Control		0.86 ± 0.03	0.76 ± 0.05	(17)
SCH 23390	1	0.93 ± 0.04	0.48 ± 0.08[b]	(8)
	10	0.87 ± 0.05	0.35 ± 0.03[b]	(8)
Sulpiride	1	0.93 ± 0.02	0.76 ± 0.09	(4)
	10	0.78 ± 0.05	0.77 ± 0.05	(4)

[a]Drugs were added to the superfusion fluid 20 minutes before S_2. Values represent mean ± standard error of the mean (SEM).
[b]$p < 0.05$ releative to control.

TABLE 2. Effect of Amphetamine in the Presence and Absence of Nipecotic Acid[a]

	Nipecotic Acid	B_2/B_1	S_2/S_1	(n)
Control	—	0.81 ± 0.05	0.73 ± 0.08	(4)
Amphetamine	—	1.10 ± 0.08^b	0.44 ± 0.09^b	(4)
Control	+	0.73 ± 0.05	0.71 ± 0.03	(4)
Amphetamine	+	0.85 ± 0.02	0.20 ± 0.05^b	(4)

[a] Where indicated nipecotic acid (100 μM) was present throughout the superfusion, while amphetamine (100 μM) was added 20 minutes before S_2. Values represent mean $\pm$ SEM.
[b] $p < 0.05$ relative to control.

experiments ($B_2/B_1 = 0.86 \pm 0.03$; $n = 17$). Similarly, electrically evoked overflow decreased somewhat with each successive stimulation ($S_2/S_1 = 0.76 \pm 0.05$; $n = 17$). SCH 23390 (1–10 μM) had no significant effect on the basal efflux of tritium. However, electrically-evoked GABA overflow was decreased (-37%) when SCH 23390, a D1 antagonist, was added 20 minutes before the second stimulation. at a concentration of 1 μm. An even greater decrease in evoked efflux (-54%) was observed with 10 μM SCH 23390. Sulpiride (1–10 μM), a D2 antagonist, had no effect on basal efflux or on electrically evoked release of GABA (TABLE 1).

Amphetamine (100 μM) added to the superfusion fluid 20 minutes before the second stimulation increased the basal efflux of GABA ($+33\%$), while decreasing electrically evoked GABA overflow (-40%). In order to investigate whether the impact of amphetamine on GABA release occurred indirectly via the release of DA, experiments were carried out using animals lesioned with intraventricular 6-hydroxy-dopamine. The lesion, which depleted striatal DA by 98%, completely blocked the stimulatory effect of amphetamine on basal GABA efflux, but failed to abolish the ability of amphetamine to inhibit GABA overflow. SCH 23390 (10 μM) and sulpiride (10 μM) did not modify the impact of amphetamine on either spontaneous or evoked GABA efflux. In contrast, nipecotic acid (100 μM), a GABA uptake inhibitor, prevented the effect of amphetamine on resting GABA efflux, although it potentiated the effect of amphetamine on evoked overflow (TABLE 2).

These results suggest that endogenous DA released by brief electrical field stimulation can stimulate GABA release in striatum through an activation of D1 receptors. Higher concentrations of endogenous DA released by amphetamine can stimulate resting GABA efflux by acting on the high affinity GABA transport system. Finally, amphetamine appears to have an inhibitory influence on evoked GABA overflow that is independent of DA and of the high affinity GABA transporter.

REFERENCES

1. KUBOTA, Y., S. INAGAKI, S. KITO & J.-Y. WU. 1987. Brain Res. **406:** 147–156.
2. GIRAULT, J. A., U. SPAMPINATO, J. GLOWINSKI & M. J. BESSON. 1986. Neuroscience **19:** 1109–1117.
3. STARR, M. 1987. J. Neurochem. **49:** 1042–1049.
4. BERNATH, S. & M. J. ZIGMOND. 1989. Brain Res. **476:** 373–376.

Stereochemical Effects of Mono- and Dihydroxyaporphines on Presynaptic Inhibition of Dopamine Synthesis *In Vitro*[a]

RAYMOND G. BOOTH,[b] ROSS J. BALDESSARINI,[b,d]
NORA S. KULA,[b] AND JOHN L. NEUMEYER[c]

[b]*Departments of Psychiatry and Neuroscience Program*
Harvard Medical School
and
Laboratories for Psychiatric Research
Mailman Research Center
McLean Division of Massachusetts General Hospital
Belmont, Massachusetts 02178

[c]*Section of Medicinal Chemistry*
College of Pharmacy and Allied Health Professions
Northeastern University
Boston, Massachusetts 02115

There is compelling evidence from both *in vivo* and *in vitro* experiments that dopamine (DA) agonists can modulate DA synthesis via presynaptic autoreceptor–mediated inhibition of tyrosine hydroxylase (TH), the rate-limiting step in the biosynthesis of DA.[1] We have synthesized novel enantiomeric mono- and dihydroxyaporphines, rigid molecules which structurally incorporate the active *trans* α-rotamer of DA[2] (FIGURE 1), to probe stereospecific structural requirements of DA autoreceptor function in striatal minces from normal and DA-depleted (reserpinized) rats. Effects of the $R(-)$ and $S(+)$ isomers of the noncatechol aporphine 11-hydroxy-*N*-*n*-propylnoraporphine (11-OH-NPa) and of the corresponding catecholic analogue *N*-*n*-propylnorapomorphine (NPA) on TH were assessed by measuring the formation of $^{14}CO_2$ evolved during the decarboxylation of L-dihydroxyphenylalanine (DOPA) to DA using excess [^{14}C]-L-tyrosine as substrate and minced corpus striatum from rat forebrain as a source of DA-rich nerve endings. This method is based on the marked substrate preference of endogenous aromatic L-amino acid decarboxylase for L-DOPA over L-tyrosine.

Both isomers of NPA fully inhibited TH activity (IC_{50} = 0.3 and 1.0 μM, for $R(-)$- and $S(+)$NPA, respectively; TABLE 1). Furthermore, their effects were fully blocked by the nonselective DA-receptor antagonist fluphenazine as well as by the D_2-selective antagonists spiroperidol and $(-)$sulpiride, but were not affected by the D_1 antagonist SCH-23390. These results suggest a D_2-type autoreceptor-mediated inhibition of DA synthesis with limited enantiomeric selectivity of this catechol aporphine.

The isomers of the monohydroxy analogue, $R(-)$- and $S(+)$-11-OH-NPa (IC_{50} = 42 and 87 μM, respectively; TABLE 1), were about 100-fold less potent than

[a]This work was supported by U.S. Public Health Service (National Institute of Mental Health) grants MH-14275, MH-34006, and MH-47370, and an award from the Bruce J. Anderson Foundation.

[d]Author to whom correspondence should be addressed at Mailman Research Center, McLean Hospital, 115 Mill Street, Belmont, Mass. 02178.

their dihydroxy NPA congeners in inhibiting TH activity in normal tissue. $R(-)11$-OH-NPa appeared to have a biphasic concentration-response function, with a high-potency component. This component was made clearer by depleting endogenous DA [>90% as measured by high performance liquid chromatography-electrochemical detection (HPLC-EC)] by systemic pretreatment with reserpine at 20 and 2 hours prior to sacrifice, with no evidence of up regulation of the binding of [³H]spiroperidol or [³H]SCH-23390 with this acutely reserpinized striatal tissue. Under these conditions, $R(-)11$-OH-NPa was a highly potent but *partial* agonist ($IC_{25} = 7$ nM) and fluphenazine fully blocked its effects. Evidently, actions mediated by endogenous DA contribute to the effect of high concentrations of $R(-)11$-OH-NPa to evoke a full inhibition of DA synthesis, but it also appears to exert autoreceptor-mediated effects with very high potency. Similar to its weak effect in normal tissue, $S(+)11$-OH-NPa showed almost

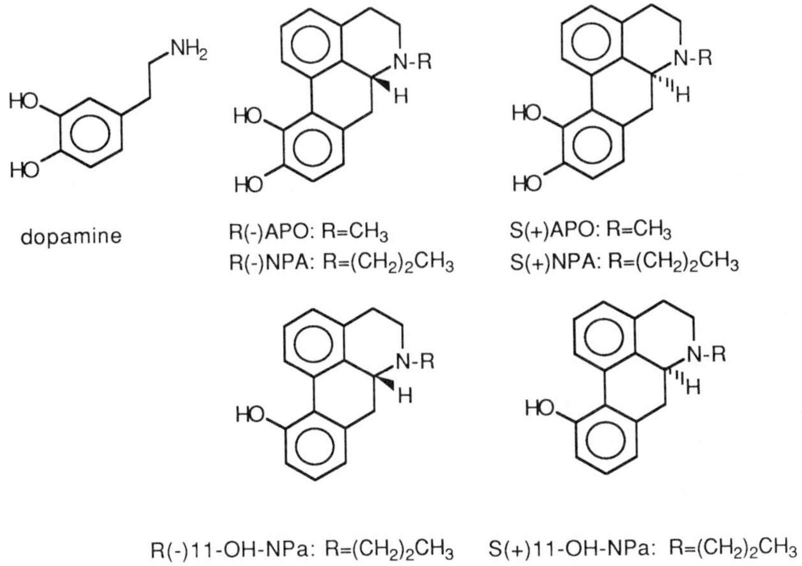

FIGURE 1. Structures of dopamine and aporphines.

no effect on TH activity in DA-depleted tissue at concentrations below 10 μM ($IC_{25} = 30\,\mu$M).

These results demonstrate that NPA acts as a full agonist to inhibit striatal DA synthesis via a presynaptic autoreceptor of the D_2 type with low (ca. threefold) stereoselectivity. This observation contrasts sharply with previous studies of behavioral[3] and neurophysiological[4] expression of central dopaminergic activities as well as with results of *in vitro* striatal binding studies[2] with NPA where the $S(+)$ enantiomer was 100 to 1000 times less potent than $R(-)$NPA, and may even act behaviorally as a DA antagonist.[3] However, the present *in vitro* results are consistent with our previous work assessing *in vivo* DA autoreceptor function as a decrease in formation of DOPA, where the $R:S$ enantiomeric ratio for NPA was ca. fivefold.[5] These findings suggest that, in contrast to its $R(-)$ isomer, $S(+)$NPA may have a greater efficacy at striatal D_2 nerve terminal autoreceptors than at postsynaptic or somatodendritic sites[4] and that $S(+)$NPA might be a selective, full, striatal autoreceptor agonist of moderate potency.

While the apparent loss of efficacy of $R(-)$11-OH-NPa in DA-depleted tissue from reserpinized animals suggests that DA contributes to the effect of high concentrations of this compound, the 2000-fold increase in potency at lower concentrations in reserpinized tissue (with no effect on D_1 or D_2 receptor populations) indicates that $R(-)$11-OH-NPa has high affinity and partial intrinsic activity at the striatal DA autoreceptor as well as very high stereoselectivity. Further support for this conclusion includes the ability of fluphenazine to block fully the high-potency effect of $R(-)$11-OH-NPa on TH. The very high (4300-fold) stereoselectivity ($R > S$) between the 11-OH-NPa isomers after reserpine pretreatment also seems consistent with a receptor-mediated mechanism.

In summary, we found evidence of a D_2 striatal autoreceptor mechanism which accommodated a DA agonist of the 10,11-dihydroxyaporphine structure (NPA) with relatively little stereoselectivity, while the 11-monohydroxyaporphine analogue (11-OH-

TABLE 1. Inhibition of Tyrosine Hydroxylase Activity in Rat Striatal Tissue by Dopamine and Aporphine Isomers

Compound		R/S Potency
	IC_{50} (μM)	
Dopamine	0.2	—
$R(-)$NPA	0.3	
$S(+)$NPA	1.0	3.3
$R(-)$11-OH-NPa	42	
$S(+)$11-OH-NPa	87	2.1
	IC_{25} (μM)	
(Normal tissue)		
$R(-)$11-OH-NPa	14.0	
$S(+)$11-OH-NPa	30.0	2.1
(Reserpinized tissue)		
$R(-)$11-OH-NPa	0.007	
$S(+)$11-OH-NPa	30.0	4300

NPa) was a potent partial agonist that was much more strongly preferred in the $R(-)$ configuration. These results differ from D_2 radioreceptor assay studies with striatal homogenates where the affinity of the NPA isomers was highly stereoselective (100-fold, $R > S$) and 11-OH-NPa and NPA had about equal affinities.[2] Perhaps the structural requirements for affinity or agonist activity at striatal D_2 autoreceptors are different from postsynaptic and other D_2-type receptors. This characteristic may aid the development of autoreceptor-selective agonists.

REFERENCES

1. WOLF, M. E. & R. H. ROTH. 1987. Dopamine autoreceptors. *In* Dopamine Receptors. I. Creese & C. M. Fraser, Eds.: 45–96. Alan R. Liss Inc. New York, N.Y.
2. GAO, Y., R. ZONG, A. CAMPBELL, N. S. KULA, R. J. BALDESSARINI & J. L. NEUMEYER. 1988. Synthesis and dopamine agonist and antagonist effects of $R(-)$ and $S(+)$-11-hydroxy-N-n-propylnoraporphine. J. Med. Chem. **31:** 1389–1396.

3. CAMPBELL, A., R. J. BALDESSARINI, M. J. TEICHER & J. L. NEUMEYER. 1986. Behavioral effects of apomorphine isomers in the rat: selective locomotor-inhibitory effects of $S(+)N$-n-propylnorapomorphine. Psychopharmacology **88:** 158–164.

4. COX, R. F., J. L. NEUMEYER & B. L. WASZCZAK. 1988. Effects of N-n-propylnorapomorphine enantiomers on single unit activity of substantia nigra pars compacta and ventral tegmental area dopamine neurons. J. Pharmacol. Exp. Ther. **247:** 355–362.

5. BALDESSARINI, R. J., E. R. MARSH, N. S. KULA, R. ZONG, Y. GAO & J. L. NEUMEYER. Effects of isomers of hydroxyaporphines on dopamines metabolism in rat brain regions. Biochem. Pharmacol. (In press.)

Regulation of Nerve Impulse Frequency and Transmitter Release by Serotonergic Autoreceptor Agonists

P. A. BRODERICK[a] AND M. F. PIERCEY[b]

[a]*Department of Pharmacology*
City University of New York Medical School
Convent Ave and West 138th Street
New York, New York 10031

[b]*CNS Research*
The Upjohn Company
Kalamazoo, Michigan 49001

Somatodendritic autoreceptors for serotonergic cells belong to the 5-HT_{1A} receptor subtype.[1] The nonbenzodiazepine (NBZD) anxiolytic drugs buspirone and ipsapirone bind with high affinity to the 5-HT_{1A} receptor.[2] In addition, these anxiolytic drugs depress 5-HT nerve impulse frequencies by activation of 5-HT_{1A} autoreceptors.[3] Benzodiazepine anxiolytics also depress nerve impulse generation in serotonergic cells.[4] However, benzodiazepine anxiolytics also depress nerve impulse generation in norepinephrine (NE) cells.[5] In contrast, buspirone has been reported to increase NE cell impulse frequency.[5] The present report evaluates the effects of the NBZD anxiolytics buspirone and ipsapirone on 5-HT and NE nerve impulse generation and transmitter release.

Therefore, we studied the effects of buspirone and ipsapirone on 5-HT nerve impulse frequencies in dorsal raphe and NE nerve impulse frequencies in locus ceruleus, electrophysiologically. We also studied serotonin and norepinephrine release in CA_1 region of hippocampus, voltammetrically. All studies were done *in vivo,* in the chloral hydrate anesthetized animal, species *Rattus norvegicus,* male, Sprague-Dawley (300–400 g). Impulse activity was measured extracellularly with glass micro-electrodes filled with pontamine sky blue in 2 M NaCl, after intravenous administration of buspirone and ipsapirone. Neurotransmitter release was measured extracellularly with stearate electrodes in conjunction with semidifferential voltammetry, after subcutaneous administration of buspirone and ipsapirone (1 mg/kg). Norepinephrine and serotonin voltammetric signals were selectively measured. Dopamine, 3-4-dihydroxyphenylacetic acid (DOPAC), ascorbic acid, 5-hydroxyindoleacetic acid (5-HIAA), and uric acid are not detected at the same oxidation potentials used for the detection of NE and 5-HT. The details for the electrophysiological[3,6] and voltammetric[7,8] techniques are described elsewhere.

Both anxiolytics depressed firing rates of 5-HT neurons; complete inhibition occurred at the highest doses. The ED_{50} for the buspirone effect was 15 μg/kg; the ED_{50} for the ipsapirone effect was 16 μg/kg. Buspirone was just as potent in exciting NE cells ($ED_{25} = 26$ μg/kg) as it was in depressing 5-HT cells. Ipsapirone, however, was not as potent as buspirone in its effect on nerve impulse frequency, i.e., the ED_{25} for excitement of NE cells was 586 μg/kg. Buspirone and ipsapirone depressed 5-HT release in hippocampal CA_1 region. The percent decrease from basal (control) values one hour after buspirone administration was 60%; and one hour after ipsapirone administration, 23%. Both 5-HT_{1A} anxiolytics depressed hippocampal CA_1 norepineph-

"

rine release (buspirone = 98%; ipsapirone = 61%). Depression of norepinephrine release may be mediated at NE nerve terminals, thus bypassing the stimulant effects on NE impulse generation. It is possible that the depression in NE release could be of some importance to the therapeutic and/or side effects of these anxiolytic drugs. Nonetheless, a depression in 5-HT release, coupled with a depression in nerve impulse generation, is compatible with an anxiolytic role for 5-HT$_{1A}$ autoreceptors.

REFERENCES

1. SPROUSE, J. S. & G. K. AGHAJANIAN. 1987. Synapse **1**: 3–9.
2. PEROUTKA, S. J. 1985. Biol. Psychiatry **20**: 971–979.
3. LUM, J. T. & M. F. PIERCEY. 1988. Eur. J. Pharmacol. **149**: 9–15.
4. GALLAGHER, D. W. 1978. Eur. J. Pharmacol. **49**: 133–143.
5. SANGHERA, M. K. & D. W. GERMAN. 1983. J. Neural Transm. **57**: 267–279.
6. CEDARBAUM, J. M. & G. K. AGHAJANIAN. 1976. Brain Res. **112**: 413–419.
7. BRODERICK, P. A. 1988. Neurosci. Lett. **95**: 275–280.
8. BRODERICK, P. A. 1989. Brain Res. **495(1)**: 115–121.

GABA$_B$ Receptors Control GABA Release of Neocortical Neurones

R. A. DEISZ AND W. ZIEGLGÄNSBERGER

Clinical Neuropharmacology
Max-Planck-Institute for Psychiatry
Kraepelinstrasse 2
8000 Munich 40, Federal Republic of Germany

A wealth of evidence from neurochemical studies suggests that GABA (γ-aminobutyric acid) analogues may modulate the release of GABA. Most of these investigations used fairly crude procedures such as high K$^+$ induced transmitter release (see Reference 13 for review), but whether this phenomenon occurs under physiological conditions remains unclear. A conclusive link to the well-established reduction of inhibitory postsynaptic potentials (IPSPs) at higher frequencies of stimulation in hippocampal pyramidal cells[2,11] is lacking. Establishment of this connection is hampered by the problem of separating the relative contribution of several conceivable mechanisms to IPSP depression.

For instance, any decline of inhibition may be brought about by a postsynaptic desensitization of GABA receptors as demonstrated in isolated hippocampal neurones.[12] In addition, shifts of the Cl$^-$ equilibrium potential reduced the postsynaptic response even before the onset of desensitization.[10] In neurones of the hippocampal slice the decline of IPSPs at higher frequencies of stimulation was associated with a significant elevation of extracellular K$^+$ concentration,[11] which may indirectly affect inhibition. Increases of extracellular K$^+$ concentration were shown to diminish the inhibitory driving force,[5] as recently confirmed in neocortical neurones.[14] Furthermore, the reduction of inhibition may be related to GABA uptake. In some hippocampal neurones nipecotic acid, a blocker of GABA uptake, slightly prolonged IPSPs.[8] But nipecotic acid virtually abolished IPSPs in the majority of neurones.[8] Similar observations had led to the proposal of a negative feedback control of GABA release.[1] Here we briefly review electrophysiological data of IPSPs from neocortical neurones that provide evidence for a negative feedback of GABA on its own release.

Orthodromic stimulation of the neocortex *in vitro* evokes a complex synaptic pattern.[4,7,9] Two temporally distinct inhibitory postsynaptic potentials (IPSP$_A$ and IPSP$_B$) were analyzed from current voltage relationships, revealing different conductances of about 200 nS and 24 nS and reversal potentials of -73 mV and -89 mV, respectively (see References 7 and 9 for details of the methods). The conductances of IPSP$_A$ and IPSP$_B$ were fairly stable at a stimulus frequency of 0.1 Hz. At higher frequencies both IPSPs were rapidly attenuated, and after about 5 stimuli a steady state was reached. The average decrease in synaptic conductance between 0.1 and 1 Hz was 80% for the IPSP$_A$ and 60% for the IPSP$_B$. At these frequencies the reversal potentials decreased marginally. Membrane potential (E_m) and input resistance (R_{in}) were not affected.[7] Application of nipecotic acid (50 to 500 μM) reduced the conductance of both IPSPs evoked at low frequencies, but had no effects at higher stimulation rates. E_m and R_{in} were not consistently affected by nipecotic acid.

Local application of GABA in the presence of nipecotic acid revealed a slightly enhanced postsynaptic response.[7] The behavior of the conductance versus frequency plot in the presence of nipecotic acid (see FIGURE 1) is most readily explained by a prolonged time course of GABA in the synaptic cleft following the reduction of GABA uptake. GABA application also reduced the conductance of IPSPs by up to 50% for tens of seconds after postsynaptically detectable effects of GABA had dissipated. Application of baclofen, a selective agonist for GABA_B receptors,[3] elicited a small hyperpolarization and conductance increase. In addition, baclofen markedly and reversibly reduced EPSPs and both components of the IPSP.[7,9] The attenuation of IPSPs by baclofen in the face of unaltered postsynaptic GABA responses[9,15] indicates that GABA release was affected.

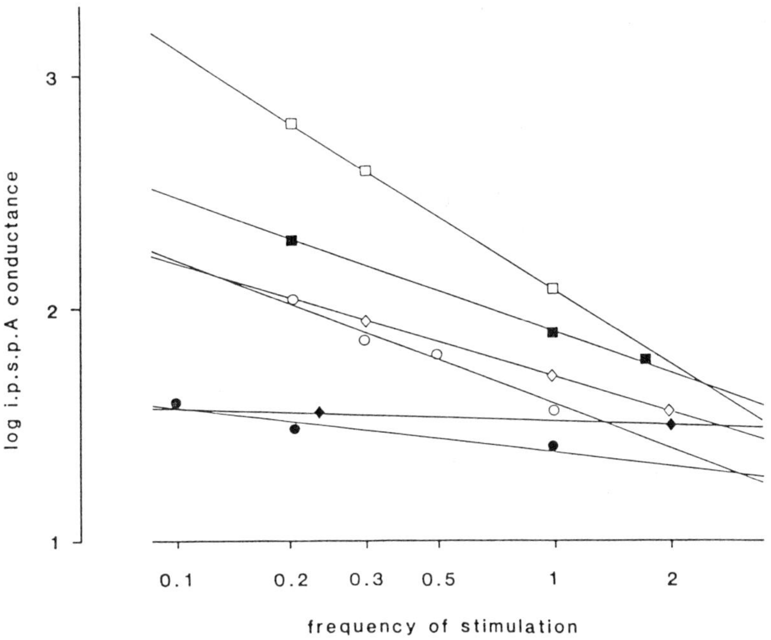

FIGURE 1. Effects of nipecotic acid on frequency dependence of IPSP_A. The logarithm of IPSP_A conductance has been plotted versus the logarithm of the frequency of stimulation for three neurones before (open symbols) and after (filled symbols) bath application of nipecotic acid. The different symbols denote three individual neurones. Note the intersection near about 2 Hz, indicating that the effects of nipecotic acid are smaller at higher frequencies of stimulation. (From Reference 7 with permission.)

In conclusion, several lines of evidence indicate that the attenuation of IPSPs by higher frequency of stimulation is located at the inhibitory synapse per se and related to a feedback control of GABA release.[7] Depression of IPSPs by nipecotic acid at low frequencies, in the face of enhanced postsynaptic responses to exogeneous GABA, indicates that the GABA transient plays a key role.[7] The prolonged presence of GABA in the synaptic cleft may cause extended presynaptic effects. The reduction of IPSPs by baclofen suggests that a GABA_B receptor is involved, perhaps modulating Ca^{2+} currents.[6] A similar conclusion was recently reached for hippocampal neurones.[15]

REFERENCES

1. AICKIN, C. C. & R. A. DEISZ. 1981. Pentobarbitone interference with inhibitory synaptic transmission in crayfish stretch receptor. J. Physiol. **315:** 175–187.
2. BEN-ARI, Y., K. KRNJEVIC & W. REINHARDT. 1979. Hippocampal seizures and failure of inhibition. Can. J. Physiol. Pharmacol. **57:** 1462–1466.
3. BOWERY, N. G., D. R. HILL, A. L. HUDSON, A. DOBLE, D. N. MIDDLEMISS, J. SHAW & M. TURNBULL. 1980. (−)Baclofen decreases neurotransmitter release in the mammalian CNS by an action at a novel GABA receptor. Nature **283:** 92–94.
4. CONNORS, B. W., M. J. GUTNICK & D. A. PRINCE. 1982. Electrophysiological properties of neocortical neurons in vitro. J. Neurophysiol. **48:** 1302–1320.
5. DEISZ, R. A. & H. D. LUX. 1982. The role of intracellular chloride in post-synaptic inhibition of crayfish stretch receptor neurones. J. Physiol. **326:** 123–138.
6. DEISZ, R. A. & H. D. LUX. 1985. γ-Aminobutyric acid–induced depression of calcium currents of chick sensory neurons. Neurosci. Lett. **56:** 205–210.
7. DEISZ, R. A. & D. A. PRINCE. 1989. Frequency-dependent depression of inhibition in guinea-pig neocortex *in vitro* by GABA$_B$ receptor feed-back on GABA release. J. Physiol. **412:** 513–541.
8. DINGLEDINE, R. & S. J. KORN. 1985. γ-Aminobutyric acid uptake and the termination of inhibitory synaptic potentials in the rat hippocampal slice. J. Physiol. **366:** 387–409.
9. HOWE, J. R., B. SUTOR & W. ZIEGLGÄNSBERGER. 1987. Baclofen reduces post-synaptic potentials of rat neocortical neurones by an action other than its hyperpolarizing action. J. Physiol. **384:** 539–569.
10. HUGUENARD, J. R. & B. E. ALGER. 1986. Whole-cell voltage-clamp study of the fading of GABA-activated currents in acutely dissociated hippocampal neurons. J. Neurophysiol. **56:** 1–18.
11. MCCARREN, M. & B. E. ALGER. 1985. Use-dependent depression of IPSPs in rat hippocampal pyramidal cells in vitro. J. Neurophysiol. **53:** 557–571.
12. NUMANN, R. E. & R. K. S. WONG. 1984. Voltage-clamp study on GABA response desensitization in single pyramidal cells dissociated from the hippocampus of adult guinea pigs. Neurosci. Lett. **47:** 289–294.
13. STARKE, K., M. GÖTHERT & H. KILBINGER. 1989. Modulation of neurotransmitter release by presynaptic autoreceptors. Physiol. Rev. **69:** 864–989.
14. THOMPSON, S. M., R. A. DEISZ & D. A. PRINCE. 1988. Relative contributions of passive equilibrium and active transport to the distribution of chloride in mammalian cortical neurons. J. Neurophysiol. **60:** 105–124.
15. THOMPSON, S. M. & B. H. GÄHWILER. 1989. Activity-dependent disinhibition III. Desensitization and GABA$_B$ receptor mediated presynaptic inhibition in the hippocampus in vitro. J. Neurophysiol. **61:** 524–533.

An α-Adrenoceptor Facilitating the Stimulation-Evoked Acetylcholine Release in the Rat Perfused Heart

HERMANN FUDER AND IRENE T. BOGNAR

Department of Pharmacology
University of Mainz
Obere Zahlbacher Strasse 67
6500 Mainz, Federal Republic of Germany

An α_1-adrenoceptor is known to inhibit the tritiated acetylcholine overflow evoked by field stimulation or high potassium solution from rat isolated atria.[1,2] In order to investigate more in detail the prejunctional modulation of acetylcholine release from the whole rat heart, we set up a preparation to allow us a selective electrical stimulation of the extrinsic intact right and left vagal nerves outside the rat heart perfused according to Langendorff with Tyrode solution at 35°C. The acetylcholine stores of the cholinergic nerves were selectively labeled by infusion of ^{14}C-choline (2-4 μCi/heart, 0.36–0.72 μmol/l, for 13 minutes) during intermittent stimulation of cholinergic nerves with 120 trains of 40 pulses at 20 Hz at 5 second intervals. The evoked ^{14}C-actylcholine/choline overflow was suppressed by tetrodotoxin 0.3 μmol/l, abolished in calcium-free Tyrode solution, and reduced by about 80% by hexamethonium 500 μmol/l. Total ^{14}C-activity was largely accounted for by choline activity in the absence of vagal nerve stimulation, but upon stimulation with 10 Hz (2 minutes), both choline and acetylcholine were increased when investigated by paper chromatography as described by Muscholl and Muth.[3]

($-$)-Noradrenaline 10 μmol/l inhibited the ^{14}C-acetylcholine/choline overflow evoked by vagal nerve stimulation (3 Hz, 24 trains of 30 pulses at intervals of 13 seconds) in the presence of propranolol 0.2 and yohimbine 0.1 μmol/l (FIGURE 1). Surprisingly, high oxymetazoline and xylometazoline concentrations enhanced the stimulation-evoked overflow about twofold above controls in a concentration-dependent (FIGURE 1) and reversible manner.

Neither clonidine nor methoxamine affected the evoked ^{14}C-acetylcholine/choline overflow (FIGURE 1). None of the agonists changed the basal overflow. The oxymetazoline-induced increase in evoked overflow was significantly reduced or abolished by some of the selective α-adrenoceptor antagonists investigated. None of the antagonists affected the basal overflow or the stimulation-evoked overflow in the absence of oxymetazoline. Whereas the high concentrations of α_2-selective (idazoxan 5, rauwolscine 1, yohimbine 5 μmol/l, TABLE 1) and low concentrations of α_1-selective antagonists {(S)-WB 4101 [(S)-2-(2,6-dimethoxyphenoxyethyl)aminomethyl-1,4-benzodioxane] 0.01, corynanthine 3 μmol/l} failed to affect the oxymetazoline-induced increase, oxymetazoline was less active or inactive in the presence of prazosin 0.03, (S)-WB 4101 0.1 (all not shown), SK&F 104078 {6-chloro-9-[(3-methyl-2-butenyl)oxyl]-3-methyl-1H-2,3,4,5-tetrahydro-3-benzazepine} 3, and corynanthine 20 μmol/l (TABLE 1).

The results are compatible with the idea that cholinergic nerves in the rat heart are endowed with α-adrenoceptors which enhance the exocytotic acetylcholine release. The receptors are activated by oxymetazoline (and possibly xylometazoline), and

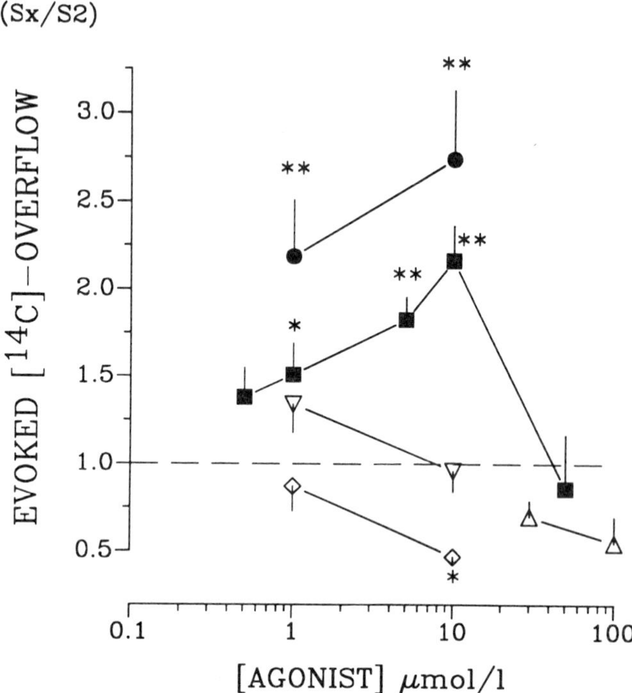

FIGURE 1. The fractional overflow evoked by vagal nerve stimulation of ^{14}C-acetylcholine/choline in the presence of α-adrenoceptor agonists (log abscissa scale, μmol/l). The overflow is expressed relative to a control stimulation period in the absence of agonist (S_x/S_2). Filled circles, xylometazoline; filled squares, oxymetazoline; inverted triangles, clonidine; triangles, methoxamine; diamonds, noradrenaline. Significant differences compared to controls in the absence of agonists, * and **, $p < 0.05$ and 0.01, U-test modified according to Holm.

TABLE 1. The Increase Induced by Oxymetazoline (OXY) of Fractional Overflow of ^{14}C-Acetylcholine/Choline Evoked by Vagal Nerve Stimulation on the Rat Perfused Heart in the Absence and Presence of α-Adrenoceptor Antagonists[a]

Antagonist (μmol/l)	n	OXY (0.5 μmol/l)	OXY (5 μmol/l)
	8–11	1.38 ± 0.17	1.83 ± 0.13
Idazoxan (5)	4	1.11 ± 0.19	1.41 ± 0.14
Rauwolscine (1)	4	1.28 ± 0.29	1.60 ± 0.17
Yohimbine (5)	4	0.88 ± 0.25	1.30 ± 0.35
SK&F 104078 (3)	4	0.88 ± 0.12	0.82 ± 0.10[c]
Corynanthine (20)	4	1.04 ± 0.13	1.25 ± 0.29[b]

[a]The overflow from n hearts is expressed relative to a control stimulation period in the absence of agonist (S_x/S_2). Significant difference compared to no antagonist: [b]$p < 0.05$; [c]$p < 0.01$; Student's t-test.

blocked by the α_1-selective prazosin, (S)-WB 4101, and corynanthine, but also by SK&F 104078 which is known to block postjunctional α_1- and α_2-adrenoceptors, but not prejunctional α_2-adrenoceptors.[4] It is not known whether this response may have a physiological or pharmacological role in prejunctional modulation of acetylcholine release.

REFERENCES

1. BENKIRANE, S., et al. 1986. Naunyn Schmiedebergs Arch. Pharmacol. **334:** 149–155.
2. McDONOUGH, P. M., et al. 1986. J. Pharmacol. Exp. Ther. **238:** 612–617.
3. MUSCHOLL, E. & A. MUTH. 1982. Naunyn Schmiedebergs Arch. Pharmacol. **320:** 60–169.
4. DALY, R. N., et al. 1988. J. Pharmacol. Exp. Ther. **247:** 122–128.

Modification of Neuronal Second Messenger Physiology

Expression of the Calcium Calmodulin Protein Kinase II Gene and the Yeast Adenylate Cyclase Gene in Neurons from HSV-1 Vectors

ALFRED I. GELLER[a] AND RACHAEL NEVE[b]

[a]*Dana Farber Cancer Institute*
44 Binney Street
Boston, Massachusetts 02115

[b]*Department of Psychobiology*
University of California
Irvine, California 92717

We are studying the regulation of neurotransmitter release from sympathetic neurons by second messenger systems. Our experimental strategy is to use defective HSV-1 vectors[1-3] to express altered second messenger enzymes[4] that are not regulated, instead they are always active; the increased activity of a second messenger enzyme may clarify its role in regulating neurotransmitter release from sympathetic neurons.

pHSVlac, our prototype HSV-1 vector, expresses the *E. coli Lac Z* gene from the HSV-1 immediate early 4/5 promoter. Expression of β-galactosidase has been observed in cultured rat neurons from throughout the nervous system, including superior cervical ganglia, dorsal root ganglia, spinal cord, cerebellum, thalamus, striatum, hippocampus, occipital cortex, temporal cortex, and frontal cortex.[1,2] Furthermore, expression of β-galactosidase is obtained following stereotaxic injection of purified pHSVlac virus into the adult rat brain.[3] Expression is observed in cells surrounding the injection site and in neurons that project to the injection site. Expression is stable for at least six weeks; pHSVlac virus does not spread through the brain.

To study neurotransmitter release we are using HSV-1 vectors to express the segment of the yeast adenylate cyclase gene[5] or the calcium calmodulin protein kinase II gene[6] encoding the catalytic domain of each enzyme; this results in enzymes that are always active. Both vectors are properly packaged into HSV-1 virus particles as determined by Southern analysis of the virus stocks. *In situ* hybridization demonstrated that both genes are expressed. Expression of the catalytic fragment of adenylate cyclase in PC12 rat pheochromocytoma cells results in a 20-fold increase in cAMP concentration, as determined with a radioimmunoassay for cAMP. In addition, using a rabbit anti-cAMP antibody in an immunofluorescent assay, we found some cells containing elevated cAMP levels. Furthermore, metabolic labeling of cells with ^{32}P PO_4 demonstrated an increase in protein phosphorylation similar to that observed following addition of dibutyryl cAMP. Expression of the catalytic fragment of calcium calmodulin protein kinase II results in an increase in protein phosphorylation in PC12 cells as determined by metabolic labeling of cells with ^{32}P PO_4. We are studying the effects of these enzymes on neurotransmitter release and searching for correlations with changes in second messenger levels, protein phosphorylation, and gene expression.

REFERENCES

1. GELLER, A. I. & X. O. BREAKEFIELD. 1988. A defective HSV-1 vector expresses *Escherichia coli* β-galactosidase in cultured peripheral neurons. Science **241:** 1667–1669.
2. GELLER, A. I. & A. FREESE. 1990. A defective HSV-1 vector expresses *E. coli* β-galactosidase in cultured CNS neurons. Proc. Nat. Acad. Sci. USA. **87:** 1149–1153.
3. GELLER, A. I., A. FREESE, L. H. HEMMENDINGER & B. SABEL. 1989. Gene transfer into neurons of the adult rat brain. Soc. Neurosci. Abstr. **15:** 8.4.
4. GELLER, A. I. & R. NEVE. 1989. Expression of the calcium calmodulin protein kinase II gene and the yeast adenylate cyclase gene in neurons from HSV-1 vectors. Soc. Neurosci. Abstr. **15:** 336.11.
5. KATAOKA, T., D. BROEK & M. WIGLER. 1985. DNA sequence and characterization of the *S. cerivisiae* gene encoding adenylate cyclase. Cell **43:** 493–505.
6. BULLEIT, R. F., M. K. BENNETT, S. S. MOLLOY, J. B. HURLEY & M. B. KENNEDY. 1988. Conserved and variable regions in the subunits of brain type II Ca^{2+} calmodulin-dependent protein kinase. Neuron **1:** 63–72.

Regulation of Striatal Acetylcholine Release by Dopamine after Nigrostriatal Bundle Injury

DENISE JACKSON AND MICHAEL J. ZIGMOND

Department of Behavioral Neuroscience
Center for Neuroscience
University of Pittsburgh
Pittsburgh, Pennsylvania 15260

Lesions of the dopaminergic component of the nigrostriatal bundle (NSB) are associated with marked neurological deficits. These deficits are due in part to the disruption of the normal presynaptic inhibition of cholinergic activity in striatum by dopamine (DA). Thus, the deficits can be alleviated by pharmacological treatments that either increase dopaminergic activity (e.g., L-DOPA) or block cholinergic activity (e.g., atropine). Most cases of Parkinson's disease are associated with progressive NSB degeneration and accompanying neurological dysfunctions. However, under conditions in which the degeneration is not progressive, both patients and experimental animals may show a recovery of function (for review see Reference 1). We have previously observed that such recovery is associated with an increase in the capacity of DA to inhibit acetylcholine (ACh) release as measured *in vitro*.[2] The purpose of the present study was to investigate the mechanism of that recovery of function.

Striatal DA depletions in adult male rats were produced by administration of 6-hydroxydopamine (6-HDA, 8 μg) into the NSB, thereby causing the permanent loss of striatal DA (mean depletion, 94%; range 89% to 98%). These lesions were associated with marked deficits in motor function which abated within 1–2 months. Release studies were conducted with striatal slices from control rats and lesioned rats 2–5 days postoperative (neurological deficits) or 1–2 months postoperative (recovered). Striatal slices were prepared and superfused as previously described.[3] The efflux of tritium after prelabeling with [^{3}H]choline was used as an index of ACh release. DA in superfusates and tissue slices was measured by high performance liquid chromatography (HPLC) with electrochemical detection, and the DA content in superfusates was expressed either as absolute release (pg DA per mg protein) or fractional DA release (pg DA per ng tissue DA). Release was evoked with electrical field depolarization as described below.

Normally the D_2 antagonist sulpiride elevates electrically evoked ACh release from striatal slices. In confirmation of previous results from our laboratory,[2] we observed that immediately after 6-HDA-induced lesions the effectiveness of sulpiride was greatly reduced, but that considerable recovery could be detected 1–2 months later. Sulpiride (1 μM) increased evoked ACh overflow by 93 ± 20% in slices from control rats, only 26 ± 2% in slices from unrecovered rats ($p < 0.05$), and 63 ± 10% in slices from recovered rats. This recovery of the impact of DA on ACh release might be due to an increase in DA overflow from residual DA neurons and/or an increase in the responsiveness of cholinergic neurons to DA.

We wished to differentiate between these two hypotheses. To examine the first, DA efflux was measured from slices prepared from control and 6-HDA-lesioned animals (TABLE 1). NSB lesions reduced basal and evoked DA release in both groups of

TABLE 1. Effects of 6-HDA on DA Release[a]

Group	(*n*)	Absolute DA Release (pg DA per mg protein)		Fractional DA Release (pg DA per ng tissue DA)	
		Basal	Overflow	Basal	Overflow
Control	(4)	274 ± 68	9709 ± 1690	4 ± 1	131 ± 20
2–5 days post 6-HDA	(9)	128 ± 11[b]	326 ± 93[b]	83 ± 15[b]	164 ± 52
1–2 months post 6-HDA	(10)	188 ± 39[b]	515 ± 83[b]	73 ± 12[b]	266 ± 70[b]

[a]Slices were superfused with modified Krebs bicarbonate buffer at 100 μl/minute and stimulated with bipolar pulses (18 mA) delivered at 8 Hz for 1 minute. Five-minute fractions were collected and endogenous DA efflux was determined with HPLC. Basal DA release is the average of the three 5-minute fractions preceding stimulation. Overflow represents the amount of DA exceeding basal levels during the 20 minutes immediately following the stimulation.

[b]$p < 0.05$ in comparison with control efflux.

lesioned rats to below control levels. However, evoked overflow in slices from recovered rats was higher than in slices from unrecovered rats. Moreover, when expressed as fractional efflux, overflow from slices prepared from recovered rats was twofold higher than in the control or unrecovered conditions, confirming our previous observations.[4,5] These observations suggest the gradual onset of compensatory mechanisms which

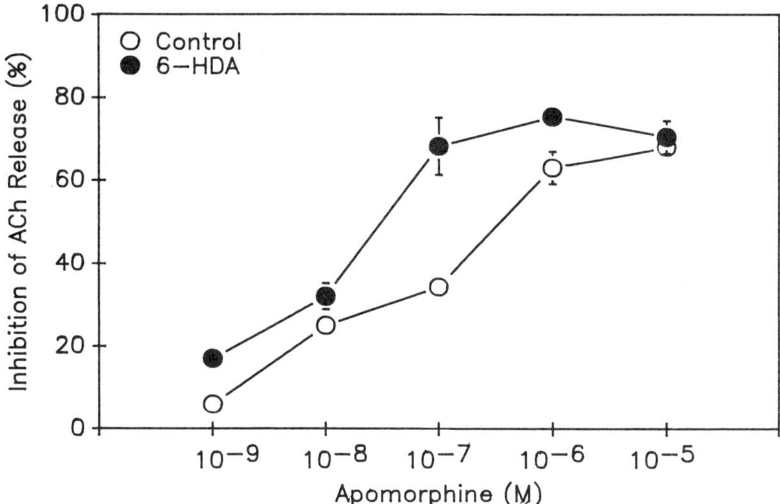

FIGURE 1. Apomorphine-induced inhibition of ACh release after 6-HDA. The effects of apomorphine (1 nM–10 μM) on electrically evoked ACh release from striatal slices of control and recovered 6-HDA-lesioned rats are expressed as percent decrease from evoked release in the drug-free condition. Slices were treated as described in TABLE 1 except that a less intense stimulation paradigm was used (1 Hz, 2 minutes 10 mA) so that the release of endogenous DA would not obscure the impact of apomorphine. In addition, for these experiments, slices were stimulated twice, 80 minutes (S_1) and 125 minutes (S_2) after the beginning of superfusion. Apomorphine was added to the superfusion buffer beginning 25 minutes prior to the second stimulation. The S_2/S_1 ratio then was determined and expressed as a percentage of the S_2/S_1 ratio observed in the absence of apomorphine. This provided a measure of the inhibition of ACh release by apomorphine. Each point represents mean ± standard error of the mean from 3–4 experiments.

increase striatal DA release from those nerve terminals that are spared by 6-HDA. To examine changes in sensitivity to DA, we measured the ability of apomorphine (1 nM–10 μM) to inhibit ACh overflow. Slices prepared from 6-HDA-lesioned animals 1–2 months postoperatively showed an approximately 10-fold shift to the left in the dose response curve for apomorphine-induced inhibition of ACh release (FIGURE 1). This result is consistent with previous observations,[2,6] and suggests an increase in the sensitivity of cholinergic neurons to DA.

In summary, after NSB lesions there is a recovery of the capacity of endogenous DA to inhibit ACh release which parallels recovery of behavioral function. These events may be due in part to an increase in the capacity of residual DA terminals to release DA and an increase in the effectiveness of that DA to inhibit cholinergic neurons.

REFERENCES

1. ZIGMOND, M. J. & E. M. STRICKER. 1989. Int. Rev. Neurobiol. **31:** 1–79.
2. MACKENZIE, R. G., M. K. STACHOWIAK & M. J. ZIGMOND. 1989. Eur. J. Pharmacol. **168:** 43–52.
3. JACKSON, D., M. K. STACHOWIAK, J. P. BRUNO & M. J. ZIGMOND. 1988. Brain Res. **457:** 259–266.
4. STACHOWIAK, M. K., R. W. KELLER, JR., E. M. STRICKER & M. J. ZIGMOND. 1987. J. Neurosci. **7:** 1648–1654.
5. SNYDER, G. L., M. K. STACHOWIAK, R. W. KELLER, JR., E. M. STRICKER & M. J. ZIGMOND. 1986. Soc. Neurosci. Abstr. **12:** 136.
6. STOOF, J. C., R. E. THIEME, M. C. VRIJMOED-DE VRIES & A. H. MULDER. 1979. Arch. Pharmacol. **309:** 119.

Autoreceptor- and Heteroreceptor-Mediated Regulation of Monoamine Release in Spinal Cord Synaptosomes[a]

I. MATSUMOTO, M. COMBS, S. BRANNAN,
AND D. J. JONES

Departments of Anesthesiology and Pharmacology
The University of Texas Health Science Center at San Antonio
7703 Floyd Curl Drive
San Antonio, Texas 78284

The spinal cord contains descending norepinephrine (NE) and serotonin (5HT) neuronal systems, the terminals of which overlap in their anatomical distribution in the spinal gray. A great deal of interest has focussed on both these neuronal systems in mediating nociception in the rat[1] and the possible interaction between these two systems. Previously studies in regional brain areas had demonstrated that adrenergic receptors located presynaptically on 5HT nerve terminals modulate the release of [^{3}H]5HT.[2] Pharmacological studies have characterized these "heteroreceptors" as alpha$_2$-receptors[3] and have also shown in the hippocampus that endogenous NE can modulate 5HT release via these receptors.[4] Early studies by McGrath[5] and more recently by Feuerstein and Hertting[6] have demonstrated that 5HT can also regulate NE release from both peripheral and central tissues. However, the 5HT receptor mediating this response has not been clearly characterized. Since interactions between NE vs. 5HT neurons have been shown in supraspinal areas, the purpose of the present study was to investigate the possible integration of these two neuronal systems in spinal cord by determining the effects of NE and alpha$_2$ agonists on [^{3}H]5HT release and 5HT and 5HT agonists on [^{3}H]NE release from spinal cord synaptosomes.

TABLE 1. IC$_{50}$ Values for Alpha Agonist-Induced Inhibition of [^{3}H]5HT Release in Spinal Cord Synaptosomes

Additions	IC$_{50}$ (percent releasable [^{3}H]5HT)
NE	5.0
Alpha methyl NE	3.9
Clonidine	1.0
Guanabenz	0.9
BHT 920	6.4
UK 14304	50
BHT 933	>1000
Methoxamine	>1000
Phenylephrine	>1000

[a]Supported by National Science Foundation grant BNS 8820008.

The K^+-stimulated release of both [³H]5HT and [³H]NE was performed according to Stauderman and Jones[7] using the crude synaptosomal preparation (P_2) from adult rat spinal cords. The synaptosomes were preloaded with either [³H]5HT or [³H]NE (0.05 μM final concentration) and superfused continuously with oxygenated Krebs-Ringer buffer. Drugs were added 10 minutes before superfusion with 15 mM K^+ buffer following 30 minutes of washing. Data are expressed as the difference in percent total releasable [³H]5HT or [³H]NE under basal conditions minus peak K^+-stimulated release.

Our results demonstrate that both autoreceptors and heterorecptors regulate the K^+-induced release of monoamines from spinal cord synaptosomes. Based on pharmacological studies, the autoreceptor regulating 5HT release was demonstrated to have characteristics similar to the $5HT_{1B}$ receptor. The autoreceptor regulating NE release demonstrated characteristics similar to the alpha$_2$-receptor. Alpha$_2$ receptors also regulate the release of 5HT via apparent heteroreceptors located on 5HT terminals (TABLE 1). The blockade of NE autoreceptors with alpha$_2$-receptor antagonists was shown to increase K^+-stimulated [³H]NE release and presumably endogenous NE release. Alpha$_2$ antagonists reduced [³H]5HT release possibly through this mechanism of increased availability of NE to the alpha$_2$-heteroreceptor. Thus a potential *in vivo*

TABLE 2. Comparative Effects of 5HT Agonists on K^+-Induced [³H]NE Release in Spinal Cord Synaptosomes[a]

Additions	Percent Releasable [³H]NE
K^+ (15 mM)	2.72 ± 0.25
K^+ + 5HT (0.01 μM)	1.88 ± 0.09
K^+ + 5CT (0.1 μM)	1.93 ± 0.19
K^+ + RU 24969 (0.1 μM)	1.61 ± 0.23
K^+ + 5MEODMT (0.1 μM)	4.35 ± 0.59
K^+ + 8-OHDPAT (0.1 μM)	2.51 ± 0.31

[a]Abbreviations: 5 CT, 5-carboxamidotryptamine; RU 24969 (5-methoxy-3-[1,2,3,6-tetrahydro-4-pyridinyl]1H indole); 5 MEO DMT, 5-methoxydimethyltryptamine; 8-OH DPAT, 8-hydroxy-2-diproylaminotetralin.

mechanism for interregulation between these two monoaminergic systems is possible. At concentrations of 0.1 μM and less, 5HT decreased NE release and other 5HT agonists such as RU 24969 and 5CT also reduced NE release (TABLE 2). At concentrations greater than 1 μM, 5HT enhanced [³H]NE release possibly due to displacement of [³H]NE from the terminal. The $5HT_{1A}$ agonist 8 OH DPAT did not alter release. It is unclear what type of 5HT receptor mediates the inhibition of NE release since metitepine only partially attenuates the effects of 5HT on [³H]NE release. These studies provide evidence for possible integration of 5HT and NE neuronal activity in descending pathways in the spinal cord. Since both monoamines have been demonstrated to modulate spinal cardiovascular, motor, and pain reflexes, the interaction of presynaptic mechanisms may be a part of the overall balance in reflex activity.

REFERENCES

1. ARCHER, T., G. JONSSON, B. G. MINOR & C. POST. 1988. Eur. J. Pharmacol. **120:** 295–307.
2. BENKIRANE, S., S. ARBILLA & S. Z. LANGER. 1986. Naunyn Schmiedebergs Arch. Pharmacol. **334:** 149–155.

3. GOTHERT, M., H. HUTH & E. SCHLICKER. 1981. Naunyn Schmiedebergs Arch. Pharmacol. **317:** 199–203.
4. FEUERSTEIN, T. J., G. HERTTING & R. JACKISCH. 1985. Naunyn Schmiedebergs Arch. Pharmacol. **329:** 216–221.
5. MCGRATH, M. A. 1977. Circ. Res. **41:** 428–435.
6. FEUERSTEIN, T. J. & G. HERTTING. 1987. Naunyn Schmiedebergs Arch. Pharmacol. **333:** 191–197.
7. STAUDERMAN, K. A. & D. J. JONES. 1986. Eur. J. Pharmacol. **120:** 107–109.

In Vivo Microdialysis Measurement of the Extracellular Concentration of the Dopaminergic D1 Agonist SKF 38393 in the Rat Striatum

JOHN BROCK AND J. B. JUSTICE, JR.

Department of Chemistry
Emory University
Atlanta, Georgia 30322

SKF 38393 is a dopaminergic agonist which binds to D1 and D2 dopamine (DA) receptors. The selectivity of this binding depends on the concentration of drug in the medium around the receptors. The binding constants for SKF 38393 are 1.1 nM for D1 receptors and 157 nM for D2 receptors.[1] Quantitation of *in vivo* brain concentrations of SKF 38393 is useful for behavioral or neurochemical studies involving these subpopulations of DA receptors.

Therefore an *in vivo* method to sample, separate, and quantitate SKF 38393 from the extracellular fluid of the rat striatum was developed. This method consists of microdialysis sampling and quantitation by off-line high pressure liquid chromatography with electrochemical detection. The construction of the microdialysis probe has been discussed previously.[2] The perfusion flow rate was 0.3 μl/minute. The chromatographic separation of SKF 38393 was accomplished via a smallbore column (0.5 mm $\times$ 10 cm) packed with 5 μm Spherisorb ODS2 stationary phase (Alltech Associ-

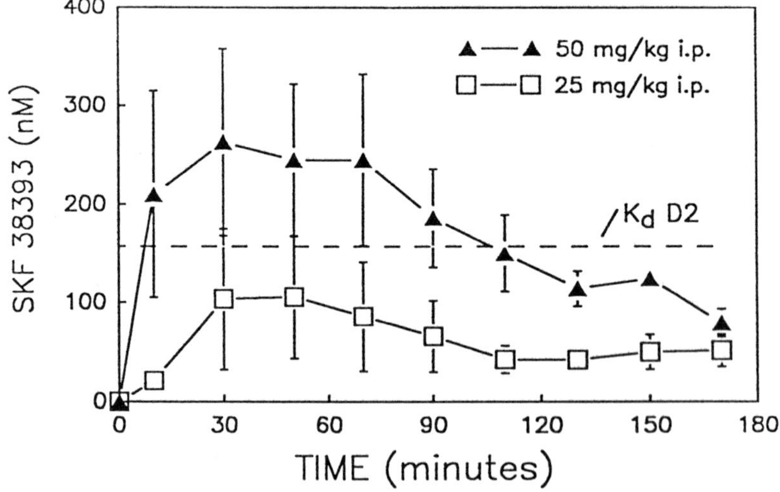

FIGURE 1. Pharmacokinetics of SKF 38393 in the rat striatum after ip injection of two different doses; $n = 4$ for each dose. Error bars are standard error of the mean.

ates). The mobile phase consisted of 62.5 mM sodium phosphate, 0.27 mM disodium ethylene diamine tetraacetate, 28.7 mM triethylamine, 0.73 mM sodium hexyl sulfate in distilled water with 15% methanol (pH = 3.0). An Isco LC 5000 syringe pump delivered the mobile phase at 33 μl/minute. SKF 38393 eluted at about 10 minutes and was detected electrochemically with a PAR EC 400 detector. The glassy carbon working electrode (BAS Inc.) was set at +0.7 volts versus Ag/AgCl.

Overall method calibration accuracy was evaluated. Any differences in diffusion between the calibration technique and the actual rat brain were quantitated by measuring transport characteristics for microdialysis probes under *in vitro* and *in vivo* conditions. Inflow to and outflow from the probe were approximately equal under *in vitro* conditions representing free diffusion conditions. In the rat brain only outflow was monitored and found to be slightly lower than *in vitro* conditions. These results agree well with findings by other researchers.[3]

This method was used to generate striatal pharmacokinetic curves for two different intraperitoneal (ip) doses of the drug (25 and 50 mg/kg); 50 mg/kg SKF 38393 ip injections gave a broad maximum to 263 nM between 30 and 60 minutes. The 25 mg/kg injections yielded a 105 nM maximum. Striatal extracellular SKF 38393 concentrations remained elevated for 3 hours (see FIGURE 1).

REFERENCES

1. SEEMAN, P. & H. B. NIZNICK. 1988. *In* ISI Atlas of Science: Pharmacology: 61–70. Institute for Scientific Information. Philadelphia, Pa.
2. CHURCH, W. & J. B. JUSTICE, JR. 1987. Anal. Chem. **59:** 712–716.
3. ALEXANDER, G. M., J. R. GROTHUSEN & R. J. SCHWARTZMAN. 1988. Life Sci. **43:** 595–601.

Excitatory Amino Acid Receptor Involvement in the Regulation of Striatal Extracellular Dopamine

KRISTEN A. KEEFE, MICHAEL J. ZIGMOND, AND
ELIZABETH D. ABERCROMBIE

Departments of Behavioral Neuroscience and Psychiatry
Center for Neuroscience
University of Pittsburgh
Pittsburgh, Pennsylvania 15260

Both in vitro and in vivo pharmacological studies, as well as studies in which the corticostriatal pathway is stimulated, have shown that excitatory amino acids (EAAs) or EAA agonists can evoke the release of dopamine (DA) in the striatum via both N-methyl-d-aspartate (NMDA) and non-NMDA receptor subtypes.[1-7] It has been suggested that EAAs regulate this DA release in the striatum at the level of the DA terminal either directly or via intrinsic striatal neurons. Although it is apparent from such studies that the direct application of EAAs or EAA agonists or the stimulation of excitatory afferent input can evoke the release of DA in the striatum, it is not clear to what extent endogenous EAAs regulate DA release under basal conditions in the intact animal. Therefore, to more fully investigate the role of endogenous EAAs in the regulation of basal DA release in the striatum, we have begun to examine the effects of local application of EAA antagonists on extracellular DA levels in striata of unanesthetized rats using in vivo microdialysis.

Loop-style microdialysis probes were implanted into striata of male Sprague-Dawley rats (AP + 0.5 mm, ML ± 2.5 mm both from bregma, DV − 7.0 mm from dura). The dialysis probes had an active dializing length of 4 mm and were perfused at a rate of 1.5 μl/minute with a modified Ringer's solution (in mM: NaCl 117; KCl 4.7; $CaCl_2$ 1.25; $MgCl_2$ 1.2; $NaHCO_3$ 25; NaH_2PO_4 1.2; pH = 7.4). The in vitro recovery of the dialysis probes for DA was 40%. At least 12 hours after the probe was implanted, basal levels of DA were measured in 15-minute dialysis samples with high performance liquid chromatography and electrochemical detection (HPLC-EC). The applied potential was +0.60 V (Waters 460 detector). Once stable baseline levels of DA in the extracellular fluid were obtained (4 consecutive samples with <10% variability), the perfusion medium was switched to one containing an EAA antagonist in Ringer's (pH = 7.4). The nonspecific antagonist kynurenic acid (10 mM) and the specific antagonists DL-2-amino-5-phosphonovaleric acid (APV; .75 mM) and 6-cyano-7-nitroquinoxaline-2,3-dione (CNQX; 1 mM) were used to antagonize NMDA and non-NMDA receptor subtypes, respectively.

As shown in FIGURE 1, the antagonists increased the level of DA in the extracellular fluid of the striatum. The effects of kynurenic acid and CNQX were consistent across all animals. The effect of APV administration, however, was more variable. In two cases, APV produced an 80% increase in extracellular DA, whereas a much smaller, delayed increase was observed in three additional cases.

To determine the effectiveness of the receptor blockade produced by APV and CNQX, we examined the ability of these compounds to antagonize NMDA (1 mM) and kainic acid (100 μM) induced increases in extracellular DA. Coinfusion of the

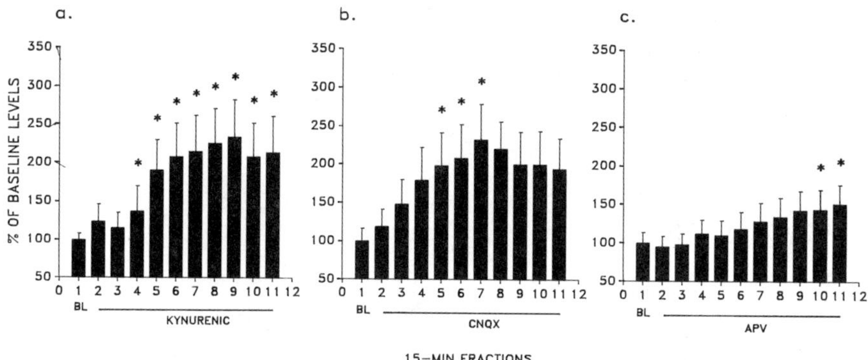

FIGURE 1. Mean [± standard error of the mean (SEM)] extracellular DA levels in the striatum in response to local infusion of (a) 10 mM kynurenic acid ($n = 4$); (b) 1 mM CNQX ($n = 4$); and (c) 0.75 mM APV ($n = 5$). *$p \leq 0.05$.

antagonists effectively prevented the increases in extracellular DA levels seen with the EAA agonists alone (FIGURE 2).

The present results with EAA antagonists provide evidence suggesting that endogenous EAAs are involved in the regulation of basal extracellular DA level in the striatum and that this regulation may be inhibitory in nature. Such an inhibitory effect of EAAs presumably must be mediated indirectly via another neuron. Under basal conditions, therefore, EAAs may excite GABAergic (γ-aminobutyric acid) interneurons or GABAergic projection neurons which then inhibit DA release at the level of the DA terminal or via a feedback loop through the substantia nigra, respectively. The fact that the local application of either EAA agonists or antagonists increases striatal extracellular DA levels may be due to an action of EAAs or agonists acting on EAA receptors on DA terminals that do not typically receive EAA input under basal

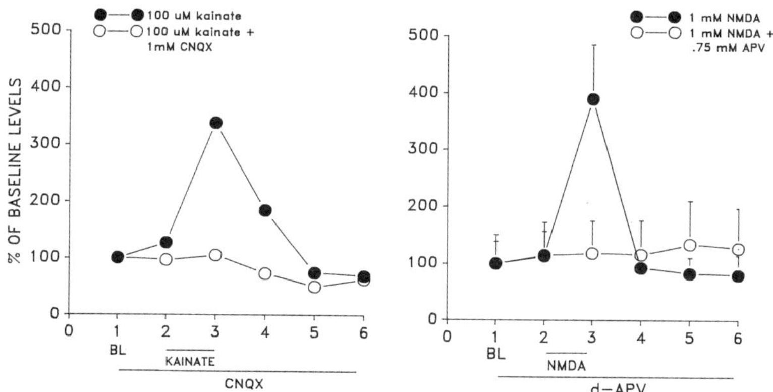

FIGURE 2. Effects of CNQX (1 mM; $n = 1$) and APV (0.75 mM; mean ± SEM; $n = 3$) infusion on DA release in the striatum stimulated by a local, 20-minute application of kainic acid (100 μM) and NMDA (1 mM), respectively. Basal DA values (BL) represent the mean of 4 consecutive samples after 1.5 hours of CNQX or APV infusion.

conditions. Under conditions in which the concentration of the ligand is high, however, such as exogenous application or electrical stimulation of an afferent pathway, the concentration of EAA in the extracellular fluid might be great enough to reach and stimulate these receptors. Alternatively, it is conceivable that EAA receptors located on DA terminals and on neuronal cell bodies in the striatum are differentially sensitive to EAAs and EAA agonists. In this case, exogenously applied EAAs might have a direct excitatory effect on DA release at the level of the terminal, whereas the lower concentrations of EAAs encountered in vivo under basal conditions might preferentially act upon GABAergic neurons in the striatum.

REFERENCES

1. CARTER, C. J., R. L'HEUREUX & B. SCATTON. 1988. J. Neurochem. **51:** 462–468.
2. CLOW, D. W. & K. JHAMANDAS. 1989. J. Pharmacol. Exp. Ther. **248:** 722–728.
3. JHAMANDAS, K. & M. MARIEN. 1987. Br. J. Pharmacol. **90:** 641–650.
4. NIEOULLON, A., A. CHERAMY & J. GLOWINSKI. 1978. Brain Res. **145:** 69–83.
5. ROBERTS, P. J. & N. A. SHARIF. 1978. Brain Res. **157:** 391–395.
6. ROMO, R., A. CHERAMY, G. GODEHEU & J. GLOWINSKI. 1986. Neuroscience **19:** 1067–1079.
7. SNELL, L. D. & K. M. JOHNSON. 1986. J. Pharmacol. Exp. Ther. **238:** 939–946.

Immunocytochemical Study of Peptide Receptors

DAVID J. BERLOVE, GREGORY J. MICHAEL,
DIANE T. PIEKUT, AND KARL M. KNIGGE

Neuroendocrine Unit
University of Rochester School of Medicine and Dentistry
601 Elmwood Avenue
Rochester, New York 14642

INTRODUCTION

Demonstration of receptors by immunocytochemical procedures may have some advantages over present autoradiographic techniques including ease of preparation, decreased incubation times, opportunity for immunostaining of multiple receptors or transmitter-receptor combinations and capability of resolution at the electron microscope (EM) level. Receptor-specific antibodies might also provide the ability to direct a variety of pharmacologic or toxic agents to specific binding sites.

Immunocytochemical study of a receptor requires a highly specific antibody directed against that receptor. One means of obtaining an antibody directed against a particular receptor is the generation of antiidiotypic antibodies. The network theory of Jerne[1] suggests that the set of determinants (idiotype) expressed in the variable portion of an antibody can provoke an immune response and stimulate the production of complementary antiidiotypic antibodies. In addition to binding at the ligand-binding site present in the idiotype, the antiidiotypic antibody may also bind to receptor sites specific for this ligand. Antiidiotypic binding to peptide receptors is well established and has been reported for a number of neuroactive peptides. The receptor binding characteristics of an antiidiotype allow the immunocytochemical identification and localization of receptors. We have generated antiidiotypes against vasopressin (AVP) and corticotropin releasing factor (CRF) and have employed the antibodies to immunolabel putative receptors in the rat brain.

The affinity of an antiidiotype for a receptor reflects its mimicry of a peptide bound by the idiotype. It is therefore possible that a relationship exists between the affinity of a peptide for its receptor and the binding of the peptide's antiidiotype to that receptor. If the native peptide binds to several pharmacologically distinct receptor sites and analogues are available that selectively bind to one receptor type, antiidiotypes to these analogues may be selective in their receptor binding. Two vasopressin receptors outside of the central nervous system have been well characterized. Vasopressin receptors in brain are less well characterized, and multiple subtypes have been described. Using an antiidiotype to a selective vasopressin analogue, we have attempted to immunocytochemically label a specific vasopressin receptor subtype in brain.

ANTIBODY GENERATION

A primary vasopressin antiserum was generated in a rabbit immunized with synthetic vasopressin. The antiserum immunocytochemically stained vasopressin cells intensely at a dilution of 1:8000, and preabsorption with synthetic AVP completely eliminated staining, whereas preabsorption with oxytocin or other peptides did not alter staining. Immunoglobulin G (IgG) from this antiserum was used to immunize another female rabbit. The generation of antiidiotypic antibody was measured by a precipitin test in which the antiidiotypic antiserum was coincubated with antivasopressin IgG in a wide range of dilutions. The antiserum was also coincubated with a rat neural membrane preparation.[2,3] A concentration-dependent reduction in vasopressin binding to this receptor preparation was observed.

A primary antiserum to the selective V_1 receptor antagonist [1-(β-mercapto-β,β-cyclopentamethylene-propionic acid)2-(O-Methyl)Tyr] arginine vasopressin (β-mercapto-AVP) was generated in rabbit. Immunostaining was detected by both immunoblot and immunocytochemistry. IgG from this anti-β-mercapto-vasopressin antiserum was used to generate an antiidiotypic antiserum in another rabbit as described in detail elsewhere.[4]

The antiidiotypic antibody related to CRF was made in a similar manner to the vasopressin antiidiotype. Immunization was performed using IgG purified from an antiserum made against the rat/human sequence of CRF (rCRF). The specificity of the CRF antiserum was confirmed in absorption studies using peptides including rCRF, oCRF, VIP, and PHI. This antiserum specifically recognizes the rat sequence of CRF with epitopes toward the carboxyl terminus responsible for immunocytochemical staining. Following immunization of the second rabbit, an antibody response was induced which yielded an antiserum useful in immunocytochemical labeling.[5]

IMMUNOCYTOCHEMICAL PROCEDURES

Sprague-Dawley and homozygous Brattleboro rats were perfused with physiologic saline followed by 4% paraformaldahyde fixative containing 0.2% picric acid in 0.025M phosphate-buffered saline buffer, pH 8.0. Immunocytochemistry was performed on 40 μm vibratome-cut frontal tissue sections of brain. After 48 hour primary incubation, sections were subsequently immunolabeled with the avidin-biotin complex (ABC) method using the protocol and reagents of the Vectastain ABC kit as described in Berlove and Piekut.[6] Controls included the deletion of primary antisera, the use of preimmune rabbit serum in the incubation, and the use of primary idiotypic antisera preincubated for 24 hours with the peptide it was generated against.

RESULTS

Sprague-Dawley rats display antivasopressin immunoreactivity in the magnocellular neurons of the supraoptic and paraventricular (FIGURE 1A) nuclei (SON and PVN) of hypothalamus and in their fibers that course ventrally through the hypothalamus and zona interna of the median eminence into the neurohypophysis. Parvocellular vasopressin immunoreactive neurons are also seen in the suprachiasmatic nucleus (SCN). Tissue sections from the Brattleboro brains are devoid of vasopressin immunoreactivity.

Immunolabeling with the vasopressin antiidiotype revealed immunoreactivity in the magnocellular neurons of PVN (FIGURE 1B) and SON in both Sprague-Dawley and Brattleboro rats. Antiidiotype immunoreactivity was not present in the parvocellular vasopressinergic neurons of the SCN. A marked difference was apparent in the appearance of the reaction product of these two antisera. Vasopressin (idiotype)

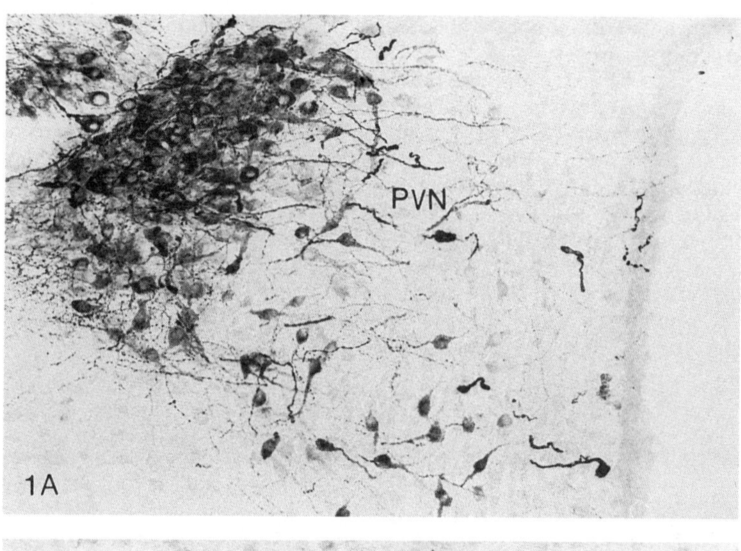

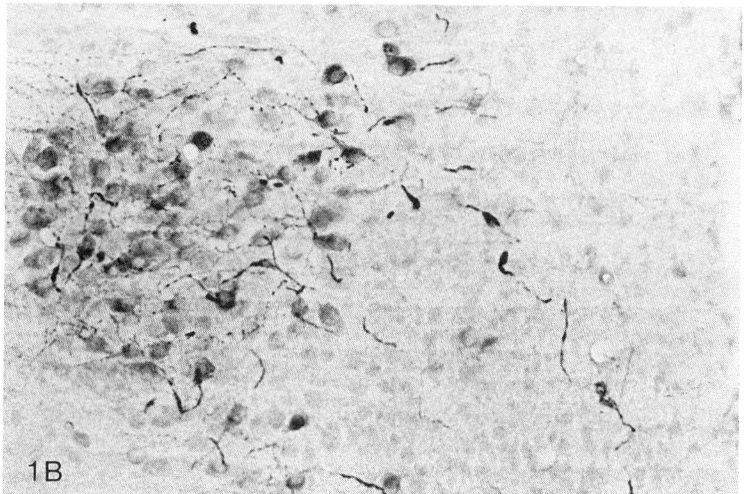

FIGURE 1. (A) Magnocellular neurons in the paraventricular nucleus of hypothalamus (PVN) immunolabeled with antivasopressin antiserum. ×125. (B) Magnocellular neurons of PVN immunolabeled with vasopressin antiidiotypic antiserum. ×125.

immunoreaction product was observed in the form of large coarse granules in immuno-labeled neurons. In contrast, the antiidiotype reaction product presented a more delicate punctate appearance.

The distribution of β-mercapto-AVP antiidiotype immunolabeling was similar to that seen with AVP antiidiotype. Immunoreactive neurons were present in the magno-cellular neurons of SON and dorsolateral PVN but not in SCN. Antiidiotype immuno-label was observed as fine points of immunoreactive product in contrast to the large granular reaction product seen with the idiotype. Immunoblots confirm that this antiidiotypic antiserum is not immunoreactive against arginine vasopressin, oxytocin, or β-mercapto-vasopressin.

Immunocytochemistry utilizing the CRF antiidiotypic antiserum yielded labeling of neurons and their processes in a large number of brain regions including the cerebral cortex, striatum, nuclei of the diagonal band of Broca, interpeduncular nucleus, laterodorsal tegmentum, pedunculopontine tegmentum, and lamina X of the spinal cord (FIGURE 2). Lightly labeled cells were also observed within the paraventricular, dorsomedial, and lateral hypothalamic nuclei, dorsal lateral periaqueductal grey, superior and inferior colliculus, nucleus of the solitary tract and nucleus of the spinal tract of V. CRF immunoreactive neurons were unlabeled for the most part by the CRF antiidiotypic antibody. Cell bodies and processes labeled with the CRF antiidiotype are identified in numerous areas of brain known to contain CRF immunoreactive fibers and/or terminals.

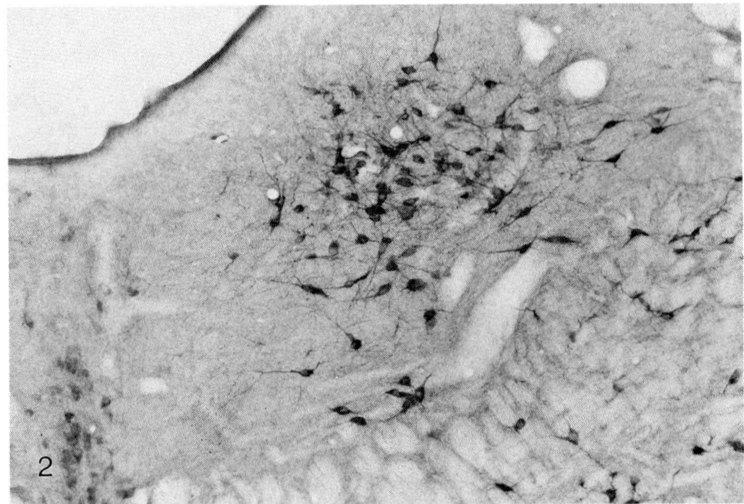

FIGURE 2. Cells in the laterodorsal tegmentum are labeled with the CRF antiidiotype. ×125.

DISCUSSION

Antiidiotypic antibodies have been used successfully to immunolabel putative vasopressin receptor sites in brain. AVP antiidiotype and β-mercapto-vasopressin antiidiotype both immunolabel neurons in SON and magnocellular portions of PVN and their fiber processes. They do not immunolabel the parvocellular vasopressinergic neurons of SCN. The AVP antiidiotype immunolabels PVN and SON neurons in the vasopressin-deficient Brattleboro rat. The presence of vasopressin receptors in these nuclei has been reported in autoradiographic studies[7] and presumably represents autoreceptors that modulate the synthesis and/or release of vasopressin from vasopressinergic neurons.

Preliminary analysis indicates that immunolabeled granules associated with cell bodies visualized using the CRF antiidiotype were not localized for the most part in the areas containing CRF immunoreactive perikarya; they were seen instead in regions where CRF immunoreactive fiber projections and terminal fields are demonstrated. The distribution of labeling obtained using the CRF antiidiotypic antibody suggests that it labels neurons that are postsynaptic to CRF-producing cells.

The immunocytochemical labeling of peptide receptors provides an important tool in studies to elucidate receptor distribution and relationship to other transmitters or receptors. The generation of antiidiotypic antisera that recognize these receptors can provide the ability to specifically immunolabel them. Immunocytochemical methods for the study of receptors offer the flexibility and resolution to characterize the presynaptic/postsynaptic cell pair and extend these observations to the electron microscopic level.

REFERENCES

1. JERNE, N. 1974. Towards a network theory of the immune system. Ann. Immunol. **125C:** 373–389.
2. JUNIG, J. & L. G. ABOOD. 1987. Solubilization and purification of the Ni-stimulated arginine-vasopressin binding site of rat brain membranes. Neurochem. Res. **12:** 809–817.
3. KNIGGE, K., D. T. PIEKUT, L. G. ABOOD, S. A. JOSEPH, G. J. MICHAEL, L. XIN & D. J. BERLOVE. 1989. Immunocytochemistry of receptors using anti-idiotypic antibodies. Methods Enzymol. **178:** 212–221.
4. BERLOVE, D. J. & D. T. PIEKUT. Anti-idiotypic antibody to V_1 receptor antagonist immunolabels magnocellular neurons in the rat hypothalamus. Exp. Brain Res. (Submitted.)
5. PIEKUT, D. T. & K. M. KNIGGE. 1989. Immunocytochemistry of putative CRF receptors in rat forebrain using a CRF anti-idiotypic antibody. Neurosci. Res. Commun. **4(3):** 167–173.
6. BERLOVE, D. J. & D. T. PIEKUT. 1989. Vasopressin receptor distribution in adrenalectomized rats using vasopressin anti-idiotype. Peptides **10:** 871–881.
7. BRINTON, R., K. GEE, J. WAMSLEY, T. DAVIS & H. YAMAMURA. 1984. Regional distribution of putative vasopressin receptors in rat brain and pituitary by quantitative autoradiography. Proc. Nat. Acad. Sci. USA **81:** 7248–7252.

GABA$_A$-Benzodiazepine Receptors Modulate Serotonin Release

In Vivo Electrophysiological Studies in the Rat Hippocampus

ALVARO LISTA,[a] PIERRE BLIER, AND
CLAUDE DE MONTIGNY

Neurobiological Psychiatry Unit
Department of Psychiatry
McGill University
Montreal, Quebec, H3A 1A1, Canada

The existence of benzodiazepine (BZ) receptors modulating serotonin (5-HT) release has been documented in the rat frontal cortex, hippocampus, and substantia nigra using the method of electrically releasing [^{3}H]serotonin from preloaded brain slices.[1] These BZ receptors are coupled to the γ-aminobutyric acid (GABA$_A$) chloride channel receptor complex.[1,2] Electrophysiological experiments, using an *in vivo* paradigm that permits the assessment of the effectiveness of 5-HT synaptic transmission, also indicate the existence of BZ receptors modulating 5-HT release in the rat dorsal hippocampus.[3,4]

The present study was undertaken to investigate the pharmacological properties of the BZ receptors that modulate *in vivo* the stimulation-induced release of 5-HT in the rat hippocampus. Male Sprague-Dawley rats (250–300 g) were anesthetized with chloral hydrate [400 mg/kg, intraperitoneally (ip)] and extracellular unitary recordings were obtained from CA$_3$ dorsal hippocampus pyramidal neurons with five-barreled micropipettes. 5-HT and GABA were applied by microiontophoresis in order to assess the responsiveness of postsynaptic pyramidal neurons to these neurotransmitters. The ascending 5-HT pathway was electrically stimulated in the ventromedial tegmentum with square pulses of 0.5 msecond, at a frequency of 1 Hz and an intensity of 300 μA. The period of suppression of the firing activity of pyramidal neurons (SIL values) produced by the stimulation was determined from peristimulus time histograms before (S_1) and after (S_2) the intravenous administration of drugs.[5] The effect of drugs on the effectiveness of the 5-HT pathway stimulation was measured as the S_2/S_1 ratio. A cumulative-dose procedure was used to generate dose-response curves. Data were analyzed with either the paired or the unpaired two-tailed Student's t-test, using the Dunnett's t distribution for multiple comparisons.

The acute administration of diazepam [0.05–2 mg/kg, intravenously (iv)], a nonselective BZR agonist, dose dependently enhanced the effectiveness of the electrical stimulation of the ascending 5-HT pathway (FIGURE 1). Similar effects were observed with lorazepam (0.05–2 mg/kg) and with the selective type I BZ receptor agonist CL 218872 (1–10 mg/kg; data not shown). In contrast, FG 7142 (0.01–1 mg/kg), a β-carboline inverse agonist, dose dependently reduced the effectiveness of the stimulation (FIGURE 1). Similar results were obtained with DMCM (0.05–1

[a]Present affiliation: Center for Pharmacological and Therapeutic Investigation, T. Gomez 1565, CP 11600, Montevideo, Uruguay.

mg/kg), another β-carboline inverse agonist (data not shown). Both the effect of diazepam and that of FG 7142 were completely blocked by the specific BZ receptor antagonist flumazenil (1 mg/kg, FIGURE 1).

In order to determine if these BZ receptors modulating 5-HT transmission were coupled to GABA receptors, the effect of the GABA$_A$ chloride channel blocker picrotoxine and that of the GABA$_A$ competitive antagonist bicuculline were studied.

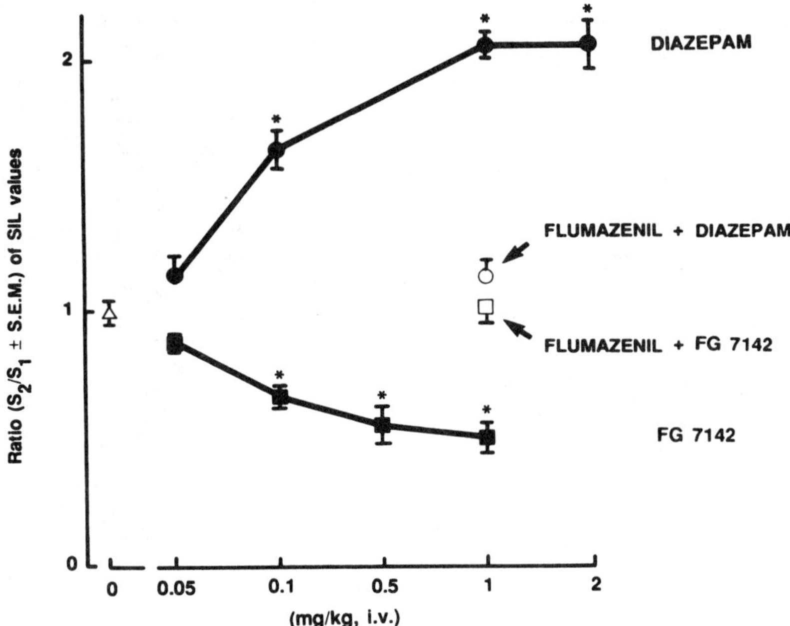

FIGURE 1. Ratio of the effectiveness of the electrical stimulation of the ascending 5-HT pathway in suppressing the firing activity of CA$_3$ dorsal hippocampus pyramidal neurons before (S$_1$) and after (S$_2$) the administration of successive intravenous doses of diazepam and FG 7142. The open circle represents the ratio following 1 mg/kg of diazepam injected immediately after the administration of flumazenil (1 mg/kg, iv). The open square represents the ratio following 1 mg/kg of FG 7142 injected immediately after the administration of flumazenil (1 mg/kg, iv). The control value (open triangle) represents the ratio of the effectiveness of the stimulation before and after the injection of the vehicle (0.1 ml, iv). The mean duration [± standard error of the mean (SEM)] of the suppression induced by the stimulation before injecting the vehicle was of 37 ± 3 mseconds. Each point was calculated from 4 experiments. *$p < 0.001$ compared with control value, using the two-tailed Student's t-test with the Dunnett's t distribution for multiple comparisons.

Both compounds reduced, in a dose-dependent manner, the effectiveness of the stimulation (FIGURE 2). The maximal effects of picrotoxine and bicuculline in reducing the stimulation were reversed by diazepam (FIGURE 2).

Neither the inhibitory effect of 5-HT nor that of GABA, applied by microiontophoresis onto the same hippocampus neurons, was modified by diazepam, FG 7142, picrotoxine, or bicuculline (data not shown).

The present electrophysiological data provide evidence that 5-HT synaptic transmission in the rat dorsal hippocampus can be modulated *in vivo* by $GABA_A/BZ$ receptors. These appear to be of the central type I, as suggested by the enhancing effect of CL 218872 on the stimulation of the 5-HT pathway. The present results also show that these BZ receptors possess both agonist and inverse agonist sites. This was evidenced by the following observations: (1) diazepam enhanced the effect of the

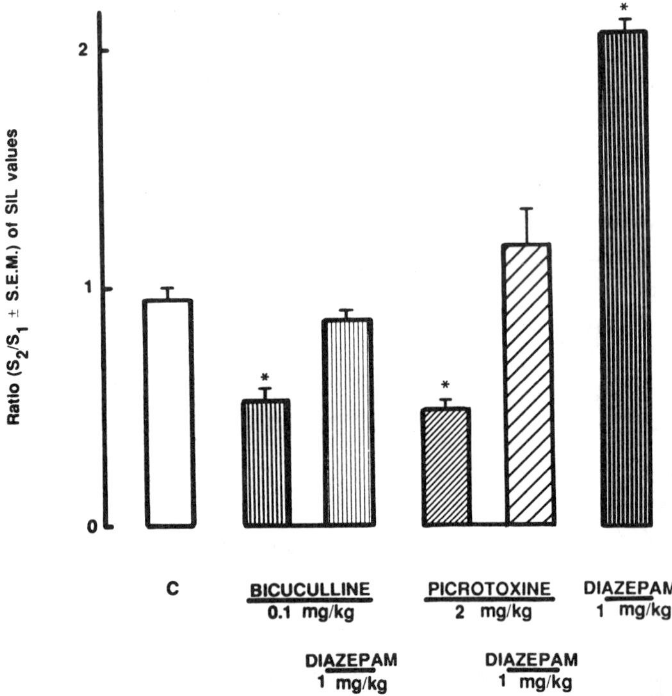

FIGURE 2. Ratio of the effectiveness of the electrical stimulation of the ascending 5-HT pathway in suppressing the firing activity of CA_3 dorsal hippocampus pyramidal neurons depicting the effects of the intravenous administration of bicuculline and picrotoxine, and, their subsequent reversals by an injection of diazepam (1 mg/kg, iv). The column on the extreme right represents the effect of diazepam (1 mg/kg, iv) when administered alone in a separate experimental series. The control value (C) represents the ratio of the effectiveness of the stimulation before and after the injection of vehicle (0.1 ml, iv). The mean duration ($\pm$SEM) of the suppression induced by the stimulation before injecting the vehicle was of 39 ± 4 mseconds. Each value was calculated from 4 experiments. *$p < 0.001$ when compared with the control value, using the paired Student's t-test.

stimulation, while the inverse agonists DMCM and FG 7142 decreased it; (2) both of these effects were blocked by the BZ receptor antagonist flumazenil. These BZ receptors are coupled to $GABA_A$ receptors and, hence, most likely exert their effects by modulating that of endogenous GABA on 5-HT transmission. This is suggested by the activity of GABA antagonists, which by themselves reduced by 50% the effectiveness of 5-HT transmission. It is noteworthy that the effects of the GABA antagonists were reversed by diazepam.

The modulation of 5-HT transmission by GABA$_A$/BZR in our model is functionally linked to the presynaptic serotonergic endings, as suggested by the unchanged postsynaptic responsiveness of GABA and 5-HT receptors following the administration of diazepam at a dose that produced the maximal effect on the efficacy of 5-HT pathway stimulation.

In conclusion, the activation of type I BZR coupled to GABA$_A$ receptors modulates the effectiveness of 5-HT transmission in the rat dorsal hippocampus via a presynaptic mechanism regulating the release of endogenous 5-HT.

REFERENCES

1. LISTA, A., S. ARBILLA & S. Z. LANGER. 1988. J. Neurochem. **51:** 1414–1421.
2. LISTA, A., S. ARBILLA & S. Z. LANGER. 1988. Eur. J. Pharmacol. **152:** 195–196.
3. LISTA, A., P. BLIER & C. DE MONTIGNY. 1989. Soc. Neurosci. Abstr. **15:** 199.1.
4. LISTA, A., P. BLIER & C. DE MONTIGNY. 1989. Eur. J. Pharmacol. **171:** 229–231.
5. CHAPUT, Y., P. BLIER & C. DE MONTIGNY. 1986. J. Neurosci. **6:** 2796–2801.

L-Glutamic Acid Evokes Ca^{2+}-Independent Release of Dopamine from Rat Striatum via Dopamine Uptake System[a]

GYORGY LONART AND MICHAEL J. ZIGMOND

Department of Behavioral Neuroscience and Psychiatry
Center for Neuroscience
University of Pittsburgh
Pittsburgh, Pennsylvania 15260

Although the glutamatergic modulation of dopamine (DA) release has been intensively investigated in the striatum,[1] the majority of *in vitro* experiments have been carried out in Mg^{2+}-free medium. This is because Mg^{2+} is known to block the actions of L-glutamate (L-Glu) mediated through receptors of the *N*-methyl-D-aspartate (NMDA) type. However, changes in available Mg^{2+} could be expected to exert considerable influence on flux rates through numerous metabolic pathways and on the activity of Na^+,K^+-ATPase[2] which has been shown to modulate biochemical responses[3] mediated by NMDA receptors. Thus we have examined the effects of L-Glu on striatal DA in the presence of a physiological concentration of Mg^{2+}.

Slices (350 μm) were prepared from the striatum of adult Sprague Dawley rats. Slices (5 per chamber) were superfused (100 μl/minute) with a Krebs bicarbonate buffer consisting of the following (unless stated otherwise): 117 mM NaCl, 4.7 mM KCl, 1.2 mM $MgCl_2$, 1.25 mM $CaCl_2$ 1.2 mM NaH_2PO_4, 25 mM $NaHCO_3$, 0.015% ascorbic acid, 11.5 mM dextrose, and 10 μM tyrosine, which was constantly bubbled with gas containing 5% CO_2 and 95% O_2. Endogenous DA was measured by high performance liquid chromatography (HPLC) and expressed as ng DA per mg protein in the slices per 5 minute superfusion.

L-Glu incrased basal DA release. This effect was highly significant in the presence of 3 mM (5.9-fold increase) and 10 mM (9.3-fold increase) L-Glu (FIGURE 1), but was not significant at lower L-Glu concentrations. Since L-Glu has a potent neurotoxic effect, the integrity of the slices was examined during exposure to these high concentrations of the amino acid by measuring the release of the enzyme lactic acid dehydrogenase, a cytosolic marker. Although superfusion with H_2O produced a large increase in lactic acid dehydrogenase release ($+352 \pm 159\%$), L-Glu (10 mM) produced no significant change in this marker ($+48 \pm 35$). The failure of L-Glu to produce a cytotoxic response even at such high concentrations may be due to the presence of a highly effective acidic amino acid uptake system.

The tetrodotoxin (TTX) insensitivity of the L-Glu stimulation (TABLE 1) suggests that L-Glu can evoke the release of DA via a direct effect that is independent of action potentials. The Ca^{2+} independence and reserpine insensitivity of the response (TABLE 1) indicates that this L-Glu-stimulated release of DA is from the cytoplasmic pool. This DA release can be blocked by 10 μM nomifensine, an inhibitor of high affinity DA uptake, or by Na^+ deprivation (117 mM NaCl replaced by 234 mM sucrose). Both of these treatments suggest that the release is mediated via the DA transporter. The

[a]This work was supported in part by U. S. Public Health Service grants NS 19608, MH 43947, and MH 00058.

effects of L-Glu and nomifensine were not additive which suggests that L-Glu increases the DA efflux by reversing the DA transport system rather than blocking DA uptake.

Omitting the Mg^{2+} from the medium potentiated the DA release evoked by lower concentrations of L-Glu (0.03–1 mM). However, raising Mg^{2+} above the normal concentration of this ion (to 3 mM) increased the L-Glu (10 mM) stimulated DA release (TABLE 1), probably due to the ability of excess Mg^{2+} to potentiate neurotransmitter release via reversal of the Na^+-dependent uptake system.[2]

Selective NMDA receptor antagonists 2-amino-5-phosphonovaleric acid (AP5, 100 μM) and MK 801 (1 μM) reduced the L-Glu-evoked DA release by 66% (TABLE 1). The extent of inhibition was not further increased by higher antagonist concentra-

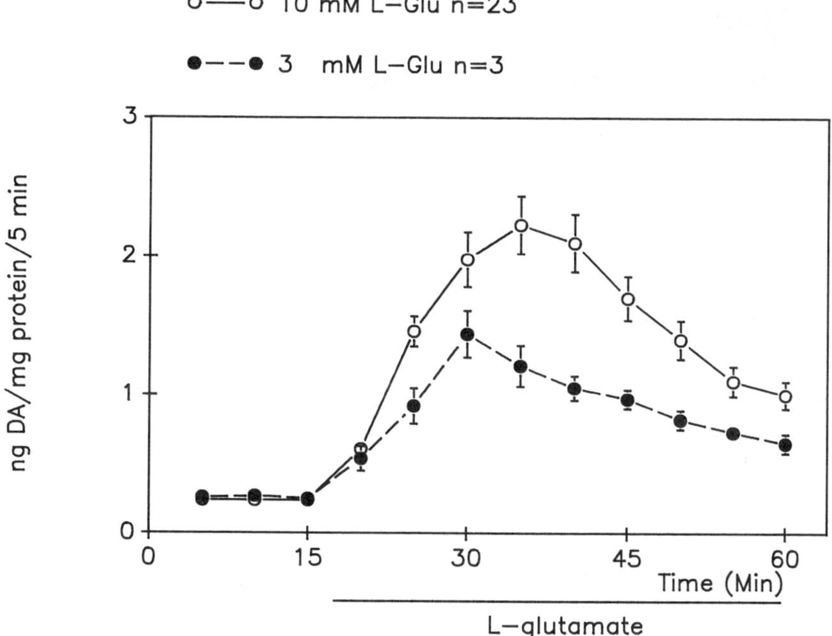

FIGURE 1. Effect of L-Glu on DA efflux. Each point represents the mean ± standard error of the mean.

tions. To determine whether there was another excitatory amino acid receptor mediated effect, we applied 10 μM CNQX 20 minutes before the L-Glu stimulation. CNQX had no significant effect on the L-Glu-stimulated DA release (TABLE 1). It is therefore possible that the non-NMDA receptor mediated component of the evoked DA release is due to the electrogenic uptake of L-Glu. The electrogenic uptake of L-Glu via the acidic amino acid uptake system might depolarize the neurons[4] and reduce the activity of other uptake systems. Due to the relative low potency of the available acidic amino acid uptake blockers (dihydrokainic acid, and DL-threo-β-hydroxyaspartate) and the high L-Glu concentration applied, we could not obtain any reliable data on the influence of L-Glu uptake on the DA release.

In summary, these data suggest that high L-Glu concentration evokes a Ca^{2+}-

TABLE 1. Effects of TTX, Ca^{2+}-Free Medium, Reserpine Pretreatment, Nomifensine, Na^+ Deprivation, and Excitatory Amino Acid Antagonists on the Basal and L-Glu-Stimulated DA Efflux

	DA Efflux (ng DA/mg protein per 5 minutes)[a]	
	Basal	10 mM L-Glu
Control	0.29 + 0.03 (3)	2.28 + 0.14 (3)
TTX (0.5 μM)	0.30 + 0.04 (3)	2.03 + 0.14 (3)
Control	0.19 ± 0.03 (3)	2.94 ± 0.43 (3)
Ca^{2+} free[b]	0.29 ± 0.04 (5)	2.18 ± 0.47 (5)
Control	0.27 + 0.04 (4)	1.85 + 0.45 (4)
Reserpine (5 mg/kg)[c]	0.15 + 0.03 (4)	1.91 + 0.69 (4)
Control	0.22 + 0.02 (3)	2.01 + 0.41 (3)
Nomifensine (10 μM)	0.39 + 0.04[e] (5)	0.65 + 0.12[e] (5)
Control	0.34 ± 0.04 (4)	2.22 ± 0.39 (4)
Na^+ deprivation	1.76 ± 0.21[e] (4)	1.31 ± 0.13[e] (4)
Control	0.25 + 0.03 (3)	2.00 + 0.12 (3)
AP5 (100 μM)[d]	0.19 + 0.02 (3)	0.78 + 0.13[e] (3)
Control	0.48 + 0.03 (4)	2.28 + 0.61 (4)
MK 801 (1 μM)	0.36 + 0.07 (4)	0.97 + 0.15[e] (3)
Control	0.18 + 0.06 (3)	1.63 + 0.37 (3)
CNQX (1 μM)	0.19 + 0.02 (3)	1.18 + 0.26 (3)
Control	0.21 + 0.05 (3)	1.86 + 0.32 (3)
Mg^{2+} free	0.27 + 0.02 (3)	2.75 + 0.69 (3)
Control	0.26 + 0.08 (3)	2.5 + 0.32 (3)
3 mM Mg^{2+}	0.32 + 0.10 (3)	7.2 + 1.54[e] (3)

[a]Shown is the amount of DA overflow (ng/mg protein per 5 minutes) during the fourth 5-minute period of L-Glu exposure, the time of maximal response. Values represent mean ± standard error of the mean. The number of experiments is shown in parentheses.

[b]Samples superfused with Ca^{2+}-free medium containing EGTA (1 mM).

[c]Animals received reserpine (5 mg/kg, subcutaneously) 24 hours prior to the experiment.

[d]Antagonists were applied 20 minutes before L-Glu stimulation.

[e]Significant difference from the control (Mann-Whitney U-test) $p < 0.05$.

independent release via the DA transport system. Such a mechanism could permit impulse-independent release of DA as a result of a local presynaptic influence. Further experimentation will be required to determine if such a process occurs under physiological conditions.

ACKNOWLEDGMENTS

We thank Drs. Jon W. Johnson, Alan M. Palmer, and Ian J. Reynolds for helpful discussions.

REFERENCES

1. CLOW, D. W. & K. JHAMADAS. 1989. J. Pharm. Exp. Ther. **248:** 722–728.
2. SACHS, J. R. 1988. J. Physiol. **400:** 575–591.
3. CARTER, C. J., J. GUEUGNON & B. SCATTON. 1988. J. Neurochem. **51:** 944–949.
4. MCMAHON, H.T., A. P. BARRIE, M. LOVE & D. G. NICHOLLS. 1989. J. Neurochem. **53:** 71–79.

Dopamine D$_2$ Receptor Agonists Inhibit the Calcium-Evoked Release of Endogenous Dopamine from Striatal Synaptosomes[a]

JOSEPH M. MASSERANO, JOHN F. BOWYER,
AND GREG A. GERHARDT

Department of Pharmacology (C-236)
University of Colorado Health Sciences Center
4200 East Ninth Avenue
Denver, Colorado 80262

Dopamine (DA) neurons in the central nervous system regulate the quantity of dopamine in the synapse through the process of autoregulation. Autoregulation occurs when dopamine released from a neuron acts upon presynaptic dopamine D$_2$ receptors located on these same neurons (autoreceptors) and inhibits the further release of dopamine. A number of authors have reported an inhibition of dopamine release from striatal slices by selective dopamine agonists.[1] However, this autoreceptor regulation of dopamine release has not been obtained in synaptosomal preparations of the striatum.[2,3] Recently, Bowyer and coworkers found that after incubation of striatal synaptosomes in Krebs-Ringer-bicarbonate buffer containing no calcium, a robust release of ^{3}H-dopamine could be obtained by the addition of 1.25 mM calcium.[4,5] This calcium-evoked release of ^{3}H-dopamine was inhibited by selective D$_2$ receptor agonists and could be reversed by selective D$_2$ receptor antagonists. We have extended these findings by examining the D$_2$ receptor regulation of calcium-evoked release of endogenous dopamine from synaptosomes made from the striatum and nucleus accumbens of rats.

Rats were decapitated under halothane anesthesia and the striatum and nucleus accumbens were dissected out. A P2 synaptosomal pellet was obtained as previously described.[6] The P2 pellet was resuspended in a Krebs-Ringer-HEPES (KRH) buffer (pH 7.4) and aliquots placed into 25 mm Millipore polypropylene disc filter holders. The synaptosomes were superfused (0.75 ml/minute) with the KRH buffer containing no calcium +0.1 mM EGTA for a washout period of 20 minutes. After the 20 minute superfusion, two samples were collected (1½ ml each) to determine baseline release. Next the synaptosomes were superfused with 1.25 mM calcium containing buffer for 10 minutes in the presence or absence of the D$_2$ receptor agonist pergolide (10^{-7} M). For the potassium-release experiments, the same procedures were followed except 15 mM KCl was used to depolarize the synaptosomes and 1.25 mM calcium was present at all times. At the end of each experiment, the Whatman filter discs containing the synaptosomes were removed and placed into a citrate/acetate (pH 4.0) mobile phase buffer overnight for extraction of dopamine from the synaptosomes. Dopamine released from the synaptosomes and dopamine remaining in the synaptosomal tissue were measured by high pressure liquid chromatography coupled with coulometric

[a]This work was supported by U.S. Public Health Service grants MH41551 and AG06434.

electrochemical detection (HPLC-EC). The data are presented as either percent fractional release or as picograms of dopamine released per nanogram of tissue dopamine content. Detection sensitivity of dopamine on our HPLC-EC system approaches 0.05 to 0.1 picogram.

FIGURE 1 illustrates the effects of pergolide on endogenous dopamine release from striatal synaptosomes depolarized with 15 mM potassium. There was a sevenfold increase in the total endogenous release of dopamine following 15 mM potassium superfusion. The inclusion of pergolide in the superfusion buffer had no effect on the potassium-evoked release of dopamine. Therefore, our results confirm the work of Raiteri *et al.* indicating that potassium-evoked release of endogenous dopamine, like that of ^{3}H-dopamine, is not affected by D_2 receptor agonists in synaptosomes.[2,3]

FIGURE 2 illustrates the effects of pergolide on the endogenous release of dopamine from striatal synaptosomes treated with 1.25 mM calcium. The incubation of the striatal synaptosomes in a no-calcium buffer plus 0.1 mM EGTA followed by the inclusion of 1.25 calcium into the perfusate markedly increased the release of endogenous dopamine. The inclusion of pergolide (10^{-7} M) in the perfusion buffer six minutes prior to the addition of calcium produced profound inhibition of dopamine release throughout the entire period of calcium exposure. As shown in the bar graph in the upper right portion of the figure, endogenous dopamine release was inhibited 73% by 10^{-7} M pergolide and 78% by 10^{-6} M pergolide. The inhibition of endogenous dopamine release produced by 10^{-7} M pergolide was completely reversed by inclusion of the D_2 selective antagonist sulpiride (10^{-6} M). We have found similar results using the D_2 receptor agonist quinpirole in both striatal and nucleus accumbens synaptosomes. We propose that functional potassium channels regulate calcium-evoked re-

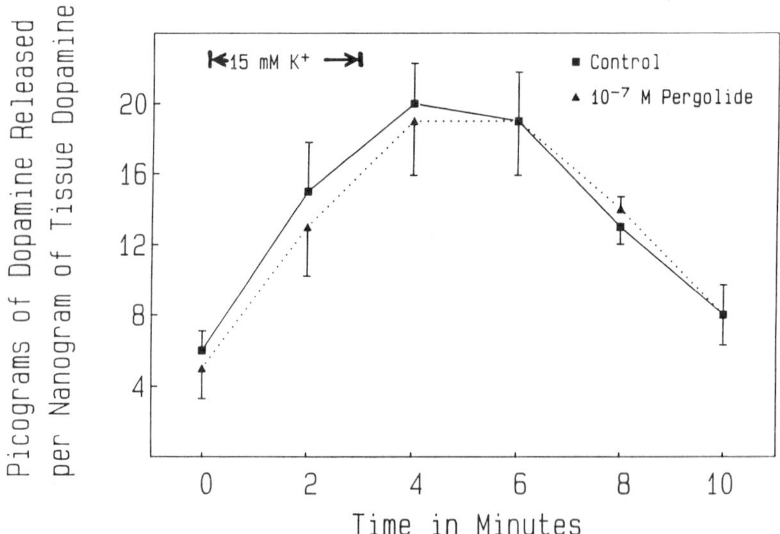

FIGURE 1. The effects of pergolide on potassium-evoked endogenous dopamine release from striatal synaptosomes. Data represent the means ± standard errors of the mean (SEM) for 5 determinations.

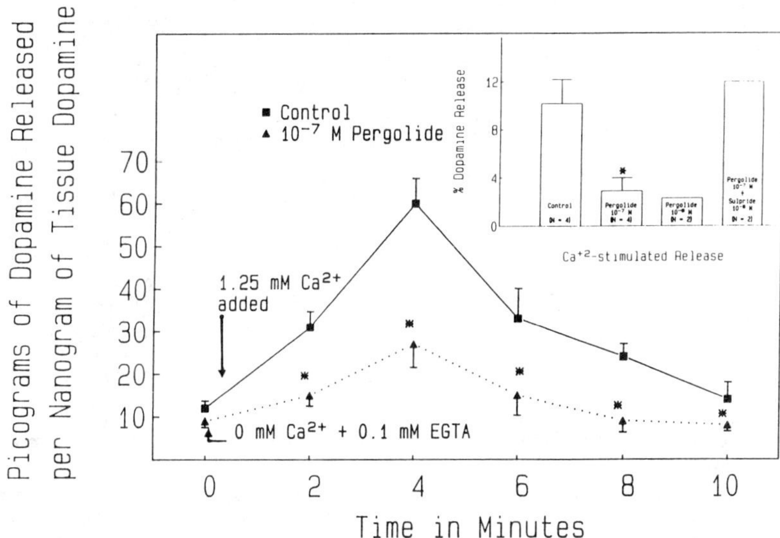

FIGURE 2. The effects of pergolide on the calcium-evoked release of endogenous dopamine from striatal synaptosomes. The data represent the means ± SEM for 5 determinations. *Significantly different from control release, $p < 0.01$.

lease of endogenous dopamine and may be necessary for the dopamine D₂ receptor modulation of this release.

These data indicate that the regulation of dopamine release by dopamine D₂ autoreceptors occurs directly on the presynaptic dopamine terminal and is not dependent upon interneurons and glial cells that are present in tissue slice preparations.

REFERENCES

1. CHESSELET, M.-F. 1984. Presynaptic regulation of neurotransmitter release in the brain: facts and hypothesis. Neuroscience **12:** 346–375.
2. RAITERI, M., F. CERRITO, G. CASAZZA & G. LEVI. 1980. Presynaptic dopamine receptors in striatal nerve endings: absence of haloperidol-induced supersensitivity. Adv. Biochem. Psychopharmacol. **24:** 37–43.
3. RAITERI, M., A. M. CERVONI & R. DEL CARMINE. 1978. Do presynaptic autoreceptors control dopamine release. Nature **274:** 706–708.
4. BOWYER, J. F. & N. WEINER. 1987. Modulation of the Ca²⁺-evoked release of (³H) dopamine from striatal synaptosomes by dopamine (D2) agonists and antagonists. J. Pharmacol. Exp. Ther. **241:** 27–33.
5. BOWYER, J. F. & N. WEINER. 1989. K⁺ channel and adenylate cyclase involvement in regulation of Ca²⁺-evoked release of (³H)dopamine from synaptosomes. J. Pharmacol. Exp. Ther. **248:** 514–520.
6. BOWYER, J. F., J. M. MASSERANO & N. WEINER. 1987. Inhibitory effects of amphetamine on potassium stimulated release of (³H)dopamine from striatal slices and synaptosomes. J. Pharmacol. Exp. Ther. **240:** 177–186.

Presynaptic Modulation of Oxytocin and Vasopressin Secretion from Isolated Nerve Terminals of the Rat Neural Lobe

KEMAL PAYZA AND JAMES T. RUSSELL

Laboratory of Developmental Neurobiology
Building 36, Room B-316
National Institute of Child Health and Human Development
National Institutes of Health
Bethesda, Maryland 20892

Isolated nerve endings (neurosecretosomes, NSS) of the rat neural lobe respond to depolarizing stimulus with calcium-dependent secretion of the hormones oxytocin (OT) and vasopressin (VP).[1] Using the NSS as a model system, we investigate the role of dynorphin and somatostatin, peptides coreleased from the neural lobe,[2,3] in the presynaptic modulation of OT and VP secretion.

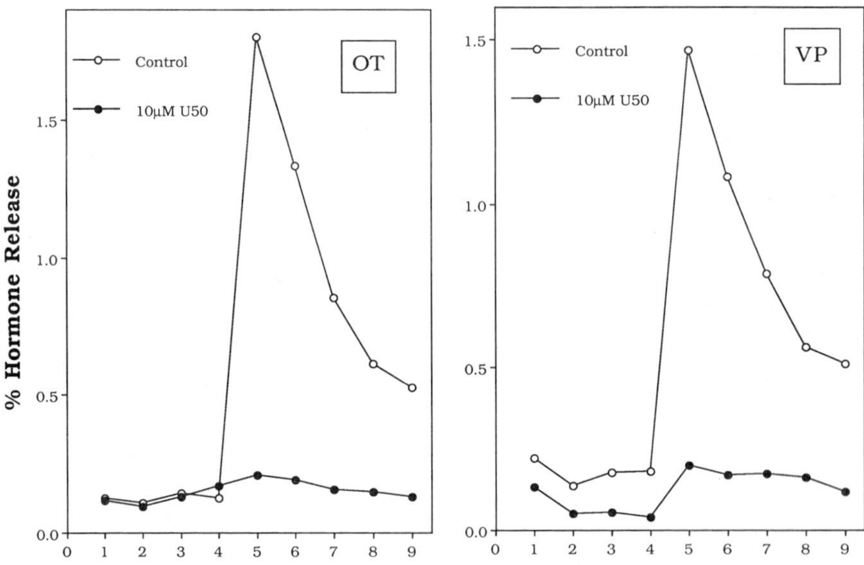

Collection Period (5 minutes)

FIGURE 1. The kappa opiate agonist U-50488H inhibits hormone secretion. The NSS were perfused in duplicate with either 0 (control) or 10 μM U-50488H (U50) in 0 Na Ringer. In periods 5–9, the $[K^+]_o$ was elevated to 25 mM.

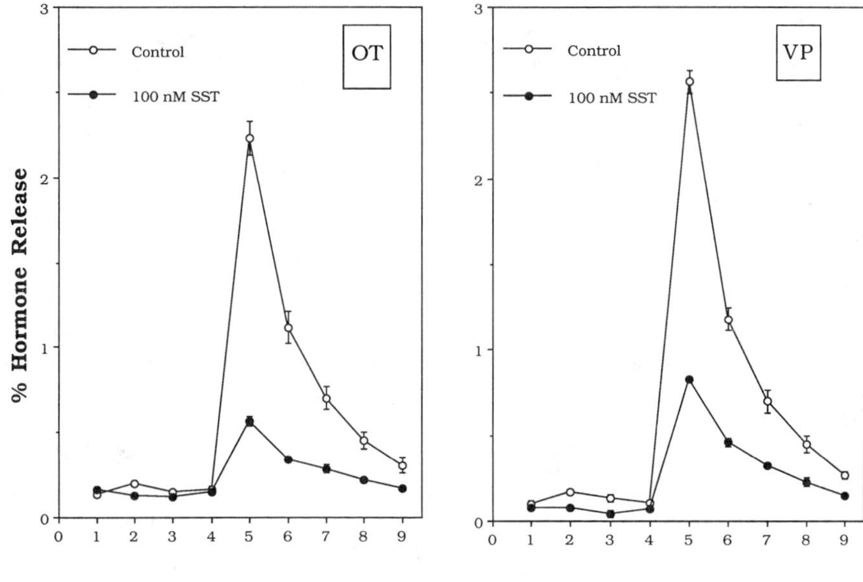

Collection Period (5 minutes)

FIGURE 2. Somatostatin inhibits hormone secretion. The NSS were perfused in quadruplicate with either 0 (control) or 100 nM somatostatin (SST) in 0 Na Ringer. In periods 5–9, the $[K^+]_o$ was elevated to 35 mM. Stimulated control secretion was 4.73% ± 0.35 (OT) and 5.07% ± 0.26 (VP); with 100 nM SST these values were reduced to 1.57% ± 0.05 (OT) and 1.98% ± 0.05 (VP).

METHOD

The "150 Na" Ringer was of the following composition (mM): NaCl, 150; KCl, 4.5; CaCl$_2$, 2.0; MgCl$_2$, 1.5; HEPES, 10 (pH 7.3 with Tris); dextrose, 10; bovine serum albumin (BSA), 0.15%. In some Ringers (0 Na and 10 Na) sodium was reduced (to 0 or 10 mM) and replaced on an equimolar basis with N-methyl-D-glucamine (NMDG). In cases of elevated potassium, NMDG was reduced to maintain the osmolality constant. ADP ribosylations were performed by Dr. Jane Halpern.

The NSS were prepared essentially as described,[1] loaded onto filters (pore size 0.45 μm), and perfused with 150 Na Ringer at 200 or 250 μl/minute. Following a wash period of 10 minutes, the NSS were perfused for another 30 minutes with 0 Na or 10 Na Ringer as indicated. The perfusate was then sampled continuously, and duplicate aliquots were taken for radioimmunoassay. The hormone released into the medium per fraction was expressed as percentage of total in the NSS. Data shown are means; the standard error of the mean (SEM) is shown for $n = 4$.

RESULTS

The kappa opiate agonist U-50488H inhibited the high-K^+-induced secretion of OT and VP by 98% and 84%, respectively (FIGURE 1). At 1 μM in 10 Na Ringer, dynorphin A (1–8) also inhibited 70% of OT and 50% of VP secretion induced by 35 mM K^+. In the presence of the kappa antagonist norbinaltorphimine (1 μM), these inhibitory effects of dynorphin were reversed by about one-half (reduced to 39% and 23% inhibition of OT and VP, respectively). Incubation of the NSS overnight in pertussis toxin (PTX, 1 μg/ml) did not affect hormone secretion nor did it block the inhibition by U50488H or dynorphin. We find that the NSS do contain a G protein substrate for ADP ribosylation by PTX. Somatostatin inhibited secretion of OT (67%) and VP (61%) (FIGURE 2).

DISCUSSION

We find that somatostatin and kappa opiate agonists inhibit high-K^+-induced secretion of both OT and VP from the NSS. This is the first report of a modulatory role for somatostatin in the neural lobe. Although our opiate data are in agreement with those of Zhao *et al.*,[4] other work shows only OT subject to inhibition.[5,6] We find that the inhibition is partially reversible with 1μM norbinaltorphimine, a specific kappa antagonist. Higher concentrations may be required for complete reversal. These data, along with the localization of dynorphin and its receptors[7] in the neural lobe, support a role for this neuropeptide as an endogenously released presynaptic inhibitor of one or both primary neurohypophysial hormones. Kappa opiate receptors may inhibit calcium channels directly *via* a G-protein,[8] or indirectly through inhibition of adenylate cyclase.[9] In either case, PTX should inactivate the G protein and block the kappa opiate effect. The failure of PTX may be due to resistance of the G protein to inactivation or of the NSS to PTX entry.

REFERENCES

1. CAZALIS, M., G. DAYANITHI & J. J. NORDMANN. 1987. J. Physiol. **390:** 55–70.
2. WHITNALL, M. H., H. GAINER, B. M. COX & C. J. MOLINEAUX. 1983. Science **222:** 1137–1139.
3. PATEL, Y. C., H. H. ZINGG & J. J. DREIFUS. 1972. Nature **267:** 852–853.
4. ZHAO, B.-G., C. CHAPMAN & R. J. BICKNELL. 1988. Brain Res. **462:** 62–66.
5. BONDY, C. A., H. GAINER & J. T. RUSSELL. 1988. Endocrinology **122:** 1321–1327.
6. FALKE, N. 1988. Neuropeptides **11:** 163–167.
7. HERKENHAM, M., R. C. RICE, R. E. JACOBSEN & R. B. ROTHMAN. 1986. Brain Res. **382:**365–371.
8. BROWN, A. M., A. YATANI A., Y. IMOTO, J. CODINA, R. METTERA & L. BIRNBAUMER. 1989. Ann. N. Y. Acad. Sci. **560:** 373–386.
9. ATTALI, B., D. SAYA & Z. VOGEL. 1989. J. Neurochem. **52:** 360–369.

Rapid Kinetic Analysis of Presynaptic Receptor Modulation of Acetylcholine Release[a]

L. BRUCE PEARCE AND EMIL ADAMEC

Department of Pharmacology
Boston University School of Medicine
80 East Concord Street
Boston, Massachusetts 02118

INTRODUCTION

Presynaptic receptor mechanisms modulating neurotransmission have been under scrutiny since the 1950s with evidence of the widespread existence of these receptors supported by numerous investigations. Despite the general acceptance of autoreceptors our understanding of the function of presynaptic receptors on the time scale of signaling events within neuronal systems remains more a matter of enlightened speculation than scientific fact. Accordingly, the focus of our research is to establish the time course of presynaptic receptor mediated effects on transmitter release. Our approach to this question has been application of the technique of rapid superfusion[1,2] to rapid kinetic analysis of potassium-evoked release of [^{3}H]acetylcholine ([^{3}H]ACh) and presynaptic autoreceptor modulation of release.

RESULTS AND DISCUSSION

Density gradient purified whole rat brain synaptosomes[3] (150μg) prelabeled with 60 μCi [^{3}H]choline (80 Ci/mmol) were superfused at 0.35 or 2.0 ml/second with Krebs buffer pH 7.4 and serial fractions were collected for a total of 67 and 1.5 seconds, respectively. Subsecond kinetic analysis of the exponential washout of tracer radiolabel revealed the $t_{1/2}$ of mixing to be 137 and 49 mseconds. Lysis of synaptosomes by rapid introduction of distilled water revealed that the synaptosomes are superfused with >90% efficiency. Superfusion with isoosmotic Krebs solutions containing 4.7, 10, 20, 50, 75, and 100 mM KCl produced concentration-dependent evoked release of radiolabel. All (100%) evoked release at 20 mM KCl and a small but significant (by BMDP2V ANOVA p < 0.0001) fraction of basal release (4.7mM KCl) were calcium dependent. Evoked release (20 mM KCl) data fit (by BMDP3R) the sum of two exponentials suggesting that evoked release measured over 48 seconds occurs in two distinct phases. Superfusion with 100 mM of the agonists oxotremorine and adenosine resulted in 40 and 65% inhibition of KCl-evoked release. Simultaneous addition of 500 mM oxotremorine and evoking agent resulted in 52% inhibition of release indicating that presynaptic muscarinic autoreceptor activation leads to exceedingly rapid inhibition of release (FIGURE 1). The ability to examine real time release of transmitter using

[a]This work was supported by a PMA Foundation Starter Grant and Eppley Foundation Grant.

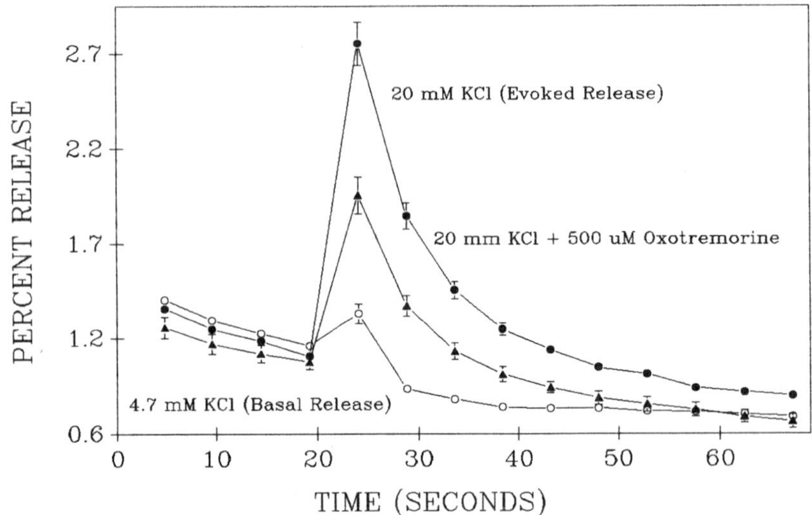

FIGURE 1. Rapid kinetics of autoreceptor-mediated inhibition. Purified whole rat brain synaptosomes (2.25 mg) prepared on discontinuous Percoll gradients were preincubated for 30 minutes under 1 atmosphere of 95% O_2/5% CO_2 and in the presence of 5 μM (60 μCi) [^{3}H]choline to label the releasble pool of ACh. Aliquots (150 μg) of purified synaptosomes were superfused at 0.35 ml/second with Krebs buffer pH 7.4 containing 4.7 mM KCl for 30 seconds to remove excess radiolabel and to determine individual flow rates. Collection of serial 4.8 second fractions was begun immediately, and at the onset of fraction 5 superfusion was switched to a source containing either buffer with 4.7 mM KCl or isoosmotic buffer with 20 mM KCl and 500 μM of the muscarinic agonist oxotremorine. Simultaneous exposure to oxotremorine and potassium-evoked depolarization resulted in 52% inhibition of release with marked inhibition observed within the first 4.8 seconds. Data are expressed as percent of total radiolabel with total radiolabel equal to the amount remaining associated with the synaptosomes after superfusion plus the amount released during the last 48 seconds of superfusion. Error bars represent standard errors of the mean.

rapid superfusion (RS) has provided evidence that brain muscarinic autoreceptors have the ability to produce very rapid inhibition of evoked release, strongly suggesting that these receptors function on the physiological time scale of signaling within the central nervous system. Studies are currently under way to extend these observations to the subsecond actions of presynaptic receptor activation on potassium-evoked release and the rapid kinetics of receptor activated second messengers in superfused synaptosomes.

REFERENCES

1. PEARCE, L. B., R. D. CALHOON, P. BURNS, A. VINCENT & S. M. GOLDIN. 1988. Two functionally distinct forms of guanosine cyclic 3',5'-phosphate stimulated cation channels in a bovine rod photoreceptor disk preparation. Biochemistry **27:** 4396–4406.
2. TURNER, T. J., L. B. PEARCE & S. M. GOLDIN. 1989. A superfusion system designed to measure release of radiolabeled neurotransmitter on a subsecond time scale. Anal. Biochem. **178:** 8–16.
3. NAGY, A. & V. DELGADO-ESCUETA. 1984. A rapid preparation of synaptosomes from mammalian brain using nontoxic iso-osmotic gradient material (Percoll). J. Neurochem. **43:** 1114–1123.

Neurotransmitter Release and Impulse Frequency Studies of Dopaminergic Autoreceptor Agonists

M. F. PIERCEY, P. A. BRODERICK,[a] W. E. HOFFMANN,
AND G. D. VOGELSANG

CNS Research
The Upjohn Company
Kalamazoo, Michigan 49001

[a]*Department of Pharmacology*
City University of New York Medical School
Convent Avenue and West 138th Street
New York, New York 10031

Antipsychotic medications are thought to act therapeutically by depressing dopaminergic neurotransmission postsynaptically.[1] The inherent limitation of a postsynaptic mechanism of action is the well-known resultant increased impulse activity and neurotransmitter release in dopaminergic cell bodies and terminals respectively, at least after acute administration.[2] It is entirely possible that stimulation of dopaminergic autoreceptors can overcome this limitation by depressing dopamine release presynaptically. This paper presents data on the autoreceptor properties of two dopaminergic agonist dihydrophenalenes, U-66444B and its active enantiomer U-68553B, synthesized by Dr. J. Szmuszkovicz, the Upjohn Company. The data are compared with data gleaned from the prototypical dopaminergic agonist ($-$)-apomorphine.

The effects of the dihydrophenalenes on impulse frequency and neurotransmitter release from dopaminergic neurons were studied electrophysiologically and voltammetrically. Impulse frequency studies took place in dopaminergic cell bodies in substantia nigra pars compacta (SNPC) and ventral tegmental area (VTA); voltammetric studies took place in dopaminergic nerve terminals located in striatum and nucleus accumbens. All studies were done in the male, chloral hydrate anesthetized Sprague-Dawley rat (300–400 g). Supplemental chloral hydrate injections were provided as needed. Procedural techniques for electrophysiology[3] and voltammetry[4,5] are previously described.

The results showed that U-66444B depressed dopaminergic neurons in SNPC and VTA with a potency three times that for apomorphine. The ED_{50} for U-66444B inhibiting SNPC neurons was 3.0 ± 0.94 μg/kg intravenously (iv) [mean $\pm$ standard error of the mean (SEM), $n = 9$]. The ED_{50} for SNPC inhibition by apomorphine was 9.6 ± 3.2 μg/kg iv ($n = 6$). Activity was found to reside principally in the ($+$) stereoisomer, U-68553B. The dopaminergic cells in the VTA responded similarly to U-66444B. Haloperidol reversed the dihydrophenalene-induced and the apomorphine-induced inhibition of SNPC and VTA cells. Animals chronically treated with 0.2 ml/day vehicle or with 6 mg/kg U-66444B for 14 days still responded to the potent dopaminergic autoreceptor agonist effects of U-66444B. Therefore, the propensity to produce tolerance was weak.

In vivo voltammetric studies showed that U-68553B suppressed striatal dopamine release by more than 25% for longer than 50 minutes after injection [100 μg/

kg intraperitoneally (ip)] ($n = 4$). By contrast, apomorphine (500 μg/kg ip) effects persisted for only 10 minutes after injection ($n = 4$). The maximal change in the dopamine signal induced by U-68553B was 44% compared to only 20% for apomorphine.[6] U-68553B, (100 μg/kg ip) also depressed dopamine release in the nucleus accumbens ($n = 4$), Following a transient (10-minute) but statistically significant ($p < 0.05$) increase in the accumbens dopamine signal, there was a dramatic 75% decrease in dopamine release, which persisted even after one hour after injection.

It is concluded that U-68553B is a structurally novel, very potent dopamine autoreceptor agonist that, because it is more potent, longer acting, possibly more specific, and less likely to induce tolerance, could have enhanced probability for success when compared to previous autoreceptor agonists in treating schizophrenia.

REFERENCES

1. CARLSSON, A. & M. LINDQUIST. 1963. Acta Pharmacol. **20:** 140–144.
2. BUNNEY, B. S. & G. K. AGHAJANIAN. 1978. *In* Psychopharmacology: a Generation of Progress. M. A. Lipton, A. DiMascio & K. F. Killam, Eds.: 159–169. Raven Press. New York, N.Y.
3. VOGELSANG, G. D. & M. F. PIERCEY. 1985. Eur. J. Pharmacol. **110:** 267–269.
4. BRODERICK, P. A. 1987. Neuropeptides **10(4):** 369–386.
5. BRODERICK, P. A. 1989. Brain Res. **475(1):** 115–121.
6. BRODERICK, P. A. 1985. Ann. N.Y. Acad. Sci. **473:** 506–508.

Prazosin-Sensitive Prejunctional α_2-Adrenergic Receptor Modulates Release of the Adrenergic Transmitter Elicited by Periarterial Nerve Stimulation in the Rat Kidney

DEAN D. SCHWARTZ AND K. U. MALIK

Department of Pharmacology
University of Tennessee, Memphis
Memphis, Tennessee 38163

Prejunctional α-adrenergic autoreceptors have been traditionally considered to be of the α_2-adrenergic receptor subtype (see Reference 1). Recently, however, using selective α-adrenergic receptor agonists and antagonists, prejunctional inhibitory α_1-adrenergic receptors have been reported.[2-4] We have recently reported that in the isolated perfused rat kidney, the preferential α_1-adrenergic receptor antagonists prazosin and corynanthine enhanced stimulus-induced fractional [³H]norepinephrine overflow.[5] The purpose of this study was to determine the subtype of α-adrenergic receptor responsible for the enhanced fractional overflow. The results of this study suggest that prazosin enhanced stimulus-induced fractional overflow by blockade of prejunctional α_2, presumably α_{2B} subtype of adrenergic receptors.

Male Sprague-Dawley rats were anesthetized with ether and the left kidney and renal artery isolated and perfused at 6 ml/minute with Tyrode's solution warmed to 37°C and gassed with 95% O_2/5% CO_2. Bipolar electrodes were placed around the renal artery for periarterial nerve stimulation. Following stabilization for 30 minutes, 25 μCi of tritiated norepinephrine ([³H]NE), (1.67 pmol, 14.9 Ci/mmol) was infused into the renal artery to label endogenous neurotransmitter stores. The kidney was then washed with [³H]NE-free Tyrode's solution for 1 hour. Basal overflow was collected for one minute before each stimulation. Stimulus-induced tritium overflow was collected for one minute during each period of stimulation. Stimulations were carried out at 0.5–4 Hz for 30 seconds at supramaximal voltage, 1 msecond duration every 10 minutes. At the end of each experiment, kidney tritium content was determined. Neurotransmitter release was expressed in these experiments as fractional overflow. Fractional overflow = [(stimulus-induced overflow) − (basal overflow)/tissue tritium content at the time of stimulation] × 100.

FIGURE 1 depicts the effect of the preferential α_1-adrenergic receptor antagonists prazosin and corynanthine and the preferential α_2-adrenergic receptor antagonist rauwolscine, on fractional overflow elicited by periarterial nerve stimulation (0.5–4.0 Hz). Prazosin and corynanthine both enhanced stimulation-induced fractional overflow at 2 and 4 Hz. Rauwolscine enhanced fractional overflow at 1, 2, and 4 Hz. Combination of prazosin (0.1 μM) and rauwolscine (0.1 μM) did not have any additive effect on the enhancement of fractional overflow caused by nerve stimulation (data not shown). Stimulus-induced fractional overflow at 0.5 Hz was not enhanced by any of the α-adrenergic receptor antagonists suggesting negligible feedback modulation at this frequency. Therefore, the remaining experiments were carried out at 0.5 Hz. At 0.5 Hz, the α_2-adrenergic receptor agonists UK-14,304 (1–10 nM) and clonidine (1–10 nM) dose dependently inhibited fractional overflow wheras the α_1-adrenergic receptor agonists cirazoline (1–10 nM) and methoxamine (1–10 nM) were ineffective. The inhibition of fractional overflow elicited by UK-14,304 (10 nM) was completely abolished by 10 nM rauwolscine confirming that UK-14,304 exerted its inhibitory effect through the activation of prejunctional α_2-adrenergic receptors. FIGURE 2 depicts the effect of prazosin and corynanthine on the inhibitory effect of UK-14,304 on tritium overflow. Both prazosin and corynanthine dose dependently attenuated the inhibition of fractional overflow produced by UK-14,304. Prazosin completely reversed the effect of UK-14,304 at the concentration of 1.0 μM, whereas corynanthine completely abolished the effect at 10 μM.

The selectivity of prazosin as an α_1-adrenergic receptor antagonist has led to the wide use of this agent in characterizing α_1-adrenergic receptor–mediated responses. Ligand binding studies have demonstrated, however, that prazosin can also bind with high affinity to [^{3}H]rauwolscine binding sites in rodent kidney and with a lower affinity to α_2-adrenergic receptors in human kidney.[6] This apparent heterogeneity in α_2-adrenergic receptors has led to the proposal that the low affinity prazosin binding site in

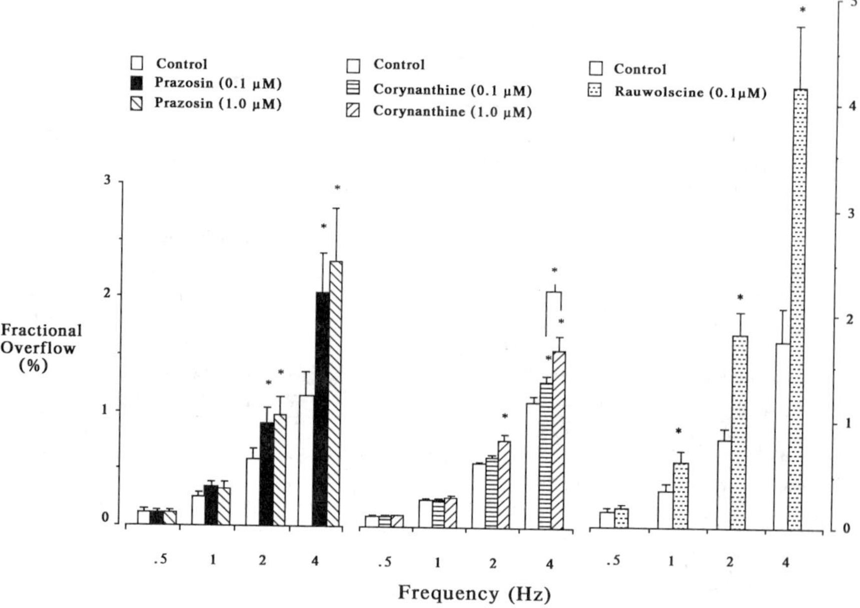

FIGURE 1. Effects of prazosin (0.1–1.0 μM), corynanthine (0.1–1.0 μM), and rauwolscine (0.1 μM) on fractional overflow, elicited by periarterial nerve stimulation (0.5–4.0 Hz), in the isolated perfused rat kidney. Control stimulations were followed by stimulations in the presence of increasing concentrations of prazosin ($n = 6$), corynanthine ($n = 6$), and rauwolscine ($n = 4$). Bars and vertical lines represent the mean and standard error of the mean (SEM), respectively. An asterisk over the bar indicates significant difference from control for the same frequency. An asterisk over the bracket indicates a significant difference between the two concentrations of the antagonist for the same frequency of stimulation (*$p < 0.05$).

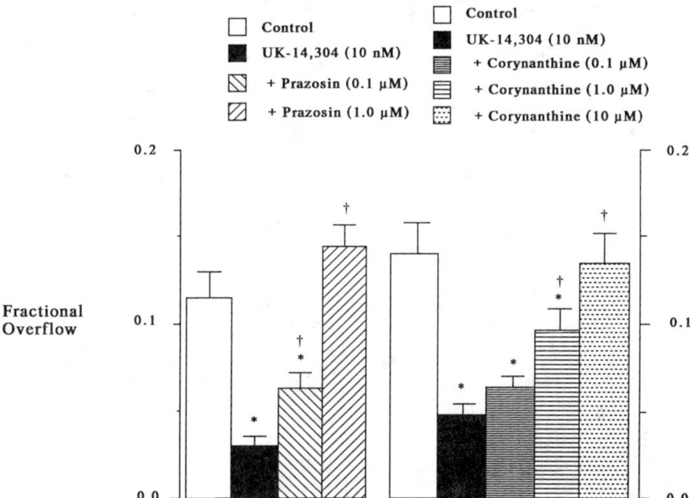

FIGURE 2. Effect of prazosin (0.1–1.0 μM) and corynanthine (0.1–10 μM) on the inhibition of fractional overflow elicited by UK-14,304 (10 nM) at 0.5 Hz in the rat kidney. Control stimulations were followed by stimulations first in the presence of UK-14,304 (10 nM) alone and then in the presence of increasing concentrations of prazosin or corynanthine ($n = 5$). Each concentration of the drug was infused for two stimulation periods and the average fractional overflow was used to determine the percent of the control response. Bars and vertical lines represent the mean and SEM, respectively. An asterisk over the bar indicates significant difference from control for the same frequency (*$p < 0.05$). A dagger over the bar indicates significant difference from UK-14,304 (†$p < 0.05$).

human platelets be termed α_{2A}- whereas the high affinity prazosin binding site in neonatal rat lung be termed α_{2B}-adrenergic receptors.[7] In tissues with predominantly α_{2A}-adrenergic receptors, corynanthine has a higher affinity for α_2-adrenergic receptors than prazosin (HT-29 cells, Ki-nM 144 ± 14 vs. 340 ± 20; human platelets, Ki-nM 91 ± 11 vs. 267 ± 28, respectively) whereas in tissues with predominantly α_{2B}-adrenergic receptors, corynanthine is less potent than prazosin (rat lung, Ki-nM 100 ± 19 vs. 5.4 ± 0.5; NG108–15 cells, Ki-nM 43 ± 14 vs. 3.7 ± 0.7, respectively).[8] In the present study, prazosin was 10 times more potent than corynanthine in both enhancing stimulus-induced fractional overflow and in reversing the inhibitory effect of UK-14,304 suggesting that prazosin enhances fractional overflow by blocking prejunctional α_2-adrenergic receptors. Furthermore, the prazosin-sensitive α_2-adrenergic receptor modulating neurotransmitter release in the rat kidney is of the α_{2B}-adrenergic receptor subtype.

REFERENCES

1. STARKE, K. 1977. Rev. Physiol. Biochem. Pharmacol. **77:** 1–124.
2. RUMP, L. C. & H. MAJEWSKI. 1987. J. Cardiovasc. Pharmacol. **9:** 500–507.
3. DEJONGE, A., G. VAN DEN BERG, J. Q. QIAN, B. WILFFERT, M. J. M. C. THOOLEN, P. B. M. W. M. TIMMERMANS & P. A. VAN ZWIETEN. 1986. J. Pharmacol. Exp. Ther. **236:** 500–504.

4. STORY, D. F., C. A. STANFORD-STARR & M. J. RAND. 1985. Clin. Sci. **68**(Suppl. 10): 111s–115s.
5. SCHWARTZ, D. D. & K. U. MALIK. 1989. J. Pharmacol. Exp. Ther. **250**: 764–771.
6. NEYLON, C. B. & R. J. SUMMERS. 1985. Br. J. Pharmacol. **85**: 349–359.
7. BYLUND, D. B. 1985. Pharmacol. Biochem. Behav. **22**: 835–843.
8. BYLUND, D. B., C. RAY-PRENGER & T. J. MURPHY. 1988. J. Pharmacol. Exp. Ther. **245**: 600–607.

Pharmacological and Biochemical Evaluation of Hippocampal Muscarinic Cholinergic Autoreceptors

THOMAS W. VICKROY

Department of Physiological Sciences
University of Florida
Box J-144 JHMHC
Gainesville, Florida 32610-0144

While release-inhibiting autoreceptors have been described for many putative central nervous system (CNS) neurotransmitters, the signal transduction events that mediate autoreceptor control of impulse-dependent neurotransmitter release are poorly delineated. In CNS cholinergic neurons, it is well recognized that activation of presynaptic muscarinic receptors (mAChR) effectively reduces depolarization-dependent acetylcholine (ACh) release from nerve terminals. In view of the strong evidence for pharmacological and molecular heterogeneity among mAChR, receptor-selective drugs have been used to characterize muscarinic autoreceptors as belonging to the M_2 mAChR subclass.[1,2] Since the primary signal transduction pathway for M_2 mAChR appears to be inhibition of membrane-bound adenylate cyclase via interactions with an inhibitory guanine nucleotide binding protein (G_i), the studies described in this report were undertaken in order to evaluate the potential cause-effect relationship between inhibition of adenylate cyclase (reduced cyclic AMP formation) and muscarinic autoreceptor inhibition of ACh release. Studies were carried out *in vitro* using [³H]choline-prelabeled hippocampal slices from adult male albino Sprague-Dawley rats. [³H]ACh release was stimulated by electrical field depolarization under assay conditions in which mAChR agonists maximally inhibited [³H]ACh release by 80% while mAChR antagonists had no effect other than blockade of the response to agonists.[3] Preliminary characterization of the structure-activity profile for a series of mAChR agonists and antagonists demonstrated an M_2 profile (methylatropine = atropine > scopolamine ≫ pirenzepine) although the M_1-selective antagonist pirenzepine was exceptionally weak.

In order to establish evidence for a cause-effect relationship between adenylate cyclase inhibition and autoreceptor reduction of [³H]ACh release, it was decided that three criteria must be satisfied. These included: (1) elevation of intraneuronal cyclic AMP should facilitate [³H]ACh release; (2) elevation of intraneuronal cyclic AMP should reverse autoreceptor-mediated inhibition of [³H]ACh release; and (3) uncoupling mAChR from adenylate cyclase should eliminate autoreceptor control over [³H]ACh release. As shown in TABLE 1, elevation of intraneuronal cyclic AMP by forskolin (an adenylate cyclase stimulator) or the cell-permeant analogue 8-bromocyclic AMP does facilitate stimulated (but not basal) [³H]ACh release. However, neither of these agents attenuates autoreceptor control over [³H]ACh release as demonstrated by the slightly improved inhibitory effect of the mAChR agonist carbachol (TABLE 1). Even stronger evidence is provided by results with *N*-ethylmaleimide (NEM) and pertussis toxin (PT). Both NEM, a sulfhydryl alkylating agent, and PT disrupt the inhibitory coupling of cell surface receptors with adenyl cyclase via

TABLE 1. Insensitivity of Muscarinic Autoreceptor-Mediated Inhibition of [3H]Acetylcholine Release to Forskolin and 8-Bromo Cyclic AMP[a]

Additions (mM)	S_2/S_1 Ratio	Percent Inhibition
0.1% Ethanol	0.72 ± 0.09	—
+ carbachol (0.3)	0.33 ± 0.05	54
+carbachol (3)	0.24 ± 0.05	67
Forskolin (0.001)	0.92 ± 0.07^b	—
+ carbachol (0.3)	0.40 ± 0.04	57
+ carbachol (3)	0.26 ± 0.03	72
Forskolin (0.01)	1.14 ± 0.06^c	—
+ carbachol (0.3)	0.46 ± 0.05	60
+ carbachol (3)	0.23 ± 0.04	80
Buffer	0.68 ± 0.03	—
+ carbachol (0.3)	0.33 ± 0.02	52
+ carbachol (3)	0.22 ± 0.03	68
8-Bromo cyclic AMP (0.03)	0.93 ± 0.06	—
+ carbachol (0.3)	0.42 ± 0.04	55
+ carbachol (3)	0.21 ± 0.03	67
8-Bromo cyclic AMP (0.3)	1.01 ± 0.08^b	—
+ carbachol (0.3)	0.34 ± 0.02	66
+ carbachol (3)	0.29 ± 0.08	71

[a]Data are the mean ± standard error of the mean (SEM) from 4 to 25 individual chambers for each treatment group. Ethanol (0.1%) and buffer were vehicles for forskolin and 8-bromo cyclic AMP, respectively. [b]$p < 0.05$, [c]$p < 0.02$ vs. respective control groups (Mann-Whitney U test).

chemical modification of the G_i protein family. While pretreatment of hippocampal slices with PT or NEM effectively eliminates mAChR inhibition of cyclic AMP formation in washed membranes, neither treatment altered autoreceptor control over stimulated [3H]ACh release (TABLE 2). Taken together, these data argue strongly against the involvement of adenyl cyclase (or intraneuronal cyclic AMP) as a mediator of muscarinic autoreceptor function in hippocampal cholinergic neurons.

TABLE 2. Muscarinic Receptor–Mediated Inhibition of Adenyl Cyclase and [3H]Acetylcholine Release: Differential Sensitivity to Alteration of Guanine Nucleotide Binding Proteins

Slice Pretreatment[a]	Percent Inhibition by 3 mM Carbachol	
	Forskolin-Activated Adenyl Cyclase[b]	Stimulated [3H]ACh Release
Vehicle (water)	$20 \pm 2.4\%^c$	76 ± 3.8^c
30 μM N-ethylmaleimide	$2.3 \pm 0.4\%$	$81 \pm 4.1\%^c$
Vehicle (buffer)	$31 \pm 3.8\%^c$	$61 \pm 5.0\%^c$
1 μg/ml pertussis toxin	$-8.3 \pm 4.1\%$	$63 \pm 3.2\%^c$

[a]Hippocampal slices were preincubated (30 min @ 37°C) in superfusion buffer containing water (NEM vehicle), NEM, 50 mM NaH_2PO_4 in 100 mM NaCl (pertussis toxin vehicle) or pertussis toxin (all diluted 1:100).
[b]Values are percent inhibition of forskolin (10 μM) stimulated formation of cyclic AMP (above baseline values) in hippocampal membranes.
[c]Indicates a significant ($p < 0.02$) inhibitory effect by carbachol.

REFERENCES

1. MARCHI, M. & M. RAITERI. 1985. On the presence in the cerebral cortex of muscarinic receptor subtypes which differ in neuronal localization, function and pharmacological properties. J. Pharmacol. Exp. Ther. **235:** 230–233.
2. MEYER, E. M. & D. H. OTERO. 1985. Pharmacological and ionic characterizations of the muscarinic receptors modulating [^{3}H]acetylcholine release from rat cortical synaptosomes. J. Neurosci. **5:** 1202–1207.
3. VICKROY, T. W. & E. D. CADMAN. 1989. Dissociation between muscarinic receptor–mediated inhibition of adenyl cyclase and autoreceptor inhibition of [^{3}H]acetylcholine release in rat hippocampus. J. Pharmacol. Exp. Ther. **251:** 1039–1044.

C5a Complement Peptide Releases Catecholamines from Rat Brain Synaptosomes via a Novel Receptor

CURTIS WILLIAMS, FRANCESCA NOVARESE,
ELLEN MILLER, REBECA LEGGIERE,
AND ROZELLE CORDA

Division of Natural Sciences
State University of New York at Purchase
Purchase, New York 10577

INTRODUCTION

Anaphylatoxins C3a and C5a are pharmacologically active inflammatory peptides, 77 and 74 amino acids respectively, cleaved from C3 and C5 during the serum complement cascade. The complement cascade is activated by immune complexes and by microbial cell wall products. Among the biological activities of the anaphylatoxins is the calcium-dependent release of histamine from mast cells and basophils. It has also been shown that this activity is inhibited by pertussis toxin (PT) suggesting linkage to a G protein of the Ni type.[1]

Our research has shown that human C3a and C5a in the rat perifornical hypothalamus mimic dopamine (DA) and norepinephrine (NE) respectively in their effects on eating and drinking behaviors, and that these effects are blocked by catecholamine antagonists.[2,3] The activity of immune-complex formation at this site also mimics NE and has been shown to be complement dependent.[4] There are at least two different binding sites for C3a in the rat brain.[5] C5a binding is also complex, but the site mediating the behavioral responses we assay appears to be presynaptic, since the C5a activity is dependent on endogenous catecholamines.[6] We have determined that C5a, but not C3a, is a releaser of preloaded radioactive DA and NE from rat forebrain P2 synaptosomes in a concentration-dependent manner (0.3–2.0 μM), and most interesting is that C5a releases no radioactive derivatives from synaptosomes preloaded with ^{3}H-choline.[7]

This report begins to characterize C5a as a catecholamine releaser in the central nervous system (CNS) and to compare the properties of its receptors to those in the periphery.

RESULTS

TABLE 1 shows that DA release by C5a is significantly reduced in calcium-free synaptosome suspensions, about the same reduction seen in release by potassium ion depolarization. We cannot demonstrate calcium dependence of C5a-induced NE release (data not shown). The column heading X/C expresses the increase of experimental over control release. The column heading (X-C)/(T-C) expresses the fraction of the total releasable pool actually released.

TABLE 1. Calcium Dependence of DA Release by Anaphylatoxin C5a[a]

| Release Condition | Standard GBS | | | Calcium-Free GBS | | |
	Picograms per Milligram Protein	$\frac{X}{C}$	$\frac{X-C}{T-C}$	Picomoles per Milligram Protein	$\frac{X}{C}$	$\frac{X-C}{T-C}$
T H$_2$O lysis	117			120		
C GBS control	38			37		
X KCl 70 mM	72	1.89	0.42	57	1.54	0.23
X C5a 1 uM	57[b]	1.50	0.24	46[c]	1.24	0.14

[a]Crude forebrain synaptosomes (P2) are prepared by homogenization in 9 vol/wt 0.32 M sucrose, sedimentation of the 700 $\times$ g supernatant (S1) at 8500 $\times$ g and again at 10500 $\times$ g. Resuspended in PO$_4$-buffered balanced salts and glucose (GBS), uptake of ^{14}C-DA by P2 proceeds for 30 minutes at 37°C in oxygenated GBS. Washed, preloaded synaptosomes are resuspended to ¼ S1 volume and triplicate 25 μl aliquots are added to 25 μl of releasing cocktail (or 225 μl H$_2$O for lysis) in microfuge tubes for a 10 minute incubation at 37°C. Release supernatants are assayed by liquid scintillation after stopping with iced GBS and centrifugation 5 minutes at 12000 $\times$ g. Protein assays are made with Bio Rad Reagent. Data are analyzed by Student t-tests of corrected counts per minute to compare groups using the same synaptosome preparations, or of picograms/milligram protein calculations to compare preparations treated differently.
[b]C5a vs. control, $p < 0.05$.
[c]CF-C5a vs. CF-control, $p < 0.05$; CF-C5a vs. standard C5a, $p < 0.05$.

TABLE 2 suggests, in the (X-C)/(T-C) columns, that pretreatment of synaptosomes with PT does not affect C5a release of DA. The X/C calculations suggest, however, that release is enhanced in both C5a and high potassium conditions. No calculations suggest an inhibition of the C5a release mechanism. Primary radioactivity data are presented because the unexplained high DA uptake by the PT-treated synaptosomes makes other than intragroup data analysis impossible. Other G protein reactive substances are being tested.

CONCLUSIONS

Our research to date demonstrates that anaphylatoxin C5a is a selective catecholamine releaser. The results we report here lead us to think that the C5a-sensitive

TABLE 2. Effect of Pertussus Toxin on DA Release by Anaphylatoxin C5a[a]

| Release Condition | Standard GBS | | | 10 μg PT/ml GBS | | |
	Mean Counts per Minute	$\frac{X}{C}$	$\frac{X-C}{T-C}$	Mean Counts per Minute	$\frac{X}{C}$	$\frac{X-C}{T-C}$
T H$_2$O lysis	3607			8712		
C GBS control	1545			2446		
X KCl 70 mM	2100	1.36	0.27	3900	1.59	0.23
X C5a 1 μM	1834[b]	1.19	0.14	3476[c]	1.42	0.16

[a]General procedures were as described in TABLE 1. P2 suspensions were incubated 60 minutes at 37°C in oxygenated GBS or PT in GBS and washed twice in GBS before ^{14}C-DA uptake. The PT concentration was 10 times that reported effective for basophils.
[b]C5a vs. control, $p < 0.05$.
[c]C5a vs. control, $p < 0.001$.

receptor at CNS nerve terminals is unlike those described in the periphery, or that its signal is tranduced by a different mechanism.

Sites like these in the human brain, however, could contribute to the neurological and neuropsychiatric disorders associated with certain autoimmune syndromes, especially those characterized as immune-complex diseases.

REFERENCES

1. WARNER, J. A., K. B. YANCY & D. W. MacGLASHAN, JR. 1987. J. Immunol. **139:** 161–165.
2. SCHUPF, N., C. A. WILLIAMS, T. E. HUGLI & J. COX. 1983. J. Neuroimmunol. **5:** 305–316.
3. WILLIAMS, C. A., N. SCHUPF & T. E. HUGLI. 1985. J. Neuroimmunol. **9:** 29–40.
4. SCHUPF, N. & C. A. WILLIAMS. 1987. J. Neuroimmunol. **13:** 293–303.
5. WILLIAMS, C. A., N. SCHUPF, C. L. REILLY & J. WAGNER. 1988. Drug Dev. Res. **15:** 175–187.
6. SCHUPF, N., C. A. WILLIAMS, A. BERKMAN, W. S. CATTELL & L. KERPER. 1989. Brain Behav. Immun. **3:** 28–38.
7. WILLIAMS, C. A. & N. SCHUPF. 1987. Ann. N.Y. Acad. Sci. **496:** 250–263.

Stimulation of D2 Dopamine Receptors Decreases Intracellular Calcium Levels in Rat Anterior Pituitary Cells but Not Striatal Synaptosomes

A Flow Cytometric Study Using Indo-1

MARINA E. WOLF[a] AND GREGORY KAPATOS

Center for Cell Biology
Sinai Hospital of Detroit
6767 West Outer Drive
Detroit, Michigan 48235-2899

Pharmacologically similar D2 dopamine (DA) receptors are found on rat anterior pituitary cells (AP cells) and striatal nerve terminals. In fact, the anterior pituitary D2 DA receptor has long been considered a model for central nervous system (CNS) D2 DA receptors. Stimulation of D2 DA receptors on AP cells has been shown to inhibit Ca^{2+} entry and decrease free intracellular calcium levels ($[Ca^{2+}]_i$).[1] If D2 DA receptors employ similar signal transduction mechanisms regardless of the type of cell on which they are located, striatal D2 DA receptors might also be expected to modulate Ca^{2+} entry.

The purpose of the present study was to compare, under identical experimental conditions, rat AP cells and striatal synaptosomes with respect to the mechanism of depolarization-induced Ca^{2+} entry and the effect of D2 DA receptor stimulation on $[Ca^{2+}]_i$. A fluorescence-activated cell sorter (FACS), in combination with the fluorescent Ca^{2+} indicator indo-1, was used to monitor $[Ca^{2+}]_i$ in both preparations. Fluorescent voltage-sensitive dyes were used to monitor changes in membrane potential. FACS analysis enables cells or synaptosomes to be analyzed one at a time at a rate of several thousand per second. This makes it possible to define the properties of a population of cells or synaptosomes without obscuring the presence of subpopulations which may exhibit important differences in responsiveness. Methods for the FACS analysis of synaptosomes have been described previously.[2,3]

In AP cells, we found that stimulus-induced increases in $[Ca^{2+}]_i$ were mediated largely by L-type Ca^{2+} channels which are insensitive to ω-conotoxin. In striatal synaptosomes, however, the Na^+ gradient was found to be critical in maintaining $[Ca^{2+}]_i$ under resting conditions and for generating increases in $[Ca^{2+}]_i$ in response to depolarization. This suggests that stimulus-induced increases in synaptosomal $[Ca^{2+}]_i$ are mediated primarily by reversal of Na^+/Ca^{2+} exchange on the time scale of this study (30 seconds–2 minutes after depolarization). Pharmacological studies supported this conclusion.

Stimulation of D2 DA receptors decreased resting and stimulated levels of $[Ca^{2+}]_i$ in AP cells. The DA agonist-induced decrease in $[Ca^{2+}]_i$ appeared to reflect hyperpolarization of the cells and a resultant elimination of spontaneous Ca^{2+}-dependent

---

[a]Present affiliation: Department of Psychiatry, Wayne State University School of Medicine, Lafayette Clinic, 951 E. Lafayette, Detroit, Mich. 48207.

action potentials. This is consistent with other studies demonstrating that DA agonists increase the K^+ conductance of AP cells.[4] In contrast, stimulation of D2 DA receptors hyperpolarized a subpopulation of striatal synaptosomes but had no apparent effect on $[Ca^{2+}]_i$. It is possible that the DA receptor itself functions similarly in both systems, but that the ability of receptor occupation to influence $[Ca^{2+}]_i$ depends on the mechanisms that regulate $[Ca^{2+}]_i$ in that cell and whether those mechanisms are sensitive to the effects transduced by the receptor. Alternatively, cells may differ with respect to coupling molecules (e.g., G proteins) necessary for the expression of receptor function. These findings have been described in detail elsewhere.[5]

REFERENCES

1. MALGAROLI, A., L. VALLAR, F. R. ELAHI, T. POZZAN, A. SPADA & J. MELDOLESI. 1987. J. Biol. Chem. **263:** 13920–13927.
2. WOLF, M. E. & G. KAPATOS. 1989. J. Neurosci. **9:** 94–105.
3. WOLF, M. E. & G. KAPATOS. 1989. J. Neurosci. **9:** 106–114.
4. CASTELLETTI, L., M. MEMO, C. MISSALE, P. F. SPANO & A. VALERIO. 1989. J. Physiol. **410:** 251–265.
5. WOLF, M. E. & G. KAPATOS. 1989. Synapse **4:** 353–370.

Index of Contributors